TIME SERIES ANALYSIS

WILEY SERIES IN PROBABILITY AND STATISTICS

Established by WALTER A. SHEWHART and SAMUEL S. WILKS

A complete list of the titles in this series appears at the end of this volume.

TIME SERIES ANALYSIS

Wilfredo Palma
Pontificia Universidad Católica de Chile

WILEY

Published by John Wiley & Sons, Inc., Hoboken, New Jersey
Published simultaneously in Canada

Library of Congress Cataloging-in-Publication Data:

Palma, Wilfredo, 1963-
Time series analysis / Wilfredo Palma, Pontificia Universidad Católica de Chile.
pages cm
Includes bibliographical references and indexes.
ISBN 978-1-118-63432-5 (cloth)
1. Time-series analysis. I. Title.
QA280.P354 2016
519.5′5–dc23

2015024282

Printed in the United States of America

10 9 8 7 6 5 4 3 2 1

CONTENTS

Preface xiii

Acknowledgments xvii

Acronyms xix

About the Companion Website xxv

1 Introduction **1**

1.1 Time Series Data 2

1.1.1 Financial Data 2

1.1.2 Economic Data 6

1.1.3 Hydrological Data 6

1.1.4 Air Pollution 7

1.1.5 Transportation Data 9

1.1.6 Biomedical Time Series 9

1.1.7 Sociological Data 10

1.1.8 Energy Data 11

1.1.9 Climatic Data 12

1.2 Random Variables and Statistical Modeling 16

1.3 Discrete-Time Models 22

1.4 Serial Dependence 22
1.5 Nonstationarity 25
1.6 Whiteness Testing 32
1.7 Parametric and Nonparametric Modeling 36
1.8 Forecasting 38
1.9 Time Series Modeling 38
1.10 Bibliographic Notes 39
Problems 40

2 Linear Processes 43

2.1 Definition 44
2.2 Stationarity 44
2.3 Invertibility 46
2.4 Causality 46
2.5 Representations of Linear Processes 46
2.5.1 Wold Decomposition 47
2.5.2 Autoregressive Representation 48
2.5.3 State Space Systems 48
2.6 Weak and Strong Dependence 49
2.7 ARMA Models 51
2.7.1 Invertibility of ARMA Processes 52
2.7.2 Simulated ARMA Processes 52
2.8 Autocovariance Function 56
2.9 ACF and Partial ACF Functions 57
2.9.1 Sample ACF 60
2.9.2 Partial ACF 63
2.10 ARFIMA Processes 64
2.10.1 Long-Memory Processes 64
2.10.2 Linear Representations 65
2.10.3 Autocovariance Function 66
2.10.4 Sample Mean 67
2.10.5 Partial Autocorrelations 67
2.10.6 Illustrations 68
2.11 Fractional Gaussian Noise 71
2.11.1 Sample Mean 72
2.12 Bibliographic Notes 72
Problems 72

3 State Space Models 89

3.1 Introduction 90
3.2 Linear Dynamical Systems 92
3.2.1 Stability 92
3.2.2 Hankel Matrix 93
3.2.3 Observability 94
3.2.4 Controllability 94
3.2.5 Minimality 95
3.3 State space Modeling of Linear Processes 96
3.3.1 State Space Form to Wold Decomposition 96
3.3.2 Wold Decomposition to State Space Form 96
3.3.3 Hankel Matrix to State Space Form 96
3.4 State Estimation 97
3.4.1 State Predictor 98
3.4.2 State Filter 98
3.4.3 State Smoother 99
3.4.4 Missing Values 99
3.4.5 Additive Noise 105
3.4.6 Structural Models 110
3.4.7 Estimation of Future States 111
3.5 Exogenous Variables 113
3.6 Bibliographic Notes 114
Problems 114

4 Spectral Analysis 121

4.1 Time and Frequency Domains 122
4.2 Linear Filters 122
4.3 Spectral Density 123
4.4 Periodogram 125
4.5 Smoothed Periodogram 128
4.6 Examples 130
4.7 Wavelets 136
4.8 Spectral Representation 138
4.9 Time-Varying Spectrum 140
4.10 Bibliographic Notes 145
Problems 145

5 Estimation Methods **151**

5.1 Model Building 152
5.2 Parsimony 152
5.3 Akaike and Schwartz Information Criteria 153
5.4 Estimation of the Mean 153
5.5 Estimation of Autocovariances 154
5.6 Moment Estimation 155
5.7 Maximum-Likelihood Estimation 156
5.7.1 Cholesky Decomposition Method 156
5.7.2 Durbin-Levinson Algorithm 157
5.8 Whittle Estimation 157
5.9 State Space Estimation 160
5.10 Estimation of Long-Memory Processes 161
5.10.1 Autoregressive Approximations 162
5.10.2 Haslett-Raftery Method 162
5.10.3 A State Space Method 164
5.10.4 Moving-Average Approximations 165
5.10.5 Semiparametric Approach 168
5.10.6 Periodogram Regression 169
5.10.7 Rescaled Range Method 170
5.10.8 Variance Plots 171
5.10.9 Detrended Fluctuation Analysis 171
5.10.10 A Wavelet-Based Method 174
5.10.11 Computation of Autocovariances 177
5.11 Numerical Experiments 178
5.12 Bayesian Estimation 180
5.12.1 Markov Chain Monte Carlo Methods 181
5.12.2 Metropolis-Hastings Algorithm 181
5.12.3 Gibbs Sampler 182
5.13 Statistical Inference 184
5.14 Illustrations 189
5.15 Bibliographic Notes 193
Problems 194

6 Nonlinear Time Series **209**

6.1 Introduction 210
6.2 Testing for Linearity 211
6.3 Heteroskedastic Data 212

6.4 ARCH Models 213
6.5 GARCH Models 216
6.6 ARFIMA-GARCH Models 218
6.7 ARCH(∞) Models 220
6.8 APARCH Models 222
6.9 Stochastic Volatility 222
6.10 Numerical Experiments 223
6.11 Data Applications 225
6.11.1 SP500 Data 225
6.11.2 Gold Data 226
6.11.3 Copper Data 231
6.12 Value at Risk 236
6.13 Autocorrelation of Squares 241
6.13.1 Squares of Gaussian Processes 241
6.13.2 Autocorrelation of Squares 242
6.13.3 Illustrations 243
6.14 Threshold autoregressive models 247
6.15 Bibliographic Notes 252
Problems 253

7 Prediction 267

7.1 Optimal Prediction 268
7.2 One-Step Ahead Predictors 268
7.2.1 Infinite Past 268
7.2.2 Finite Past 269
7.2.3 Innovations Algorithm 269
7.2.4 An Approximate Predictor 273
7.3 Multistep Ahead Predictors 275
7.3.1 Infinite Past 275
7.3.2 Finite Past 275
7.4 Heteroskedastic Models 276
7.4.1 Prediction of Returns 278
7.4.2 Prediction of Volatility 278
7.5 Prediction Bands 281
7.6 Data Application 287
7.7 Bibliographic Notes 289
Problems 289

8 Nonstationary Processes 295

8.1 Introduction 296

8.1.1 Deterministic Trends 296

8.1.2 Stochastic Trends 296

8.2 Unit Root Testing 297

8.3 ARIMA Processes 298

8.4 Locally Stationary Processes 301

8.4.1 State-Space Representations 308

8.4.2 Whittle Estimation 310

8.4.3 State Space Estimation 311

8.4.4 Asymptotic Variance 316

8.4.5 Monte Carlo Experiments 319

8.4.6 Data Application 320

8.5 Structural Breaks 326

8.6 Bibliographic Notes 331

Problems 332

9 Seasonality 337

9.1 SARIMA Models 338

9.1.1 Spectral Density 341

9.1.2 Several Seasonal Components 343

9.1.3 Estimation 343

9.1.4 Estimator Performance 343

9.1.5 Heating Degree Day Data Application 345

9.2 SARFIMA Models 351

9.3 GARMA Models 353

9.4 Calculation of the Asymptotic Variance 355

9.5 Autocovariance Function 355

9.6 Monte Carlo Studies 359

9.7 Illustration 362

9.8 Bibliographic Notes 364

Problems 365

10 Time Series Regression 369

10.1 Motivation 370

10.2 Definitions 373

10.2.1 Grenander Conditions 373

10.3 Properties of the LSE 375
10.3.1 Asymptotic Variance 375
10.3.2 Asymptotic Normality 376
10.4 Properties of the BLUE 376
10.4.1 Efficiency of the LSE Relative to the BLUE 376
10.5 Estimation of the Mean 378
10.5.1 Asymptotic Variance 379
10.5.2 Relative Efficiency 381
10.6 Polynomial Trend 382
10.6.1 Consistency 384
10.6.2 Asymptotic Variance 384
10.6.3 Relative Efficiency 385
10.7 Harmonic Regression 385
10.7.1 Consistency 386
10.7.2 Asymptotic Variance 387
10.7.3 Efficiency 387
10.8 Illustration: Air Pollution Data 388
10.9 Bibliographic Notes 392
Problems 392

11 Missing Values and Outliers 399

11.1 Introduction 400
11.2 Likelihood Function with Missing Values 401
11.2.1 Integration 402
11.2.2 Maximization 402
11.2.3 Calculation of the Likelihood Function 404
11.2.4 Kalman Filter with Missing Observations 404
11.3 Effects of Missing Values on ML Estimates 405
11.3.1 Monte Carlo Experiments 407
11.4 Effects of Missing Values on Prediction 407
11.5 Interpolation of Missing Data 410
11.6 Spectral Estimation with Missing Values 418
11.7 Outliers and Intervention Analysis 421
11.7.1 Methodology 424
11.7.2 Known time of the event 424
11.7.3 Unknown time of the event 426
11.8 Bibliographic Notes 434
Problems 435

12 Non-Gaussian Time Series 441

12.1 Data Driven Models 442
12.1.1 INAR Processes 442
12.1.2 Conditional Distribution Models 446
12.2 Parameter Driven Models 452
12.3 Estimation 453
12.3.1 Monte Carlo Experiments 461
12.3.2 Diagnostics 465
12.3.3 Prediction 465
12.4 Data Illustrations 466
12.4.1 IBM Trading Volume 466
12.4.2 Glacial Varves 469
12.4.3 Voting Intentions 473
12.5 Zero-Inflated Models 477
12.6 Bibliographic Notes 483
Problems 483

Appendix A: Complements 487

A.1 Projection Theorem 488
A.2 Wold Decomposition 490
A.2.1 Singularity and Regularity 490
A.2.2 Wold Decomposition 490
A.2.3 Causality 492
A.2.4 Invertibility 493
A.2.5 Best Linear Predictor 493
A.2.6 Szegö-Kolmogorov Formula 494
A.2.7 Ergodicity 494
A.2.8 Fractional Brownian Motion 496
A.3 Bibliographic Notes 497

Appendix B: Solutions to Selected Problems 499

Appendix C: Data and Codes 557

References 559

Topic Index 573

Author Index 577

PREFACE

This book aims to provide an overview of time series analysis to a wide audience of students, practitioners, and scientists from different fields. It is intended as an introductory text on the vast time series subject. Consequently, it focuses on methodologies and techniques rather than theoretical results. This book strives to provide a working knowledge of the practical applications of time series methods. However, it does not attempt to cover all of the relevant topics in this field.

Consistent with this objective, the first chapter reviews the main features of a number of real-life time series arising in different fields, including finance, hydrology, meteorology, sociology, and politics, among others. At the same time, this chapter refreshes some basic knowledge on statistical distributions. Furthermore, Chapter 1 provides an overview of the time series modeling fundamentals by taking a first look to concepts such as stationarity, non-stationarity, parametric and nonparametric approaches, and whiteness tests. Further readings are suggested in a bibliographical notes section. This chapter ends with a number of proposed exercises.

Chapter 2 addresses linear processes, one of the fundamental concepts of time series analysis. It reviews the different representations of linear time series and discusses essential topics such as stationarity, invertibility, and causality. One interesting feature of this chapter is that it covers both short and long-memory linear processes.

State space models are discussed in Chapter 3. Apart from being another representation of linear processes, state space systems provide several practical tools for estimating, smoothing, and predicting time series models. Moreover, they are useful for handling nonstationarity and missing data problems. In particular, we discuss applications of state space techniques to parameter estimation in Chapter 4, to nonstationary processes in Chapter 8, to seasonal models in Chapter 9, and to missing values in Chapter 11.

As discussed in Chapter 4, time series analysis can be carried out from a time-domain or from a spectral domain. However, in practice one usually combine both approaches. For example, spectral analysis is fundamental for modeling time series exhibiting seasonal patterns. On the other hand, Chapter 5 provides an overview of the realm of methodologies for estimating time series models. It begins with some essential notions about specifying appropriate statistical models, including the concepts of parsimony and information criteria. Afterwards, it proceeds to discuss several estimation techniques such as maximum likelihood, Whittle approach, Bayesian estimation, along with an extensive list of specific methods developed for long-memory models. This chapter also reviews techniques for carrying out statistical inferences about the fitted models. A number of simulations and practical applications complete this chapter.

Nonlinear time series are addressed in Chapter 6. This is an important subject that provides tools for modeling time series data which do not fit into the linear processes category discussed in Chapter 2. For example, most of financial time series are better described by heteroskedastic nonlinear models. This chapter also discusses techniques for assessing financial risk and modeling time series with complex structure. For instance, these data can be modeled via threshold processes that allow for the treatment of time series undergoing regime changes.

The fundamental topic of forecasting time series is discussed in Chapter 7. Optimal prediction with finite and infinite past is reviewed in the context of linear and nonlinear processes. Calculating one-step and multistep predictors and procedures for establishing prediction bands are described.

Chapter 8 examines the subject of nonstationary time series models. Since many real-life time series do not exhibit stationary behavior, this chapter discusses methods for handling nonstationarity, including autoregressive integrated moving-average models and locally stationary processes. While the former assume that the series results from the integration of an stationary process, the later assume that the parameters of the model change smoothly over time. In order to account for abrupt changes, this chapter also discusses methods for treating structural breaks.

Seasonal patterns are present in time series data as diverse as Internet traffic, sales revenues, and transportation. Methods for analyzing these time series are described in Chapter 10, including models based on seasonal differentiation. This chapter illustrates the finite sample performance of these techniques via Monte Carlo simulations and a real-life data application.

Time series regression methods are reviewed in Chapter 9. These techniques allow for the modeling of time series data affected by some underlying trend or exogenous variables. For example, these trends can be described by a polynomial structure. Additionally, harmonic regression can be a useful tool for handling seasonality.

Data gaps and outliers are frequent problems in time series analysis. These topics are reviewed in Chapter 11. The effects of incomplete data on parameter estimates and predictors is studied in this chapter. The problem of defining an appropriate likelihood function in the context of incomplete data is discussed along with state space techniques for obtaining estimates. On the other hand, methods to account for outliers are also addressed in this chapter.

Most time series models assume that the observations are normally distributed. In practice, however, many real-life time series do not fit this assumption. For example, the time series may correspond to count data or positive observations. Consequently, Chapter 12 provides an overview of several methods for estimating and predicting non-normal time series.

Some basic knowledge of calculus is required for understanding most methods discussed in this book. Apart from this, the text intends to be self-contained in terms of other more advanced concepts. In particular, Appendix A provides some specific technical details about fundamental concepts in time series analysis. On the other hand, Appendix B contains solutions to some of the proposed problems. Finally, Appendix C provides information about the data and the computing codes used in this book. It is worth noting that similarly to Chapter 1, every chapter of this book ends with a bibliographical notes section and a list of proposed problems. Supplementary material for this book can be also found on the Book Companion Site through the books page on wiley.com.

W. PALMA

Santiago, Chile
January, 2015

ACKNOWLEDGMENTS

I wish to express my gratitude to Steve Quigley for encouraging me to write this book and for many valuable comments on its contents. I would like to thank Jon Gurstelle and Sari Friedman, and the editorial staff at John Wiley & Sons for their continuous support and for making the publication of this book possible. I am also indebted to many coauthors and colleagues, some of the results described in this text reflect part of that fruitful collaboration. I am also grateful of the support from the Department of Statistics and the Faculty of Mathematics at the Pontificia Universidad Católica de Chile. Several chapters of this book evolved from lecture notes for undergraduate and graduate courses on time series analysis. I would like to thank many students for constructive remarks on the text and for trying out the proposed exercises. The financial support by the Ministry of Economy, Development, and Tourism's Millennium Science Initiative through grant IC120009, awarded to The Millennium Institute of Astrophysics, MAS, and from Fondecyt Grant 1120758 are gratefully acknowledged.

W.P.

ACRONYMS

ACF	autocorrelation function
AIC	Akaike's information criterion
ANOVA	analysis of variance
APARCH	asymmetric power autoregressive conditionally heteroskedastic
AO	additive outlier
AR	autoregressive
ARCH	autoregressive conditionally heteroskedastic
ARFIMA	autoregressive fractionally integrated moving-average
ARMA	autoregressive moving-average
BLUE	best linear unbiased estimator
DFA	detrended fluctuation analysis
DWT	discrete wavelet transform
EGARCH	exponential generalized autoregressive conditionally heteroskedastic
fBm	fractional Brownian motion

FFT	fast Fourier transform
fGn	fractional Gaussian noise
FI	fractionally integrated
FIGARCH	fractionally integrated generalized autoregressive conditionally heteroskedastic
FIEGARCH	fractionally integrated exponential generalized autoregressive conditionally heteroskedastic
FN	fractional noise
GARCH	generalized autoregressive conditionally heteroskedastic
GARMA	Gegenbauer autoregressive moving-average
HDD	heating degree day
HTTP	hyper text transfer protocol
IG	inverse gamma distribution
i.i.d.	independent identically distributed
IM	intermediate memory
IO	innovative outlier
LM	long memory
LMGARCH	long-memory generalized autoregressive conditionally heteroskedastic
LMSV	long-memory stochastic volatility
LS	level shift
LSARFIMA	locally stationary autoregressive fractionally integrated moving-average
LSE	least squares estimator
MA	moving-average
MCMC	Markov chain Monte Carlo algorithm
ML	maximum likelihood
MLE	maximum-likelihood estimator
MSE	mean-squared error
MSPE	mean-squared prediction error
PACF	partial autocorrelation function
PM	particulate matterl
QMLE	quasi-maximum-likelihood estimator
R/S	rescaled range statistics
SARFIMA	seasonal autoregressive fractionally integrated moving-average

SD	standard deviation
SETAR	self-exciting threshold autoregressive
SM	short memory
SV	stochastic volatility
TAR	threshold autoregressive
TC	temporary change
VaR	value at risk
WN	white noise
ZIM	zero-inflated model
ZINB	zero-inflated Negative Binomial model

ABOUT THE COMPANION WEBSITE

This book is accompanied by both Instructor and Student companion websites, which are available via the book's page on www.wiley.com.

The Instructor and Student websites includes:

- Data Sets
- R script

CHAPTER 1

INTRODUCTION

A time series is a collection of observations taken sequentially in time. The nature of these observations can be as diverse as numbers, labels, colors, and many others. On the other hand, the times at which the observations were taken can be regularly or irregularly spaced. Moreover, time can be continuous or discrete. In this text, we focus primarily on describing methods for handling numeric time series observed at regular intervals of time. Note, however, that many nonnumeric data can be readily transformed to numeric. For instance, data concerning an election between candidate A and candidate B can be described by a numeric variable taking the value 0 for candidate A and 1 for candidate B. However, data observed at irregular time intervals are more difficult to handle. In this case, one may approximate the actual time to the closest integer value and still use the methodologies for handling regularly spaced series. If this approach does not provide adequate results, there are a number of more advanced techniques to treat those types of data. Another common problem in time series analysis is missing observations. In this case, the collected data display irregularly spaced observation times. There are special techniques for handling this problem and some of them are discussed in Chapter 11.

Time Series Analysis. First Edition. Wilfredo Palma.

This introductory chapter presents a number of real-life time series data examples as well as provides a general overview of some essential concepts of the statistical analysis of time series, such as random variable, stochastic process, probability distribution and autocorrelation, among others. These notions are fundamental for the statistical modeling of serially dependent data.

Since this text attempts to reach a large audience interested in time series analysis, many of the more technical concepts are explained in a rigorous but simple manner. Readers interested in extending their knowledge of some particular concept in time series analysis will find an extensive list of references and a selected bibliographical discussion at the end of each chapter.

1.1 TIME SERIES DATA

Let us denote by $\{y_t\}$ a time series where t denotes the time at which the observation was taken. Usually, $t \in \mathbb{Z}$, where $\mathbb{Z} = \{\ldots, -2, -1, 0, 1, 2, \ldots\}$ is the set of positive and negative integer values. In practice, however, only a finite stretch of data is available. In such situations, we can write the time series as $\{y_1, y_2, \ldots, y_n\}$. A time series $\{y_t\}$ corresponds to a *stochastic process* which in turn is composed of *random variables* observed across time. Both concepts are explained in detail later in this chapter.

Several examples of real-life time series data are presented in the following subsections. These data illustrations come from fields as diverse as, finance, economic, sociology, energy, medicine, climatology, and transport, among many others. Apart from exhibiting the time series, we describe their main features and some basic data transformations that help uncovering these characteristics.

1.1.1 Financial Data

Finance is a field where time series arises naturally from the evolution of indexes and prices. In what follows, we present two basic examples, the evolution of a well-known stock index and its volume of stock transactions.

Standard & Poor's Stock Index. Figure 1.1 shows the logarithm of the S&P500 daily stock index for the period from January 1950 to January 2014. Note that this index seems to increase with time, but there are some downward periods commonly denoted as *bear markets*. In order to study these indices, it is customary in finance to consider the logarithm return, which is defined as

$$r_t = \log \frac{P_t}{P_{t-1}} = \log P_t - \log P_{t-1},$$

where P_t denotes the price or the index value at time t. These returns are displayed in Figure 1.2. Observe the great drop in returns experienced on October 1987 and the abrupt changes or great *volatility* during 2009.

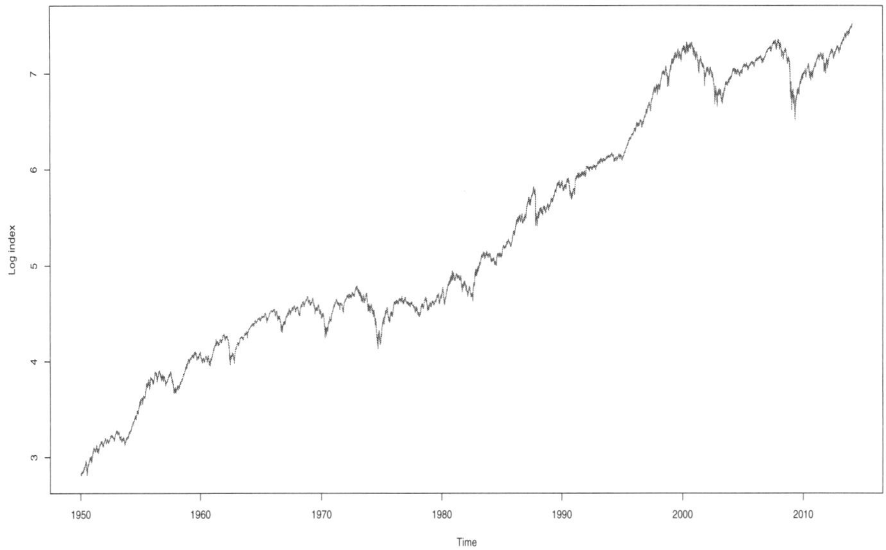

Figure 1.1 *S&P500 daily stock log index, January 1950 to January 2014.*

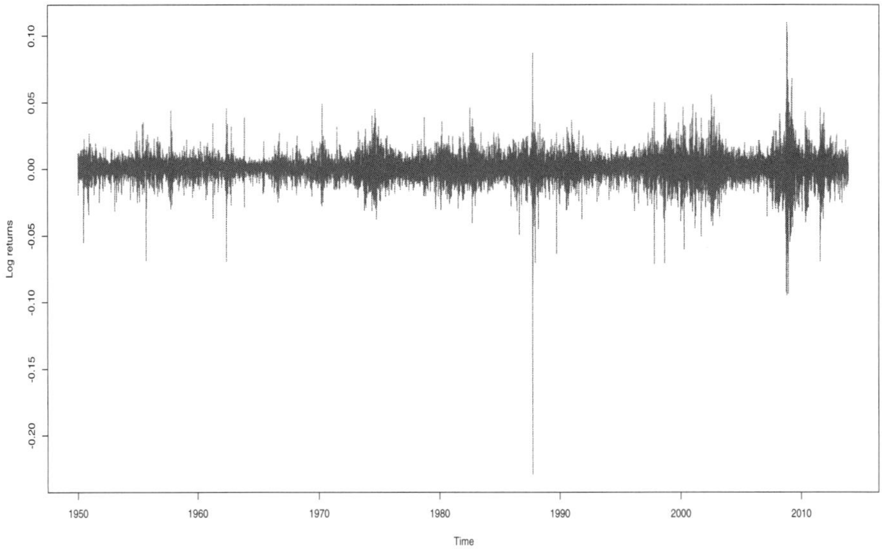

Figure 1.2 *S&P500 daily log returns, January 1950 to January 2014.*

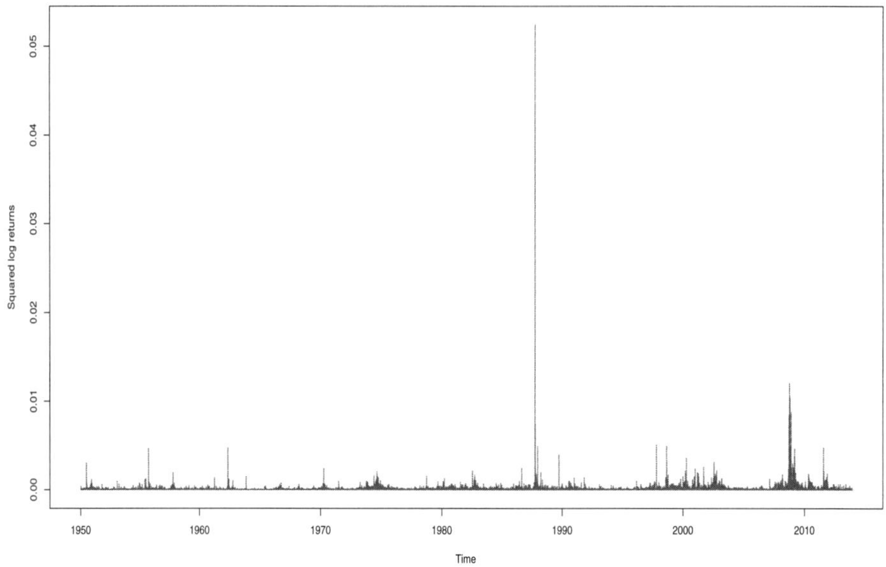

Figure 1.3 *S&P500 daily square log returns, January 1950 to January 2014.*

Another look at the volatility is shown in Figure 1.3 where the squared returns, r_t^2, are plotted. From this graph, the high volatility of this stock index is evident during these periods.

Financial time series possess specific features, such as those indicated above. Consequently, Chapter 6 describes methodologies for handling this type of data. These time series can be analyzed by means of the so-called conditionally heteroskedastic processes or stochastic volatility models, among others.

Volume of Transactions. As another example of financial data, the daily volume of transactions of the S&P500 stocks is displayed in Figure 1.4. Observe that this series exhibits an upward trend up to 2009.

On the other hand, Figure 1.5 depicts a logarithm transformation of the above time series. Note that the variance of the data across time is now more stabilized, emerging a seemingly overall upward trend, excepting the values after 2009 and some other periods.

These transaction volume data can be considered as an example of a non-Gaussian time series. In particular, these observations are positive counts. Specific methods for modeling and predicting non-Gaussian data are described in Chapter 12.

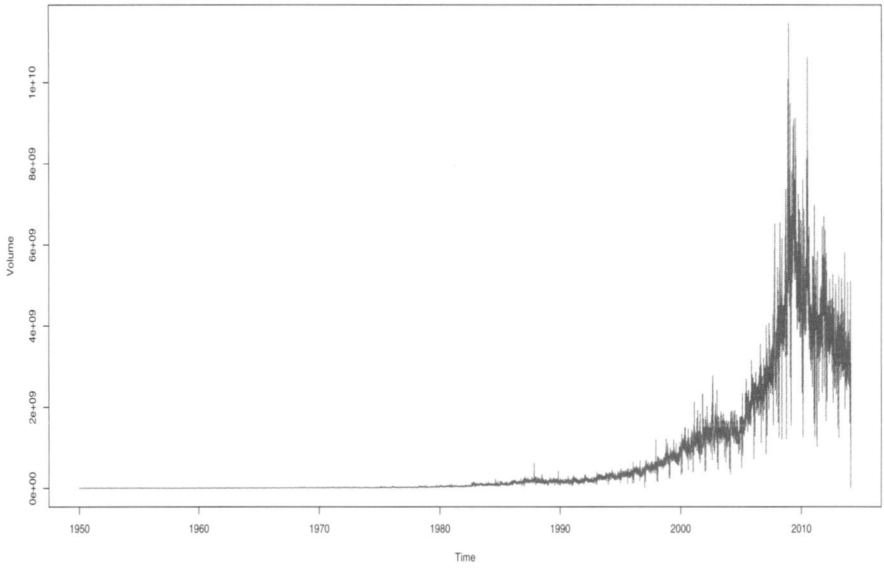

Figure 1.4 *S&P500 daily volume of transactions, January 1950 to January 2014.*

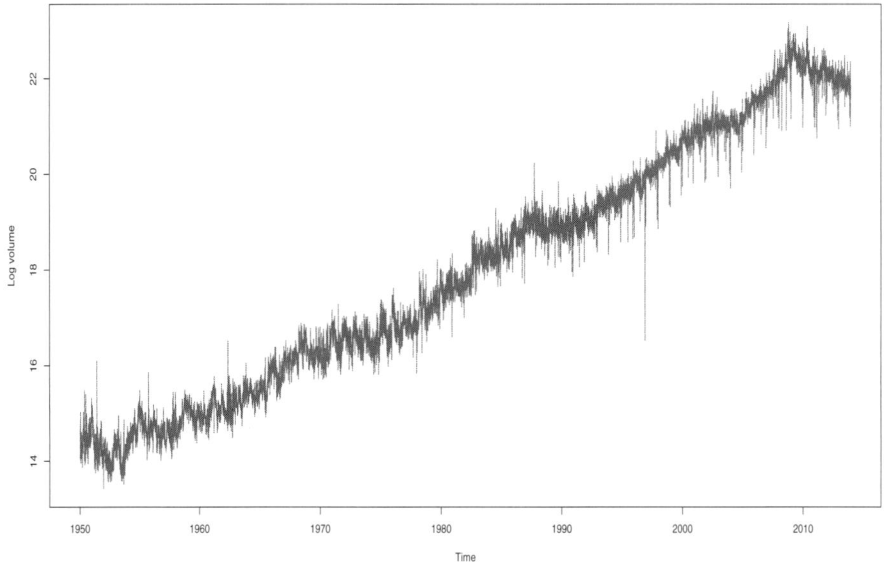

Figure 1.5 *S&P500 daily log volume of transactions, January 1950 to January 2014.*

1.1.2 Economic Data

Figure 1.6(a) exhibits the monthly US employment in the arts, entertainment and recreation section for the January 1990 to December 2012, measured in thousands of persons. On the other hand, Figure 1.6(b) shows the logarithm transformation of these data. Notice that this data transformation seems to stabilize the variance of the series across time. On both panels, however, a seasonal pattern and an upper trend are evident.

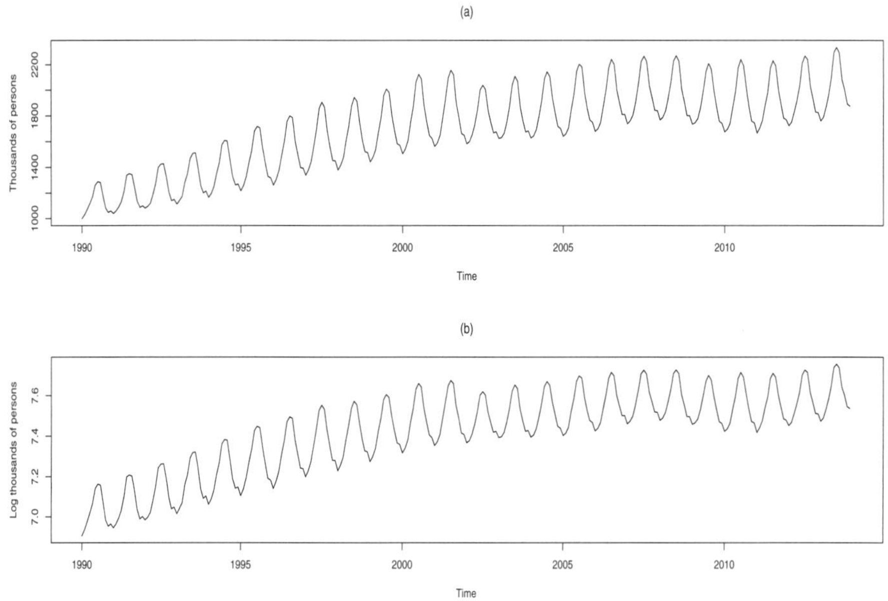

Figure 1.6 *US employment arts, entertainment, and recreation, January 1990 to December 2012.*

1.1.3 Hydrological Data

In hydrology, time series data is usually related to the collection of river flows observations though the years. For example, the yearly minimum water levels of the Nile river measured at the Roda gauge is a well-known time series exhibiting high levels of serial dependency. These measurements, available from Statlib, www.stat.cmu.edu, are displayed in Figure 1.7 spanning a time period from 622 A.D. to 1921 A.D.

Notice that there are several blocks of seemingly repeated values. That is, consecutive years having exactly the same minimum water level. Since the observations are specified by four digits, these repetitions are probably the result of a lack of new information. The analysis of this time series data

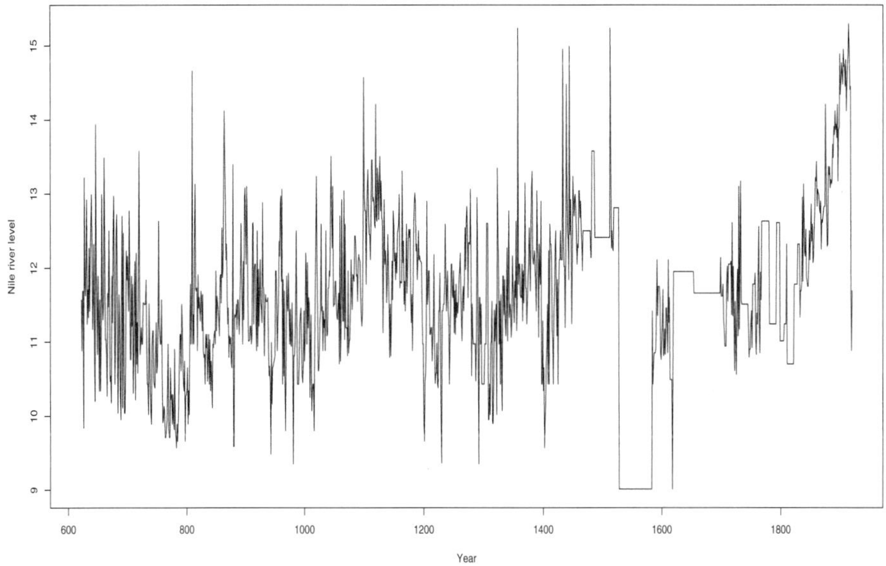

Figure 1.7 *Nile river yearly minimum level at the Roda gauge, from 622 A.D. to 1921 A.D.*

also indicates that the first 100 observations seem to have a different serial dependence structure, suffering a structural break phenomenon.

A detailed analysis of this historically important hydrological time series is proved in Chapter 8, where the changes in the serial dependence structure of these observations are modeled. The analysis of these hydrological data was crucial in the formal study of the so-called long-memory processes reviewed in Chapter 2.

1.1.4 Air Pollution

Figure 1.8 exhibits a daily index that measures the particulate matter of diameter less than 2.5 μ in Santiago, Chile, for the period 1989–1999, commonly referred to as PM2.5. A log-transformed data is shown in Figure 1.9.

These measurements indicate the level of air pollution in certain city or region. Note the seasonal behavior of this series, due to the effects of climate conditions across the year. In winter, the PM2.5 level increases dramatically. On the other hand, it appears that there is downward trend in the series, indicating an improvement of the air quality during that period. In order to stabilize the variance exhibited by this data, a logarithmic transformation has been made and the resulting series is shown in Figure 1.9. A possible downward trend is now more clear in the transformed data.

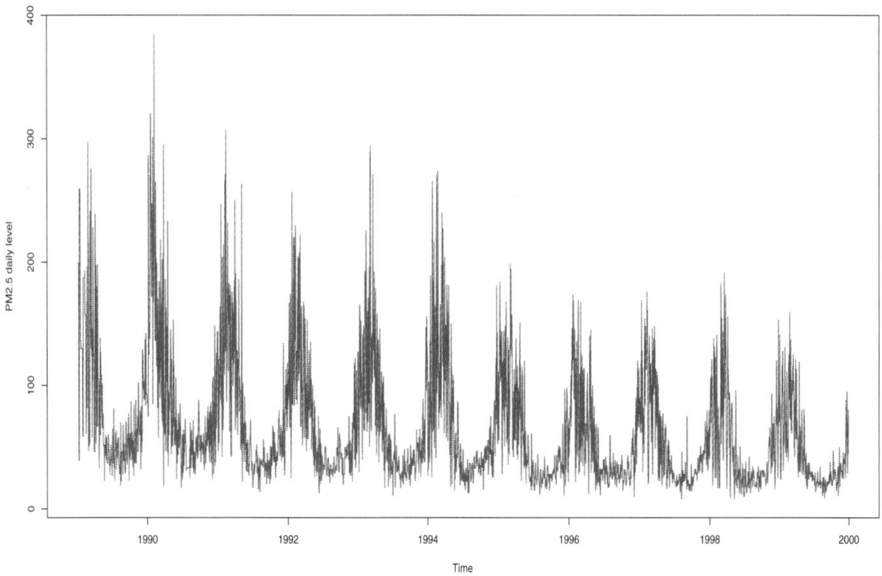

Figure 1.8 *Air pollution data: daily PM2.5 measurements at Santiago, Chile, 1989 - 1999.*

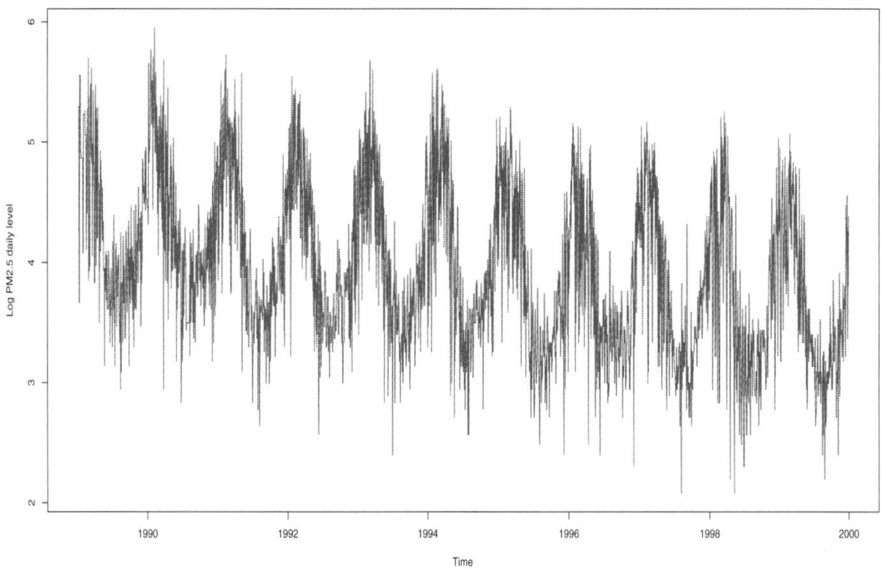

Figure 1.9 *Air pollution data: log daily PM2.5 measurements at Santiago, Chile, 1989 - 1999.*

1.1.5 Transportation Data

Figure 1.10 shows the number of monthly passenger enplanements for the period from January 2004 to December 2013. These observations correspond to the number of passenger boarding an airplane in the United States in a given month. Note the seasonal behavior of this series derived from the annual cycle of winter and summer seasons. Besides, it seems that there was a drop on passenger enplanements around 2009, revealing a plausible effect of the financial crisis of that year. In this situation, it is possible that the process was affected by a structural break or structural change. Methodologies for handling these situations will be discussed in Chapter 8.

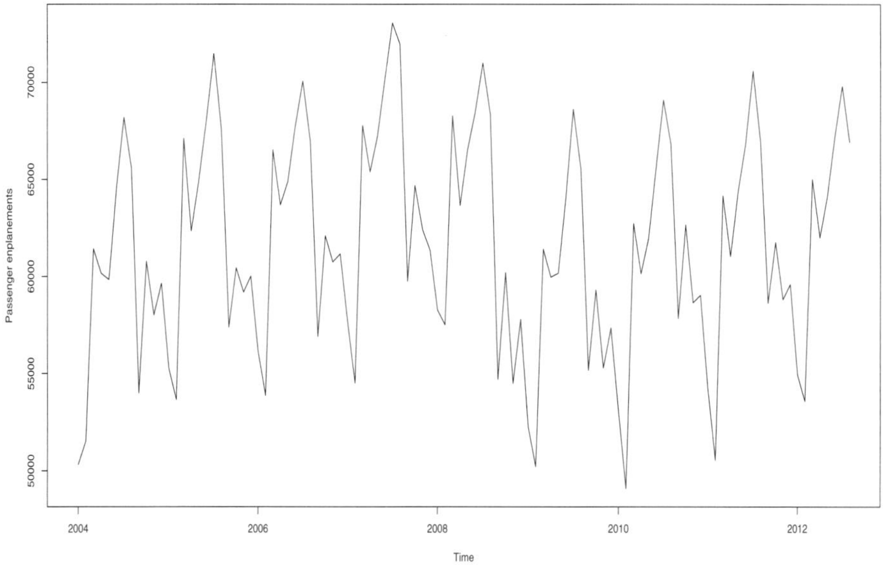

Figure 1.10 *Passenger enplanements from January 2004 to January 2013.*

1.1.6 Biomedical Time Series

The annual global number of dengue cases during the period 1955 - 2008 is depicted in Figure 1.11. In order to emphasize that these data correspond to counts, the values are shown as bars. Note the increase of cases by the end of this period, reaching very high values around the year 2000 and then a sharp decay by the end of that decade. Apart from the analysis of the evolution of diseases, there are several others applications of time series methods to biomedical studies. Techniques for modeling count data is described in Chapter 12.

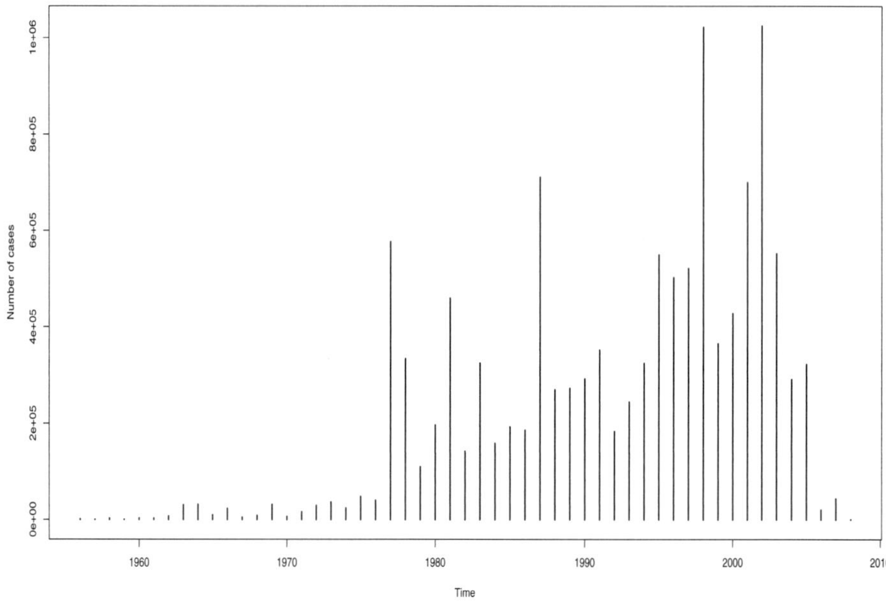

Figure 1.11 *Annual global number of dengue cases during the period 1955 - 2008.*

1.1.7 Sociological Data

The results from a series of monthly United Kingdom voting intention surveys for the period June 1984 to March 2012 are shown in Figure 1.12. For simplicity, we have only plotted the vote intentions for the Conservative Party and the Labor Party.

The heavy line indicates the Labor Party voting intention, while the dotted line corresponds to the voting intention of the Conservative Party. Observe the seemingly mirror effect in the behavior of these two series given that these two political parties historically concentrate a large percentage of vote intentions.

Furthermore, for the period from 1993 to 2005, there is large distance between the voting intentions of the two parties. During this period, the Labor Party shows a higher level of voting intention than the Conservative Party. The opposite is true for the following period, from 2006 to 2010.

These time series are additional examples of cases where the data is not necessarily Gaussian and specific methods must be developed for handling them. Some of these technique are reviewed in Chapter 12, including, for instance, the conditional distribution models.

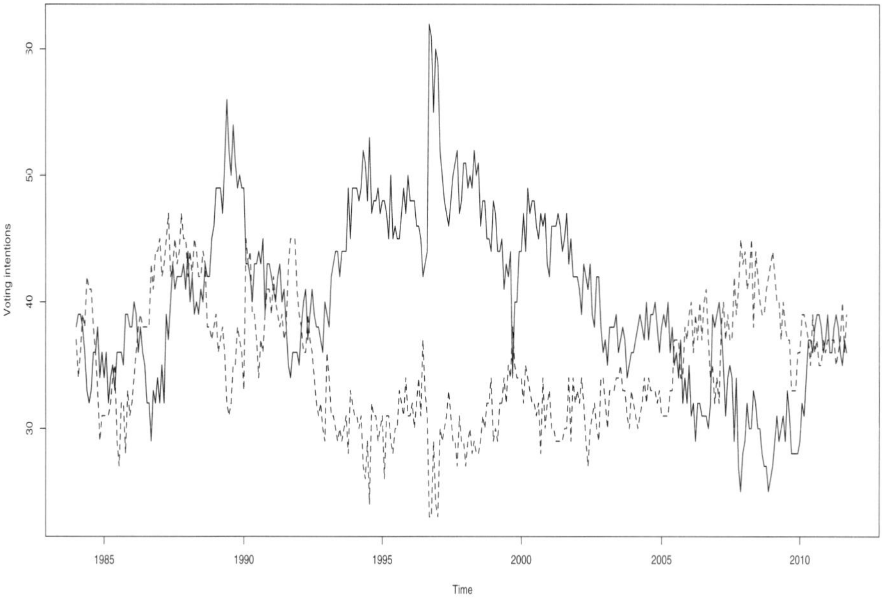

Figure 1.12 *Monthly United Kingdom voting intentions, June 1984 to March 2012. Heavy line, Labor Party; dotted line: Conservative Party.*

1.1.8 Energy Data

A time series consisting of monthly heating degree day (HDD) measurements is shown in Figure 1.13. HDD values are indicative of the amount of energy required to heat a building, and they result from from measurements of outside air temperature. The heating requirements for a particular structure at a specific location are usually considered to be directly proportional to the number of HDD at that location.

These series are important in the analysis of energy demand, and they can be measured at different specific locations such as cities or at a county level.

In this case, the data correspond to the global measurements for 27 European Union countries from January 1980 to May 2010. As in other series previously presented, this time series displays a clear seasonal behavior, expected from the different heating requirements within a year.

This series displays a clear seasonal behavior and no upper or downward trends are evident, as compared to the previous example of employment data. In Chapter 9, we discuss methodologies for modeling this type of seasonal time series and apply the techniques to the monthly HDD measurements.

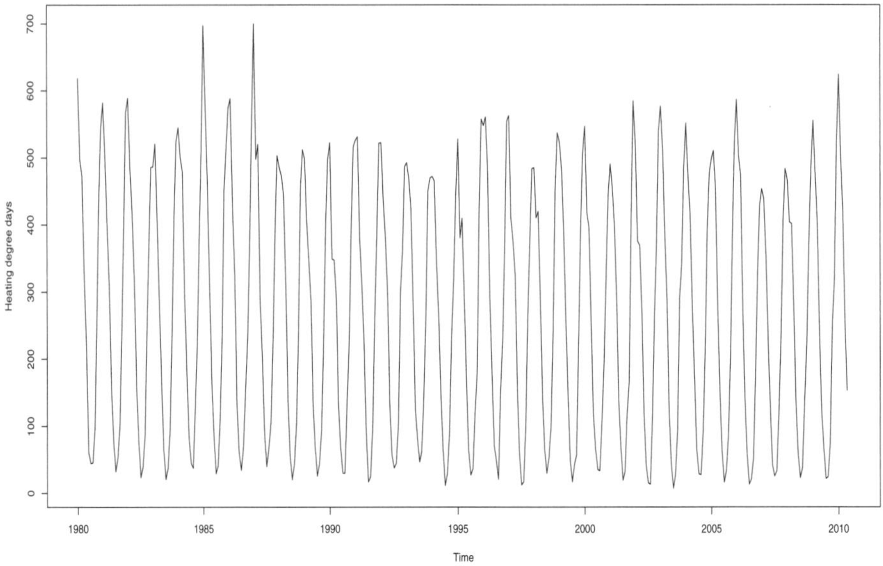

Figure 1.13 *Monthly heating degree days (HDDs) measurements for 27 countries of the European Union, for the the period January 1980 to May 2010.*

1.1.9 Climatic Data

Time series play an important role in the study of climate, allowing to model actually observed values in the historical records or reconstructing unobserved data. For example, paleoclimatic studies usually rely on the reconstruction of climatic conditions by means of tree rings, mineral sediments, and other related data.

Temperature Reconstruction. Figure 1.14 exhibits a reconstruction of annual temperatures in north-central China from 1470 A.D. to 2002 A.D., based on drought/flood index and tree-ring records. Observe the apparent nonstationary behavior of the data, exhibiting heteroskedasticity and periods with upward and downward trends. However, detailed studies must be conducted on these data to distinguish whether there are *deterministic* trends or they are just random or stochastic. On the other hand, reconstructions of precipitation conditions for the same region and period of time are displayed in Figure 1.15. Note the increasing variability of the series.

Mineral Deposits. Figure 1.16 displays a 2,650-year centered time series of speleothem cave deposit data. This series is composed by stalagmite layer thickness observations taken at Shihua Cave, Beijing, China, from 665 B.C. to 1985 A.D., see Appendix C for details about these data. A logarithm transformation of these data is exhibited in Figure 1.17.

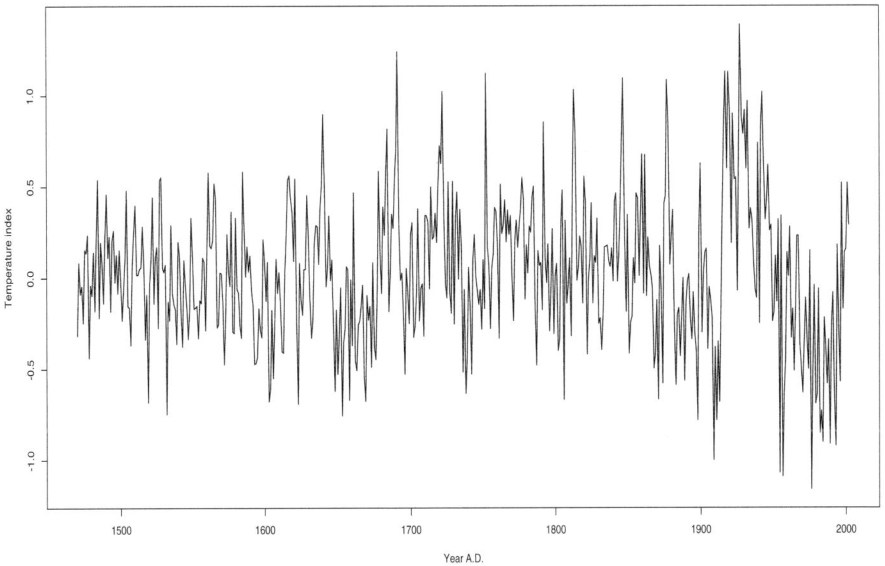

Figure 1.14 *North-Central China temperature, 1470 A.D. to 2002 A.D.*

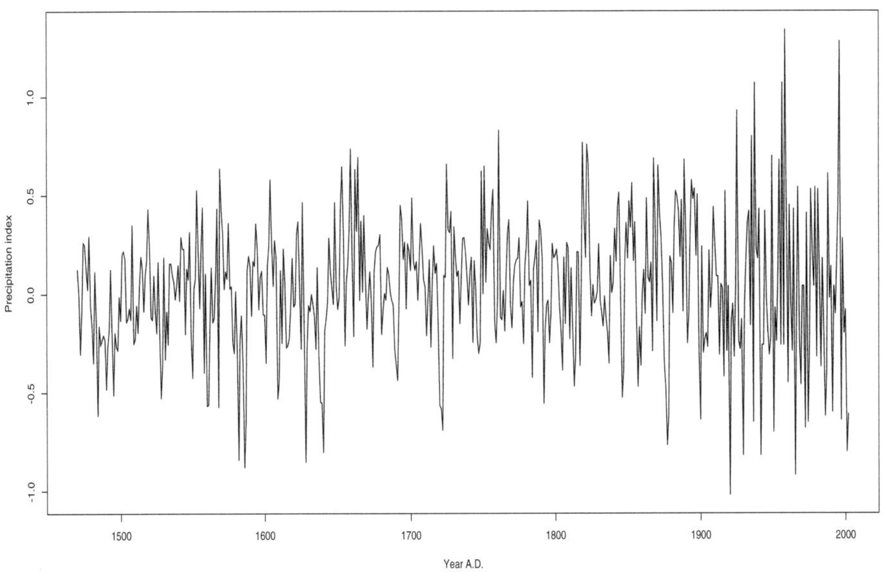

Figure 1.15 *North-Central China precipitation, 1470 A.D. to 2002 A.D.*

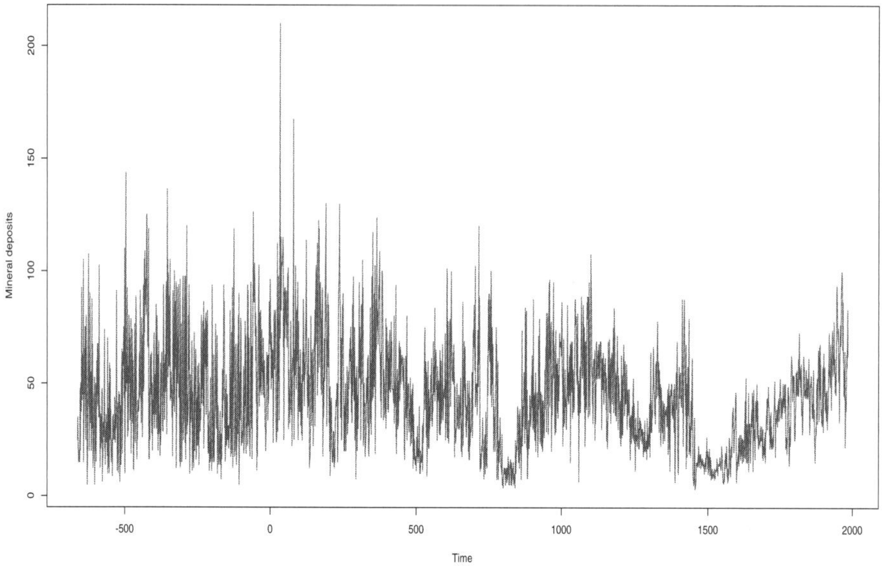

Figure 1.16 *Speleothem cave deposit data.*

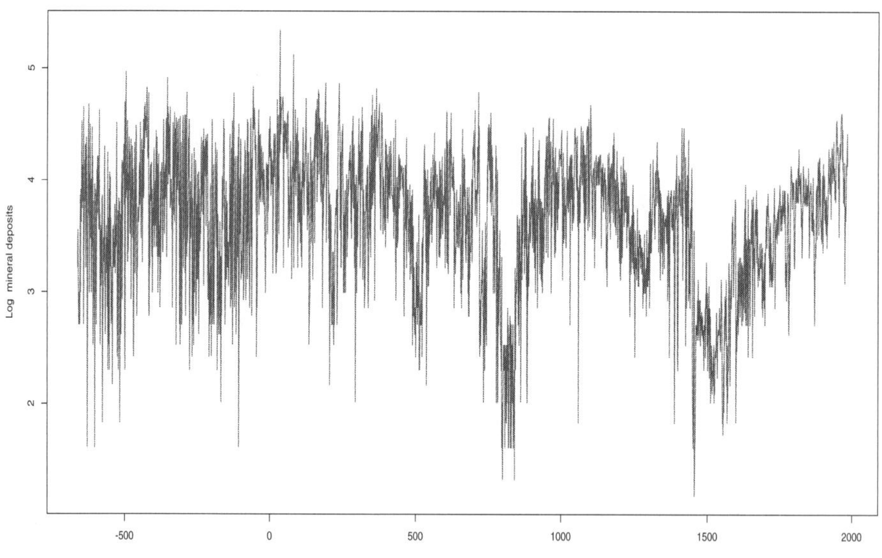

Figure 1.17 *Log speleothem cave deposit data.*

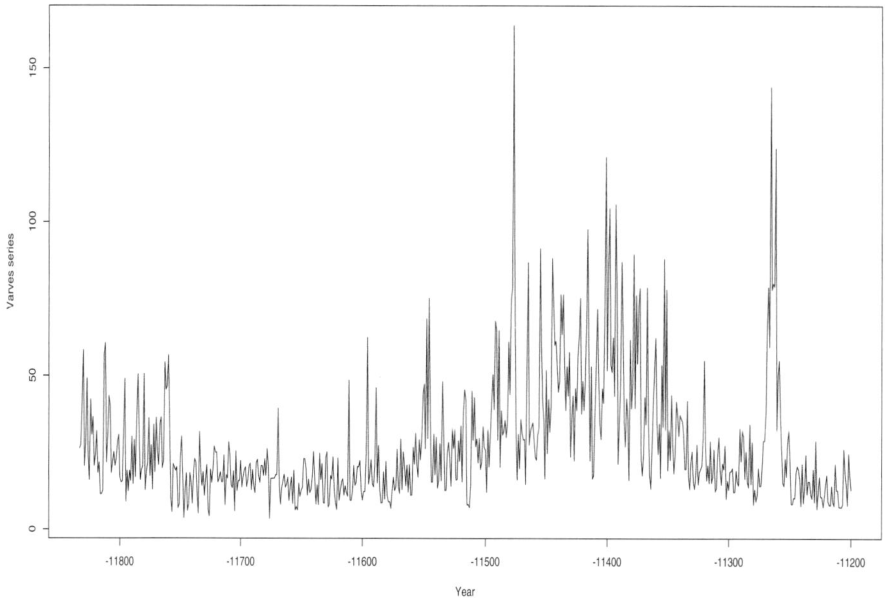

Figure 1.18 *Glacial varves time series data.*

Glacial Varves Data. A *varve* is an annual layer of sediment or sedimentary rock. This word derives from the Swedish word *varv* which means layers or circles. Figure 1.18 exhibits the thicknesses of the yearly varves at one location in Massachusetts for the period 11,833 B.C. to 11,200 B.C., see Appendix C for further details about these data.

Tree Ring Data. Figure 1.19 displays a time series consisting of annual *Pinus longaeva* tree ring width measurements at Mammoth Creek, Utah, from 0 A.D. to 1989 A.D. An analysis of these data is described in Section 8.4.6.

As discussed above, real-life time series may exhibit several features such as trends, seasonality, and heteroskedasticity, among others. These aspects can be considered as *nonstationary* behavior, where *stationarity* is associated to, for example, constant mean and variance across time.

Given that most time series methodologies are mainly focussed on stationary data, a number of techniques have been developed to transform a real-life time series into a stationary one. However, before reviewing some of these procedures, it is necessary to introduce some fundamental probabilistic and statistical concepts such as random variable, probability distribution, and autocorrelation, among others.

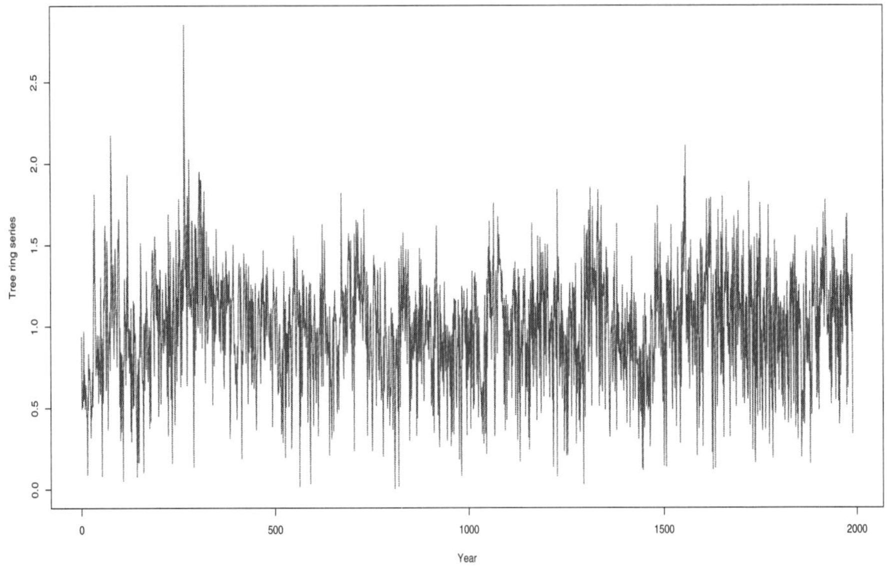

Figure 1.19 *Tree ring data. Mammoth Creek, Utah, from 0 A.D. to 1989 A.D.*

1.2 RANDOM VARIABLES AND STATISTICAL MODELING

Time series are stochastic processes which, in turn, correspond to a sequence of random variables. Loosely speaking, a random variable is a function between a *sample space* Ω containing all possible outcomes and the set of real numbers denoted by $\mathbb{R}$. A number of examples of real-valued random variables are presented next. In some cases, the random variables take values on a discrete set such as $\{0, 1\}$, and in other cases, the random variable is continuous, taking values on $\mathbb{R}$ or $\mathbb{R}_+$, the set of positive real numbers.

■ EXAMPLE 1.1

A very simple illustration of random variable is the tossing of a coin. In this case, the two possible outputs are ω_1 = Heads and ω_2 = Tails. Thus, the sample space is composed of the events ω_1 and ω_2, that is, $\Omega = \{\omega_1, \omega_2\}$. We can write the random variable $x : \Omega \to \{0, 1\}$, where $x(\omega_1) = 1$ and $x(\omega_2) = 0$. Now, under the assumption of a *fair coin*, we can establish the probability distribution of the random variable x as $\mathbb{P}(x = 1) = \frac{1}{2}$ and $\mathbb{P}(x = 0) = \frac{1}{2}$. More generally, the situation described may represent the choice between two options A and B, where the probability assigned to each case can be specified as $\mathbb{P}(x = A) = p_A$ and $\mathbb{P}(x = B) = p_B$. This is called the *Bernoulli* distribution.

EXAMPLE 1.2

A useful extension of the previous example is the repetition of the coin tossing or considering multiple selection of options A and B. If n denotes the number of coin tosses or the number of selections between options A and B, then the sample space to describe this situation is the *product space* $\Omega_n = \Omega \times \cdots \times \Omega = \Omega^n$. The random variable now can be denoted as $x_n : \Omega_n \to \{0,1\}^n$, which is called *Binomial* distribution. Consider $n = 3$, a particular case could be, for instance, $x_3(\omega_1, \omega_2, \omega_3) = (1,0,1)$. The probability of this event is $\mathbb{P}[x_3 = (1,0,1)] = p_A^2 p_B$. More generally, the probability distribution of this random variable is given by

$$\mathbb{P}(x = k) = \binom{n}{k} p^k (1-p)^{n-k},$$

for $k = 0, 1, 2, \ldots, n$.

EXAMPLE 1.3

In many situations, the observations that we are interested in are *continuous*. For example, consider the returns from a financial instrument. A basic question that one may be interested is: What is the probability that my investment returns at least 8.5% annually. In this situation, if the random variable x represents the annual percentage return, the event of interest is $\omega = \{x \geq 8.5\}$. The probability of this event can be expressed as $\mathbb{P}(x \geq 8.5) = 1 - \mathbb{P}(x < 8.5) = 1 - F(8.5)$, where F is called the *distribution function* of x.

$$F(x) = \int_{-\infty}^{x} f(u)du,$$

where, f is the *density* of the distribution function F. For instance, if we assume that the returns are normally distributed, then the density is given by

$$f(x) = \frac{1}{\sqrt{2\pi}\sigma} e^{-\frac{1}{2}(x-\mu)^2/\sigma^2},$$

with $\mu = E(x)$ and $\sigma^2 = \text{Var}(x)$.

Given that financial returns commonly do not follow a Gaussian distribution, see, for example, Figures 1.20 and 1.21, some extensions of the normal random variables have been considered in the literature. One of these generalizations is the so-called *exponential power* family of random variables, which has density

$$f(x) = \frac{\beta}{2\alpha\Gamma(1/\beta)} e^{-\frac{1}{2}|x-\mu|^\beta/\sigma^\beta},$$

where the mean of x is μ and the variance is now

$$\mathrm{Var}(x) = \alpha^2 \frac{\Gamma(3/\beta)}{\Gamma(1/\beta)},$$

with Γ denoting the *gamma function*. In Figure 1.20, observe that the returns of the S&P500 index are more concentrated around zero and more dispersed towards the tails than a Gaussian random variable. This behavior is further evidenced by Figure 1.21, which shows that the quantile of the returns are far from those corresponding to a normal distribution, cf. 1.21(a).

■ EXAMPLE 1.4

An important distribution in time series analysis is the *multivariate normal distribution*. Given the series $\{y_1, y_2, \ldots, y_n\}$, we can write the random vector $X = (y_1, y_2, \ldots, y_n)$. It is said that X is a multivariate Gaussian random variable, with mean μ and variance-covariance matrix Σ, denoted as $N(\mu, \Sigma)$, if the density function is

$$f(X) = (2\pi)^{-n/2} |\Sigma|^{-n/2} e^{-\frac{1}{2}(X-\mu)'\Sigma^{-1}(X-\mu)}.$$

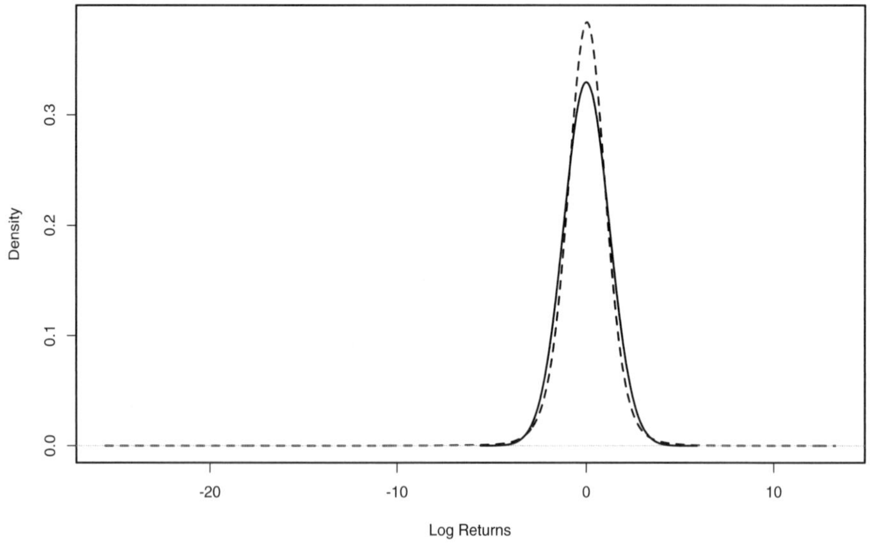

Figure 1.20 *Distribution of S&P500 daily log returns, January 1950 to January 2014. Dotted line, data distribution; heavy line, normal distribution.*

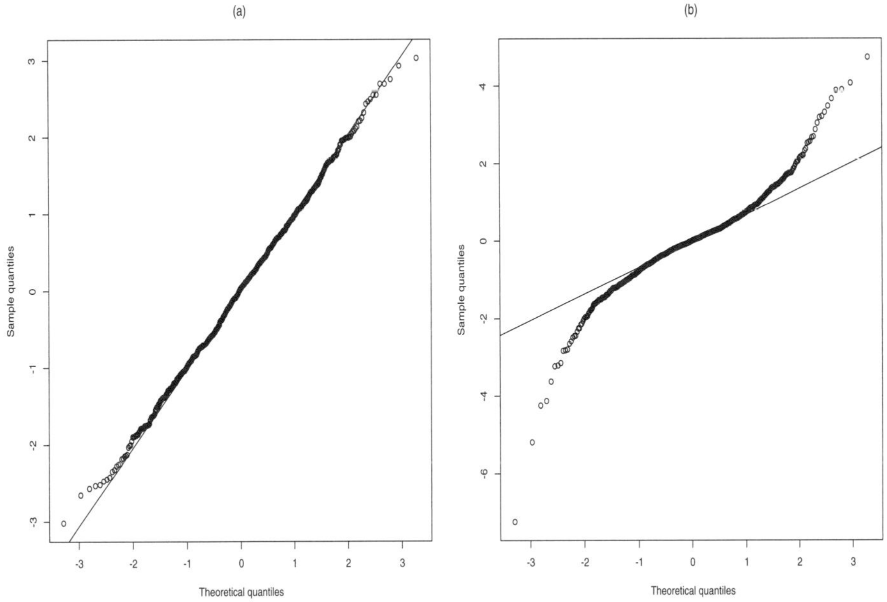

Figure 1.21 *Quantiles of S&P500 daily log returns, January 1950 to January 2014. (a) Normal sample quantile-quantile plot. (b) Data quantile-quantile plot.*

Observe that the elements of Σ correspond to the covariances among the components of the random vector X, that is, $\Sigma_{t,s} = \text{Cov}(y_t, y_s)$. If the time series is stationary, there exists a function $\gamma(\cdot)$ such that $\gamma(|t-s|) = \text{Cov}(y_t, y_s)$. In this case, the variance-covariance matrix can be written as

$$\Sigma = \begin{bmatrix} \gamma(0) & \gamma(1) & \gamma(2) & \cdots & \gamma(n-1) \\ \gamma(1) & \gamma(0) & \gamma(1) & \cdots & \gamma(n-2) \\ \vdots & \vdots & \vdots & & \vdots \\ \gamma(n-1) & \gamma(n-2) & \gamma(n-3) & \cdots & \gamma(0) \end{bmatrix}.$$

Note that if $y = AX$, where A is an $m \times n$ matrix and X is a multivariate normal random variable with mean μ and variance-covariance matrix Σ, then y corresponds to a multivariate normal random variable with mean $A\mu$ and variance-covariance matrix $A\Sigma A'$.

■ EXAMPLE 1.5

The *Chi-squared* or χ^2_ν distribution is quite commonly found in statistics. Here the parameter ν indicates the degree of freedom of the distribution. If x follows a standard normal distribution, then its square corresponds to a χ^2_1 random variable. Besides, if $x_1, x_2, \ldots, x_n$ is a sequence of inde-

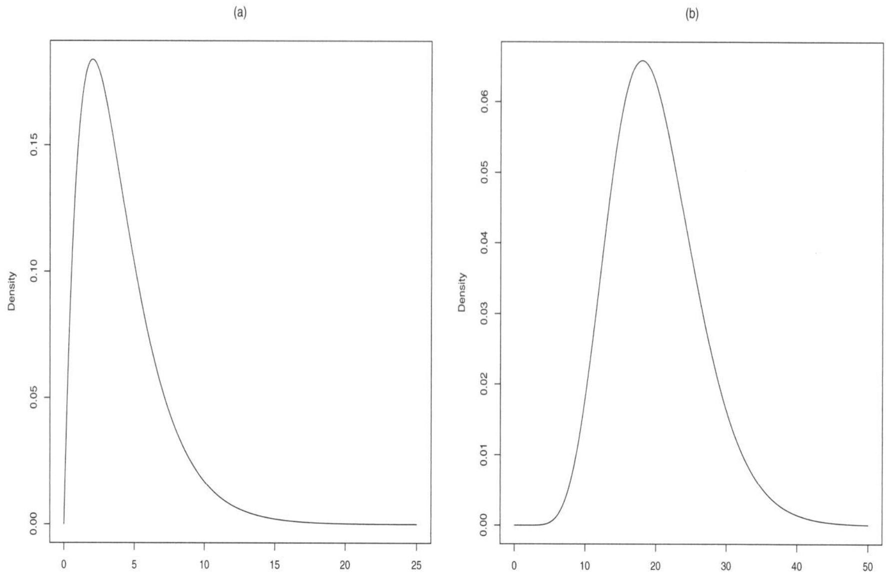

Figure 1.22 *Density functions of χ^2_ν distributions, where ν are the degrees of freedom. (a) $\nu = 4$. (b) $\nu = 20$.*

pendent standard Gaussian random variables, then $\sum_1^n x_t^2$ follows a χ^2_n distribution. The expected value of a χ^2_ν is equal to ν.

$$f(x) = \frac{2^{-\nu/2}}{\Gamma(\nu/2)} x^{\nu/2-1} e^{-x/2},$$

Figure 1.22 exhibits the density functions of two χ^2_ν distributions with $\nu = 4$ and $\nu = 20$ degrees of freedom, respectively.

■ EXAMPLE 1.6

Consider a time series of counts such as the number of calls received by a telephone call center within a given interval of time. In this case, a random variable that could describe the data observed at time t can be specified as $x : \chi \to \mathbb{N}$. An example of such random variable is the so-called *Poisson distribution* given by

$$\mathbb{P}(x = k) = e^{-\lambda} \frac{\lambda^k}{k!},$$

for $k = 0, 1, 2, \ldots$, where the parameter λ is the expected value of x, $E(x) = \lambda$. In addition, $\text{Var}(x) = \lambda$. A simulated series of 365 Poisson counts is shown in Figure 1.23 with $\lambda = 4$. On the other hand, Figure 1.24 exhibits an histogram of these data.

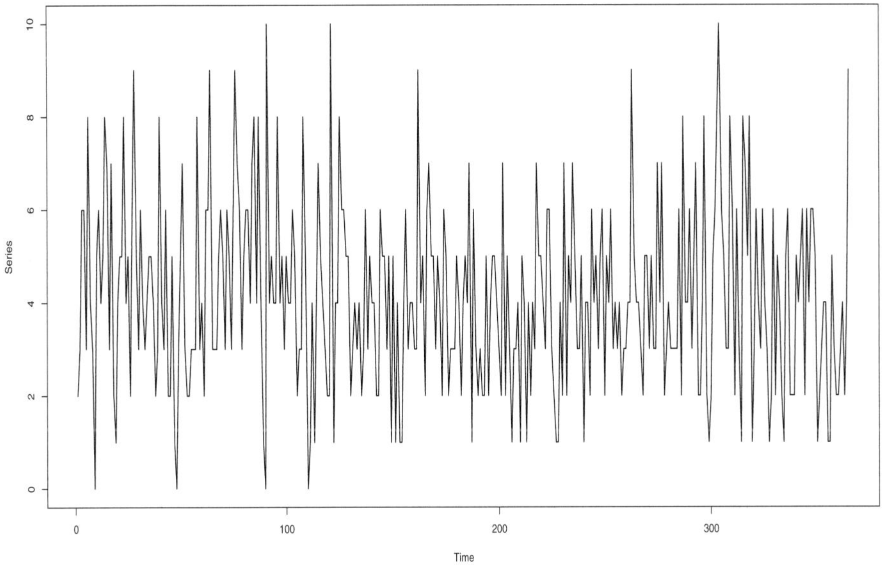

Figure 1.23 *Time series of Poisson counts.*

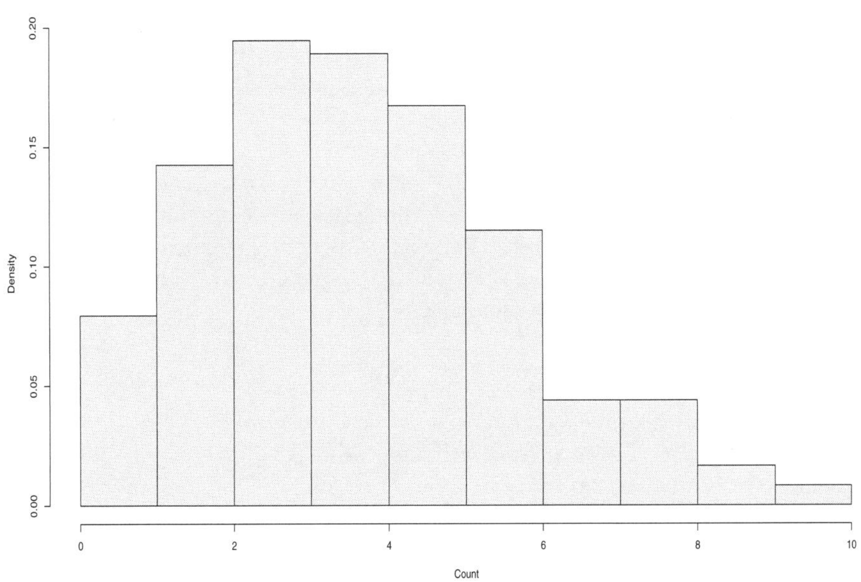

Figure 1.24 *Histogram of the time series of Poisson counts.*

Poisson distribution is widely used for modeling data from different fields, including number of people in a queue, number of patients having a particular disease and number of shoppers arriving at a store, among many other examples.

1.3 DISCRETE-TIME MODELS

This book focuses primarily on stochastic processes that are observed at discrete-times $\dots, t_0, t_1, t_2, \dots$, as opposed to continuous time, meaning that the process has been observed at *all* times in a given interval, for example $t \in (0, T)$. Note that most of the models developed in the time series literature are concerned with equally spaced times. In this case, the observations can be written as $\{y_t : t \in \mathbb{Z}\}$ for $t \in \{\dots, -2, -1, 0, 1, 2, \dots\}$. There are models for treating unequally spaced times, but they are usually more complex to specify and study. In this context, we can also mention the missing data problem, where the series is not observed for some values of t. Methods for dealing with this situation are discussed in Chapter 11.

1.4 SERIAL DEPENDENCE

Consider the stochastic process $\{y_t\}$ and suppose that its mean is $\mu_t = E(y_t)$. If this process is Gaussian, then we can decompose additively it as $y_t = \mu_t + \eta_t$, where η_t is zero-mean stochastic process. In order to specify the process $\{y_t\}$, one may give μ_t some particular structure. For instance, for a *stationary* process, the mean is assumed constant across time so that $\mu_t = \mu$, for all t. More generally, the mean can be specified by a linear model that depends upon time $\mu_t = \beta_0 + \beta_1 t + \cdots + \beta_p t^p$ or depends on other covariates $\mu_t = \beta_0 + \beta_1 x_{1t} + \cdots + \beta_p x_{pt}$.

Stationarity, which is formally discussed in Chapter 2, means that the statistical characteristics of the time series are preserved across time. In particular, the mean and variance of the series are constant and that the relative dependence of an observation with respect to past values remains the same, regardless of the moment at which it is evaluated. That is, suppose that there exists a function γ such that

$$\gamma(h) = \text{Cov}(y_t, y_{t+h}).$$

The existence of this function, denoted as the *autocovariance function*, means that the covariance between observations y_t and y_{t+h} does not depend on t. Stationarity is a key assumption in time series analysis for carrying statistical inferences and prediction. The *autocorrelation function*, ACF hereafter, is then defined as

$$\rho(h) = \frac{\gamma(h)}{\gamma(0)}.$$

Empirical estimates of the ACF are given by the so-called *moment estimators*

$$\widehat{\rho}_k = \frac{\widehat{\gamma}(k)}{\widehat{\gamma}(0)}, \tag{1.1}$$

with

$$\widehat{\gamma}(k) = \frac{1}{n}\sum_{t=1}^{n-k}(y_t - \bar{y})(y_{t+k} - \bar{y}).$$

Examples of the calculation of the sample ACF for some of the time series reviewed earlier in this chapter are presented next. The sample ACF of the returns from the S&P500 index is displayed in Figure 1.25(a) while the sample ACF of the squared returns is exhibited in Figure 1.25(b). Note that the returns show a very low level of autocorrelation, but the squared returns display a large level of autocorrelation.

On the other hand, Figure 1.26 shows the sample ACF of the passenger enplanements data. Observe the clear seasonal pattern that emerges from this graph. In this case, the period is 12 months, showing the annual cycle of the airplane traffic.

The sample ACF of the HDDs series is exhibited in Figure 1.27. As in the case of the passenger enplanements, note the seasonal behavior of the autocorrelation, reflecting the summer/winter effects on the heating requirements.

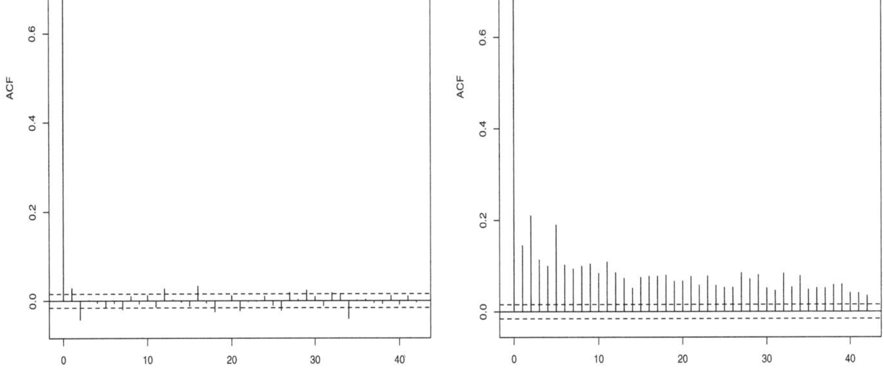

Figure 1.25 *S&P500 data. (a) Sample ACF of log returns. (b) Sample ACF of squared log returns.*

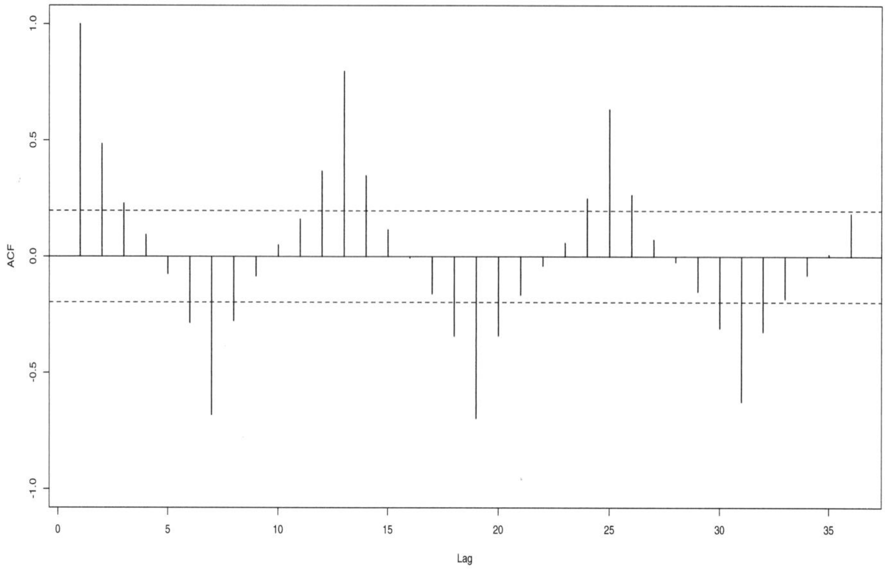

Figure 1.26 *Sample ACF of passenger enplanements from January 2004 to January 2013.*

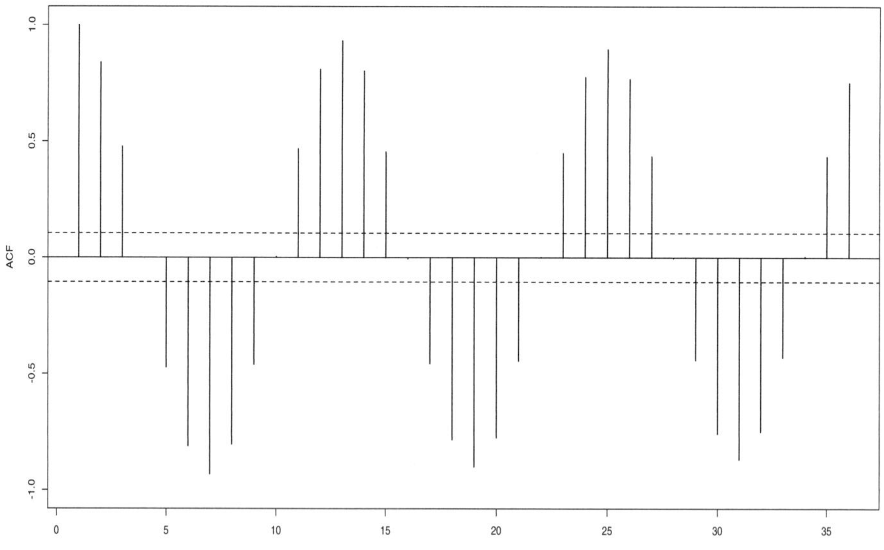

Figure 1.27 *Heating degree days, European Union, January 1980 to May 2010.*

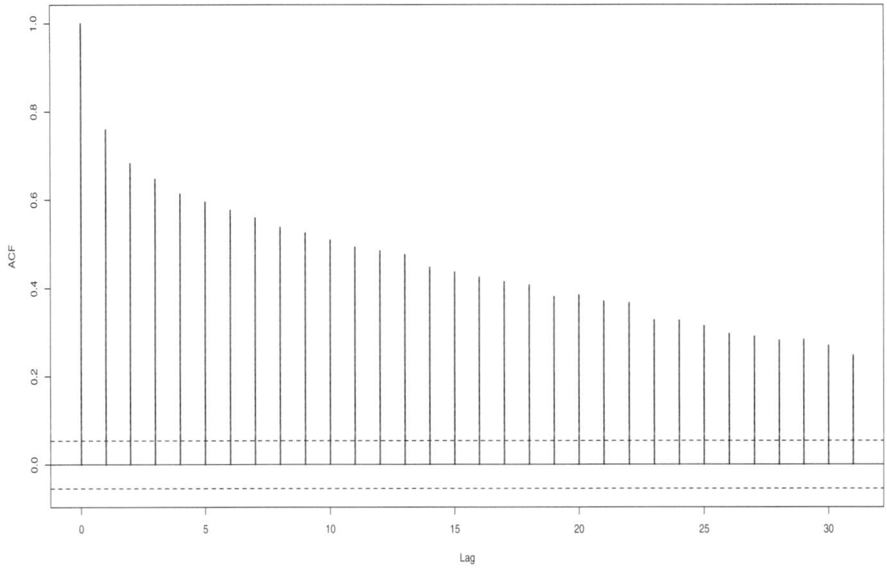

Figure 1.28 *Sample ACF of the Nile river yearly minimum level at the Roda gauge, from 622 A.D. to 1921 A.D.*

A typical sample ACF of a time series exhibiting long-range serial dependence structure is provided in Figure 1.28. Notice that the ACF decays very slowly. This behavior is usual for time series satisfying a long-memory model, such as the autoregressive fractionally integrated moving-average (ARFIMA) models described in Chapter 2.

1.5 NONSTATIONARITY

As seen in the examples discussed in Section 1.1, many real-life time series display nonstationary features such as trends or seasonalities. Given that most of the methodologies for analyzing time series rely on the *stationarity* assumption, there are a number of techniques developed for transforming nonstationary data into stationary. Among these approaches, variance stabilization, trend estimation through linear regression and differentiation of the series are often used.

Variance stabilization is usually achieved by a *Box-Cox transformation* of the data. If the original data is denoted as x_t, the transformed series y_t is given by

$$y_t = \begin{cases} \alpha^{-1}(x_t^\alpha - 1) & \text{if} \quad \alpha \neq 0, \\ \log x_t & \text{if} \quad \alpha = 0. \end{cases}$$

Linear models are tools for removing a *deterministic trend* from the data. This regression model typically includes a polynomial in t, an harmonic component, or may contain other covariates. Thus, the model may be written as

$$\begin{aligned} y_t &= \beta_0 + \beta_1 x_{1t} + \cdots + \beta_p x_{pt} + \eta_t, \\ &= \boldsymbol{X\beta} + \boldsymbol{\eta}, \end{aligned}$$

where the matrix $\boldsymbol{X} = (1, x_{1t}, \ldots, x_{pt})$ are the covariates and the vector $\boldsymbol{\eta} = (\eta_1, \eta_2, \ldots, \eta_n)$ represents non-systematic errors. The coefficients $(\beta_0, \beta_1, \ldots, \beta_p)$ can be obtained, for instance, by *least squares estimates* (LSE), which are studied in Chapter 10. The LSE in this case is given by

$$\widehat{\boldsymbol{\beta}} = (\boldsymbol{X}'\boldsymbol{X})^{-1}\boldsymbol{X}'y.$$

After estimating the regression parameters, the *detrended series* is obtained by removing the regression part from the series $\{y_t\}$, $e_t = y_t - \widehat{\beta}_0 - \widehat{\beta}_1 x_{1t} - \cdots - \widehat{\beta}_p x_{pt}$. Afterwards, the time series methods can be applied to this resulting sequence. In many applications, the regressors are either polynomials or harmonic function, such as in the case of seasonal behavior,

$$y_t = \sum_{j=1}^{m} [\alpha_j \sin(\omega_j t) + \beta_j \cos(\omega_j t)] + \eta_t,$$

where the coefficients α_j and β_j are unknown but the frequencies ω_j are usually considered known or obtained from the spectral analysis, as discussed in Chapter 3.

■ EXAMPLE 1.7

To illustrate the versatility of regression methods for modeling time series data, consider the following simple examples. Figure 1.29 exhibits a simulated process

$$y_t = \beta_0 + \beta_1 t + \eta_t,$$

with $\beta_0 = 0$, $\beta_1 = 0.01$ and η_t is a Gaussian white noise with zero-mean and unit variance, for $t = 1, \ldots, n$ and $n = 300$.

A more complex data structure is shown in Figure 1.30. This graph exhibits a set of 300 simulated observations drawn from the *trend break* regression model

$$y_t = \begin{cases} \beta_0 + \beta_1 t + \eta_t, & t \leq T \\ \beta_2 + \beta_3 t + \eta_t, & t > T, \end{cases} \tag{1.2}$$

where $\beta_0 = 0$, $\beta_1 = 0.01$, $\beta_2 = 0.5$, $\beta_3 = 0.0067$, and $T = 150$. Observe that in this example, there is a change in the slope of the linear trend, but not a discontinuity of the trend.

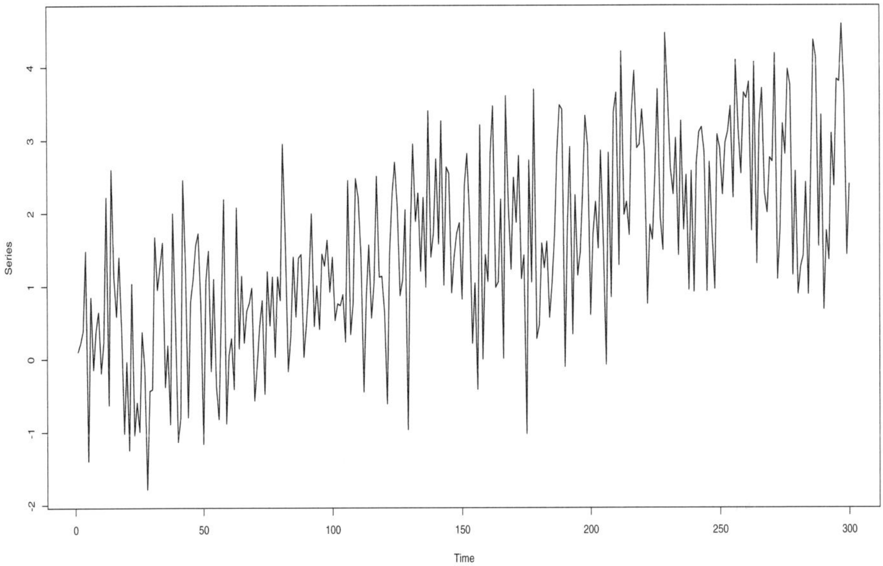

Figure 1.29 *Simulated linear regression time series.*

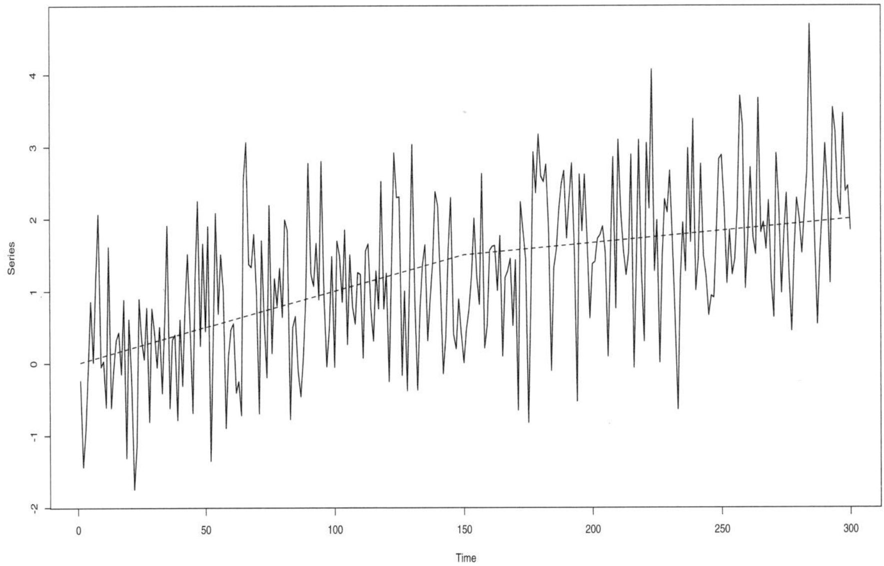

Figure 1.30 *Structural break regression model with change in the slope of the trend at time* $T = 150$.

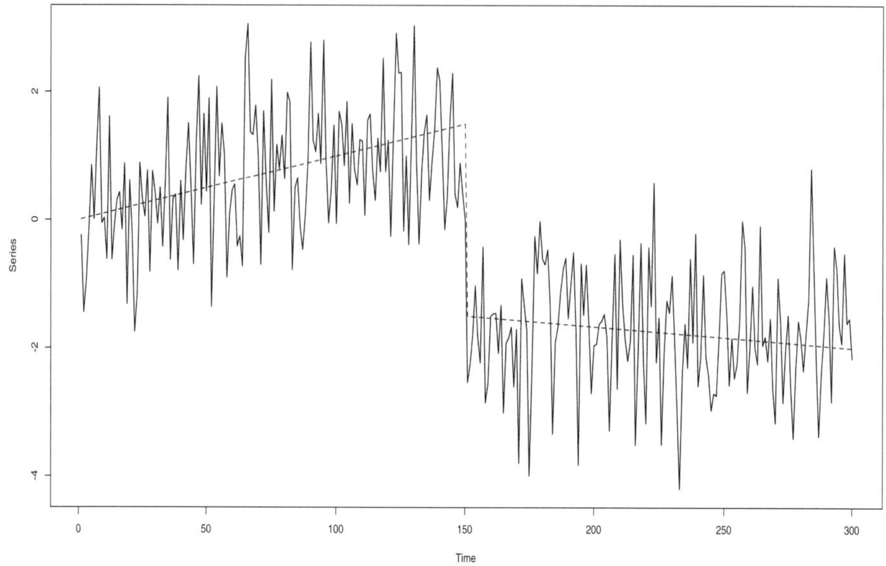

Figure 1.31 *Structural break regression model with change in the slope and the level of the trend at the point $T = 150$.*

Figure 1.31 shows another example of the structural break model described by (1.2), with $\beta_0 = 0$, $\beta_1 = 0.01$, $\beta_2 = -0.0033$, and $\beta_3 = -0.0033$. In this case, there are two changes involved since both the slope of the linear trends and the intercepts vary at time $T = 150$.

Another common situation is described in Figure 1.32, which displays 300 observations from the *instrumental variable* model

$$y_t = \beta_0 + \beta_1 t + \beta_2 D_t + \eta_t,$$

where

$$D_t = \begin{cases} 1, & T_1 \leq t \leq T_2 \\ 0, & \text{otherwise.} \end{cases} \tag{1.3}$$

In this illustration, $\beta_0 = 0$, $\beta_1 = 0.01$, $\beta_1 = 10$, $T_1 = 150$, and $T_2 = 160$.

On the other hand, harmonic regressions allow us to model a great variety of seasonal patterns. Figure 1.33 shows a series of 400 observations from the harmonic model

$$y_t = \beta \cos(\omega t) + \eta_t,$$

with $\beta = 10$ and $\omega = 0.05$.

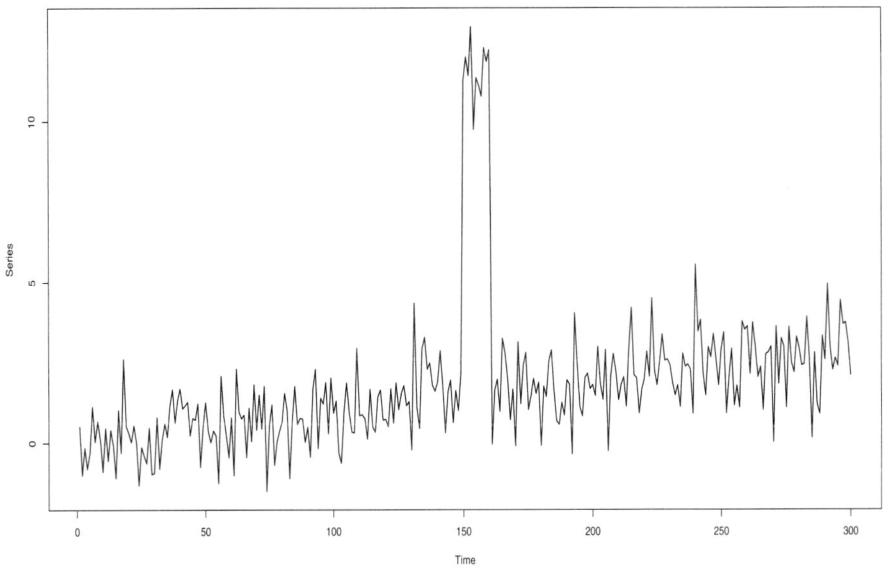

Figure 1.32 *Structural break regression model with instrumental variable for period* $T_1 = 150$ *to* $T_2 = 160$.

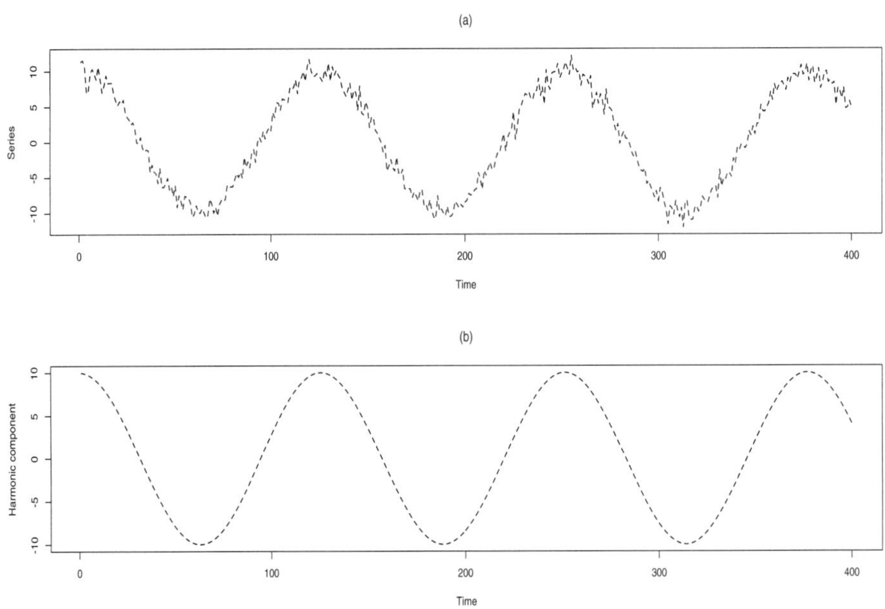

Figure 1.33 *Harmonic regression time series with one frequency. (a) Simulated data. (b) Underlying harmonic trend.*

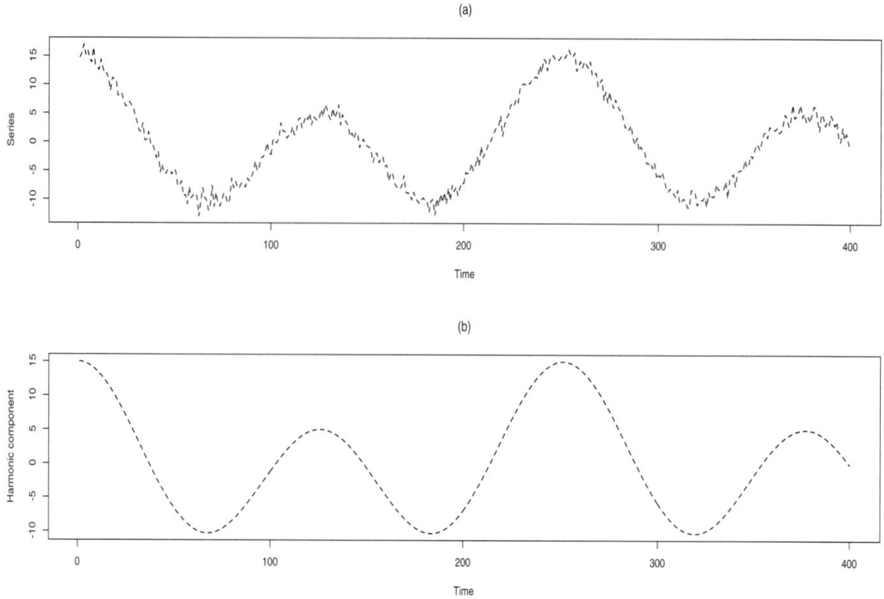

Figure 1.34 *Harmonic regression time series with two frequencies. (a) Simulated data. (b) Underlying harmonic trend.*

An example of an harmonic time series with two frequencies is displayed in Figure 1.34, for the model

$$y_t = \beta_1 \cos(\omega_1 t) + \beta_2 \cos(\omega_2 t) + \eta_t,$$

with $\beta_1 = 10$, $\beta_2 = 5$, $\omega_1 = 0.05$ and $\omega_2 = 0.025$.

Figure 1.35 shows a series of 800 observations from the three-component harmonic model

$$y_t = \beta_1 \cos(\omega_1 t) + \beta_2 \cos(\omega_2 t) + \beta_3 \cos(\omega_3 t) + \eta_t,$$

with $\beta_1 = \beta_2 = \beta_3 = 10,$, $\omega_1 = 0.075$, $\omega_2 = 0.0325$ and $\omega_3 = 0.0125$.

Differentiation. Another approach for removing a trend in the data is differentiation. In this case, however, the underlying trend is assumed to be *nondeterministic* or *stochastic*. Under this framework, the data is assumed to be generated by some time aggregation process, for example,

$$y_t \quad = \quad \sum_{k=1}^{t} \eta_k, \tag{1.4}$$

where η_t is a zero-mean, constant variance stationary process, which may be sequentially correlated. Thus, by *differencing* $\{y_t\}$, we obtain the series $\{z_t\}$

$$z_t = y_t - y_{t-1} = \eta_t,$$

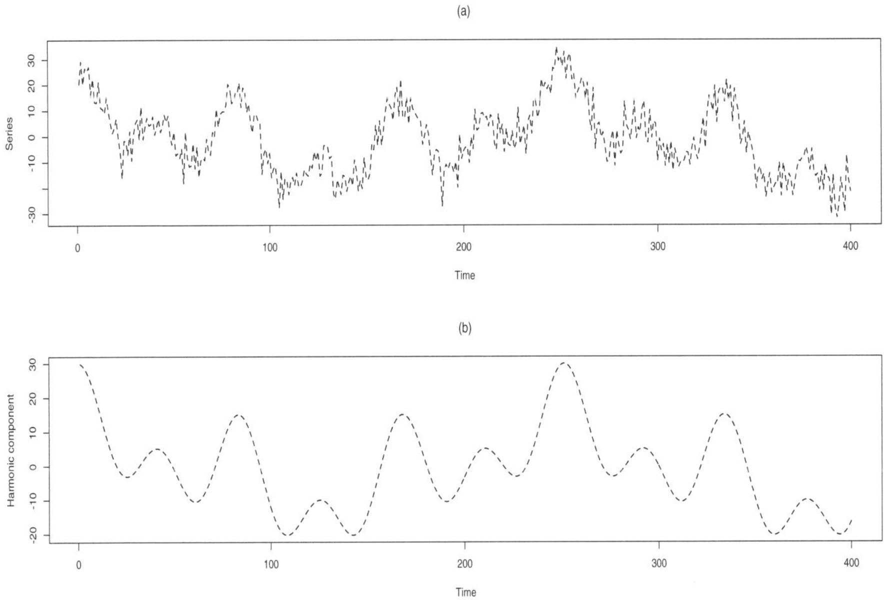

Figure 1.35 *Harmonic regression time series with three frequencies. (a) Simulated data. (b) Underlying harmonic trend.*

which shares the same *stationary* properties of $\{\eta_t\}$. A common problem with this technique is to decide when to stop differencing. Two basic aspects should be taken into account. First, the differenced series should look stationary and second, its variance should be no greater than the original series. A disproportionate variance increase in the resulting series could indicate over-differentiation.

Another usual dilemma is choosing between regression or differentiation. Even though there is no general guidances about this, one can apply any of these techniques and see whether they produce adequate results or not. It can be shown that applying a regression method to an integrated model will produce heteroskedastic errors. On the other hand, differencing a deterministic trend will generate artificially correlated errors. These two aspects can be explained as follows.

Assume that the series corresponds to an integrated model, as in (1.4) with white noise sequence η_t, and a linear model $y_t = \beta_0 + \beta_1 t + \varepsilon_t$ is fitted. Thus, $y_t = y_{t-1} + \beta_1 + \varepsilon_t - \varepsilon_{t-1}$ and $\varepsilon_t = \sum_{k=1}^{t} \eta_k - \beta_1 t$. Clearly, this error variable has a time-dependent mean $E(\varepsilon_t) = \beta_1 t$ and heteroskedasticity $E(\varepsilon_t) = t\sigma_\eta^2$. Therefore, the regression model does not satisfy the basic assumptions about the non-systematic error term.

Conversely, if the process y_t satisfies the regression model $y_t = \beta_0 + \beta_1 t + \varepsilon_t$ with white noise ε_t and it is differenced, we obtain $z_t = y_t - y_{t-1} = \beta_1 +$

$\varepsilon_t - \varepsilon_{t-1}$. The variance of resulting series z_t is twice the variance of the original data, $\text{Var}(z_t) = 2\sigma_\varepsilon^2$, which is an indication of a wrong differentiation procedure.

Apart from the two techniques discussed above, there are many other transformation methods for achieving stationarity. On the other hand, there are methodologies that allows for the direct treatment of nonstationary data, without transformation. One example of these methods is the so-called *locally stationary* models, described in Chapter 5.

1.6 WHITENESS TESTING

A white noise process is a sequence of zero-mean uncorrelated random variables. If this sequence is Gaussian, then the process is also independent. A fundamental procedure in time series analysis is testing for whiteness. That is, testing whether the series is white noise or it processes some more complex mean or dependence structure. Given the sequence $y_1, \ldots, y_n$, the null hypothesis is $H_0 : \{y_t\}$ is white noise versus $H_1 : \{y_t\}$ is not white noise. Observe that H_0 may fail due to many causes. For example, the mean of the process is not constant, its variance is not constant, its observations are correlated, or combinations of these aspects.

Whiteness testing procedures usually do not involve checking for independence, unless the series is assumed to be Gaussian. It is important to emphasize at this point that the definition of white noise refers only to an uncorrelated sequence. In particular, this means that a sequence with correlated squares is still white noise, according to this definition. This is usually the case for financial time series: returns are often uncorrelated but volatilities or squared returns are just often correlated. Typically, a white noise test takes into consideration the estimated autocorrelations $r_1, \ldots, r_L$ with

$$r_k = \widehat{\rho}_k,$$

where $\widehat{\rho}_k$ is given by (1.1). The Box-Ljung test, a well-known procedure for checking whether a sequence is white noise or not, can be written as

$$Q_L = n(n+2) \sum_{h=1}^{L} \frac{r_k^2}{n-h},$$

and it can be shown that the statistic Q_L follows, approximately, a χ^2 distribution with L degrees of freedom.

■ EXAMPLE 1.8

Figure 1.36 shows 500 observations of a Gaussian white noise sequence with zero-mean and unit variance, while Figure 1.37 exhibits the sample ACF and the results from a Ljung-Box test with $L = 10$. Note that, as expected, the series complies with the test at the 5% significance level.

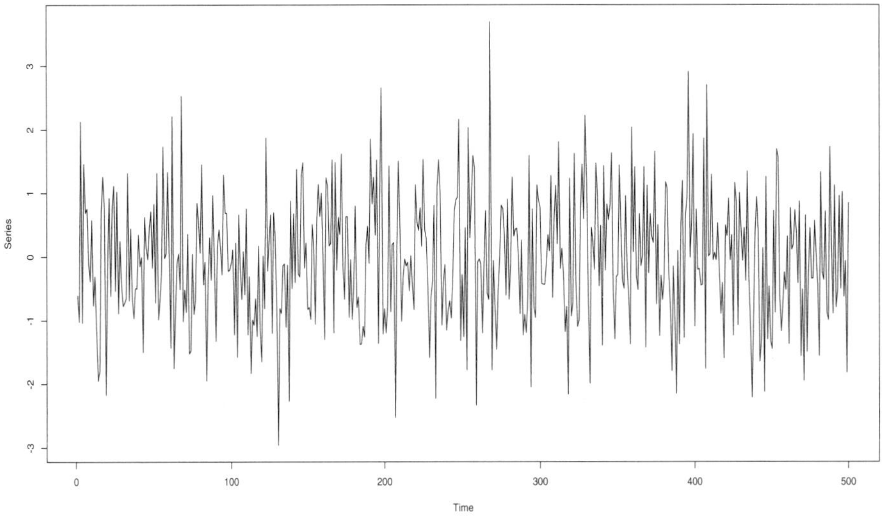

Figure 1.36 *Simulated Gaussian white noise sequence with zero-mean, unit variance, and 500 observations.*

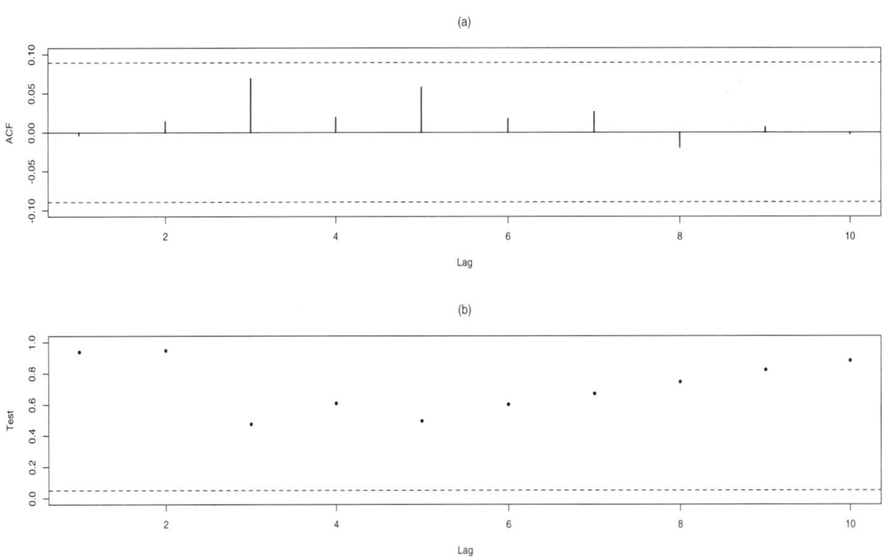

Figure 1.37 *Sample ACF (a) and Box-Ljung test (b) for the white noise sequence.*

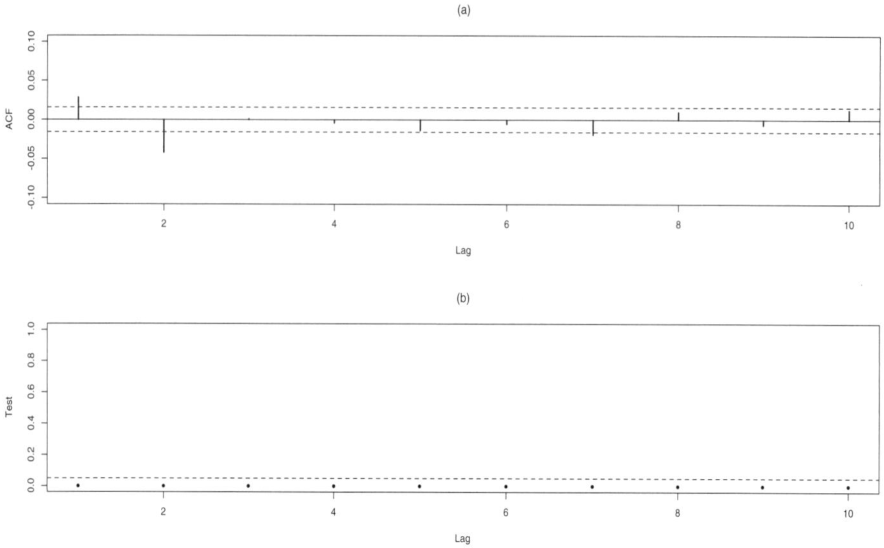

Figure 1.38 *Sample ACF (a) and Box-Ljung test (b) for the S&P500 log returns.*

An application of the Box-Ljung test to the S&P500 log returns is shown in Figure 1.38. Note that in this case, there is evidence that the series is not white noise, given that the null hypothesis is rejected by the Box-Ljung test at the 5% significance level for all values of L.

■ EXAMPLE 1.9

The ability of the Box-Ljung to detect deviations from the white noise assumption can be illustrated as follows. Consider the moving-average model defined in (1.5). Figure 1.39 exhibits the percentage of rejection of the white noise hypothesis for different values of the parameter ψ in the case of a time series composed of 1000 observations and considering a Box-Ljung test with 10 lags. Notice that the percentage of rejection gets close to the 100% for very low values of $|\psi|$, in this case, for values just above 0.2. On the other hand, as ψ get close to zero, as expected, the rejection rate decays very sharply.

As a second illustration, consider the first-order auto-regressive model (1.9). Figure 1.40 exhibits the percentage of rejection of the white noise hypothesis for different values of the parameter ϕ in the case of a time series composed of 1000 observations and considering a Box-Ljung test with 10 lags. Similar to the AR(1) case, the percentage of rejection gets close to the 100% for very low values of $|\phi|$.

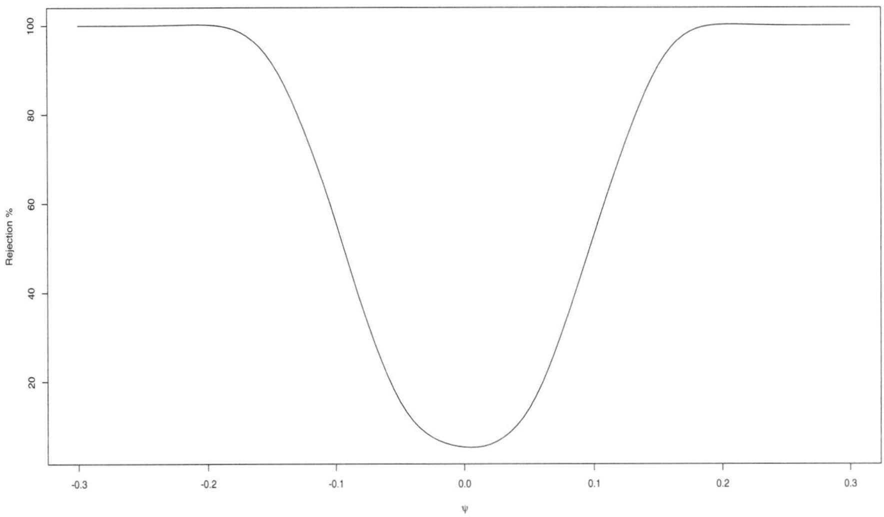

Figure 1.39 *Percentage of rejection of the white noise hypothesis for different values of the parameter ψ in the case of a time series composed of 1000 observations and considering a Box-Ljung test with 10 lags.*

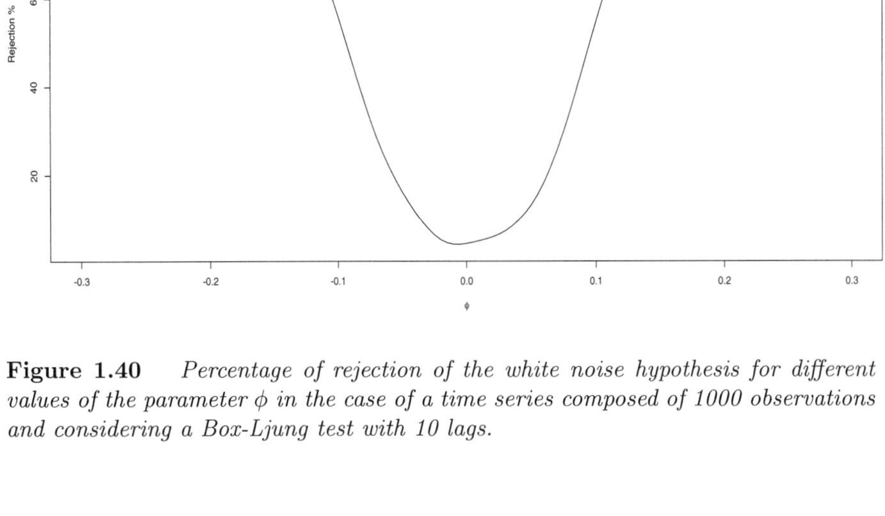

Figure 1.40 *Percentage of rejection of the white noise hypothesis for different values of the parameter ϕ in the case of a time series composed of 1000 observations and considering a Box-Ljung test with 10 lags.*

1.7 PARAMETRIC AND NONPARAMETRIC MODELING

An important distinction among statistical procedures is related to parametric versus non parametric modeling. Before entering in further technical details, one can think of the data as coming from a model, usually unknown, which is specified by a number of coefficients or *parameters*. This vision is referred to as *Fisher paradigm*. In this context, the statistical analysis is essentially guessing which are the parameters of the model generating the observed data. In order to accomplish this goal, one must select the model and estimate the corresponding parameters. Examples of this procedures are autoregressive moving-average (ARMA) models, which are defined in Chapter 2. For specifying a parametric model, one can provide a *parameter vector* $\theta = (\theta_1, \ldots, \theta_p)$ and a *parameter space* Θ, such that $\theta \in \Theta$. Observe that the dimension of this parameter is finite, p. As an illustration, consider the simple model

$$y_t = \varepsilon_t + \psi\varepsilon_{t-1}, \tag{1.5}$$

where $\{\varepsilon_t\}$ is a white noise sequence and ψ is a one-dimensional parameter. This is an example of the so-called moving-average models studied in Chapter 2. In this case, the model can be specified by the two-dimensional vector $\theta = (\psi, \sigma)$, where σ is the white noise standard deviation. Besides, the parameter space is given by $\Theta = (\mathbb{R}, \mathbb{R}_+)$, where $\mathbb{R}_+$ denotes the positive real numbers.

A generalization of the simple model (1.5) is considering several parameters

$$y_t = \varepsilon_t + \psi_1\varepsilon_{t-1} + \psi_2\varepsilon_{t-2} + \cdots + \psi_q\varepsilon_{t-q}, \tag{1.6}$$

which will be denoted as moving-average model MA(q) in Chapter 2. In this case, the parameter vector is $\theta = (\psi_1, \psi_2, \ldots, \psi_q, \sigma)$. Another extension is allowing the coefficients ψ_j to depend on a specific finite-dimensional parameter vector, $\psi_j(\theta)$ and write

$$\begin{aligned} y_t &= \varepsilon_t + \psi_1(\theta)\varepsilon_{t-1} + \psi_2(\theta)\varepsilon_{t-2} + \cdots && (1.7) \\ &= \sum_{j=0}^{\infty} \psi_j(\theta)\varepsilon_{t-j}. && (1.8) \end{aligned}$$

In this case, even though there are infinite coefficients $\psi_j(\theta)$, model (1.7) is still parametric since they depend on a *finite-dimensional* parameter θ.

EXAMPLE 1.10

Consider the first-order autoregressive model

$$y_t = \phi y_{t-1} + \varepsilon_t. \tag{1.9}$$

If $|\phi| < 1$, this model can be also written as

$$\begin{aligned} y_t &= \varepsilon_t + \phi\varepsilon_{t-1} + \phi^2\varepsilon_{t-2} + \cdots \\ &= \sum_{j=0}^{\infty} \psi_j(\theta)\varepsilon_{t-j}, \end{aligned}$$

where $\psi_j(\theta) = \phi^{j-1}$ with $\theta = (\phi, \sigma)$.

The models previously discussed are called *linear* in the sense they are linear combinations of the noise sequence $\{\varepsilon_t\}$. Other parametric models, called *nonlinear*, contain multiplicative error terms. For example, consider the model

$$y_t = \varepsilon_t + \psi\varepsilon_{t-1}\varepsilon_{t-2}.$$

This process in nonlinear since it includes the product of two lagged noise values, ε_{t-1} and ε_{t-2}. Note, however, that this model is still parametric, since it is defined by the bivariate vector $\theta = (\psi, \sigma)$. More general expressions for the nonlinear parametric can be provided. For instance,

$$y_t = f_\theta(\varepsilon_t, \varepsilon_{t-1}, \varepsilon_{t-2}, \dots),$$

where f_θ is a measurable function defined by the finite-dimensional parameter θ.

On the other hand, one may not want to specify a finite parameter model but rather consider the data as generated by an unspecified mechanism. In this context, the statistical analysis focuses on finding a general function that describes the data well. Examples of this approach are *kernel smoothing methods*, *neural networks* and *regression trees*, among others.

Contrary to the parametric setup, one may consider a nonparametric model as specified by an infinite-dimensional parameter space. For instance, the observation y_t can be specified by

$$y_t = f(\varepsilon_t, \varepsilon_{t-1}, \varepsilon_{t-2}, \dots),$$

where $\{\varepsilon_t\}$ is an input sequence and $f(\cdot)$ is a function. In this case, the *model* is specified by $f \in \mathcal{F}$, where $\mathcal{F}$ is a space of functions.

Still, there is a third approach, the *semiparametric modeling* which combines some of the parametric and nonparametric aspects. For example, one may partially specify the behavior of the spectrum around a given frequency, leaving unspecified the behavior of the spectrum at other frequencies, see its definition in Chapter 3. For instance, if $f_y(\lambda)$ denotes the spectral density of the process $\{y_t\}$, then one may specify the behavior of the spectrum in a neighborhood of the origin as

$$f_y(\lambda) \sim C|\lambda|^\alpha,$$

for small λ, where C is a positive constant and α is the parameter of interest, that is, the rate at which the spectrum converges to zero in the case that α is positive or diverges in the case that α is negative.

1.8 FORECASTING

Regardless of the modeling approach that one may consider, parametric, semiparametric, or nonparametric, in time series analysis, one is usually interested in predicting future values of the process. Given observations up to time t, one may want to forecast h-steps ahead, the *prediction horizon*, $\widehat{y}_{t+h}$. As shown later in this text, the best linear predictor is given by the conditional expectation,

$$\widehat{y}_{t+h} = E[y_{t+h}|y_t, y_{t-1}, y_{t-2}, \dots].$$

The best predictor may not necessarily be linear, so that the optimal forecast may have a different expression that will depend on the relationship between y_t and its past. For a Gaussian process, the *best predictor* is precisely the *best linear predictor*, independently of the model linking y_{t+h} with $y_t, y_{t-1}, y_{t-2}, \dots$. Besides, note that in practice only a finite stretch of data is available, $\{y_1, y_2, \dots, y_n\}$, say. In this case, the *finite past* predictor is given by,

$$\widehat{y}_{n+h} = E[y_{n+h}|y_n, y_{n-1}, y_{n-2}, \dots, y_1].$$

In some other situations, the interest is focused on estimating past values of the process, the so-called *backcasting* procedure. In this case, one may want to estimate, for instance, y_0, given $\{y_1, y_2, \dots, y_n\}$,

$$\widehat{y}_0 = E[y_0|y_1, y_2, \dots, y_n].$$

Another procedure, called *smoothing* is concerned with estimating the value of the process at a particular time $t \in \{1, 2, \dots, n\}$ given the remaining observations,

$$\widehat{y}_t = E[y_t|y_1, \dots, y_{t-1}, y_{t+1}, \dots, y_n].$$

Let $\{e_t\}$ be the prediction error sequence, that is, $e_t = y_t - \widehat{y}_t$, for $t = 1, 2, \dots, n$. One basic criterion for goodness of fit in time series analysis is that $e_1, e_2, \dots, e_n$ is a white noise sequence. This hypothesis can be formally tested by the Box-Ljung procedure or another technique. If the resulting sequence of prediction errors is not white noise, then the model may not be appropriate for the data.

As in other areas of statistics, when fitting a time series model one set some observations aside, so that we can assess the out-of-sample performance of the model.

1.9 TIME SERIES MODELING

This section provides an overview of the process involved in time series *parametric* modeling and prediction. These procedures take into account the topics discussed in the previous sections of this chapter. As real-life time series data usually appear nonstationary, there are techniques that transform the data into a more *stationary* sequence. Next, the basic distributional features of the data are analyzed, establishing constant mean, variance, and checking for the presence of outliers or nonnormality. Once these properties have been found consistent with stationarity, the focus is shifted toward empirically investigating the autocorrelation structure of the observations. Based on this analysis, a model can be proposed, for example, an ARMA model as defined in Chapter 2. The model is usually selected by considering an information criterion such as Akaike's information criterion (AIC) or Bayesian information criterion (BIC), see Chapter 4 for definitions of these concepts. The main idea behind this model selection process is that while adding more parameters could improve the fitting ability of a particular model, it also diminishes its degree of freedom to evaluate the fitting and out-of-sample prediction quality. In this sense, criteria such as AIC or BIC help to strike a balance between fitting ability and model complexity.

After selecting the model, a battery of tests can be applied to determine the parameter significance as well as the goodness of fit. Moreover, a residual analysis can be performed to test the whiteness of the model errors. If the model is appropriate, the residuals should be white noise.

Finally, once the model is judged to be adequate for the data, it can be used for producing forecasts of future values. In many situations, part of the data is set aside from the fitting procedure, so that it can be used for an out-of-sample assessment of the prediction quality. This step is important also for the *nonparametric approach*. In this case, a methodology such as neural network or regression tree can be applied to a portion of the time series, the so-called *training data*, and the remaining part of the series is left for evaluating the out-of-sample forecasting capacity of the nonparametric method.

1.10 BIBLIOGRAPHIC NOTES

There is a pleayade of books in time series analysis. The monograph by Brockwell and Davis (2002) is good introductory text to the subject. The book by Shumway and Stoffer (2011) provides another excellent treatment of the fundamental time series techniques. Diggle (1990) offered an overview of the subject from a biostatistics standpoint. Kedem and Fokianos (2002) covered a number of regression time series methodologies, including models for count data. The books by Tsay (2005) and Tsay (2013) provide excellent discussions about financial time series and a number of techniques for fitting and predicting heteroskedastic data. Hamilton (1994) is another interesting text on financial time series covering several topics. More advanced texts in the subject are, for example, Brockwell and Davis (1991) and Fuller (1996).

Problems

1.1 Explain the following concepts: (a) Stationarity, (b) Seasonality, (c) Deterministic trend, (d) Stochastic trend, (e) White noise, and (f) Structural change.

1.2 What does it mean a *sample realization* of a time series?

1.3 Why tree rings and mineral sediments time series are important for climatic and paleoclimatic studies?

1.4 Explain the concept and utility of the *autocovariance function* and the *autocorrelation function*. Write down the corresponding equations that define these concepts.

1.5 Suppose that you have a sample realization of n observations given by $\{x_1, \dots, x_n\}$. Explain how you could estimate the autocovariance and sample autocorrelation functions.

1.6 Explain the concept of *strict stationarity* of a time series.

1.7 What are sufficient conditions for weak stationarity of a time series? Is a weakly stationary time series necessarily strictly stationary?

1.8 How is a white noise defined?

1.9 Suppose you have a series y_t defined as $y_t = \sum_{j=0}^{p} \beta_j t^j + \omega_t$. Calculate the expected value and the autocovariance function of y_t.

1.10 Regarding the previous question, propose two weak stationarity-inducing transformations of y_t.

1.11 Let $\{y_1, \dots, y_n\}$ be a sample from a stationary process. Suggest a procedure to determine if the stochastic process is the sum of a constant and a white noise?

1.12 Let $\{\varepsilon_1, \dots, \varepsilon_n\}$ be Gaussian white noise with zero-mean and unit variance and suppose that we are interested in simulating a Gaussian stationary process $\{y_1, y_2, \dots, y_n\}$ with autocovariance function γ and mean μ. Show that this process can be simulated by generating a sample of the white noise $\varepsilon = \{\varepsilon_1, \dots, \varepsilon_n\}$ and then obtaining $y = \mu + A\varepsilon$, where A is a square matrix satisfying $AA' = \Gamma$ with $\Gamma_{i,j} = \gamma(i - j)$.

1.13 Let $x_1, x_2, \dots, x_n$ be a sample of χ_1^2 independent random variables. By taking into account that the fourth moment of a standard normal distribution is $E(x^4) = 3$, calculate the variance of the random variable $y = x_1 + x_2 + \cdots + x_n$.

1.14 Suppose that the model $y_t = \beta_0 + \beta_1 t + \epsilon_t$ with $t = 1, ..., n$, $E(\epsilon_t) = 0$, $E(\epsilon_t^2) = \sigma^2$ and for $t \neq s$, $E(\epsilon_t \epsilon_s) = 0$. Define $w_t = \frac{1}{3} \sum_{j=-1}^{1} y_{t+j}$.

(a) Find the expected value of the time series w_t.

(b) Calculate $\text{Cov}(w_{t+k}, w_t)$ and show that this covariance does not depend on t. Is the sequence $\{w_t\}$ stationary?

1.15 Explain why it is sometimes necessary to apply a functional transformation of the data, such as a logarithmic or a Box-Cox transformation.

1.16 Suppose that the price of a stock at time t is denoted as P_t and that the sequence of prices is described by the equation

$$P_t = P_{t-1}(1 + r_t),$$

where r_t is the return at time t. By taking logarithms and using the approximation $\log(1 + x) \sim x$ for small values of x, show that

$$r_t = \Delta \log P_t = \log P_t - \log P_{t-1}.$$

1.17 Explain the following concepts related to the analysis of financial time series, (a) Return, (b) Volatility, (c) Heteroskedasticity, and (d) Risk.

1.18 Let $P_k(t)$ be a k order polynomial defined by

$$P_k(t) = a_0 + a_1 t + a_2 t^2 + \cdots + a_k t^k.$$

Show that $(1 - B)^k P_k(t) = c$, where c is a constant.

CHAPTER 2

LINEAR PROCESSES

The concept of linear process is fundamental in time series analysis. As in the time series examples discussed in the previous chapter, many social, physical and economic phenomena can be analyzed and described by this class of models. A scientist or a practitioner knows that a phenomenon under study may be highly complex, but often a linear approach offers a first good description of the data. On the basis of a linear process, more complex models can be built afterwards. A linear process contains three basic components: an *input noise*, a *linear filter*, and the output *observed* data. In practice, one only has a finite set of observations, but one can still imagine or assume that the available time series comes from a linearly filtered noise. Even though this approach seems to oversimplify the data generating mechanism, it usually provides a powerful tool for modeling a wide range of time series data. In this chapter, we review the foundations of the linear processes and study some of their applications. Three basic representations of a linear process are described, the Wold expansion, the autoregressive expansion and the state space systems. Stationarity, invertibility, and causality are also reviewed. Another important aspect to consider in the analysis of a time series is whether the dependence structure of its observations is weak or strong. This topic is discussed when describing

Time Series Analysis. First Edition. Wilfredo Palma.

autoregressive moving-average (ARMA) and autoregressive fractionally integrated moving-average (ARFIMA) models. This chapter also analyzes the autocovariance structure of these processes, providing methods for computing autocorrelation (ACF) and partial autocorrelation (PACF).

2.1 DEFINITION

A linear process can be written as,

$$y_t = \psi(B)\varepsilon_t, \tag{2.1}$$

where y_t are the observed values, $\psi(B)$ is a linear filter on the backshift operator B, and ε_t are the input noise. Recall that the effect of B is lagging an observation, that is, $By_t = y_{t-1}$.

The filter can be written as

$$\psi(B) = \sum_{j=-\infty}^{\infty} \psi_j B^j, \tag{2.2}$$

where $\sum_{j=-\infty}^{\infty} \psi_j^2 < \infty$. Consequently, we can write the observed time series as,

$$y_t = \sum_{j=-\infty}^{\infty} \psi_j \varepsilon_{t-j}. \tag{2.3}$$

Note that this filter is said to be *linear* since it does not contain mixed input noise terms such as $\varepsilon_{t-i}\varepsilon_{t-j}$. Furthermore, this expression tell us that the observed value at time t, y_t, depends on past, present, and future values of the input noise, that is, $\{\ldots, \varepsilon_{-3}, \varepsilon_{-2}, \varepsilon_{-1}, \varepsilon_0, \varepsilon_1, \varepsilon_2, \varepsilon_3, \ldots\}$.

2.2 STATIONARITY

This is an essential concept in time series analysis. Generally speaking, we can distinguish two definitions of stationarity. The first definition is focused on the joint distribution of the process, while the second focuses on the second order structure of the time series model.

Strict Stationarity. Let $y_h(\omega) = \{y_{t_1+h}(\omega), \ldots, y_{t_n+h}(\omega)\}$ be a trajectory of the process $\{y_t\}$ with $t_1 + h, \ldots, t_n + h \in \mathbb{Z}$. The process is said to be *strictly stationary* if and only if the distribution of y_h is the same regardless of h.

Weak Stationarity. A process y_t is said to be *weakly stationary* or *second-order stationary* if (a) it has a constant mean, (b) it has finite and constant second moment, and (c) there is a function $\gamma(\cdot)$ such that $\gamma(k) = \mathrm{Cov}(y_t, y_{t+|k|})$ for any t, k. For example, for the linear process (2.3), we have (a) $E(y_t) =$

$\sum_{j=-\infty}^{\infty} \psi_j \, E(\varepsilon_{t-j}) = 0$. On the other hand, given that $\{\varepsilon_t\}$ are uncorrelated,

$$E(y_t^2) = \sum_{j=-\infty}^{\infty} \psi_j^2 \, E(\varepsilon_{t-j}^2) = \sigma^2 \sum_{j=-\infty}^{\infty} \psi_j^2 < \infty. \tag{2.4}$$

Furthermore, we can write the autocovariance function $\gamma(k) = E(y_t y_{t+h})$ as

$$\begin{aligned} \gamma(h) &= \sum_{i=-\infty}^{\infty} \sum_{j=-\infty}^{\infty} \psi_i \psi_j \, E(\varepsilon_{t-i} \varepsilon_{t+h-j}) \\ &= \sigma^2 \sum_{i=-\infty}^{\infty} \sum_{j=-\infty}^{\infty} \psi_i \psi_j \delta(h+i-j) \\ &= \sigma^2 \sum_{j=-\infty}^{\infty} \psi_j \psi_{j+h}. \end{aligned} \tag{2.5}$$

Stationarity is an important concept in time series analysis. Loosely speaking, it means that the statistical properties of the process remain constant over time. In practice, this implies that all values of the process are comparable, no matter at what time they were observed. In turn, comparability of the observations allows us to draw statistical conclusions about the whole process.

A process can be strictly stationary but not necessarily weakly stationary and vice versa. For instance, the process $\{y_t : t \in \mathbb{Z}\}$ where, y_t are independent and identically distributed Cauchy random variables, is strictly stationary but not weakly stationary since the first and the second moments do not exist. A more sophisticated example of this is the *fractionally integrated generalized autoregressive conditionally heteroskedastic* (FIGARCH) model introduced in Chapter 6. Conversely, let $\{\varepsilon_t\}$ be a sequence of independent and identically distributed normal random variables with zero-mean and unit variance, and let $\{\eta_t\}$ be a sequence of independent and identically distributed exponential random variables with rate 1. Then the process generated by $y_t = \varepsilon_t[t/2] + (\eta_t - 1)[(t+1)/2]$, where $[\cdot]$ denotes the integer part function, is weakly stationary but not strictly stationary. Nevertheless, these two concepts are equivalent for Gaussian processes.

A *strict white noise* process is a sequence of independent and identically distributed random variables, while a *weak white noise* process is a sequence of uncorrelated random variables with zero-mean and constant finite variance, that is, with an autocovariance function satisfying $\gamma(0) < \infty$ and $\gamma(h) = 0$ for all $h \neq 0$.

2.3 INVERTIBILITY

The linear process (2.3) is *invertible* if there exists a filter $\pi(B)$ such that we can write

$$\pi(B)y_t = \sum_{j=-\infty}^{\infty} \pi_j y_{t-j} = \varepsilon_t.$$

The filter $\pi(B)$ can be considered as the *inverse* of the filter $\psi(B)$, that is, $\pi(B)\psi(B) = 1$. Note that an invertible time series y_t can be expressed as

$$y_t = \sum_{j=-\infty}^{-1} \pi_j y_{t-j} + \sum_{j=1}^{\infty} \pi_j y_{t-j} + \varepsilon_t.$$

2.4 CAUSALITY

One way to describe a discrete-time stochastic process is writing it as the result of the filtering of a white noise sequence $\{\varepsilon_t\}$,

$$y_t = \varphi(\dots, \varepsilon_{-2}, \varepsilon_{-1}, \varepsilon_0, \varepsilon_1, \varepsilon_2, \dots),$$

where $\varphi(\cdot)$ is a measurable function, that is, the resulting sequence $\{y_t\}$ is a well-defined random process. Assume now that the noise sequence is generated simultaneously as the observed process $\{y_t\}$, so that at any instant t, the generated noise sequence is $\dots, \varepsilon_{-2}, \varepsilon_{-1}, \varepsilon_0, \varepsilon_1, \varepsilon_2, \dots, \varepsilon_t$ and the observed process is given by $\dots, y_{-2}, y_{-1}, y_0, y_1, y_2, \dots, y_t$. In this context, the process $\{y_t\}$ is *causal* and it can be written as

$$y_t = \varphi(\dots, \varepsilon_{-2}, \varepsilon_{-1}, \varepsilon_0, \varepsilon_1, \varepsilon_2, \dots, \varepsilon_t).$$

Thus, a causal process depends only on past and present noise, and does not depend on *future* values of the noise. This is an important feature of the process $\{y_t\}$, meaning that only past or present shocks can affect it. If the process is not causal, for example, $y_t = \varepsilon_{t+1} - \theta\varepsilon_t$, then a *future* event ε_{t+1} can affect its *present* value. Even though this process is not causal, we can still predict it. As we will see later, the best linear predictor of y_t is given by $\widehat{y}_t = E[y_t | y_{t-1}, y_{t-2}, \dots]$. Note that for $|\theta| < 1$, we can write $\varepsilon_t = \sum_{j=1}^{\infty} \theta^{j-1} y_{t-j}$, so that the predictor can be expressed as $\widehat{y}_t = E[y_t | \varepsilon_t, \varepsilon_{t-1}, \dots] = -\theta\varepsilon_t$. In this case, the prediction error $e_t = y_t - \widehat{y}_t$ variance is $\mathrm{Var}(e_t) = \sigma_\varepsilon^2$.

2.5 REPRESENTATIONS OF LINEAR PROCESSES

A linear process may be represented in many different forms, for instance, as the Wold decomposition, an autoregressive expansion, or a state space system. To a large extent, these representations are equivalent and a key issue is how to pass from one representation to another. We have to keep in mind, however, that in most cases, these representations are not unique.

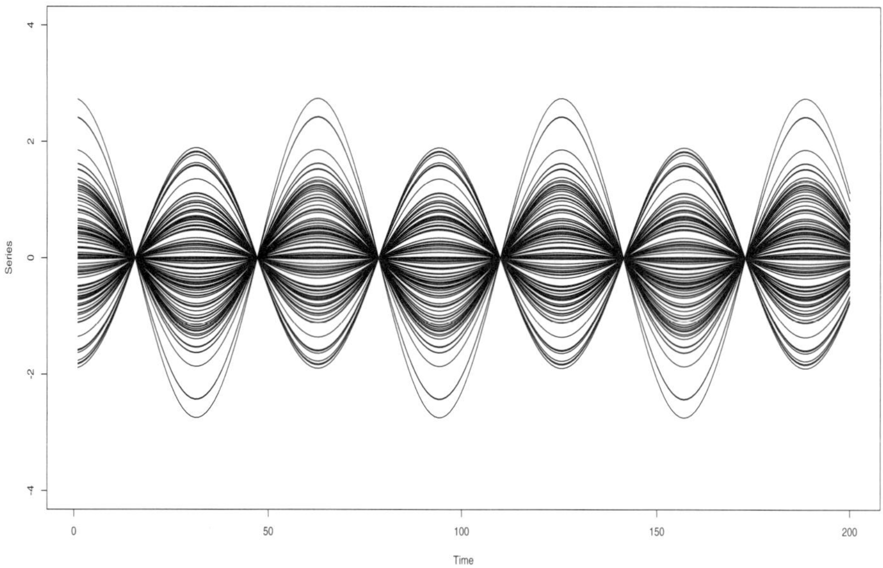

Figure 2.1 *Several paths of the singular process singular with zero-mean and unit variance Gaussian random variable Z. Real axis paths of 200 observations each, with* $\omega = 0.1$.

2.5.1 Wold Decomposition

A model described by (2.3) is usually referred to as *regular* process. Consider now the process

$$y_t = Ze^{\imath\,\omega\,t} = Z\cos(\omega t) + Z\imath\cos(\omega t), \tag{2.6}$$

where Z is a zero-mean random variable with variance σ^2 and ω is a known frequency. This time series satisfies $E(y_t) = 0$, $\mathrm{Var}(y_t) = \sigma^2$, and there is a function $\gamma(\cdot)$ such that $\mathrm{Cov}(y_t, y_s) = \gamma(t-s) = \sigma^2 e^{\imath\omega(t-s)}$. Thus, according to the definition discussed in the previous section, the process $\{y_t\}$ is weakly stationary. Note that this process has a different representation as the one described in (2.3). This sequence is an example of the so-called *singular* processes. See Figure 2.1 and Appendix A for further details about these definitions.

Even though both a regular process and a singular process may be stationary, the fundamental difference between them is that the former cannot be totally predicted, while the latter can be perfectly predicted. Generally speaking, as we accumulate more and more observations of a regular process, we can improve our forecasting ability, reducing, for example, the mean square prediction error. But there is always a certain level of prediction error. On the contrary, for the case of a singular process, it suffices to have only one observation in order to obtain a perfect predictor of any future value. In our example (2.6), any future value of the process can be written in terms of the observation y_t as

$$y_{t+h} = Ze^{\imath\,\omega\,(t+h)} = y_t e^{\imath \omega h}.$$

Thus, given that y_t and ω are assumed to be known, we know the exact value of y_{t+h}.

In this context, an important result called the *Wold decomposition* establishes that *any* stationary process can be written as the sum of a *regular* and a *singular* process and that decomposition is unique. Put it simply, any stationary process may contain a perfectly predictable or deterministic part and another non deterministic component.

Given that in practice, we usually have to deal with processes that cannot be fully predictable, in the remaining of this book we will only consider regular processes. However, singular processes are of interest in some fields such as electrical engineering where signal can be described by a particular frequency and a random amplitude.

2.5.2 Autoregressive Representation

Assuming that the linear process $\{y_t\}$ is invertible and causal, it can be expressed as

$$y_t = \sum_{j=1}^{\infty} \pi_j y_{t-j} + \varepsilon_t.$$

This is often called the autoregressive representation of the time series since the present value y_t is written in terms of its past values plus a noise.

2.5.3 State Space Systems

The linear processes introduced in (2.3) were described by the Wold decomposition. However, these processes can also be expressed in terms of a *state space linear system.*

A linear state space system may be described by the discrete-time equations

$$x_{t+1} = Fx_t + H\varepsilon_t, \tag{2.7}$$

$$y_t = Gx_t + \varepsilon_t, \tag{2.8}$$

where x_t is the *state* vector for all time t, $y_t \in \mathbb{R}$ is the *observation* sequence, $F : \mathcal{H} \to \mathcal{H}$ is the *state transition matrix* or the *state matrix*, G is the *observation matrix*, H is a state noise vector, and $\{\varepsilon_t\}$ is the *state white noise* sequence with variance σ^2; (2.7) is called the *state equation* while (2.8) is called the *observation equation*.

EXAMPLE 2.1

As an illustration, consider the following ARMA(1,1) model described by the equation

$$y_t - \phi y_{t-1} = \varepsilon_t - \theta\varepsilon_{t-1},$$

where $|\phi| < 1$, $|\theta| < 1$ and ε_t is a white noise sequence. Thus, in order to obtain the Wold expansion for this process, we can write

$$(1 - \phi B)y_t = (1 - \theta B)\varepsilon_t$$

and then

$$y_t = (1 - \phi B)^{-1}(1 - \theta B)\varepsilon_t.$$

Therefore, the Wold expansion is given by

$$y_t = \varepsilon_t + (\phi - \theta)\varepsilon_{t-1} + \phi(\phi - \theta)\varepsilon_{t-2} + \phi^2(\phi - \theta)\varepsilon_{t-3} + \cdots$$

On the other hand, by inverting $\theta(B)$, we obtain the infinite AR expansion. Therefore, the infinite AR expansion is given by

$$y_t = \varepsilon_t + (\theta - \phi)y_{t-1} + \theta(\theta - \phi)y_{t-2} + \theta^2(\theta - \phi)y_{t-3} + \cdots$$

A state space representation of this ARMA process is given by

$$\begin{aligned} x_{t+1} &= \phi x_t + (\phi - \theta)\varepsilon_t \\ y_t &= x_t + \varepsilon_t. \end{aligned}$$

Observe that according to the state equation, we can write

$$\begin{aligned} x_{t+1} &= (1 - \phi B)^{-1}(1 - \theta B)x_t \\ &= x_t + (\phi - \theta)x_{t-1} + \phi(\phi - \theta)x_{t-2} + \phi^2(\phi - \theta)x_{t-3}\cdots \end{aligned}$$

Finally, by replacing this expression in the observation equation, we obtain the Wold expansion for the model.

2.6 WEAK AND STRONG DEPENDENCE

Consider a stationary process $\{y_t\}$ with mean μ. An important and fundamental problem in time series is finding estimates of μ. Given a trajectory $y_1, y_2, \ldots, y_n$ of this process, a simple estimator of μ is the sample mean

$\widehat{\mu}_n = \bar{y}_n$. The variance of this estimator is

$$\begin{aligned}
\mathrm{Var}(\widehat{\mu}_n) &= \mathrm{Var}\left(\frac{1}{n}\sum_{t=1}^{n} y_t\right) \\
&= \frac{1}{n^2}\sum_{t=1}^{n}\sum_{s=1}^{n}\mathrm{Cov}(y_t, y_s) \\
&= \frac{1}{n^2}\sum_{t=1}^{n}\sum_{s=1}^{n}\gamma(t-s) \\
&= \frac{1}{n^2}\sum_{h=1-n}^{n-1}(n-|h|)\gamma(h).
\end{aligned}$$

Therefore, an upper bound for the variance of $\widehat{\mu}_n$ is given by

$$\mathrm{Var}(\widehat{\mu}_n) \leq \frac{2}{n}\sum_{h=0}^{n-1}|\gamma(h)|.$$

Consequently, as the sample size gets larger, we conclude that

$$\lim_{n\to\infty}\mathrm{Var}(\widehat{\mu}_n) \leq \lim_{n\to\infty}\frac{2}{n}\sum_{h=0}^{n-1}|\gamma(h)| = 2\lim_{n\to\infty}|\gamma(n)|.$$

From the above expression, if $\lim_{n\to\infty}\gamma(n) = 0$, then the variance of the estimator tends to zero, as the sample size tends to infinity. Thus, by an application of the Chebyshev inequality, we can conclude that $\lim_{n\to\infty}\gamma(n) = 0$ implies that the sample mean is a consistent estimator of μ.

At this point, one may ask about other important properties of $\widehat{\mu}_n$ such as rate of convergence of $\mathrm{Var}(\widehat{\mu}_n)$ to zero, asymptotic normality, and efficiency. In order to study these properties, it is necessary to incorporate to the analysis the *rate* at which $\gamma(n)$ tends to zero as $n\to\infty$. If this rate is sufficiently fast so that $\sum_{h=0}^{\infty}|\gamma(h)| < \infty$, then it can be shown that

$$\mathrm{Var}(\widehat{\mu}_n) = \sigma_n^2 \sim \frac{2}{n}\sum_{h=0}^{\infty}|\gamma(h)|,$$

and that $\widehat{\mu}_n$ satisfies a central limit theorem and is efficient,

$$\widehat{\mu}_n \sim \mathrm{N}\left(\mu, \sigma_n^2\right).$$

Time series satisfying the condition $\sum_{h=0}^{\infty}|\gamma(h)| < \infty$, are usually referred to as *short-memory* or *weakly dependent* processes.

On the other hand, if the rate of decaying of $\gamma(n)$ is slow enough so that $\sum_{h=0}^{\infty}|\gamma(h)| = \infty$, the analysis of the properties of $\widehat{\mu}_n$ is more complex. Consider, for example, that the ACF of a process satisfies the condition $\gamma(h) \sim$

$C|h|^{2d-1}$ for large h. Naturally, in order to satisfy that $\lim_{n\to\infty}\gamma(n) = 0$, the parameter d must satisfy $d < \frac{1}{2}$. Furthermore, note that for $d \in (0, \frac{1}{2})$, $\sum_{h=0}^{\infty}|\gamma(h)| = \infty$. In this case, we can calculate the variance of the sample mean for large n as follows:

$$\begin{aligned}\mathrm{Var}(\widehat{\mu}_n) &= \frac{1}{n^2}\sum_{h=1-n}^{n-1}(n-|h|)\gamma(h) \sim \frac{1}{n}\sum_{h=1-n}^{n-1}\gamma(h) \sim \frac{1}{n}\sum_{h=1-n}^{n-1}C|h|^{2d-1} \\ &= \frac{2}{n}\sum_{h=1}^{n-1}C\left|\frac{h}{n}\right|^{2d-1}n^{2d-1} = 2n^{2d-1}\sum_{h=1}^{n-1}C\left|\frac{h}{n}\right|^{2d-1}\frac{1}{n} \\ &\sim 2Cn^{2d-1}\int_0^1|x|^{2d-1}\,dx \sim C_1 n^{2d-1}.\end{aligned}$$

Observe that in this case, the variance of the sample mean converges to zero at a *slower* rate, as compared to the short-memory case. Therefore, stationary time series satisfying $\sum_{h=0}^{\infty}|\gamma(h)| = \infty$, are usually referred to as *long-memory* or *strongly dependent* processes.

2.7 ARMA MODELS

ARMA models are fundamental tools for analyzing short-memory time series. It can be shown that this class of models approximate arbitrarily well any linear stationary process with continuous spectral density defined in Chapter 3. Besides, there is a large number of numerical and computational tools for fitting, diagnosing, and forecasting ARMA models. They have been very useful for modeling a large number of time series exhibiting weak dependence. On the other hand, autoregressive fractionally integrated moving-average (ARFIMA) processes have been widely used for fitting time series data exhibiting long-range dependence. These two classes of models are discussed next.

An ARMA(p,q) process $\{y_t\}$ can be specified by the discrete-time equation,

$$\phi(B)y_t = \theta(B)\varepsilon_t, \tag{2.9}$$

where $\phi(B) = 1 - \phi_1 B - \cdots - \phi_p B^p$ is an autoregressive polynomial on the backshift operator B, $\theta(B) = 1 + \theta_1 B + \cdots + \theta_q B^q$ is a moving-average polynomial, with roots different from those of $\phi(B)$ and $\{\varepsilon_t\}$ is a white noise sequence with zero-mean and variance σ^2. Before studying the properties of this class of models, consider the following three examples. Note that an autoregressive AR(p) process corresponds to an ARMA$(p,0)$ model. On the other hand, a moving-average MA(q) is the special case ARMA$(0,q)$.

EXAMPLE 2.2

According to (2.9) an AR(1) model can be expressed as $y_t = \phi y_{t-1} + \varepsilon_t$, where ε_t is a white noise sequence with variance σ^2. Note that the mean of y_t must satisfy $E(y_t) = \phi\, E(y_{t-1})$ since $E(\varepsilon_t) = 0$. For $\phi \neq 0$, the stationarity condition implies that $E(y_t) = 0$. On the other hand, the variance of this process must satisfy $\mathrm{Var}(y_t) = \phi^2\, \mathrm{Var}(y_{t-1}) + \sigma^2$. Under stationarity, we must have $(1 - \phi^2)\,\mathrm{Var}(y_t) = \sigma^2$. Consequently, given that variances must be positive, we conclude that $|\phi| < 1$ and that $\mathrm{Var}(y_t) = \sigma^2/(1-\phi^2)$. With the above condition on the parameter ϕ, the process y_t may be expressed as the Wold expansion $y_t = \sum_{j=0}^{\infty} \phi^j\, \varepsilon_{t-j}$. The autocovariance function of this process can be obtained as follows. Consider $h > 1$, so that

$$y_{t+h} = \phi^h y_t + \phi^{h-1}\varepsilon_t + \phi^{h-2}\varepsilon_{t+1} + \cdots + \varepsilon_{t+h-1}.$$

Thus,

$$\begin{aligned}
\gamma(h) &= E(y_t y_{t+h}) \\
&= E[y_t(\phi^h y_t + \phi^{h-1}\varepsilon_t + \phi^{h-2}\varepsilon_{t+1} + \cdots + \varepsilon_{t+h-1})] \\
&= \phi^h\, E\, y_t^2 + \phi^{h-1}\, E\, y_t\varepsilon_t + \cdots + E\, y_t\varepsilon_{t+h-1} \\
&= \phi^h\, E\, y_t^2 = \frac{\sigma^2\, \phi^h}{1-\phi^2}.
\end{aligned}$$

2.7.1 Invertibility of ARMA Processes

An ARMA(p, q) process is said to be *invertible* if all the roots z of the polynomial $\Theta(z) = 1 + \theta_1 z + \cdots + \theta_q z^q$ satisfy $|z| > 1$. This means that we can write the following expression for the noise sequence

$$\varepsilon_t = \theta(B)^{-1}\phi(B) y_t = \Pi(B) y_t = \sum_{j=0}^{\infty} \pi_j y_{t-j}.$$

2.7.2 Simulated ARMA Processes

In order to gain an insight about how a trajectory of an ARMA time series looks like in what follows, we present several simulated processes.

■ **EXAMPLE 2.3**

As a first example, Figure 2.2 shows a simulated trajectory of 1000 observations from an AR(1) process with parameter $\phi = 0.5$. Note that due to the dependence of the observations, the series seems to have some trends. An even stronger effect of the dependence can be observed in Figure 2.3, which shows an AR(1) process with autoregressive parameter $\phi = 0.9$. In this case, the trends seem much more relevant. Nevertheless, since these two processes are stationary, there are no *true* trends in these series.

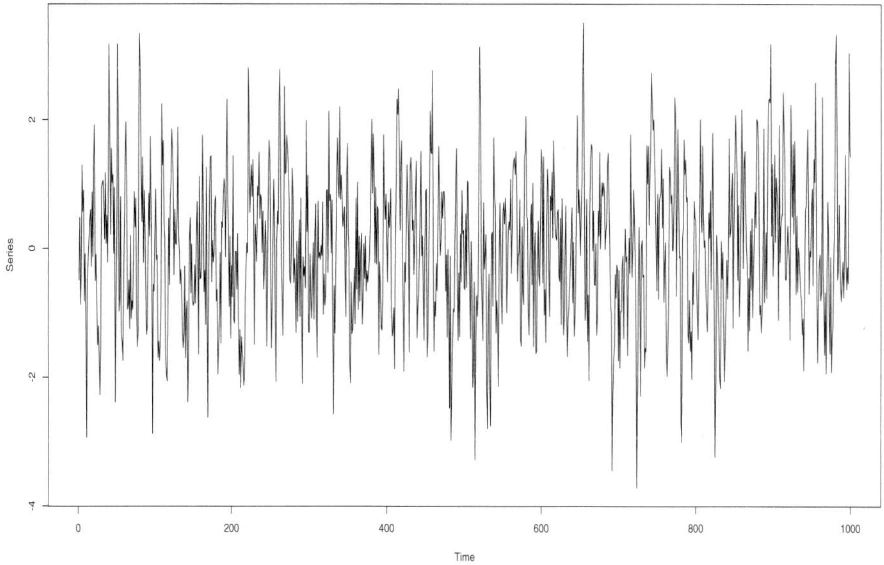

Figure 2.2 *Simulated $AR(1)$ time series of length 1000 with $\phi = 0.5$.*

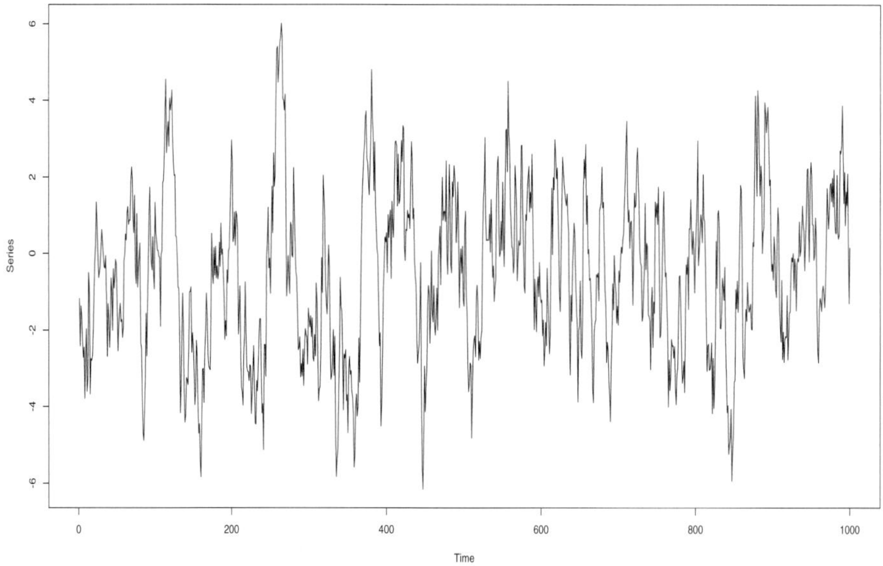

Figure 2.3 *Simulated $AR(1)$ time series of length 1000 with $\phi = 0.9$.*

EXAMPLE 2.4

The behavior of moving-average models is explored in Figures 2.4 and 2.5. Both series are of length 1000 observations. Figure 2.4 displays the trajectory of an MA(2) process with parameters $\theta = (0.7, -0.6)$, while Figure 2.5 shows an MA(4) process with moving-average parameters $\theta = (0.3, -0.4, 0.7, 0.5)$. Similar to the previous AR examples, in this case, we can also observe the effect of the dependence in the trajectory of the series.

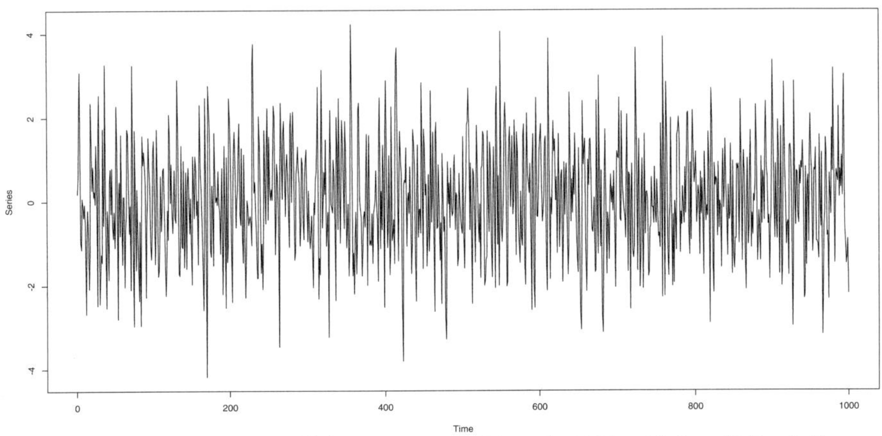

Figure 2.4 *Simulated MA(2) time series of length 1000 with* $\theta = (0.7, -0.6)$.

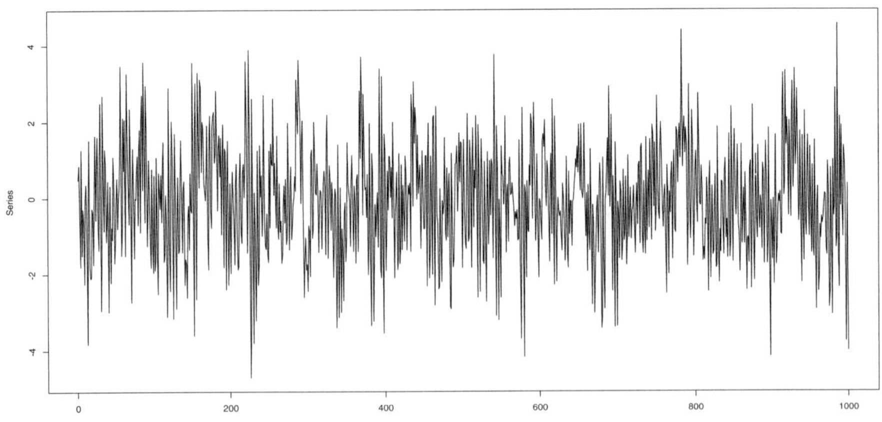

Figure 2.5 *Simulated MA(4) time series of length 1000 with* $\theta = (0.3, -0.4, 0.7, 0.5)$.

■ EXAMPLE 2.5

Finally, Figures 2.6 and 2.7 exhibit two simulated ARMA(1, 1) processes, with parameters $\phi = 0.5$ and $\theta = 0.7$, and $\phi = 0.5$ and $\theta = -0.7$, respectively. Notice the variety of trajectories generated by these models. In particular, observe that by just changing the sign of the moving-average parameter, we obtain a much different time series.

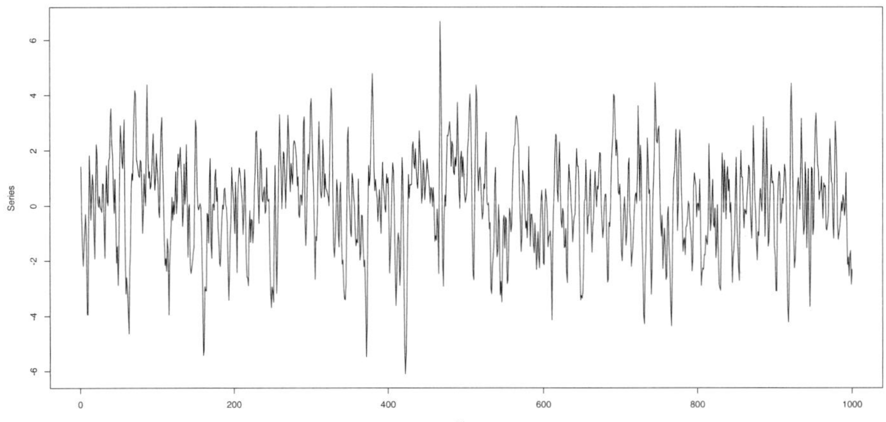

Figure 2.6 *Simulated ARMA(1, 1) time series of length 1000 with* $\phi = 0.5$ *and* $\theta = 0.7$.

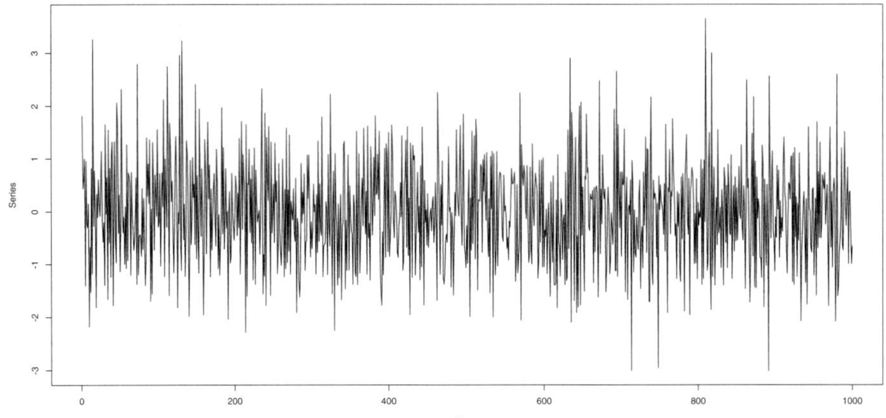

Figure 2.7 *Simulated ARMA(1, 1) time series of length 1000 with* $\phi = 0.5$ *and* $\theta = -0.7$.

2.8 AUTOCOVARIANCE FUNCTION

As defined in Chapter 1, a second-order stationary time series possesses an autocovariance function $\gamma(\cdot)$, which measures the level of dependence of the value at a certain time t and another value at time $t+h$. For a stationary process, this statistical dependence indicated by the covariance between both values does not depend on t but only on the lag h. Several examples of calculations of the ACF for ARMA processes are presented next.

Consider the moving-average model MA(q) given by

$$y_t = \varepsilon_t + \theta_1 \varepsilon_{t-1} + \cdots + \theta_q \varepsilon_{t-q}. \tag{2.10}$$

In this case, the ACF is

$$\gamma(h) = \begin{cases} \sigma^2(\theta_1^2 + \cdots + \theta_q^2) & \text{for} \quad h = 0 \\ \sigma^2 \sum_{j=0}^{q-|h|} \theta_j \theta_{j+|h|} & \text{for} \quad |h| = 1, \ldots, q \\ 0 & \text{for} \quad |h| > q. \end{cases} \tag{2.11}$$

Based on this expression, a moving-average process can be identified from its empirical ACF. If the sample ACF is not significant after a lag q, this is an indication that a MA(q) process may model the data adequately.

■ EXAMPLE 2.6

Consider the ARMA(1, 1) satisfying the discrete-time equation

$$y_t - \phi y_{t-1} = \varepsilon_t + \theta \varepsilon_{t-1}. \tag{2.12}$$

There are at least two ways to calculate its autocovariance function.

First Method: From expression (2.5), it suffices to calculate the coefficients ψ_j and then calculate the sum. To this end, we can write

$$\begin{aligned} y_t &= (1-\phi B)^{-1}(1+\theta B)\varepsilon_t \\ &= [1 + (\phi+\theta)B + \phi(\phi+\theta)B^2 + \phi^2(\phi+\theta)B^3 + \cdots]\varepsilon_t \\ &= \varepsilon_t + (\phi+\theta)\varepsilon_{t-1} + \phi(\phi+\theta)\varepsilon_{t-2} + \phi^2(\phi+\theta)\varepsilon_{t-3} + \cdots. \end{aligned} \tag{2.13}$$

Thus, $\psi_0 = 1$, $\psi_j = \phi^{j-1}(\phi+\theta)$ for $j \geq 1$. Consequently,

$$\gamma(h) = \sigma^2 \left[\psi_{|h|} + \frac{(\phi+\theta)^2 \phi^{|h|}}{1-\phi^2} \right]. \tag{2.14}$$

Second Method: Another way to compute the autocovariance function is as follows. Multiplying both sides of (2.12) by y_{t-h}, we get

$$y_t y_{t-h} - \phi y_{t-1} y_{t-h} = \varepsilon_t y_{t-h} + \theta \varepsilon_{t-1} y_{t-h}.$$

Now, taking expected values, we obtain

$$E(y_t y_{t-h}) - \phi\, E(y_{t-1} y_{t-h}) = E((\varepsilon_t y_{t-h}) + \theta\, E(\varepsilon_{t-1} y_{t-h}).$$

From (2.13), we have that $E\,\varepsilon_t y_t = \sigma^2$ and $E(\varepsilon_{t-1} y_{t-h}) = \psi_h \sigma^2$. That is,

$$\gamma(h) - \phi\gamma(h-1) = E(\varepsilon_t y_{t-h}) + \theta\, E(\varepsilon_{t-1} y_{t-h}).$$

For $h = 0$, we have that

$$\gamma(0) - \phi\gamma(1) = \sigma^2(1 + \theta\psi_1), \tag{2.15}$$

and for $h = 1$, we get

$$\gamma(1) - \phi\gamma(0) = \sigma^2\theta. \tag{2.16}$$

Given that ε_t and ε_{t-1} are uncorrelated with y_{t-h} for $h > 2$, we get

$$\gamma(h) = \phi\gamma(h-1). \tag{2.17}$$

Note that from (2.15) and (2.16), we conclude that

$$\gamma(0) = \sigma^2 \frac{1 + 2\phi\theta + \theta^2}{1 - \phi^2},$$

$$\gamma(1) = \sigma^2 \frac{(\phi + \theta)(1 + \phi\theta)}{1 - \phi^2},$$

and from (2.17),

$$\gamma(h) = \phi^{|h|}\gamma(0),$$

for $|h| > 1$.

2.9 ACF AND PARTIAL ACF FUNCTIONS

As discussed previously, the ACF is a *standardized* measure of the dependence of two observations y_t and y_{t+h}, corresponding to the autocovariance function divided by the variance of the process.

$$\rho(h) = \frac{\gamma(h)}{\gamma(0)}. \tag{2.18}$$

From the previous section, we can write the ACF of a moving-average process MA(q) as follows:

$$\rho(h) = \begin{cases} 1 & \text{for} \quad h = 0 \\ \frac{\sum_{j=0}^{q-|h|} \theta_j \theta_{j+|h|}}{\theta_1^2 + \cdots + \theta_q^2} & \text{for} \quad |h| = 1, \ldots, q \\ 0 & \text{for} \quad |h| > q. \end{cases}$$

In what follows, we present a number of examples of the behavior of the ACF for different ARMA models. Figure 2.8 exhibits the exact ACF for two AR(1) processes. Figure 2.8(a) shows the ACF for a model with $\phi = 0.5$

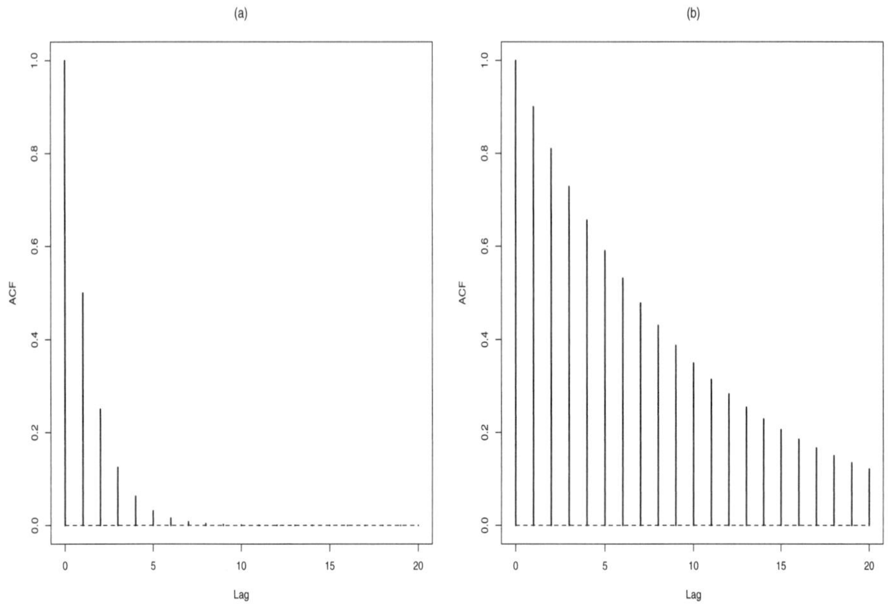

Figure 2.8 *Exact autocorrelation function (ACF) of an AR(1) model. (a)* $\phi = 0.5$, *(b)* $\phi = 0.9$.

while Figure 2.8(b) depicts the ACF for $\phi = 0.9$. Note the different decaying rates of these ACFs. As ϕ increases, the ACF decays more slowly. Recall, however, that a necessary condition for ensuring the stationarity of these AR(1) models is that $|\phi| < 1$. In this situation, as ϕ approaches the upper limit 1, the convergence rate to zero is extremely slow, as shown in Figure 2.9 (a) and (b) which depicts the ACF for $\phi = 0.95$ and $\phi = 0.99$, respectively.

The theoretical ACF of MA(q) time series models is exhibited in Figure 2.10. The ACF on Figure 2.10(a) corresponds to an MA(2) model with $\theta = (0.7, -0.6)$ while the ACF on Figure 2.10(b) is for an MA(4) process with $\theta = (0.3, -0.4, 0.7, 0.5)$. As expected, the ACF vanishes for lags greater than the order q of the model.

Figure 2.11 shows the exact ACF for two ARMA$(1,1)$ models. The ACF on Figure 2.11(a) corresponds to a model with parameters $\phi = 0.5$ and $\theta = 0.7$ while the ACF on Figure 2.11(b) corresponds to an ARMA$(1,1)$ time series model with parameters $\phi = 0.5$ and $\theta = -0.7$. Note that the decaying rate of the ACF is governed by $\phi^{|h|}$. Thus, it is expected that in both cases, the ACF converges rapidly to zero as the lag increases. But the sign of the ACF is governed in this case for the moving-average parameter θ, which is clearly reflected on Figure 2.11(b) which corresponds to a negative value of this parameter.

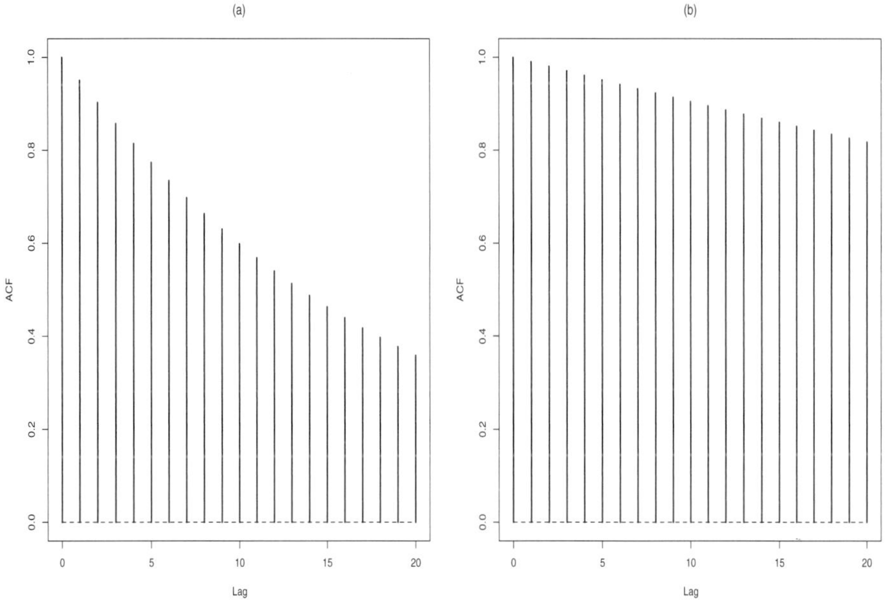

Figure 2.9 *Exact autocorrelation function (ACF) of an AR(1) model. (a)* $\phi = 0.95$, *(b)* $\phi = 0.99$.

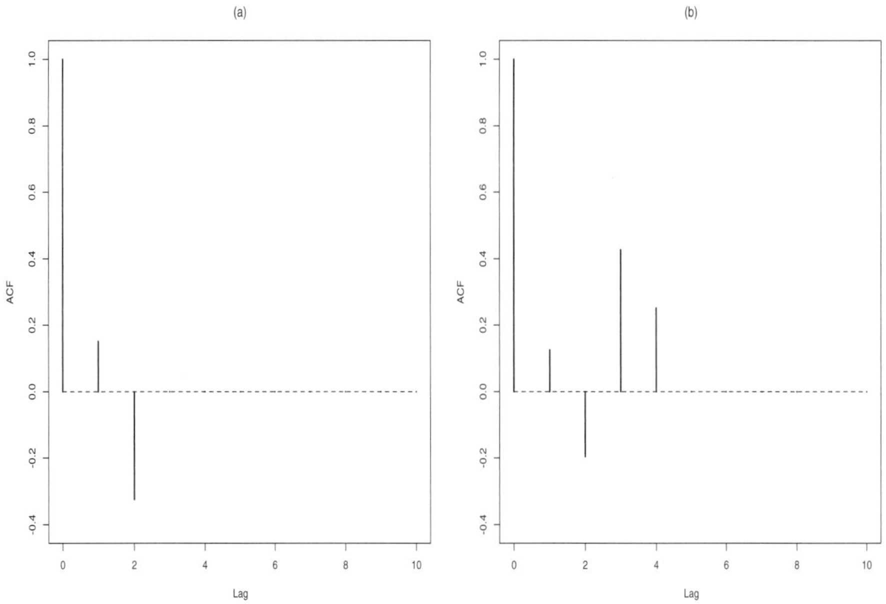

Figure 2.10 *Exact autocorrelation function (ACF) of a MA(q) model, for* $q = 2$ *and* $q = 4$. *(a)* $\theta = (0.7, -0.6)$, *(b)* $\theta = (0.3, -0.4, 0.7, 0.5)$.

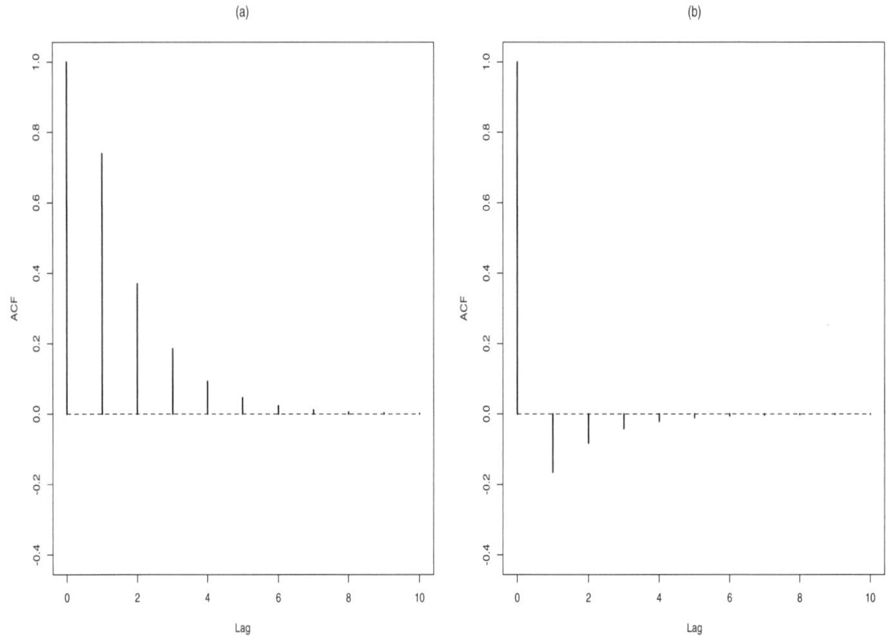

Figure 2.11 *Exact autocorrelation function (ACF) of an ARMA$(1, 1)$ model. (a) $\phi = 0.5$ and $\theta = 0.7$, (b) $\phi = 0.5$ and $\theta = -0.7$.*

2.9.1 Sample ACF

The sample counterparts of the ACF are shown in the following figures. All these illustrations are based on 1000 simulated observations of the respective processes.

For instance, Figure 2.12 depicts the sample ACF of an AR(1) time series model with autoregressive parameter $\phi = 0.5$. Note that the sample ACF decays as expected from Figure 2.8(a).

Similarly, Figure 2.13 exhibits the sample version of the ACF for an AR(1) model with autoregressive parameter $\phi = 0.9$. This empirical ACF can be compared to its theoretical counterpart depicted on Figure 2.8(b).

The behavior of the sample ACF for two ARMA(1,1) models is exhibited in Figures 2.14 and 2.15. The model in Figure 2.14 has parameters $\phi = 0.5$ and $\theta = 0.7$, while the model in Figure 2.15 has parameters $\phi = 0.5$ and $\theta = -0.7$. Observe that these two plots are close to their theoretical versions shown in Figure 2.11.

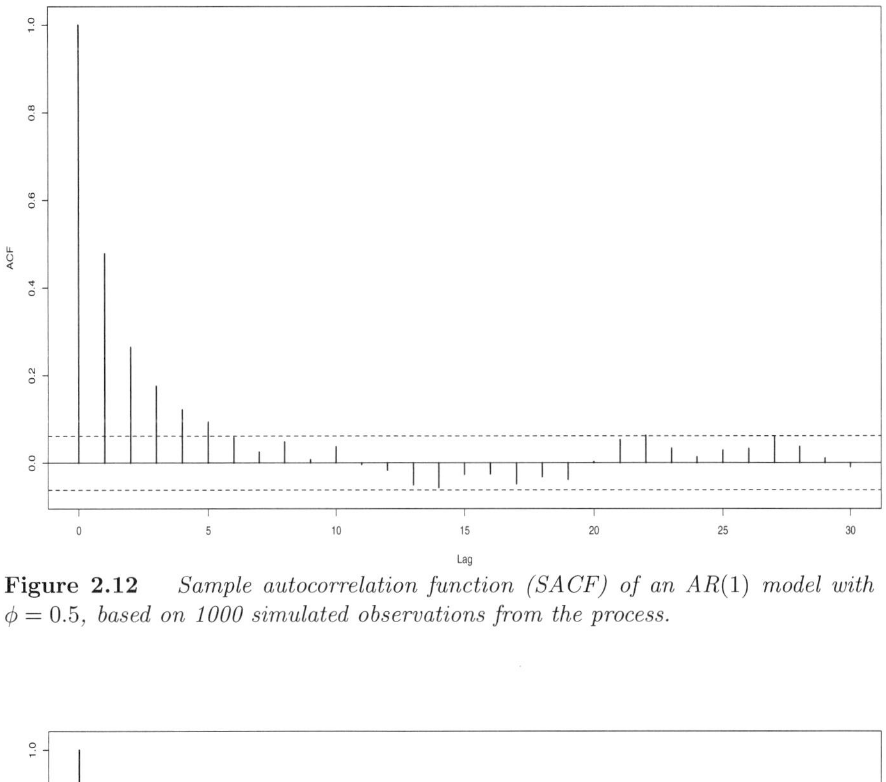

Figure 2.12 *Sample autocorrelation function (SACF) of an* $AR(1)$ *model with* $\phi = 0.5$*, based on 1000 simulated observations from the process.*

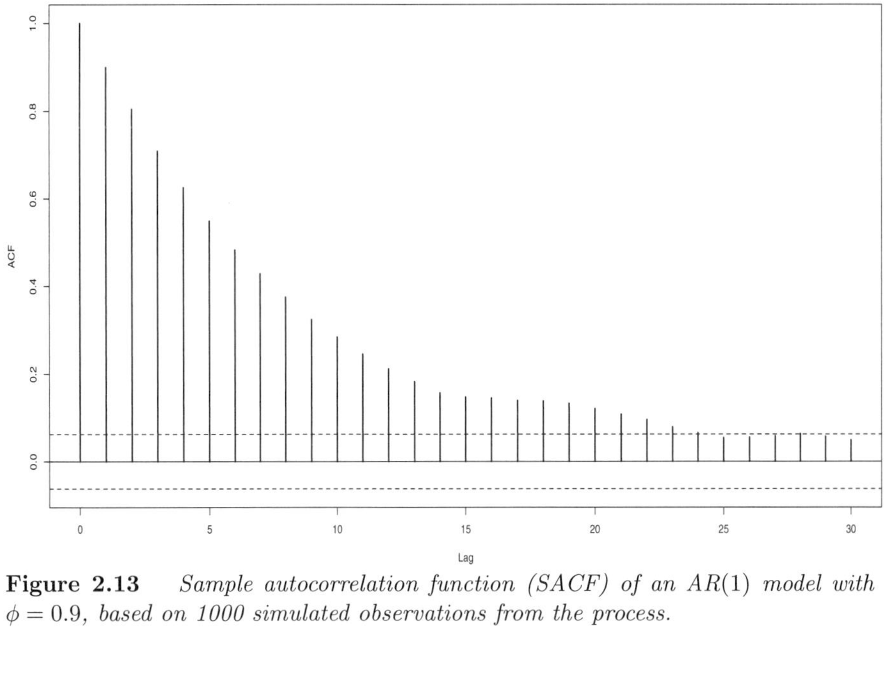

Figure 2.13 *Sample autocorrelation function (SACF) of an* $AR(1)$ *model with* $\phi = 0.9$*, based on 1000 simulated observations from the process.*

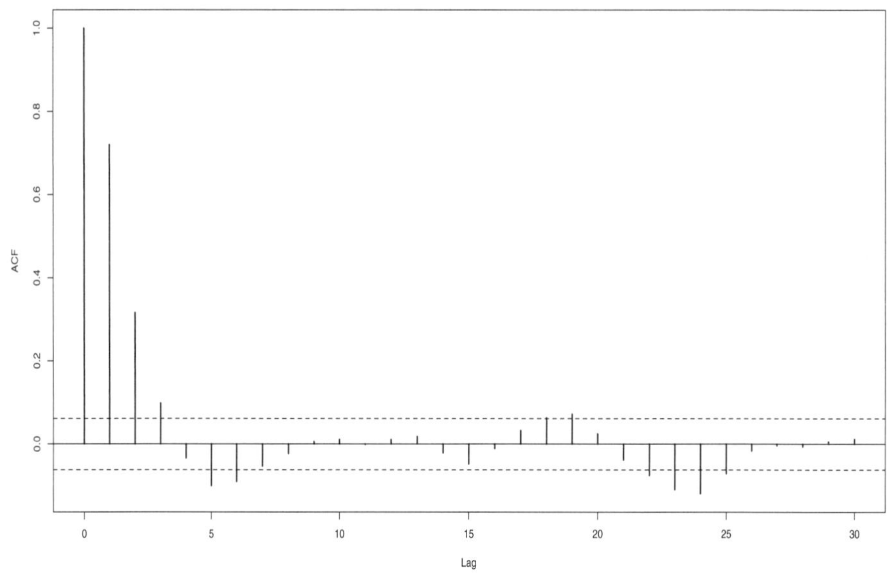

Figure 2.14 *Sample autocorrelation function (SACF) of an ARMA(1,1) model with $\phi = 0.5$ and $\theta = 0.7$.*

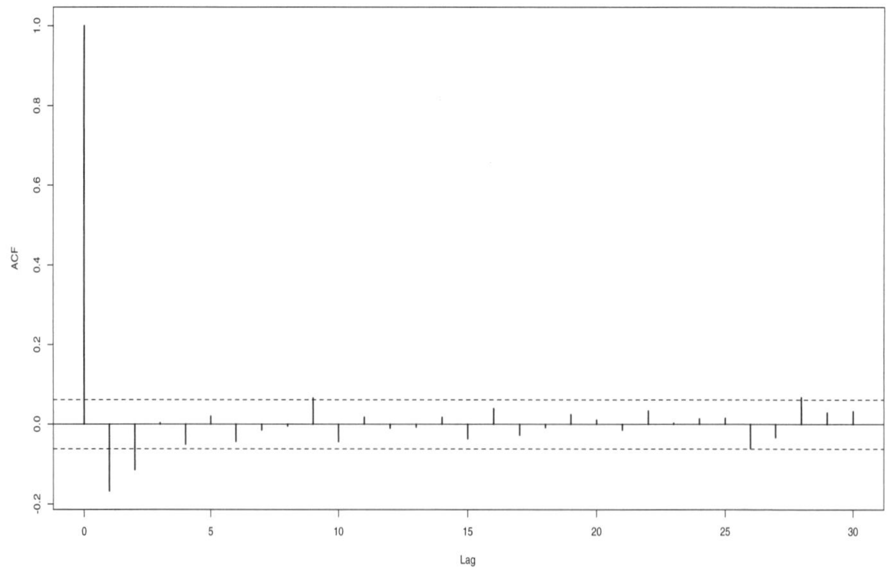

Figure 2.15 *Sample autocorrelation function (SACF) of an ARMA(1,1) model with $\phi = 0.5$ and $\theta = -0.7$.*

2.9.2 Partial ACF

Consider the series $y_1, y_2, \ldots, y_n$ and the predictors

$$\begin{aligned} \widehat{y}_1 &= E(y_1|y_2, \ldots, y_n), \\ \widehat{y}_{n+1} &= E(y_{n+1}|y_2, \ldots, y_n), \end{aligned}$$

for $n \geq 2$ and $\widehat{y}_1 = \widehat{y}_2 = 0$. The PACF is the correlation between the errors $e_1 = y_1 - \widehat{y}_1$ and $e_{n+1} = y_{n+1} - \widehat{y}_{n+1}$, that is,

$$\alpha(n) = \frac{\operatorname{Cov}(e_{n+1}, e_1)}{\sqrt{\operatorname{Var}(e_{n+1})\operatorname{Var}(e_1)}}.$$

Observe that, based on the definition of the predictors $\widehat{y}_1$ and $\widehat{y}_0$, we have that the prediction errors are $e_1 = y_1$ and $e_2 = y_2$ and then $\alpha(1) = \rho(1)$.

Even though this correlation may appear as difficult to interpret, for an AR(p) $\alpha(h) = 0$ for $h > p$. Thus, it allows for the easy identification of an autoregressive process. Figure 2.16 depicts the PACF of an AR(2) model with parameters $\phi_1 = 0.4$ and $\phi_2 = 0.2$. Figure 2.16(a) shows the exact PACF, while Figure 2.16(b) exhibits the ample PACF based on a series of 1000 observations. Notice that the theoretical PACF vanishes after the second lag. Furthermore, its sample version shows a similar behavior when the confidence bands are taken into account.

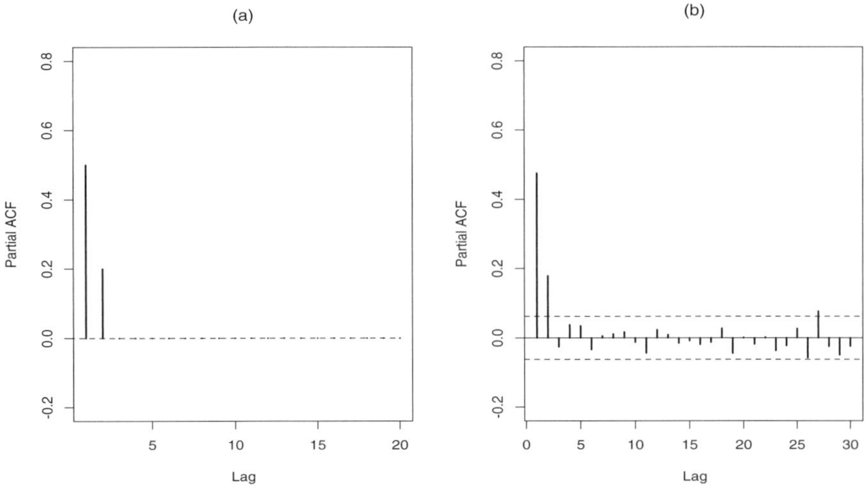

Figure 2.16 *Partial autocorrelation function (PACF) of an AR(2) model with* $\phi_1 = 0.4$ *and* $\phi_2 = 0.2$. *(a) Exact PACF, (b) Sample PACF based on a series of 1000 observations.*

2.10 ARFIMA PROCESSES

In this section, we focus our attention on a particular class of linear time series called *long-memory* or *long-range-dependent* processes. There are several definitions of this type of time series in the literature. One fundamentals aspect is related to the estimation of the mean of a process.

If the autocovariance function of a stationary process is summable, then the sample mean is root-n consistent, where n is the sample size. This is the case, for instance, for sequences of independent and identically distributed random variables or Markovian processes. Generally speaking, these processes are said to have *short memory*. On the contrary, a process has *long memory* if its autocovariances are not absolutely summable.

In what follows, we provide a brief overview of these class of time series models.

2.10.1 Long-Memory Processes

Let $\gamma(h) = \langle y_t, y_{t+h} \rangle$ be the autocovariance function at lag h of the stationary process $\{y_t : t \in \mathbb{Z}\}$. A usual definition of long memory is that

$$\sum_{h=-\infty}^{\infty} |\gamma(h)| = \infty. \tag{2.19}$$

However, there are alternative definitions. In particular, long memory can be defined by specifying a hyperbolic decay of the autocovariances

$$\gamma(h) \sim h^{2d-1} \ell_1(h), \tag{2.20}$$

as $h \to \infty$, where d is the so-called *long-memory parameter* and $\ell_1(\cdot)$ is a slowly varying function. Recall that a positive measurable function defined on some neighborhood $[a, \infty)$ of infinity is said to be *slowly varying* if for any $c > 0$, $\ell(cx)/\ell(x)$ converges to 1 as x tends to infinity. Examples of slowly varying functions are $\ell(x) = \log(x)$ and $\ell(x) = b$, where b is a positive constant. Hereafter, the notation $x_n \sim y_n$ means that $x_n/y_n \to 1$ as $n \to \infty$, unless specified otherwise.

A well-known class of long-memory models is the autoregressive fractionally integrated moving-average (ARFIMA) processes. An ARFIMA process $\{y_t\}$ may be defined by

$$\phi(B) y_t = \theta(B)(1 - B)^{-d} \varepsilon_t, \tag{2.21}$$

where $\phi(B) = 1 + \phi_1 B + \cdots + \phi_p B^p$ and $\theta(B) = 1 + \theta_1 B + \cdots + \theta_q B^q$ are the autoregressive and moving-average operators, respectively; $\phi(B)$ and $\theta(B)$ have no common roots, $(1 - B)^{-d}$ is a fractional differencing operator defined by the binomial expansion

$$(1 - B)^{-d} = \sum_{j=0}^{\infty} \eta_j B^j = \eta(B),$$

where

$$\eta_j = \frac{\Gamma(j+d)}{\Gamma(j+1)\Gamma(d)}, \tag{2.22}$$

for $d < \frac{1}{2}$, $d \neq 0, -1, -2, \ldots$, and $\{\varepsilon_t\}$ is a white noise sequence with finite variance.

Consider the ARFIMA process defined by (2.21). Assume that the polynomials $\phi(\cdot)$ and $\theta(\cdot)$ have no common zeros and that $d \in (-1, \frac{1}{2})$. Then, the stationarity, causality, and invertibility of an ARFIMA model can be established as follows.

(a) If the zeros of $\phi(\cdot)$ lie outside the unit circle $\{z\colon |z| = 1\}$, then there is a unique stationary solution of (2.21) given by

$$y_t = \sum_{j=-\infty}^{\infty} \psi_j \varepsilon_{t-j},$$

where $\psi(z) = (1-z)^{-d}\theta(z)/\phi(z)$.

(b) If the zeros of $\phi(\cdot)$ lie outside the closed unit disk $\{z\colon |z| \leq 1\}$, then the solution $\{y_t\}$ is causal.

(c) If the zeros of $\theta(\cdot)$ lie outside the closed unit disk $\{z\colon |z| \leq 1\}$, then the solution $\{y_t\}$ is invertible.

2.10.2 Linear Representations

Infinite AR and MA expansions for an ARFIMA process can be described as follows. Under the assumption that the roots of the polynomials $\phi(B)$ and $\theta(B)$ are outside the closed unit disk $\{z\colon |z| \leq 1\}$ and $d \in (-1, \frac{1}{2})$, the ARFIMA(p, d, q) process is stationary, causal, and invertible. In this case we can write

$$y_t = (1-B)^{-d}\phi(B)^{-1}\theta(B)\varepsilon_t = \psi(B)\varepsilon_t,$$

and

$$\varepsilon_t = (1-B)^{d}\phi(B)\theta(B)^{-1}y_t = \pi(B)y_t.$$

The MA(∞) coefficients, ψ_j, and AR(∞) coefficients, π_j, satisfy the following asymptotic relationships,

$$\psi_j \sim \frac{\theta(1)}{\phi(1)} \frac{j^{d-1}}{\Gamma(d)}, \tag{2.23}$$

$$\pi_j \sim \frac{\phi(1)}{\theta(1)} \frac{j^{-d-1}}{\Gamma(-d)}, \tag{2.24}$$

as $j \to \infty$.

For a fractional noise process with long-memory parameter d, these coefficients are given by

$$\begin{aligned} \psi_j &= \prod_{t=1}^{j} \frac{t-1+d}{t} = \frac{\Gamma(j+d)}{\Gamma(d)\Gamma(j+1)}, \\ \pi_j &= \prod_{t=1}^{j} \frac{t-1-d}{t} = \frac{\Gamma(j-d)}{\Gamma(-d)\Gamma(j+1)}, \end{aligned}$$

for $j \geq 1$ and $\psi_0 = \pi_0 = 1$.

2.10.3 Autocovariance Function

The autocovariance function of the ARFIMA$(0, d, 0)$ process is given by

$$\gamma_0(h) = \sigma^2 \frac{\Gamma(1-2d)}{\Gamma(1-d)\Gamma(d)} \frac{\Gamma(h+d)}{\Gamma(1+h-d)}, \tag{2.25}$$

where $\Gamma(\cdot)$ is the gamma function and the ACF is

$$\rho_0(h) = \frac{\Gamma(1-d)}{\Gamma(d)} \frac{\Gamma(h+d)}{\Gamma(1+h-d)}.$$

For the general ARFIMA(p, d, q) process, observe that the polynomial $\phi(B)$ in (2.21) may be written as

$$\phi(B) = \prod_{i=1}^{p} (1 - \rho_i B).$$

Assuming that all the roots of $\phi(B)$ have multiplicity one, it can be deduced that

$$\gamma(h) = \sigma^2 \sum_{i=-q}^{q} \sum_{j=1}^{p} \psi(i) \xi_j C(d, p+i-h, \rho_j), \tag{2.26}$$

with

$$\psi(i) = \sum_{k=\max(0,i)}^{\min(q,q+i)} \theta_k \theta_{k-i},$$

$$\xi_j = \left[\rho_j \prod_{i=1}^{p} (1 - \rho_i \rho_j) \prod_{m \neq j} (\rho_j - \rho_m) \right]^{-1},$$

and

$$C(d,h,\rho) = \frac{\gamma_0(h)}{\sigma^2}[\rho^{2p}\beta(h) + \beta(-h) - 1], \tag{2.27}$$

where $\beta(h) = F(d+h, 1, 1-d+h, \rho)$ and $F(a,b,c,x)$ is the Gaussian hypergeometric function

$$F(a,b,c,x) = 1 + \frac{a \cdot b}{\gamma \cdot 1}x + \frac{a \cdot (a+1) \cdot b \cdot (b+1)}{\gamma \cdot (\gamma+1) \cdot 1 \cdot 2}x^2 + \cdots .$$

It can be shown that

$$\gamma(h) \sim c_\gamma |h|^{2d-1}, \tag{2.28}$$

as $|h| \to \infty$, where

$$c_\gamma = \frac{\sigma^2}{\pi} \frac{|\theta(1)|^2}{|\phi(1)|^2} \Gamma(1-2d)\sin(\pi d).$$

2.10.4 Sample Mean

Let $y_1, y_2, \ldots, y_n$ be a sample from an ARFIMA(p,d,q) process, and let $\bar{y}$ be the sample mean. The variance of $\bar{y}$ is given by

$$\mathrm{Var}(\bar{y}) = \frac{1}{n}\left[2\sum_{j=1}^{n-1}\left(1 - \frac{j}{n}\right)\gamma(j) + \gamma(0)\right].$$

By formula (2.28), $\gamma(j) \sim c_\gamma j^{2d-1}$ for large j. Hence, for large n, we have

$$\begin{aligned}
\mathrm{Var}(\bar{y}) &\sim 2c_\gamma n^{2d-1} \sum_{j=1}^{n-1}\left(1 - \frac{j}{n}\right)\left(\frac{j}{n}\right)^{2d-1}\frac{1}{n} \\
&\sim 2c_\gamma n^{2d-1} \int_0^1 (1-t)t^{2d-1}dt \\
&\sim \frac{c_\gamma}{d(2d+1)}n^{2d-1}.
\end{aligned} \tag{2.29}$$

2.10.5 Partial Autocorrelations

Explicit expressions for the PACF for the general ARFIMA model are difficult to find. In the particular case of a fractional noise process FN(d), the coefficients of the best linear predictor

$$\widehat{y}_{n+1} = \phi_{n1}y_n + \cdots + \phi_{nn}y_1$$

are given by

$$\phi_{nj} = -\binom{n}{j}\frac{\Gamma(j-d)\Gamma(n-d-j+1)}{\Gamma(-d)\Gamma(n-d+1)},$$

for $j = 1, \ldots, n$. Thus, the partial autocorrelations are simply

$$\phi_{nn} = \frac{d}{n-d} \tag{2.30}$$

and then $\phi_{nn} \sim d/n$ for large n.

2.10.6 Illustrations

In the following two simulation examples, we illustrate some of the concepts discussed in the previous sections about ARFIMA processes. For simplicity, consider the family of ARFIMA$(1, d, 1)$ models

$$(1 + \phi B)y_t = (1 + \theta B)(1 - B)^{-d}\varepsilon_t,$$

where the white noise sequence satisfies $\{\varepsilon_t\} \sim N(0, 1)$. In these examples, the sample autocorrelation function has been calculated by first estimating the autocovariance function by means of the expression

$$\widehat{\gamma}(h) = \frac{1}{n}\sum_{t=1}^{n-h}(y_t - \bar{y})(y_{t+h} - \bar{y}),$$

for $h = 0, \ldots, n-1$, where $\bar{y}$ is the sample mean and then defining the autocorrelation function estimate as:

$$\widehat{\rho}(h) = \frac{\widehat{\gamma}(h)}{\widehat{\gamma}(0)}.$$

The theoretical values of the ACF were calculated as follows. According to formula (2.26), the autocovariance function for the process $\{y_t\}$ is given by

$$\gamma(h) = \frac{\theta C(d, -h, -\phi) + (1 + \theta^2)C(d, 1-h, -\phi) + \theta C(d, 2-h, -\phi)}{\phi(\phi^2 - 1)},$$

where the function $C(\cdot, \cdot, \cdot)$ is defined in (2.27). Hence, we have $\rho(h) = \gamma(h)/\gamma(0)$. On the other hand, the theoretical PACF of a fractional noise process is given by (2.30) while the theoretical PACF of the ARFIMA$(1, d, 1)$ model can be computed by means of the Durbin-Levinson algorithm; see Chapter 5 for further details.

■ **EXAMPLE 2.7**

Figure 2.17 shows 1000 simulated values from an ARFIMA$(0, d, 0)$ process with $d = 0.4$. The theoretical and the empirical ACF are shown in Figure 2.18.

From Figure 2.17, note that this time series exhibits a persistence in its values, they tend to stay at a certain level for a while and then

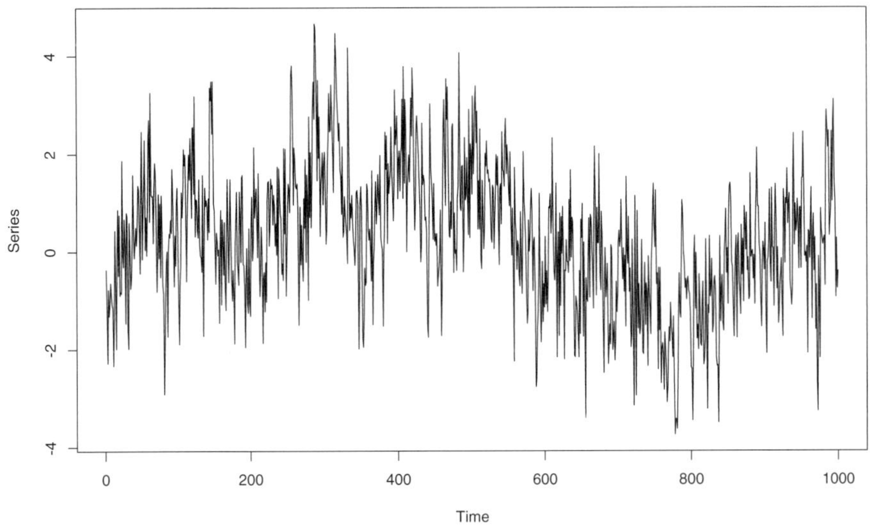

Figure 2.17 *Simulated ARFIMA$(0,d,0)$ time series with 1000 observations with $d = 0.4$.*

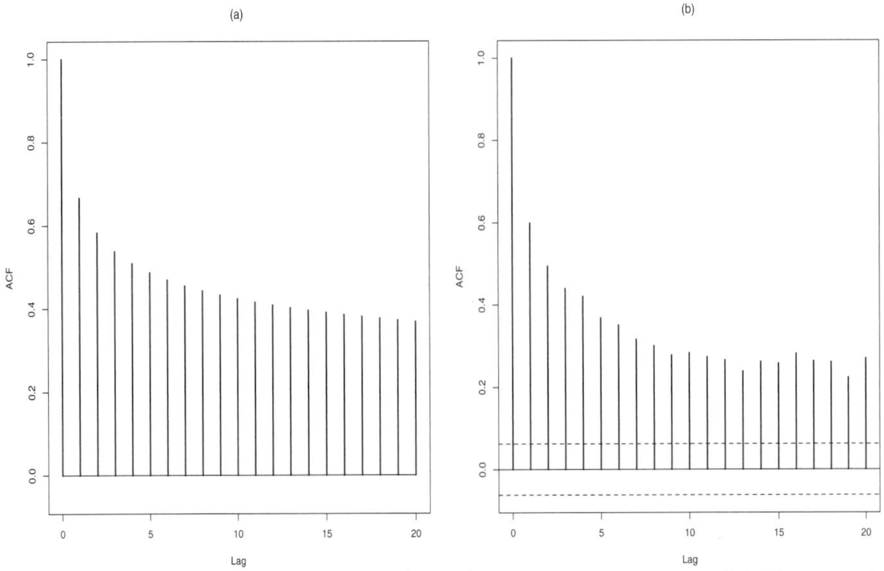

Figure 2.18 *ACF of an ARFIMA$(0,d,0)$ time series $d = 0.4$. (a) Theoretical ACF, (b) Sample ACF*

jump to another level. The first behavior is known as the *Joseph effect* while the second is known as the *Noah effect*, that is, abrupt events that change dramatically the level of the observations.

On the other hand, notice from Figure 2.18 that the sample autocorrelations are significant, even after a large number of lags. This behavior is expected from Figure 2.18(a), which shows its theoretical counterpart.

EXAMPLE 2.8

Figure 2.19 exhibits 1000 observations from an ARFIMA$(1, d, 1)$ process with $d = 0.3$, $\phi = 0.6$, and $\theta = -0.2$. Notice the persistence of the values and the apparent long-term trend.

On the other hand, Figure 2.20 displays both the theoretical and a sample ACF of this model. Observe the slowly decaying of both ACF and the similarity between the exact and the empirical values. It is clear that this time series displays significant empirical ACF levels even beyond lag 20.

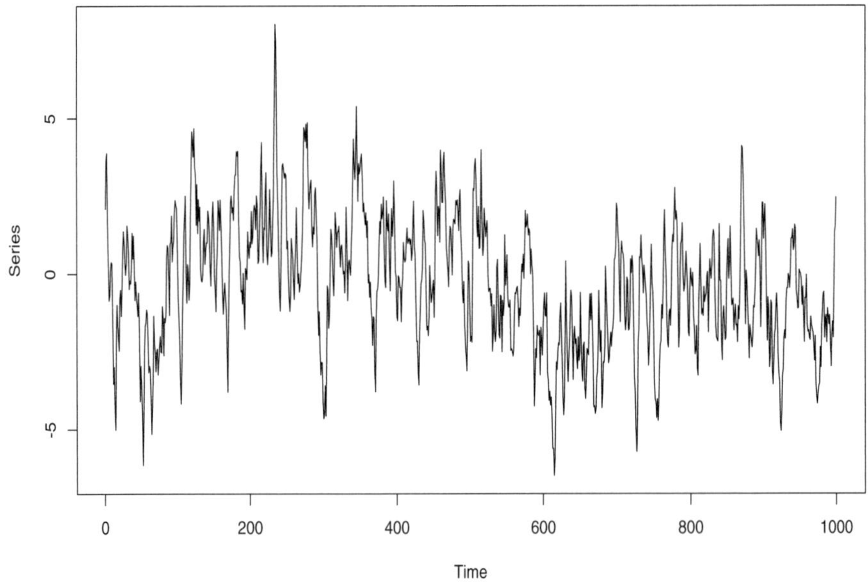

Figure 2.19 *Simulated ARFIMA$(1, d, 1)$ time series of 1000 observations with* $d = 0.3$, $\phi = 0.6$, *and* $\theta = -0.2$.

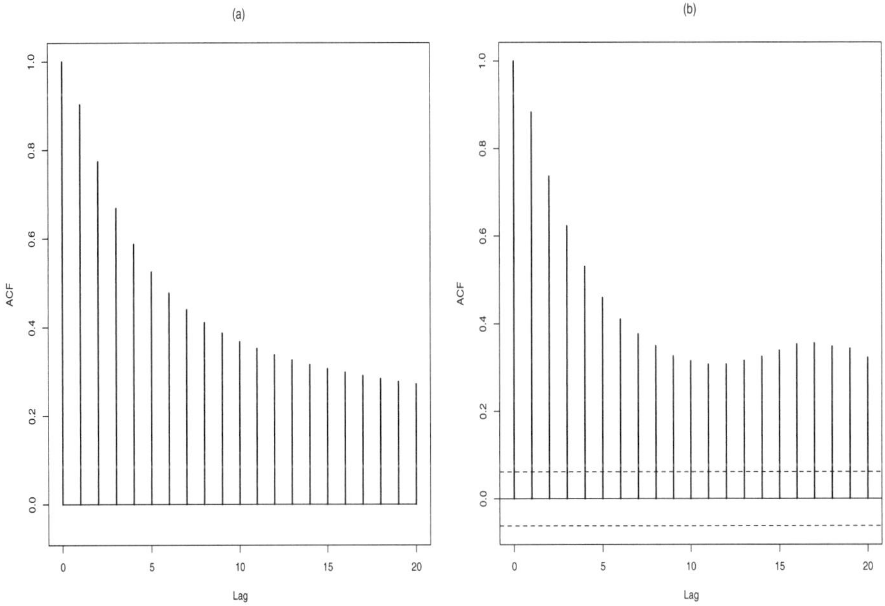

Figure 2.20 *ACF of an ARFIMA$(1, d, 1)$ time series with $d = 0.3$, $\phi = 0.6$, and $\theta = -0.2$. (a) Theoretical ACF, (b) Sample ACF.*

2.11 FRACTIONAL GAUSSIAN NOISE

Another well-known long-range-dependent process is the so-called *fractional Gaussian noise* (fGn). This process may be defined as follows. Consider the fractional Brownian motion $B_d(t)$ defined in the Section A.2.8, and let $\{y_t : t \in \mathbb{Z}\}$ be defined by the increments of $B_d(t)$:

$$y_t = B_d(t+1) - B_d(t). \tag{2.31}$$

The discrete-time process $\{y_t : t \in \mathbb{Z}\}$ is called *fractional Gaussian noise* (fGn). Let $\{y_t : t \in \mathbb{Z}\}$ be the process defined by (2.31). Then

(a) $\{y_t : t \in \mathbb{Z}\}$ is stationary for $d \in (-\frac{1}{2}, \frac{1}{2})$.

(b) $E(y_t) = 0$.

(c) $E(y_t^2) = E[B(1)^2]$.

(d) The autocovariance function of $\{y_t : t \in \mathbb{Z}\}$ is

$$\gamma(h) = \frac{\sigma^2}{2} \left(|h+1|^{2d+1} - 2|h|^{2d+1} + |h-1|^{2d+1} \right),$$

where $\sigma^2 = \mathrm{Var}(y_t)$.

(e) For $d \neq 0$, the asymptotic behavior of the ACF is given by

$$\gamma(h) \sim \sigma^2 d(2d+1)\,|h|^{2d-1},$$

as $|h| \to \infty$.

2.11.1 Sample Mean

Let $\bar{y}_n$ be the sample mean of an fGn described by (2.31). Then, by telescopic sum, we have

$$\bar{y}_n = \frac{1}{n}[B_d(n+1) - B_d(1)].$$

Thus, an application of formula (A.11) yields

$$\mathrm{Var}(\bar{y}_n) = \sigma^2 n^{2d-1}.$$

Consequently, since the process $\{B_d(t)\}$ is Gaussian, we conclude that

$$\bar{y}_n \sim N(0, \sigma^2 n^{2d-1}),$$

for all $n \in \mathbb{N}$.

2.12 BIBLIOGRAPHIC NOTES

Several textbooks address the concept of linear processes and ARMA models; see, for example, Box, Jenkins, and Reinsel (1994) and Brockwell and Davis (2002). The book by Shumway and Stoffer (2011) provides another excellent treatment of the fundamental time series techniques. ARFIMA models are discussed, for instance, in Doukhan, Oppenheim, and Taqqu (2003), Rangarajan and Ding (2003), Teyssière and Kirman (2007), and Palma (2007), among others. Definitions of long-memory processes have been extensively discussed in the literature; see, for instance, Chapter 3 of Palma (2007). The articles by Cox (1984) and Hall (1997) give overviews about different definitions of long-range dependence. On the other hand, the paper by Hosking (1981) discusses several properties of ARFIMA models, including results about stationarity, invertibility, autocorrelations, and the like. Formulas for the exact autocovariance function of an ARFIMA process were established by Sowell (1992). A nice review of fractional Gaussian noise processes and their properties is given in Taqqu (2003).

Problems

2.1 Let $y_t = \varepsilon_t + X\,\varepsilon_{t-1}$ be a sequence of independent identically distributed (i.i.d.) random variables, where X is a random variable with mean μ and variance σ_X^2, $\{\varepsilon_t\}$ is a sequence of i.i.d. random variables with zero-mean and variance σ_ε^2. Furthermore, X and $\{\varepsilon_t\}$ are independent. Is $\{y_t\}$ a weakly stationary process? If this is true, calculate the autocovariance function of $\{y_t\}$.

2.2 Let $\{\varepsilon_t\}$ be a sequence of independent identically distributed random variables with zero-mean and variance σ^2. Is the sequence $\{y_t : t \in \mathbb{Z}\}$ defined by $y_t = \varepsilon_t + \varepsilon_{t-1} + \varepsilon_{t-2}\,\varepsilon_{t-3}$ a linear process? Is the process $y_t = \varepsilon_t + \varepsilon_{t-1} + \varepsilon_{t-3}$ causal?

2.3 Can a stationary process be non-causal? Discuss.

2.4 Let the process $\{y_t : t \in \mathbb{Z}\}$ be defined by $y_t = \phi\, y_{t-1} + \varepsilon_t$ where $\{\varepsilon_t\}$ is a sequence of identically distributed *Student* t random variables with $\nu = 2$ degrees of freedom. Is this process second order stationary?

2.5 Consider the ARMA$(1,1)$ model

$$y_t - \phi y_{t-1} = \varepsilon_t + \theta \varepsilon_{t-1},$$

where $\varepsilon_t \sim \mathrm{WN}(0, \sigma^2)$ with $|\phi| < 1$, $|\theta| < 1$

(a) Let $\Psi(z) = (1 - \phi z)^{-1}(1 + \theta z) = \sum_{j=0}^{\infty} \psi_j z^j$. Verify that $\psi_0 = 1$, $\psi_j = \phi^{j-1}(\phi + \theta)$ $j \geq 1$

(b) Starting from the fact that for $k \geq 0$

$$\gamma(k) = \sigma^2 \sum_{j=0}^{\infty} \psi_j \psi_{j+k},$$

find $\rho(k) = \frac{\gamma(k)}{\gamma(0)}$ for $k \geq 1$.

2.6 Consider the ARMA$(2,2)$ model,

$$y_t - 0.2 y_{t-1} + 0.5 y_{t-2} = \varepsilon_t + 2\varepsilon_{t-1} - \varepsilon_{t-2}.$$

Verify if this process is stationary and invertible.

2.7 Consider the ARMA(2,1) process

$$y_t = 1.3\, y_{t-1} - 0.4\, y_{t-2} + z_t + z_{t-1}.$$

(a) Specify if the process is stationary and invertible.

(b) Calculate the autocovariance function of y_t.

2.8 Let y_t be a process satisfying

$$y_t = \alpha + \beta t + \eta_t$$
$$\eta_t = \phi \eta_{t-1} + \epsilon_t,$$

where ϵ_t is white noise $(0, \sigma_\epsilon^2)$.

(a) Assume that the parameter ϕ is known and equal to 1. What would you do for modeling y_t?
(b) Assume now that you only know that $|\phi| < 1$. What would you do for modeling y_t?

2.9 Consider the following ARMA(2,1) model

$$y_t - y_{t-1} + 0.1y_{t-2} = \epsilon_t - 0.3\epsilon_t,$$

where ϵ_t is white noise $(0, \sigma_\epsilon^2)$.
(a) Show that this process is causal and invertible.
(b) Calculate the coefficients of the Wold expansion.
(c) Calculate the ACF of y_t.

2.10 Consider the process $\{x_t\}$ given

$$x_t = \phi x_{t-s} + z_t,$$

with $\{z_t\} \sim \text{WN}(0, \sigma^2)$, $|\phi| < 1$ and $s \in \mathbb{N}$. Determine the partial autocorrelation function of x_t.

2.11 The autocorrelation and partial autocorrelation functions are key tools for identifying ARMA processes. To assess their ability to recognize ARMA we simulated six processes and plotted their autocorrelations, see Figure 2.21 to Figure 2.26. Identify the following processes justifying your decision:
(a) $(1 - 0.5\,B)\,y_t = (1 + 0.5\,B)\,z_t$.
(b) $(1 - 0.6\,B)\,y_t = z_t$.
(c) $y_t = (1 + 1.3\,B + 0.4\,B^2)\,z_t$.
(d) $(1 - 0.6\,B + 0.05\,B^2)\,y_t = (1 + 0.7\,B)\,z_t$.

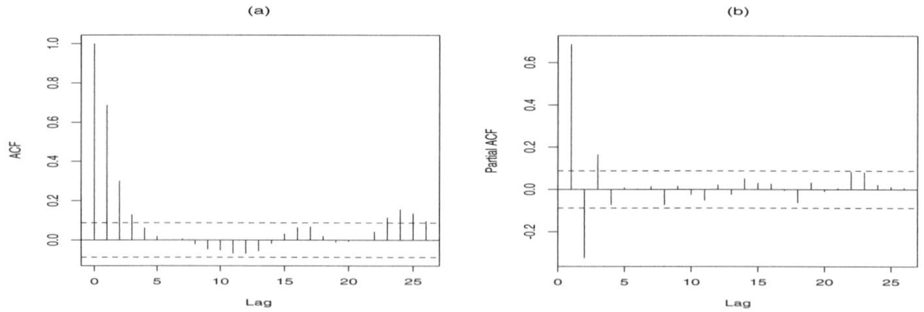

Figure 2.21 Time series I, (a) Sample ACF, (b) Sample PACF.

2.12 Explain the following concepts:
(a) Strict stationarity.
(b) Second order stationarity.

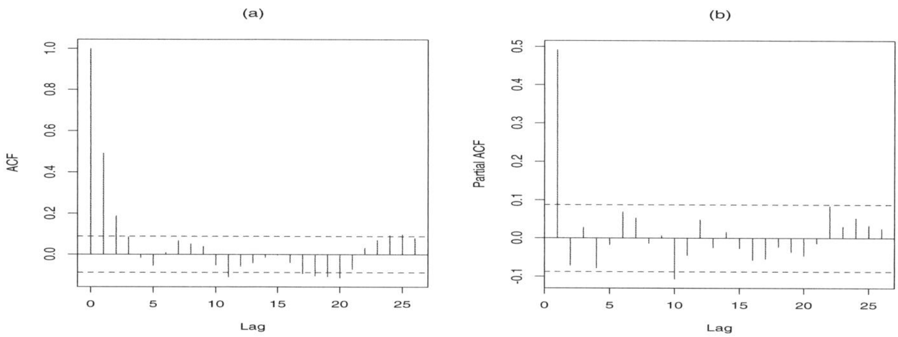

Figure 2.22 Time series II, (a) Sample ACF, (b) Sample PACF.

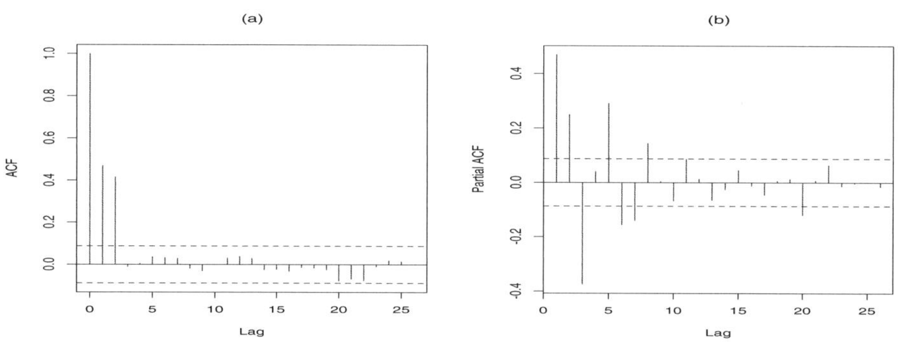

Figure 2.23 Time series III, (a) Sample ACF, (b) Sample PACF.

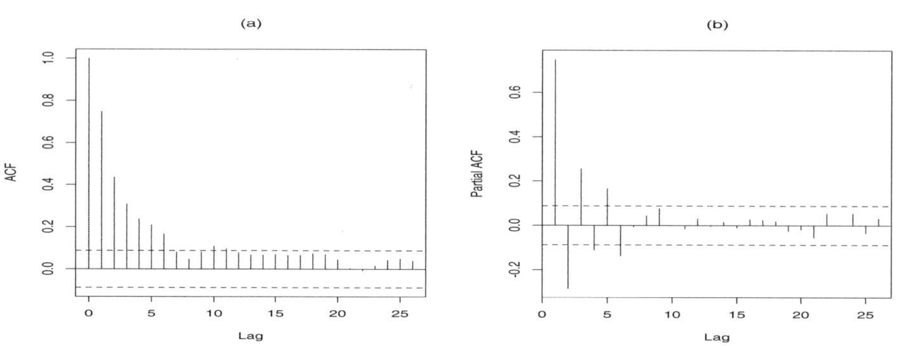

Figure 2.24 Time series IV, (a) Sample ACF, (b) Sample PACF.

(c) How is the decaying of the ACF of an ARMA or an ARFIMA process?

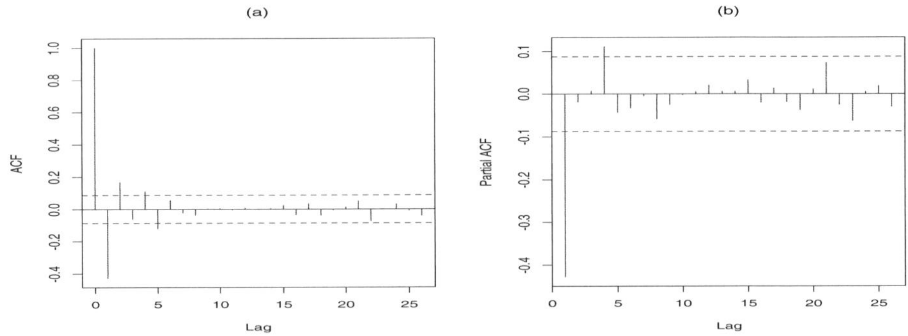

Figure 2.25 Time series V, (a) Sample ACF, (b) Sample PACF.

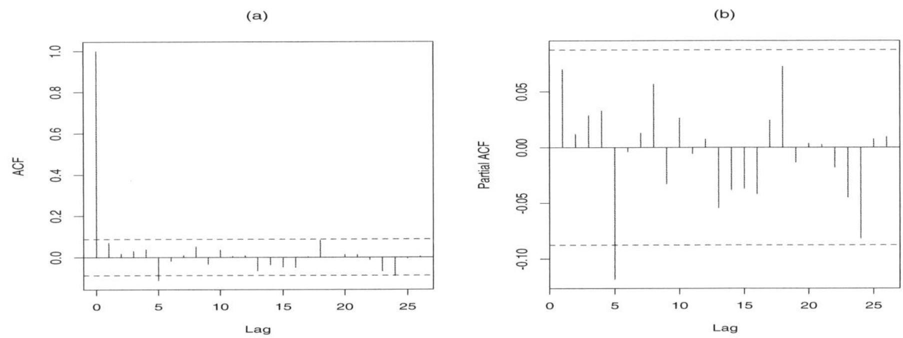

Figure 2.26 Time series VI, (a) Sample ACF, (b) Sample PACF.

2.13 Show that if $\{y_t\}$ is stationary and $|\theta| < 1$, then for each t, $\sum_{j=1}^{m} \theta^j y_{t+1-j}$ converges in mean square error as $m \to \infty$.

2.14 Consider the process ARMA(2,1)

$$y_t = 1.3y_{t-1} - 0.4y_{t-2} + z_t + z_{t-1}.$$

(a) Analyze if this model is stationary and invertible.
(b) Calculate the ACF of this process.

2.15 Consider the following ARMA(2,1) model,

$$y_t - y_{t-1} + 0.1y_{t-2} = \epsilon_t - 0.3\epsilon_t,$$

where ϵ_t is a white noise sequence with $(0, \sigma_\epsilon^2)$.

(a) Show that this time series is causal and invertible.
(b) Calculate the coefficients of its Wold expansion.
(c) Calculate the ACF of y_t.

2.16 Consider the ARMA(1,1) time series y_t defined by

$$y_t - \phi\, y_{t-1} = \varepsilon_t + \theta\, \varepsilon_{t-1},$$

with $\sigma^2 = \text{Var}(\varepsilon_t) = 1$ and let $\eta = (\phi,\, \psi)$.

(a) Show that a state space representation of y_t is given by

$$\begin{aligned} x_{t+1} &= \phi\, x_t + \psi\, \varepsilon_t \\ y_t &= x_t + \varepsilon_t, \end{aligned}$$

and find ψ.

(b) For which values of η the observed process is stationary?

(c) Plot a trajectory of y_t for different values of η.

2.17 Verify if the time series

$$y_t = 0.20 y_{t-1} + 0.63 y_{t-2} + \epsilon_t + 2.3\epsilon_{t-1} - 0.50\epsilon_{t-2}$$

where $\epsilon_t \sim \text{WN}(0, \sigma^2)$ is invertible and stationary.

2.18 Consider the following processes and verify if they are causal, stationary and invertible,

(a) $y_t - 0.4 y_{t-1} + 0.03 y_{t-2} = \varepsilon_t$, where ε_t are independent random variables with Student t distribution of 6 degree of freedom.

(b) $y_t = \varepsilon_{t+1} + 0.45\varepsilon_t$, where ε_t is a white noise sequence with distribution $N(0, 1)$.

(c) $y_t = \varepsilon_t + (\alpha - \beta) \sum_{j=1}^{\infty} \alpha^{j-1} \varepsilon_{t-j}$, where ε_t is a white noise sequence with variance σ_ε^2, $|\alpha| < 1$ and $|\beta| < 1$.

2.19 Let y_t be an ARMA(1,2) time series where $\text{Var}\, \varepsilon_t = \sigma_\varepsilon^2$ and

$$y_t - 0.5 y_{t-1} = \varepsilon_t - \varepsilon_{t-1} + 0.24\varepsilon_{t-2}.$$

(a) Verify if this model is causal, stationary and invertible.

(b) Find the first five terms of the Wold expansion of this process, $\psi_1, \ldots, \psi_5$.

(c) Find the ACF.

2.20 Show that an ARFIMA model satisfies the definition of a long-memory process described in this chapter and discuss alternative definitions of strongly dependent time series models.

2.21 Let $y_t = \varepsilon_t\, y_{t-1} + \varepsilon_{t-1}$ where ε_t is a white noise with zero-mean and unit variance. Is $\{y_t : t \in \mathbb{Z}\}$ a lineal process? Is it causal? Calculate $E(y_t)$.

2.22 The ACF of an ARFIMA$(0, d, 0)$ process is given by

$$\gamma(h) = \sigma^2 \frac{\Gamma(1 - 2d)}{\Gamma(1 - d)\Gamma(d)} \frac{\Gamma(h + d)}{\Gamma(1 + h - d)},$$

where $\Gamma(\cdot)$ is the gamma function. Besides, the PACF is

$$\phi_{nn} = -\frac{\Gamma(n-d)\Gamma(1-d)}{\Gamma(-d)\Gamma(n-d+1)}.$$

(a) Show that as $h \to \infty$, we have

$$\gamma(h) \sim \sigma^2 \frac{\Gamma(1-2d)}{\Gamma(1-d)\Gamma(d)} h^{2d-1}.$$

(b) Verify that as n tends to infinity, the PACF behaves as

$$\phi_{nn} \sim \frac{d}{n}.$$

Hint: $\Gamma(x+1) = x\,\Gamma(x)$.

2.23 Prove that

$$f(\lambda) \sim \frac{\sigma^2}{2\pi} \frac{|\theta(1)|^2}{|\phi(1)|^2} |\lambda|^{-2d},$$

for $|\lambda| \to 0$. **Hint**: $\sin(x) \sim x$ for $|x| \to 0$.

2.24 Calculate explicitly the autocovariance function of an ARFIMA$(1, d, 0)$ and ARFIMA$(0, d, 1)$.

2.25 Use Stirling's approximation

$$\Gamma(x) \sim \sqrt{2\pi} e^{1-x} (x-1)^{x-1/2},$$

as $x \to \infty$, to show that

$$\frac{\Gamma(n+\alpha)}{\Gamma(n+\beta)} \sim n^{\alpha-\beta}, \qquad (2.32)$$

as $n \to \infty$.

2.26 Applying Stirling's approximation, show directly that for an ARFIMA$(0, d, 0)$ process the following asymptotic expressions hold

$$\begin{aligned} \psi_k &\sim \frac{k^{d-1}}{\Gamma(d)} \\ \pi_k &\sim \frac{k^{-d-1}}{\Gamma(-d)} \\ \rho(k) &\sim k^{2d-1} \frac{\Gamma(1-d)}{\Gamma(d)} \end{aligned}$$

as $k \to \infty$.

2.27 Show that the AR(∞) and MA(∞) coefficients of an ARFIMA(p,d,q) process satisfy

$$\begin{aligned}\psi_j &\sim \frac{\theta(1)}{\phi(1)}\frac{j^{d-1}}{\Gamma(d)}\\ \pi_j &\sim \frac{\phi(1)}{\theta(1)}\frac{j^{-d-1}}{\Gamma(-d)}\end{aligned}$$

as $j \to \infty$.

2.28 Using the following formula involving the gamma function

$$\Gamma(1-d)\Gamma(d) = \frac{\pi}{\sin(\pi d)},$$

show that the constant c_γ appearing in expression (2.28) may be written as

$$c_\gamma = \sigma^2 \frac{|\theta(1)|^2}{|\phi(1)|^2}\frac{\Gamma(1-2d)}{\Gamma(1-d)\Gamma(d)}.$$

2.29 Show that the autocovariance function of a fractional Gaussian noise is positive for $d \in (0, \frac{1}{2})$ and negative for $d \in (-\frac{1}{2}, 0)$.

2.30 Let η_j be the MA(∞) coefficients of an ARFIMA$(0,d,0)$ process with $d \in (0,\frac{1}{2})$ and define $\varphi_0 = 1$ and $\varphi_j = \eta_j - \eta_{j-1}$ for $j = 1, 2, \ldots$.

(a) Verify that $\varphi_j = \frac{d-1}{j}\eta_{j-1}$.

(b) Show that $\varphi_j \sim \frac{j^{d-2}}{\Gamma(d-1)}$ as $j \to \infty$.

(c) Prove that

$$\sum_{j=0}^{\infty} \varphi_j^2 = \frac{\Gamma(3-2d)}{\Gamma(2-d)^2}.$$

2.31 Consider the linear process $\{y_t\}$ with Wold expansion

$$y_t = \varepsilon_t + \sum_{j=1}^{\infty} \frac{\varepsilon_{t-j}}{j},$$

where $\{\varepsilon_t\}$ is a white noise sequence with unit variance.

(a) Show that the autocovariance function of this process is

$$\gamma(h) = \frac{1}{|h|} + \frac{1}{|h|}\sum_{j=1}^{|h|}\frac{1}{j},$$

for $|h| > 0$ and

$$\gamma(0) = 1 + \frac{\pi^2}{6}.$$

(b) Show that

$$\gamma(h) \sim \frac{\log h}{h},$$

as $h \to \infty$.

(c) Verify that for any $m > n > 0$,

$$\sum_{h=n}^{m} \gamma(h) > \sum_{h=n}^{m} \frac{1}{h}.$$

(d) Is $\{y_t\}$ a long-memory process?
Hint: The following formulas may be useful,

$$\sum_{j=1}^{\infty} \frac{1}{j(j+h)} = \frac{1}{h} \sum_{j=1}^{h} \frac{1}{j},$$

for $h > 0$,

$$\sum_{j=1}^{h} \frac{1}{j} = C + \log h + \mathcal{O}\left(\frac{1}{h}\right),$$

where C is the Euler' constant, $C = 0.5772\cdots$

$$\sum_{j=1}^{\infty} \frac{1}{j^2} = \frac{\pi^2}{6}.$$

2.32 Let γ_j be the autocovariance function of a fractional noise FN(d) with $d < \frac{1}{4}$ and white noise variance σ^2.

(a) Show that

$$\sum_{j=-\infty}^{\infty} \gamma_j^2 = \sigma^4 \frac{\Gamma(1-4d)}{\Gamma^2(1-2d)}.$$

(b) Prove that

$$\sum_{j=0}^{\infty} \gamma_j \gamma_{j+1} = \sigma^4 \frac{d\,\Gamma(1-4d)}{(1-2d)\Gamma^2(1-2d)}.$$

2.33 Show that for $d \in (-\frac{1}{2}, \frac{1}{2})$,

$$(1-\phi)^{-d} = \sum_{j=0}^{\infty} \psi_j \phi^j,$$

where

$$\psi_j = \frac{\Gamma(j+d)}{\Gamma(j+1)\Gamma(d)}.$$

2.34 Prove that for $d \in (-\frac{1}{2}, \frac{1}{2})$,

$$\sum_{j=1}^{\infty} j\psi_j \phi^{j-1} = \frac{\partial}{\partial \phi}(1-\phi)^{-d} = \frac{d}{(1-\phi)^{1+d}}.$$

2.35 Show that the two following MA(1) processes

$$x_t = z_t + \theta z_{t-1}, \ \{z_t\} \sim \text{WN}(0, \sigma^2)$$
$$y_t = \tilde{Z}_t + \frac{1}{\theta}\tilde{Z}_{t-1}, \ \{\tilde{Z}_t\} \sim \text{WN}(0, \sigma^2\theta^2)$$

where $0 < |\theta| < 1$, have the same autocovariance function.

2.36 Show that

$$\int_{-\pi}^{\pi} e^{i(k-h)\lambda} d\lambda = \begin{cases} 2\pi, & \text{if } \ k = h, \\ 0, & \text{if } \ k \neq h. \end{cases}$$

2.37 Consider the autoregressive process of order 1 defined by

$$x_t - \mu = \alpha(x_{t-1} - \mu) + \varepsilon_t$$

where $\{\varepsilon_t\}$ is white noise and $-1 < \alpha < 1$.

(a) Show that given n observations $x_1, \ldots, x_n$ the least squares estimator that minimizes

$$S = \sum_{t=2}^{n} [x_t - \mu - \alpha(x_{t-1} - \mu)]^2$$

is given by

$$\hat{\mu} = \frac{\overline{x}_{(2)} - \hat{\alpha}\,\overline{x}_{(1)}}{1 - \hat{\alpha}}$$

and

$$\hat{\alpha} = \frac{\sum_{t=1}^{n-1}(x_t - \hat{\mu})(x_{t+1} - \hat{\mu})}{\sum_{t=1}^{n-1}(x_t - \hat{\mu})^2}$$

where $\overline{x}_{(1)}, \overline{x}_{(2)}$ are the averages of the first and last $n-1$ observations, respectively.

(b) Show that if we can make the approximation

$$\overline{x}_{(1)} \cong \overline{x}_{(2)} \cong \overline{x}$$

then

$$\hat{\mu} = \overline{x}$$

and

$$\hat{\alpha} = \frac{\sum_{t=1}^{N-1}(x_t - \overline{x})(x_{t+1} - \overline{x})}{\sum_{t=1}^{n-1}(x_t - \overline{x})^2}$$

(c) Find the autocorrelation function for an AR(1) process. In view of this, why $\hat{\alpha}$ given in part (b) is a reasonable estimate of α?

(d) Show that the spectral density of this process is given by

$$f(\omega) = \frac{\sigma_x^2(1-\alpha^2)}{\pi(1-2\alpha\cos(\omega)+\alpha^2)}.$$

2.38 Suppose that the y_t series follows the model $y_t = y_{t-1} + \varepsilon_t$ with $\varepsilon_t \sim \text{WN}(\sigma_\varepsilon^2)$. Is y_t stationary? If this sequence is not stationary, propose a transformation to induce stationarity.

2.39 Suppose that the series $\{y_t\}$ is modeled by $y_t = \mu_t + x_t$, where $\mu_t = \mu_{t-1} + \varepsilon_t$ is not observable and $x_t = \omega_t - \theta\omega_{t-1}$. Assume that $\varepsilon_t \sim \text{WN}(\sigma_\varepsilon^2)$, $\omega_t \sim \text{WN}(\sigma_\omega^2)$, and $\text{Cov}(\varepsilon_t, \omega_s) = 0$ for all (t, s). Please answer the following questions:

(a) What is the mean and the variance of y_t?

(b) Is y_t stationary?

(c) If part (b) is true, what is the autocovariance function of y_t?

(d) If part (b) is false, suggest an appropriate transformation to induce stationarity in y_t. Then, calculate the mean and autocovariance function of the transformed stationary series.

2.40 Let $\{x_t\}$ a sequence of independent random variables distributed as

$$x_t \sim \begin{cases} \exp(1), & t\,\text{even}, \\ \text{N}(0,1), & t\,\text{odd}. \end{cases}$$

(a) Is $\{x_t\}$ stationary?
(b) Is $\{x_t\}$ strictly stationary?

2.41 Let $\{z_t\}$ be a sequence of i.i.d. Bernoulli random variables with parameter $\frac{1}{2}$. Decide on the stationarity of the following series:

(a) $\{x_t; t \in \{0, \pm1, \pm2, \ldots\}\}$ where if t is odd, x_t is the value of a normal observation with mean $\frac{1}{2}$ and variance $\frac{1}{4}$, while for t even, $x_t = z_t$.
(b) $x_0 = c_0$; $x_t = 0.6x_{t-1} + \epsilon_t$ where ϵ_t is a sequence of independent identically distributed random variables with zero-mean.

2.42 Which of the following processes is weakly stationary for $T = \{0, 1, 2, \ldots\}$ where ϵ_t is a sequence of random errors with zero-mean and variance 1 and a_1 and a_2 are real constants

(a) $\epsilon_1 + \epsilon_2 \cos(t)$

(b) $\epsilon_1 + \epsilon_2 \cos(t) + \epsilon_3 \cos(t)$

(c) $a_1 + \epsilon_1 \cos(t)$

(d) $a_1 + \epsilon_1 a_1^t + \epsilon_2$

2.43 Using a normal random number generator, generate 100 observations of the following series, plot and discuss the differences.

(a) AR(1), $\phi = 0.6$

(b) MA(1), $\theta = -0.6$

(c) ARMA(1,1) con $\phi = 0.6$ and $\theta = -0.6$

2.44 Consider a series modeled by the following process:

$$(1 - 0.82B + 0.22B^2 + 0.28B^4)[\log(z) - \mu] = \varepsilon_t,$$

where ε_t is white noise sequence.

(a) Factorize the autoregressive operator, and explain the aspects that the factorization reveals regarding the autocorrelation function and the periodic components of this series.

(b) What is the formula which allows the forecasting of this series?

2.45 Let $\{\epsilon_t\}_{t\geq 1}$ be an i.i.d. sequence of random variables $N(\mu, \sigma^2)$ and θ real parameter. Consider the sequence $\{x_t\}_{t\geq 1}$ defined by:

$$x_1 = \epsilon_1, \quad x_t = \theta x_{t-1} + \epsilon_t \qquad (t \geq 2).$$

In what follows, consider $\mu = 0$.

(a) Calculate $V(x_t)$

(b) Calculate $\operatorname{Cov}(x_t, x_{t-k})$, $0 \leq k \leq t$

(c) What is the distribution of x_t?

(d) For what values of θ, (x_t) converges in distribution?

(e) What is the distribution of $(x_1, x_2, \ldots, x_n)$? Calculate its probability density.

(f) Is this process stationary?

2.46 Let z_1, z_2 two random variables such that $E[z_1] = \mu_1$; $E[z_2] = \mu_2$; $\operatorname{Var}[z_1] = \sigma_{11}$; $\operatorname{Var}[z_2] = \sigma_{22}$; $\operatorname{Cov}[z_1, z_2] = \sigma_{12}$, let the process $x(t, \omega)$ be defined by $x(t, \omega) = z_1(w) I_{\mathbb{R}^- \cup 0}(t) + z_2(w) I_{\mathbb{R}^+}(t)$.

(a) Describe the trajectories of x.

(b) What should be necessary to make the process stationary?

(c) Calculate $\mu_x(t)$ and $\gamma_x(t_1, t_2)$.

(d) Find necessary and sufficient conditions on μ_1, μ_2, σ_1, σ_2 and σ_{12} so that the process x is second order stationary. In this case, find autocovariance and autocorrelation functions.

2.47 The sequence $W(t), t \in \mathbb{R}$ is called a Wiener process if it satisfies

1. W(0)=0.
2. $W(t_2) - W(t_1) \sim N(0, \sigma^2(t_2 - t_1))$, $t_2 > t_1$.
3. If $t_0 \leq t_1 \leq \cdots \leq t_n$, then $W(t_1) - W(t_0), W(t_2) - W(t_1), \ldots, W(t_n) - W(t_{n-1})$ are independent.

(a) Calculate the mean $\mu_W(t)$ and $\text{Cov}(W_{t_1}, W_{t_2})$.
(b) Is $W(t)$ a second-order stationary process?

2.48 Find the autocorrelation function of the MA(2) process given by

$$x_t = \epsilon_t + 0.4\,\epsilon_{t-1} - 0.2\,\epsilon_{t-2}.$$

2.49 Calculate the autocorrelation function of the MA(n) process given by

$$x_t = \sum_{k=0}^{n} \frac{\epsilon_{t-k}}{n+1}.$$

2.50 Find the ACF of the AR(1) process defined by $x_t = 0.7\,x_{t-1} + \epsilon_t$ and plot $\rho_x(k)$ for lags $k = \pm 6, \pm 5, \pm 4, \pm 3, \pm 2, \pm 1$.

2.51 Let x_t be a process given by $x_t = \mu + \epsilon_t + \beta\epsilon_{t-1}$, where $\mu \in \mathbb{R}$ is a constant. Show that the ACF does not depend on μ.

2.52 Find the values of $\lambda_1,\ \lambda_2 \in \mathbb{R}$ such that the AR(2) process $x_t = \lambda_1\,x_{t-1} + \lambda_2\,x_{t-2} + \epsilon_t$ is stationary.

2.53 Assume that $\gamma_x(k)$ is the ACF of the stationary process $\{x_t, t \in \mathbb{Z}\}$. Calculate the ACF of the process $\nabla x_t = x_t - x_{t-1}$ in terms of $\gamma_x(k)$. Find $\gamma_{\nabla x}(k)$ if $\gamma_x(k) = \lambda^{-k}$.

2.54 Find the MA(∞) representation of the process $x_t = 0.3\,x_{t-1} + \epsilon_t$.

2.55 Suppose that the process $\{x_t, t \in \mathbb{Z}\}$ can be represented as $\epsilon_t = \sum_{j=0}^{\infty} c_j x_{t-j}$ as well as $x_t = \sum_{j=0}^{\infty} b_j \epsilon_{t-j}$.

(a) If $c_0 = b_0 = 1$, $A(s) = \sum_{j=0}^{\infty} c_j s^j$, and $B(s) = \sum_{j=0}^{\infty} b_j s^j$, show that $A(s) \cdot B(s) = 1$.
(b) If $R(s) = \sum_{k=-\infty}^{\infty} r_X(k) s^k$, show that

$$R(s) = \sigma_\epsilon^2 B(s)B(1/s) = \sigma_\epsilon^2 [A(s)A(1/s)]^{-1}.$$

2.56 Consider the MA(2) process given by $x_t = -1.7 + \epsilon_t - 0.6\epsilon_{t-1} + 0.3\epsilon_{t-2}$.

(a) Is this a stationary process?
(b) Find μ_x, $\gamma_x(0)$, $\gamma_x(1)$, $\gamma_x(2)$, $\gamma_x(3)$, $\gamma_x(23)$, $\rho_x(1)$, $\rho_x(2)$, $\rho_x(3)$, and $\rho_x(23)$.

2.57 Consider the ARMA(1,1) process given by: $x_t = 0.4x_{t-1} + \epsilon_t - 0.8\epsilon_{t-1}$.

(a) Is this a linear process?

(b) Is this a stationary process?

(c) Is this an invertible process?

(d) Find μ_X, $\gamma_X(0)$, $\gamma_X(1)$, $\gamma_X(2)$, $\gamma_X(3)$, $\rho_X(0)$, $\rho_X(1)$, $\rho_X(2)$, and $\rho_X(3)$.

(e) If $x_{71} = -5$, what is the expected value of x_{72}?

(f) Write the process in an AR form.

(g) Write the process in an MA expansion.

2.58 Let $\{x_t, t \in \mathbb{Z}\}$ be an MA(1) process given by $x_t = \epsilon_t + \alpha a_{t-1}$. Show that $|\rho_x(1)| \leq 1$.

2.59 Let $x_t = \sum_{j=0}^{\infty} \epsilon_{t-j}$ be a linear process.

(a) Is this process stationary?

(b) Is ∇x_t a stationary process?

(c) Show that the best linear predictor of x_{t+k} based on $\{x_u : u \leq t\}$ is x_t.

2.60 Consider the AR(2) process $x_t = \phi_1 x_{t-1} + \phi_2 x_{t-2} + \epsilon_t$. Calculate $\rho_X(k)$ based on the roots of the model, under the following conditions:

(a) $\rho_x(0) = 1, \rho_x(1) = \frac{\phi_1}{1-\phi_2}$

(b) $\rho_x(0) = 1, \rho_x(1) = \rho_x(-1)$.

Compare and discuss the two results.

2.61 Consider the following stochastic processes

1.- $x_t = 0.5x_{t-1} + \epsilon_t$.

2.- $x_t = 1.5x_{t-1} - 0.5x_{t-2} + \epsilon_t$.

(a) In both cases express x_t as an MA(∞) and an AR(∞).

(b) Calculate $\rho_x(k)$ and ϕ_{kk}.

2.62 Let x_t be an ARMA(p_2, q_2) process and let y_t be an ARMA(p_2, q_2) process, such that x_t, y_t are independent. Define $z_t = x_t + y_t$. Verify that z_t is an ARMA(p, q) process such that $p \leq p_1 + p_2$ and $q \leq \max\{p_1 + q_2, q_1 + p_2\}$.

2.63 If $\{x_t, t \in \mathbb{Z}\}$ and $\{y_t, t \in \mathbb{Z}\}$ are stationary, is $\{ax_t + by_t, t \in \mathbb{Z}\}$ also stationary?

2.64 If an ARMA(p, q) model with $p > 0$ and $q > 0$ is stationary, then provide conditions such that

(a) It is also invertible.

(b) It can be written as an infinite-order MA model.

(c) It can be written as a finite-order MA model.

(d) It can be written as a finite-order AR model.

2.65 If $W_{1,t} = (1 - \theta_1 B)\varepsilon_{1,t}$ and $W_{2,t} = (1 - \theta_2 B)\varepsilon_{2,t}$, where $\varepsilon_{1,t}$ and $\varepsilon_{2,t}$ are two independent white noise sequences, show that $W_{3,t} = W_{1,t} + W_{2,t}$ can be written as $W_{3,t} = (1 - \theta_3 B)\varepsilon_{3,t}$. Find expressions for θ_3 and $\sigma_{3,t}^2$ in terms of the corresponding parameters of the other two processes.

2.66 Consider ARMA(1,1) process

$$y_t = 10 + 0.8y_{t-1} + \varepsilon_t - 0.5\varepsilon_{t-1}$$

(a) Is this a stationary and an invertible process?

(b) Calculate the mean of y_t.

(c) Calculate the autocovariance and autocorrelation functions.

(d) If possible, find the AR(∞) and MA(∞) representation.

2.67 Let $\{y_t\}$ be an ARMA time series plus a noise defined by

$$y_t = x_t + W_t,$$

where $\{W_t\} \sim \text{WN}(0, \sigma_w^2)$, $\{x_t\}$ is the ARMA(p,q) process satisfying $\Phi(B)x_t = \Theta(B)\varepsilon_t$, $\{\varepsilon_t\} \sim \text{WN}(0, \sigma_\varepsilon^2)$ and $E(W_s z_t) = 0$ for all s and t.

(a) Show that $\{x_t\}$ is stationary and find its autocovariance function in terms of σ_w^2 and the autocovariance function of $\{x_t\}$.

(b) Show that $U_t := \Phi(B)y_t$, is r-correlated where $r = \max(p, q)$. Conclude that $\{y_t\}$ is an ARMA(p, r) process.

2.68 Suppose that $\{x_t\}$ is a non-invertible MA(1) process

$$x_t = \varepsilon_t + \theta\varepsilon_{t-1}$$

$$\{z_t\} \sim \text{WN}(0, \sigma^2)$$

where $|\theta| > 1$. Define a new process $\{W_t\}$ as

$$W_t = \sum_{j=0}^{\infty} (-\theta)^{-j} x_{t-j}$$

and show that $\{W_t\} \sim \text{WN}(0, \sigma_w^2)$. Express σ_w^2 in terms of θ and σ^2 and show that $\{x_t\}$ has representation *invertible* (in terms of $\{W_t\}$)

$$x_t = W_t + \frac{1}{\theta} W_{t-1}.$$

2.69 If $\{x_t\}$ denotes the unique stationary solution of the autoregressive equations

$$x_t = \phi x_{t-1} + \varepsilon_t, \qquad t = 0, \pm 1, \ldots$$

where $\{\varepsilon_t\} \sim WN(0, \sigma^2)$ and $|\phi| > 1$. Define the new sequence

$$W_t = x_t - \frac{1}{\phi} x_{t-1}$$

show that $\{W_t\} \sim WN(0, \sigma_W^2)$ and express σ_W^2 in terms of σ^2 and ϕ. Show that $\{x_t\}$ is the (stationary only) solution of the equations

$$x_t = \frac{1}{\phi} x_{t-1} + W_t, \quad t = 0, \pm 1, \ldots$$

2.70 Let $\{B_0(t)\}$ be a fractional Brownian motion with $d = 0$. Verify that

$$\mathrm{Cov}[B_0(t), B_0(s)] = \min\{|t|, |s|\}.$$

2.71 Let $\{B_d(t)\}$ be a fractional Brownian motion with $d \in (-\frac{1}{2}, \frac{1}{2})$. Show that this process has *stationary increments*, that is,

$$B_d(t+h) - B_d(t) \sim B_d(h) - B_d(0),$$

for all $t, h \in \mathbb{R}$.

2.72 Let $\{B_d(t)\}$ be a fractional Brownian motion with $d \in (-\frac{1}{2}, \frac{1}{2})$. Prove that for any $p > 0$,

$$E[B_d(t+h) - B_d(t)]^p = |h|^{p(d+1/2)}\, E[B(1)^p],$$

and

$$E\left[\frac{B_d(t+h) - B_d(t)}{h}\right]^2 = |h|^{2d-1}.$$

2.73 Let $\delta \in (1,2)$, $n \geq 1$. Show that

$$\sum_{j=1}^{\infty} [j(j+n)]^{-\delta} = \mathcal{O}(n^{-\delta}).$$

2.74 Show that for $a < 1$,

$$\lim_{m \to \infty} m^{2\beta} \sum_{i=1}^{m-1} a^i (m-i)^{-2\beta} = \frac{a}{1-a}.$$

2.75 Assume that $\epsilon_t \sim t_\nu$ with $\nu > 4$. Given that for even n

$$E(\epsilon_t^n) = \frac{\Gamma\left(\frac{n+1}{2}\right)\Gamma\left(\frac{\nu-n}{2}\right)}{\sqrt{\pi}\,\Gamma\left(\frac{\nu}{2}\right)} \nu^{n/2},$$

prove the following results:

(a) The kurtosis of ϵ_t is

$$\eta_\nu = 3\left(\frac{\nu-2}{\nu-4}\right),$$

and $\eta_\nu > 3$ for any fixed ν.

(b) The kurtosis of the process y_t defined by the Wold expansion (6.24) is

$$\kappa_\nu = \frac{6}{\nu-4}\left(\sum_{j=0}^{\infty}\psi_j^2\right)^{-2}\left(\sum_{j=0}^{\infty}\psi_j^4\right) + 3,$$

and $\kappa_\nu \to 3$ as $\nu \to \infty$.

Hint: Recall that $x\Gamma(x) = \Gamma(x+1)$ and $\Gamma(\frac{1}{2}) = \sqrt{\pi}$.

CHAPTER 3

STATE SPACE MODELS

The linear processes introduced in the previous chapter were described in terms of the Wold expansion or an infinite moving-average representation. However, these processes can also be expressed in terms of a *state space linear system*. This chapter is devoted to describe these systems and investigate some of the relationships between Wold expansions and state space representations. State space models are very useful for calculating estimators, predictors, and interpolators. Consequently, Wold expansions and state space representations of linear time series will be extensively used throughout this book.

This chapter begins with a motivating example of state spaces models in the context of air pollution. Section 3.2 introduces the state space linear systems and discusses a number of fundamental concepts such as stability, observability, controllability, and minimality. Additionally, three equivalent representations of a linear process including the Wold decomposition, state space systems, and the Hankel matrix are analyzed in Section 3.3. Section 3.4 describes the Kalman filter equations to calculate recursively state estimates, forecasts, and smoothers along with their variances. This section also discusses techniques for handling missing values and predicting future observations. Some extensions of these procedures to incorporate exogenous variables are

Time Series Analysis. First Edition. Wilfredo Palma.

described in Section 3.5, and further readings on theoretical and practical issues of state space modeling are suggested in Section 3.6. A list of problems is given at the end of the chapter.

3.1 INTRODUCTION

State space models are useful representations of linear processes. However, these systems can also be helpful to model practical situations. Suppose that the air pollution is generated by two sources, $x_{t,1}$ and $x_{t,2}$. But, these components are not directly observed. The instruments measure a combination of these two components plus an instrumental error, that is, $y_t = \alpha_1 x_{t,1} + \alpha_2 x_{t,2} + \varepsilon_t$, where ε_t is a white noise sequence with zero-mean and variance σ. An important problem is estimating the parameters α_1, α_2, σ and the magnitude of these sources from the observed data. For simplicity, suppose that we know from the physical dynamic of the air pollution that the state variables satisfy the equation

$$x_{t+1} = \begin{bmatrix} 0.6 & 0.1 \\ 0.2 & 0.7 \end{bmatrix} x_t + \begin{bmatrix} \sigma & 0 \\ 0 & \sigma \end{bmatrix} v_t,$$

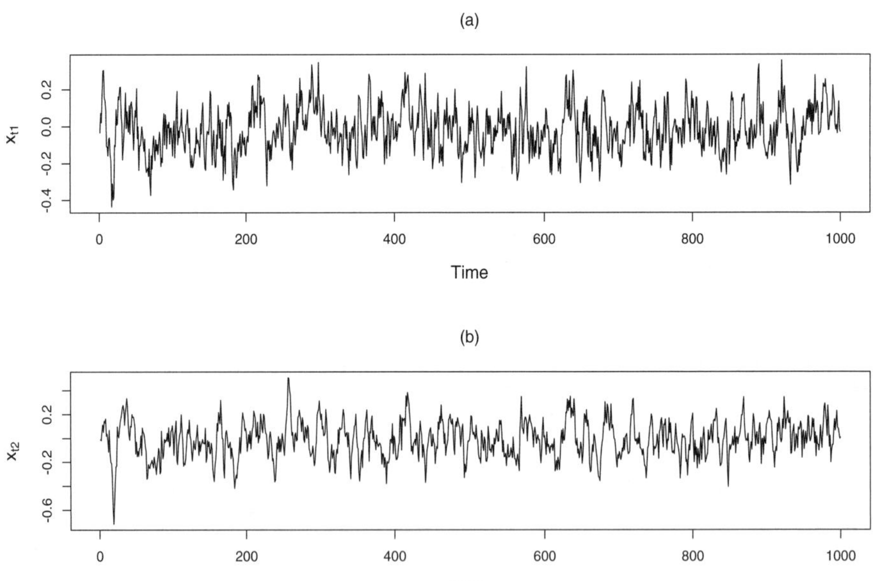

Figure 3.1 *State space model example. State values x_t, $t = 1, \ldots, 1000$. (a) first component $x_{t,1}$, (b) second component $x_{t,2}$.*

where v_t is a state noise two-dimensional vector with zero-mean and variance I_2. On the other hand, suppose that the instrumental measurements satisfy

$$y_t = \begin{bmatrix} \alpha_1 & \alpha_2 \end{bmatrix} x_t + \sigma\, w_t,$$

where w_t is a normal white noise with zero-mean and unit variance, independent of the state noise v_t. Figure 3.1 shows the two components of the state vector for a simulated process with $\alpha_1 = 0.7$, $\alpha_2 = 0.3$, $\sigma = 0.1$ and 1000 observations. The variance of the state vector is given by

$$\mathrm{Var}(x_t) = \begin{pmatrix} 0.0172 & 0.0064 \\ 0.0064 & 0.0237 \end{pmatrix},$$

and the variance of the observation sequence is $\mathrm{Var}(y_t) = 0.0243$ and standard deviation $\mathrm{sd}(y_t) = 0.1560$. On the other hand, Figure 3.2 displays the values of the observed process y_t. The sample ACF and sample PACF of this series is exhibited in Figure 3.3. Notice the level of dependence of the observation y_t on its past values. Later in this chapter we will come back to this illustrative example to show the states and parameter estimates.

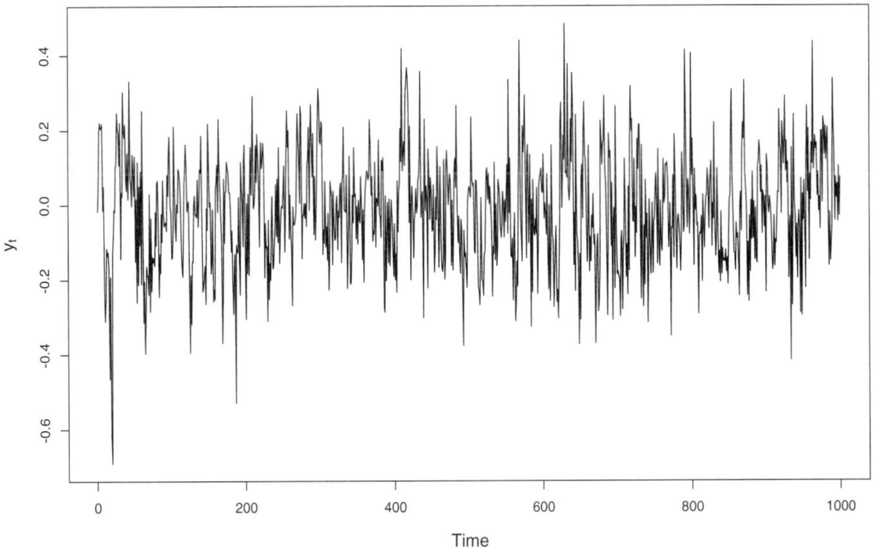

Figure 3.2 *State space model example. Observations y_t, $t = 1, \ldots, 1000$.*

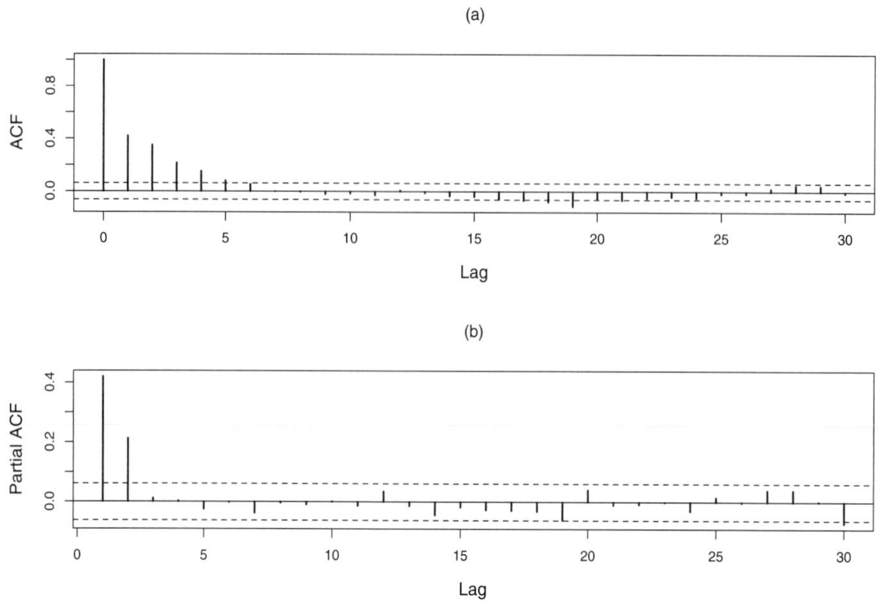

Figure 3.3 *State space model example. (a) Sample ACF, and (b) sample PACF of y_t, $t = 1, \ldots, 1000$.*

3.2 LINEAR DYNAMICAL SYSTEMS

A more general expression for the linear state space system described in the previous section is may be described by the discrete-time equations

$$x_{t+1} = F x_t + v_t, \tag{3.1}$$

$$y_t = G x_t + w_t, \tag{3.2}$$

where x_t is the *state* vector for all time t, y_t is the *observation* sequence, F is the *state matrix*, G is the *observation matrix*, $\{v_t\}$ is the *state white noise* sequence with zero-mean variance σ_v^2 and $\{w_t\}$ is the observation error sequence with zero-mean variance σ_w^2. Equation (3.1) is called the *state equation* while (3.2) is the *observation equation*.

3.2.1 Stability

A state space system (3.1)–(3.2) is said to be *stable* if F^n converges to zero as n tends to ∞. The system is said to be *exponentially stable* if there exist positive constants c and α such that $\|F^n\| \leq ce^{-\alpha n}$. The stability of a state space model means that the state vector does not explode as time increases and that the effect of the initial value of the state vanishes as time progresses. In what follows we will assume that the system is stable unless stated otherwise.

EXAMPLE 3.1

Consider the state space of Section 3.1. In this case,

$$F = \begin{bmatrix} 0.6 & 0.1 \\ 0.2 & 0.7 \end{bmatrix}.$$

Note that if λ is an eigenvalue of F associated to the eigenvector x, then $F^n x = \lambda^n x$. Thus, if the eigenvalues of F satisfy $|\lambda| < 1$ then $\lambda^n \to 0$ as n increases. Consequently, $F^n x$ also converges to zero as $n \to \infty$. In this case the eigenvalues of F are 0.8 and 0.5 so they satisfy the convergence condition.

3.2.2 Hankel Matrix

Suppose that $\psi_0 = 1$ and $\psi_j = GF^{j-1} \in \mathbb{R}$ for all $j > 0$ such that $\sum_{j=0}^{\infty} \psi_j^2 < \infty$. Then from (3.1)-(3.2), the process $\{y_t\}$ may be written as the Wold expansion

$$y_t = \sum_{j=0}^{\infty} \psi_j \varepsilon_{t-j}. \tag{3.3}$$

This linear process can be characterized by the *Hankel matrix* given by

$$H = \begin{pmatrix} \psi_1 & \psi_2 & \psi_3 & \cdots \\ \psi_2 & \psi_3 & \psi_4 & \cdots \\ \psi_3 & \psi_4 & \psi_5 & \cdots \\ \vdots & \vdots & \vdots & \end{pmatrix}.$$

Note that this Hankel matrix specifies the Wold expansion (3.3) and vice versa. Furthermore, the dimensionality of the state space system (3.1)-(3.2) is closely related to the dimensionality of the matrix $\mathcal{H}$ and to the rationality of the spectral density of the process (3.3). Specifically, the rank of $\mathcal{H}$ is finite if and only if the spectral density of (3.3) is rational.

The class of *autoregressive moving-average* (ARMA) processes have rational spectrum, hence the rank of $\mathcal{H}$ is finite for these models. In turn, as we will see later, this means that any state space system representing an ARMA process is finite-dimensional. On the contrary, the class of long-memory processes [e.g., *autoregressive fractionally integrated moving-average* (ARFIMA) models] does not have rational spectrum. Consequently, all state space systems representing such models are infinite-dimensional.

Since the state space representation of a linear regular process is not necessarily unique, one may ask which is the minimal dimension of the state vector.

In order to answer this question it is necessary to introduce the concepts of *observability* and *controllability*.

3.2.3 Observability

Let $\mathcal{O} = (G', F'G', F'^2G', \dots)'$ be the *observability matrix*. The system (3.1)–(3.2) is said to be *observable* if and only if $\mathcal{O}$ is full rank or, equivalently, $\mathcal{O}'\mathcal{O}$ is invertible.

The definition of observability is related to the problem of determining the value of the *unobserved* initial state x_0 from a trajectory of the *observed* process $\{y_0, y_1, \dots\}$ in the absence of state or observational noise. Consider, for example, the deterministic state space system

$$\begin{aligned} x_{t+1} &= Fx_t, \\ y_t &= Gx_t, \end{aligned}$$

and let $Y = (y_0, y_1, \dots)'$ be a trajectory of the process. Since

$$\begin{aligned} y_0 &= Gx_0, \\ y_1 &= GFx_0, \\ y_3 &= GF^2x_0, \\ &\vdots \end{aligned}$$

we may write $Y = \mathcal{O}x_0$. Now, if $\mathcal{O}'\mathcal{O}$ is full rank, we can determine the value of the initial state explicitly as $x_0 = (\mathcal{O}'\mathcal{O})^{-1}\mathcal{O}'Y$.

3.2.4 Controllability

Consider the case where the state error is written in terms of the observation so that $vt = H\,w_t$ and the state space model can be expressed as

$$x_{t+1} = F\,x_t + H\,w_t, \tag{3.4}$$

$$y_t = G\,x_t + w_t. \tag{3.5}$$

Let $\mathcal{C} = (H, FH, F^2H, \dots)$ be the *controllability matrix*. The system (3.4)-(3.5) is *controllable* if $\mathcal{C}$ is full rank or $\mathcal{C}'\mathcal{C}$ is invertible.

The key idea behind the concept of controllability of a system is as follows. Let $\mathcal{E}_{t-1} = (\dots, w_{t-2}, w_{t-1})'$ be the history of the state noise process at time t and suppose that we want the state to reach a particular value x_t. The question now is whether we can choose an adequate sequence $\mathcal{E}_{t-1}$ to achieve that goal. In order to answer this question, we may write the state at time t as

$$x_t = Hw_{t-1} + FHw_{t-2} + F^2Hw_{t-3} + \cdots = \mathcal{C}\mathcal{E}_{t-1}.$$

Thus, if the system is controllable, $\mathcal{C}'\mathcal{C}$ is full rank and we may write

$$\mathcal{E}_{t-1} = (\mathcal{C}'\mathcal{C})^{-1}\mathcal{C}'x_t.$$

For a finite-dimensional state space system the observability and the controllability matrices may be written as

$$\begin{aligned} \mathcal{O} &= (G', F'G', \dots, F'^{n-1}G')', \\ \mathcal{C} &= (H, FH, \dots, F^{n-1}H), \end{aligned}$$

where $n = \text{rank}(\mathcal{O}) = \text{rank}(\mathcal{C})$.

EXAMPLE 3.2

For the state space of Section 3.1 we have

$$\mathcal{O} = \begin{bmatrix} 0.70 & 0.45 \\ 0.30 & 0.35 \end{bmatrix}.$$

Since the rank of this matrox is 2, the system is observable. On the other hand, suppose that we can write

$$v_t = \begin{bmatrix} 1 \\ 2 \end{bmatrix} w_t,$$

where w_t is a white noise sequence with zero-mean and unit variance. Thus, the controllability matrix is given by

$$\mathcal{C} = \begin{bmatrix} 1 & 0.8 \\ 2 & 1.6 \end{bmatrix}.$$

Therefore, the system is not controllable. However, if

$$v_t = \begin{bmatrix} 1 \\ 1 \end{bmatrix} w_t,$$

then

$$\mathcal{C} = \begin{bmatrix} 1 & 0.7 \\ 1 & 0.9 \end{bmatrix}.$$

So, that this model is controllable.

3.2.5 Minimality

A state space system is *minimal* if F is of minimal dimension among all representations of the linear process (3.3). The problem of finding minimal representations for finite-dimensional systems minimality is highly relevant since a state space representation with the smallest dimension may be easier to interpret or easier to handle numerically. A state space system is minimal if and only if it is observable and controllable.

3.3 STATE SPACE MODELING OF LINEAR PROCESSES

A linear process may be represented in many different forms, for instance, as a state space system, a Wold decomposition, or by its Hankel matrix. To a large extent, these representations are equivalent and a key issue is how to pass from one representation to another. We have to keep in mind, however, that in most cases these representations are not unique.

3.3.1 State Space Form to Wold Decomposition

Given a state space system (3.4)-(3.5) with the condition that the sequence $\{GF^jH\}$ is square summable, we may find the Wold representation (3.3) by defining the coefficients $\psi_j = GF^{j-1}H$. Observe that the stability of F is not sufficient for assuring the above condition. However, square summability is guaranteed if F is exponentially stable.

3.3.2 Wold Decomposition to State Space Form

A state space representation of the process (3.3) can be specified by the state

$$x_t = [\, y(t|t-1) \quad y(t+1|t-1) \quad y(t+2|t-1) \quad \cdots]', \tag{3.6}$$

where $y(t+j|t-1) = E[y_{t+j}|y_{t-1}, y_{t-2}, \ldots]$ and

$$F = \begin{bmatrix} 0 & 1 & 0 & 0 & 0 & \cdots \\ 0 & 0 & 1 & 0 & 0 & \cdots \\ 0 & 0 & 0 & 1 & 0 & \\ \vdots & \vdots & & & \ddots & \end{bmatrix}, \tag{3.7}$$

$$H = [\psi_1 \quad \psi_2 \quad \psi_3 \quad \cdots]', \tag{3.8}$$

$$G = [1 \quad 0 \quad 0 \quad \cdots], \tag{3.9}$$

$$y_t = Gx_t + \varepsilon_t. \tag{3.10}$$

3.3.3 Hankel Matrix to State Space Form

Let A be a linear operator that selects rows of $\mathcal{H}$ such that $\mathcal{H}_0 = A\mathcal{H}$ consists of the basis rows of $\mathcal{H}$. Thus, $\mathcal{H}_0$ is full rank and consequently $\mathcal{H}_0\mathcal{H}_0'$ is invertible. Given the Hankel representation, a state space system can be specified by the state vector

$$x_t = \mathcal{H}_0(\varepsilon_{t-1}, \varepsilon_{t-2}, \ldots)',$$

and the system matrices

$$F = A \begin{bmatrix} \psi_2 & \psi_3 & \psi_4 & \cdots \\ \psi_3 & \psi_4 & \psi_5 & \cdots \\ \psi_4 & \psi_5 & \psi_6 & \cdots \\ \vdots & \vdots & \vdots & \end{bmatrix} \mathcal{H}_0'(\mathcal{H}_0\mathcal{H}_0')^{-1},$$

$$H = A(\psi_1, \psi_2, \dots)',$$

and

$$G = (\psi_1, \psi_2, \dots)\mathcal{H}_0'(\mathcal{H}_0\mathcal{H}_0')^{-1}.$$

Let $e_j = (0, \dots, 0, 1, 0, 0, \dots)$ where the 1 is located at the jth position. Observe that by induction we can prove that $F^{j-1}\mathcal{H}_0 e_1 = \mathcal{H}_0 e_j$. For $j = 1$ is trivial. Suppose that the assertion is valid for j, we will prove that the formula holds for $j + 1$:

$$\begin{aligned} F^j\mathcal{H}_0 e_1 &= F\mathcal{H}_0 e_j = A \begin{bmatrix} \psi_2 & \psi_3 & \psi_4 & \cdots \\ \psi_3 & \psi_4 & \psi_5 & \cdots \\ \psi_4 & \psi_5 & \psi_6 & \cdots \\ \vdots & \vdots & \vdots & \end{bmatrix} e_j \\ &= A \begin{bmatrix} \psi_{j+1} \\ \psi_{j+2} \\ \psi_{j+3} \\ \vdots \end{bmatrix} = A\mathcal{H}e_{j+1} = \mathcal{H}_0 e_{j+1}. \end{aligned}$$

Therefore,

$$\begin{aligned} GF^{j-1}H &= (\psi_1, \psi_2, \dots)\mathcal{H}_0'(\mathcal{H}_0\mathcal{H}_0')^{-1}F^{j-1}\mathcal{H}_0 e_1 \\ &= (\psi_1, \psi_2, \dots)\mathcal{H}_0'(\mathcal{H}_0\mathcal{H}_0')^{-1}\mathcal{H}_0 e_j. \end{aligned}$$

On the other hand, since $(\psi_1, \psi_2, \dots)$ belongs to span of $\mathcal{H}$, it may be written as $(\psi_1, \psi_2, \dots) = b\mathcal{H}_0$. Thus,

$$GF^{j-1}H = b\mathcal{H}_0 e_j = (\psi_1, \psi_2, \dots)e_j = \psi_j.$$

3.4 STATE ESTIMATION

The state space model described by (3.1)-(3.2) can be readily extended to handle time-varying state and observation matrices. This makes the system versatile enough to model, for instance, nonstationary processes. This more general state space model can be written as,

$$x_{t+1} = F_t x_t + v_t, \tag{3.11}$$

$$y_t = G_t x_t + w_t \tag{3.12}$$

where $\operatorname{Var}(v_t) = \Sigma_t$, $\operatorname{Var}(w_t) = \sigma_t$ and $\operatorname{Cov}(v_t, w_t) = \Gamma_t$. As discussed previously, in many practical situations the state vector of the model is not directly observed. We have indirect information of the state x_t from the observed sequence y_t. Based on what information is available for estimating the state, we may consider the following three cases. If only the past of the process $(\ldots, y_{t-2}, y_{t-1})$ is available the we proceed to the *prediction* of the state x_t. If the past and the present of the process $(\ldots, y_{t-1}, y_t)$ is available, then we proceed to *filtering* the state x_t. Finally, if the full trajectory of the process is available $(\ldots, y_{t-1}, y_t, y_{t+1}, \ldots)$, then we consider *smoothing* of the state x_t.

In what follows we summarize the Kalman recursive equations to find state predictors, filters, and smoothers. The main purpose of these equations is to simplify the numerical calculation of the estimators. Since in practice we usually have only a finite stretch of data, we focus our attention on projections onto subspaces generated by a finite trajectories of the process $\{y_t : t \in \mathbb{Z}\}$.

3.4.1 State Predictor

Let $\widehat{x}_t$ be the predictor of the state x_t based on $\{y_s : 1 \le s \le t-1\}$ and let $\Omega_t = E[(x_t - \widehat{x}_t)(x_t - \widehat{x}_t)']$ be the state error variance, with $\widehat{x}_1 = 0$ and $\Omega_1 = E[x_1 x_1']$. Then, the state predictor $\widehat{x}_t$ is given by the following recursive equations for $t \ge 1$:

$$
\begin{aligned}
\Delta_t &= G_t\,\Omega_t\,G_t' + \sigma_t^2, && (3.13)\\
K_t &= (F_t\,\Omega_t\,G_t' + \Gamma_t)\Delta_t^{-1}, && (3.14)\\
\Omega_{t+1} &= F_t\,\Omega_t\,F_t' + \Sigma_t - \Delta_t\,K_t\,K_t', && (3.15)\\
\nu_t &= y_t - G_t\,\widehat{x}_t, && (3.16)\\
\widehat{x}_{t+1} &= F_t\,\widehat{x}_t + K_t\,\nu_t. && (3.17)
\end{aligned}
$$

The sequence K_t is the *Kalman gain* and $\{\nu_t\}$ is the *innovation sequence* that represents the part of the observation y_t which cannot be predicted from its past.

3.4.2 State Filter

Define $\widehat{x}_{t|t}$ as the conditional expectation of x_t based on $\{y_s : 1 \le s \le t\}$ and let $\Omega_{t|t} = E[(x_t - \widehat{x}_{t|t})(x_t - \widehat{x}_{t|t})']$ be its error variance, with $\widehat{x}_{1|1} = 0$. Then the state filter $\widehat{x}_{t|t}$ is given by the following recursive equations for $t \ge 1$:

$$
\begin{aligned}
\Omega_{t|t} &= \Omega_t - \Omega_t\,G_t'\,\Delta_t^{-1}\,G_t\,\Omega_t,\\
\widehat{x}_{t|t} &= \widehat{x}_t + \Omega_t\,G_t'\,\Delta_t^{-1}\,\nu_t.
\end{aligned}
$$

3.4.3 State Smoother

Let $\widehat{x}_{t|s}$ be conditional expectation of x_t based on $\{y_j : 1 \leq j \leq s\}$. The state smoother $\widehat{x}_{t|n}$ is given by the following recursive equations for $s \geq t$:

$$\begin{aligned} \widehat{x}_{t|s} &= \widehat{x}_{t|s-1} + A_{t,s}\, G_t'\, \Delta_s^{-1}\, \nu_s, \\ A_{t,s+1} &= A_{t,s}\, (F_t - K_s\, \Delta_s^{-1}\, G_t)'. \end{aligned}$$

The state smoother error variance is obtained from the equation

$$\Omega_{t|s} = \Omega_{t|s-1} - A_{t,s} G_t'\, \Delta_s^{-1}\, G_t\, A_{t,s}',$$

with initial conditions $A_{t,t} = \Omega_{t|t-1} = \Omega_t$ and $\widehat{x}_{t|t-1} = \widehat{x}_t$ from (3.15) and (3.17), respectively.

3.4.4 Missing Values

When the series has missing observations , the Kalman prediction equations (3.15)–(3.17) must be modified as follows. If the observation y_t is missing, then

$$\begin{aligned} \Omega_{t+1} &= F_t\, \Omega_t\, F_t' + \Sigma_t, \\ \nu_t &= 0, \\ \widehat{x}_{t+1} &= F_t \widehat{x}_t. \end{aligned}$$

As a consequence, the missing value y_t affects the estimation of the state at time $t+1$ making the innovation term zero and increasing the state prediction error variance with respect to the observed y_t case since the subtracting term $\Delta_t\, K_t\, K_t'$ appearing in (3.15) is absent in the modified equations.

Furthermore, when y_t is missing, the modified Kalman filtering equations are

$$\begin{aligned} \Omega_{t|t} &= \Omega_t, \\ \widehat{x}_{t|t} &= \widehat{x}_t, \end{aligned}$$

and the state smoother equations become

$$\begin{aligned} \widehat{x}_{t|s} &= \widehat{x}_{t|s-1}, \\ A_{t,s+1} &= A_{t,s}\, F_t', \\ \Omega_{t|s} &= \Omega_{t|s-1}. \end{aligned}$$

Additional details about this modifications can be found in Subsection 11.2.4.

EXAMPLE 3.3

In order to illustrate the application of the Kalman recursions consider the air pollution state space model discussed in Section 3.1.

Let $\theta = c(\alpha_1, \alpha_2, \sigma)$ the parameter describing this model. The R package FKF provides a fast implementation of the Kalman recursions. With the help of this package we obtain the following output:

```
> fit
$par
       ar1        ar2      sigma
0.68476597 0.33517491 0.09911715

$value
[1] -577.6894

$counts
function gradient
     178       NA

$convergence
[1] 0

$message
NULL

$hessian
            ar1       ar2      sigma
ar1    444.0878  337.6504   8584.973
ar2    337.6504  322.5501   5345.319
sigma 8584.9733 5345.3187 203590.721
```

The approximated variance-covariance matrix of the parameter estimate $\widehat{\theta}$ is given by

$$\operatorname{Var}(\widehat{\theta}) = \begin{pmatrix} 0.1396 & -0.0307 & 0.0838 \\ 1.2299 & 0.7011 & 0.1144 \end{pmatrix}.$$

From this matrix we obtain 95% confidence intervals for the parameters, $(0.1396, 1.2299)$ for α_1, $(-0.0307, 0.7011)$ for α_2 and $(0.0838, 0.1144)$ for σ.

Figure 3.4 and Figure 3.5 show the first and second components of the state predictors, respectively. Note that the predictors follows the trajectory of the state vectors closely.

The standard deviations of these state predictors are exhibited in Figure 3.6. The first standard deviations were no plotted since they

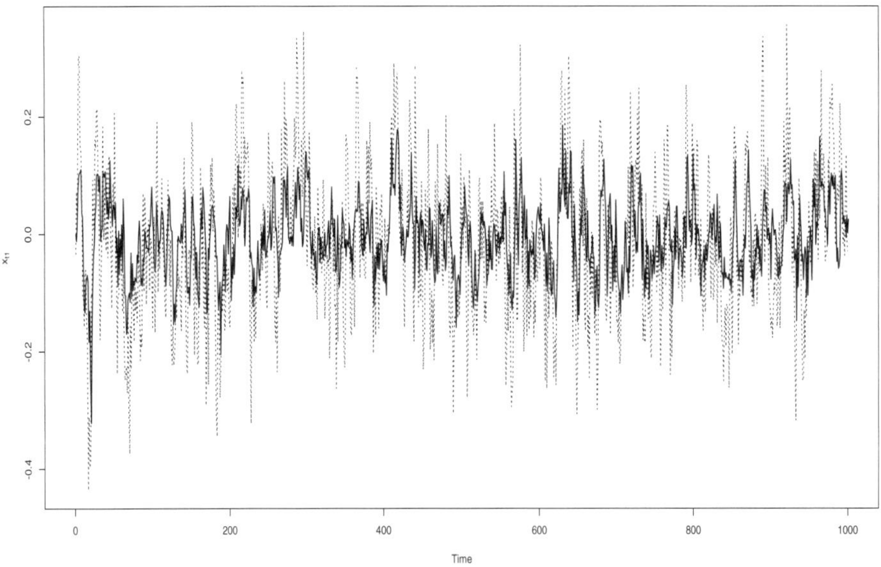

Figure 3.4 *State space model example. State predictors, $t = 1, \ldots, 1000$. Heavy line, first component predictors. Dotted line, first component state values.*

represent an initial guess, which is usually a very high value, 100 in this case. Note that the standard deviations decay quite rapidly after a few steps and converge to 0.1143 and 0.1298, respectively.

On the other hand, Figure 3.7 and Figure 3.8 exhibit the filtered state vectors. Observe that the estimated states are closer to their true counterparts than the predicted states.

The standard deviations of the filtered states is depicted in Figure 3.9. Notice that for both components these standard deviations converge very quickly to 0.093 and 0.1200, respectively.

Figure 3.10 display the innovation sequence of this fitted state space model along with the sample ACF and their standard deviations. Note that the innovations seem to be uncorrelated.

Finally, the Kalman gain vectors are shown in Figure 3.11. Notice that both components converge fast to 0.4891 and 0.3669, respectively. Summarizing, this simple example illustrate the versatility of a state space system to model complex practical situations where the variables of interest are not directly observed. In this case, the model can identify the relative contribution of each pollution source and can estimate them from the observed data.

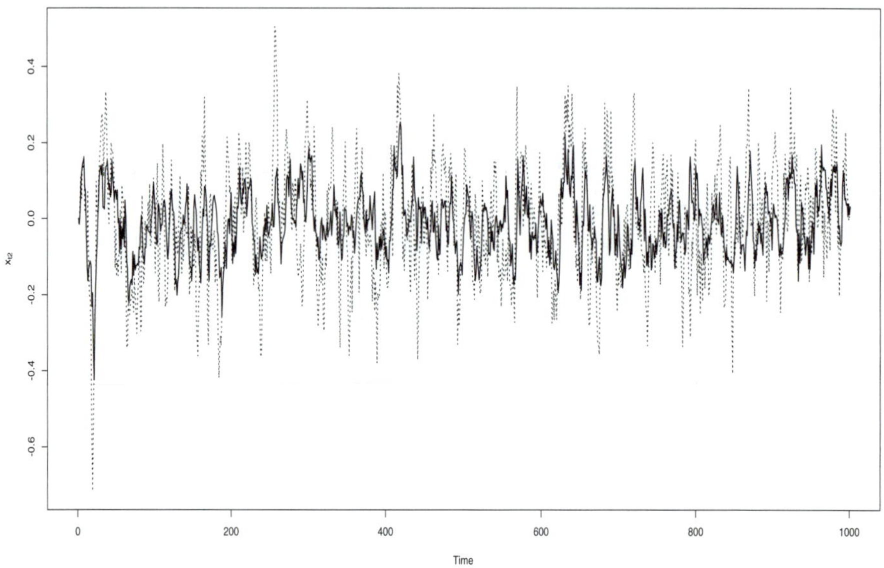

Figure 3.5 *State space model example. State predictors, $t = 1, \ldots, 1000$. Heavy line, second component predictors. Dotted line, second component state values.*

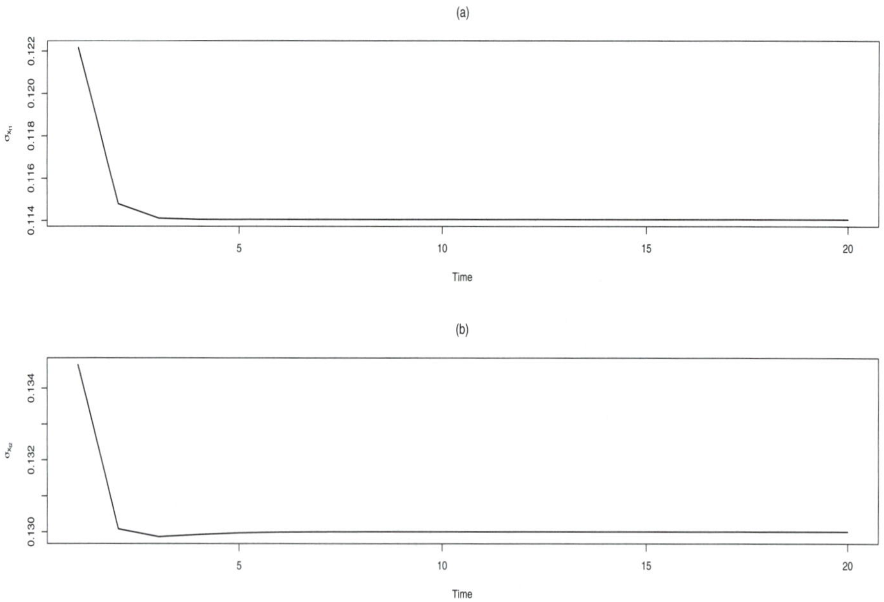

Figure 3.6 *State space model example. Standard deviations of state predictors, $t = 2, \ldots, 20$. (a) first state component, (b) second state component $x_{t,2}$.*

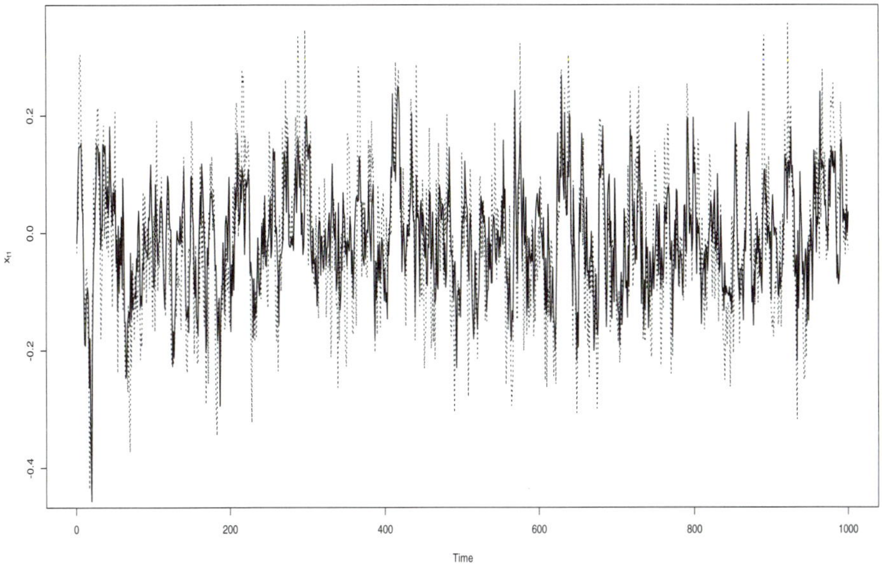

Figure 3.7 *State space model example. States filters, $t = 1, \ldots, 1000$. Heavy line, filtered first component. Dotted line, first component state values.*

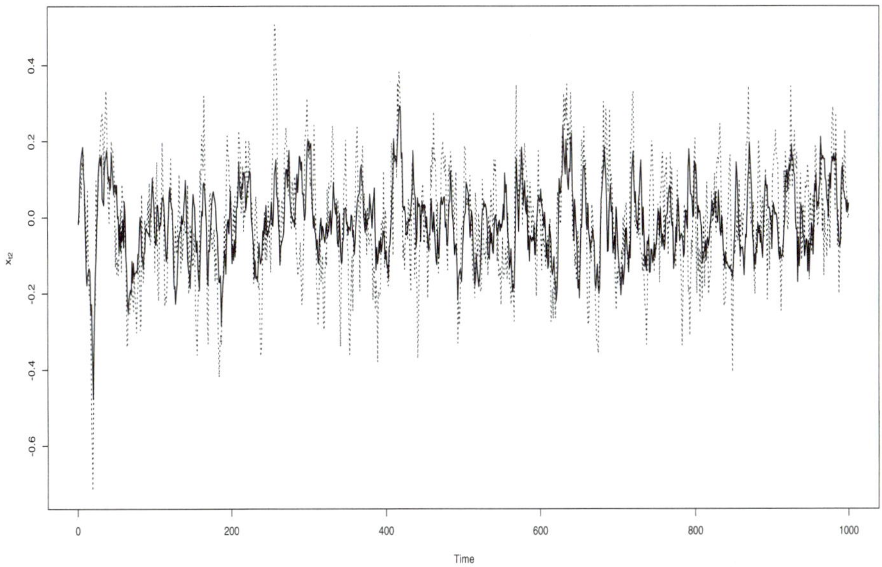

Figure 3.8 *State space model example. States filters, $t = 1, \ldots, 1000$. Heavy line, filtered second component. Dotted line, second component state values.*

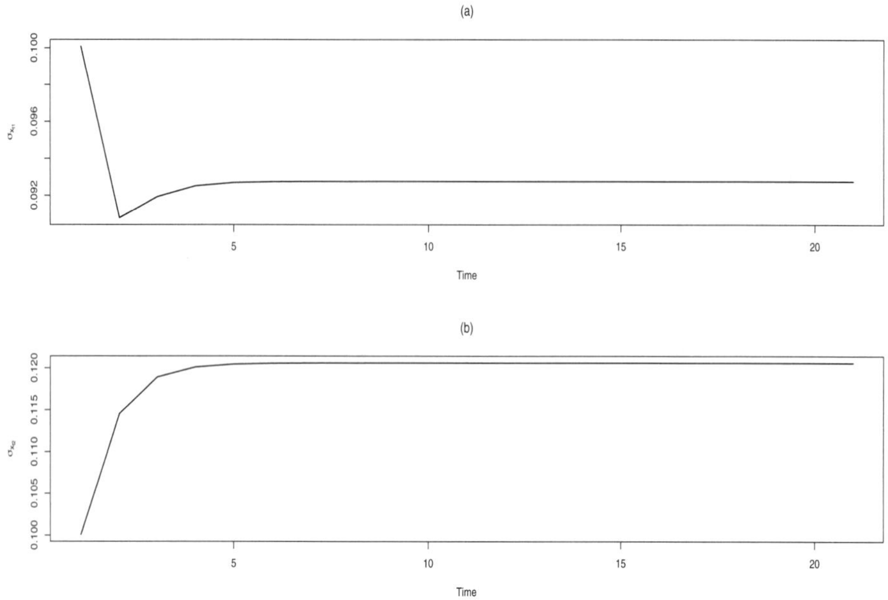

Figure 3.9 *State space model example. Standard deviations of state filters,* $t = 1, \ldots, 20$. *(a) first state component. (b) second state component* $x_{t,2}$.

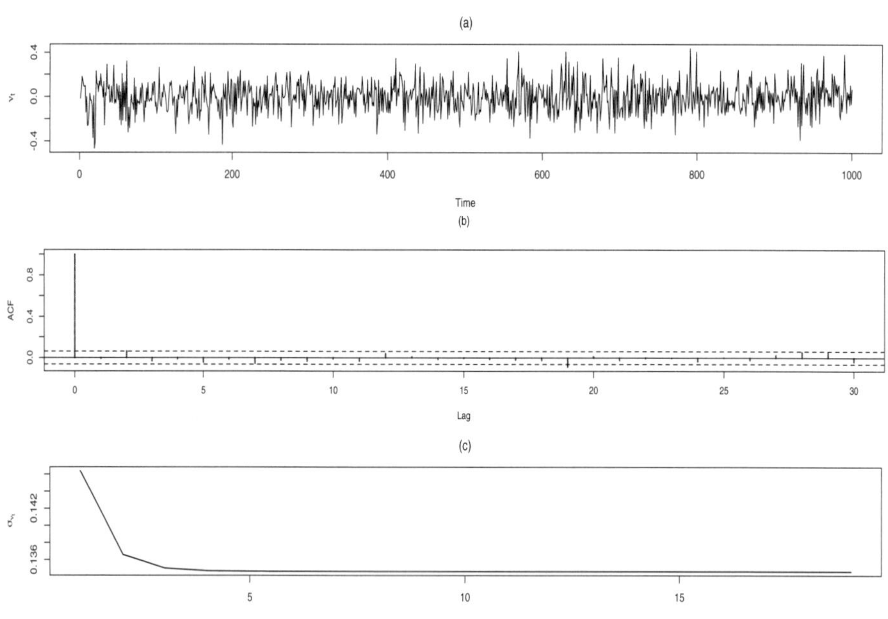

Figure 3.10 *State space model example. (a) Innovations, (b) sample ACF, and (c) standard deviations for* $t = 2, \ldots, 20$.

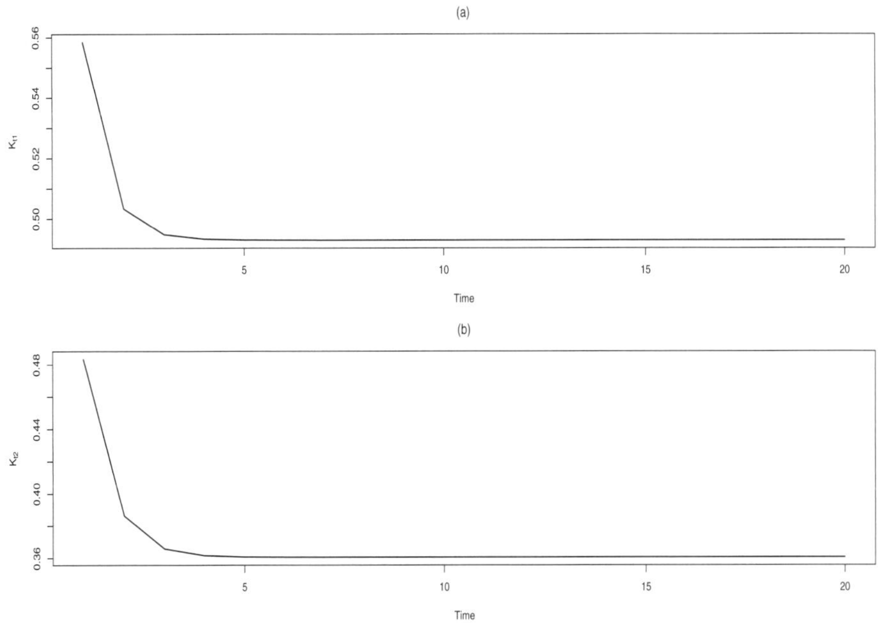

Figure 3.11 *State space model example. Kalman gain, $t = 1, \ldots, 20$. (a) first component $K_{t,1}$, (b) second component $K_{t,2}$.*

3.4.5 Additive Noise

State space models allows the handling of ARMA or ARFIMA processes observed with error. As an illustration consider the AR(1) model with additive noise described by

$$\begin{aligned} x_{t+1} &= \phi x_t + v_t, \\ y_t &= x_t + w_t, \end{aligned}$$

where x_t is an AR(1) underlying process and y_t is the observed process, with error w_t which is a white noise sequence. Suppose that $\mathrm{Var}(v_t) = \sigma_v^2$ and $\mathrm{Var}(w_t) = \sigma_w^2$ are the state and observational noise variances, respectively. Assume also that the state and observational noises are uncorrelated. Notice that the variance of the state and observed process are

$$\begin{aligned} \mathrm{Var}(x_t) &= \frac{\sigma_v^2}{1-\phi^2}, \\ \mathrm{Var}(y_t) &= \frac{\sigma_v^2}{1-\phi^2} + \sigma_w^2. \end{aligned}$$

Thus, that the variance ratio is

$$\frac{\mathrm{Var}(y_t)}{\mathrm{Var}(x_t)} = 1 + \frac{\sigma_w^2}{\sigma_v^2}(1-\phi^2),$$

showing that the increase in variance from x_t to y_t depend on both the noise variance ratio and the value of the autoregressive parameter.

The Kalman equations in this case are give by

$$\begin{aligned}
\Delta_t &= \Omega_t + \sigma_w^2, \\
K_t &= \phi\,\Omega_t\,\Delta_t^{-1}, \\
\Omega_{t+1} &= \phi\,\Omega_t + \sigma_v^2 - \Delta_t\,K_t^2, \\
\nu_t &= y_t - \widehat{x}_t, \\
\widehat{x}_{t+1} &= \phi\,\widehat{x}_t + K_t\,\nu_t.
\end{aligned}$$

These equations can be simplified as follows,

$$\begin{aligned}
\Delta_t &= \Omega_t + \sigma_w^2, \\
K_t &= \frac{\phi\,\Omega_t}{\Omega_t + \sigma_w^2}, \\
\Omega_{t+1} &= \frac{(\phi^2\sigma_w^2 + \sigma_v^2)\,\Omega_t + \sigma_v^2\sigma_w^2}{\Omega_t + \sigma_w^2}, \\
\nu_t &= y_t - \widehat{x}_t, \\
\widehat{x}_{t+1} &= \phi\,\widehat{x}_t + \frac{\phi\,\Omega_t}{\Omega_t + \sigma_w^2}\,\nu_t.
\end{aligned}$$

Figure 3.12 displays 100 values of the state and observations for a simulated AR(1) plus noise process with $\phi = 0.9$, $\sigma_v^2 = 1$, $\sigma_w^2 = 2$. The sample ACF and PACF of the process y_t is exhibited in Figure 3.13

The evolutions of Δ_t and Ω_t for the simulated process is shown in Figure 3.14. Notice that both sequences converge rapidly to their asymptotic values. On the other hand, Figure 3.15 displays the evolutions of the Kalman gain K_t as well as the innovation sequence ν_t. Similarly to the case of the sequences Δ_t and Ω_t, the Kalman gains converges fast to its limit. Moreover, the sample ACF and PACF of the innovations ν_t, see Figure 3.16 suggest that this residual sequence seems to be white noise. This hypothesis is formally tested by means of the Box-Ljung test,

```
> Box.test(nu,lag=10,type="Ljung")

Box-Ljung test

data:  nu
X-squared = 7.0116, df = 10, p-value = 0.7243
```

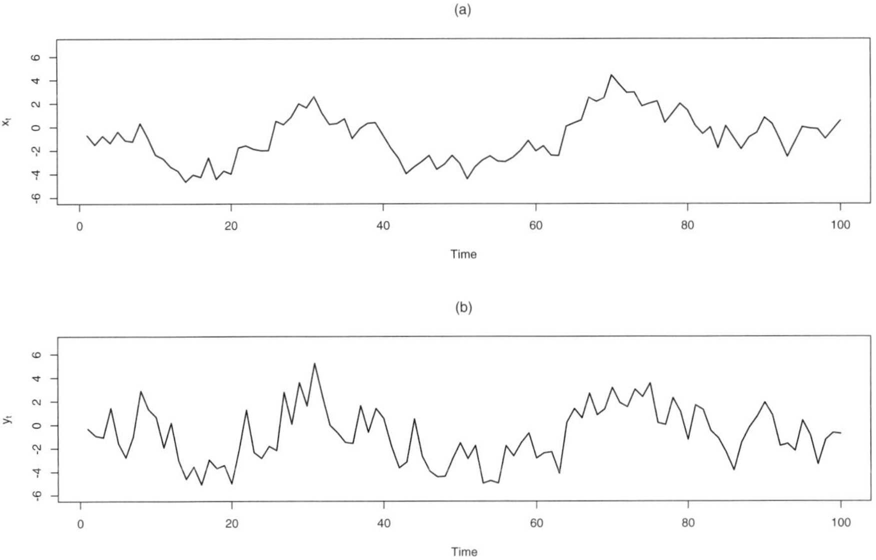

Figure 3.12 *Additive noise state space model. Simulated process with $\phi = 0.9$, $\sigma_v^2 = 1$, $\sigma_w^2 = 2$, $n = 100$. (a) state x_t, (b) observation y_t.*

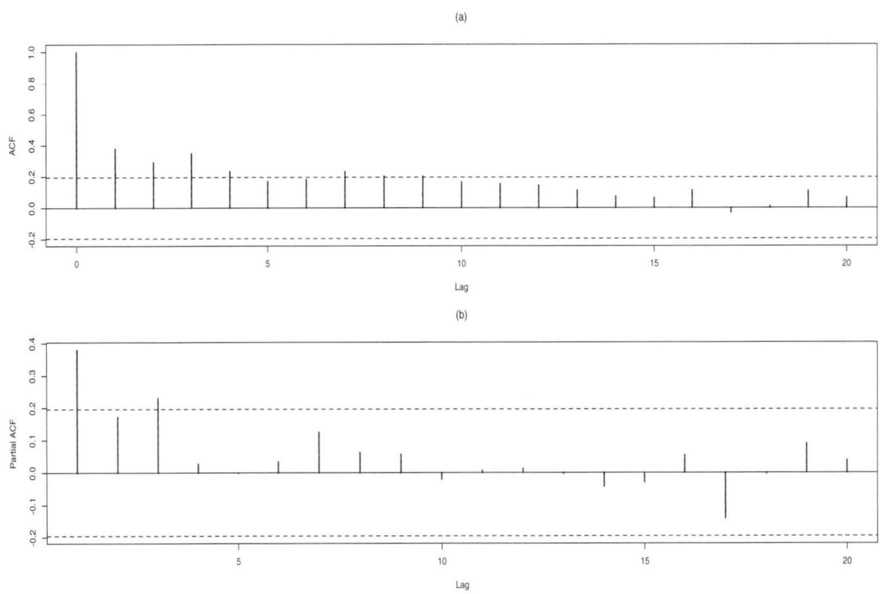

Figure 3.13 *Additive noise state space model. Sample ACF and PACF of the simulated process with $\phi = 0.9$, $\sigma_v^2 = 1$, $\sigma_w^2 = 2$, $n = 100$. (a) sample ACF of y_t, (b) sample PACF of y_t.*

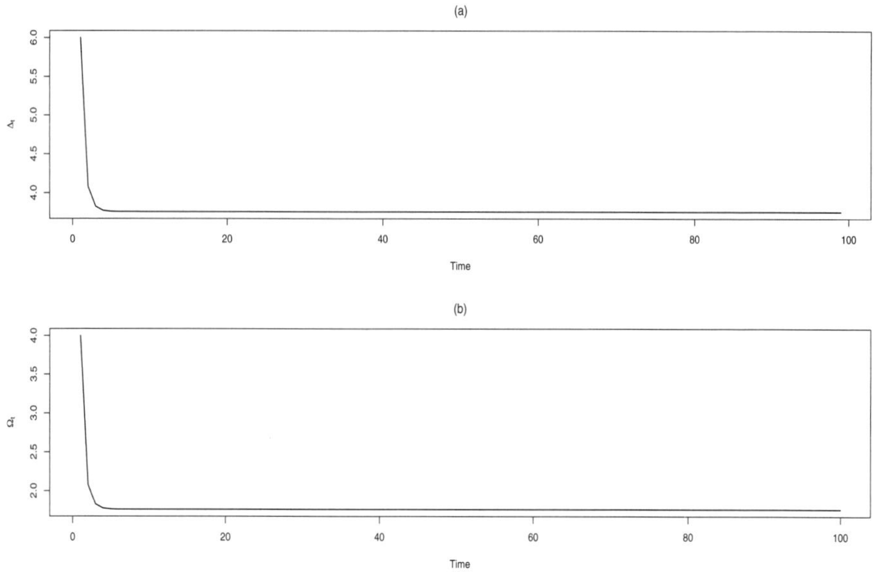

Figure 3.14 *Additive noise state space model. Evolution of Δ_t and Ω_t for the simulated process with $\phi = 0.9$, $\sigma_v^2 = 1$, $\sigma_w^2 = 2$, $n = 100$. (a) state x_t, (b) observation y_t.*

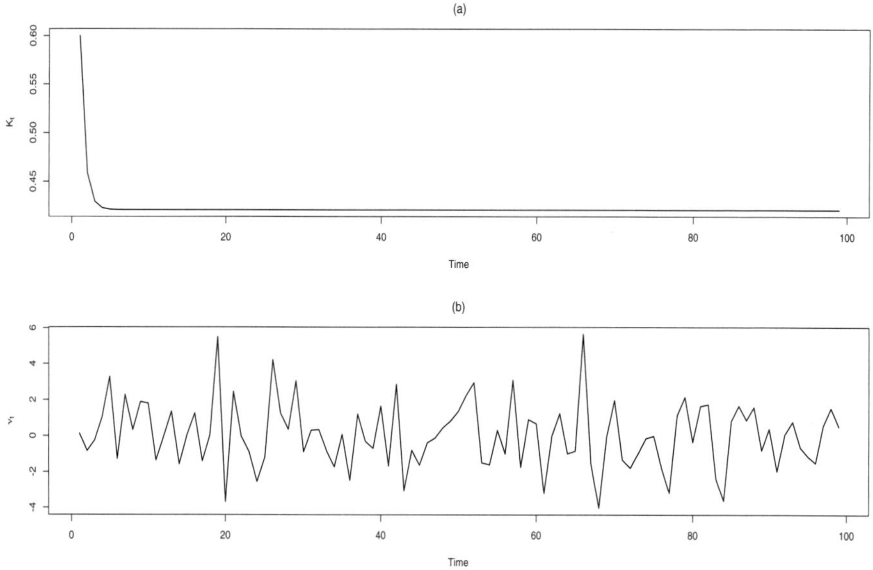

Figure 3.15 *Additive noise state space model. Evolution of K_t and ν_t for the simulated process with $\phi = 0.9$, $\sigma_v^2 = 1$, $\sigma_w^2 = 2$, $n = 100$. (a) state x_t, (b) observation y_t.*

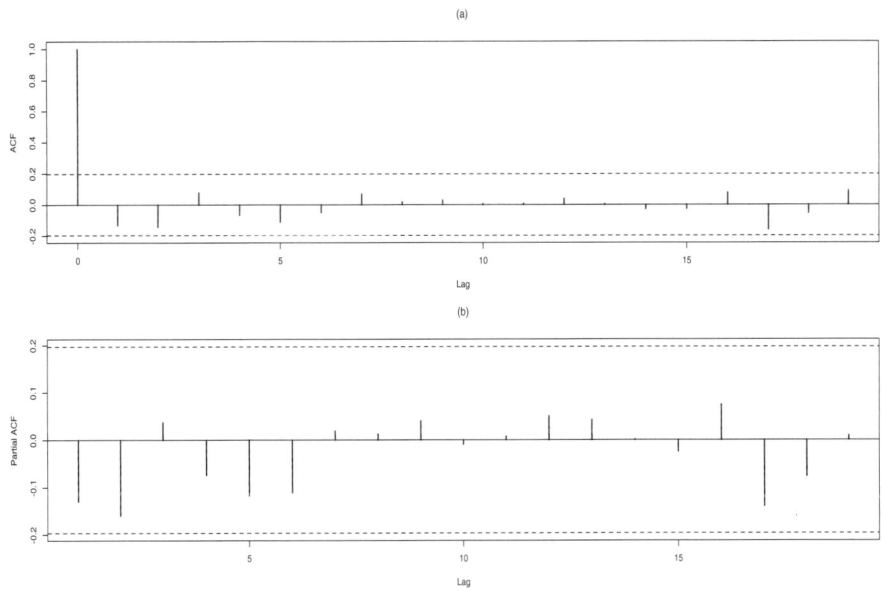

Figure 3.16 *Additive noise state space model. Sample ACF and sample PACF of ν_t for the simulated process with $\phi = 0.9$, $\sigma_v^2 = 1$, $\sigma_w^2 = 2$, $n = 100$. (a) state x_t, (b) observation y_t.*

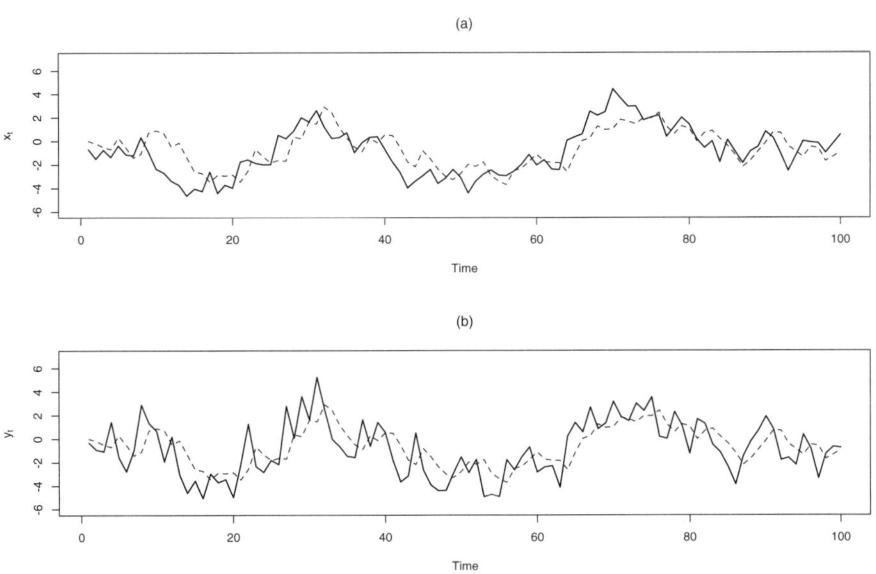

Figure 3.17 *Additive noise state space model. Predictions of the simulated process with $\phi = 0.9$, $\sigma_v^2 = 1$, $\sigma_w^2 = 2$, $n = 100$. (a) state predictions $\widehat{x}_t$, (b) observation predictions $\widehat{y}_t$.*

This procedure indicates that the white noise hypothesis is not rejected at, for instance, the 5% significance level.

Finally, the state and observation predictions are exhibited in Figure 3.17. Notice that the predicted values follows closely theirs true counterparts.

3.4.6 Structural Models

A very useful class of processes in econometrics are the so-called *structural models*, described by

$$\begin{aligned} x_{t+1} &= x_t + v_t, \\ y_t &= x_t + w_t, \end{aligned}$$

where the state x_t is an unobservable variable that governs a economical or financial phenomenon and y_t is an observable process. Notice that this state space model is a particular case of the process studied in the previous subsection. In this situation, the state follows a random walk process with noise standard deviation σ_v. In addition, the observed values correspond to the underlying random walk plus an observational noise w_t with standard deviation σ_w.

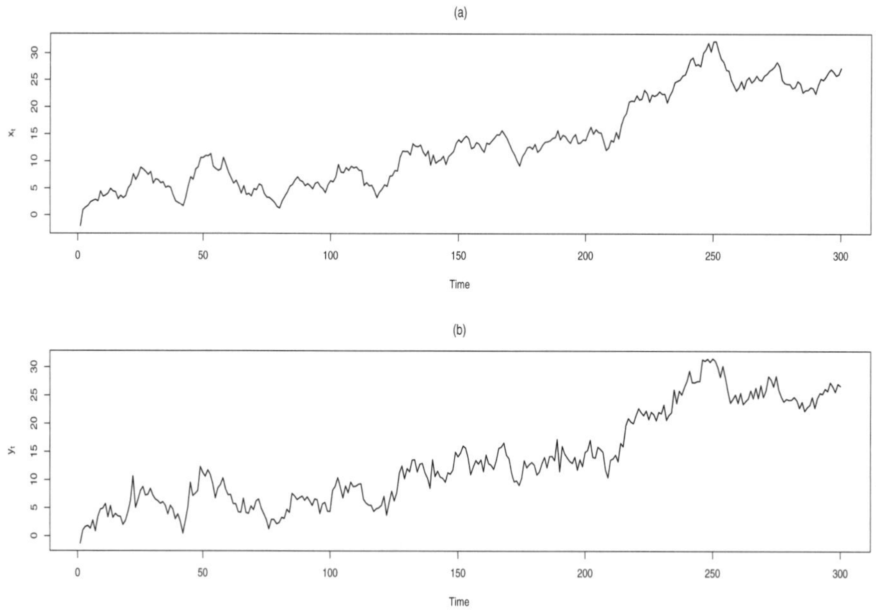

Figure 3.18 *Structural model. Simulated process with $\sigma_v^2 = 1$, $\sigma_w^2 = 1$, $n = 300$. (a) state x_t, (b) observation y_t.*

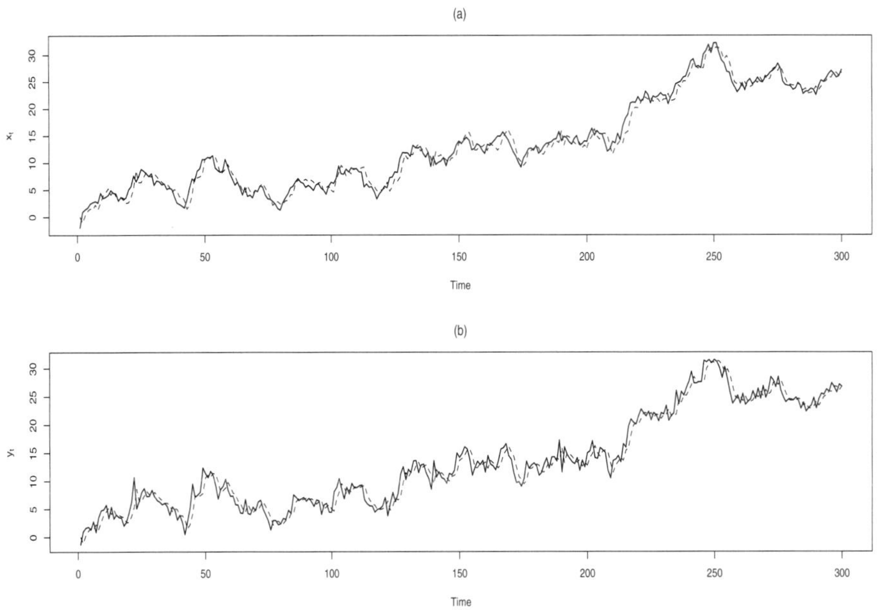

Figure 3.19 *Structural Model. Predictions of the simulated process* $\sigma_v^2 = 1$, $\sigma_w^2 = 1$, $n = 300$. *(a) state predictions* $\widehat{x}_t$, *(b) observation predictions* $\widehat{y}_t$.

Figure 3.18 shows a trajectory of this time series structural model with 300 values and parameters $\sigma_v^2 = 1$ and $\sigma_w^2 = 1$. Observe that the series y_t is, as expected, more noisy than the state sequence x_t.

On the other hand, Figure 3.19 exhibits the predictions of the state and observations obtained from de Kalman recursions. Notice that in both cases the predicted values are close to their theoretical counterparts.

3.4.7 Estimation of Future States

Future values of the state vector can be estimated from the Kalman recursions. Let $h \geq 0$ and define $\mathcal{F}_h = F_t \, F_{t+1} \, \cdots \, F_{t+h}$, the h-step forward state predictor is given by

$$\widehat{x}_{t+h} = \mathcal{F}_h \, \widehat{x}_t,$$

with error variance

$$\Omega_t^h = \mathcal{F}_h \, \Omega_t \, \mathcal{F}_h' + \sum_{j=0}^{h-1} \mathcal{F}_j \, \Sigma_j \, \mathcal{F}_h'.$$

Consequently, the h-step predictor of observation y_{t+h}, for $h \geq 0$, given its finite past $y_1, \ldots, y_{t-1}$ is readily obtained from the state predictor $\widehat{x}_{t+h}$ as

$$\widehat{y}_{t+h} = G_{t+h}\,\widehat{x}_{t+h},$$

since the sequence $\varepsilon_{t+1}, \varepsilon_{t+2}, \ldots$ is orthogonal to the past observations $\{y_1, \ldots, y_{t-1}\}$. Consequently, we conclude that

$$\widehat{y}_{t+h} = G_{t+h}\,\mathcal{F}_h \widehat{x}_t,$$

with h-step prediction error variance

$$\mathrm{Var}(y_{t+h} - \widehat{y}_{t+h}) = \Delta_{t+h} = G_{t+h}\,\Omega_t^h\,G_{t+h}' + \sigma_{t+h}^2.$$

■ EXAMPLE 3.4

Figure 3.20 and Figure 3.21 exhibit the estimated state components for $t = 1000$ and $h = 1, \ldots, 300$, along with their estimated standard deviations. Notice that both states estimates rapidly converge to zero as the horizon h increases. Additionally, the standard deviations plots show that they decrease after a few steps starting at time $t = 1$, they remain steady and then start to grow after $t = 1,000$ reaching the state standard deviations $\mathrm{sd}(x_{t,1}) = 0.1312$ and $\mathrm{sd}(x_{t,2}) = 0.1539$, respectively.

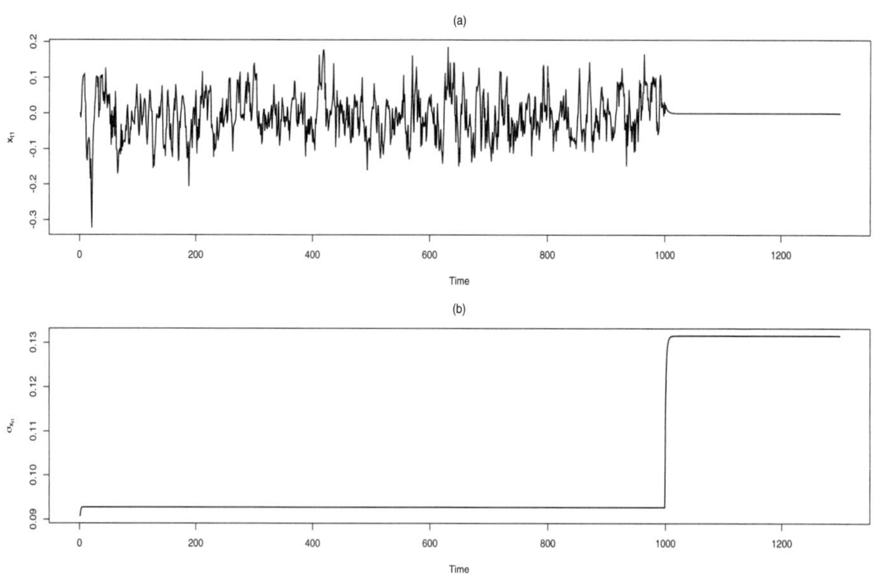

Figure 3.20 *State space model example. Estimated future states, $t = 1000$, $h = 1, \ldots, 300$. (a) estimated first state component, (b) estimated standard deviations.*

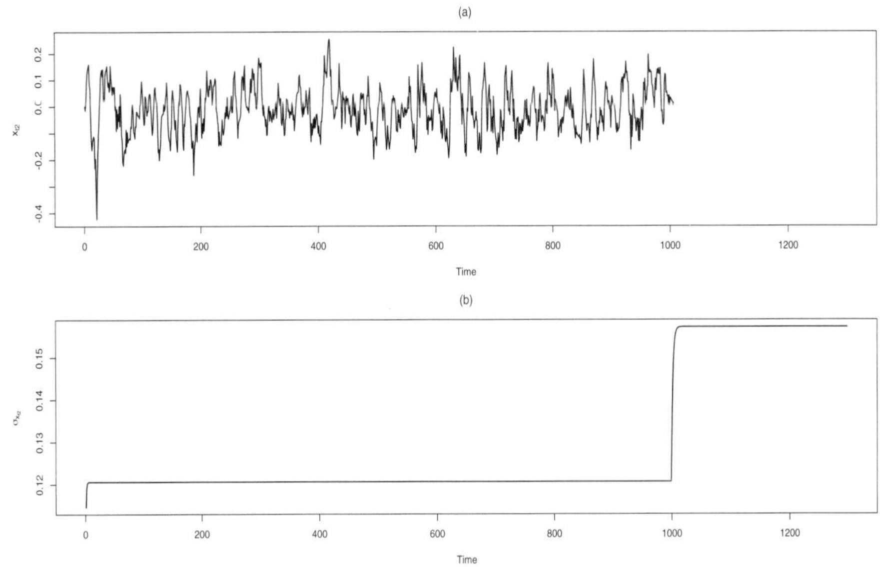

Figure 3.21 *State space model example. Estimated future states,* $t = 1000$, $h = 1, \ldots, 300$. *(a) estimated second state component., (b) estimated standard deviations.*

3.5 EXOGENOUS VARIABLES

An additional versatility onstate space models is that they can be readily extended to incorporate an exogenous process $\{z_t\}$. This sequence is usually a deterministic component such as a trend or cycle. An extended state space system may be written as

$$\widetilde{x}_{t+1} = F\,\widetilde{x}_t + L\,z_t + v_t, \quad (3.18)$$

$$\widetilde{y}_t = G\,\widetilde{x}_t + w_t. \quad (3.19)$$

The extended model described by equations (3.18)–(3.19) can be modified to fit into the simpler structure (3.1)–(3.2) as follows. Let us define the variable

$$r_t = \sum_{t=1}^{t-1} F^{j-1}\,L\,z_{t-j},$$

for $t \geq 1$ and $r_0 = 0$. With this definition we have that

$$r_{t+1} = F\,r_t + L\,z_t. \quad (3.20)$$

Let $x_t = \widetilde{x}_t - r_t$ be the modified state vector and $y_t = \widetilde{y}_t - G\,r_t$ be the modified observation at time t. Then, from (3.18)–(3.19) and (3.20) we conclude that the state x_t and the observations y_t satisfy the system (3.1)–(3.2).

3.6 BIBLIOGRAPHIC NOTES

The monographs by Anderson and Moore (1979), Harvey (1989), Aoki (1990), and Durbin and Koopman (2001) offer an excellent overview of both methodological and applied aspects of state space modeling. Besides, Chapter 12 of Brockwell and Davis (1991) gives a very good introduction to state space systems in a finite-dimensional context, including descriptions of Kalman recursions and treatment of missing values.

The Kalman filter equations were introduced by Kalman (1961) and Kalman and Bucy (1961). Applications of state space systems to the analysis of time series data are reported in fields as diverse as aeronautics [e.g., Kobayashi and Simon (2003)] and oceanography [e.g., Bennett (1992)].

In some texts, the term *reachability* is used instead of *controllability*. Several definitions of stability of state space systems in a very general context of infinite-dimensional systems is given in Curtain and Zwart (1995).

Fitting time series models with missing data has been extensively discussed in the state space context. For ARMA and ARIMA models, see, for instance, Jones (1980), Ansley and Kohn (1983), Kohn and Ansley (1986), and Bell and Hillmer (1991). On the other hand, for ARFIMA models see, for example, Palma and Chan (1997), and Ray and Tsay (2002).

Finally, the book by Hannan and Deistler (1988) gives an excellent theoretical treatment of the linear systems. In particular, the relationships among the different representations of these processes are analyzed in full detail.

Problems

3.1 Consider the following structural model

$$\begin{aligned} y_t &= \frac{\mu_t}{2} + \varepsilon_t \\ \mu_{t+1} &= \mu_t + \nu_t + \eta_t \\ \nu_{t+1} &= \nu_t + \omega_t, \end{aligned}$$

where ε_t, η_t and ω_t are uncorrelated white noise sequences.

(a) Write this model in terms of a state space representation, identifying all its components.

(b) Is the state sequence stationary?

(c) Is the observed process y_t stationary?

(d) Write down the recursive Kalman equations for this state space model.

3.2 Consider ARMA(3,2) process $y_t = \phi_1 y_{t-1} + \phi_2 y_{t-2} + \phi_3 y_{t-3} + \varepsilon_t - \theta_1 \varepsilon_{t-1} - \theta_2 \varepsilon_{t-2}$, where ϵ_t is white noise $(0, \sigma^2)$. Find a state space system representation of y_t.

3.3 Consider the following state space model:

$$\begin{aligned} x_{t+1} &= \begin{bmatrix} 3 & 1 & 0 \\ 2 & 0 & 0 \\ 0 & 1 & 2 \end{bmatrix} x_t + \begin{bmatrix} 1 \\ 0 \\ 2 \end{bmatrix} \varepsilon_t, \\ y_t &= \begin{bmatrix} 1 & 1 & 0 \end{bmatrix} x_t. \end{aligned}$$

(a) Is this state space model stable?
(b) Verify whether this state space model is observable.
(c) Verify whether this state space model is controllable.
(d) What can you conclude about the model?

3.4 Find a minimal state space representation of the process $y_t = \varepsilon_t + \theta\varepsilon_{t-1}$ where $|\theta| < 1$ and ε_t is white noise.

3.5 Consider the linear process $y_t = \phi y_{t-1} + \varepsilon_t$ where $|\phi| < 1$ and ε_t is white noise.

(a) Find a minimal state space representation of the process y_t.
(b) Verify that the system is stable.
(c) Find the Hankel matrix representing this process.
(d) What is the rank of this Hankel matrix?

3.6 Consider the state space system

$$\begin{aligned} x_{t+1} &= \begin{bmatrix} \theta_1 & 0 & 0 \\ 0 & \theta_2 & 0 \\ 0 & 0 & \theta_3 \end{bmatrix} x_t + \begin{bmatrix} 1 \\ 0 \\ 0 \end{bmatrix} \varepsilon_t, \\ y_t &= \begin{bmatrix} 1 & 1 & 1 \end{bmatrix} x_t. \end{aligned}$$

(a) For which values of the parameter $\boldsymbol{\theta} = (\theta_1, \theta_2, \theta_3)$ is this system stable?
(b) Assume that ε_t is an independent and identically distributed sequence $N(0, 1)$. Simulate several trajectories of the system for a sample size $n = 1000$ and different parameters $\boldsymbol{\theta}$.
(c) Implement computationally the Kalman recursions for this state space system.

3.7 Provide another state space representation based on the infinite autoregressive expansion of an ARFIMA process. Discuss the advantages or disadvantages of this AR(∞) with respect to the MA(∞) representation.

3.8 Consider following the state transition matrix associated to the AR(m) state space approximation:

$$F = \begin{pmatrix} \pi_1 & \pi_2 & \cdots & \pi_{m-1} & \pi_m \\ 1 & 0 & \cdots & 0 & 0 \\ 0 & 1 & \cdots & 0 & 0 \\ \vdots & & & & \vdots \\ 0 & 0 & \cdots & 0 & 0 \\ 0 & 0 & \cdots & 1 & 0 \end{pmatrix}.$$

(a) Show that the eigenvalues of this matrix are the roots of the polynomial

$$\lambda^m - \pi_1\lambda^{m-1} - \cdots - \pi_{m-1}\lambda - \pi_m = 0.$$

(b) Verify that space of eigenvectors is given by the $\{(\lambda^{m-1}, \ldots, \lambda, 1)'\}$.

3.9 Consider the following state space model:

$$\begin{aligned} x_{t+1} &= \begin{bmatrix} 1 & 0 & 0 \\ 1 & 0 & 0 \\ 0 & 1 & 0 \end{bmatrix} x_t + \begin{bmatrix} 1 \\ 0 \\ 0 \end{bmatrix} \varepsilon_t, \\ y_t &= \begin{bmatrix} 1 & 0 & 0 \end{bmatrix} x_t. \end{aligned}$$

(a) Is this state space model stable?
(b) Verify whether this state space model is observable.
(c) Verify whether this state space model is controllable.
(d) What can you conclude about the model?

3.10 Consider a finite-dimensional state space system where $x_t \in \mathbb{R}^n$. Write a computer program implementing the Kalman recursion equations (3.13)–(3.17).

3.11 Given a sample $\{y_1, \ldots, y_n\}$, verify that $\Omega_{t|n} \leq \Omega_{t|t} \leq \Omega_t$ for all $t \leq n$, where the matrix inequality $A \leq B$ means that $x'(B - A)x \geq 0$ for all x.

3.12 Consider the following state space system:

$$\begin{aligned} x_{t+1} &= \phi x_t + \varepsilon_t, \\ y_t &= \theta x_t + \varepsilon_t, \end{aligned}$$

where ε_t is white noise with unit variance.

(a) For which values of the parameter $\boldsymbol{\theta} = (\phi, \theta)$ is this system stable?
(b) For which values of $\boldsymbol{\theta}$ is the system observable or controllable?
(c) For which values of $\boldsymbol{\theta}$ are the Kalman recursions stable?
(d) Assume that ε_t is an independent and identically distributed sequence $N(0, 1)$. Simulate several trajectories of the system for a sample size $n = 1000$ and different parameters $\boldsymbol{\theta}$.

3.13 Consider the following state space system for $x_t \in \mathbb{R}^n$ with $n \geq 2$:

$$\begin{aligned} x_{t+1} &= Fx_t + H\varepsilon_t, \\ y_t &= Gx_t + \varepsilon_t, \end{aligned}$$

where

$$\begin{aligned} F &= \begin{bmatrix} 0 & 1 & 0 & 0 & \cdots & 0 \\ 0 & 0 & 1 & 0 & \cdots & 0 \\ 0 & 0 & 0 & 1 & \cdots & 0 \\ \vdots & \vdots & & & \ddots & \\ 0 & 0 & 0 & \cdots & 0 & 1 \\ 0 & 0 & 0 & \cdots & 0 & 0 \end{bmatrix}, \\ G &= [\psi_n \quad \psi_{n-1} \quad \psi_{n-2} \quad \cdots], \\ H &= [1 \ 0 \ 0 \ \cdots \ 0]', \end{aligned}$$

and the coefficients ψ_j are given by the expansion (3.3).

(a) Verify that this system is stable.

(b) Verify that $\|F\| = \sup_x \|Fx\|/\|x\| = 1$.

(c) Does $\|F\|^n$ converges to zero at an exponential rate?

(d) Find the observability matrix for this system $\mathcal{O}$ and verify that it is of full rank if and only if $\psi_n \neq 0$.

3.14 Consider the AR(p) process given by the equation

$$y_t - \phi_1 y_{t-1} \cdots - \phi_p y_{t-p} = \varepsilon_t.$$

(a) Show that by defining the state vector $x_t = (y_{t+1-p}, \ldots, y_t)'$, the AR($p$) process may be written in terms of the following state space representation:

$$\begin{aligned} x_{t+1} &= Fx_t + H\varepsilon_{t+1}, \\ y_t &= Gx_t, \end{aligned}$$

where

$$\begin{aligned} F &= \begin{bmatrix} 0 & 1 & 0 & 0 & \cdots & 0 \\ 0 & 0 & 1 & 0 & \cdots & 0 \\ 0 & 0 & 0 & 1 & \cdots & 0 \\ \vdots & \vdots & & & \ddots & \\ 0 & 0 & 0 & \cdots & 0 & 1 \\ \phi_p & \phi_{p-1} & \phi_{p-2} & \cdots & \phi_2 & \phi_1 \end{bmatrix}, \\ G &= [0 \ 0 \ 0 \ \cdots \ 0 \ 1], \\ H &= [0 \ 0 \ 0 \ \cdots \ 0 \ 1]'. \end{aligned}$$

(b) Write down the observation matrix $\mathcal{O}$ for this state space model. Is this system observable?

(c) Find the controllability matrix $\mathcal{C}$ for this state space representation and check whether this system is controllable.

(d) Verify whether this system converges to its steady state or not. If yes, write down the steady state system equations.

3.15 Given the state space system

$$\begin{aligned} x_{t+1} &= Fx_t + H\varepsilon_t, \\ y_t &= Gx_t + \varepsilon_t, \end{aligned}$$

where $\{\varepsilon_t\}$ is a white noise sequence, show that the product of the observation matrix and the controllability matrix yields

$$\mathcal{O}\mathcal{C} = \begin{pmatrix} GH & GFH & GF^2H & GF^3H & \cdots \\ GFH & GF^2H & GF^3H & GF^4H & \cdots \\ GF^2H & GF^3H & GF^4H & GF^5H & \cdots \\ \vdots & \vdots & \vdots & & \end{pmatrix}.$$

Is this a Hankel matrix?

3.16 Suppose that the state transition matrix F satisfies $\|F^j\| \leq ce^{-\alpha j}$ so that the corresponding state space system is stable.

(a) Verify that in this case,

$$\sum_{j=0}^{n} F^j z^j,$$

converges for all $|z| \leq 1$.

(b) Show that

$$(I - zF)^{-1} = \sum_{j=0}^{\infty} F^j z^j,$$

for all $|z| \leq 1$.

3.17 Consider the following state space model:

$$\begin{aligned} x_{t+1} &= Fx_t + H\varepsilon_t, \\ y_t &= Gx_t + \varepsilon_t, \end{aligned}$$

and let $\psi(z)$ be the operator $\psi(z) = 1 + \psi_1 z + \psi_2 z^2 + \cdots$, where $\psi_j = GF^{j-1}H$ and $|z| \leq 1$.

(a) Prove that $\psi(z)$ may be written as

$$\psi(z) = 1 + G(I - zF)^{-1}Hz.$$

(b) Assume that $F = 0$. Show that if $|GH| \leq 1$, then $\psi(z)$ is invertible for $|z| \leq 1$.

(c) If $F = 0$ and $|GH| \leq 1$, verify that the state x_t may be expressed as

$$x_t = Hy_{t-1} + H \sum_{j=1}^{\infty} (-GH)^j y_{t-1-j}.$$

3.18 Consider the linear state space model where the transition matrix depends on time

$$\begin{aligned} x_{t+1} &= F_t x_t + H\varepsilon_t, \\ y_t &= Gx_t + \varepsilon_t, \end{aligned}$$

for $t \geq 1$, ε_t is a white noise sequence and x_0 is the initial state.

(a) Show that the state at time $t + 1$ may be written as

$$x_{t+1} = \left(\prod_{j=1}^{t} F_j \right) x_0 + \sum_{j=0}^{t} \varphi_j H \varepsilon_{t-j},$$

and find the coefficients φ_j.

(b) Let $z_n = \sum_{j=1}^{n} \log \|F_j\|$ and assume that the limit

$$\gamma = \lim_{n \to \infty} z_n / n$$

exists. Prove that if $\gamma < 0$, then

$$\lim_{t \to \infty} \left(\prod_{j=1}^{t} F_j \right) x_0 = 0, \tag{3.21}$$

in probability.

(c) Suppose that $\|F_t\| \leq e^{-\alpha t}$ where α is a positive constant. Show that (3.21) holds in this situation.

(d) Assume that $\|F_t\| \leq t^{-\beta}$ where β is a positive constant. Prove that the limit (3.21) holds under these circumstances.

CHAPTER 4

SPECTRAL ANALYSIS

Some fundamental concepts about spectral analysis are introduced in this chapter. A time series can be analyzed by studying its *time domain* related characteristics such as mean, variance and autocovariances. However, it can be also described by its *frequency domain* related properties such as spectral density or Cramer representation. In what follows we describe briefly these two ways of describing and modeling time series data. As described in this chapter, the frequency domain is particularly appropriate for analyzing time series exhibiting periodic or seasonal patterns. These patterns can be deterministic as in a an harmonic regression or stochastic as in seasonal autoregressive model. However, the application of the spectral analysis is also important for estimating and forecasting time series. As an example of such application we can mention the Whittle likelihood function which allows for the efficient calculation of quasi maximum likelihood estimators. A detailed account of these spectral based parameter estimation methods is provided in Chapter 5.

Time Series Analysis. First Edition. Wilfredo Palma.

4.1 TIME AND FREQUENCY DOMAINS

In the previous chapters we have discussed some basic concepts in time series analysis such as moments, autocovariances and autocorrelations. Time lags have played and essential role in these definitions. Starting from these concepts, more elaborated techniques can be implemented such as maximum likelihood estimation which in the Gaussian case is based on the mean and covariance structure of the observations. Generally speaking, these methods belong to the so-called *time domain*. On the other hand, a time series can be analyzed by taking into account it spectral features, including its spectral distribution, its spectral density or its Cramer representation. These techniques are intimately related to the study of data exhibiting seasonal o cyclical behavior and they are usually denoted as belonging to the so-called *frequency domain*. For example, in many engineering contexts, it is natural to study the frequency at which a signal is propagated. Nevertheless, as we will see later, there is a deep relationship between time and frequency domains.

4.2 LINEAR FILTERS

In linear filtering theory, one usually assumes that a signal is the result of the application of a linear filter to a white noise input sequence. This simple idea has proven to be quite powerful and useful in practice. As in the previous chapters, the data can be written as

$$y_t = \psi(B)\varepsilon_t = \sum_{j=-\infty}^{\infty} \psi_j \varepsilon_{t-j}. \tag{4.1}$$

where $\sum_{j=-\infty}^{\infty} \psi_j^2 < \infty$. Observe that this condition guarantees that the process $\{y_t : t \in \mathbb{Z}\}$ possesses finite variance since

$$\operatorname{Var}(y_t) = \operatorname{Var}(\sum_{j=-\infty}^{\infty} \psi_j \varepsilon_{t-j}) = \sigma^2 \sum_{j=-\infty}^{\infty} \psi_j^2 < \infty.$$

In some cases, processes with infinite variance are also of interest. However the analysis of these processes are beyond the scope of this book.

When the filter ψ is invertible, the process $\{y_t : t \in \mathbb{Z}\}$ satisfies the discrete-time equation

$$\psi(B)^{-1} y_t = \pi(B) y_t = \sum_{j=-\infty}^{\infty} \pi_j y_{t-j} = \varepsilon_t. \tag{4.2}$$

For simplicity, in this chapter we will consider linear processes satisfying the summability condition $\sum_{h=-\infty}^{\infty} |h\gamma(h)| < \infty$. Most the results discussed next can be extended to more general autocovariance structures such as, for instance, long-memory processes. Nevertheless, the above condition greatly simplify the exposition.

■ **EXAMPLE 4.1**

The autoregressive process introduced in the previous chapter can be readily represented as the result of a linear filtering. For example, an AR(p) process

$$y_t = \phi_1 y_{t-1} + \phi_2 y_{t-2} + \cdots + \phi_p y_{t-p} + \varepsilon_t,$$

can be readily expressed in terms of (4.2) by setting $\pi_j = 0$ for $j < 0$ and $j > p$, and $\pi_j = -\phi_j$ for $j = 1, \ldots, p$.

■ **EXAMPLE 4.2**

Consider the periodic process $y_t = A\cos(\omega t + \phi)$ where A is a zero-mean random variable with finite variance. Then, this process satisfies the equation

$$y_t - 2\cos(\omega) y_{t-1} + y_{t-2} = 0.$$

Thus, this periodic process could fit equation (4.2) with $\pi_j = 0$ for $j < 0$, $\pi_0 = 1$ $\pi_1 = -2\cos(\omega)$, $\pi_2 = 1$ and ε_t a sequence of zeroes. The ACF of this process is given by

$$\gamma(h) = \sigma_A^2 \cos(\omega h),$$

where σ_A^2 is the variance of the random variable A.

4.3 SPECTRAL DENSITY

The spectral density of the process defined by the linear filter (4.1) is given by

$$f(\omega) = \frac{\sigma^2}{2\pi} |\psi(e^{i\omega})|^2.$$

It can be readily shown that an alternative definition of the spectral density of the process is provided by

$$f(\omega) = \frac{1}{2\pi} \sum_{h=-\infty}^{\infty} \gamma(h) e^{i\omega h}. \tag{4.3}$$

This expression is obtained as follows,

$$\begin{aligned} f(\omega) &= \frac{\sigma^2}{2\pi} \sum_{h=-\infty}^{\infty} \sum_{j=-\infty}^{\infty} \psi_j \psi_{j+h} e^{i\omega h} \\ &= \frac{\sigma^2}{2\pi} \sum_{h=-\infty}^{\infty} \sum_{j=-\infty}^{\infty} \psi_j e^{-i\omega j} \psi_{j+h} e^{i\omega(j+h)} \\ &= \frac{\sigma^2}{2\pi} \sum_{j=-\infty}^{\infty} \psi_j e^{-i\omega j} \sum_{h=-\infty}^{\infty} \psi_{j+h} e^{i\omega(j+h)} \end{aligned}$$

Now by changing the index from $j+h$ to k in the second sum above we have

$$\begin{aligned} f(\omega) &= \frac{\sigma^2}{2\pi} \sum_{j=-\infty}^{\infty} \psi_j e^{-i\omega j} \sum_{k=-\infty}^{\infty} \psi_k e^{i\omega k} \\ &= \frac{\sigma^2}{2\pi} \left| \sum_{j=-\infty}^{\infty} \psi_j e^{i\omega j} \right|^2 . \end{aligned}$$

Note that we can also obtain an expression for $\gamma(h)$ based on the spectral density of the process. Multiplying (4.3) by $e^{i\omega - h}$ on both sides we get

$$f(\omega) e^{i\omega - h} = \frac{1}{2\pi} \sum_{j=-\infty}^{\infty} \gamma(h) e^{i\omega(j-h)},$$

and by integrating both sides with respect to ω

$$\int_{-\pi}^{\pi} f(\omega) e^{-i\omega h} d\omega = \int_{-\pi}^{\pi} \frac{1}{2\pi} \sum_{j=-\infty}^{\infty} \gamma(h) e^{i\omega(j-h)} d\omega.$$

Recalling that the sequence $\{\gamma(h)\}$ is absolutely summable, we obtain

$$\int_{-\pi}^{\pi} f(\omega) e^{i\omega - h} d\omega = \frac{1}{2\pi} \sum_{j=-\infty}^{\infty} \gamma(h) \int_{-\pi}^{\pi} e^{i\omega(j-h)} d\omega$$

But, since

$$\int_{-\pi}^{\pi} e^{i(j-h)\lambda} d\lambda = \begin{cases} 2\pi & \text{if } j = h, \\ 0 & \text{if } j \neq h, \end{cases}$$

see Problem 4.8, we conclude that

$$\gamma(h) = \int_{-\pi}^{\pi} f(\omega) e^{i\omega - h} d\omega.$$

EXAMPLE 4.3

Consider the ARMA(p, q) model

$$\pi(B)y_t = \theta(B)\varepsilon_t, \tag{4.4}$$

where $\pi(B) = 1 - \pi_1 B - \cdots - \pi_p B^p$, $\theta(B) = 1 - \theta_1 B - \cdots - \theta_q B^q$ and $\{\varepsilon_t\}$ is a white noise sequence with variance σ^2. In this case, the spectral density is given by

$$f(\omega) = \frac{\sigma^2}{2\pi} \frac{|\theta(e^{i\omega})|^2}{|\pi(e^{i\omega})|^2} = \frac{\sigma^2}{2\pi} \frac{|1 - \theta_1 e^{i\omega} - \cdots - \theta_q e^{i\omega q}|^2}{|1 - \pi_1 e^{i\omega} - \cdots - \pi_p e^{i\omega p}|^2}.$$

EXAMPLE 4.4

Consider the fractional noise model ARFIMA$(0, d, 0)$ model

$$y_t = (1 - B)^{-d}\varepsilon_t, \tag{4.5}$$

where $(1 - B)^{-d}$ fractional difference operator and $\{\varepsilon_t\}$ is a white noise sequence with variance σ^2. In this case, the spectral density is given by

$$f(\omega) = \frac{\sigma^2}{2\pi} |1 - e^{i\omega}|^{-2d}.$$

4.4 PERIODOGRAM

The periodogram is an estimator of the spectral density. Given the sample $\{y_1, y_2, \dots, y_n\}$, its *periodogram* is defined as

$$I(\omega) = \frac{1}{2n} \left| \sum_{t=1}^{n} (y_t - \bar{y}) e^{i\omega t} \right|^2. \tag{4.6}$$

This expression can be rewritten as

$$\begin{aligned} I(\omega) &= \frac{1}{2\pi n} \sum_{t=1}^{n} \sum_{s=1}^{n} (y_t - \bar{y})(y_s - \bar{y}) e^{i\omega(t-s)} \\ &= \frac{1}{2\pi n} \sum_{h=1-n}^{n-1} \sum_{t=1}^{n-|h|} (y_t - \bar{y})(y_{t+|h|} - \bar{y}) e^{i\omega h} \\ &= \frac{1}{2\pi} \sum_{h=1-n}^{n-1} \widehat{\gamma}(h) e^{i\omega h} \end{aligned}$$

Thus, we conclude that the spectral density of the process can be written as

$$I(\omega) = \frac{1}{2\pi} \sum_{h=1-n}^{n-1} \widehat{\gamma}(h) e^{i\omega h}.$$

Notice that the periodogram is an asymptotically unbiased estimator of the spectral density, as shown next,

$$\begin{aligned} E[I(\omega)] &= \frac{1}{2\pi} E\left(\sum_{h=1-n}^{n-1} \widehat{\gamma}(h) e^{i\omega h} \right) \\ &= \frac{1}{2\pi} \sum_{h=1-n}^{n-1} E\left[\widehat{\gamma}(h)\right] e^{i\omega h}. \end{aligned}$$

However,

$$\begin{aligned} E\left[\widehat{\gamma}(h)\right] &= \frac{1}{n} E[\sum_{t=1}^{n-|h|} (y_t - \bar{y})(y_{t+|h|} - \bar{y})] \\ &= \frac{1}{n} \sum_{t=1}^{n-|h|} E[(y_t - \bar{y})(y_{t+|h|} - \bar{y})] \\ &= \frac{1}{n} \sum_{t=1}^{n-|h|} E[(y_t - \mu)(y_{t+|h|} - \mu)] - E(\mu - \bar{y})^2 \\ &= \frac{1}{n} \sum_{t=1}^{n-|h|} \gamma(h) - \operatorname{Var}(\bar{y}) \\ &= \frac{n - |h|}{n} \gamma(h) - \operatorname{Var}(\bar{y}) \\ &= \gamma(h) - \frac{|h|}{n} \gamma(h) - \operatorname{Var}(\bar{y}), \end{aligned}$$

and then,

$$E\left[\widehat{\gamma}(h)\right] = \gamma(h) - \frac{|h|}{n} \gamma(h) - \sigma_n^2, \tag{4.7}$$

where $\sigma_n^2 = \operatorname{Var}(\bar{y})$. Notice from this expression that as the sample size increases, for a fixed lag h we have

$$\lim_{n\to\infty} E\left[\widehat{\gamma}(h)\right] = \gamma(h) - \lim_{n\to\infty} \frac{|h|}{n} \gamma(h) - \lim_{n\to\infty} \sigma_n^2 = \gamma(h).$$

Thus, $\widehat{\gamma}(h)$ is asymptotically unbiased. Now, by virtue of (4.7) we have

$$\begin{aligned} E\,I(\omega) &= \frac{1}{2\pi}\sum_{h=1-n}^{n-1} E\left[\widehat{\gamma}(h)\right] e^{i\omega h} \\ &= \frac{1}{2\pi}\sum_{h=1-n}^{n-1}\left[\gamma(h) - \frac{|h|}{n}\gamma(h) - \sigma_n^2\right] e^{i\omega h} \\ &= \frac{1}{2\pi}\sum_{h=1-n}^{n-1}\gamma(h)e^{i\omega h} - \frac{1}{2\pi n}\sum_{h=1-n}^{n-1}|h|\gamma(h)e^{i\omega h} - \frac{\sigma_n^2}{2\pi}\sum_{h=1-n}^{n-1} e^{i\omega h}. \end{aligned}$$

Observe that the last two terms of the expression above vanish as the sample size tends to infinity. In the case of the second term, since the ACF are summable,

$$\lim_{n\to\infty}\frac{1}{2\pi n}\Big|\sum_{h=1-n}^{n-1}|h|\gamma(h)e^{i\omega h}\Big| \le \lim_{n\to\infty}\frac{1}{2\pi n}\sum_{h=1-n}^{n-1}|h\gamma(h)| = 0.$$

For the third term, we have the identity

$$\sum_{h=1-n}^{n-1} e^{i\omega h} = \frac{e^{i\omega n}-1}{e^{i\omega}-1} + \frac{e^{-i\omega n}-1}{e^{-i\omega}-1} - 1.$$

Therefore,

$$\lim_{n\to\infty}\frac{1}{2\pi n}\sum_{h=1-n}^{n-1} e^{i\omega h} = \lim_{n\to\infty}\frac{1}{2\pi n}\left(\frac{e^{i\omega n}-1}{e^{i\omega}-1} + \frac{e^{-i\omega n}-1}{e^{-i\omega}-1} - 1\right) = 0.$$

Consequently,

$$\lim_{n\to\infty} E[I(\omega)] = \frac{1}{2\pi}\sum_{h=-\infty}^{\infty}\gamma(h)e^{i\omega h} = f(\omega).$$

■ EXAMPLE 4.5

As an illustration consider the *white noise* concept introduced in Chapter 1. From a time-domain point of view, this process consists of uncorrelated random variables with zero-mean and constant variance. On the other hand, from a frequency-domain, a white noise sequence can be characterized by a flat spectrum. Figure 4.1(a) shows the raw periodogram and theoretical spectral density of a Gaussian white noise sequence with zero-mean and unit variance. By comparison, Figure 4.1(b) exhibits the raw periodogram and theoretical spectral density of a Gaussian colored noise sequence. In this case, this colored noise corresponds to an MA(1) model with $\theta = 0.5$.

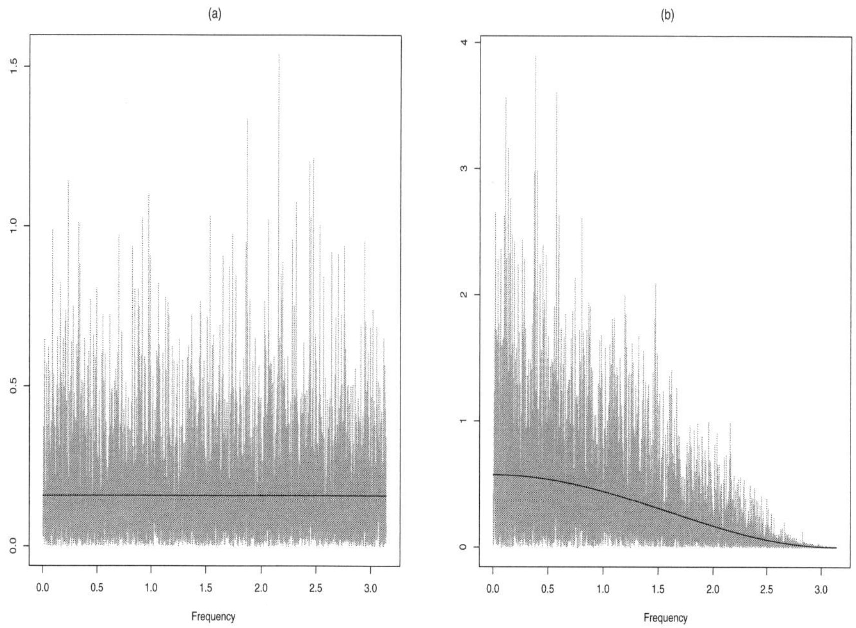

Figure 4.1 *(a) White and (b) colored noise estimated (gray line) and theoretical spectral densities (black line).*

It is interesting to mention that the name white noise comes from the white light of the sun. From a physical standpoint, this light is a composition of light of different colors, but this composition is such that every frequency is equally represented. Thus, the spectrum of white light is flat, that is, every frequency makes the same contribution. On the contrary, in the case of colored noise, some frequencies have a stronger presence in the spectrum.

4.5 SMOOTHED PERIODOGRAM

The raw periodogram discussed in the previous section can be smoothed via different techniques. One of these methods is weighting the raw periodogram around a Fourier frequency ω_j as follows

$$\widehat{f}(\omega_j) = \frac{1}{2\pi} \sum_{h=-m}^{m} W(h) I(\omega_{j+h}).$$

The weighting function $W(\cdot)$ usually is symmetric, such that $W(h) = W(-h)$ and $\sum_{h=-m}^{m} W(h) = 1$. One of the most well-known weighting functions is the so-called *Daniell window* defined by

$$W(\omega) = \begin{cases} \frac{r}{2\pi}, & -\frac{\pi}{r} \leq \omega \leq \frac{\pi}{r}, \\ 0 & \text{otherwise} \end{cases}$$

It is straightforward to verify that

$$\int_{-\pi}^{\pi} W(\omega) d\omega = 1.$$

As an illustration consider the following MA(1) process

$$y_t = \varepsilon_t + \theta\, \varepsilon_{t-1},$$

where ε_t is a normal white noise with zero-mean and unit variance. Given that $\sigma = 1$, the spectral density of this model is

$$f(\omega) = \frac{1}{2\pi}(1 + \theta^2 + 2\,\theta\, \cos\omega).$$

Figure 4.2 displays the theoretical spectral density (heavy line), the raw periodogram (gray line) and the Daniell smoothed periodogram (dotted line). Notice that the raw periodogram exhibits high variability while the smoothed periodogram follows closely the theoretical spectral density of the model.

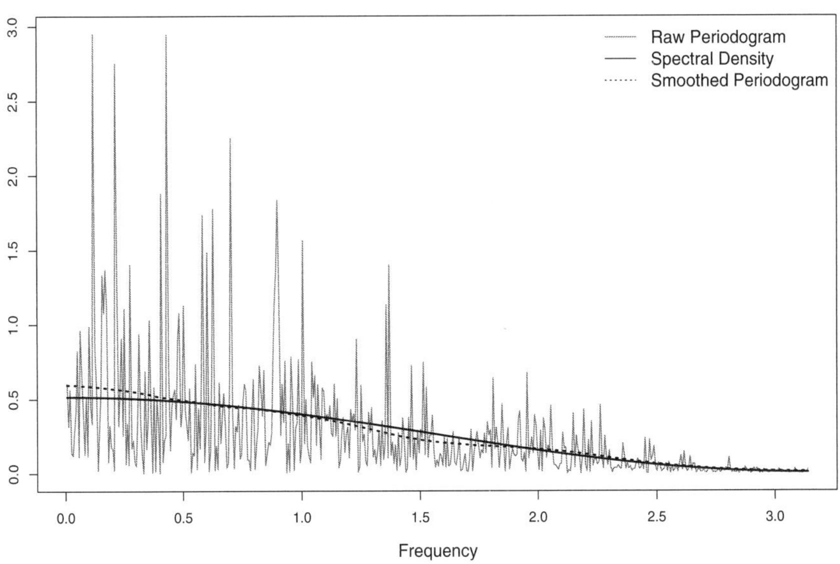

Figure 4.2 *Estimation of the spectral density of a MA(1) process with $\theta = 0.8$ and $\sigma = 1$.*

4.6 EXAMPLES

As a first illustrative example consider the following deterministic harmonic series, with two frequencies ω_1 and ω_2,

$$y_t = \alpha_1 \sin(\omega_1 t) + \alpha_2 \sin(\omega_2 t). \tag{4.8}$$

A sample of this time series is depicted in Figure 4.3 with 512 observations, $\omega_1 = \pi/16$, $\omega_2 = \pi/32$, $\alpha_1 = 2$, and $\alpha_2 = 1$. The periodogram of this series is displayed in Figure 4.4. Observe that the two frequencies are clearly detected by the peaks in the estimated spectral density displayed in Figure 4.4.

Consider now the harmonic time series of the previous example, but with added white noise, as described by equation (4.9). A sample of 512 observations from this series is exhibited in Figure 4.5.

$$y_t = \alpha_1 \sin(\omega_1 t) + \alpha_2 \sin(\omega_2 t) + \varepsilon_t, \tag{4.9}$$

The periodogram of this time series is plotted in Figure 4.6. Notice that it is very similar to the periodogram shown in Figure 4.4. Thus, it seems that the periodogram is not greatly affected by the presence of noise in the harmonic series.

The third example illustrates the estimation of the spectral density in the case of an AR(1) time series with parameter $\phi = 0.5$. Figure 4.7 shows a

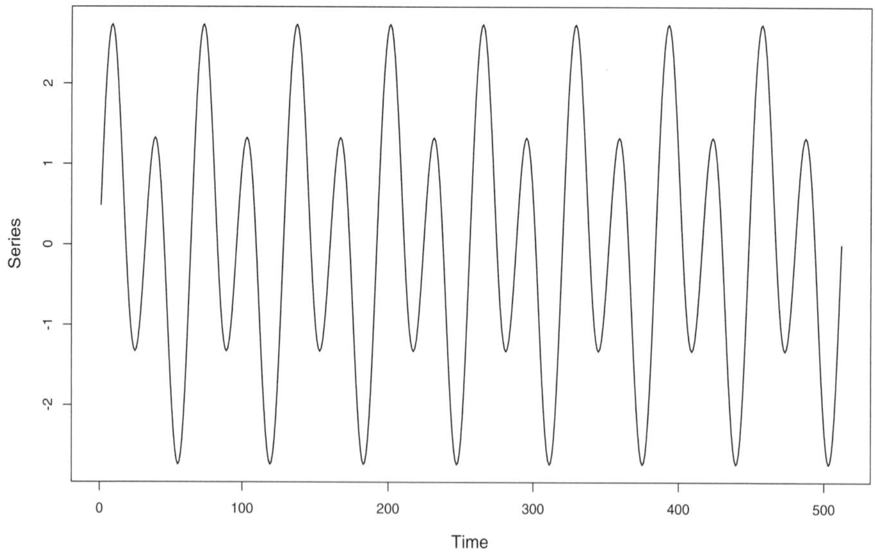

Figure 4.3 *Simulated harmonic time series model (4.8) with 512 observations with $\omega_1 = \pi/16$ and $\omega_2 = \pi/32$.*

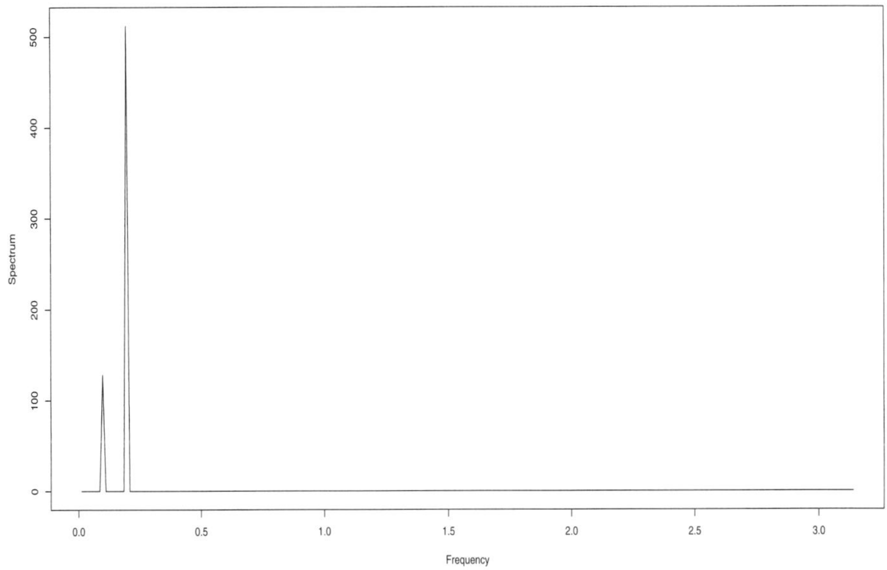

Figure 4.4 *Periodogram of simulated* (4.8) *with 512 observations with* $\omega_1 = \pi/16$ *and* $\omega_2 = \pi/32$.

sample of this process, with length $n = 512$. The periodogram is plotted in Figure 4.8 along with the theoretical spectral density.

As shown in Figure 4.8, the periodogram is usually ragged and it is sometime difficult to pinpoint which are the key features of the spectrum. One way to smooth the periodogram is applying a kernel function to the raw periodogram as described in Section 4.5. For instance, Figure 4.9 show the periodogram of the AR(1), along with the theoretical and the smoothed periodogram. Observe that the smoother version of the periodogram is very close to its theoretical counterpart.

The following example discusses the estimation of the spectral density in the context of a long-memory process. It considers an ARFIMA$(0, d, 0)$ time series with long-memory parameter $d = 0.4$. Figure 4.10 shows one realization of this process, with $n = 512$ observations. The periodogram of this series is displayed in Figure 4.11 along with its theoretical counterpart. Observe that both lines indicate the presence of a peak near the origin, which is expected in the case of a strongly dependent process.

Finally, Figure 4.12 shows the heating degree days data introduced in Chapter 1 along with its periodogram. Note that this time series exhibits a annual seasonal component due to the different winter and summer heating requirements in Europe. Consequently, in the periodogram we can observe a seasonal frequency of $\omega = 2\pi/12 = \pi/6$.

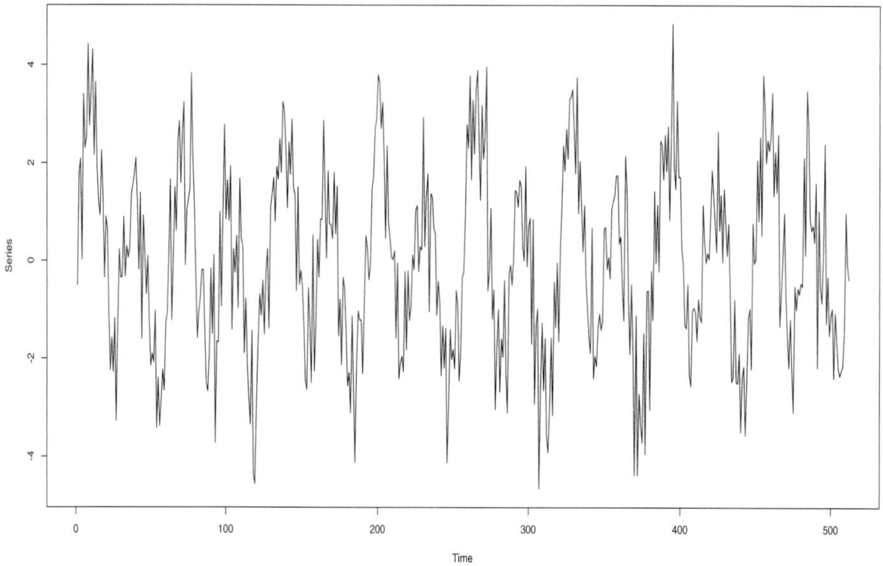

Figure 4.5 *Simulated harmonic time series model* (4.9) *with 512 observations with* $\omega_1 = \pi/16$ *and* $\omega_2 = \pi/32$.

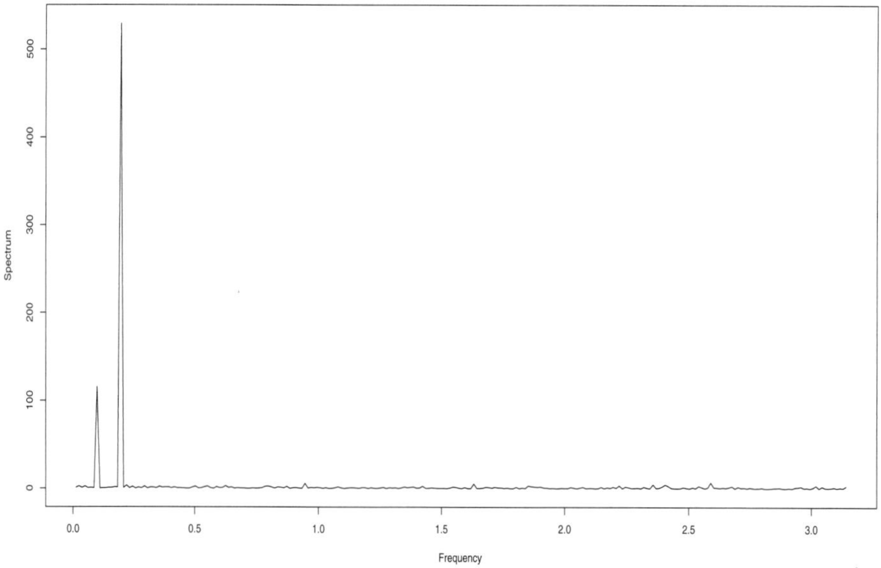

Figure 4.6 *Periodogram of simulated* (4.9).

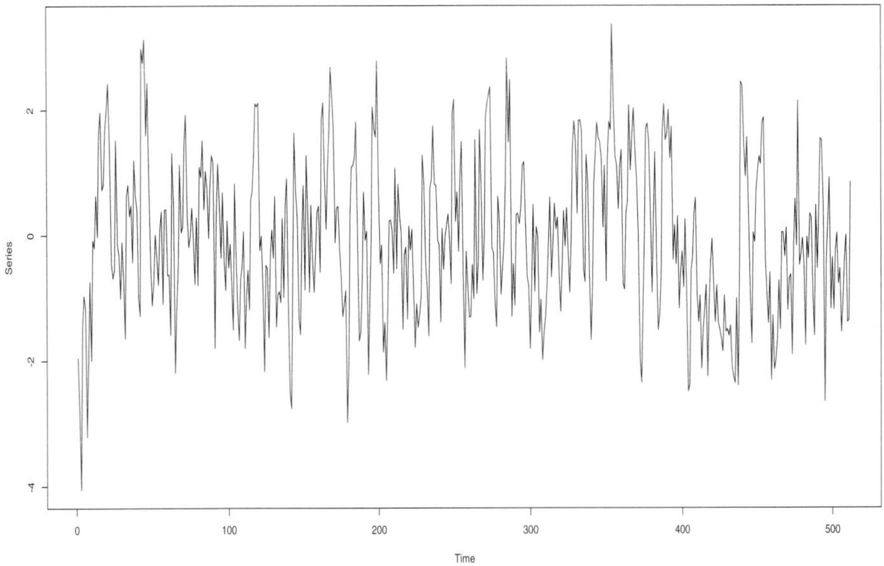

Figure 4.7 *Simulated AR(1) time series with 512 observations with* $\phi = 0.5$.

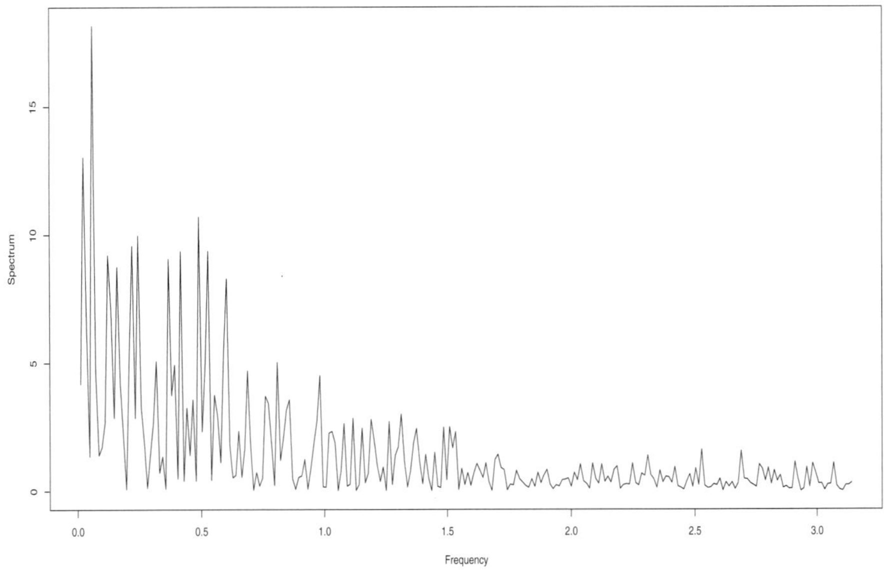

Figure 4.8 *Periodogram of simulated AR(1) time series with 512 observations with* $\phi = 0.5$.

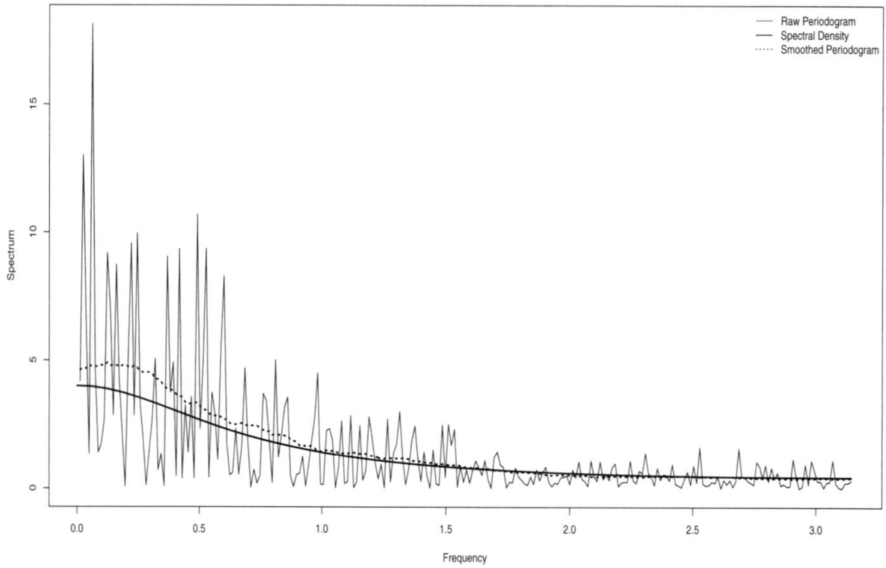

Figure 4.9 *Raw periodogram of simulated $AR(1)$ time series with 512 observations with $\phi = 0.5$, along with its smoothed version (broken line) and the theoretical spectral density (heavy line).*

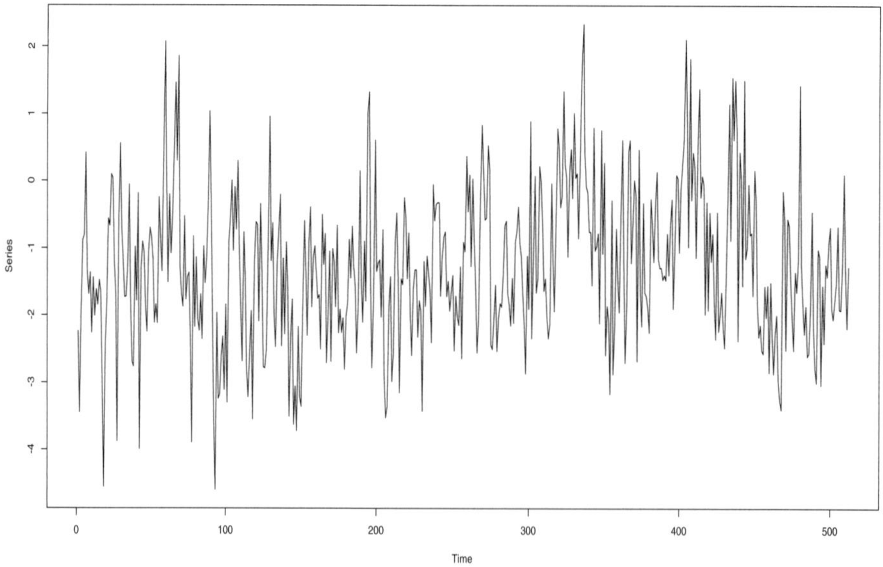

Figure 4.10 *Simulated $ARFIMA(0, d, 0)$ time series with 512 observations with $d = 0.4$.*

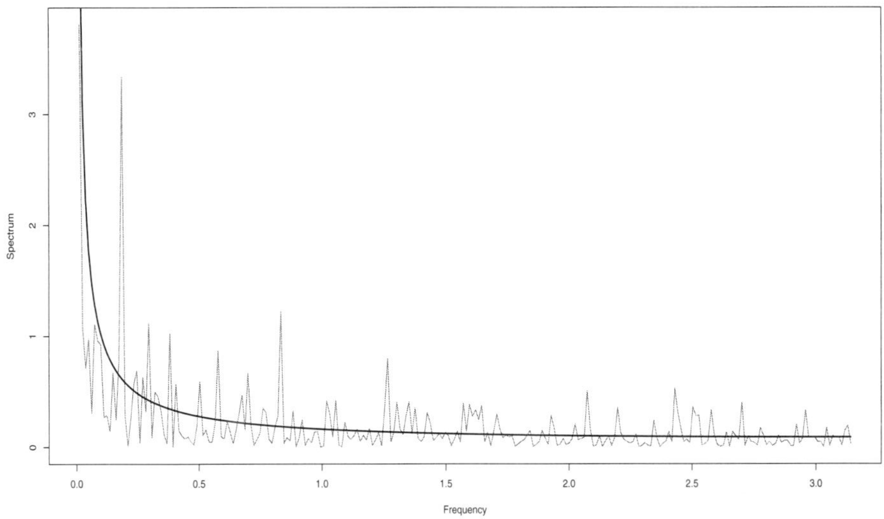

Figure 4.11 *Periodogram of simulated ARFIMA$(0, d, 0)$ time series with 512 observations with $d = 0.4$.*

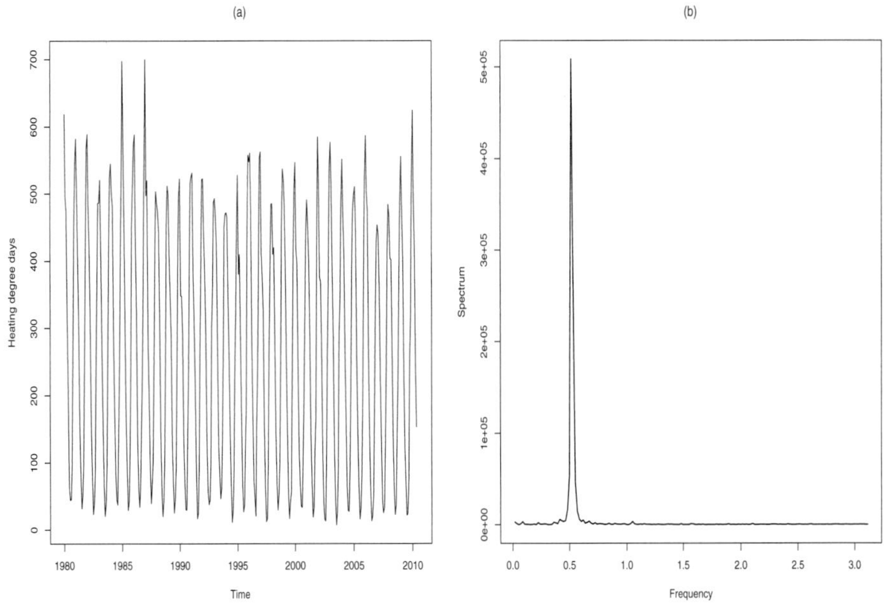

Figure 4.12 *Spectral analysis of the Heating Degree Day data. (a) Time Series. (b) Periodogram.*

4.7 WAVELETS

The spectral analysis described in the previous sections correspond to decompose the time series in terms of a sequence of harmonic functions. Nevertheless, there are other decompositions that may be very useful for analyzing a time series. One important example of these decomposition are the so-called wavelets which we briefly describe next.

A *wavelet* is a real-valued integrable function $\psi(t)$ satisfying

$$\int \psi(t)\, dt = 0. \tag{4.10}$$

A wavelet has n vanishing moments if

$$\int t^p \psi(t)\, dt = 0,$$

for $p = 0, 1, \ldots, n-1$.

Consider the following family of *dilations* and *translations* of the wavelet function ψ defined by

$$\psi_{jk}(t) = 2^{-j/2}\psi(2^{-j}t - k),$$

for $j, k \in \mathbb{Z}$. In this context, the terms j and 2^j are usually called the *octave* and the *scale*, respectively. It can be shown that (see Problem 4.12)

$$\int \psi_{jk}^2(t)\, dt = \int \psi^2(t)\, dt.$$

The *discrete wavelet transform* (DWT) of a process $\{y(t)\}$ is then defined by

$$d_{jk} = \int y(t)\psi_{jk}(t)\, dt,$$

for $j, k \in \mathbb{Z}$.

Provided that the family $\{\psi_{jk}(t)\}$ forms an orthogonal basis, that is,

$$\int \psi_{ij}(t)\psi_{k\ell}(t)\, dt = 0,$$

for all i, j, k, ℓ, excepting $i = j = k = \ell$, we obtain the following representation of the process $\{y(t)\}$:

$$y(t) = \sum_{j=-\infty}^{\infty} \sum_{k=-\infty}^{\infty} d_{jk}\psi_{jk}(t).$$

■ **EXAMPLE 4.6**

The *Haar wavelet system*

$$\psi(t) = \begin{cases} 1 & \text{if } t \in [0, \frac{1}{2}) \\ -1 & \text{if } t \in [\frac{1}{2}, 1) \\ 0 & \text{otherwise,} \end{cases}$$

is a simple example of a function satisfying (4.10). Observe that

$$\int t^p \psi(t)\, dt = \frac{1}{p+1} \left\{ \frac{1}{2^p} - 1 \right\}.$$

Therefore, this wavelet has a vanishing moment only for $p = 0$.

■ **EXAMPLE 4.7**

The so-called *Daubechies wavelets* are a family of wavelets that extends the previous example achieving a greater number of vanishing moments. This family forms an orthogonal wavelet basis and it is built in terms of the *multiresolution analysis*. Only a hint of this procedure is given here and further references are provided in the bibliographic section.

Starting from a *scaling function* ϕ that satisfies

$$\phi\left(\tfrac{t}{2}\right) = \sqrt{2} \sum_j u_j \phi(t - j),$$

we obtain the *mother wavelet* ψ by defining

$$\psi(t) = \sqrt{2} \sum_j v_j \phi(2t - j).$$

Thus, the Haar wavelet described in the previous example is obtained by setting $\phi(t) = \mathbf{1}_{[0,1]}(t)$, $u_0 = u_1 = v_0 = -v_1 = 1/\sqrt{2}$ and $u_j = v_j = 0$ for $j \neq 0, 1$, where $\mathbf{1}_A$ denotes the indicator function of the set A, that is,

$$\mathbf{1}_A(t) = \begin{cases} 1 & \text{if} \quad t \in A, \\ 0 & \text{if} \quad t \in A^c. \end{cases}$$

■ **EXAMPLE 4.8**

In order to illustrate the application of the wavelet analysis to real-life data, consider the heating degree day data introduced in Chapter 1. The R package wavelets allows for the calculation of the discrete wavelet transform for univariate and multivariate time series. The lower panel of Figure 4.13 shows the heating degree data while the upper panel exhibits the coefficients of the wavelet transform up to 3 levels. Observe the cyclical regularity of the components.

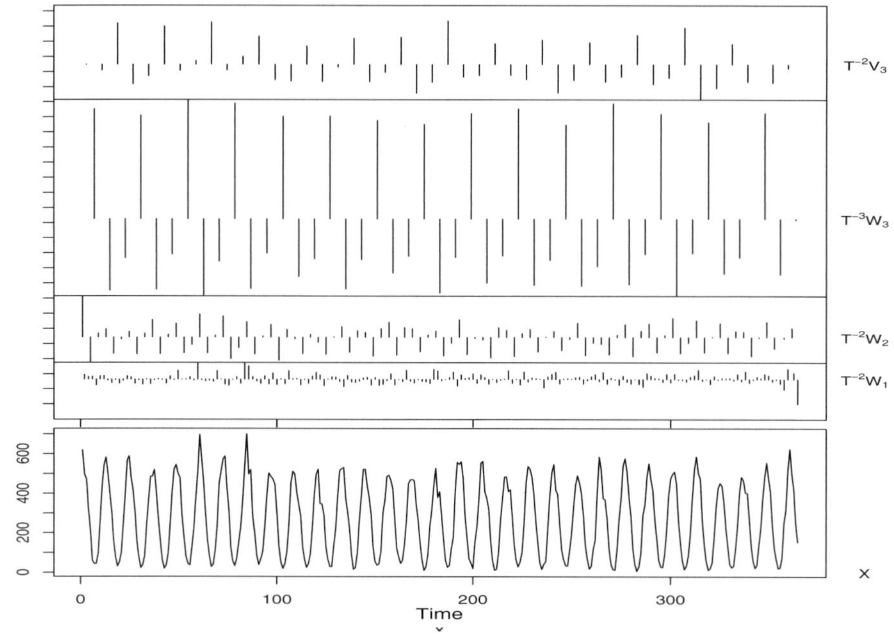

Figure 4.13 *Discrete wavelet transform for the HDD data.*

4.8 SPECTRAL REPRESENTATION

A linear stationary process y_t can be also written in terms of the so-called spectral representation or Cramer representation,

$$y_t = \int_{-\pi}^{\pi} A(\lambda)\, e^{i\lambda t}\, dB(\lambda), \tag{4.11}$$

where $A(\lambda)$ is a transfer function and $B(\lambda)$ is a orthogonal increments process on $[-\pi, \pi]$ such that

$$\text{Cov}[B(\lambda), B(\omega)] = \frac{\sigma^2}{2\pi}\delta(\lambda - \omega)d\lambda d\omega.$$

According to these equations, we have that

$$\begin{aligned}
\text{Cov}(y_t, y_s) &= \text{Cov}\left[\int_{-\pi}^{\pi} A(\lambda)\, e^{i\lambda t}\, dB(\lambda), \int_{-\pi}^{\pi} A(\omega)\, e^{-i\omega s}\, dB(\omega)\right] \\
&= \int_{-\pi}^{\pi}\int_{-\pi}^{\pi} A(\lambda)\, \overline{A(\omega)}\, e^{i\lambda t - i\omega s}\, \text{Cov}[dB(\lambda), dB(\omega)]
\end{aligned}$$

Thus,

$$\begin{aligned}\mathrm{Cov}(y_t, y_s) &= \frac{\sigma^2}{2\pi}\int_{-\pi}^{\pi}\int_{-\pi}^{\pi} A(\lambda)\,\overline{A(\omega)}\,e^{i\lambda t - i\omega s}\,\delta(\lambda-\omega)d\lambda d\omega \\ &= \frac{\sigma^2}{2\pi}\int_{-\pi}^{\pi} |A(\lambda)|^2\,e^{i\lambda(t-s)}\,d\lambda\end{aligned}$$

Therefore, we can write the spectral density of the process as

$$f(\lambda) = \frac{\sigma^2\,|A(\lambda)|^2}{2\pi}.$$

EXAMPLE 4.9

Define the sequence of random variables

$$\varepsilon_t = \int_{-\pi}^{\pi} e^{i\lambda t}\,dB(\lambda).$$

Thus, we have that

$$\begin{aligned}\mathrm{Cov}(\varepsilon_t, \varepsilon_s) &= \mathrm{Cov}\left[\int_{-\pi}^{\pi} e^{i\lambda t}\,dB(\lambda), \int_{-\pi}^{\pi} e^{-i\omega s}\,dB(\omega)\right] \\ &= \int_{-\pi}^{\pi}\int_{-\pi}^{\pi} e^{i\lambda t - i\omega s}\,\mathrm{Cov}[dB(\lambda), dB(\omega)] \\ &= \frac{\sigma^2}{2\pi}\int_{-\pi}^{\pi} e^{i\lambda(t-s)}\,d\lambda = \sigma^2\delta(t-s).\end{aligned}$$

Consequently, ε_t is a white noise sequence. Therefore, we can write the MA(1) model

$$y_t = \varepsilon_t - \theta\,\varepsilon_{t-1}.$$

Replacing ε_t by its definition we get

$$\begin{aligned}y_t &= \int_{-\pi}^{\pi} e^{i\lambda t}\,dB(\lambda) - \theta\int_{-\pi}^{\pi} e^{i\lambda(t-1)}\,dB(\lambda) \\ &= \int_{-\pi}^{\pi} e^{i\lambda t} - \theta\,e^{i\lambda(t-1)}\,dB(\lambda) \\ &= \int_{-\pi}^{\pi} [1-\theta\,e^{-i\lambda}]e^{i\lambda t}\,dB(\lambda).\end{aligned}$$

Thus, by defining the transfer function

$$A(\lambda) = 1 - \theta\,e^{-i\lambda}$$

we obtain a spectral representation of the process y_t.

4.9 TIME-VARYING SPECTRUM

As discussed in the context of locally stationary processes, the spectral density can be defined in terms of both frequency and time. In the case of non stationary processes, in some cases it is possible to extend the definition of the spectral density to the so-called time-varying spectral density. To illustrate this extension, consider the a class of LSARMA(1,1) processes defined by

$$y_t - \phi\left(\frac{t}{T}\right) y_{t-1} = \varepsilon_t + \theta\left(\frac{t}{T}\right) \varepsilon_{t-1},$$

for $t = 1, \ldots, T$ where ε_t is a white noise sequence with zero-mean and variance σ^2. Rescaling the time to the unit interval we can write with $t = [u\,T]$

$$y_t - \phi(u) y_{t-1} = \varepsilon_t + \theta(u)\, \varepsilon_{t-1}.$$

The limiting time-varying spectral density of this non stationary process is

$$\begin{aligned} f(\lambda, u) &= \frac{\sigma^2}{2\pi} \left| \frac{1 + \theta(u)\, e^{i\,\lambda}}{1 - \phi(u)\, e^{i\,\lambda}} \right|^2 \\ &= \frac{\sigma^2}{2\pi} \frac{1 + \theta(u)^2 + 2\,\theta(u)\,\cos\lambda}{1 + \phi(u)^2 - 2\,\phi(u)\,\cos\lambda}. \end{aligned}$$

■ **EXAMPLE 4.10**

As a first illustration of the analysis of time-varying spectrum, consider the LS-MA(1) model with a first-order moving-average parameter evolving according to the linear equation,

$$\theta(u) = 0.1 + 0.7\,u.$$

The spectral density of this process is displayed in Figure 4.14. On the other hand, a simulated time series from this model is exhibited in Figure 4.15.

Observe that the variance of the series seems to increase. This is expected from the fact that the variance of this time series evolves in an increasing manner

$$\begin{aligned} \operatorname{Var} y_t &= \sigma^2[1 + \theta(u)^2] \\ &= \sigma^2(1.01 + 0.14\,u + 0.49\,u^2). \end{aligned}$$

The time-varying periodogram of these data is displayed in Figure 4.16. Notice that this periodogram looks similar to its theoretical counterpart shown in Figure 4.14.

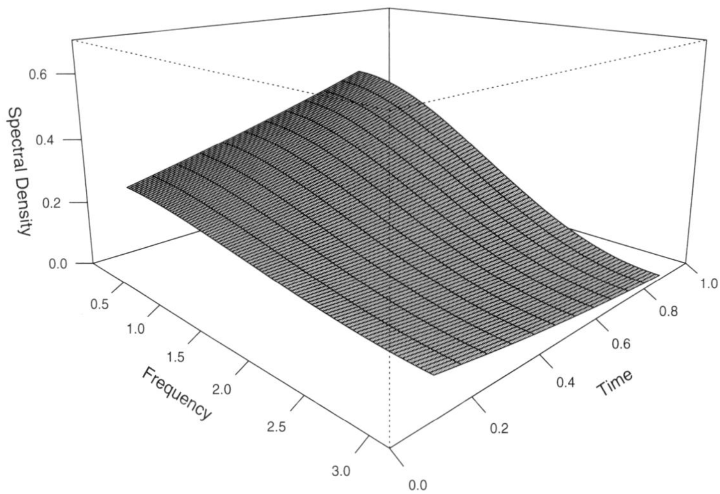

Figure 4.14 *Time-varying spectral density of a LS-MA(1) model with* $\theta(u) = 0.1 + 0.7\,u$.

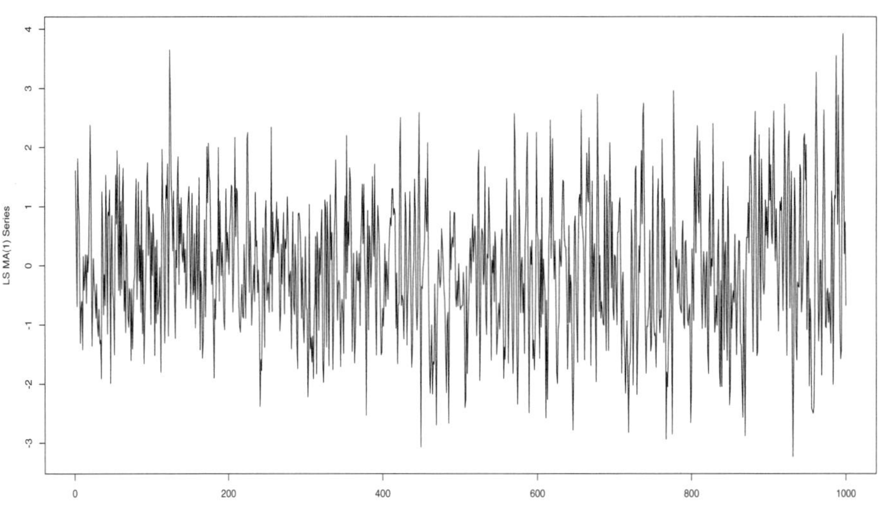

Figure 4.15 *Simulated LS-MA(1) model with* $\theta(u) = 0.1 + 0.7\,u$.

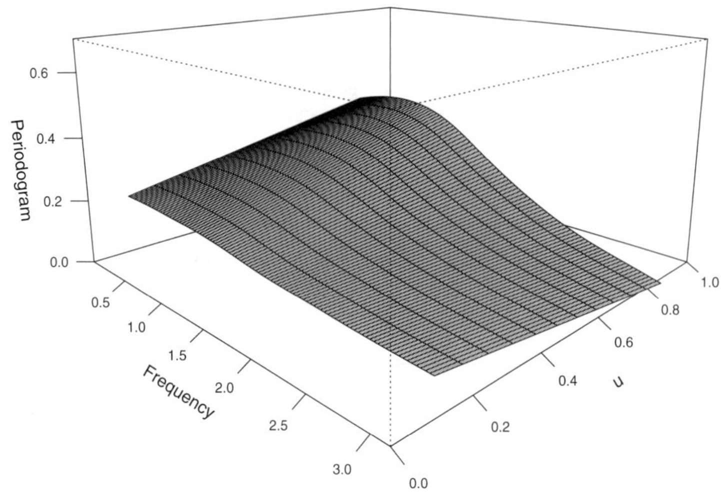

Figure 4.16 *Time-varying periodogram of a LS-MA(1) model with* $\theta(u) = 0.1 + 0.7\,u$.

■ EXAMPLE 4.11

A second illustration of the spectral analysis of non stationary time series is provided by the LS-AR(1) process with a first-order autoregressive parameter defined by $\phi(u) = 0.9 - 0.8\,u^2$. Figure 4.17 show the evolution of this parameter for scaled time values of $u \in [0, 1]$. Observe that in this case, the parameter $\phi(u)$ decays at a quadratic rate from $\phi(0) = 0.9$ to $\phi(1) = 0.1$.

A simulated trajectory of this model with 1000 observations is displayed in Figure 4.18.

Note that the variance of this series seems to decrease, as expected from the asymptotic formula for the variance of the process given by

$$\begin{aligned} \operatorname{Var} y_t &= \frac{\sigma^2}{1 - \phi(u)^2} \\ &= \frac{\sigma^2}{0.19 + 1.44\,u^2 - 0.62\,u^4}. \end{aligned}$$

The spectral density of this process is plotted in Figure 4.19 and its periodogram is exhibited in Figure 4.20. Notice that both graphs have similar shape in terms of both the frequency axis and the scaled time axis.

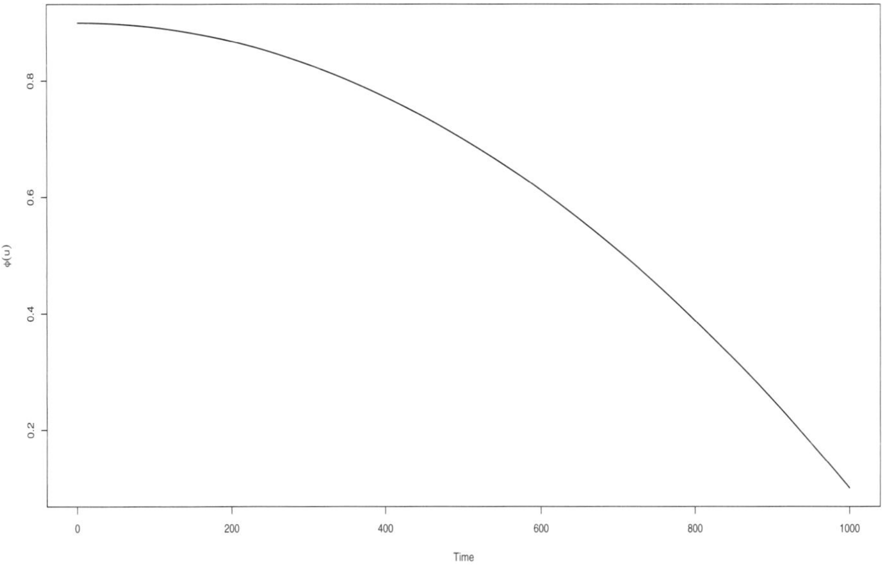

Figure 4.17 *Evolution of the time-varying first-order autoregressive parameter of a LS-AR(1) model with* $\phi(u) = 0.9 - 0.8\,u^2$.

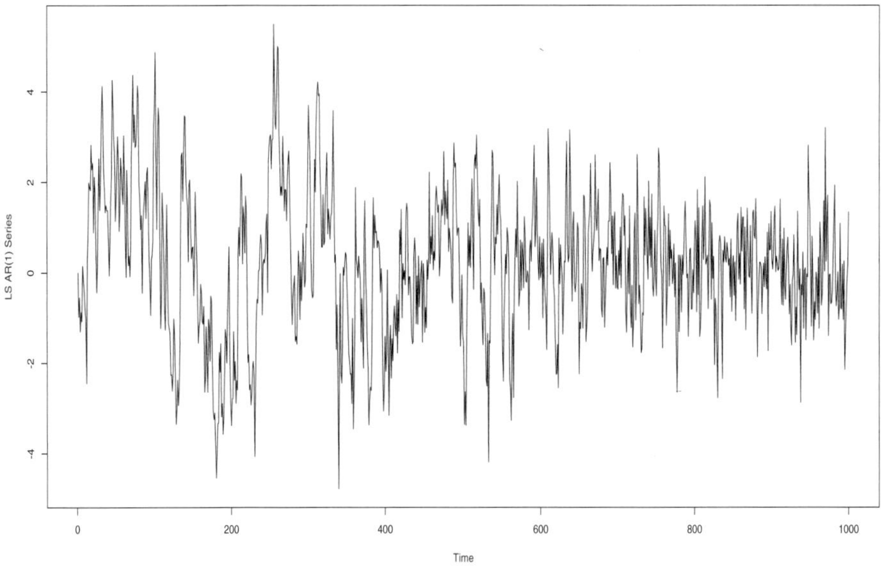

Figure 4.18 *Simulated LS-AR(1) model with* $\phi(u) = 0.9 - 0.8\,u^2$.

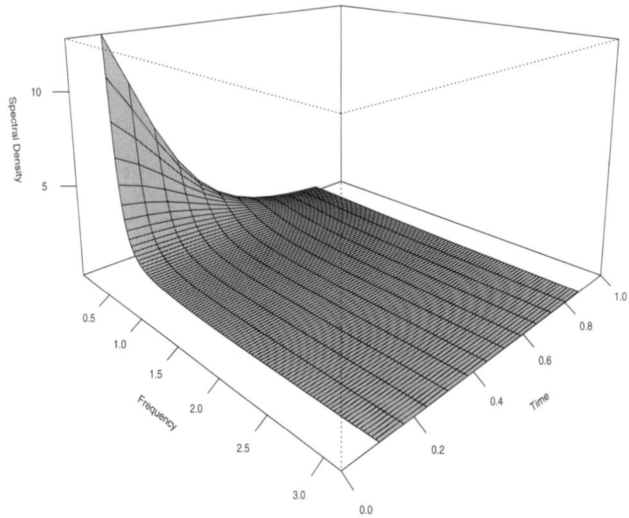

Figure 4.19 *Time-varying spectral density of a LS-AR(1) model with* $\phi(u) = 0.9 - 0.8\,u^2$.

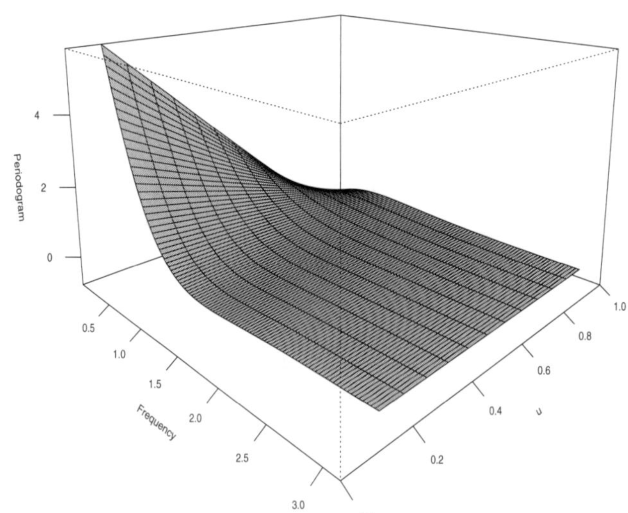

Figure 4.20 *Time-varying periodogram of a LS-AR(1) model with* $\phi(u) = 0.9 - 0.8\,u^2$.

4.10 BIBLIOGRAPHIC NOTES

Frequency-domain time series analysis has a vast literature in statistics, signal processing, engineering and many other fields. Three classical book on this subject are Priestley (1981a,b) and Brillinger and Krishnaiah (1983). Extensions of spectral techniques to nonlinear time series can be found in the monograph Subba Rao and Gabr (1984). Chapter 11 of Press, Teukolsky, Vetterling, and Flannery (2007) provides an excellent account of computational aspects of the periodogram calculation by means of the fast Fourier transform. Fourier analysis is reviewed by Körner (1989, 1993) while some wavelets fundamental concepts are discussed by Flandrin (1999). Other useful references in spectral analysis are Carmona, Hwang, and Torresani (1998), Koopmans (1995) and Zurbenko (1986). Several parametric and nonparametric methods in the frequency domain are addressed by Castanié (2006, 2011) while hypothesis testing in this context is reviewed by Dzhaparidze (1986) Haykin (1979) Jenkins and Watts (1968). Applications of the spectral analysis techniques to economics are discussed in Granger (1964). Wavelets have been vastly discussed in the literature. A nice recent revision, especially related to long-range dependence, is the article by Abry, Flandrin, Taqqu, and Veitch (2003) while Chapter 9 of Percival and Walden (2006) offers a comprehensive revision of wavelet methods for stationary processes. Additionally, Chapter 2 of Flandrin (1999) and Chapter 13 of Press, Teukolsky, Vetterling, and Flannery (1992) provide overviews about this topic, including several details about the Daubechies wavelets.

Problems

4.1 Let $\{x_t\}$ and $\{y_t\}$ be two stationary satisfying

$$\begin{aligned} x_t - \alpha x_{t-1} &= w_t, \\ y_t - \alpha y_{t-1} &= x_t + z_t, \end{aligned}$$

where $\{w_t\}$ and $\{z_t\}$ are two uncorrelated white noise sequences $(0, \sigma^2)$. Find the spectral density of $\{y_t\}$.

4.2 If $\{x_t\}$ and $\{y_t\}$ are two uncorrelated stationary processes with autocovariance functions $\gamma_X(\cdot)$ and $\gamma_Y(\cdot)$ and spectral densities $f_X(\cdot)$ and $f_Y(\cdot)$, respectively. Show that the process $\{z_t\} = \{x_t + y_t\}$ is stationary with autocovariance function $\gamma_Z = \gamma_X(\cdot) + \gamma_Y(\cdot)$ and spectral density $f_Z = f_X + f_Y$.

4.3 Consider the periodogram defined as

$$I(\lambda) = \frac{1}{2\pi n} \left| \sum_{j=1}^{n} y_j e^{i\lambda j} \right|^2 .$$

(a) Show that the periodogram satisfies

$$\int_{-\pi}^{\pi} e^{ik\lambda} I(\lambda)\, d\lambda = \begin{cases} w(k,n), & |k| < n, \\ 0, & |k| > n, \end{cases}$$

where

$$w(k,n) = \frac{1}{n} \sum_{t=1}^{n-k} (y_t - \bar{y}_n)(y_{t+k} - \bar{y}_n).$$

(b) Prove that if the process y_t is stationary with mean μ we have that

$$\lim_{n\to\infty} E\, w(k,n) = \gamma(k),$$

where $\gamma(k)$ is the autocovariance at lag k.

4.4 Consider the following inner product of two functions f and g given by

$$\langle f, g \rangle = \frac{1}{2\pi} \int_{-\pi}^{\pi} f(\lambda)\overline{g}(\lambda)\, d\lambda.$$

(a) Let $e_j(\lambda) = e^{i\lambda j}$ for $j \in \mathbb{Z}$. Show that the functions $\{e_j(\lambda)\}$ are orthonormal, that is,

$$\langle e_t, e_s \rangle = 0,$$

for all $s, t \in \mathbb{Z}$, $t \neq s$, and

$$\langle e_t, e_t \rangle = 1,$$

for all $t \in \mathbb{Z}$.

4.5 Suppose that a function f is defined by

$$f(\lambda) = \sum_{j=-m}^{m} \alpha_j e_j(\lambda),$$

where e_j are the functions defined in Problem 4.4.

(a) Verify that the coefficients α_j are given by

$$\alpha_j = \frac{1}{2\pi} \int_{-\pi}^{\pi} f(\lambda) e^{-i\lambda j}\, d\lambda.$$

The coefficients α_j are called the *Fourier coefficients* of the function $f(\cdot)$.

(b) Prove that

$$\|f\| = \sqrt{\sum_{j=-m}^{m} \alpha_j^2}.$$

4.6 Show that if the spectral density of a stationary process satisfies $f_\theta(\lambda) > c$ for all $\lambda \in (-\pi, \pi]$ where c is a positive constant, then $x'\Gamma_\theta x > 0$ for all $x \in \mathbb{R}^n$, $x \neq 0$.

4.7 Suppose that $\{y_1, \ldots, y_n\}$ follows an ARFIMA model and $I_n(\lambda)$ is its periodogram. Prove that if $f(\lambda)$ is the spectral density of the ARFIMA process and $g(\lambda)$ is a continuous function on $[-\pi, \pi]$ then

$$\lim_{n\to\infty} \int_{-\pi}^{\pi} g(\lambda) I_n(\lambda)\, d\lambda = \int_{-\pi}^{\pi} g(\lambda) f(\lambda)\, d\lambda.$$

4.8 Show that

$$\int_{-\pi}^{\pi} e^{i(k-h)\lambda} d\lambda = \begin{cases} 2\pi & \text{if } k = h, \\ 0 & \text{if } k \neq h \end{cases}$$

4.9 Assume that $\{x_t\}$ is a non-causal and non-invertible ARMA$(1, 1)$ process that satisfies

$$x_t - \phi x_{t-1} = z_t + \theta z_{t-1}, \quad z_t \sim RB(0, \sigma^2)$$

where $|\phi| > 1$, $|\theta| > 1$. Define $\tilde{\phi}(B) = 1 - \frac{1}{\phi}$ y $\tilde{\theta}(B) = 1 + \frac{1}{\theta}$, and let $\{W_t\}$ be a process given by

$$W_t = \tilde{\theta}^{-1}(B)\tilde{\phi}(B)x_t.$$

(a) Verify that the process $\{W_t\}$ possesses constant spectral density.

(b) Deduce that $\{W_t\} \sim RB(0, \sigma_w^2)$ and provide an explicit expression for σ_w^2 in terms of ϕ, θ and σ^2.

(c) Deduce that $\tilde{\phi}(B)x_t = \tilde{\theta}(B)W_t$, such that $\{x_t\}$ is a causal and invertible ARMA$(1, 1)$ process relative to the white noise sequence$\{W_t\}$.

4.10 Prove that if $\{y_t : t \in \mathbb{Z}\}$ is a stationary process such that $\sum_{k=0}^{\infty} |\gamma(k)| < \infty$, then its spectral density may be written as

$$f(\lambda) = \frac{1}{2\pi} \sum_{h=-\infty}^{\infty} \gamma(h) e^{-i\lambda h},$$

and this function is symmetric and positive.

4.11 Let $\Gamma_n = [\gamma(i - j)]_{i,j=1,\ldots,n}$ be the variance-covariance matrix of a linear process with

$$\gamma(h) = \int_{-\pi}^{\pi} f(\lambda) e^{-i\lambda h} d\lambda.$$

Show that the variance of the sample mean of $y_1, \dots, y_n$ is given by

$$\mathrm{Var}(\bar{y}_n) = \frac{1}{n^2} \int_{-\pi}^{\pi} f(\lambda) \left| \sum_{h=1}^{n} e^{-i\lambda h} \right|^2 d\lambda.$$

4.12 Verify that the $\mathcal{L}_2$ norm of ψ and ψ_{ik} are identical for all $j, k \in \mathbb{Z}$, that is,

$$\int \psi_{jk}^2(t)\, dt = \int \psi^2(t)\, dt.$$

4.13 Show that the Haar system generates an orthonormal basis, that is,

$$\int \psi_{ij}(t)\psi_{k\ell}(t)\, dt = \delta_{ij}\delta_{k\ell}.$$

4.14 Consider the so-called Littlewood-Paley decomposition

$$\psi(t) = \begin{cases} 1 & \text{if } |t| \in [1/2, 1), \\ 0 & \text{otherwise.} \end{cases}$$

Prove that the Littlewood-Paley decomposition generates an orthonormal basis.

4.15 Consider the following processes:

$$\begin{aligned} x_t &= \phi\, x_{t-s} + z_t \\ y_t &= W_t + \theta\, W_{t-s} \end{aligned}$$

with $|\phi| < 1$, $|\theta| < 1$, $s \in \mathbb{N}$, $\{z_t\} \sim \mathrm{WN}(0, \sigma^2)$, $\{W_t\} \sim \mathrm{WN}(0, \sigma^2)$ and $\mathrm{Cov}(z_k, W_h) = 0$ for all k and h. Calculate the spectral density of the process $v_t = x_t + y_t$.

4.16 Consider the periodogram defined in (4.6).

(a) Show that the periodogram satisfies

$$\int_{-\pi}^{\pi} e^{ik\lambda} I(\lambda)\, d\lambda = \begin{cases} w(k, n), & |k| < n, \\ 0, & |k| > n, \end{cases}$$

where

$$w(k, n) = \frac{1}{n} \sum_{t=1}^{n-k} (y_t - \bar{y}_n)(y_{t+k} - \bar{y}_n).$$

(b) Prove that if the process y_t is stationary and ergodic with mean μ we have that

$$\lim_{n\to\infty} w(k, n) = \lim_{n\to\infty} \frac{1}{n} \sum_{t=1}^{n-k} (y_t - \mu)(y_{t+k} - \mu) = \gamma(k),$$

where $\gamma(k)$ is the autocovariance at lag k.

4.17 Suppose that $\{y_1, \ldots, y_n\}$ follows an ARFIMA model and $I_n(\lambda)$ is its periodogram. Based on the previous problem, prove that if $f(\lambda)$ is the spectral density of the ARFIMA process and $g(\lambda)$ is a continuous function on $[-\pi, \pi]$ then

$$\lim_{n\to\infty} \int_{-\pi}^{\pi} g(\lambda) I_n(\lambda)\, d\lambda = \int_{-\pi}^{\pi} g(\lambda) f(\lambda)\, d\lambda.$$

CHAPTER 5

ESTIMATION METHODS

This chapter reviews several methodologies for estimating time series models. There are a number of well-known techniques such as the maximum likelihood estimation and its different computational formulations such as Cholesky decomposition or state space equations. On the other hand, there are approximate maximum likelihood methods including for example the Whittle approach, moving-average and autoregressive approximations. However, before reviewing these specific techniques this chapter begins with an overview of model building and specification. This is a necessary step in the analysis of time series before applying a specific estimation technique. In the model building and specification stage, we investigate the dependence structure of the data and decide which class of models may fit them more adequately. For example, decide whether an ARMA or an ARFIMA model would be more appropriate. In turn, in order to carry out this first stage of the time series analysis it is necessary to find estimates of the mean and the autocorrelation function of the series. Thus, after discussing general aspects of model building we focus our attention on estimating the mean and the ACF, which are essential tools for specifying the model.

Time Series Analysis. First Edition. Wilfredo Palma.

5.1 MODEL BUILDING

As discussed in previous chapters, real-life time series exhibit a number of distinctive features. Among these, it is relevant to decide whether the series will be treated as stationary or not. In the first case, we can proceed to the modeling stage by looking at the autocorrelation structure shown by the sample ACF and PACF. Based on these moment estimates, a stationary model such an ARMA or ARFIMA model can be proposed and fitted. On the other hand, if the series displays a number of nonstationary characteristics, we can apply transformations procedures to obtain a stationary series. Among these procedures we have previously revised the detrending of the data by means of regression techniques and the differentiation of the time series data. If a regression model is fitted, we can treat the residuals as the series to be analyzed. If differentiation is taken place, we can apply an ARIMA model. If the series displays seasonal behavior, then we can use harmonic regression or SARIMA models.

The specification of a model usually is concerned with selecting a class of processes such as ARIMA(p, d, q) or SARIMA$(p, d, q) \times (P, D, Q)_s$. The orders of these models can be selected from the sample ACF o PACF. It is common to consider a nested family of models and then estimate all the model in the class. For example, ARMA(p, q) with orders $p, q = 0, 1, 2, 3$. Since the models are nested, we can use an information criterion such as AIC or BIC to select appropriate values of p and q.

5.2 PARSIMONY

In theory, if a linear process has continuous spectrum, then we can always find values of p and q such that an ARMA(p, q) approximates it arbitrarily well. In other words, in practice we can always rely on this class of processes to model a linear time series. However, this general mathematical result does not guarantee that the values of the p and q are small. In fact, they eventually could be quite large.

Having an ARMA model with large autoregressive and moving-average orders could be cumbersome from both numerically and statistical perspectives. Let us say that $p = 45$ and $q = 37$. This model could fit well a data set but it requires the numerical calculation of 83 parameters and checking if the fitted model is stationary and invertible.

In this context, it is usually desirable that the fitted model be parsimonious, that is, the orders are relatively small. In this sense, there is a trade off between the approximation quality of the model, which usually requires larger values of p and q, and the simplicity of the model or parsimony, which strive for a small number of parameters.

This balance is commonly stricken by means of a criterion information that penalizes the model with the number of parameter estimated. These criteria are described next.

5.3 AKAIKE AND SCHWARTZ INFORMATION CRITERIA

The Akaike information criterion or AIC for short is defined as Akaike's information criterion (AIC)

$$\text{AIC} = -2\log \mathcal{L}(\widehat{\boldsymbol{\theta}}) + 2r,$$

where $\mathcal{L}(\boldsymbol{\theta})$ is the likelihood of the data, $\widehat{\boldsymbol{\theta}}$ is the maximum likelihood estimate, and r is the number of estimated parameters of the model. For example, for and ARMA(p,q) model $r = p + q + 1$ since we have to add the estimation of noise standard deviation parameter σ.

The Schwartz information criterion or Bayesian information criterion (BIC) is defined by

$$\text{BIC} = -2\log \mathcal{L}(\widehat{\boldsymbol{\theta}}) + r\log n,$$

Note that for a sample size $n > 8$, BIC penalizes more strongly the increment of the number of parameters in the model, as compared to the AIC.

5.4 ESTIMATION OF THE MEAN

Estimating the mean of a stationary process is a fundamental stage of the time series analysis. Even though there are several estimators of the mean, the sample mean and the BLUE are the most commonly considered.

Given the sample $Y_n = (y_1, y_2, \ldots, y_n)'$ from a stationary process with mean μ and variance Γ, the sample mean is defined as

$$\widehat{\mu} = \frac{1}{n}\sum_{t=1}^{n} y_t = \frac{1}{n}\mathbf{1}'Y_n,$$

where $\mathbf{1} = (1, 1, \ldots, 1)'$ and the BLUE is

$$\widetilde{\mu} = (\mathbf{1}'\Gamma^{-1}\mathbf{1})^{-1}\mathbf{1}'\Gamma^{-1}Y_n.$$

The large sample behavior of these two well-known estimators depend critically on the memory of the process. For a short memory process such as an ARMA model, the sample mean and the BLUE converge to the true mean at rate $\mathcal{O}(n^{-1})$. Furthermore, their asymptotic variance are similar. In this sense, the sample mean is an efficient estimator.

On the other hand, for a strongly dependent process with long memory parameter d, the convergence rate of both estimators is $\mathcal{O}(n^{2d-1})$. Note that since $d > 0$, this convergence rate is slower than for the short memory case. Besides, the asymptotic variances of the sample mean and the BLUE are different, implying that the sample mean is not an efficient estimator.

Specifically, the following expressions can be established

$$\sqrt{n}\,(\,\bar{y}_n - \mu\,) \to N(0, v),$$

where $v = 2\pi f(0) = \sum_{-\infty}^{\infty} \gamma(h)$. Analogously for the BLUE we have

$$\sqrt{n}\,(\,\tilde{\mu}_n - \mu\,) \to N(0, v).$$

Notice that for an ARMA(p, q) process we have that $v = \sigma^2 \frac{|\theta(1)|^2}{|\phi(1)|^2}$.

For long memory processes,

$$n^{1/2-d}\,(\,\bar{y}_n - \mu\,) \to N(0, w),$$

with $w = \sigma^2 \frac{|\theta(1)|^2}{|\phi(1)|^2} \frac{\Gamma(1-2d)}{d(1+2d)\Gamma(d)\Gamma(1-d)}$, and for the BLUE we have

$$n^{1/2-d}\,(\,\tilde{\mu}_n - \mu\,) \to N(0, w),$$

with $w = \sigma^2 \frac{|\theta(1)|^2}{|\phi(1)|^2} \frac{\Gamma(1-2d)\Gamma(2-2d)}{\Gamma(1-d)^2}$.

5.5 ESTIMATION OF AUTOCOVARIANCES

For a stationary process, autocovariances are commonly estimated by means of moment methods. That is, given the sample $\{y_1, y_2, \ldots, y_n\}$, the usual estimator of the autovariance at a lag $h \geq 0$, $\gamma(h)$ is given by

$$\widehat{\gamma}_n(h) = \frac{1}{n} \sum_{t=1}^{n-h} (y_t - \bar{y})(y_{t+h} - \bar{y}).$$

It can be shown that for a fixed h, this estimate of $\gamma(h)$ is asymptotically unbiased, i.e.,

$$\lim_{n \to \infty} \widehat{\gamma}_n(h) = \gamma(h).$$

Additionally, assuming that the input noise of the process satisfies $E\,\varepsilon_t^4 = \eta\sigma^2 < \infty$. if the autocovariances of the process are absolutely summable, then

$$\sqrt{n}\,[\,\widehat{\gamma}_n(h) - \gamma(h)\,] \to N(0, v),$$

where $v = (\eta - 3)\gamma(h)^2 + \sum_{j=-\infty}^{\infty} [\gamma(j)^2 + \gamma(j-h)\gamma(j+h)]$.

Similar expressions can be found for the asymptotic behavior of the sample ACF. In this case, we have

$$\sqrt{n}\,[\,\widehat{\rho}_n(h) - \rho(h)\,] \to N(0, w),$$

where w is giveb by the *Barlett formula*

$$w = \sum_{j=-\infty}^{\infty} \{[1 + 2\rho(h)^2]\rho(j)^2 + \rho(j)\rho(j+2h) - 4\rho(h)\rho(j)\rho(j+h)\}.$$

EXAMPLE 5.1

As an illustration, consider a white noise sequence with $\rho(0) = 1$ and $\rho(h) = 0$ for all $h \neq 0$. In this situation, the Barlett formula indicates that $w = 1$ and we obtain the usual Barlett confidence bands for the sample ACF, $(-\frac{2}{n}, \frac{2}{n})$.

5.6 MOMENT ESTIMATION

A simple estimation approach is based on comparing the sample ACF with their theoretical counterparts. Given the sample $\{y_1, y_2, \dots, y_n\}$ and the specification of a time series model with ACF $\gamma_\theta(h)$, we write the moment equations

$$\gamma_\theta(h) = \widehat{\gamma}(h),$$

for different values of h. The solution of this equation system, $\widehat{\theta}$ is the moment estimator of θ. Analogously, we can write these equations in terms of the sample autocorrelations

$$\rho_\theta(h) = \widehat{\rho}(h).$$

As an example, consider the AR(1) model $y_t = \phi\, y_{t-1} + \varepsilon_t$. The ACF is given by $\rho(h) = \phi^{|h|}$. Thus, we can find an estimate for ϕ based on the equation

$$\rho_\phi(1) = \widehat{\rho}(1),$$

obtaining $\widehat{\phi} = \widehat{\rho}(1)$. Furthermore, an estimate of the variance noise σ^2 can be obtained as follows. Given that

$$\gamma(0) = \frac{\sigma^2}{1 - \phi^2},$$

we can write

$$\sigma^2 = \gamma(0)\,(1 - \phi^2).$$

Therefore, a moment estimate of the noise variance is given by

$$\widehat{\sigma}^2 = \widehat{\gamma}(0)\,(1 - \widehat{\phi}^2).$$

On the other hand, for and AR(2) model, the Yule Walker estimate of the process

$$y_t = \phi_1 y_{t-1} + \phi_2 y_{t-2} + \varepsilon_t, \tag{5.1}$$

are given by

$$\begin{pmatrix} \widehat{\phi}_1 \\ \widehat{\phi}_2 \end{pmatrix} = \begin{bmatrix} \widehat{\gamma}(0) & \widehat{\gamma}(1) \\ \widehat{\gamma}(1) & \widehat{\gamma}(0) \end{bmatrix}^{-1} \begin{pmatrix} \widehat{\gamma}(1) \\ \widehat{\gamma}(2) \end{pmatrix}.$$

The idea of comparing theoretical and sample moments can be extended to ARMA or more complex models. For example, consider the MA(1) model

$$y_t = \varepsilon_t + \theta\varepsilon_{t-1}.$$

In this case the moment estimate $\widehat{\theta}$ is given by

$$\widehat{\theta} = \frac{1}{2\widehat{\rho}(1)} \pm \sqrt{\left(\frac{1}{2\widehat{\rho}(1)}\right)^2 - 1}$$

Note that are two possible solutions for $\widehat{\theta}$. However, the adequate value corresponds to the one that satisfies the invertibility condition $|\widehat{\theta}| < 1$.

5.7 MAXIMUM-LIKELIHOOD ESTIMATION

Assume that $\{y_t\}$ is a zero-mean stationary Gaussian process. Then, the log-likelihood function of this process is given by

$$\mathcal{L}(\theta) = -\tfrac{1}{2}\log\det\Gamma_\theta - \tfrac{1}{2}Y'\Gamma_\theta^{-1}Y, \tag{5.2}$$

where $Y = (y_1, \ldots, y_n)'$, $\Gamma_\theta = \mathrm{Var}(Y)$, and θ is the parameter vector. Consequently, the *maximum-likelihood* (ML) estimate $\widehat{\theta}$ is obtained by maximizing $\mathcal{L}(\theta)$. The log-likelihood function (5.2) requires the calculation of the determinant and the inverse of the variance-covariance matrix Γ_θ. However, these calculations can be conducted by means of the Cholesky decomposition method. In the following subsections, we review this and other procedures for computing the function (5.2) such as the Durbin-Levinson algorithm and state space techniques.

5.7.1 Cholesky Decomposition Method

Given that the matrix Γ_θ is symmetric positive definite, it can be written as

$$\Gamma_\theta = U'U,$$

where U is an upper triangular matrix. According to this Cholesky decomposition, the determinant of Γ_θ is given by $\det\Gamma_\theta = (\det U)^2 = \prod_{j=1}^n u_{jj}^2$, where u_{jj} denotes the jth diagonal element of the matrix U. Besides, the inverse of Γ_θ can be obtained as $\Gamma_\theta^{-1} = U^{-1}(U^{-1})'$, where the inverse of U can be computed by means of a very simple procedure.

5.7.2 Durbin-Levinson Algorithm

The Cholesky decomposition could be inefficient for long time series. Thus, faster methods for calculating the log-likelihood function (5.2) have been developed. One of these algorithms designed to exploit the Toeplitz structure of the variance-covariance matrix Γ_θ, is known as the Durbin-Levinson algorithm.

Suppose that $\widehat{y}_1 = 0$ and $\widehat{y}_{t+1} = \phi_{t1}y_t + \cdots + \phi_{tt}y_1$, $t = 1, \ldots, n-1$, are the one-step ahead forecasts of the process $\{y_t\}$ based on the finite past $\{y_1, \ldots, y_{t-1}\}$, where the regression coefficients ϕ_{tj} are given by the equations

$$\begin{aligned}
\phi_{tt} &= [\nu_{t-1}]^{-1}\left[\gamma(t) - \sum_{i=1}^{t-1}\phi_{t-1,i}\gamma(t-i)\right], \\
\phi_{tj} &= \phi_{t-1,j} - \phi_{tt}\phi_{t-1,t-j}, \qquad j = 1, \ldots, t-1, \\
\nu_0 &= \gamma(0), \\
\nu_t &= \nu_{t-1}[1 - \phi_{tt}^2], \qquad j = 1, \ldots, t-1.
\end{aligned}$$

Furthermore, if $e_t = y_t - \widehat{y}_t$ is the prediction error and $\boldsymbol{e} = (e_1, \ldots, e_n)'$, then $\boldsymbol{e} = LY$ where L is the lower triangular matrix:

$$L = \begin{pmatrix}
1 & & & & & \\
-\phi_{11} & 1 & & & & \\
-\phi_{22} & -\phi_{21} & 1 & & & \\
-\phi_{33} & -\phi_{32} & -\phi_{31} & & & \\
\vdots & \vdots & & & 1 & \\
-\phi_{n-1,n-1} & -\phi_{n-1,n-2} & -\phi_{n-1,n-3} & \cdots & -\phi_{n-1,1} & 1
\end{pmatrix}.$$

Hence, Γ_θ may be decomposed as $\Gamma_\theta = LDL'$, where $D = \mathrm{diag}(\nu_0, \ldots, \nu_{n-1})$. Therefore, $\det \Gamma_\theta = \prod_{j=1}^{n} \nu_{j-1}$ and $Y'\Gamma_\theta^{-1}Y = \boldsymbol{e}'D^{-1}\boldsymbol{e}$. As a result, the log-likelihood function (5.2) may be expressed as

$$\mathcal{L}(\theta) = -\frac{1}{2}\sum_{t=1}^{n}\log \nu_{t-1} - \frac{1}{2}\sum_{t=1}^{n}\frac{e_t^2}{\nu_{t-1}}.$$

The numerical complexity of this algorithm is $\mathcal{O}(n^2)$ for a linear stationary process. Nevertheless, for some Markovian processes such as the family of ARMA models, the Durbin-Levinson algorithm can be implemented in only $\mathcal{O}(n)$ operations. Unfortunately, this reduction in operations count does not apply to ARFIMA models since they are not Markovian.

5.8 WHITTLE ESTIMATION

A well-known methodology to obtain approximate maximum-likelihood estimates is based on the calculation of the periodogram—see equation (4.6)—by

Table 5.1 Maximum Likelihood Estimation of ARMA models

p	q	ϕ_1	ϕ_2	θ_1	θ_2	μ	AIC	t_{ϕ_1}	t_{ϕ_2}	t_{θ_1}	t_{θ_2}	LB
1	0	0.65				0.25	1051.45	15.75			1.36	
2	0	0.37	0.43			0.24	984.56	7.5	8.77		0.83	0.47
0	1			0.37		0.25	1150.14			10.3	2.47	
1	1	0.88		-0.4		0.24	1010.48	27.36		-7.59	0.85	
2	1	0.14	0.58	0.28		0.24	981.02	1.53	9.06	2.54	0.91	0.95
0	2			0.44	0.51	0.25	1040.55			8.43	2.03	
1	2	0.77		-0.33	0.26	0.24	992.9	12.6		-3.95	1.02	0.05
2	2	0.15	0.57	0.28	0.01	0.24	983	1.32	5.21	2.17	0.92	0.95

means of the *fast Fourier transform* (FFT) and the use of the so-called *Whittle approximation* of the Gaussian log-likelihood function. Since the calculation of the FFT has a numerical complexity of order $\mathcal{O}[n \log_2(n)]$, this approach produces very fast algorithms for computing parameter estimates.

Suppose that the sample vector $Y = (y_1, \ldots, y_n)'$ is normally distributed with zero-mean and variance Γ_θ. Then, the log-likelihood function divided by the sample size is given by

$$\mathcal{L}(\theta) = -\frac{1}{2n} \log \det \Gamma_\theta - \frac{1}{2n} Y' \Gamma_\theta^{-1} Y. \tag{5.3}$$

Notice that the variance-covariance matrix Γ_θ may be expressed in terms of the spectral density of the process $f_\theta(\cdot)$ as follows:

$$(\Gamma_\theta)_{ij} = \gamma_\theta(i - j),$$

where

$$\gamma_\theta(k) = \int_{-\pi}^{\pi} f_\theta(\lambda) \exp(i\lambda k)\, d\lambda.$$

In order to obtain the Whittle method, two approximations are made. Since

$$\frac{1}{n} \log \det \Gamma_\theta \to \frac{1}{2\pi} \int_{-\pi}^{\pi} \log[2\pi f_\theta(\lambda)]\, d\lambda,$$

as $n \to \infty$, the first term in (5.3) is approximated by

$$\frac{1}{2n} \log \det \Gamma_\theta \approx \frac{1}{4\pi} \int_{-\pi}^{\pi} \log[2\pi f_\theta(\lambda)]\, d\lambda.$$

On the other hand, the second term in (5.3) is approximated by

$$\begin{aligned}
\frac{1}{2n} Y' \Gamma_\theta^{-1} Y &\approx \sum_{\ell=1}^{n} \sum_{j=1}^{n} y_\ell \left\{ \frac{1}{8\pi^2 n} \int_{-\pi}^{\pi} f_\theta^{-1}(\lambda) \exp[i\lambda(\ell - j)]\, d\lambda \right\} y_j \\
&= \frac{1}{8\pi^2 n} \int_{-\pi}^{\pi} f_\theta^{-1}(\lambda) \sum_{\ell=1}^{n} \sum_{j=1}^{n} y_\ell y_j \exp[i\lambda(\ell - j)]\, d\lambda \\
&= \frac{1}{8\pi^2 n} \int_{-\pi}^{\pi} f_\theta^{-1}(\lambda) \left| \sum_{j=1}^{n} y_j \exp(i\lambda j) \right|^2 d\lambda \\
&= \frac{1}{4\pi} \int_{-\pi}^{\pi} \frac{I(\lambda)}{f_\theta(\lambda)}\, d\lambda,
\end{aligned}$$

where

$$I(\lambda) = \frac{1}{2\pi n} \left| \sum_{j=1}^{n} y_j e^{i\lambda j} \right|^2$$

is the *periodogram* of the series $\{y_t\}$ defined in equation (4.6).

Thus, the log-likelihood function is approximated, up to a constant, by

$$\mathcal{L}_3(\theta) = -\frac{1}{4\pi} \left[\int_{-\pi}^{\pi} \log f_\theta(\lambda)\, d\lambda + \int_{-\pi}^{\pi} \frac{I(\lambda)}{f_\theta(\lambda)}\, d\lambda \right]. \tag{5.4}$$

The evaluation of the log-likelihood function (5.4) requires the calculation of integrals. To simplify this computation, the integrals can be substituted by Riemann sums as follows:

$$\int_{-\pi}^{\pi} \log f_\theta(\lambda)\, d\lambda \approx \frac{2\pi}{n} \sum_{j=1}^{n} \log f_\theta(\lambda_j),$$

and

$$\int_{-\pi}^{\pi} \frac{I(\lambda)}{f_\theta(\lambda)} d\lambda \approx \frac{2\pi}{n} \sum_{j=1}^{n} \frac{I(\lambda_j)}{f_\theta(\lambda_j)},$$

where $\lambda_j = 2\pi j/n$ are the Fourier frequencies. Thus, a discrete version of the log-likelihood function (5.4) is

$$\mathcal{L}_4(\theta) = -\frac{1}{2n} \left[\sum_{j=1}^{n} \log f_\theta(\lambda_j) + \sum_{j=1}^{n} \frac{I(\lambda_j)}{f_\theta(\lambda_j)} \right].$$

Other versions of the Whittle likelihood function are obtained by making additional assumptions. For instance, if the spectral density is normalized as

$$\int_{-\pi}^{\pi} \log f_\theta(\lambda)\, d\lambda = 0, \tag{5.5}$$

then the Whittle log-likelihood function is reduced to

$$\mathcal{L}_5(\theta) = -\frac{1}{4\pi}\int_{-\pi}^{\pi}\frac{I(\lambda)}{f_\theta(\lambda)}\,d\lambda,$$

with the corresponding discrete version

$$\mathcal{L}_6(\theta) = -\frac{1}{2n}\sum_{j=1}^{n}\frac{I(\lambda_j)}{f_\theta(\lambda_j)}.$$

5.9 STATE SPACE ESTIMATION

The state space methodology may be also used for handling autoregressive approximations. For instance, starting from the AR(m) truncation (5.6) and dropping θ from the coefficients π_j and the tilde from ε_t, we have

$$y_t = \pi_1 y_{t-1} + \cdots + \pi_m y_{t-m} + \varepsilon_t.$$

Thus, we may write the following state space system:

$$\begin{aligned} x_{t+1} &= F x_t + H\varepsilon_{t+1}, \\ y_t &= G x_t, \end{aligned}$$

where the state is given by $x_t = [y_t \;\; y_{t-1} \;\; \cdots \;\; y_{t-m+2} \;\; y_{t-m+1}]'$, the state transition matrix is

$$F = \begin{pmatrix} \pi_1 & \pi_2 & \cdots & \pi_{m-1} & \pi_m \\ 1 & 0 & \cdots & 0 & 0 \\ 0 & 1 & \cdots & 0 & 0 \\ \vdots & & & & \vdots \\ 0 & 0 & \cdots & 0 & 0 \\ 0 & 0 & \cdots & 1 & 0 \end{pmatrix},$$

the observation matrix is

$$G = [1\ 0\ 0 \cdots\ 0],$$

and the state noise matrix is given by

$$H = [1\ 0\ 0 \cdots\ 0]'.$$

The variance of the observation noise is $R = 0$, the covariance between the state noise and the observation noise is $S = 0$ and the state noise variance-covariance matrix is given by

$$Q = \sigma^2 \begin{pmatrix} 1 & 0 & \cdots & 0 \\ 0 & 0 & \cdots & 0 \\ \vdots & & & \vdots \\ 0 & 0 & \cdots & 0 \\ 0 & 0 & \cdots & 0 \end{pmatrix}.$$

The Kalman recursions for this system are

$$\begin{aligned}
\Delta_t &= \omega_t(1,1), \\
\Theta_t &= \left[\sum_{i=1}^{m} \pi_i \omega_t(i,1) \;\; \omega_t(1,1) \;\; \cdots \;\; \omega_t(m-1,1)\right]', \\
\omega_{t+1}(1,1) &= \sigma^2 + \sum_{i,j=1}^{m} \pi_i \omega_t(i,j)\pi_j - \left[\sum_{i=1}^{m} \pi_i \omega_t(i,1)\right]^2 / \omega_t(1,1), \\
\omega_{t+1}(1,j) &= \sum_{i=1}^{m} \pi_i \omega_t(i,j) - \left[\sum_{i=1}^{m} \pi_i \omega_t(i,1)\right] \frac{\omega_t(1,j)}{\omega_t(1,1)} \;\text{ for } j \geq 2, \\
\omega_{t+1}(i,j) &= \omega_t(i-1,j-1) - \omega_t(i-1,1)\omega_t(j-1,1)/\omega_t(1,1) \;\text{ for } i,j \geq 2, \\
\widehat{x}_{t+1}(1) &= \sum_{i=1}^{m} \pi_i \widehat{x}_t(i) + \left[\sum_{i=1}^{m} \pi_i \omega_t(i,1)\right] \frac{y_t - \widehat{x}_t(1)}{\omega_t(1,1)}, \\
\widehat{x}_{t+1}(j) &= \widehat{x}_t(j-1) + \omega_t(j-1,1)\, \frac{y_t - \widehat{x}_t(1)}{\omega_t(1,1)} \;\text{ for } j \geq 2, \\
\widehat{y}_{t+1} &= \widehat{x}_{t+1}(1).
\end{aligned}$$

The initial conditions for these iteration may be

$$\widehat{x}_0 = 0,$$

and

$$\Omega_0 = \begin{pmatrix} \gamma(0) & \gamma(1) & \cdots & \gamma(m-1) \\ \gamma(1) & \gamma(0) & \cdots & \gamma(m-2) \\ \vdots & \vdots & & \vdots \\ \gamma(m-2) & \gamma(m-3) & \cdots & \gamma(1) \\ \gamma(m-1) & \gamma(m-2) & \cdots & \gamma(0) \end{pmatrix}.$$

The calculation of the log-likelihood proceeds analogously to the previous full dimension case. This representation is particularly useful for interpolation of missing values since the state noise is uncorrelated with the observation noise.

5.10 ESTIMATION OF LONG-MEMORY PROCESSES

This section discusses some specific estimation methods developed to deal with the estimation of long-range dependent time series. Among these techniques we consider maximum-likelihood procedures based on AR and MA approximations,a log-periodogram regression, the so-called *rescaled range statistic* (R/S), the variance plots, the *detrended fluctuation analysis*, and a wavelet-based approach.

5.10.1 Autoregressive Approximations

Given that the computation of exact ML estimates is computationally demanding, many authors have considered the use of autoregressive approximations to speed up the calculation of parameter estimates. Let $\{y_t : t \in \mathbb{Z}\}$ be a long-memory process defined by the autoregressive expansion

$$y_t = \varepsilon_t + \pi_1(\theta)y_{t-1} + \pi_2(\theta)y_{t-2} + \pi_3(\theta)y_{t-3} + \cdots,$$

where $\pi_j(\theta)$ are the coefficients of $\phi(B)\theta^{-1}(B)(1-B)^d$. Since in practice only a finite number of observations is available, $\{y_1, \ldots, y_n\}$, the following truncated model is considered

$$y_t = \widetilde{\varepsilon}_t + \pi_1(\theta)y_{t-1} + \pi_2(\theta)y_{t-2} + \cdots + \pi_m(\theta)y_{t-m}, \tag{5.6}$$

for $m < t \leq n$. Then, the approximate maximum-likelihood estimate $\widehat{\theta}_n$ is obtained by minimizing the function

$$\mathcal{L}_1(\theta) = \sum_{t=m+1}^{n} [y_t - \pi_1(\theta)y_{t-1} - \pi_2(\theta)y_{t-2} - \cdots - \pi_m(\theta)y_{t-m}]^2. \tag{5.7}$$

Many improvements can be made on this basic framework to obtain better estimates. In the following subsections, we describe some of these refinements. For simplicity, an estimator produced by the maximization of an approximation of the Gaussian likelihood function will be called *quasi-maximum-likelihood estimate* (QMLE).

5.10.2 Haslett-Raftery Method

Consider an ARFIMA process. An approximate one-step forecast of y_t is given by

$$\widehat{y}_t = \phi(B)\theta(B)^{-1}\sum_{j=1}^{t-1}\phi_{tj}y_{t-j}, \tag{5.8}$$

with prediction error variance

$$v_t = \operatorname{Var}(y_t - \widehat{y}_t) = \sigma_y^2\kappa\prod_{j=1}^{t-1}(1-\phi_{jj}^2),$$

where $\sigma_y^2 = \operatorname{Var}(y_t)$, κ is the ratio of the innovations variance to the variance of the ARMA(p,q) process

$$\phi_{tj} = -\binom{t}{j}\frac{\Gamma(j-d)\Gamma(t-d-j+1)}{\Gamma(-d)\Gamma(t-d+1)}, \tag{5.9}$$

for $j = 1, \ldots, t$.

To avoid the computation of a large number of coefficients ϕ_{tj}, the last term of the predictor (5.8) is approximated by

$$\sum_{j=1}^{t-1} \phi_{tj} y_{t-j} \approx \sum_{j=1}^{M} \phi_{tj} y_{t-j} - \sum_{j=M+1}^{t-1} \pi_j y_{t-j}, \tag{5.10}$$

since $\phi_{tj} \sim -\pi_j$ for large j, where for simplicity π_j denotes $\pi_j(\theta)$.

An additional approximation is made to the second term on the right-hand side of (5.10):

$$\sum_{j=M+1}^{t-1} \pi_j y_{t-j} \approx M \pi_M d^{-1} \left[1 - \left(\frac{M}{t} \right)^d \right] \bar{y}_{M+1,t-1-M},$$

where $\bar{y}_{M+1,t-1-M} = \frac{1}{t-1-2M} \sum_{j=M+1}^{t-1-M} y_j$. Hence, a QMLE $\widehat{\theta}_n$ is obtained by maximizing

$$\mathcal{L}_2(\theta) = \text{constant} - \tfrac{1}{2} n \log[\widehat{\sigma}^2(\theta)],$$

with

$$\widehat{\sigma}^2(\theta) = \frac{1}{n} \sum_{t=1}^{n} \frac{(y_t - \widehat{y}_t)^2}{\nu_t}.$$

The Haslett-Raftery algorithm has numeric complexity of order $\mathcal{O}(nM)$. Therefore, if the truncation parameter M is fixed, then this method is order $\mathcal{O}(n)$. Thus, it is usually faster than the Cholesky Decomposition and the Durbin-Levinson method. It has been suggested that $M = 100$ works fine in most applications. Besides, by setting $M = n$ we get the exact ML estimator for the fractional noise process. But, the numerical complexity in situation is order $\mathcal{O}(n^2)$.

Another autoregressive approximation method is described next. Consider the following Gaussian innovation sequence:

$$\varepsilon_t = y_t - \sum_{j=1}^{\infty} \pi_j(\theta) y_{t-j}.$$

Since the values $\{y_t, t \leq 0\}$ are not observed, an approximate innovation sequence $\{u_t\}$ may be obtained by assuming that $y_t = 0$ for $t \leq 0$,

$$u_t = y_t - \sum_{j=1}^{t-1} \pi_j(\theta) y_{t-j},$$

for $j = 2, \ldots, n$. Let $r_t(\theta) = u_t(\theta)/\sigma$ and $\theta = (\sigma, \phi_1, \ldots, \phi_p, \theta_1, \ldots, \theta_q, d)$. Then, a QMLE for θ is provided by the minimization of

$$\mathcal{L}_2(\theta) = 2n \log(\sigma) + \sum_{t=2}^{n} r_t^2(\theta).$$

Now, by taking partial derivatives with respect to θ, the minimization problem is equivalent to solving the nonlinear equations

$$\sum_{t=2}^{n} \{r_t(\theta)\dot{r}_t(\theta) - E[r_t(\theta)\dot{r}_t(\theta)]\} = 0, \tag{5.11}$$

where $\dot{r}_t(\theta) = \left(\frac{\partial r_t(\theta)}{\partial \theta_1}, \ldots, \frac{\partial r_t(\theta)}{\partial \theta_r}\right)'$.

5.10.3 A State Space Method

The state space methodology may be also used for handling autoregressive approximations. For instance, starting from the AR(m) truncation (5.6) and dropping θ from the coefficients π_j and the tilde from ε_t, we have

$$y_t = \pi_1 y_{t-1} + \cdots + \pi_m y_{t-m} + \varepsilon_t.$$

Thus, we may write the following state space system:

$$\begin{aligned} x_{t+1} &= F x_t + H \varepsilon_{t+1}, \\ y_t &= G x_t, \end{aligned}$$

where the state is given by $x_t = [y_t \;\; y_{t-1} \;\; \cdots \;\; y_{t-m+2} \;\; y_{t-m+1}]'$, the state transition matrix is

$$F = \begin{pmatrix} \pi_1 & \pi_2 & \cdots & \pi_{m-1} & \pi_m \\ 1 & 0 & \cdots & 0 & 0 \\ 0 & 1 & \cdots & 0 & 0 \\ \vdots & & & & \vdots \\ 0 & 0 & \cdots & 0 & 0 \\ 0 & 0 & \cdots & 1 & 0 \end{pmatrix},$$

the observation matrix is

$$G = [1 \; 0 \; 0 \cdots \; 0],$$

and the state noise matrix is given by

$$H = [1 \; 0 \; 0 \cdots \; 0]'.$$

The variance of the observation noise is $R = 0$, the covariance between the state noise and the observation noise is $S = 0$ and the state noise variance-covariance matrix is given by

$$Q = \sigma^2 \begin{pmatrix} 1 & 0 & \cdots & 0 \\ 0 & 0 & \cdots & 0 \\ \vdots & & & \vdots \\ 0 & 0 & \cdots & 0 \\ 0 & 0 & \cdots & 0 \end{pmatrix}.$$

The Kalman recursions for this system are

$$
\begin{aligned}
\Delta_t &= \omega_t(1,1), \\
\Theta_t &= \left[\sum_{i=1}^{m} \pi_i \omega_t(i,1) \;\; \omega_t(1,1) \;\; \cdots \;\; \omega_t(m-1,1) \right]', \\
\omega_{t+1}(1,1) &= \sigma^2 + \sum_{i,j=1}^{m} \pi_i \omega_t(i,j) \pi_j - \left[\sum_{i=1}^{m} \pi_i \omega_t(i,1) \right]^2 / \omega_t(1,1), \\
\omega_{t+1}(1,j) &= \sum_{i=1}^{m} \pi_i \omega_t(i,j) - \left[\sum_{i=1}^{m} \pi_i \omega_t(i,1) \right] \frac{\omega_t(1,j)}{\omega_t(1,1)} \quad \text{for } j \geq 2, \\
\omega_{t+1}(i,j) &= \omega_t(i-1,j-1) - \omega_t(i-1,1)\omega_t(j-1,1)/\omega_t(1,1) \quad \text{for } i,j \geq 2, \\
\widehat{x}_{t+1}(1) &= \sum_{i=1}^{m} \pi_i \widehat{x}_t(i) + \left[\sum_{i=1}^{m} \pi_i \omega_t(i,1) \right] \frac{y_t - \widehat{x}_t(1)}{\omega_t(1,1)}, \\
\widehat{x}_{t+1}(j) &= \widehat{x}_t(j-1) + \omega_t(j-1,1) \frac{y_t - \widehat{x}_t(1)}{\omega_t(1,1)} \quad \text{for } j \geq 2, \\
\widehat{y}_{t+1} &= \widehat{x}_{t+1}(1).
\end{aligned}
$$

The initial conditions for these iteration may be

$$\widehat{x}_0 = 0,$$

and

$$
\Omega_0 = \begin{pmatrix}
\gamma(0) & \gamma(1) & \cdots & \gamma(m-1) \\
\gamma(1) & \gamma(0) & \cdots & \gamma(m-2) \\
\vdots & \vdots & & \vdots \\
\gamma(m-2) & \gamma(m-3) & \cdots & \gamma(1) \\
\gamma(m-1) & \gamma(m-2) & \cdots & \gamma(0)
\end{pmatrix}.
$$

5.10.4 Moving-Average Approximations

An alternative methodology to autoregressive approximations is the truncation of the Wold expansion of a long-memory process. Two advantages of this approach are the easy implementation of the Kalman filter recursions and the simplicity of the analysis of the theoretical properties of the ML estimates. Besides, if the long-memory time series is differenced, then the resulting moving-average truncation has smaller error variance than the autoregressive approximation.

A causal representation of an ARFIMA(p,d,q) process $\{y_t\}$ is given by

$$y_t = \sum_{j=0}^{\infty} \psi_j \varepsilon_{t-j}, \tag{5.12}$$

and we may consider an approximate model for (5.12) given by

$$y_t = \sum_{j=0}^{m} \psi_j \varepsilon_{t-j}, \tag{5.13}$$

which corresponds to a MA(m) process in contrast to the MA(∞) process (5.12). A canonical state space representation of the MA(m) model (5.13) is given by

$$\begin{aligned} x_{t+1} &= F x_t + H \varepsilon_t, \\ y_t &= G x_t + \varepsilon_t, \end{aligned}$$

with

$$x_t = [y(t|t-1)\ y(t+1|t-1)\ \cdots\ y(t+m-1|t-1)]',$$

where $y(t+j|t-1) = E[y_{t+j}|y_{t-1}, y_{t-2}, \cdots]$ and system matrices

$$F = \begin{bmatrix} 0 & I_{m-1} \\ 0 & 0 \end{bmatrix}, \qquad G = \begin{bmatrix} 1 & 0 & \cdots & 0 \end{bmatrix}, \quad H = [\psi_1 \cdots \psi_m\ ]'.$$

The approximate representation of a causal ARFIMA(p, d, q) has computational advantages over the exact one. In particular, the order of the MLE algorithm is reduced from $\mathcal{O}(n^3)$ to $\mathcal{O}(n)$.

The log-likelihood function, excepting a constant, is given by

$$\mathcal{L}(\theta) = -\frac{1}{2}\left\{ \sum_{t=1}^{n} \log \Delta_t(\theta) + \sum_{t=1}^{n} \frac{[y_t - \widehat{y}_t(\theta)]^2}{\Delta_t(\theta)} \right\},$$

where $\theta = (\phi_1, \dots, \phi_p, \theta_1, \dots, \theta_q, d, \sigma^2)$ is the parameter vector associated to the ARFIMA representation (5.2).

In order to evaluate the log-likelihood function $\mathcal{L}(\theta)$ we may choose the initial conditions $\widehat{x}_1 = E[x_1] = 0$ and $\Omega_1 = E[x_1 x_1'] = [\omega(i,j)]_{i,j=1,2,\ldots}$ where $\omega(i,j) = \sum_{k=0}^{\infty} \psi_{i+k}\psi_{j+k}$.

The evolution of the state estimation and its variance, Ω_t, is given by the following recursive equations. Let $\delta_i = 1$ if $i \in \{0, 1, \dots, m-1\}$ and $\delta_i = 0$ otherwise. Furthermore, let $\delta_{ij} = \delta_i \delta_j$. Then, the elements of Ω_{t+1} and x_{t+1} are as follows:

$$\Delta_t = \omega_t(1,1) + 1, \tag{5.14}$$

$$\omega_{t+1}(i,j) = \omega_t(i+1, j+1)\delta_{ij} + \psi_i\psi_j - \frac{[\omega_t(i+1,1)\delta_i + \psi_i][\omega_t(j+1,1)\delta_j + \psi_j]}{\omega_t(1,1)+1}, \tag{5.15}$$

the state estimation is

$$\widehat{x}_{t+1}(i) = \widehat{x}_t(i+1)\delta_i + \frac{[\omega_t(i+1,1)\delta_i + \psi_i][y_t - \widehat{x}_t(1)]}{\omega_t(1,1)+1}, \tag{5.16}$$

and the observation predictor is given by

$$\widehat{y}_t = G\widehat{x}_t = \widehat{x}_t(1).$$

A faster version of the previous algorithm can be obtained by differencing the series $\{y_t\}$ since the infinite MA representation of the differenced series converges more rapidly than the MA expansion of the original process. To illustrate this approach, consider the differenced process

$$z_t = (1 - B)y_t = \sum_{j=0}^{\infty} \varphi_j \varepsilon_{t-j}, \tag{5.17}$$

where $\varphi_j = \psi_j - \psi_{j-1}$.

Remark 5.1. It is worth noting that the process $\{z_t\}$ is stationary and invertible for any $d \in (0, \frac{1}{2})$, provided that the AR(p) and MA(q) polynomials do not have common roots and all their roots are outside the closed unit disk.

By truncating the MA(∞) expansion (5.17) after m components, we get the approximate model

$$z_t = \sum_{j=0}^{m} \varphi_j \varepsilon_{t-j}. \tag{5.18}$$

An advantage of this approach is that, as shown in Problem 2.30, the coefficients φ_j converge faster to zero than the coefficients ψ_j. Consequently, a smaller truncation parameter m is necessary to achieve a good approximation level.

The truncated model (5.18) can be represented in terms of a state space system as

$$\begin{aligned} x_{t+1} &= \begin{bmatrix} 0 & I_{m-1} \\ 0 & 0 \end{bmatrix} x_t + \begin{bmatrix} \varphi_1 \\ \vdots \\ \varphi_m \end{bmatrix} \varepsilon_t, \\ z_t &= \begin{bmatrix} 1 & 0 & \cdots & 0 \end{bmatrix} x_t + \varepsilon_t. \end{aligned}$$

Under normality, the log-likelihood function of the truncated model (5.18) may be written as

$$\mathcal{L}_n(\theta) = -\frac{1}{2n} \log \det T_{n,m}(\theta) - \frac{1}{2n} \boldsymbol{z}' T_{n,m}(\theta)^{-1} \boldsymbol{z}, \tag{5.19}$$

where $[T_{n,m}(\theta)]_{r,s=1,\ldots,n} = \int_{-\pi}^{\pi} f_{m,\theta}(\lambda) e^{i\lambda(r-s)} d\lambda$, is the covariance matrix of $\boldsymbol{z} = (z_1, ..., z_n)'$ with $f_{m,\theta}(\lambda) = (2\pi)^{-1}\sigma^2 |\varphi_m(e^{i\lambda})|^2$ and the polynomial $\varphi_m(\cdot)$ is given by

$$\varphi_m(e^{i\lambda}) = 1 + \varphi_1 e^{i\lambda} + \cdots + \varphi_m e^{mi\lambda}.$$

The matrices involved in the truncated Kalman equations are of size $m \times m$. Thus, only $\mathcal{O}(m^2)$ evaluations are required for each iteration and the algorithm has an order $\mathcal{O}(n \times m^2)$. For a fixed truncation parameter m, the calculation of the likelihood function is only of order $\mathcal{O}(n)$ for the approximate ML method. Therefore, for very large samples, it may be desirable to consider truncating the Kalman recursive equations after m components. With this truncation, the number of operations required for a single evaluation of the log-likelihood function is reduced to $\mathcal{O}(n)$.

It is worth noting that the autoregressive, AR(m), and the moving-average, MA(m), approximations produce algorithms with numerical complexity of order $\mathcal{O}(n)$, where n is the sample size. Nevertheless, the quality of these approximations is governed by the truncation parameter m. The variance of the truncation error for an AR(m) approximation is of order $\mathcal{O}(1/m)$ while this variance is of order $\mathcal{O}(m^{2d-1})$ in the MA(m) case. On the other hand, the truncation error variance is of order $\mathcal{O}(m^{2d-3})$ for the differenced approach.

The methods reviewed so far apply to Gaussian processes. However, if this assumption is dropped, we still can find well-behaved Whittle estimates. For example, let $\{y_t\}$ be a stationary process with Wold decomposition:

$$y_t = \sum_{j=0}^{\infty} \psi_j(\theta)\varepsilon_{t-j},$$

where ε_t is an independent and identically distributed sequence with finite four cumulant and $\sum_{j=0}^{\infty} \psi_j^2(\theta) < \infty$. The following result establishes the consistency and the asymptotic normality of the Whittle estimate under these circumstances. Let $\widehat{\theta}_n$ be the value that maximizes the log-likelihood function $\mathcal{L}_5(\theta)$. Then, under some regularity conditions, $\widehat{\theta}_n$ is consistent and $\sqrt{n}(\widehat{\theta}_n - \theta_0) \to N[0, \Gamma(\theta_0)^{-1}]$ as $n \to \infty$, where $\Gamma(\theta_0)$ is the matrix defined in (5.27). It is important to emphasize that this result does not assume the normality of the process.

5.10.5 Semiparametric Approach

In this subsection we analyze a generalization of the Whittle approach called the *Gaussian semiparametric* estimation method. This technique does not require the specification of a parametric model for the data. It only relies on the specification of the shape of the spectral density of the time series.

Assume that $\{y_t\}$ is a stationary process with spectral density satisfying

$$f(\lambda) \sim G\lambda^{1-2H},$$

as $\lambda \to 0+$, with $G \in (0, \infty)$ and $H \in (0, 1)$. Observe that for an ARFIMA model, the terms G and H correspond to $\sigma^2\theta(1)^2/[2\pi\phi(1)^2]$ and $\frac{1}{2} + d$, respectively. Let us define $Q(G, H)$ as the objective function

$$Q(G, H) = \frac{1}{m}\sum_{j=1}^{m}\left[\log G\lambda_j^{1-2H} + \frac{\lambda_j^{2H-1}}{G}I(\lambda_j)\right],$$

where m is an integer satisfying $m < n/2$. If $(\widehat{G}, \widehat{H})$ is the value that minimizes $Q(G, H)$, then under some regularity conditions such as

$$\frac{1}{m} + \frac{m}{n} \to 0,$$

as $n \to \infty$, the estimator $\widehat{H}$ is consistent and $\sqrt{m}(\widehat{H} - H_0) \to N\left(0, \frac{1}{4}\right)$ as $n \to \infty$.

5.10.6 Periodogram Regression

Under the assumption that the spectral density of a stationary process may be written as

$$f(\lambda) = f_0(\lambda)[2\sin(\lambda/2)]^{-2d}, \tag{5.20}$$

we may consider the following regression method for parameter estimation.

Taking logarithms on both sides of (5.20) and evaluating the spectral density at the Fourier frequencies $\lambda_j = 2\pi j/n$, we have that

$$\log f(\lambda_j) = \log f_0(0) - d\log\left[2\sin\frac{\lambda_j}{2}\right]^2 + \log\left[\frac{f_0(\lambda_j)}{f_0(0)}\right]. \tag{5.21}$$

On the other hand, the logarithm of the periodogram $I(\lambda_j)$ may be written as

$$\log I(\lambda_j) = \log\left[\frac{I(\lambda_j)}{f(\lambda_j)}\right] + \log f(\lambda_j). \tag{5.22}$$

Now, combining (5.21) and (5.22) we have

$$\log I(\lambda_j) = \log f_0(0) - d\log\left[2\sin\frac{\lambda_j}{2}\right]^2 + \log\left\{\frac{I(\lambda_j)[2\sin(\lambda/2)]^{2d}}{f_0(0)}\right\}.$$

By defining $y_j = \log I(\lambda_j)$, $\alpha = \log f_0(0)$, $\beta = -d$, $x_j = \log[2\sin(\lambda_j/2)]^2$, and

$$\varepsilon_j = \log\left\{\frac{I(\lambda_j)[2\sin(\lambda/2)]^{2d}}{f_0(0)}\right\},$$

we obtain the regression equation

$$y_j = \alpha + \beta x_j + \varepsilon_j.$$

In theory, one could expect that for frequencies near zero (that is, for $j = 1, \ldots, m$ with $m \ll n$)

$$f(\lambda_j) \sim f_0(0)[2\sin(\lambda_j/2)]^{-2d},$$

so that

$$\varepsilon_j \sim \log\left[\frac{I(\lambda_j)}{f(\lambda_j)}\right].$$

The least squares estimate of the long-memory parameter d is given by

$$\widehat{d}_m = -\frac{\sum_{j=1}^m (x_j - \bar{x})(y_j - \bar{y})}{\sum_{j=1}^m (x_j - \bar{x})^2},$$

where $\bar{x} = \sum_{j=1}^m x_j/m$ and $\bar{y} = \sum_{j=1}^m y_j/m$.

5.10.7 Rescaled Range Method

Consider the sample $\{y_1, \ldots, y_n\}$ from a stationary long-memory process and let x_t be the partial sums of $\{y_t\}$, that is, $x_t = \sum_{j=1}^t y_j$ for $t = 1, \ldots, n$ and let $s_n^2 = \sum_{t=1}^n (y_t - \bar{y})^2/(n-1)$ be the sample variance where $\bar{y} = x_n/n$.

The *rescaled range statistic* (R/S) is defined by

$$R_n = \frac{1}{s_n}\left[\max_{1\le t\le n}\left(x_t - \frac{t}{n}x_n\right) - \min_{1\le t\le n}\left(x_t - \frac{t}{n}x_n\right)\right].$$

This statistic satisfies the following asymptotic property. Let $\{y_t : t \in \mathbb{Z}\}$ be a zero-mean stationary process such that y_t^2 is ergodic and

$$n^{-\frac{1}{2}-d}x_{[tn]} \to B_d(t),$$

in distribution, as $n \to \infty$, where $B_d(t)$ is the fractional Brownian motion defined in Subsection A.2.8. Define $Q_n = n^{-1/2-d}R_n$, then

$$Q_n \to Q,$$

in distribution, as $n \to \infty$, where

$$Q = \sup_{0\le t\le 1}[B_d(t) - tB_d(1)] - \inf_{0\le t\le 1}[B_d(t) - tB_d(1)].$$

Note that $\log R_n = E\,Q_n + (d + \frac{1}{2})\log n + (\log Q_n - E\,Q_n)$, so that we can obtain an estimator of the long-memory parameter d by a least squares technique similar to the one studied in Subsection 5.10.6. For instance, if $R_{t,k}$ is the R/S statistic based on the sample of size k, $\{y_t, \ldots, y_{t+k-1}\}$ for $1 \le t \le n-k+1$, then an estimator of d can be obtained by regressing $\log R_{t,k}$ on $\log k$ for $1 \le t \le n-k+1$.

5.10.8 Variance Plots

According to (2.29), the variance of the sample mean of a long-memory process based on m observations behaves like

$$\mathrm{Var}(\bar{y}_m) \sim c \ m^{2d-1},$$

for large m, where c is a positive constant. Consequently, by dividing a sample of size n, $\{y_1, \dots, y_n\}$, into k blocks of size m each with $n = k \times m$, we have

$$\log \mathrm{Var}(\bar{y}_j) \sim c + (2d-1) \log j, \tag{5.23}$$

for $j = 1, \dots, k$, where $\bar{y}_j$ is the average of the jth block, that is,

$$\bar{y}_j = \frac{1}{m} \sum_{t=(j-1)\times m+1}^{j\times m} y_t.$$

From (5.23), a heuristic least squares estimator of d is

$$\widehat{d} = \frac{1}{2} - \frac{\sum_{j=1}^{k} (\log j - a)[\log \mathrm{Var}(\bar{y}_j) - b]}{2 \sum_{j=1}^{k} (\log j - a)^2},$$

where $a = (1/k) \sum_{j=1}^{k} \log j$ and $b = (1/k) \sum_{j=1}^{k} \log \mathrm{Var}(\bar{y}_j)$.

Thus, for a short-memory process, $d = 0$, and then the slope of the line described by equation (5.23) should be -1. On the other hand, for a long-memory process with parameter d, the slope is $2d - 1$.

■ EXAMPLE 5.2

In order to illustrate the use of the variance plot technique, Figure 5.1 displays a variance plot for the Nile river data, from the period 622 A.D. to 1221 A.D. Notice that heavily line and the dotted line appears to have a very different slope. This is an indication of long-memory behavior of the data. On the contrary, in the variance plot of a Gaussian white noise sequence shown in Figure 5.2, the slopes of both lines are similar. This is expected from a serially uncorrelated time series.

5.10.9 Detrended Fluctuation Analysis

Let $\{y_1, \dots, y_n\}$ be a sample from a stationary long-memory process and let $\{x_t\}$ be the sequence of partial sums of $\{y_t\}$, that is, $x_t = \sum_{j=1}^{t} y_j$ for $t = 1, \dots, n$. The so-called *detrended fluctuation analysis* (DFA) method for estimating the long-memory parameter d of the process $\{y_t : t \in \mathbb{Z}\}$ proceeds as follows. The sample $\{y_1, \dots, y_n\}$ is divided into k nonoverlapping blocks, each containing $m = n/k$ observations. Within each block, we fit a linear

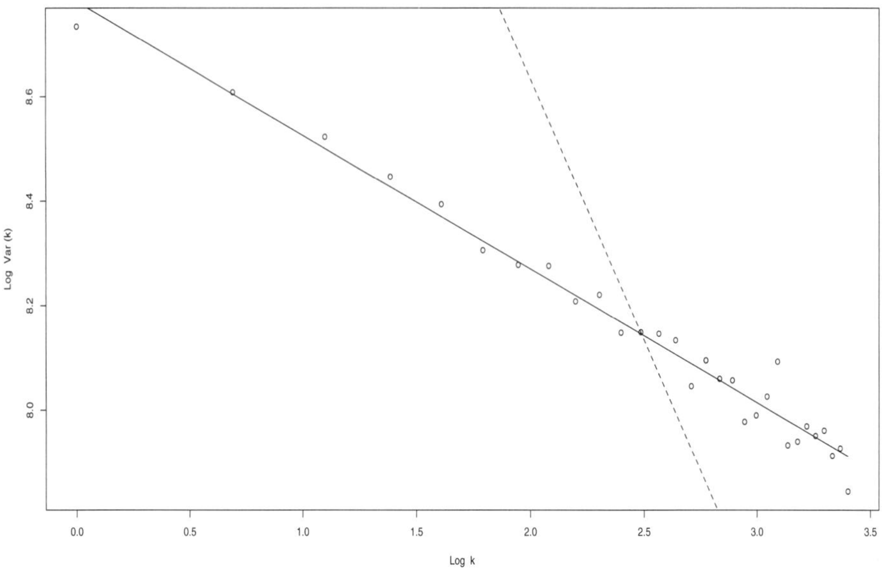

Figure 5.1 *Variance Plot for the Nile river Data.*

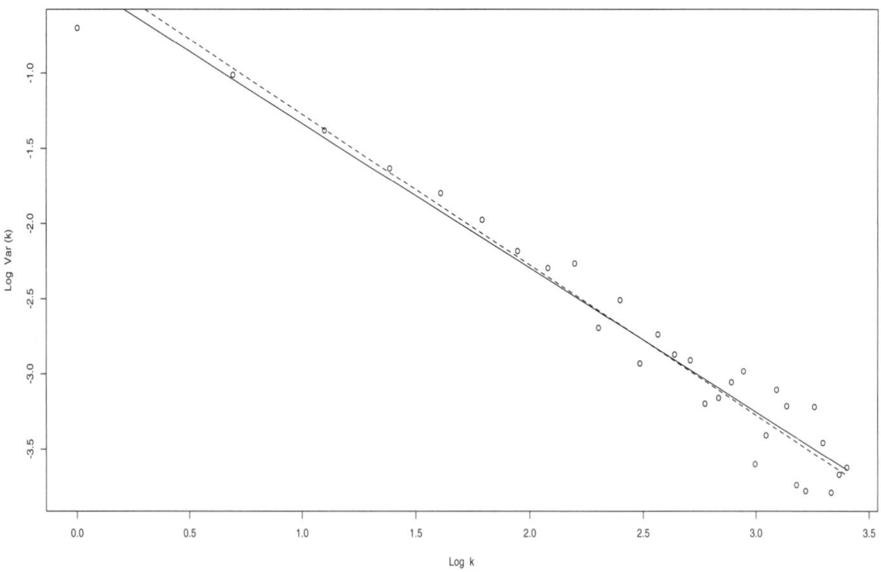

Figure 5.2 *Variance Plot for a white noise sequence.*

regression model to x_t versus $t = 1, \dots, m$. Let σ_k^2 be the estimated residual variance from the regression within block k,

$$\sigma_k^2 = \frac{1}{m} \sum_{t=1}^{m} (x_t - \widehat{\alpha}_k - \widehat{\beta}_k t)^2,$$

where $\widehat{\alpha}_k$ and $\widehat{\beta}_k$ are the least squares estimators of the intercept and the slope of the regression line.

Let $F^2(k)$ be the average of these variances

$$F^2(k) = \frac{1}{k} \sum_{j=1}^{k} \sigma_j^2.$$

For a random walk this term behaves like

$$F(k) \sim c\, k^{1/2},$$

while for a long-range sequence,

$$F(k) \sim c\, k^{d+1/2}.$$

Thus, by taking logarithms, we have

$$\log F(k) \sim \log c + (d + \tfrac{1}{2}) \log k.$$

Therefore, by fitting a least squares regression model to

$$\log F(k) = \alpha + \beta \log k + \varepsilon_k, \tag{5.24}$$

for $k \in K$, we may obtain an estimate of d as

$$\widehat{d} = \widehat{\beta} - \tfrac{1}{2},$$

where $\widehat{\beta}$ is the least squares estimator of the parameter β.

There are several ways to select the set of indexes K. If $k_0 = \min\{K\}$ and $k_1 = \max\{K\}$, then, for example, some authors choose $k_0 = 4$ and k_1 a fraction of the sample size n. Of course, for $k_0 = 2$ the regression error variance is zero since only two observations are fitted by the straight line. On the other hand, for $k_1 = n$, there is only one block and therefore the average of error variances is taken over only one sample.

This methodology derives from the following theoretical result about the behavior of the residual variances $\{\sigma_k^2\}$. Let $\{y_t\}$ be a fractional Gaussian noise process—see definition (2.31)—and let σ_k^2 be the residual variance from the least squares fitting in block k. Then,

$$E[\sigma_k^2] \sim c(d)\, m^{2d+1},$$

as $m \to \infty$ where the constant $c(d)$ is given by the formula

$$c(d) = \frac{1 - 2d}{(d+1)(2d+3)(2d+5)}.$$

5.10.10 A Wavelet-Based Method

Consider the discrete wavelet transform coefficients d_{jk} and define the statistics

$$\widehat{\mu}_j = \frac{1}{n_j} \sum_{k=1}^{n_j} \widehat{d}^2_{jk},$$

where n_j is the number of coefficients at octave j available to be calculated. As shown by Veitch and Abry (1999),

$$\widehat{\mu}_j \sim \frac{z_j}{n_j} \mathcal{X}_{n_j},$$

where $z_j = 2^{2dj}c$, $c > 0$, and $\mathcal{X}_{n_j}$ is a chi-squared random variable with n_j degrees of freedom. Thus, by taking logarithms we may write

$$\log_2 \widehat{\mu}_j \sim 2dj + \log_2 c + \log \mathcal{X}_{n_j} / \log 2 - \log_2 n_j.$$

Recall that the expected value and the variance of the random variable $\log \mathcal{X}_n$ are given by

$$\begin{aligned} E(\log \mathcal{X}_n) &= \psi(n/2) + \log 2, \\ \operatorname{Var}(\log \mathcal{X}_n) &= \zeta(2, n/2), \end{aligned}$$

where $\psi(z)$ is the psi function, $\psi(z) = d/dz \log \Gamma(z)$, and $\zeta(z, n/2)$ is the Riemann zeta function.

By defining $\varepsilon_j = \log_2 \mathcal{X}_{n_j} - \log_2 n_j - g_j$, where $g_j = \psi(n_j/2)/\log 2 - \log_2(n_j/2)$, we conclude that this sequence satisfies

$$\begin{aligned} E(\varepsilon_j) &= 0, \\ \operatorname{Var}(\varepsilon_j) &= \frac{\zeta(2, n_j/2)}{(\log 2)^2}. \end{aligned}$$

Therefore, we could write the following *heteroskedastic regression equation*:

$$y_j = \alpha + \beta x_j + \varepsilon_j,$$

where $y_j = \log_2 \widehat{\mu}_j - g_j$, $\alpha = \log_2 c$, $\beta = 2d$, and $x_j = j$. Thus, once the weighted linear regression estimate $\widehat{\beta}$ is obtained, an estimate for the long-memory parameter d is given by $\widehat{d} = \widehat{\beta}/2$. Furthermore, an estimate of the variance of $\widehat{d}$ is provided by the estimate of the variance of $\widehat{\beta}$ by means of the expression $\operatorname{Var}(\widehat{d}) = \operatorname{Var}(\widehat{\beta})/4$.

EXAMPLE 5.3

The R package fArma allows for the estimation of long-range dependent processes by several methods. The following are the results of the application of some of these fractional estimation techniques to the Nile river data. Notice that in these outputs, the Hurst parameter correspond to $H = d + \frac{1}{2}$.

```
> perFit(nile)

Title:
 Hurst Exponent from Periodgram Method

Call:
 perFit(x = nile)

Method:
 Periodogram Method

Hurst Exponent:
          H        beta
   0.9926786 -0.9853571

Hurst Exponent Diagnostic:
    Estimate  Std.Err  t-value    Pr(>|t|)
X 0.9926786 0.115791 8.573023 3.56693e-12

Parameter Settings:
      n cut.off
    663      10

> rsFit(nile)

Title:
 Hurst Exponent from R/S Method

Call:
 rsFit(x = nile)

Method:
 R/S Method

Hurst Exponent:
         H      beta
  0.8394554 0.8394554

Hurst Exponent Diagnostic:
    Estimate    Std.Err  t-value     Pr(>|t|)
X 0.8394554 0.04625034 18.15025 1.711254e-21
```

```
Parameter Settings:
       n   levels  minnpts cut.off1 cut.off2
     663       50        3        5      316
> pengFit(nile)

Title:
 Hurst Exponent from Peng Method

Call:
 pengFit(x = nile)

Method:
 Peng Method

Hurst Exponent:
         H      beta
  0.8962124 1.7924248

Hurst Exponent Diagnostic:
    Estimate    Std.Err t-value     Pr(>|t|)
X 0.8962124 0.01609048 55.6983 4.843263e-38

Parameter Settings:
       n   levels  minnpts cut.off1 cut.off2
     663       50        3        5      316
> waveletFit(nile)

Title:
 Hurst Exponent from Wavelet Estimator

Call:
 waveletFit(x = nile)

Method:
 Wavelet Method

Hurst Exponent:
         H      beta
  0.9031508 0.8063017

Hurst Exponent Diagnostic:
    Estimate    Std.Err  t-value    Pr(>|t|)
X 0.9031508 0.08329106 10.84331 0.001678205

Parameter Settings:
 length   order octave1 octave2
    512       2       2       6
```

5.10.11 Computation of Autocovariances

Precise and efficient calculation of the ACF of an ARFIMA process is a crucial aspect in the implementation of the Cholesky and the Durbin-Levinson algorithms. Recall that a closed form expression for the ACF of an ARFIMA model was discussed in previous sections.

Another approach for calculating the ACF is the so-called *splitting method.* This method is based on the decomposition of the ARFIMA model into its ARMA and its fractionally integrated (FI) parts. Let $\gamma_1(\cdot)$ be the ACF of the ARMA component and $\gamma_2(\cdot)$ be the ACF of the fractional noise given by (2.25). Then, the ACF of the corresponding ARFIMA process is given by the convolution of these two functions:

$$\gamma(h) = \sum_{j=-\infty}^{\infty} \gamma_1(j)\gamma_2(j-h).$$

If this infinite sum is truncated to m summands, then we obtain the approximation

$$\gamma(h) \approx \sum_{j=-m}^{m} \gamma_1(j)\gamma_2(j-h).$$

From this expression, the ACF $\gamma(\cdot)$ can be efficiently calculated with a great level of precision.

■ EXAMPLE 5.4

To illustrate the calculation of the ACF of a long-memory process consider the ARFIMA$(1,d,1)$ model

$$(1+\phi B)y_t = (1+\theta B)(1-B)^{-d}\varepsilon_t,$$

with $\mathrm{Var}(\varepsilon_t) = \sigma^2 = 1$.

An exact formula for the ACF of this model is given by

$$\gamma(h) = \frac{\theta C(d,-h,-\phi) + (1+\theta^2)C(d,1-h,-\phi) + \theta C(d,2-h,-\phi)}{\phi(\phi^2-1)}.$$

On the other hand, an approximated ACF is obtained by the splitting algorithm

$$\gamma(k) \approx \sum_{h=-m}^{m} \gamma_0(h)\gamma_{\mathrm{ARMA}}(k-h),$$

where

$$\gamma_0(h) = \sigma^2 \frac{\Gamma(1-2d)}{\Gamma(1-d)\Gamma(d)} \frac{\Gamma(h+d)}{\Gamma(1+h-d)},$$

Table 5.2 Calculation of the Autocorrelation Function of ARFIMA$(1, d, 1)$ Models

		ACF	
Lag	Method	$d = 0.4, \phi = 0.5, \theta = 0.2$	$d = 0.499, \phi = -0.9, \theta = -0.3$
0	Exact	1.6230971100284379	7764.0440304632230
	Approx.	1.6230971200957560	7764.0441353477199
1	Exact	0.67605709850269124	7763.5195069108622
	Approx.	0.67605707826745276	7763.5196117952073
2	Exact	0.86835879142153161	7762.8534907771409
	Approx.	0.86835880133411103	7762.8535956613778
3	Exact	0.66265875439861421	7762.0404144912191
	Approx.	0.66265877143805063	7762.0405193753304
998	Exact	0.22351300800718499	7682.7366067938428
	Approx.	0.22351301379700828	7682.7367003641175
999	Exact	0.22346824274316196	7682.7212154918925
	Approx.	0.22346824853234257	7682.7213090555442

and

$$\gamma_{\text{ARMA}}(h) = \begin{cases} \dfrac{1 - 2\phi\theta + \theta^2}{1 - \phi^2} & h = 0, \\[2ex] \dfrac{(1 - \phi\theta)(\theta - \phi)}{1 - \phi^2}(-\phi)^{|h|} & h \neq 0. \end{cases}$$

We consider two sets of parameters $d = 0.4$, $\phi = 0.5$, $\theta = 0.2$ and $d = 0.499$, $\phi = -0.9$, $\theta = -0.3$, and several lags between 0 and 999. The results are shown in Table 5.2.

Note that for the set of parameters $d = 0.4, \phi = 0.5, \theta = 0.2$, the accuracy of the splitting method is about six significant decimals while for the second set of parameters, $d = 0.499, \phi = -0.9, \theta = -0.3$, the accuracy drops to about three significant decimals for the range of lags studied.

5.11 NUMERICAL EXPERIMENTS

Table 5.3 displays the results from several simulations comparing five ML estimation methods for Gaussian processes: Exact MLE, Haslett and Raftery's

Table 5.3 Finite Sample Behavior of Maximum Likelihood Estimates

d		Exact	HR	AR	MA	Whittle
			$n = 200$			
0.40	Mean	0.3652	0.3665	0.3719	0.3670	0.3874
	SD	0.0531	0.0537	0.0654	0.0560	0.0672
0.25	Mean	0.2212	0.2219	0.2224	0.2220	0.2156
	SD	0.0612	0.0613	0.0692	0.0610	0.0706
0.10	Mean	0.0780	0.0784	0.0808	0.0798	0.0585
	SD	0.0527	0.0529	0.0561	0.0525	0.0522
			$n = 400$			
0.40	Mean	0.3799	0.3808	0.3837	0.3768	0.3993
	SD	0.0393	0.0396	0.0444	0.0402	0.0466
0.25	Mean	0.2336	0.2343	0.2330	0.2330	0.2350
	SD	0.0397	0.0397	0.0421	0.0394	0.0440
0.10	Mean	0.0862	0.0865	0.0875	0.0874	0.0753
	SD	0.0394	0.0395	0.0410	0.0390	0.0413

approach, AR(40) approximation, MA(40) approximation, and the Whittle method.

The process considered is a fractional noise ARFIMA$(0, d, 0)$ with three values of the long-memory parameter: $d = 0.1, 0.25, 0.4$, Gaussian innovations with zero-mean and unit variance, and sample sizes $n = 200$ and $n = 400$. The mean and standard deviations of the estimates are based on 1000 repetitions. All the simulations reported in Table 5.3 were carried out by means of R programs.

From Table 5.3, it seems that all estimates are somewhat downward biased for the three values of d and the two sample sizes considered. All the estimators, excepting the Whittle; seem to behave similarly in terms of bias and standard deviation. The sample standard deviations of all the methods considered are relatively close to its theoretical value 0.05513 for $n = 200$ and 0.03898 for $n = 400$. Observe that the Whittle method exhibits less bias for $d = 0.4$ but greater bias for $d = 0.1$. Besides, this procedure seems to have greater standard deviations than the other estimators, for the three values of d and the two sample sizes under study.

5.12 BAYESIAN ESTIMATION

This section discusses some applications of the Bayesian methodology to the analysis of time series data. It describes a general Bayesian framework for the analysis of ARMA and ARFIMA processes by means of the Markov chain Monte Carlo (MCMC) methodology, an important computational tool for obtaining samples from a posterior distribution. In particular, we describe applications of the Metropolis-Hastings algorithm and the Gibbs sampler in the context of long-memory processes.

The implementation of these computational procedures are illustrated with an example of Bayesian estimation of a stationary Gaussian process. Specific issues such as selection of initial values and proposal distributions are also discussed.

Consider the time series data $Y = (y_1, \ldots, y_n)'$ and a statistical model described by the parameter $\boldsymbol{\theta}$. Let $f(y|\boldsymbol{\theta})$ be the likelihood function of the model and $\pi(\boldsymbol{\theta})$ a prior distribution for the parameter. According to the Bayes theorem, the posterior distribution of $\boldsymbol{\theta}$ given the data Y is proportional to

$$\pi(\boldsymbol{\theta}|Y) \propto f(Y|\boldsymbol{\theta})\pi(\boldsymbol{\theta}).$$

More specifically, suppose that the time series follows an ARFIMA(p, d, q) model described by

$$\phi(B)(y_t - \mu) = \theta(B)(1 - B)^{-d}\varepsilon_t,$$

where the polynomials $\phi(B) = 1 + \phi_1 B + \cdots + \phi_p B^p$ and $\theta(B) = 1 + \theta_1 B + \cdots + \theta_p B^q$ do not have common roots and $\{\varepsilon_t\}$ is a white noise sequence with zero-mean and variance σ^2. Define $\mathcal{C}_d = \{d : y_t \text{ is stationary and invertible}\}$, $\mathcal{C}_\phi = \{\phi_1, \ldots, \phi_p : y_t \text{ is stationary}\}$, and $\mathcal{C}_\theta = \{\theta_1, \ldots, \theta_q : y_t \text{ is invertible}\}$.

For this model, the parameter vector may be written as

$$\boldsymbol{\theta} = (d, \phi_1, \ldots, \phi_p, \theta_1, \ldots, \theta_q, \mu, \sigma^2),$$

and the parameter space can be expressed as

$$\boldsymbol{\Theta} = \mathcal{C}_d \times \mathcal{C}_\phi \times \mathcal{C}_\theta \times \mathrm{I\!R} \times (0, \infty).$$

Sometimes, in order to simplify the specification of a prior distribution over the parameter space $\boldsymbol{\Theta}$, one may consider assigning prior distributions individually to subsets of parameters. For instance, we may assume uniform priors for d, $\phi_1, \ldots, \phi_p$, and $\theta_1 \ldots, \theta_q$, that is, $\pi(d) = U(\mathcal{C}_d)$, $\pi(\phi_1, \ldots, \phi_p) = U(\mathcal{C}_\phi)$, and $\pi(\theta_1, \ldots, \theta_q) = U(\mathcal{C}_\theta)$. Besides, we may assume an improper prior μ, $\pi(\mu) \propto 1$ and a prior $\pi(\sigma^2)$ for σ^2. With this specification, the prior distribution of $\boldsymbol{\theta}$ is simply

$$\pi(\boldsymbol{\theta}) \propto \pi(\sigma^2),$$

and the posterior distribution of $\boldsymbol{\theta}$ is given by

$$\pi(\boldsymbol{\theta}|Y) \propto f(Y|\boldsymbol{\theta})\pi(\sigma^2). \tag{5.25}$$

Apart from the calculation of this posterior distribution, we are usually interested in finding Bayes estimators for $\boldsymbol{\theta}$. For example, we may consider finding the value of $\boldsymbol{\theta}$ such that the posterior loss be minimal. That is, if $L(\boldsymbol{\theta}, Y)$ is the loss function, then

$$\widehat{\boldsymbol{\theta}} = \operatorname{argmin} \int L(\boldsymbol{\theta}, Y)\pi(\boldsymbol{\theta}|Y)\, dY.$$

As a particular case, under the quadratic loss $L(\boldsymbol{\theta}, Y) = \|\boldsymbol{\theta} - Y\|^2$ we have that the estimate of $\boldsymbol{\theta}$ is the posterior mean

$$\widehat{\boldsymbol{\theta}} = E[\boldsymbol{\theta}|Y].$$

Obtaining any of these quantities requires integration. However, in many practical situations the calculation of these integrals may be extremely difficult. To circumvent this problem, several methodologies have been proposed in the Bayesian literature, including conjugate prior distributions, numerical integration, Monte Carlo simulations, Laplace analytical approximation, and Markov chain Monte Carlo (MCMC) procedures. The analysis of all these methods is beyond the scope of this text; here we will focus on MCMC techniques.

5.12.1 Markov Chain Monte Carlo Methods

A MCMC algorithm produces a sample of a distribution of interest by a method that combines Monte Carlo techniques and Markov chains. Consider, for example, that we want to obtain a sample of the posterior distribution $\pi(\boldsymbol{\theta}|Y)$. Two well-known procedures for this purpose are the Metropolis-Hastings algorithm and the Gibbs sampler.

5.12.2 Metropolis-Hastings Algorithm

Following the Metropolis-Hastings algorithm, we start with an initial value for $\boldsymbol{\theta}$, $\boldsymbol{\theta}^{(0)}$, say. Suppose that at the stage m we have obtained the value $\boldsymbol{\theta}^{(m)}$. We update this value to $\boldsymbol{\theta}^{(m+1)}$ according to the following procedure:

1. Generate the random variable ξ from the proposal distribution $q(\xi|\boldsymbol{\theta}^{(m)})$,

$$\xi \sim q(\xi|\boldsymbol{\theta}^{(m)}).$$

2. Define

$$\alpha = \min\left\{ \frac{\pi(\xi|Y)q(\boldsymbol{\theta}^{(m)}|\xi)}{\pi(\boldsymbol{\theta}^{(m)}|Y)q(\xi|\boldsymbol{\theta}^{(m)})}, 1 \right\}.$$

3. Generate $\omega \sim \text{Ber}(\alpha)$.

4. Obtain

$$\boldsymbol{\theta}^{(m+1)} = \begin{cases} \xi & \text{if } \omega = 1, \\ \boldsymbol{\theta}^{(m)} & \text{if } \omega = 0. \end{cases}$$

The convergence of this procedure is guaranteed by the following result: Assume that the support of the proposal distribution q contains the support of the posterior distribution π. Then $\pi(\boldsymbol{\theta}^{(m)}|Y)$ converges to the unique stationary distribution of the Markov chain $\pi(\boldsymbol{\theta}|Y)$ as $m \to \infty$.

5.12.3 Gibbs Sampler

Another well-known iterative method is the Gibbs sampler. Suppose that the random variable $\boldsymbol{\theta}$ can be decomposed as $\boldsymbol{\theta} = (\theta_1, \dots, \theta_r)$ and we are able to simulate from the conditional densities

$$\theta_j|\theta_1, \dots, \theta_{j-1}, \theta_{j+1}, \dots, \theta_r \sim f_j(\theta_j|\theta_1, \dots, \theta_{j-1}, \theta_{j+1}, \dots, \theta_r),$$

for $j = 1, \dots, r$. In order to sample from the joint density of $(\theta_1, \dots, \theta_r)$ we proceed according to the following algorithm:

0. Given the sample $(\theta_1^{(m)}, \dots, \theta_r^{(m)})$, generate

1. $\theta_1^{(m+1)} \sim f_1(\theta_1|\theta_2^{(m)}, \theta_3^{(m)}, \dots, \theta_r^{(m)})$,

2. $\theta_2^{(m+1)} \sim f_2(\theta_2|\theta_1^{(m)}, \theta_3^{(m)}, \dots, \theta_r^{(m)})$,

$\vdots$

r. $\theta_r^{(m+1)} \sim f_r(\theta_r|\theta_1^{(m)}, \theta_2^{(m)}, \dots, \theta_{r-1}^{(m)})$.

The acceptance rate in this algorithm is always one, that is, all simulated values are accepted. A nice property of the Gibbs sampler is that all the simulations may be univariate. On the other hand, this algorithm *requires* that we can actually simulate samples from every conditional density f_j for $j = 1, \dots, r$. By choosing adequately these densities, it can be shown that the Gibbs sampler is a particular case of the Metropolis-Hastings algorithm.

EXAMPLE 5.5

If the process $\{y_t\}$ is Gaussian, then the likelihood function is given by

$$\begin{aligned} f(Y|\boldsymbol{\theta}) &= (2\pi\sigma^2)^{-n/2}|\Gamma(\boldsymbol{\theta})|^{-1/2} \\ &\times \exp\left\{-\frac{(Y - \mathbf{1}\mu)'\,\Gamma(\boldsymbol{\theta})^{-1}\,(Y - \mathbf{1}\mu)}{2\sigma^2}\right\}, \end{aligned} \qquad (5.26)$$

where $\Gamma(\boldsymbol{\theta}) = \text{Var}(Y)$. Hence, the posterior distribution of $\boldsymbol{\theta}$ given by (5.25) is

$$\pi(\boldsymbol{\theta}|Y) \propto (2\pi)^{-n/2}\sigma^{-n}|\Gamma(\boldsymbol{\theta})|^{-1/2} \times \exp\left\{-\frac{(Y-\mathbf{1}_{\mu})'\Gamma(\boldsymbol{\theta})^{-1}(Y-\mathbf{1}_{\mu})}{2\sigma^2}\right\}\pi(\sigma^2).$$

In this case, the MCMC method can be implemented as follows. To simplify the notation, we write the parameter $\boldsymbol{\theta}$ as $(d, \phi, \theta, \sigma^2, \mu)$ where $\phi = (\phi_1, \ldots, \phi_p)$ and $\theta = (\theta_1, \ldots, \theta_q)$.

Consider an initial sample for (d, ϕ, θ), for example,

$$\left(d^{(0)}, \phi^{(0)}, \theta^{(0)}\right) \sim N_r\left[\left(\widehat{d}, \widehat{\phi}, \widehat{\theta}\right), \Sigma\right],$$

where N_r is a multivariate Gaussian distribution with $r = p + q + 3$ and $\left(\widehat{d}, \widehat{\phi}, \widehat{\theta}\right)$ is the MLE of (d, ϕ, θ) and Σ may be obtained from the Fisher information matrix, that is,

$$\Sigma = \left[H\left(\widehat{d}, \widehat{\phi}, \widehat{\theta}\right)\right]^{-1},$$

where H is the Hessian matrix of the log-likelihood function of the data derived from (5.26).

Given the value $\left(d^{(m)}, \phi^{(m)}, \theta^{(m)}\right)$, we generate ξ from the proposal distribution $q\left(\xi|d^{(m)}, \phi^{(m)}, \theta^{(m)}\right)$ according to

$$\xi \sim N_r\left[\left(d^{(m)}, \phi^{(m)}, \theta^{(m)}\right), \Sigma\right],$$

and restricting the random variable ξ to the space $\mathcal{C}_d \times \mathcal{C}_\phi \times \mathcal{C}_\theta$ to ensure the stationarity and the invertibility of the ARFIMA process.

Now, we calculate α as

$$\alpha = \min\left\{\frac{\pi[\xi|Y]}{\pi[d^{(m)}, \phi^{(m)}, \theta^{(m)}|Y]}, 1\right\},$$

since in this case $q(\boldsymbol{\theta}|\xi) = q(\xi|\boldsymbol{\theta})$.

Then we proceed to steps 3 and 4 of the MCMC method described above. Once (d, ϕ, θ) has been updated, we update μ and σ^2.

For updating μ, one may start with an initial drawing from a normal distribution

$$\mu \sim N(\widehat{\mu}, \widehat{\sigma}_\mu^2),$$

where $\widehat{\mu}$ is an estimate of the location (e.g., the sample mean) and

$$\widehat{\sigma}_\mu^2 = \frac{\sigma^2}{\pi}\frac{|\theta(1)|^2}{|\phi(1)|^2}\frac{\Gamma(1-2d)}{d(2d+1)}\sin(\pi d)n^{2d-1},$$

cf., equation (2.28), where the values of d, $\phi(B)$, $\theta(B)$, and σ^2 are replaced by their respective MLEs. Notice, however, that we may use an *overdispersed distribution* (e.g., Student distribution), see for example Problem 12.5.

Given the sample $\mu^{(m)}$, this term may be updated to $\mu^{(m+1)}$ by generating a random variable

$$\xi \sim N(\mu^{(m)}, \widehat{\sigma}^2_{\mu^{(m)}}),$$

and then calculating

$$\alpha = \min\left\{ \frac{\pi[d^{(m)}, \phi^{(m)}, \theta^{(m)}, \xi, \widehat{\sigma}^2|Y]}{\pi[d^{(m)}, \phi^{(m)}, \theta^{(m)}, \mu^{(m)}, \widehat{\sigma}^2|Y]}, 1 \right\}.$$

Finally, for updating σ^2 one may draw samples from an inverse gamma distribution $\mathrm{IG}(\alpha, \beta)$ where the coefficients α and β can be chosen in many different ways. A simple approach is to consider that for an ARFIMA model with r parameters the MLE of σ^2, $\widehat{\sigma}^2$, satisfies approximately

$$\widehat{\sigma}^2 \sim \frac{\sigma^2}{n-r} \chi^2_{n-r},$$

for large n. Therefore, $E[\widehat{\sigma}^2] = \sigma^2$ and $\mathrm{Var}[\widehat{\sigma}^2] = 2\sigma^4/(n-r)$. Hence, by matching these moments with the coefficients α and β we have

$$\alpha = \frac{n-r+4}{2}, \qquad \beta = \frac{\widehat{\sigma}^2(n-r+2)}{2}.$$

Naturally, there are many other choices for drawing samples for σ^2, including Gamma distributions.

5.13 STATISTICAL INFERENCE

The maximum likelihood estimates of ARMA, ARIMA, ARFIMA and their seasonal counterparts are asymptotically unbiased, normally distributed and efficient. This is formally stated as follows. Let $\widehat{\theta}_n$ be the value that maximizes the exact log-likelihood where

$$\theta = (\phi_1, ..., \phi_p, \theta_1, ..., \theta_q, d)'$$

is a $p+q+1$ dimensional parameter vector and let θ_0 be the *true* parameter. Under some regularity conditions we have

(a) Consistency: $\widehat{\theta}_n \to \theta_0$ in probability as $n \to \infty$.

(b) Normality: $\sqrt{n}(\widehat{\theta}_n - \theta_0) \to N(0, \Gamma^{-1}(\theta_0))$, as $n \to \infty$, where $\Gamma(\theta) = (\Gamma_{ij}(\theta))$ with

$$\Gamma_{ij}(\theta) = \frac{1}{4\pi} \int_{-\pi}^{\pi} \left[\frac{\partial \log f_\theta(\lambda)}{\partial \theta_i} \right] \left[\frac{\partial \log f_\theta(\lambda)}{\partial \theta_j} \right] d\lambda, \tag{5.27}$$

where f_θ is the spectral density of the process.

(c) Efficiency: $\widehat{\theta}_n$ is an efficient estimator of θ_0.

In what follows we discuss the application of the previous results to the analysis of the large sample properties of MLE for some well-known time series models.

EXAMPLE 5.6

As an example, consider an ARMA(1, 1) model described by

$$y_t - \phi y_{t-1} = \varepsilon_t - \theta \varepsilon_{t-1}.$$

In this case, the parameter vector is $\boldsymbol{\theta} = (\phi, \theta)$ and the maximum likelihood estimator $\widehat{\boldsymbol{\theta}}_n = (\widehat{\phi}_n, \widehat{\theta}_n)$ satisfies the following large sample distribution

$$\sqrt{n}\left(\widehat{\boldsymbol{\theta}}_n - \boldsymbol{\theta}\right) \to N\left(0, \Gamma(\boldsymbol{\theta})^{-1}\right),$$

where

$$\Gamma(\boldsymbol{\theta})^{-1}) = \frac{1 + \phi\theta}{(\phi + \theta)^2} \left[\begin{array}{cc} (1 - \phi^2)(1 + \phi\theta) & -(1 - \theta^2)(1 - \phi^2) \\ -(1 - \theta^2)(1 - \phi^2) & (1 - \theta^2)(1 + \phi\theta) \end{array} \right],$$

EXAMPLE 5.7

For a fractional noise process with long-memory parameter d, FN(d), the maximum-likelihood estimate $\widehat{d}_n$ satisfies the following limiting distribution:

$$\sqrt{n}(\widehat{d}_n - d) \to N\left(0, \frac{6}{\pi^2}\right),$$

as $n \to \infty$. Observe that the asymptotic variance of this estimate does not depend on the value of d.

EXAMPLE 5.8

Consider the ARFIMA$(1, d, 1)$ model

$$(1+\phi B)y_t = (1+\theta B)(1-B)^{-d}\varepsilon_t,$$

where $\{\varepsilon_t\}$ is independent and identically distributed $N(0,\sigma^2)$. The parameter variance-covariance matrix $\Gamma(d,\phi,\theta)$ may be calculated as follows. The spectral density of this process is given by

$$f(\lambda) = \frac{\sigma^2}{2\pi}[2(1-\cos\lambda)]^{-d}\frac{1+\theta^2+2\theta\cos\lambda}{1+\phi^2+2\phi\cos\lambda}.$$

Hence,

$$\begin{aligned}\log f(\lambda) &= \log\left(\frac{\sigma^2}{2\pi}\right) - d\log[2(1-\cos\lambda)] \\ &\quad +\log[1+\theta^2+2\theta\cos\lambda] - \log[1+\phi^2+2\phi\cos\lambda],\end{aligned}$$

and the gradient is

$$\nabla \log f(\lambda) = \begin{bmatrix} -\log[2(1-\cos\lambda)] \\ -\dfrac{2[\phi+\cos\lambda]}{1+\phi^2+2\phi\cos\lambda} \\ \dfrac{2[\theta+\cos\lambda]}{1+\theta^2+2\theta\cos\lambda}\end{bmatrix}.$$

Thus, by dropping the parameters d, ϕ, and θ from the 3×3 matrix $\Gamma(d,\phi,\theta)$ we have

$$\Gamma_{11} = \frac{1}{4\pi}\int_{-\pi}^{\pi}\{\log[2(1-\cos\lambda)]\}^2\,d\lambda = \frac{\pi^2}{6},$$

$$\begin{aligned}\Gamma_{12} &= \frac{1}{4\pi}\int_{-\pi}^{\pi}\{\log[2(1-\cos\lambda)]\}\,\frac{2[\phi+\cos\lambda]}{1+\phi^2+2\phi\cos\lambda}d\lambda \\ &= \frac{1}{2\pi}\left\{\frac{\phi^2-1}{\phi}\int_0^{\pi}\frac{\log[2(1-\cos\lambda)]}{1+\phi^2+2\phi\cos\lambda}d\lambda\right. \\ &\qquad \left.+\frac{1}{\phi}\int_0^{\pi}\log[2(1-\cos\lambda)]d\lambda\right\}.\end{aligned}$$

Thus, we have

$$\Gamma_{12} = -\frac{\log(1+\phi)}{\phi}.$$

Analogously,

$$\Gamma_{13} = \frac{\log(1+\theta)}{\theta}.$$

In addition, for the two ARMA parameters we have

$$\begin{aligned}\Gamma_{22} &= \frac{1}{1-\phi^2},\\ \Gamma_{23} &= -\frac{1}{1-\phi\theta},\\ \Gamma_{33} &= \frac{1}{1-\theta^2}.\end{aligned}$$

Finally,

$$\Gamma(d,\phi,\theta) = \begin{bmatrix} \frac{\pi^2}{6} & -\frac{\log(1+\phi)}{\phi} & \frac{\log(1+\theta)}{\theta} \\ -\frac{\log(1+\phi)}{\phi} & \frac{1}{1-\phi^2} & -\frac{1}{1-\phi\theta} \\ \frac{\log(1+\theta)}{\theta} & -\frac{1}{1-\phi\theta} & \frac{1}{1-\theta^2} \end{bmatrix}. \tag{5.28}$$

Observe that similarly to the fractional noise case, the asymptotic variance of the MLE of an ARFIMA$(1,d,1)$ model does not depend on the value of the long-memory parameter.

■ EXAMPLE 5.9

The asymptotic variance of the MLE of ARFIMA$(1,d,0)$ and ARFIMA$(0,d,1)$ may be derived analogously to Example 5.8.

For the ARFIMA$(1,d,0)$ model we have that

$$\Gamma(d,\phi) = \begin{bmatrix} \frac{\pi^2}{6} & -\frac{\log(1+\phi)}{\phi} \\ -\frac{\log(1+\phi)}{\phi} & \frac{1}{1-\phi^2} \end{bmatrix},$$

and for the ARFIMA$(0,d,1)$ model

$$\Gamma(d,\theta) = \begin{bmatrix} \frac{\pi^2}{6} & \frac{\log(1+\theta)}{\theta} \\ \frac{\log(1+\theta)}{\theta} & \frac{1}{1-\theta^2} \end{bmatrix}.$$

From these expressions, we conclude that the asymptotic correlation between the maximum-likelihood estimates $\widehat{d}_n$ and $\widehat{\phi}_n$ of the ARFIMA$(1,d,0)$

model is

$$\lim_{n\to\infty} \operatorname{corr}(\widehat{d}_n, \widehat{\phi}_n) = \frac{\sqrt{6}}{\pi}\sqrt{1-\phi^2}\,\frac{\log(1+\phi)}{\phi}, \tag{5.29}$$

which is always positive for $\phi \in (-1, 1)$. On the other hand, the asymptotic correlation of the maximum-likelihood estimates $\widehat{d}_n$ and $\widehat{\theta}_n$ of an ARFIMA$(0, d, 1)$ model is given by

$$\lim_{n\to\infty} \operatorname{corr}(\widehat{d}_n, \widehat{\theta}_n) = -\frac{\sqrt{6}}{\pi}\sqrt{1-\theta^2}\,\frac{\log(1+\theta)}{\theta}, \tag{5.30}$$

which is always negative for $\theta \in (-1, 1)$. Observe that the asymptotic correlation formulas (5.29) and (5.30) do not depend on the value of the long-memory parameter d.

The asymptotic correlation between the maximum-likelihood estimates $\widehat{d}_n$ and $\widehat{\phi}_n$ of an ARFIMA$(1, d, 0)$ model provided by formula (5.29) is displayed in Figure 5.3 for $\phi \in (-1, 1)$. Additionally, Figure 5.4 exhibits the theoretical asymptotic correlation between the maximum-likelihood estimates $\widehat{d}_n$ and $\widehat{\theta}_n$ of an ARFIMA$(0, d, 1)$ model given by formula (5.30) for $\theta \in (-1, 1)$. Notice from these figures that the correlation between the estimators tends to 0 as $\phi \to \pm 1$ or $\theta \to \pm 1$. The maximum (minimum) value of the correlation is reached near $\phi = -0.68$ $(\theta = -0.68)$.

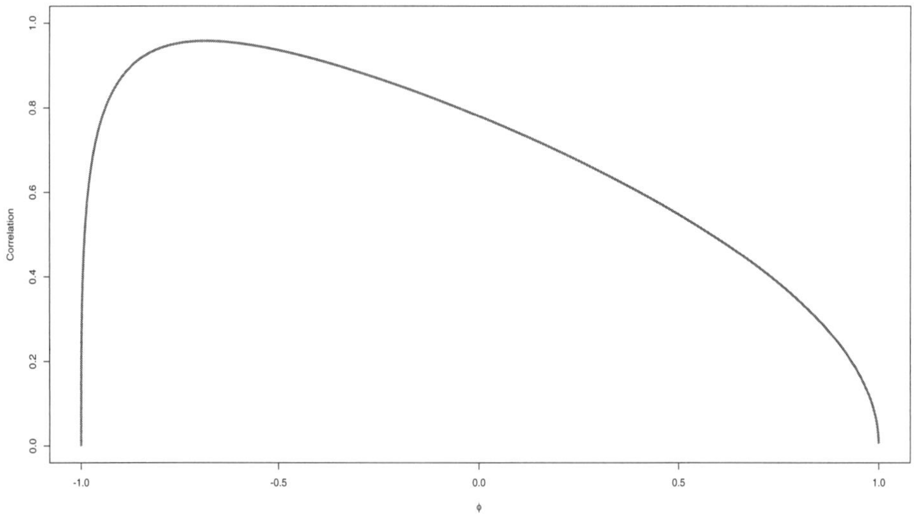

Figure 5.3 ARFIMA$(1, d, 0)$ example: Asymptotic correlation between the maximum-likelihood estimates $\widehat{d}_n$ and $\widehat{\phi}_n$.

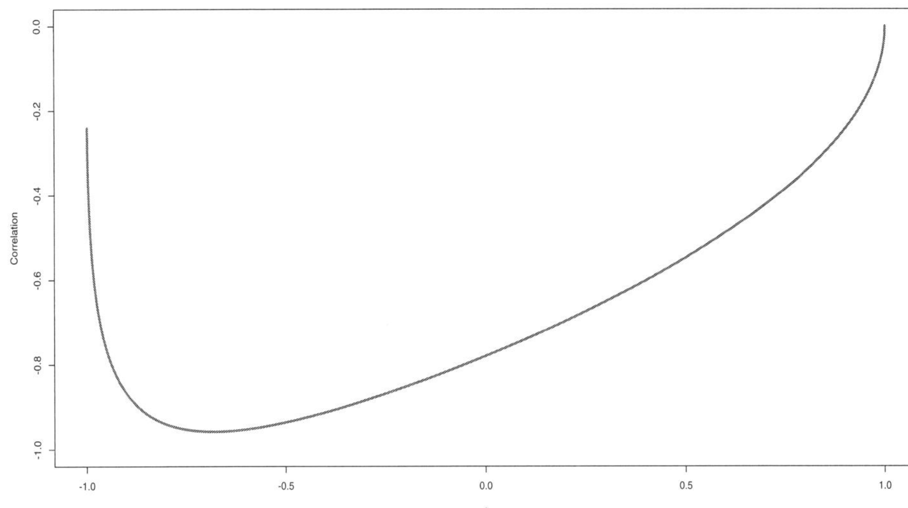

Figure 5.4 ARFIMA$(0,d,1)$ example: Asymptotic correlation between the maximum-likelihood estimates $\widehat{d}_n$ and $\widehat{\theta}_n$.

5.14 ILLUSTRATIONS

To illustrate how the finite sample performance of the maximum-likelihood estimates of ARFIMA models compare to the theoretical results revised in this chapter consider the following Monte Carlo experiments.

Table 5.4 exhibits the maximum-likelihood parameter estimations from simulated ARFIMA$(1,d,1)$ processes with sample size $n = 1000$ and parameters $d = 0.3$, $\phi = -0.5$, and $\theta = 0.2$. The results are based on 1000 replications.

Table 5.4 MLE Simulations for an ARFIMA$(1,d,1)$ Model with $d = 0.3$, $\phi = -0.5$, and $\theta = 0.2$

	d	ϕ	θ
Sample mean	0.2775	-0.5054	0.1733
Sample SD	0.0514	0.0469	0.0843
Theoretical SD	0.0487	0.0472	0.0834

Notice that the sample mean and standard deviations are close to their theoretical counterparts. The theoretical standard deviations are calculated from formula (5.28).

Figure 5.5 shows 1,164 observations from a tree ring time series dataset, see Appendix C for details. The sample ACF and sample PACF of this series is exhibited in Figure 5.6

Based on the previous plots, a family of ARMA models is proposed with orders $p, q \leq 2$ to be selected according to the AIC and the significance of the parameters. Table 5.5 reports the estimated models for all the combinations of orders. Note that the ARMA$(1, 1)$ presents the lowest AIC and botgh parameters appears to be significant at the 5% level. Consequently, this model is selected for further analysis. Figure 5.7 displays both the theoretical ACF based on the estimated parameters and the sample ACF. Observe that both plots are similar. Furthermore, a comparison of the theoretical PACF and its sample version is shown in Figure 5.8. According to these plots, the ACF and PACF of fitted model seems to be close to their theoretical counterparts.

In order to analysis the residuals of this model, Figure 5.9 exhibit its sample ACF and the corresponding Box-Ljung tests up to lag 15. Observe that the

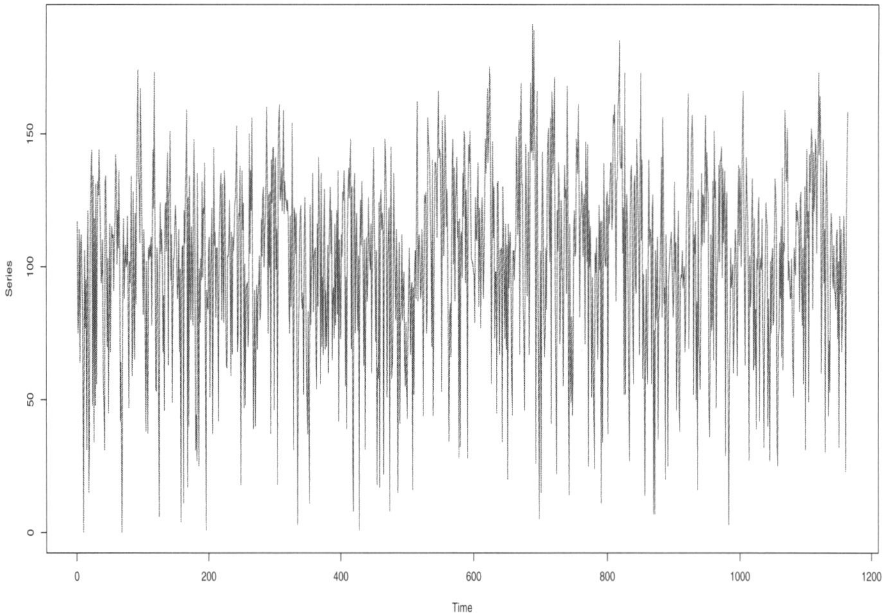

Figure 5.5 *Tree ring time series data.*

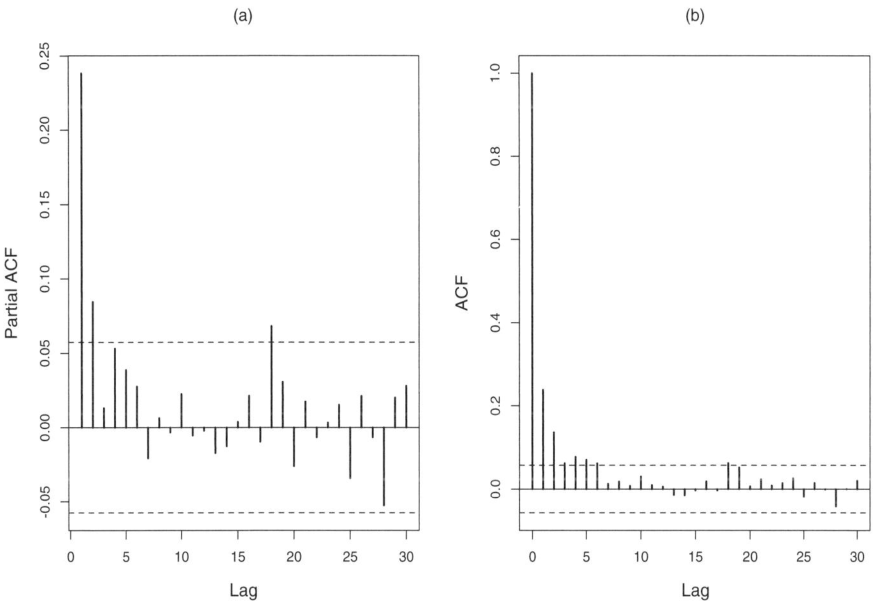

Figure 5.6 *Tree ring data: Sample ACF and PACF*

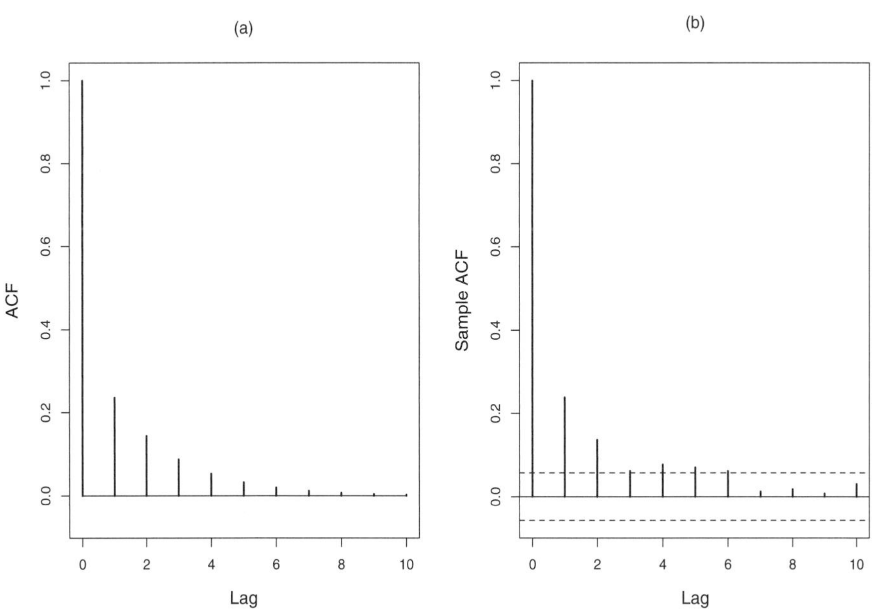

Figure 5.7 *Tree ring time series data: Theoretical and Sample ACF.*

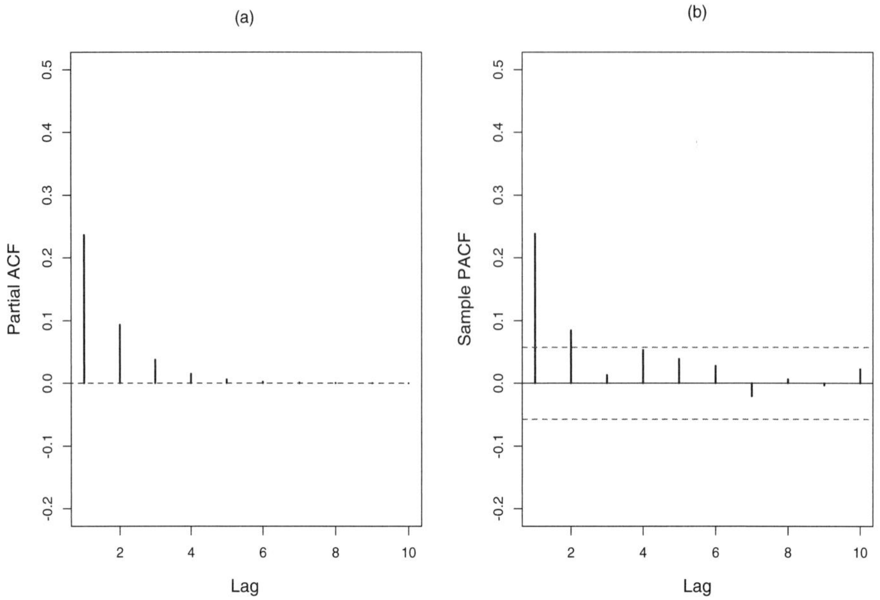

Figure 5.8 *Tree ring data: Theoretical and Sample PACF.*

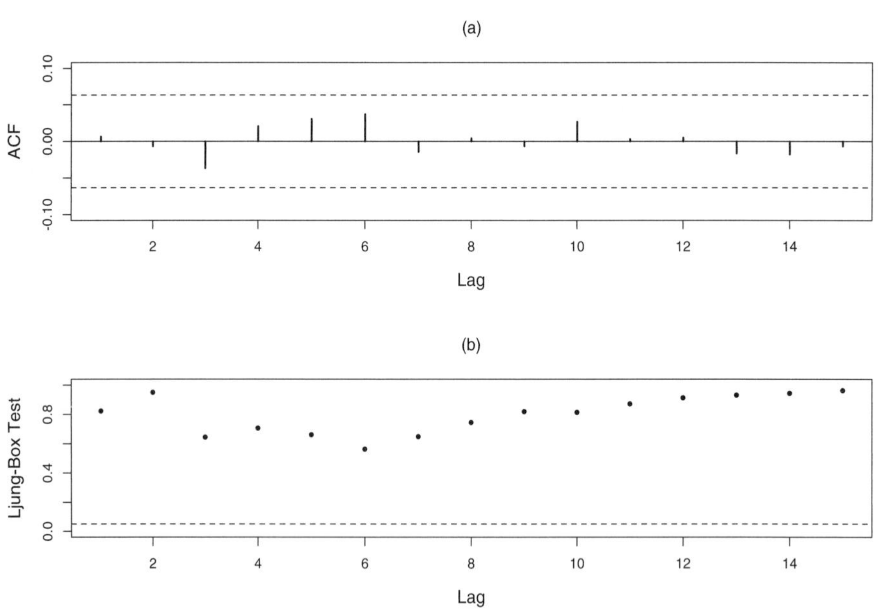

Figure 5.9 *Tree ring data: Theoretical and Sample PACF.*

Table 5.5 Maximum Likelihood Estimation of ARMA models

p	q	ϕ_1	ϕ_2	θ_1	θ_2	AIC	t_{ϕ_1}	t_{ϕ_2}	t_{θ_1}	t_{θ_2}	LB
1		0.24				11522.74	8.38			77.46	0.07
2		0.22	0.08			11516.32	7.48	2.91		71.15	0.69
	1			0.2		11534.74			7.7	84.42	0.00
1	1	0.61		-0.4		11514.73	6.16		-3.46	66.50	0.81
2	1	0.89	-0.09	-0.68		11515.44	5.23	-1.41	-4.07	61.98	0.92
	2			0.22	0.12	11519.69			7.45	76.46	0.34
1	2	0.73		-0.52	-0.05	11515.75	6.15		-4.15	63.41	0.90
2	2	0.14	0.29	0.07	-0.19	11518.73	-	-	-	66.53	0.81

null hypothesis of white noise is not rejected at the 5% level for all the lags considered.

5.15 BIBLIOGRAPHIC NOTES

Methods for estimating time series models have been extensively reviewed in the literature. A classical reference on ARMA model estimation is the book by Box, Jenkins, and Reinsel (1994). On the other hand, estimation of long-memory models have been considered by a large number of authors. An overview of the technique discussed in this chapter can be found in Palma (2007). Autoregressive approximations have been studied by Granger and Joyeux (1980), Li and McLeod (1986), Hasslett and Raftery (1989), Beran (1994), Shumway and Stoffer (2011), and Bhansali and Kokoszka (2003), among others. The Haslett-Raftery method discussed in Subsection 5.10.2 was introduced by Hasslett and Raftery (1989). State space estimation of ARFIMA and related models have been investigated by Chan and Palma (1998), Grassi and de Magistris (2014) and Dissanayake, Peiris, and Proietti (2014), among others.

The Durbin-Levinson algorithm is based on the seminal works by Levinson (1947) and Durbin (1960). The arithmetic complexity of this algorithm for a linear stationary process have been discussed, for instance, by Ammar (1998). The Durbin-Levinson algorithm can be implemented for an ARMA process in only $\mathcal{O}(n)$ operations; see, for example, Section 5.3 of Brockwell and Davis (1991). The splitting algorithm has been applied to the calculation of the ACF of long-memory processes; see, for example, the numerical experiments reported by Bertelli and Caporin (2002). Besides, several computational aspects of parameter estimation are discussed by Doornik and Ooms (2003).

The asymptotic properties of the MLE have been established by Yajima (1985) for the fractional noise process and by Dahlhaus (1989); Dahlhaus and Polonik (2006) for a general class of long-memory processes including the ARFIMA model. The so-called Whittle method was proposed by Whittle (1951). A study comparing the properties of the R/S with other estimators can be found in Giraitis, Kokoszka, Leipus, and Teyssière (2003). The large sample behavior of the periodogram of long-range-dependent processes has been extensively studied; see, for example, Fox and Taqqu (1987) and Yajima (1989), among others. Various estimators of the long-range dependence parameter including the R/S, DFA, and the Whittle methods are studied in the article by Taqqu, Teverovsky, and Willinger (1995). The books by Robert (2001) and Robert and Casella (2004) are excellent references on Bayesian methodologies. In particular, Robert (2001, Chapter 9) and Robert and Casella (2004, Chapters 6 and 7) describe several computational techniques including MCMC algorithms and the Gibbs sampler. Other good general references on Bayesian methods are the books by Box and Tiao (1992) and Press (2003). There are several versions of the MCMC algorithm. For example, the one discussed here is based on the works by Metropolis, Rosenbluth, Teller, and Teller (1953) and Hastings (1970). A comprehensive revision of Markov chain methods is provided by Tierney (1994).

Problems

5.1 Consider the AR(2) process given by: $x_t = 1.5x_{t-1} - 0.75x_{t-2} + 4.1 + \epsilon_t$.

- **(a)** Is this a stationary process?
- **(b)** Find μ_x and ρ_x.
- **(c)** Write down and solve the Yule-Walker equations. Calculate $\rho_x(3)$, $\rho_x(4)$, ..., $\rho_x(8)$.

5.2 A key tool for identifying time series processes are the ACF and the partial ACF. Figures 5.10 to 5.15 show simulated time series corresponding to the processes described below, along with their sample ACF and partial ACF. Identify which plots correspond to the models (a) to (d).

- **(a)** $y_t = 0.6\, y_{t-1} + \varepsilon_t + 0.8\, \varepsilon_{t-1}$.
- **(b)** $y_t = 0.70\, y_{t-1} - 0.12\, y_{t-2} + \varepsilon_t$.
- **(c)** $y_t = -0.4\, y_{t-1} + \varepsilon_t - \varepsilon_{t-1} + 0.21\, \varepsilon_{t-2}$.
- **(d)** $y_t = \varepsilon_t + 0.8\, \varepsilon_{t-2}$.

5.3 Consider the following MA(1) model where $\{\varepsilon_t\}$ es WN(0, 1):

$$y_t = \varepsilon_t + \theta\, \varepsilon_{t-1}.$$

- **(a)** Calculate the autocovariance function of the process.
- **(b)** Find the moment estimator of θ.
- **(c)** Show that the bias of this estimator is $-\theta/n$.
- **(d)** Assume that $y_1 = \varepsilon_1$ and define $\varepsilon = (\varepsilon_1, \ldots, \varepsilon_n)$, $y = (y_1, \ldots, y_n)$. Show that we can write $Ly = \varepsilon$ where L is a lower triangular matrix and find it.
- **(e)** Find the inverse L^{-1} and verify that the variance-covariance matrix of y can be written as $\Sigma = L^{-1}(L^{-1})'$.
- **(f)** Show that $\Sigma^{-1} = L'L$.

5.4 Figure 5.16 shows a time series of 332 observations, its sample ACF and PACF. We propose to fit an ARMA(p, q) model to these data. Based on

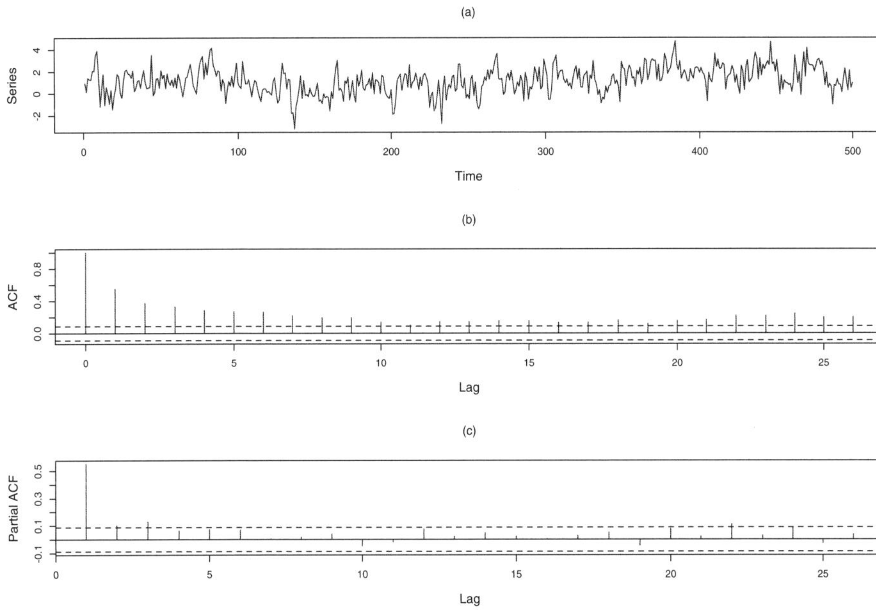

Figure 5.10 *Time series I*

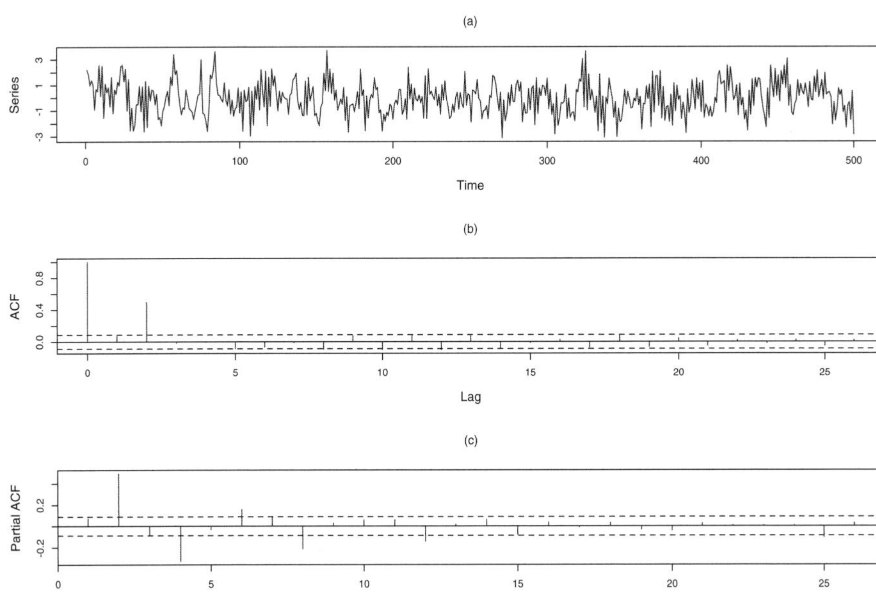

Figure 5.11 *Time series II*

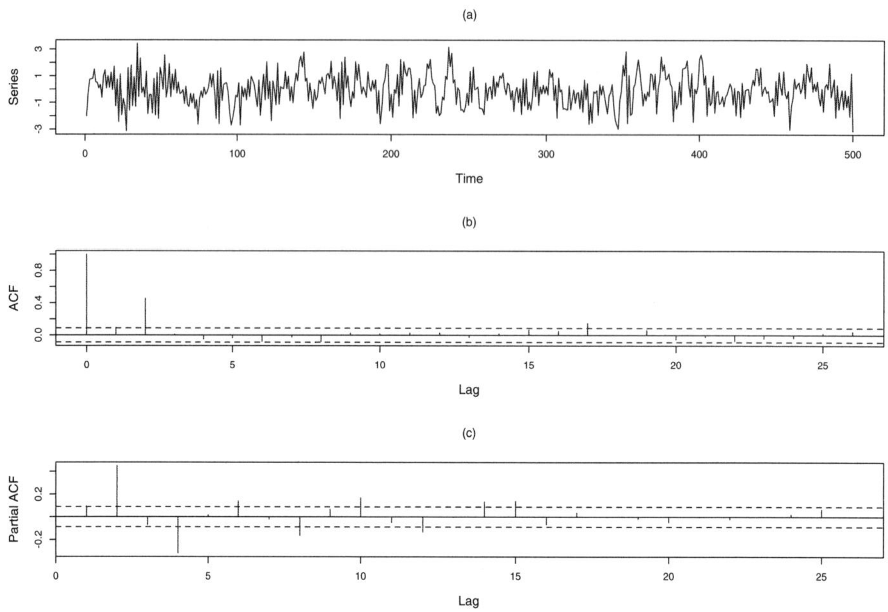

Figure 5.12 *Time series III*

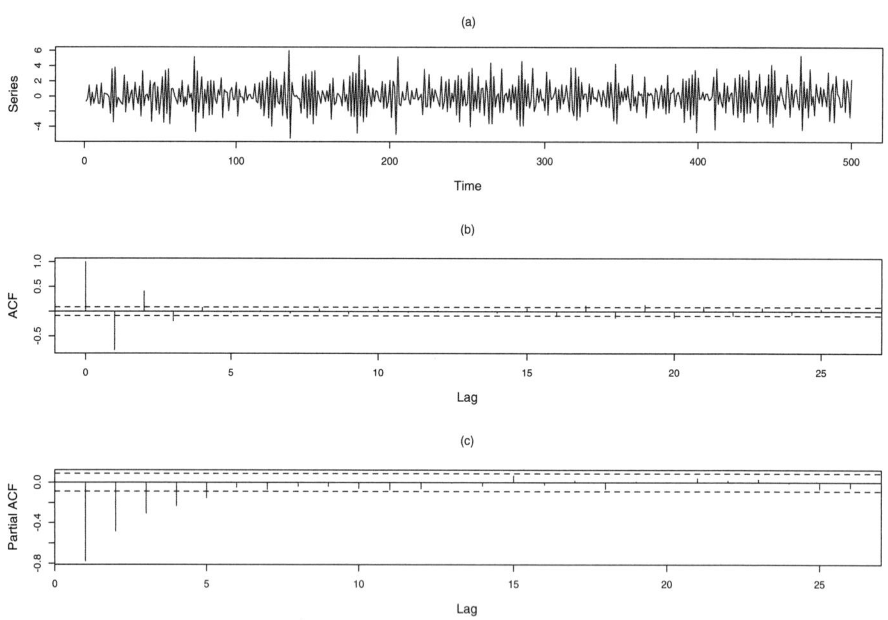

Figure 5.13 *Time series IV*

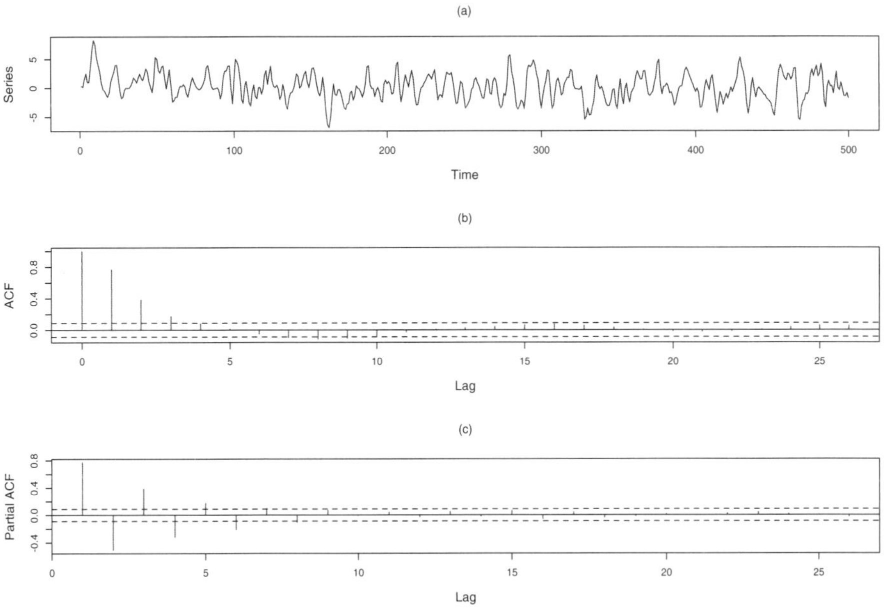

Figure 5.14 *Time series V*

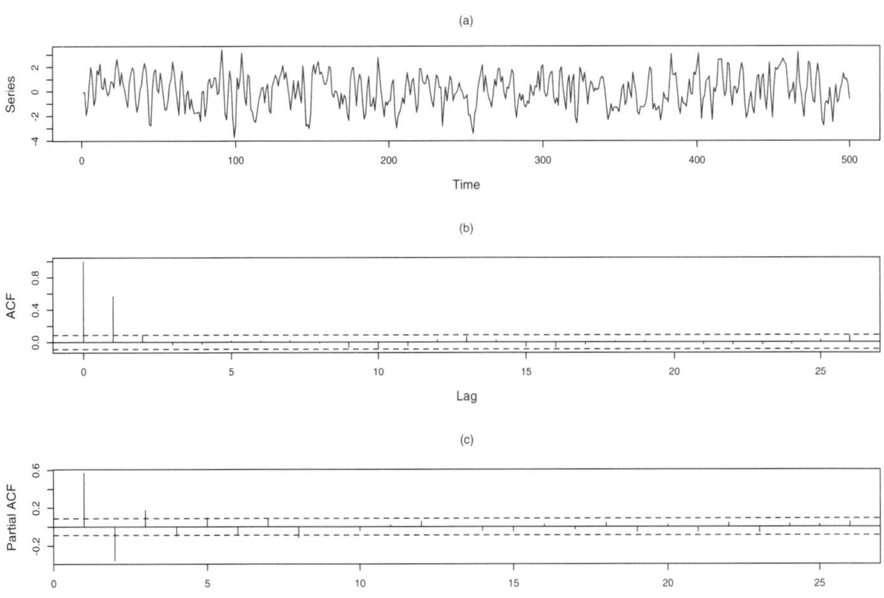

Figure 5.15 *Time series VI*

the R outputs from the fitted models presented below, information criteria, parameter significance, diagnostic plots, etc., indicate if one of these models adequately fits the time series under study. In that case, select which model seems more appropriate. Justify your answers.

```
MODEL 1

arima(x = y, order = c(2, 0, 2))

Coefficients:
         ar1     ar2     ma1     ma2  intercept
      0.2257  0.5803  0.2448  0.1128     0.0523
s.e.  0.1011  0.0975  0.1135  0.0793     0.3648

sigma^2 estimated as 0.9398:  log likelihood = -461.54,  aic = 935.08
> fit$coef
       ar1        ar2        ma1        ma2  intercept
0.22574773 0.58031755 0.24476783 0.11277715 0.05225935
> # Significance tests
> fit$coef/sqrt(diag(fit$var.coef))
      ar1       ar2       ma1       ma2 intercept
2.2331375 5.9546637 2.1559896 1.4219900 0.1432622


MODEL 2

arima(x = y, order = c(1, 0, 2))

Coefficients:
         ar1      ma1     ma2  intercept
      0.8337  -0.3546  0.3427     0.0266
s.e.  0.0398   0.0640  0.0500     0.3185

sigma^2 estimated as 0.9828:  log likelihood = -468.9,  aic = 947.8
> fit$coef
        ar1         ma1         ma2   intercept
 0.83374714 -0.35456348  0.34270842  0.02662152
> # Significance tests
> fit$coef/sqrt(diag(fit$var.coef))
       ar1        ma1        ma2  intercept
20.9293222 -5.5394974  6.8490803  0.0835905
```

```
MODEL 3

arima(x = y, order = c(2, 0, 1))

Coefficients:
         ar1     ar2     ma1  intercept
      0.1546  0.6712  0.3052     0.0653
s.e.  0.0683  0.0551  0.0863     0.3901

sigma^2 estimated as 0.9456:  log likelihood = -462.55,  aic = 935.1
> fit$coef
       ar1        ar2        ma1  intercept
0.15456221 0.67120405 0.30521334 0.06531289
> # Significance tests
> fit$coef/sqrt(diag(fit$var.coef))
       ar1        ar2        ma1  intercept
 2.2645952 12.1798059  3.5359801  0.1674072


MODEL 4

arima(x = y, order = c(1, 0, 1))

Coefficients:
         ar1      ma1  intercept
      0.9157  -0.3841     0.0703
s.e.  0.0250   0.0468     0.4104

sigma^2 estimated as 1.106:  log likelihood = -488.39,  aic = 984.79
> fit$coef
        ar1         ma1   intercept
 0.91573812 -0.38408343  0.07028509
> # Significance tests
> fit$coef/sqrt(diag(fit$var.coef))
       ar1        ma1  intercept
36.6959008 -8.2008448  0.1712706
```

5.5 Find an expression for the maximum-likelihood estimator of σ^2.

5.6 Implement computationally the convolution algorithm for estimating the ACF of an ARFIMA process. What numerical difficulties display this approach?

5.7 Using (5.9) show that $\phi_{tj} \sim -\pi_j$ for large j.

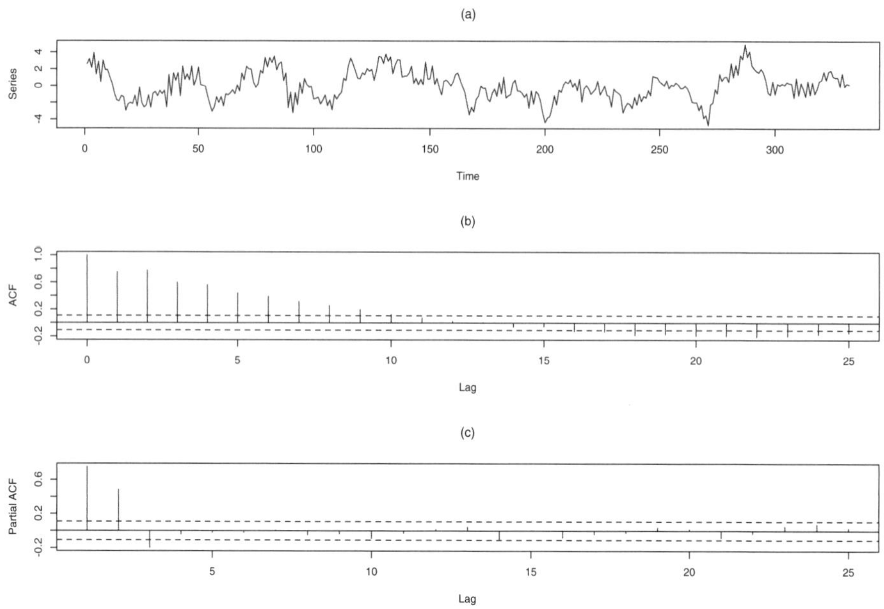

Figure 5.16 *Series, sample ACF and sample PACF*

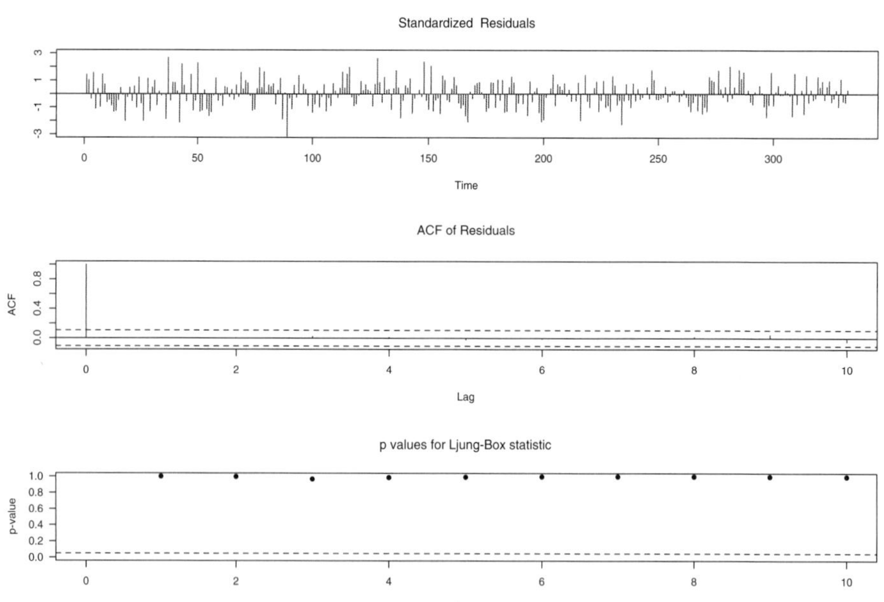

Figure 5.17 *Diagnostic graphs of Model 1.*

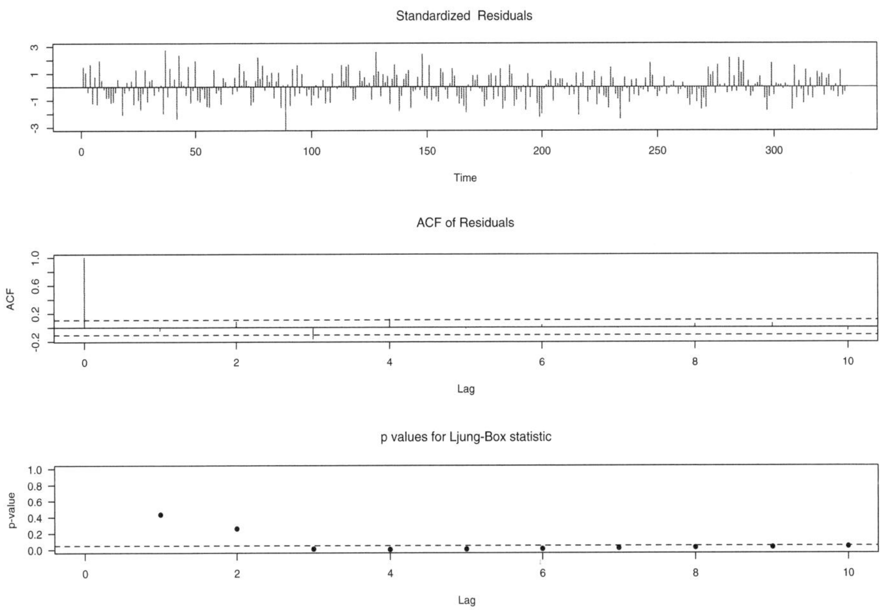

Figure 5.18 *Diagnostic graphs of Model 2.*

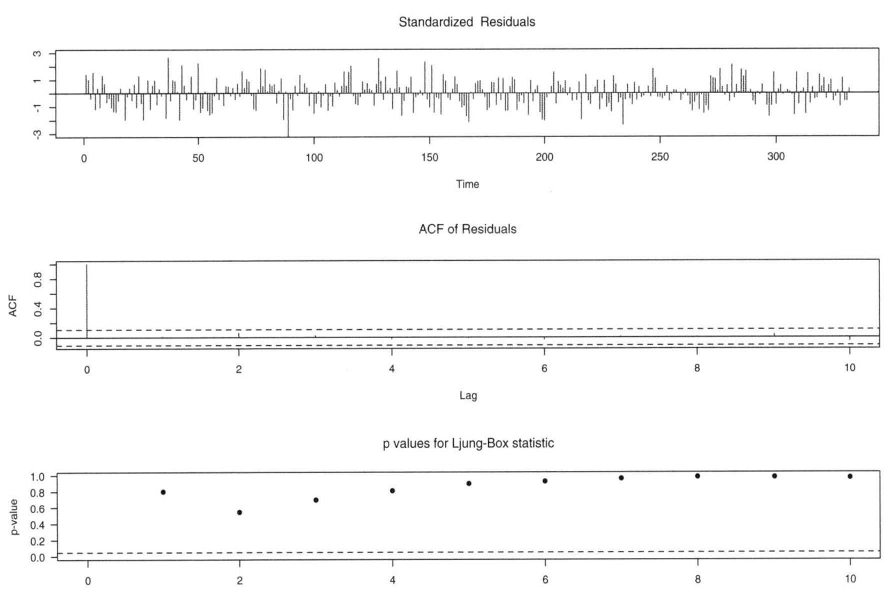

Figure 5.19 *Diagnostic graphs of Model 3.*

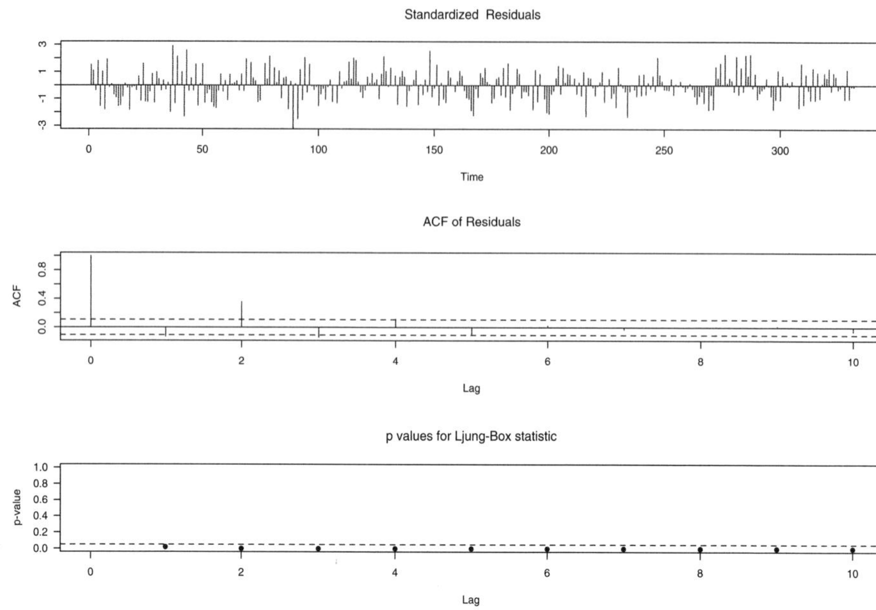

Figure 5.20 *Diagnostic graphs of Model 4.*

5.8 Explain why the Haslett-Raftery method yields exact maximum-likelihood estimates in the fractional noise case when $M = n$, where n is the sample size.

5.9 Consider the sample $y_1, ..., y_{200}$ from the ARMA(1,1) model

$$y_t - \phi\, y_{t-1} = \epsilon_t - \theta\, \epsilon_t,$$

where ϵ_t is white noise $(0, \sigma_\epsilon^2)$ and assume that the MLE of (ϕ, θ) is $(0.4, 0.7)$.

(a) Are these coefficients significative at the 5% level?

(b) Build confidence intervals for ϕ and θ at the 95% level.

(c) Find the spectral density of this process.

5.10 Consider the AR(1) process $y_t - \phi\, y_{t-1} = \epsilon_t$ with $\mathrm{Var}(\epsilon_t) = \sigma^2$. Let

$$l(\phi, \sigma^2) = -\sum_{i=1}^{n} \log(\sigma_i^2) - \sum_{i=1}^{n} e_i^2/\sigma_i^2$$

the log-likelihood function where $e_i = y_i - \hat{y}_i$ is the prediction error and $\sigma_i^2 = E(y_i - \hat{y}_i)^2$ its mean square error. Let y_1, y_2 be two observations such that $|y_1| \neq |y_2|$. Find the MLE of ϕ and σ^2.

5.11 Let y_t be a seasonal process such that $y_t = (1 + 0.2\, B)(1 - 0.8\, B^{12})\epsilon_t$, where $\sigma_\epsilon = 1$.

(a) Find the coefficients π_j of the AR(∞) expansion of the process.

(b) Plot the theoretical ACF of y_t.

(c) Plot the theoretical PACF of this process.

(d) Find the spectral density of this process and plotted it.

5.12 Let $\{x_t\}$ be the seasonal process

$$(1 - 0.7\,B^2)x_t = (1 + -.3\,B^2)z_t,$$

where $\{z_t\}$ is WN $(0, 1)$.

(a) Find the coefficients $\{\psi_j\}$ of the representation $x_t = \sum_{j=0}^{\infty} \psi_j z_{t-j}$.

(b) Find and plot the first five components of the ACF of the process $\{x_t\}$.

(c) Simulate 400 observations from this model. Plot the series and calculate the ACF and PACF.

(d) Based on the previous question, estimate the parameters of the simulated series via maximum likelihood estimation.

5.13 Explain how to select an appropriate time series model based on the following aspects:

(a) Information criteria such as AIC, BIC and others.

(b) Parsimony the model.

(c) Statistical significance of the parameters.

(d) Goodness of fit tests.

(e) Residuals whiteness testing procedures.

(f) Verification of the model assumptions, e.g. normality.

5.14 Let $Y_n = (y_1, \ldots, y_n)$ be a sequence of an ARFIMA(p, d, q) process with innovations $n(0, \sigma^2)$ and let $\boldsymbol{\theta}$ be a vector containing the ARFIMA parameters.

(a) Show that

$$f(y_{n+1}, Y_n|\boldsymbol{\theta}) = f(y_{n+1}|Y_n, \boldsymbol{\theta})f(Y_n|\boldsymbol{\theta}).$$

(b) Verify that

$$y_{n+1}|Y_n, \boldsymbol{\theta} \sim n(\mu_{n+1}, \nu_{n+1}^2\sigma^2),$$

with

$$\mu_{n+1} = \sum_{j=1}^{n} \phi_{nj} y_{n+1-j},$$

$$\nu_{n+1}^2 = \frac{\gamma(0)}{\sigma^2 \prod_{j=1}^{n}(1 - \phi_{jj}^2)},$$

where ϕ_{ij} are the partial linear regression coefficients.

5.15 Assume that the process y_t is Gaussian such that

$$Y = (y_1, \ldots, y_n)' \sim N(\boldsymbol{\theta}, \Sigma).$$

(a) Suppose that the prior distributions of $\boldsymbol{\theta}$ and Σ are

$$\pi(\boldsymbol{\theta}) \propto 1,$$

and

$$\pi(\Sigma) \propto |\Sigma|^{-(n+1)/2}.$$

Calculate the posterior distribution $\pi(\boldsymbol{\theta}, \Sigma|Y)$.

(b) Verify that the likelihood function may be written as

$$L(\boldsymbol{\theta}, \Sigma|Y) \propto |\Sigma|^{-(n/2)} e^{-(1/2)\,\mathrm{tr}[\Sigma^{-1}S(Y)]},$$

where $S(Y) = [(y_i - \theta_i)(y_j - \theta_j)]_{i,j=1,\ldots,n}$.

(c) Suppose that $\Sigma^{-1} \sim \mathrm{Wishart}_n(B^{-1}, n)$, that is,

$$\pi(\Sigma^{-1}) \propto |\Sigma|^{-(n-2)/2} e^{-(1/2)\,\mathrm{tr}[\Sigma^{-1}B]}.$$

(d) Prove that the posterior distribution of $\boldsymbol{\theta}$ and Σ given Y may be written as

$$\pi(\boldsymbol{\theta}, \Sigma|Y) \propto |\Sigma|^{-n+1} e^{-(1/2)\,\mathrm{tr}[B+S(Y)]},$$

and therefore

$$\boldsymbol{\theta}, \Sigma|Y \sim \mathrm{Wishart}_n(B + S(Y), 2n).$$

(e) Given that

$$\int |Z|^{(1/2)q-1} e^{-(1/2)\,\mathrm{tr}\,ZC} dZ = |C|^{-(1/2)(q+m-1)} 2^{(1/2)m(q+m-1)} \times \Gamma_m\left(\frac{q+m-1}{2}\right),$$

where $\Gamma_p(b)$ is the generalized gamma function

$$\Gamma_p(b) = \left[\Gamma\left(\tfrac{1}{2}\right)\right]^{(1/2)p(p-1)} \prod_{\alpha=1}^{p} \Gamma\left(b + \frac{\alpha - p}{2}\right),$$

with $b > (p-1)/2$, show that $\pi(\boldsymbol{\theta}|Y) \propto |S(Y)|^{-n/2}$.

5.16 Let $Q = (q_{ij})$ be a transition matrix of an arbitrary Markov chain on the states $0, 1, \ldots, S$ and let α_{ij} be given by

$$\alpha_{ij} = \frac{s_{ij}}{1 + \dfrac{\pi_i q_{ij}}{\pi_j q_{ji}}},$$

where $\pi_i > 0$, $i = 0, \ldots, S$, $\sum \pi_i = 1$ and s_{ij} is a symmetric function of i and j chosen such that $0 \leq \alpha_{ij} \leq 1$ for all i, j.

Consider the Markov chain on the states $0, 1, \ldots, S$ with transition matrix P given by

$$p_{ij} = q_{ij}\alpha_{ij},$$

for $i \neq j$ and

$$p_{ii} = 1 - \sum_{i \neq j} p_{ij}.$$

(a) Show that the matrix P is in fact a transition matrix.

(b) Prove that P satisfies the *reversibility* condition that

$$\pi_i p_{ij} = \pi_j p_{ji},$$

for all i and j.

(c) Verify that $\pi = (\pi_0, \ldots, \pi_S)$ is the unique stationary distribution of P, that is,

$$\pi = \pi P.$$

5.17 Consider the following two choices of the function s_{ij} on the method discussed in Problem 5.16:

$$s_{ij}^M = \begin{cases} 1 + \dfrac{\pi_i q_{ij}}{\pi_j q_{ji}} & \text{if } \dfrac{\pi_i q_{ij}}{\pi_j q_{ji}} \leq 1, \\[2ex] 1 + \dfrac{\pi_j q_{ji}}{\pi_i q_{ij}} & \text{if } \dfrac{\pi_i q_{ij}}{\pi_j q_{ji}} > 1. \end{cases}$$

and the alternative choice

$$s_{ij}^B = 1.$$

(a) Show that both choices satisfy the condition $1 \leq \alpha_{ij} \leq 1$.

(b) Suppose that the matrix Q is symmetric and consider the sampling scheme

$$x_{t+1} = \begin{cases} j & \text{with probability } \alpha_{ij}, \\ i & \text{with probability } 1 - \alpha_{ij}. \end{cases}$$

Verify that if $\pi_i = \pi_j$, then $P(x_{t+1} = j) = 1$ for the Metropolis algorithm and $P(x_{t+1} = j) = 1$ for Barker's method.

(c) According to part (b), which method is preferable?

(d) Consider the choice

$$s_{ij} = g\left[\min\left\{\frac{\pi_i q_{ij}}{\pi_j q_{ji}}, \frac{\pi_j q_{ji}}{\pi_i q_{ij}}\right\}\right],$$

where the function $g(x)$ is symmetric and satisfies $0 \leq g(x) \leq 1 + x$. Verify that for this choice, the condition $0 \leq \alpha_{ij} \leq 1$ holds.

(e) Consider the particular choice $g(x) = 1 + 2(x/2)^\gamma$ for a constant $\gamma \geq 1$. Prove that s_{ij}^M is obtained with $\gamma = 1$ and s_{ij}^B is obtained with $\gamma = \infty$.

5.18 Consider the following Poisson sampling with

$$\pi_i = e^{-\lambda} \frac{\lambda^i}{i!},$$

for $i = 0, 1, \ldots,$ and

$$q_{00} = q_{01} = \tfrac{1}{2},$$

and

$$q_{ij} = \tfrac{1}{2}, \text{ for } j = i - 1, i + 1, i \neq 0.$$

(a) Show that in this case

$$x_{t+1} = \begin{cases} i+1 & \text{with probability } \begin{cases} \dfrac{\lambda}{i+1} & \text{if } \lambda \leq i+1, \\ 1 & \text{if } \lambda > i+1, \end{cases} \\ \\ i-1 & \text{with probability } \begin{cases} \dfrac{i}{\lambda} & \text{if } \lambda \leq i, \\ 1 & \text{if } \lambda > i. \end{cases} \end{cases}$$

(b) What is the disadvantage of this method when λ is large?

5.19 Show that for an ARFIMA$(1, d, 0)$ model,

$$\text{corr}(\widehat{d}, \widehat{\phi}) \to \frac{\sqrt{6}}{\pi},$$

as $\phi \to 0$.

5.20 Show that for an ARFIMA$(0, d, 1)$ model,

$$\text{corr}(\widehat{d}, \widehat{\theta}) \to -\frac{\sqrt{6}}{\pi},$$

as $\theta \to 0$.

5.21 Let $\{y_t : t \in \mathbb{Z}\}$ be a stationary process with spectral density $f(\lambda)$ and let $Y = (y_1, y_2, \ldots, y_n)' \sim N(0, T_{\theta_0})$ where the elements of the $n \times n$ variance-covariance matrix $T_{\theta_0} = (T_{ij})$ are given by

$$T_{ij} = \int_{-\pi}^{\pi} f(\lambda) e^{i\lambda(i-j)} d\lambda.$$

Consider the function

$$\mathcal{L}_n(\theta) = \frac{1}{n} \log \det T_\theta + \frac{1}{n} Y' T_\theta^{-1} Y.$$

(a) Prove that

$$E[\mathcal{L}_n(\theta_0)] = 1 + \log \det T_{\theta_0}.$$

(b) Show that

$$\lim_{n\to\infty} E[\mathcal{L}_n(\theta_0)] = \log(2\pi) + \frac{1}{2\pi} \int_{-\pi}^{\pi} \log f(\lambda)\, d\lambda.$$

(c) Verify that

$$\lim_{n\to\infty} E[\mathcal{L}_n(\theta_0)] = 1 + \log \sigma^2.$$

(d) Verify that

$$\mathrm{Var}[\mathcal{L}_n(\theta_0)] = \frac{2}{n},$$

and prove that $\mathcal{L}_n(\theta_0)$ converges to $1 + \log \sigma^2$ in probability as n tends to infinity.

CHAPTER 6

NONLINEAR TIME SERIES

As discussed in previous chapters, linear processes are excellent tools for analyzing a great number of time series. However, they usually fail to adequately model more complicated dependence structures. For example, many real-life time series display almost no autocorrelations but exhibit strong dependence in their squares. To deal with these situations, several classes of nonlinear processes are available. Additionally, a number of testing procedures for linearity have been developed to help making a decision whether to employ a linear or a nonlinear approach. If linearity is not rejected through these procedures, then we could use some of the linear processes discussed in the previous chapters to fit the data. On the contrary, if linearity is rejected, then we could try out some of the nonlinear models discussed in this chapter. Even though nonlinear time series appear in many fields, they are commonly found in the analysis of financial instruments. Financial time series such as returns from stocks indexes exhibit almost null autocorrelation but they display an important level of dependence in their squared returns. This chapter begins defining a large class of nonlinear processes and the proceed to review some linearity testing procedures. It also discusses financial time series, an important area of applications for these models.

Time Series Analysis. First Edition. Wilfredo Palma.

6.1 INTRODUCTION

A stationary process with mean μ can be written in terms of a Volterra expansion

$$y_t = \mu + \sum_{j=-\infty}^{\infty} \psi_i \, \varepsilon_{t-i} + \sum_{i,j=-\infty}^{\infty} \psi_{ij} \, \varepsilon_{t-i}\varepsilon_{t-j} + \sum_{i,j,k=-\infty}^{\infty} \psi_{ijk} \, \varepsilon_{t-i}\varepsilon_{t-j}\varepsilon_{t-k} + \cdots,$$

where ψ_i, ψ_{ij}, $\psi_{ijk}, \ldots$ are unknown coefficients and $\{\varepsilon_t\}$ is a sequence of i.i.d. random variables with zero-mean and variance σ^2. The values ψ_i, ψ_{ij}, ψ_{ijk} are commonly referred to as linear, quadratic, cubic coefficients, respectively.

Note that when the terms ψ_{ij}, ψ_{ijk} and so on are all equal to zero, y_t reduces to a linear process. On the other hand, if the coefficients ψ_{ijk} and higher are all zero, but ψ_{ij} are non zero, the resulting process can be written as,

$$y_t = \mu + \sum_{j=-\infty}^{\infty} \psi_i \, \varepsilon_{t-i} + \sum_{i,j=-\infty}^{\infty} \psi_{ij} \, \varepsilon_{t-i}\varepsilon_{t-j}.$$

A simple example of a quadratic process is $y_t = \varepsilon_t + \theta\varepsilon_t\varepsilon_{t-1}$ where ε_t is a sequence of i.i.d. random variables with zero-mean and variance σ^2. Note

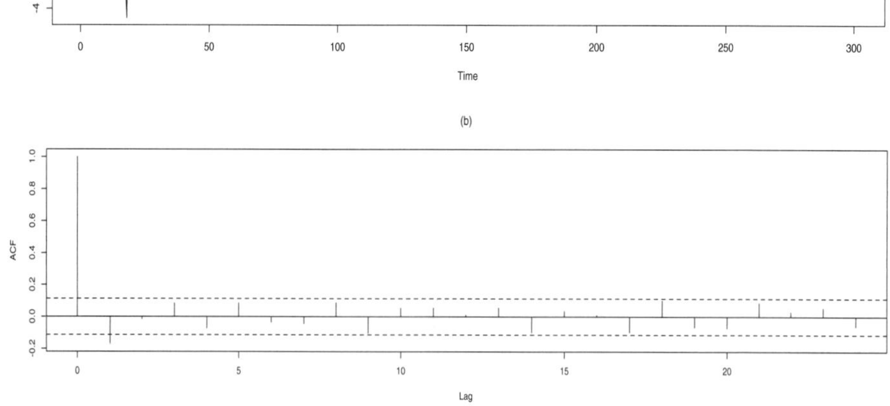

Figure 6.1 *Simulated quadratic process with* $\theta = 0.7$. *(a) Series of 300 observations. (b) Sample ACF .*

that in this case, $E\,y_t = 0$ and the autocovariance function is

$$\gamma(0) = \begin{cases} \sigma^2(1+\sigma^2\theta^2) & \text{if} \quad h = 0 \\ 0 & \text{if} \quad h \neq 0. \end{cases}$$

Thus, this quadratic process actually corresponds to a white noise sequence. A simulated sample of 300 observations from this process with $\theta = 0.7$ and $\sigma^2 = 1$ is exhibited in Figure 6.1 along with its empirical ACF.

6.2 TESTING FOR LINEARITY

There are several procedures for testing linearity of a stationary time series. For simplicity, in this section we consider a methodology based on regressing the observations on their previous values and then computing the resulting residuals.

Consider regressing the observation y_t on $1, y_{t-1}, y_{t-2}, \ldots, y_{t-p}$ for $t = p+1, \ldots, n$ where p is a previously specified autoregression order and obtain the residuals $e_{p+1}, \ldots, e_n$. Let $S_e = \sum_{j=p+1}^{n} e_t^2$ be the sum of squared residuals. Next, regress the squared process y_t^2 on $1, y_{t-1}, y_{t-2}, \ldots, y_{t-p}$ for $t = p+1, \ldots, n$ obtaining the residuals $\xi_{p+1}, \ldots, \xi_n$.

Finally, regress $e_{p+1}, \ldots, e_n$ on $\xi_{p+1}, \ldots, \xi_n$

$$e_t = \beta\xi_t + \eta_t,$$

obtaining the regression coefficient $\widehat{\beta}$ and the residual sum of squares $S_\xi = \sum_{j=p+1}^{n} \xi_t^2$. Thus, one can test for linearity by means of the statistic

$$F = \frac{\widehat{\beta}^2 S_\xi (n-2p-2)}{S_e - \widehat{\beta}^2 S_\xi}. \tag{6.1}$$

The distribution of this statistic is approximately Fisher with 1 and $n-2p-2$ degrees of freedom. As an illustration of the application of the this nonlinearity test consider the quadratic time series presented in the previous section. The R library nlts provides an implementation of this testing procedure generating the following output:

```
 order         F        df1       df2          p
1.0000    4.9697     1.0000 297.0000     0.0265

 order         F        df1       df2          p
2.000     20.333     3.0000 294.0000     0.0000

 order         F        df1       df2          p
3.0000   12.1295     6.0000 290.0000     0.0000
```

```
   order         F       df1       df2          p
  4.0000    8.1241   10.0000  285.0000     0.0000

   order         F       df1       df2          p
  5.0000    5.9962   15.0000  279.0000     0.0000
```

Based on these results, the linearity hypothesis is rejected at the 5% significance level for all the orders considered. Furthermore, this R library also offers a procedure for estimating the appropriate order. As indicated in the following output, the program suggests that $p = 3$ is an adequate order for the testing procedure.

```
The estimated order is 3 with a cross-validation error of 0.81
and Gaussian bandwidth 3 (using local polynomial with 2 degrees).

order     cv.min bw.opt          df   GCV.min GCV.bw.opt      GCV.df
1        1 1.0395           10  3.6213 0.96016        1.2      8.9953
2        2 0.8823            5 11.3163 0.79636        2.0     20.8732
3        3 0.8101            3 37.8311 0.75289        3.0     30.7917
4        4 0.8721            4 49.6511 0.77349        4.0     38.8586
5        5 0.9175           10 35.2700 0.81368        5.0     46.5358
```

6.3 HETEROSKEDASTIC DATA

Time series of returns from financial instruments usually exhibit nonlinearities. There is strong empirical evidence that a large number of time series from finance and economics show some *stylized facts* such as clusters of highly variable observations followed by clusters of observations with low variability and strong autocorrelations either in the series or its squares. In this chapter we examine some of the models proposed to account for these features. In particular, we consider *heteroskedastic* time series models where the conditional variance given the past is no longer constant.

As an illustration of the stylized facts frequently found in economic time series, consider the daily log-returns of the SP500 stock index discussed in Chapter 1. This dataset spans from January 1, 1950 to May 1, 2014. As shown in Figure 6.2 displays periods of low levels of volatility followed by periods of higher volatility. Additionally, it can be observed the presence of specific dates with exceptionally high variability. On the other hand, Figure 6.3 shows that the returns exhibit a lower level of autocorrelation as compared to the squared returns.

Another illustration of the stylized facts is provided by the daily log-returns of the copper prices from January 4, 2005 to December 31, 2014. These prices are expressed in terms of USD cents per pound at the London Metal Exchange. The series $\{r_t\}$ is displayed in panel (a) of Figure 6.4 while its squares are shown in panel (b).

From Figure 6.4(a) we note a period of high volatility during 2008–2009. Besides, Figure 6.4(b) suggests that the squared series suffers from bursts of high volatility followed by periods of low volatility. On the other hand, the sample autocorrelation function, Figure 6.5(a), shows some significant autocorrelations in the returns while the sample autocorrelation of the squares exhibits a strong level of dependence; see Figure 6.5(b).

Several models have been proposed to account for these features. Most of these models specify an ARMA or an ARFIMA process for the returns and specify some parametric model for the conditional variance of the series given its infinite past. In some cases this model resembles an ARMA in the form of a *generalized autoregressive conditionally heteroskedastic* (GARCH) process or it resembles an AR(∞) process in the form of an ARCH(∞) model.

6.4 ARCH MODELS

An ARCH(1) process is defined by the discrete-time equation

$$y_t = \sigma_t \varepsilon_t, \tag{6.2}$$

$$\sigma_t^2 = \alpha_0 + \alpha_1 y_{t-1}^2, \tag{6.3}$$

where $\sigma_t^2 = E[y_t^2 | y_{t-1}]$ is the conditional variance of the process $\{y_t\}$, the ARCH coefficients α_0, α_1 are positive, $\alpha_1 < 1$ and $\{\epsilon_t\}$ is sequence of in-

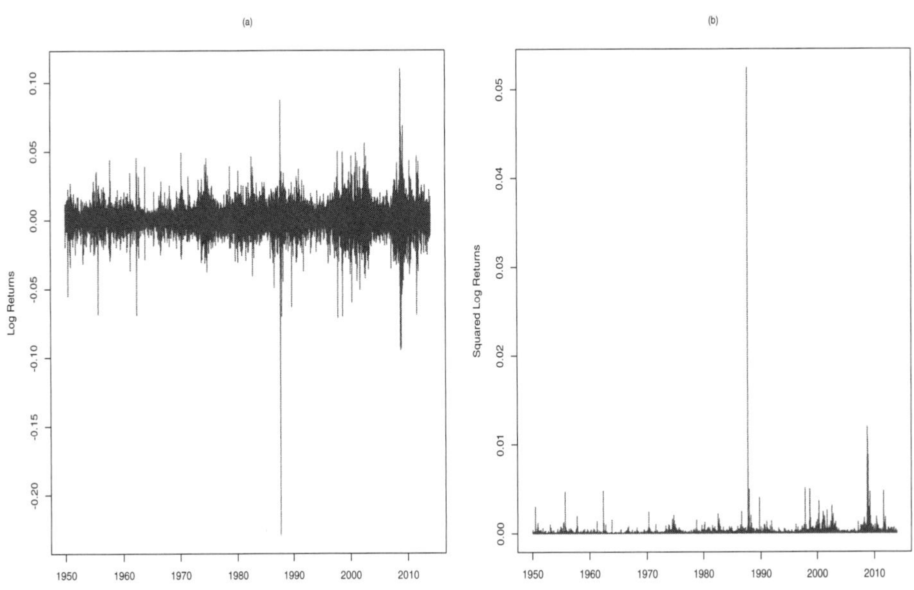

Figure 6.2 SP500 Index data (1950–2014). (a) Daily log-returns. (b) Squared daily log-returns.

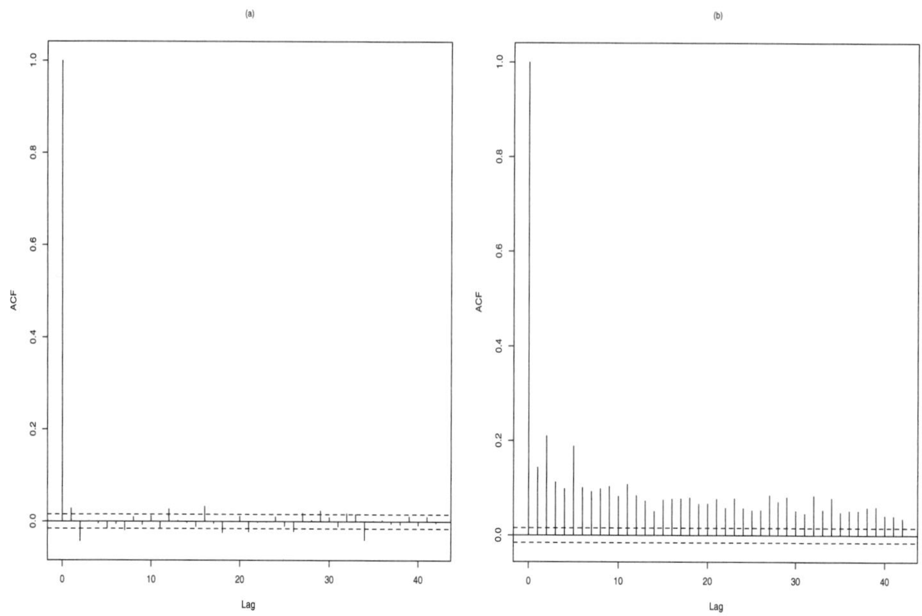

Figure 6.3 SP500 Index data (1950–2014). (a) Sample ACF of daily log-returns. (b) Sample ACF of squared daily log-returns.

dependent and identically distributed zero-mean and unit variance random variables. Although ε_t is often assumed to be Gaussian, in some cases it may be specified by a t-distribution or a double exponential distribution, among others. These distributions have a greater flexibility to accommodate a possible heavy tail behavior of some financial time series.

Observe that $E\,y_t^2 = E\,\sigma_t^2 \varepsilon_t^2$. Given that σ_t only depend on past values of the sequence $\{\varepsilon_t\}$, σ_t and ε_t are independent. Consequently,

$$E\,y_t^2 = E\,\sigma_t^2\,E\,\varepsilon_t^2 = E\,\sigma_t^2.$$

Thus,

$$E\,y_t^2 = \alpha_0 + \alpha_1\,E\,y_{t-1}^2.$$

In order to be stationary, $E\,y_t^2 = E\,y_{t-1}^2$, so that by replacing this condition in the equation above and considering that $\alpha_1 < 1$ we get

$$E\,y_t^2 = \frac{\alpha_0}{1 - \alpha_1^2}.$$

As an illustration, Figure 6.6 shows 400 observations of an ARCH(1) model with parameter $\alpha_1 = 0.9$ Additionally, Figure 6.7 exhibits the sample ACF of this series along with the sample ACF of its squares. Note that the series shows almost no autocorrelation but its squares display a high level of dependence.

This situation is due to the dependence structure of the squares of an ARCH process. In what follows we show that y_t^2 corresponds to an AR(1) process.

Let us define the sequence $\nu_t = y_t^2 - \sigma_t^2$. Thus, we can write $y_t^2 = \alpha_0 + \alpha_1 y_{t-1}^2 + \nu_t$. If μ denotes the mean of y_t^2, then $y_t^2 - \mu - \alpha_1(y_{t-1}^2 - \mu) + \nu_t$.

But ν_t is a white noise sequence. To see this, $E\,\nu_t = E\,y_t^2 - E\,\sigma_t^2 = 0$. Besides, this process is uncorrelated since for $h > 0$ we have

$$E\,\nu_t\nu_{t+h} = E\,\sigma_t^2(1 - \varepsilon_t^2)\sigma_{t+h}^2(1 - \varepsilon_{t+h}^2).$$

Given that both σ_t^2 and σ_{t+h} depend only values $\varepsilon_{t+h-1}, \varepsilon_{t+h-2}, \dots$, we conclude that

$$E\,\nu_t\nu_{t+h} = E\,\sigma_t^2(1 - \varepsilon_t^2)\sigma_{t+h}^2\,E(1 - \varepsilon_{t+h}^2) = 0.$$

Based on the previous results, the process y_t^2 satisfies an AR(1) model with autoregressive parameter α_1.

The ARCH(1) model can be readily extended to encompass a dependence of the conditional variance σ_t^2 on higher lags. The ARCH(r) process is defined by the discrete-time equation

$$y_t = \sigma_t \varepsilon_t, \tag{6.4}$$

$$\sigma_t^2 = \alpha_0 + \alpha_1 y_{t-1}^2 + \cdots + \alpha_r y_{t-r}^2. \tag{6.5}$$

In this case, the stationarity condition becomes

$$\alpha_1 + \cdots + \alpha_r < 1,$$

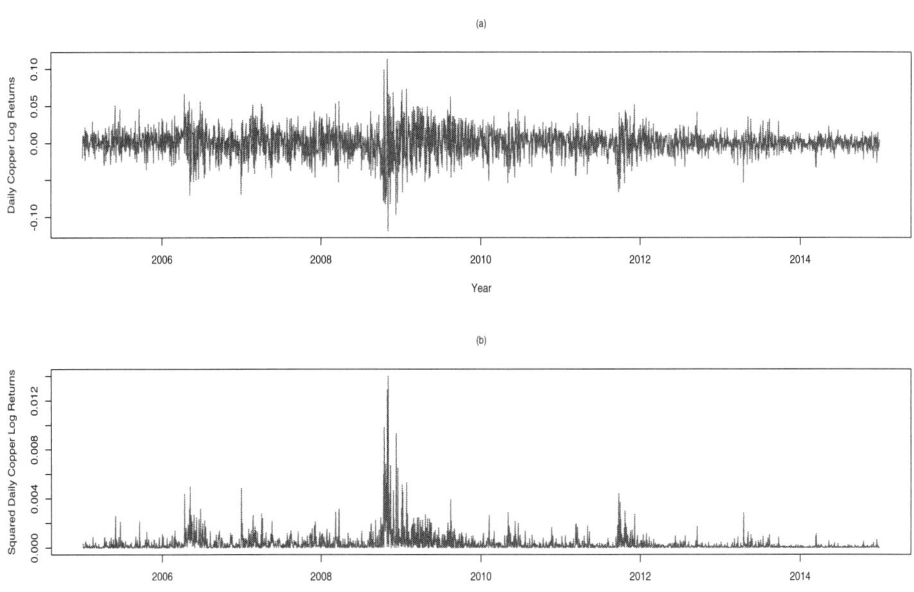

Figure 6.4 Copper price data (2005–2014): (a) Daily log-returns and (b) squared log-returns.

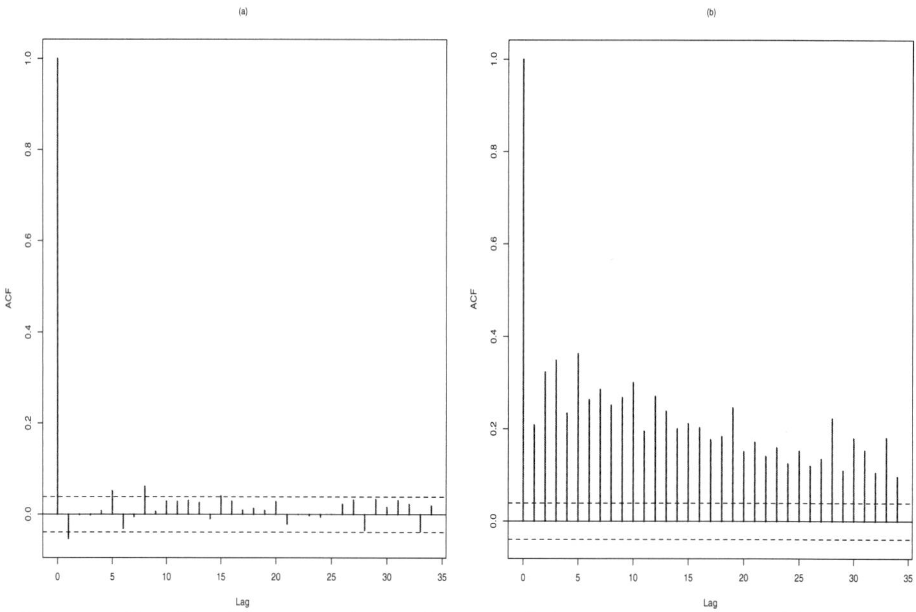

Figure 6.5 Sample autocorrelation function of the copper price data: (a) ACF of daily log-returns and (b) ACF of squared log-returns.

so that the variance of y_t is now given by

$$E\,y_t^2 = \frac{\alpha_0}{1 - \alpha_1^2 - \cdots - \alpha_r^2}.$$

Analogously to the ARCH(1) model, it can be readily shown that the squares of an ARCH(r) process y_t^2 satisfies the AR(r) model

$$y_t^2 - \mu = \alpha_1(y_{t-1}^2 - \mu) + \cdots + \alpha_r(y_{t-r}^2 - \mu) + \nu_t.$$

6.5 GARCH MODELS

A further extension of the ARCH models is the *generalized autoregressive conditionally heteroskedastic* GARCH(r, s) process is given by

$$\begin{aligned} y_t &= \sigma_t \varepsilon_t, \\ \sigma_t^2 &= \alpha_0 + \alpha_1 y_{t-1}^2 + \cdots + \alpha_r y_{t-r}^2 + \beta_1 \sigma_{t-1}^2 + \cdots + \beta_s \sigma_{t-s}^2, \end{aligned}$$

where $\sigma_t^2 = E[y_t^2 | y_{t-1}, y_{t-2}, \ldots]$ is the conditional variance of the process $\{y_t\}$, the GARCH coefficients $\alpha_0, \alpha_1, \ldots, \alpha_r$ and $\beta_1, \ldots, \beta_s$ are positive and $\{\epsilon_t\}$ is sequence of independent and identically distributed zero-mean and unit variance random variables. In order to be stationary, the coefficients of the

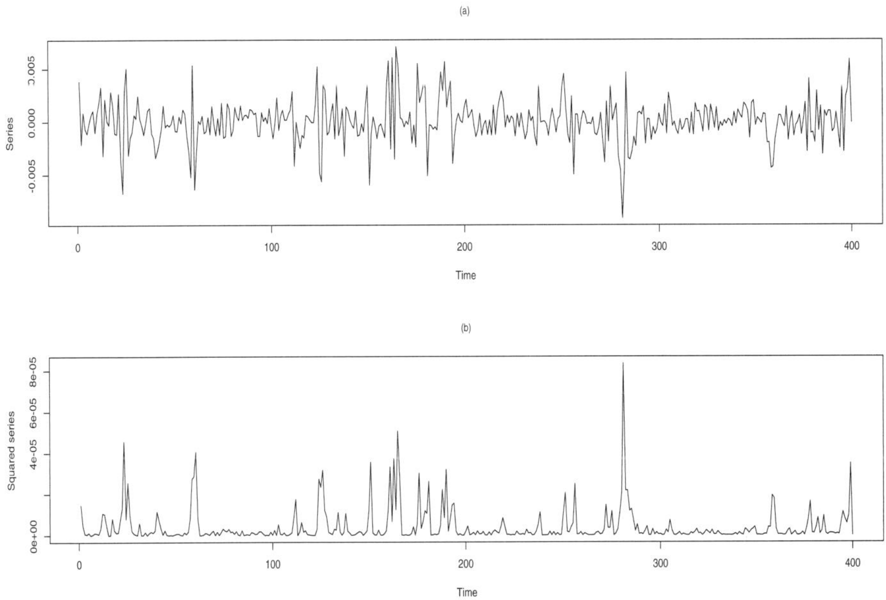

Figure 6.6 *Simulated ARCH(1) process with* $\alpha_1 = 0.9$. *(a) Series of 400 observations. (b) Squared series.*

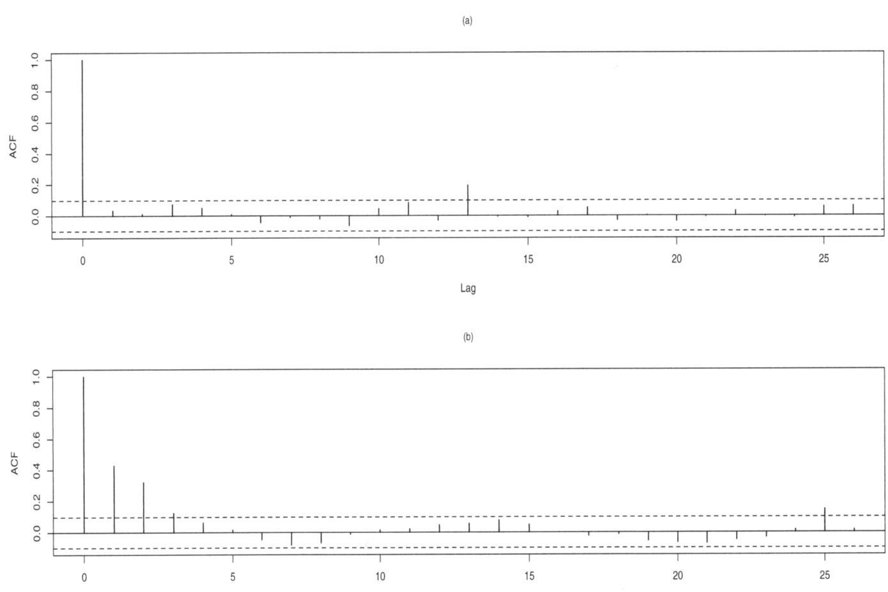

Figure 6.7 *Simulated ARCH(1) process with* $\alpha_1 = 0.9$. *(a) Sample ACF. (b) Sample ACF of the squared series.*

GARCH process must satisfy $\alpha_1 + \cdots + \alpha_r + \beta_1 + \cdots + \beta_s < 1$. Under this condition, the variance of the process can be written as

$$\operatorname{Var} y_t^2 = \frac{\alpha_0}{1 - \alpha_1^2 - \cdots - \alpha_r^2 - \beta_1^2 - \cdots - \beta_s^2}.$$

It has been observed in practice that the returns of a financial instrument may display a low level of autocorrelation and that the squared returns exhibit strong dependence. This empirical finding is usually reflected in the fact that the estimated parameters satisfy

$$\widehat{\alpha}_1^2 + \cdots + \widehat{\alpha}_r^2 + \widehat{\beta}_1^2 + \cdots + \widehat{\beta}_s^2 \sim 1,$$

indicating a near nonstationary behavior of the process due to a high level of autocorrelation of the squared returns.

In order to account for these phenomena, the GARCH models can be extended in two directions: introducing an ARMA or ARFIMA structure to model the autocorrelation of the returns and allowing a strong level of dependence in the squared returns by incorporating, for example, *integrated* GARCH processes. We begin the revision of these extension by defining the ARFIMA-GARCH models in the next section.

6.6 ARFIMA-GARCH MODELS

An ARFIMA(p, d, q)-GARCH(r, s) process is defined by the discrete-time equation

$$\phi(B) y_t = \theta(B)(1 - B)^{-d} \varepsilon_t, \tag{6.6}$$

$$\varepsilon_t = \epsilon_t \sigma_t, \tag{6.7}$$

$$\sigma_t^2 = \alpha_0 + \sum_{j=1}^{r} \alpha_j \varepsilon_{t-j}^2 + \sum_{j=1}^{s} \beta_j \sigma_{t-j}^2, \tag{6.8}$$

where $\sigma_t^2 = E[y_t^2 | y_{t-1}, y_{t-2}, \dots]$ is the conditional variance of the process $\{y_t\}$, the GARCH coefficients $\alpha_1, \dots, \alpha_r$ and $\beta_1, \dots, \beta_s$ are positive, $\sum_{j=1}^{r} \alpha_j + \sum_{j=1}^{s} \beta_j < 1$, and $\{\epsilon_t\}$ is sequence of independent and identically distributed zero-mean and unit variance random variables. Note that ϵ_t is assumed to be Gaussian, but in some cases it may be specified by a t-distribution or a double exponential distribution, among others. These distributions have a greater flexibility to accommodate a possible heavy tail behavior of some financial time series.

EXAMPLE 6.1

In order to explore the structure of the model described by (6.6)–(6.8), consider the ARFIMA(p,d,q)-GARCH$(1,1)$ process:

$$\begin{aligned} y_t &= \sum_{j=0}^{\infty} \psi_j \varepsilon_{t-j}, \\ \varepsilon_t &= \epsilon_t \sigma_t, \\ \sigma_t^2 &= \alpha_0 + \alpha_1 \varepsilon_{t-1}^2 + \beta_1 \sigma_{t-1}^2, \end{aligned}$$

where $\psi(B) = \phi(B)^{-1}\theta(B)(1-B)^{-d}$ and ϵ_t follows a standard normal distribution. Thus, we may write

$$\begin{aligned} \sigma_t^2 &= \alpha_0 + (\alpha_1 \epsilon_{t-1}^2 + \beta_1)\sigma_{t-1}^2, \\ &= \left[\prod_{k=1}^{n} (\alpha_1 \epsilon_{t-k}^2 + \beta_1)\right] \sigma_{t-1-n}^2 + \alpha_0 \left[1 + \sum_{k=0}^{n-1} \prod_{j=1}^{k+1} (\alpha_1 \epsilon_{t-j}^2 + \beta_1)\right]. \end{aligned}$$

Define the random variable $z_n = \sum_{k=1}^{n} \log(\alpha_1 \epsilon_{t-k}^2 + \beta_1)$ and let $\gamma_n = z_n/n$. By the strong law of the large numbers, $\gamma_n \to E[\log(\alpha_1 \epsilon_1^2 + \beta_1]$ almost surely as $n \to \infty$. This limit is called the *top Lyapunov exponent* of the process, γ.

If $\gamma < 0$, then we may write

$$\sigma_t^2 = \alpha_0 \left[1 + \sum_{k=0}^{\infty} \prod_{j=1}^{k+1} (\alpha_1 \epsilon_{t-j}^2 + \beta_1)\right]. \tag{6.9}$$

Consequently, the process y_t may be expressed as

$$y_t = \sqrt{\alpha_0} \sum_{j=0}^{\infty} \left\{ \psi_j \epsilon_{t-j} \left[1 + \sum_{k=0}^{\infty} \prod_{i=1}^{k+1} (\alpha_1 \epsilon_{t-j-i}^2 + \beta_1)\right]^{1/2} \right\}.$$

Thus, since $\{y_t\}$ corresponds to a transformation of the independent and identically distributed sequence $\{\epsilon_t\}$ the process $\{y_t\}$ is stationary. This result can be readily extended to the general model ARFIMA(p,d,q)-GARCH(r,s).

Observe that the conditional variance σ_t^2 specified by a GARCH(r,s) process may be expressed as an ARMA(p,r) model with $p = \max\{r,s\}$ as follows. Let $u_t = \sigma_t^2(\epsilon_t^2 - 1)$. This sequence is white noise, since $E[u_t] = 0$, $E[u_t^2] = E[\sigma_t^4(\epsilon_t^2-1)^2] = E[\sigma_t^4]\,E[(\epsilon_t^2-1)^2]$, and for $k > 0$ we have

$$E[u_t u_{t+k}] = E[E(u_t u_{t+k}|\mathcal{F}_{t+k-1})] = E[u_t \sigma_{t+k}^2\, E(\epsilon_{t+k}^2 - 1)] = 0.$$

Thus, σ_t^2 may be written as

$$(1 - \lambda_1 B - \cdots - \lambda_p B^p)\sigma_t^2 = \alpha_0 + \sum_{j=1}^{r} \alpha_j u_{t-j},$$

where $\lambda_j = \alpha_j \mathbf{1}_{\{1,\ldots,r\}}(j) + \beta_j \mathbf{1}_{\{1,\ldots,s\}}(j)$.

An approximate MLE $\widehat{\theta}$ for the ARFIMA-GARCH model is obtained by maximizing the conditional log-likelihood

$$\mathcal{L}(\theta) = -\frac{1}{2n} \sum_{t=1}^{n} \left[\log \sigma_t^2 + \frac{\varepsilon_t^2}{\sigma_t^2} \right]. \tag{6.10}$$

Let $\theta = (\theta_1, \theta_2)'$, where $\theta_1 = (\phi_1, \ldots, \phi_p, \theta_1, \ldots, \theta_q, d)'$ is the parameter vector involving the ARFIMA components and $\theta_2 = (\alpha_0, \ldots, \alpha_r, \beta_1, \ldots, \beta_s)'$ is the parameter vector containing the GARCH component. The following result establishes some asymptotic properties of this estimate: Let $\widehat{\theta}_n$ be the value that maximizes the conditional log-likelihood function (6.10). Then, under some regularity conditions, $\widehat{\theta}_n$ is a consistent estimate and $\sqrt{n}(\widehat{\theta}_n - \theta_0) \to N(0, \Omega^{-1})$, as $n \to \infty$, where $\Omega = \text{diag}(\Omega_1, \Omega_2)$ with

$$\Omega_1 = E\left[\frac{1}{\sigma_t^2} \frac{\partial \varepsilon_t}{\partial \theta_1} \frac{\partial \varepsilon_t}{\partial \theta_1'} + \frac{1}{2\sigma_t^4} \frac{\partial \sigma_t^2}{\partial \theta_1} \frac{\partial \sigma_t^2}{\partial \theta_1'} \right],$$

and

$$\Omega_2 = E\left[\frac{1}{2\sigma_t^4} \frac{\partial \sigma_t^2}{\partial \theta_2} \frac{\partial \sigma_t^2}{\partial \theta_2'} \right].$$

At this point, it is necessary to introduce the concept of *intermediate memory* which will be used in the next section. We say that a second-order stationary process has intermediate memory if for a large lag h its ACF behaves like $\gamma(h) \sim \ell(h)|h|^{2d-1}$ with $d < 0$, where $\ell(\cdot)$ is a slowly varying function. Thus, the ACF decays to zero at an hyperbolic rate but it is summable, that is,

$$\sum_{h=0}^{\infty} |\gamma(h)| < \infty.$$

6.7 ARCH(∞) MODELS

Given that the squares of many financial series have similar or greater level of autocorrelation than their returns, the memory reduction that affects the squares of an ARFIMA-GARCH process may not be adequate in practice. This circumstance leads us to explore other classes of processes to model the strong dependence of the squared returns directly. As an important example of this approach, consider the following ARCH(∞) model:

$$y_t = \sigma_t \epsilon_t, \tag{6.11}$$

$$\sigma_t^2 = \alpha_0 + \sum_{j=1}^{\infty} \alpha_j y_{t-j}^2, \tag{6.12}$$

where $\{\epsilon_t\}$ is a sequence of independent and identically distributed random variables with zero-mean and unit variance, α_0 is a positive constant, and $\alpha_j \geq 0$ for $j \geq 1$. This model can be formally written as

$$y_t^2 = \alpha_0 + \nu_t + \sum_{j=1}^{\infty} \alpha_j y_{t-j}^2, \tag{6.13}$$

where $\sigma_n^2 = E[y_t^2 | y_{t-1}, y_{t-2}, \dots]$, $\nu_t = y_t^2 - \sigma_t^2$ is a white noise sequence.

If $E[\epsilon_0^2 \sum_{j=0}^{\infty} \alpha_j] < 1$, then the conditional variance may be written in terms of a Volterra expansion

$$\sigma_t^2 = \alpha_0 \sum_{k=0}^{\infty} \sum_{j_1, \dots j_k = 1}^{\infty} \alpha_{j_1} \alpha_{j_2} \cdots \alpha_{j_k} \epsilon_{t-j_1}^2 \epsilon_{t-j_2}^2 \cdots \epsilon_{t-j_k}^2, \tag{6.14}$$

and the process $\{y_t\}$ may be expressed as

$$y_t = \epsilon_t \left[\alpha_0 \sum_{k=0}^{\infty} \sum_{j_1, \dots j_k = 1}^{\infty} \alpha_{j_1} \alpha_{j_2} \cdots \alpha_{j_k} \epsilon_{t-j_1}^2 \epsilon_{t-j_2}^2 \cdots \epsilon_{t-j_k}^2, \right]^{1/2}.$$

In particular, when the coefficients $\{\alpha_j\}$ in (6.13) are specified by an ARFIMA(p, d, q) model, the resulting expression defines the FIGARCH(p, d, q) model. If $\pi(B) = \phi(B)(1 - B)^d \theta(B)^{-1}$, then from (6.13) we get

$$\pi(B) y_t^2 = \alpha_0 + \nu_t.$$

Therefore, by multiplying both sides by $\theta(B)$ we conclude that

$$\phi(B)(1 - B)^d y_t^2 = \omega + \theta(B) \nu_t,$$

where $\omega = \theta(B)\alpha_0$. This process is strictly stationary and ergodic but not second-order stationary. On the other hand, writing this model in terms of the conditional variance as in (6.12) we have

$$\sigma_t^2 = \alpha_0 + [1 - \pi(B)] y_t^2. \tag{6.15}$$

From (6.11) we may write $\log(y_t^2) = \log(\sigma_t^2) + 2\log(|\epsilon_t|)$. Thus, by considering $\log(y_t^2)$ as the *observed* returns, it may seem natural to attempt to specify a long-memory model directly to the term $\log(\sigma_t^2)$ instead of σ_t^2. An advantage of this formulation is that $\log(\sigma_t^2)$ is allowed to be negative. Therefore, unlike the FIGARCH model, no additional conditions on the parameters are needed to ensure the positivity of σ_t^2.

An example of this type of processes is the *fractionally integrated exponential* GARCH (FIEGARCH) model specified by

$$\phi(B)(1-B)^d \log(\sigma_t^2) = \alpha + \theta(B)|\epsilon_{t-1}| + \lambda(B)\epsilon_{t-1}, \tag{6.16}$$

where $\phi(B) = 1 + \phi_1 B + \cdots + \phi_p B^p$, $\alpha \in \mathbb{R}$, $\theta(B) = \theta_1 + \cdots + \theta_q B^{q-1}$, and the polynomial $\lambda(B) = \lambda_1 + \cdots + \lambda_q B^{q-1}$ accounts for the *leverage effect*, that is, conditional variances may react distinctly to negative or positive shocks.

Consider the quasi-log-likelihood function

$$\mathcal{L}(\theta) = -\frac{1}{2}\log(2\pi) - \frac{1}{2}\sum_{t=1}^{n}\left[\log \sigma_t^2 + \frac{\varepsilon_t^2}{\sigma_t^2}\right], \tag{6.17}$$

where $\theta = (\omega, d, \phi_1, \ldots, \phi_p, , \theta_1, \ldots, \theta_q)$. A QMLE $\widehat{\theta}_n$ can be obtained by maximizing (6.17).

6.8 APARCH MODELS

Another model that incorporates asymmetry in the conditional variance is the so-called *asymmetric power autoregressive conditionally heteroskedastic* APARCH(r, s) process defined by

$$\begin{aligned} y_t &= \sigma_t \varepsilon_t, \\ \sigma_t^\delta &= \alpha_0 + \sum_{i=1}^{r} \alpha_i (|y_{t-i}| - \gamma_i y_{t-i})^\delta + \sum_{j=1}^{s} \beta_j \sigma_{t-j}^\delta. \end{aligned}$$

Notice that this model corresponds to an ARCH(r) process if $\delta = 2$, γ_i for $i = 1, \ldots, r$ and $\beta_j = 0$ for $j = 1, \ldots, s$. On the other hand, it is a GARCH(r, s) process if $\delta = 2$ and γ_i for $i = 1, \ldots, r$.

6.9 STOCHASTIC VOLATILITY

A *stochastic volatility* (SV) process is defined by the equations

$$r_t = \sigma_t \epsilon_t, \tag{6.18}$$

$$\sigma_t = \sigma \exp(v_t/2), \tag{6.19}$$

where $\{\epsilon_t\}$ is an independent and identically distributed sequence with zero-mean and unit variance, and $\{v_t\}$ is a stationary process independent of $\{\epsilon_t\}$. In particular, $\{v_t\}$ may be specified as a long-memory ARFIMA(p, d, q) process. The resulting process is called *long-memory stochastic volatility* (LMSV) model.

From (6.18), we may write

$$\begin{aligned}\log(r_t^2) &= \log(\sigma_t^2) + \log(\epsilon_t^2), \\ \log(\sigma_t^2) &= \log(\sigma^2) + v_t.\end{aligned}$$

Let $y_t = \log(r_t^2)$, $\mu = \log(\sigma^2) + E[\log(\epsilon_t^2)]$ and $\varepsilon_t = \log(\epsilon_t^2) - E[\log(\epsilon_t^2)]$. Then,

$$y_t = \mu + v_t + \varepsilon_t. \tag{6.20}$$

Consequently, the transformed process $\{y_t\}$ corresponds to a stationary long-memory process plus an additive noise. The ACF of (6.20) is given by

$$\gamma_y(h) = \gamma_v(k) + \sigma_\varepsilon^2 \delta_0(h),$$

where $\delta_0(h) = 1$ for $h = 0$ and $\delta_0(h) = 0$ otherwise. Furthermore, the spectral density of $\{y_t\}$, f_y, is given by

$$f_y(\lambda) = f_v(\lambda) + \frac{\sigma_\varepsilon^2}{2\pi},$$

where f_v is the spectral density of the long-memory process $\{v_t\}$.

In particular, if the process $\{v_t\}$ is an ARFIMA(p, d, q) model

$$\phi(B)v_t = \theta(B)(1 - B)^{-d}\eta_t, \tag{6.21}$$

and $\theta = (d, \sigma_\eta^2, \sigma_\varepsilon^2, \phi_1, \ldots, \phi_p, \theta_1, \ldots, \theta_q)'$ is the parameter vector that specifies model (6.21), then the spectral density is given by

$$f_\theta(\lambda) = \frac{\sigma_\eta^2}{2\pi} \frac{|\theta(e^{i\lambda})|^2}{|1 - e^{i\lambda}|^{2d} |\phi(e^{i\lambda})|^2} + \frac{\sigma_\varepsilon^2}{2\pi}.$$

The parameter θ can be estimated by minimizing the spectral likelihood

$$\mathcal{L}(\theta) = \frac{2\pi}{n} \sum_{j=1}^{n/2} \left[\log f_\theta(\lambda_j) + \frac{I(\lambda_j)}{f_\theta(\lambda_j)} \right]. \tag{6.22}$$

Let $\widehat{\theta}$ be the value that minimizes $\mathcal{L}(\theta)$ over the parameter space $\boldsymbol{\Theta}$. This estimator satisfies the following result: Assume that the parameter vector θ is an element of the compact parameter space $\boldsymbol{\Theta}$ and assume that $f_{\theta_1} = f_{\theta_2}$ implies that $\theta_1 = \theta_2$. Let θ_0 be the true parameter value. Then, $\widehat{\theta}_n \to \theta_0$ in probability as $n \to \infty$.

6.10 NUMERICAL EXPERIMENTS

The finite sample performance of the quasi maximum likelihood estimates based on the spectral-likelihood (6.22) is studied in this section by means of several Monte Carlo simulations.

Table 6.1 Estimation of Long-Memory Stochastic Volatility Models

d	$\widehat{d}$	$\widehat{\sigma}_\eta$	$\text{SD}(\widehat{d})$	$\text{SD}(\widehat{\sigma}_\eta)$
0.15	0.1453300	4.9880340	0.03209314	0.14602178
0.30	0.3058552	5.0067196	0.02877496	0.13827396
0.45	0.4606728	5.0364215	0.02695866	0.12792891

Table 6.2 Estimation of AR Stochastic Volatility Models

ϕ	$\widehat{\phi}$	$\widehat{\sigma}_\eta$	$\text{SD}(\widehat{\phi})$	$\text{SD}(\widehat{\sigma}_\eta)$
0.30	0.2965264	4.8788172	0.03762665	0.13678478
0.50	0.4892303	4.6579548	0.03076769	0.14218190
0.70	0.6808834	4.2779151	0.02269855	0.17127041

Table 6.3 Estimation of MA Stochastic Volatility Models

θ	$\widehat{\theta}$	$\widehat{\sigma}_\eta$	$\text{SD}(\widehat{\theta})$	$\text{SD}(\widehat{\sigma}_\eta)$
-0.30	-0.2966484	5.0985024	0.03354726	0.14166745
0.30	0.2883408	5.0785268	0.03571493	0.14252039
0.60	0.5396505	5.4969348	0.03396586	0.16060363

The models investigated are the LMSV with an ARFIMA$(0, d, 0)$ serial dependence structure, an AR(1) process and a MA(1) model. These processes have noise standard deviation $\sigma_\varepsilon = \pi/\sqrt{2}$ where ϵ_t follows a standard normal distribution, $\sigma_\eta = 5$, and the sample size is $n = 400$. Observe that $\varepsilon_t = \log \epsilon_t^2 - E[\log \epsilon_t^2]$. Thus, given that $\epsilon_t \sim N(0, 1)$ we have that $\text{Var}(\varepsilon_t) = \pi^2/2$.

The results displayed in Table 6.1 for the LMSV model, Table 6.2 for the AR(1) process and Table 6.3 for the MA(1) model correspond to the quasi maximum likelihood estimates for different values of the parameters. All the reported results are based on 1000 replications.

From these three tables, observe that the estimates of the long-memory parameter d, ϕ, θ and the scale parameter σ_η are close to their true values.

6.11 DATA APPLICATIONS

6.11.1 SP500 Data

The R library fGarch allows the fitting of GARCH models. As an illustration, the following output corresponds to the modeling of the SP500 data. The model was selected by taking into account the AIC, the significance of the parameters and the residuals diagnostics. The selected model corresponds to an AR(2) + GARCH(1, 1) process. From the output, note that the sum of the estimates α_1 and β_1 is 0.9937, that is, very close to the stationarity boundary, anticipated in Section 3 about the stylized facts in financial time series.

```
Title:
 GARCH Modeling

Call:
 garchFit(formula = ~arma(2, 0) + garch(1, 1), data = z,
       include.mean = FALSE,   trace = FALSE)

Mean and Variance Equation:
 data ~ arma(2, 0) + garch(1, 1)

Conditional Distribution:
 norm

Coefficient(s):
        ar1          ar2        omega       alpha1        beta1
 1.0405e-01  -2.3687e-02   7.7258e-07   8.1200e-02   9.1282e-01

Std. Errors:
 based on Hessian

Error Analysis:
        Estimate  Std. Error  t value Pr(>|t|)
ar1     1.040e-01   8.510e-03   12.226  < 2e-16 ***
ar2    -2.369e-02   8.371e-03   -2.829  0.00466 **
omega   7.726e-07   9.577e-08    8.067 6.66e-16 ***
alpha1  8.120e-02   4.225e-03   19.219  < 2e-16 ***
beta1   9.128e-01   4.448e-03  205.207  < 2e-16 ***
---
Signif. codes:  0 '***' 0.001 '**' 0.01 '*' 0.05 '.' 0.1 ' ' 1

Log Likelihood:
 54827.28    normalized:  3.404787
```

```
Standardised Residuals Tests:
                                Statistic p-Value
 Jarque-Bera Test   R    Chi^2  13289.56  0
 Shapiro-Wilk Test  R    W      NA        NA
 Ljung-Box Test     R    Q(10)  13.16941  0.2143535
 Ljung-Box Test     R    Q(15)  17.61137  0.2836476
 Ljung-Box Test     R    Q(20)  22.62202  0.3077331
 Ljung-Box Test     R^2  Q(10)  16.54757  0.08499406
 Ljung-Box Test     R^2  Q(15)  19.55958  0.1894863
 Ljung-Box Test     R^2  Q(20)  23.53524  0.2632833
 LM Arch Test       R    TR^2   17.63158  0.1273434

Information Criterion Statistics:
      AIC       BIC       SIC      HQIC
-6.808952 -6.806566 -6.808952 -6.808163
```

6.11.2 Gold Data

This section study the monthly gold prices for the period starting on January 1978 through September 2014. These prices correspond to US dollars per troy Oz. are exhibited in Figure 6.8. From this plot, notice the sharp rise of the prices around 2000. The log returns are displayed in Figure 6.9 while the squared returns are exhibited in Figure 6.10. In order to account for serial dependence of returns and squared returns, a class of ARMA-GARCH model is proposed for these monthly data.

The R package fGarch allows to fit these models. The selected model has ARMA$(1,1)$ dependence structure for the returns and a GARCH$(1,1)$ dependence structure for the conditional variances. Notice that all the parameters are statistically significant at the 5% level.

```
Title:
 GARCH Modeling

Call:
 garchFit(formula = ~arma(1, 1) + garch(1, 1), data = z, trace = FALSE)

Mean and Variance Equation:
 data ~ arma(1, 1) + garch(1, 1)
 [data = z]

Conditional Distribution:
 norm

Coefficient(s):
        mu         ar1          ma1       omega      alpha1       beta1
 2.1486e-03  -4.8625e-01  7.0064e-01  7.6059e-05  1.8437e-01 7.8914e-01
```

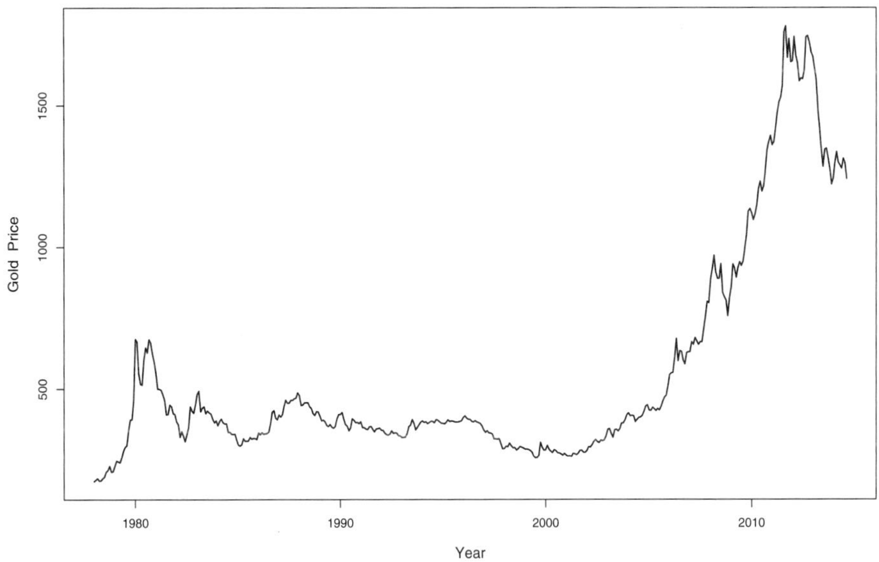

Figure 6.8 *Gold monthly prices, January 1978 to September 2014, US Dollars per Troy Oz.*

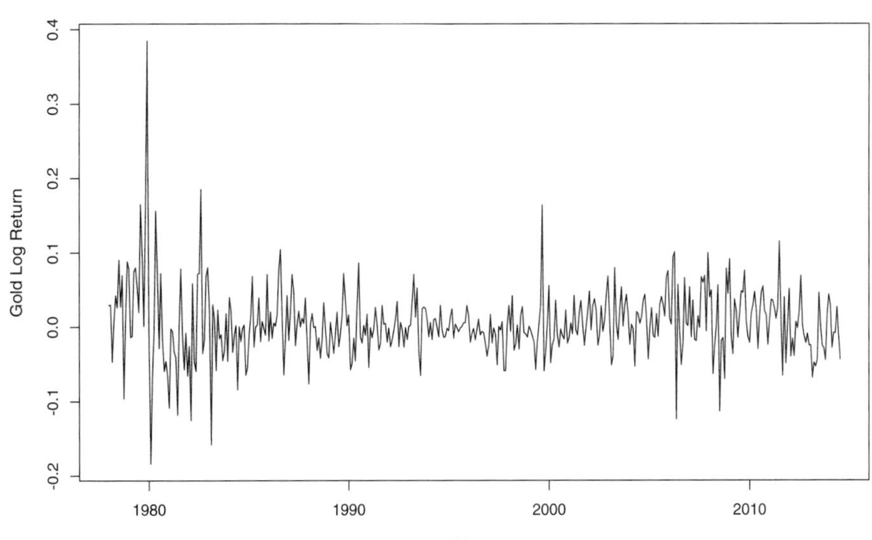

Figure 6.9 *Gold monthly log returns, January 1978 to September 2014.*

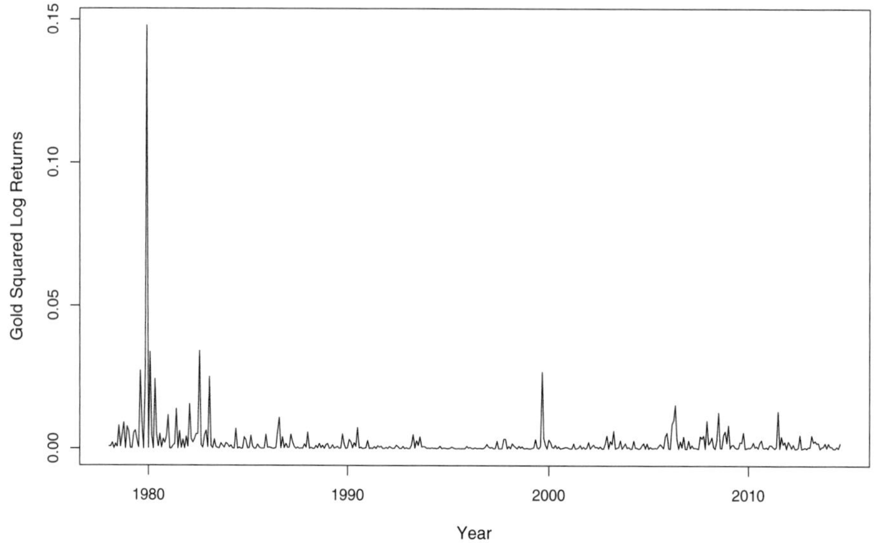

Figure 6.10 *Gold squared log returns, January 1978 to September 2014.*

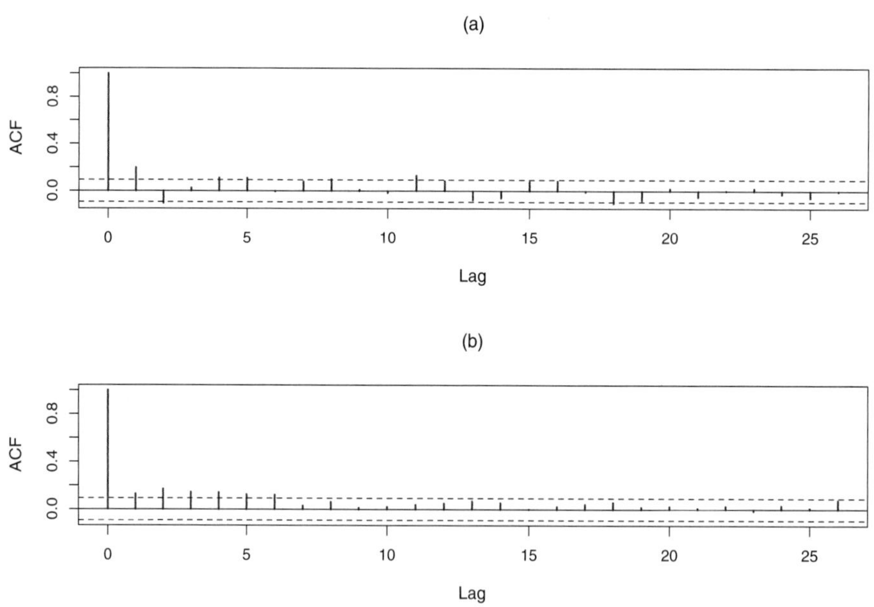

Figure 6.11 *Gold monthly log returns. (a) Sample ACF of y_t. (b) Sample ACF of y_t^2*

```
Std. Errors:
 based on Hessian

Error Analysis:
         Estimate  Std. Error  t value Pr(>|t|)
mu       2.149e-03  2.849e-03    0.754  0.45081
ar1     -4.863e-01  1.718e-01   -2.830  0.00466 **
ma1      7.006e-01  1.399e-01    5.008  5.5e-07 ***
omega    7.606e-05  3.556e-05    2.139  0.03245 *
alpha1   1.844e-01  4.817e-02    3.827  0.00013 ***
beta1    7.891e-01  4.787e-02   16.486  < 2e-16 ***
---
Signif. codes:  0 '***' 0.001 '**' 0.01 '*' 0.05 '.' 0.1 ' ' 1

Log Likelihood:
 797.8381    normalized:  1.813269

Description:
 Mon Feb  9 00:17:39 2015 by user:


Standardised Residuals Tests:
                                Statistic p-Value
 Jarque-Bera Test   R    Chi^2  104.7054  0
 Shapiro-Wilk Test  R    W      0.9744005 5.552453e-07
 Ljung-Box Test     R    Q(10)  13.33065  0.2057678
 Ljung-Box Test     R    Q(15)  23.08554  0.08234047
 Ljung-Box Test     R    Q(20)  25.57881  0.1801772
 Ljung-Box Test     R^2  Q(10)  15.77544  0.1062436
 Ljung-Box Test     R^2  Q(15)  18.89253  0.2186455
 Ljung-Box Test     R^2  Q(20)  20.06762  0.4537067
 LM Arch Test       R    TR^2   18.7434   0.09491198

Information Criterion Statistics:
      AIC       BIC       SIC      HQIC
-3.599264 -3.543536 -3.599630 -3.577279
```

On the other hand, Figure 6.12 displays the sample ACF of the standardized residuals as well as the squared standardized residuals. Notice that the serial dependence is not statistical significant at the 5% level, cf. with the Ljung -Box test for the standardized residual with p-value of 0.082 and the corresponding test for the squared standardized residuals with p-value of 0.1062 both considering 10 lags. Similar results are found when considering 20 lags.

Figure 6.13 reports two year ahead monthly forecasts and 95% predictions bands while a zoom to the last 40 predictions and prediction bands are displayed in Figure 6.14.

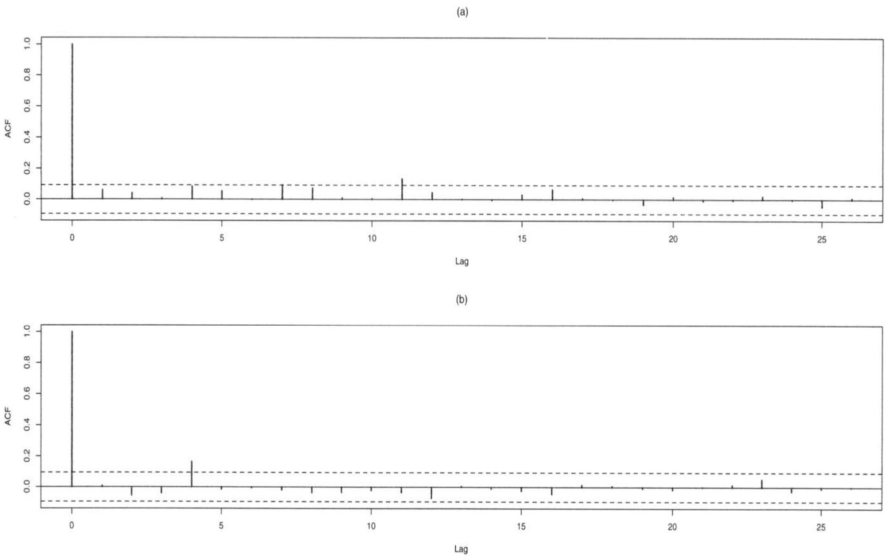

Figure 6.12 *Gold monthly log returns fitted ARMA-GARCH model. (a) Sample ACF of standardized residuals. (b) Sample ACF of squared standardized residuals.*

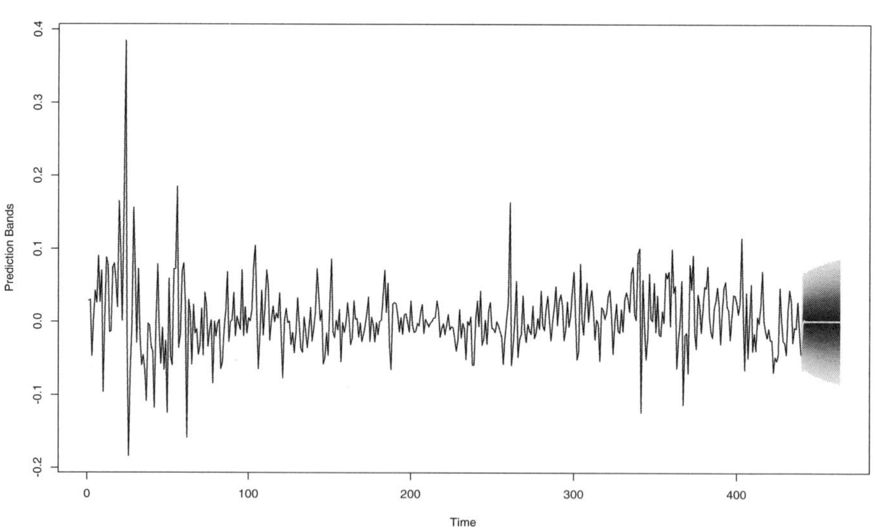

Figure 6.13 *Gold monthly log returns ARMA-GARCH model: Data and two year ahead monthly forecasts and 95% prediction bands.*

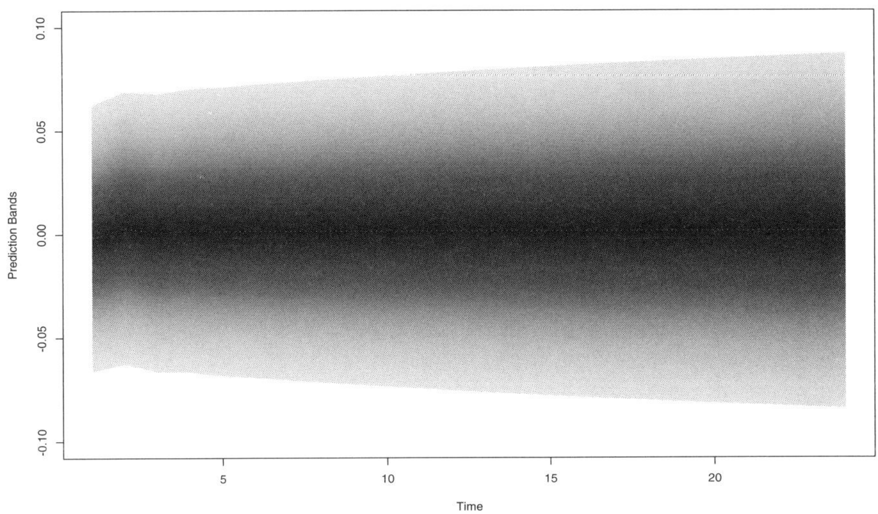

Figure 6.14 *Gold monthly log returns fitted ARMA-GARCH model: Two year ahead monthly forecasts and 95% prediction bands.*

6.11.3 Copper Data

The following example examines the daily evolution of copper prices for the ten year period from January 4, 2005 to December 31, 2014. These prices, expressed in terms of USD cents per pound, are plotted in Figure 6.15 while the corresponding log returns are exhibited in Figure 6.16. Furthermore, the evolution of the squared returns are displayed fin Figure 6.17. From these plots, notice the big drop in copper prices by the end of 2008 and their recovery starting in 2009. These big fluctuations are well represented in Figure 6.17 showing the high volatility around that period. On the other hand, the autocorrelation structure of the returns and the squared returns are exhibited in Figure 6.18.

As in the previous case of the gold data, here we use the R package fGarch to fit a family of ARMA-GARCH or ARMA-APARCH models to these daily copper prices data. We have fitted both types of models to the data to see whether a asymmetry is detected in this case. The selected models have ARMA$(1,0)$ dependence structure for the returns and a GARCH$(1,1)$ or APARCH$(1,1)$ dependence structure for the conditional variances.

From the outputs shown below, notice that leverage of the ARMA$(1,0)$-APARCH$(1,1)$ model is statically significant at the 5% level.

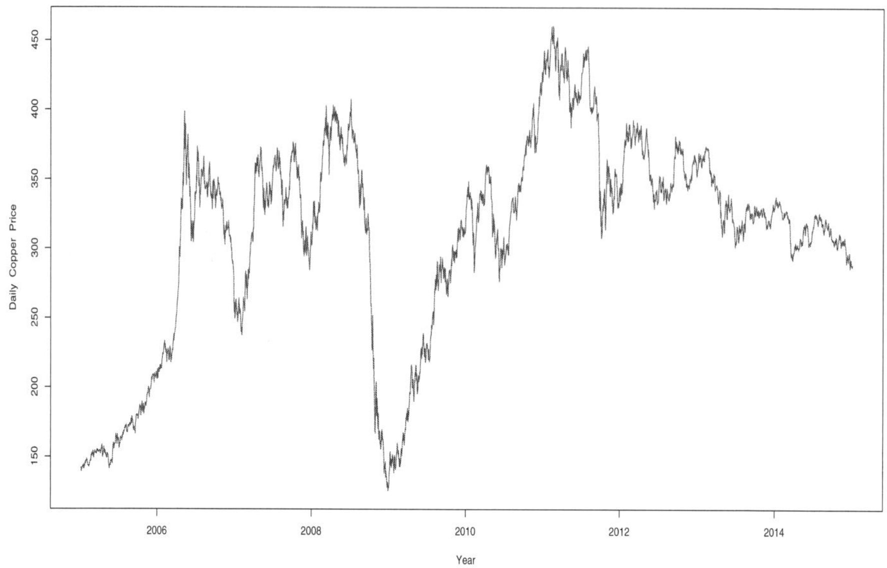

Figure 6.15 *Daily copper price, January 4, 2005 to December 31, 2014. Nominal USD cents per pound.*

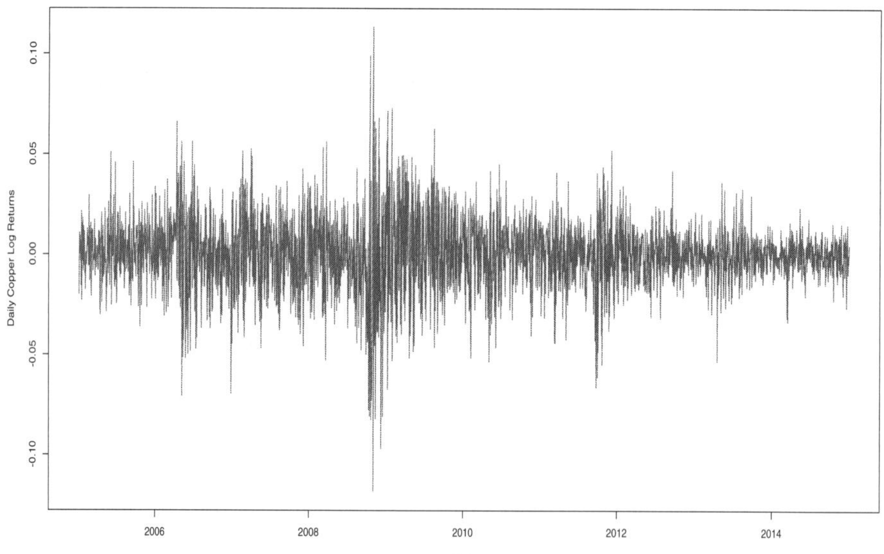

Figure 6.16 *Daily copper log returns, January 4, 2005 to December 31, 2014.*

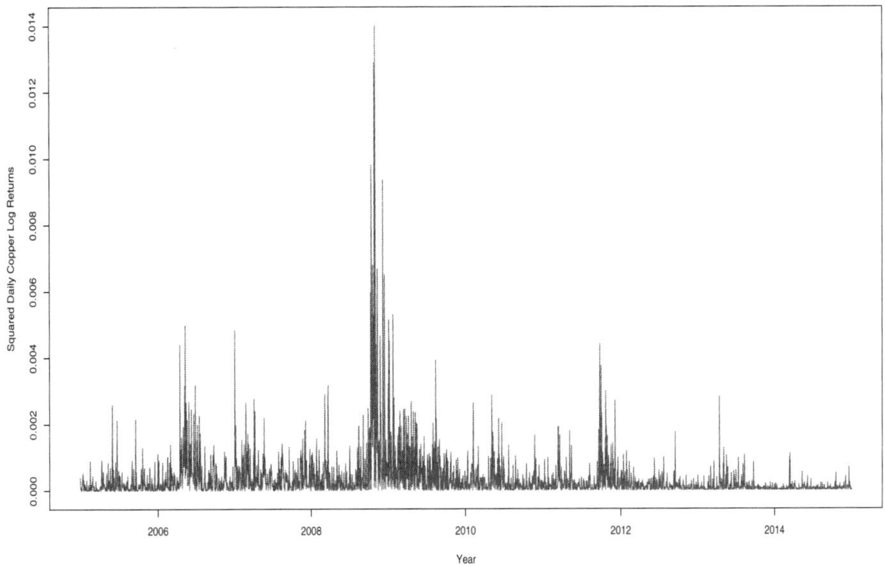

Figure 6.17 *Daily copper squared log returns, January 4, 2005 to December 31, 2014.*

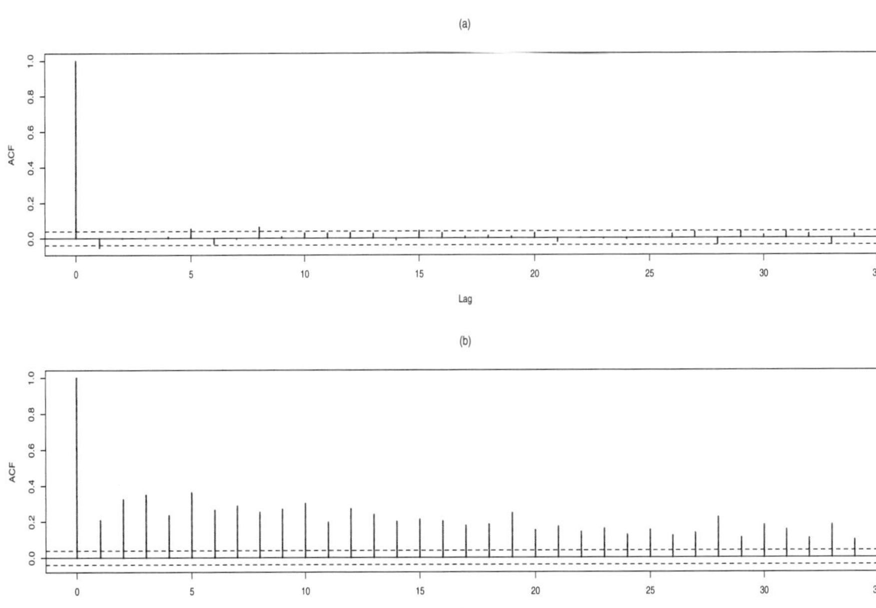

Figure 6.18 *Sample ACF. (a) Daily log returns, (b) Squared returns.*

Model without Leverage

```
Title:
 GARCH Modeling

Call:
 garchFit(formula = ~arma(1, 0) + garch(1, 1), data = z,
 trace = FALSE)

Mean and Variance Equation:
 data ~ arma(1, 0) + garch(1, 1)
 [data = z]

Conditional Distribution:
 norm

Coefficient(s):
         mu          ar1         omega        alpha1         beta1
 2.5867e-04  -6.7762e-02   1.9980e-06   7.3973e-02   9.2140e-01

Std. Errors:
 based on Hessian

Error Analysis:
         Estimate  Std. Error  t value Pr(>|t|)
mu      2.587e-04   2.727e-04    0.948  0.34288
ar1    -6.776e-02   2.065e-02   -3.282  0.00103 **
omega   1.998e-06   6.868e-07    2.909  0.00362 **
alpha1  7.397e-02   9.505e-03    7.782 7.11e-15 ***
beta1   9.214e-01   9.717e-03   94.823  < 2e-16 ***
---
Signif. codes:  0 '***' 0.001 '**' 0.01 '*' 0.05 '.' 0.1 ' ' 1

Log Likelihood:
 6857.303    normalized:  2.71684

Standardised Residuals Tests:
                                Statistic p-Value
 Jarque-Bera Test   R    Chi^2  89.50048  0
 Shapiro-Wilk Test  R    W      0.994573  5.239038e-08
 Ljung-Box Test     R    Q(10)  19.83763  0.03082641
 Ljung-Box Test     R    Q(15)  22.152    0.1038799
 Ljung-Box Test     R    Q(20)  28.05984  0.1079903
 Ljung-Box Test     R^2  Q(10)  7.296483  0.69719
 Ljung-Box Test     R^2  Q(15)  9.462268  0.8521343
 Ljung-Box Test     R^2  Q(20)  15.32643  0.7574342
 LM Arch Test       R    TR^2   8.271934  0.763535

Information Criterion Statistics:
```

```
      AIC       BIC       SIC      HQIC
-5.429717 -5.418161 -5.429725 -5.425524
```

Model with Leverage

```
Title:
 GARCH Modeling

Call:
 garchFit(formula = ~arma(1, 0) + aparch(1, 1), data = z,
 trace = FALSE)

Mean and Variance Equation:
 data ~ arma(1, 0) + aparch(1, 1)
 [data = z]

Conditional Distribution:
 norm

Coefficient(s):
         mu          ar1        omega       alpha1
 7.3223e-05  -6.6990e-02   1.5967e-06   6.7513e-02

       gamma1        beta1        delta
   1.3569e-01   9.2796e-01   2.0000e+00


Std. Errors:
 based on Hessian

Error Analysis:
         Estimate  Std. Error  t value Pr(>|t|)
mu      7.322e-05   2.767e-04    0.265  0.79131
ar1    -6.699e-02   2.072e-02   -3.234  0.00122 **
omega   1.597e-06   7.891e-07    2.023  0.04302 *
alpha1  6.751e-02   1.234e-02    5.471 4.47e-08 ***
gamma1  1.357e-01   5.034e-02    2.696  0.00702 **
beta1   9.280e-01   9.239e-03  100.435  < 2e-16 ***
delta   2.000e+00   4.080e-01    4.902 9.47e-07 ***
---
Signif. codes:  0 '***' 0.001 '**' 0.01 '*' 0.05 '.' 0.1 ' ' 1

Log Likelihood:
 6862.59    normalized:  2.718934

Standardised Residuals Tests:
                                Statistic p-Value
 Jarque-Bera Test   R    Chi^2  69.49289  7.771561e-16
 Shapiro-Wilk Test  R    W      0.9954016 4.959389e-07
 Ljung-Box Test     R    Q(10)  19.53039  0.03402065
```

```
 Ljung-Box Test     R    Q(15)  21.81076  0.1128628
 Ljung-Box Test     R    Q(20)  27.66829  0.1174872
 Ljung-Box Test     R^2  Q(10)  7.915388  0.6371015
 Ljung-Box Test     R^2  Q(15)  10.32527  0.7987958
 Ljung-Box Test     R^2  Q(20)  17.38353  0.6279475
 LM Arch Test       R    TR^2   9.140525  0.6908832

Information Criterion Statistics:
      AIC       BIC       SIC      HQIC
-5.432322 -5.416143 -5.432337 -5.426451
```

6.12 VALUE AT RISK

A usual criticism of variance or conditional variance as measures of financial risk is that they do not discriminate whether the return is positive or negative. Of course, for the investor the sign of the return makes a big difference. As a consequence, other methods for assessing investment risk have been developed. One of these techniques is the *Value at Risk*, denoted as VaR hereafter. This concept tries to measure the amount of capital that one investor can loose when exposed to a financial instrument. Let z_α be the value satisfying the equation

$$\mathbb{P}(y_t \leq z_\alpha) = \alpha,$$

where α is the probability of the left tail of the return distribution. Based on this expression, the Value at Risk of a financial instrument is given by

$$\text{VaR} = C\, z_\alpha\, \sigma_t,$$

where C is the invested capital and σ_t is the conditional standard deviation at time t.

When we are interested in evaluating the Value at Risk at a h-step horizon, the return of the financial instrument is given by

$$y_t[h] = y_{t+1} + y_{t+2} + \cdots + y_{t+h}.$$

Consequently,

$$\text{Var}(y_t[h]|\mathcal{F}_t) = \text{Var}(y_{t+1} + y_{t+2} + \cdots + y_{t+h}|\mathcal{F}_t).$$

Now, given that the sequence y_t is white noise, we conclude that

$$\text{Var}(y_t[h]|\mathcal{F}_t) = \text{Var}(y_{t+1}|\mathcal{F}_t) + \text{Var}(y_{t+2}|\mathcal{F}_t) + \cdots + \text{Var}(y_{t+h}|\mathcal{F}_t).$$

Consequently,

$$\text{Var}(y_t[h]|\mathcal{F}_t) = \sum_{j=1}^{h} \widehat{\sigma}^2_{t+j}.$$

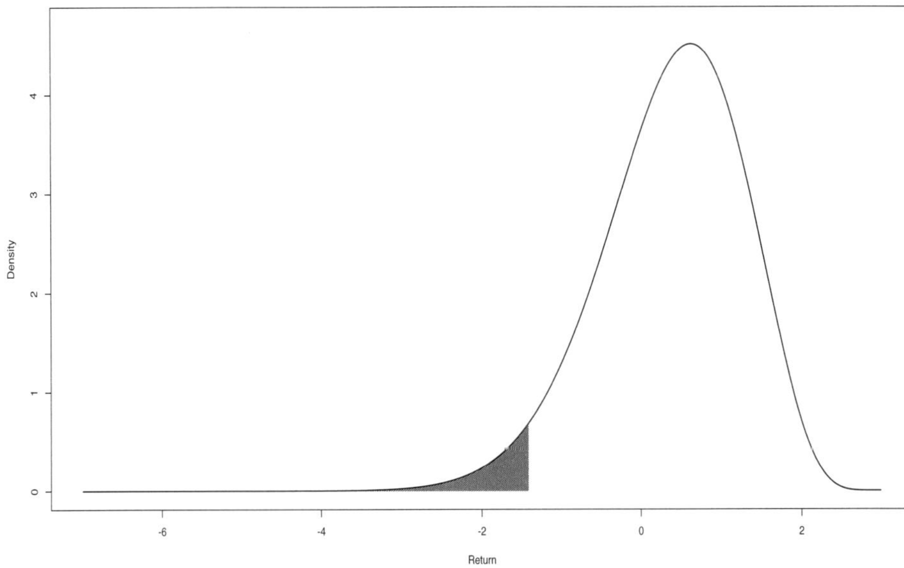

Figure 6.19 Value at risk.

Observe that for the IGARCH model $\widehat{\sigma}^2_{t+j} = \sigma^2_{t+1}$ so that

$$\mathrm{Var}(y_t[h]|\mathcal{F}_t) = h\,\sigma^2_{t+1}.$$

Therefore, the Value at Risk at horizon h is given by

$$\mathrm{VaR[h]} = C\,z_\alpha\,\sqrt{h}\,\sigma_{t+1}.$$

■ **EXAMPLE 6.2**

Consider the daily IPSA stock index, from September 1, 2002 to October 14, 2014, see Appendix C for details. In this case, the fitted model is given by the following output

```
Title:
 GARCH Modeling

Mean and Variance Equation:
 data ~ arma(0, 1) + garch(1, 1)

Conditional Distribution:
 norm
```

```
Coefficient(s):

         mu          ma1        omega       alpha1        beta1
7.1960e-04  1.8009e-01  2.8536e-06  1.4620e-01  8.2756e-01

Std. Errors:
 based on Hessian

Error Analysis:
        Estimate  Std. Error  t value Pr(>|t|)
mu     7.196e-04   1.587e-04    4.533 5.80e-06 ***
ma1    1.801e-01   1.848e-02    9.743  < 2e-16 ***
omega  2.854e-06   5.451e-07    5.235 1.65e-07 ***
alpha1 1.462e-01   1.443e-02   10.130  < 2e-16 ***
beta1  8.276e-01   1.589e-02   52.071  < 2e-16 ***
---
Signif. codes:  0 '***' 0.001 '**' 0.01 '*' 0.05 '.' 0.1 ' ' 1

Log Likelihood:
 10634.15    normalized:  3.3315

Standardised Residuals Tests:
                                Statistic p-Value
 Jarque-Bera Test   R    Chi^2  89.30602  0
 Shapiro-Wilk Test  R    W      0.9948589 3.945871e-09
 Ljung-Box Test     R    Q(10)  12.21397  0.2709916
 Ljung-Box Test     R    Q(15)  14.74627  0.4698436
 Ljung-Box Test     R    Q(20)  23.24122  0.2771061
 Ljung-Box Test     R^2  Q(10)  12.52548  0.2514255
 Ljung-Box Test     R^2  Q(15)  15.00001  0.4514162
 Ljung-Box Test     R^2  Q(20)  18.51519  0.5535096
 LM Arch Test       R    TR^2   13.70263  0.3200994

Information Criterion Statistics:
      AIC       BIC       SIC      HQIC
-6.659867 -6.650362 -6.659872 -6.656459
```

The estimated conditional standard deviation at the end of the period is $\sigma_t = 0.01030473$. For an investment of \$1,000,000, the Value at Risk for horizons $h = 1, \ldots, 40$ is shown in Figure 6.20.

Figure 6.21 shows GARCH parameters estimates based on windows of size 400 observations and 20 values shifts. The dotted line indicates β_t, the broken line corresponds to α_t while the heavy lines is the sum $\alpha_t + \beta_t$. Note that these values strongly decay around the year 2010.

As a consequence of the possible changes in the parameters of the GARCH model, the estimates of the volatility also change. Figure 6.22 exhibits this phenomenon. This plot shows the estimates of σ_t arising

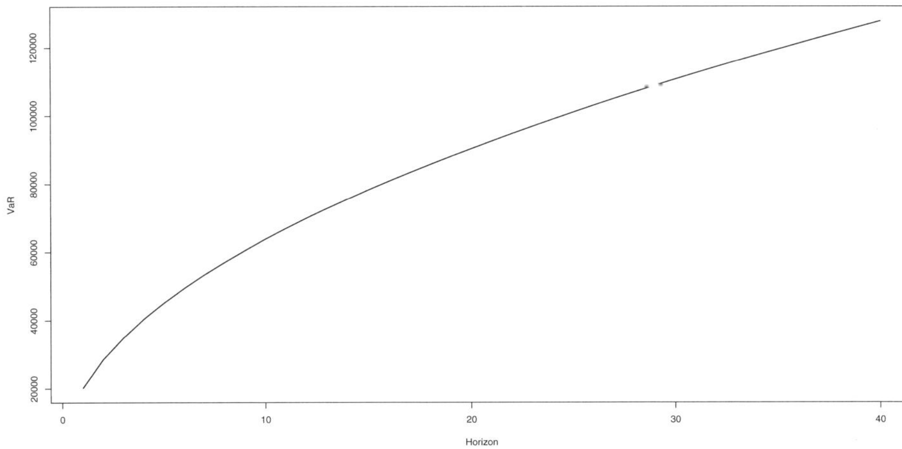

Figure 6.20 IPSA Stock Index: Value at Risk (VaR) for horizons $h = 1, \ldots, 40$.

from the sequence of windows (gray lines) and their estimates based on a single GARCH model (black line).

Furthermore, Figure 6.23 displays the associated Value at Risk estimates for both, the time-varying models indicated by the gray line as well as the fixed model denotes by the black line.

Table 6.4 IPSA Stock Index Value at Risk.

Horizon	VaR	Horizon	VaR
1	20,197.26	21	92,555.48
2	28,563.24	22	94,733.56
3	34,982.68	23	96,862.67
4	40,394.52	24	98,945.97
5	45,162.45	25	100,986.31
6	49,472.99	26	102,986.23
7	53,436.93	27	104,948.05
8	57,126.48	28	106,873.86
9	60,591.79	29	108,765.58
10	63,869.35	30	110,624.96
11	66,986.74	31	112,453.60
12	69,965.37	32	114,252.97
13	72,822.26	33	116,024.44
14	75,571.23	34	117,769.26
15	78,223.66	35	119,488.61
16	80,789.05	36	121,183.57
17	83,275.44	37	122,855.15
18	85,689.73	38	124,504.28
19	88,037.82	39	126,131.86
20	90,324.90	40	127,738.70

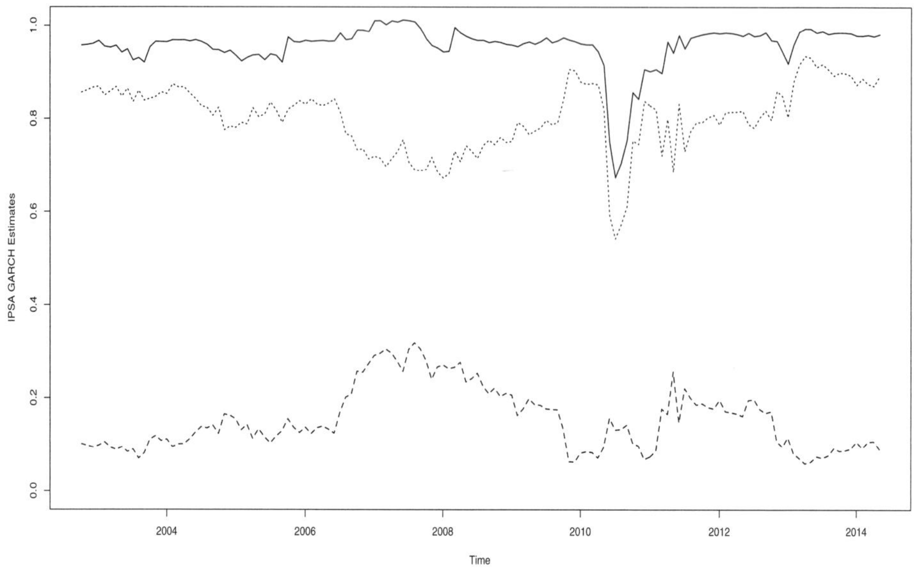

Figure 6.21 IPSA Stock Index: Parameter estimates.

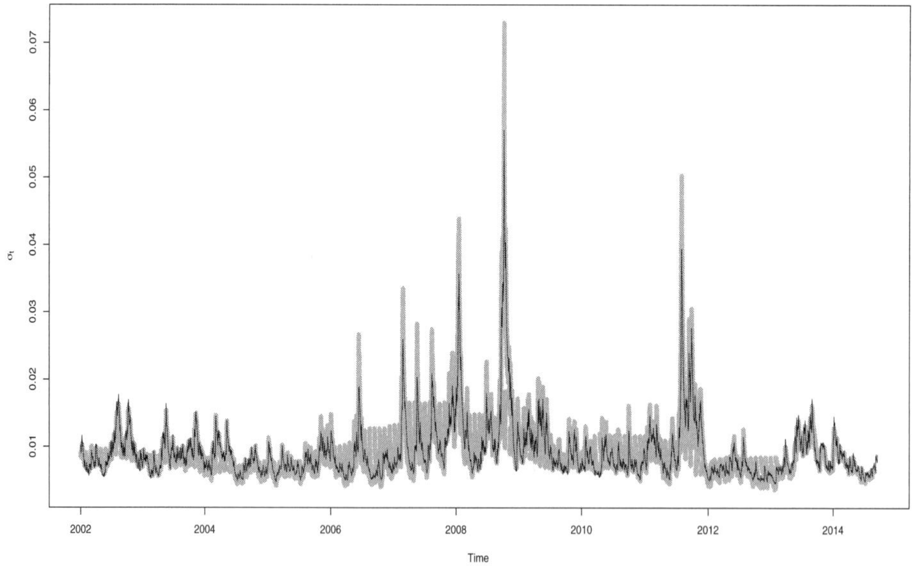

Figure 6.22 IPSA Stock Index: Volatility estimates, σ_t.

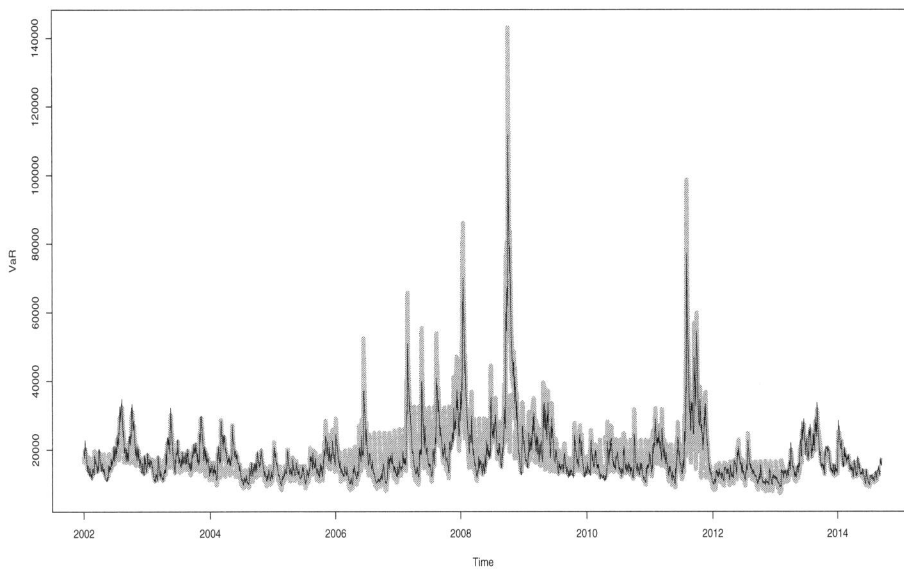

Figure 6.23 IPSA Stock Index: Value at Risk (VaR) estimates.

6.13 AUTOCORRELATION OF SQUARES

In this section we examine the autocorrelation of square transformation of a stationary process with a Wold expansion. The analysis of such transformations may give valuable clues about crucial aspects such as linearity, normality, or memory of the process. For instance, these issues are particularly important when studying the behavior of heteroskedastic processes since in this context the series usually represents the return of a financial instrument and the squared series is a rough empirical measure of its volatility. In this case, since we are interested in predicting both the returns and the volatility, we must analyze the dependence structure of a time series and its squares.

6.13.1 Squares of Gaussian Processes

Let $\{y_t : t \in \mathbb{Z}\}$ be a Gaussian process with $E[y_t] = 0$ and $\text{Var}[y_t) = 1$. For the transformation $f(y_t) = y_t^2$, the coefficients of the Hermite expansion are $\alpha_0 = 1$, $\alpha_2 = 1$, and $\alpha_j = 0$ for all $j \neq 0, 2$. Thus, we have $\langle f(y_t), f(y_s) \rangle = 1 + \rho_y^2(h)/2$. But, $E(y_t^2) = 1$ so that

$$\text{Cov}[f(y_t), f(y_s)] = \langle f(y_t), f(y_s) \rangle - 1 = \rho_y^2(h)/2,$$

and then the autocorrelation function of $f(y_t) = y_t^2$ is

$$\rho_{y^2}(h) = \rho_y^2(h). \tag{6.23}$$

From this expression we observe that since $|\rho_y| \leq 1$, $\rho_{y^2}(h)$ is smaller or equal than $\rho_y(h)$. Consequently, the autocorrelation of the squares is smaller than the autocorrelation of the original series. Actually, for a Gaussian process this reduction of the dependence is true for any transformation, as stated in the following result: Let $\{y_t : t \in \mathbb{Z}\}$ be a Gaussian process and let $\mathcal{F}$ be the class of all measurable transformations such that $E[f(y_t)] = 0$ and $E[f(y_t)^2] = 1$. Then,

$$\sup_{f \in \mathcal{F}} E[f(y_t)f(y_s)] = |\rho_y(t-s)|,$$

where $\rho_y(t-s)$ is the correlation between y_t and y_s.

As a consequence of our previous discussion, in order to account for situations where the squares exhibit more dependence than the series itself, we must abandon Gaussianity. In the next section we examine this issue in detail.

6.13.2 Autocorrelation of Squares

Consider the regular linear process $\{y_t\}$ with Wold expansion

$$y_t = \psi(B)\varepsilon_t, \tag{6.24}$$

where $\psi(B) = \sum_{i=0}^{\infty} \psi_i B^i$, $\psi_0 = 1$, and $\sum_{i=0}^{\infty} \psi_i^2 < \infty$. The *input noise* sequence $\{\varepsilon_t\}$ is assumed to be white noise.

(Linear Process) Assume that $\{\varepsilon_t\}$ are independent identically distributed random variables with zero-mean and finite kurtosis η. Then,

$$\rho_{y^2}(h) = \frac{2}{\kappa - 1}\rho_y^2(h) + \frac{\kappa - 3}{\kappa - 1}\alpha(h), \tag{6.25}$$

where κ is the kurtosis of y_t given by

$$\kappa = (\eta - 3)\left(\sum_{i=0}^{\infty} \psi_i^2\right)^{-2} \sum_{i=0}^{\infty} \psi_i^4 + 3. \tag{6.26}$$

Furthermore, if y_t is Gaussian, then $\eta = 3$ and $\kappa = 3$. Therefore, $\rho_{y^2} = \rho_y^2$, which coincides with formula (6.23).

■ **EXAMPLE 6.3**

Consider the following AR(1)-ARCH(1) process described by the equations

$$\begin{aligned} y_t &= \phi y_{t-1} + \varepsilon_t, \\ \varepsilon_t &= \epsilon_t \sigma_t, \\ \sigma_t^2 &= \alpha_0 + \beta \varepsilon_{t-1}^2, \end{aligned}$$

where ϵ_t is sequence of independent and identically distributed random variables with distribution $N(0,1)$. The autocorrelation function of $\{y_t^2\}$ is given by

$$\rho_{y^2}(h) = \phi^{2|h|}\left[1 + \frac{\eta-1}{\kappa-1}\Delta(0)[\beta^{|h|}-1]\right].$$

In this case, $\rho_{y^2}(h) = \mathcal{O}(\phi^{2|h|})$ and therefore the squared process $\{y_t^2\}$ has short memory.

6.13.3 Illustrations

Figure 6.24 shows the sample ACF of a series of 1000 observations from a Gaussian ARFIMA$(0,d,0)$ process with $d = 0.4$ and the sample ACF of the squares. Since this is a Gaussian process, from formula (6.23) we expect that $\widehat{\rho}_{y^2} \sim \widehat{\rho}_y^2$. Additionally, given that $\rho_y(h) \sim Ch^{2d-1}$, the ACF of the squared series should behave like $\rho_{y^2}(h) \sim C^2h^{2\widetilde{d}-1}$, where $\widetilde{d} = 2d - \frac{1}{2}$. In this case, $\widetilde{d} = 0.3$. Thus, the sample ACF of y_t^2 should decay a bit more rapidly than the ACF of y_t as it seems to be the case when comparing panels (a) and (b). A similar behavior occurs when $d = 0.2$; see Figure 6.25, where $\widetilde{d} = -0.1$.

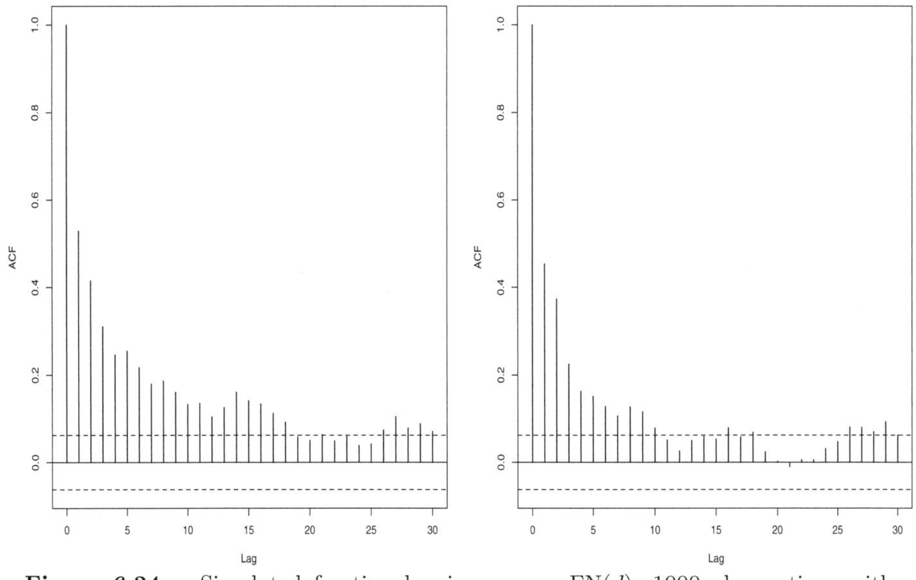

Figure 6.24 Simulated fractional noise process FN(d), 1000 observations with $d = 0.4$. (a) ACF of the series and (b) ACF of the squared series.

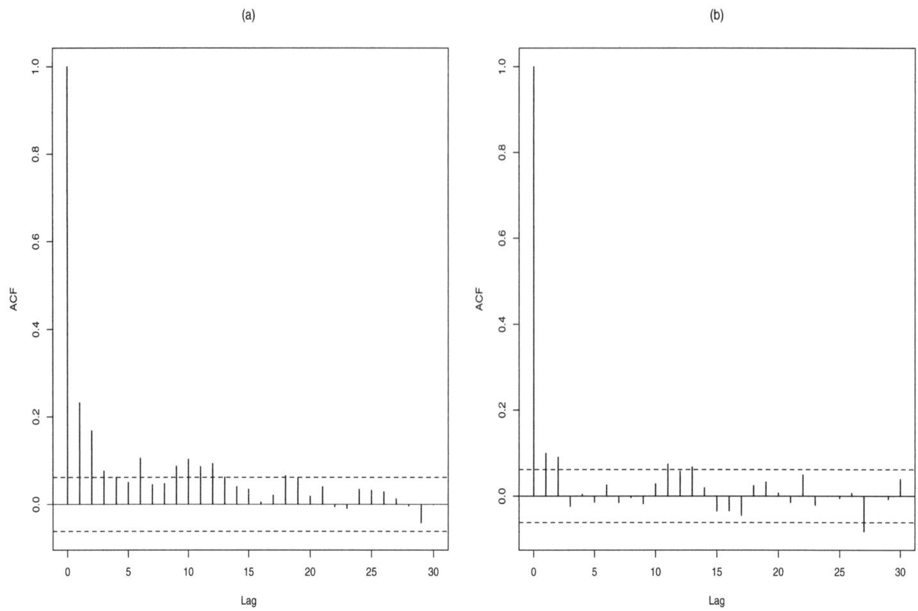

Figure 6.25 Simulated fractional noise process FN(d), 1000 observations with $d = 0.2$. (a) ACF of the series and (b) ACF of the squared series.

Figure 6.26 displays the sample ACF from 1000 simulated observations of the ARCH(1) process:

$$\begin{aligned} y_t &= \epsilon_t \, \sigma_t, \\ \sigma_t^2 &= 0.1 + 0.8 \, y_{t-1}^2, \end{aligned}$$

where ϵ_t is assumed to be a sequence of independent and identically distributed $N(0, 1)$ random variables. Note that in panel (a), as expected, the sample ACF of the series y_t shows no significant correlations. On the contrary, the sample ACF of the squared series shown in panel (b) exhibits a substantial level of autocorrelation, which decays at an exponential rate. A similar behavior of the autocorrelation is displayed by Figure 6.27, which depicts the sample ACF of 1000 observations from the following GARCH(1, 1) model:

$$\begin{aligned} y_t &= \epsilon_t \, \sigma_t, \\ \sigma_t^2 &= 0.1 + 0.7 \, y_{t-1}^2 + 0.2 \, \sigma_{t-1}^2, \end{aligned}$$

where $\{\epsilon_t\}$ is a sequence of independent and identically distributed $N(0, 1)$ random variables.

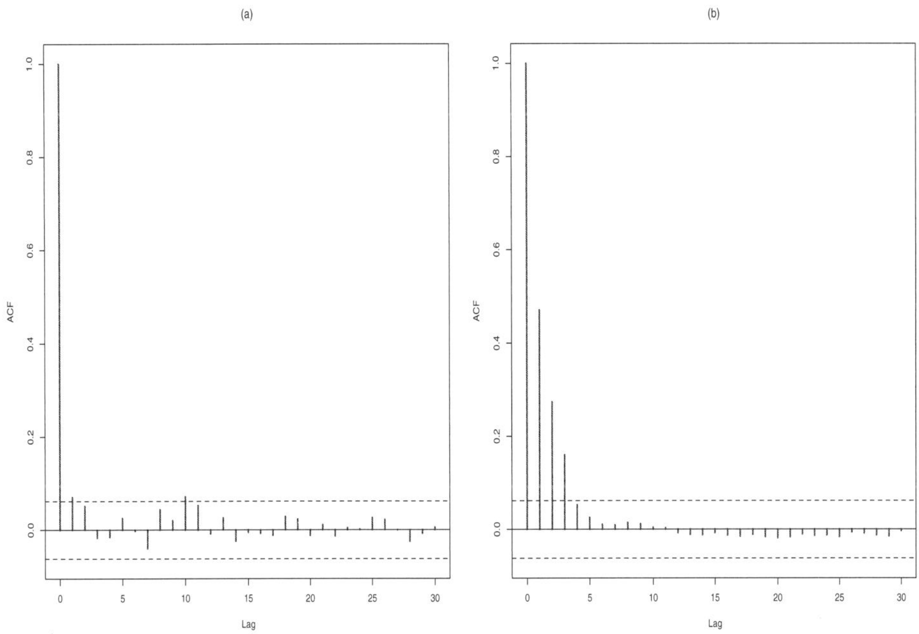

Figure 6.26 Simulated ARCH(1) process: 1000 observations with $\alpha_0 = 0.1$ and $\alpha_1 = 0.8$. (a) ACF of the series and (b) ACF of the squared series.

Figure 6.28 exhibits a trajectory of 1000 observations from the ARFIMA$(0, d, 0)$-GARCH$(1, 1)$ process:

$$\begin{aligned} y_t &= \sum_{j=0}^{\infty} \psi_j \, \varepsilon_{t-j}, \\ \varepsilon_t &= \epsilon_t \, \sigma_t, \\ \sigma_t^2 &= 0.1 + 0.7 \, \varepsilon_{t-1}^2 + 0.2 \, \sigma_{t-1}^2, \end{aligned}$$

where $d = 0.4$,

$$\psi_j = \frac{\Gamma(0.4 + j)}{\Gamma(1 + j)\Gamma(0.4)},$$

and ϵ_t is an independent and identically distributed Gaussian sequence with zero-mean and unit variance. Panel (a) shows the series while panel (b) shows the squares. On the other hand, Figure 6.29 shows the sample ACF of this series; see panel (a) and the sample ACF of the squares; see panel (b). Note that in this case both panels seem to exhibit long-memory behavior because $d \in (\frac{1}{4}, \frac{1}{2})$.

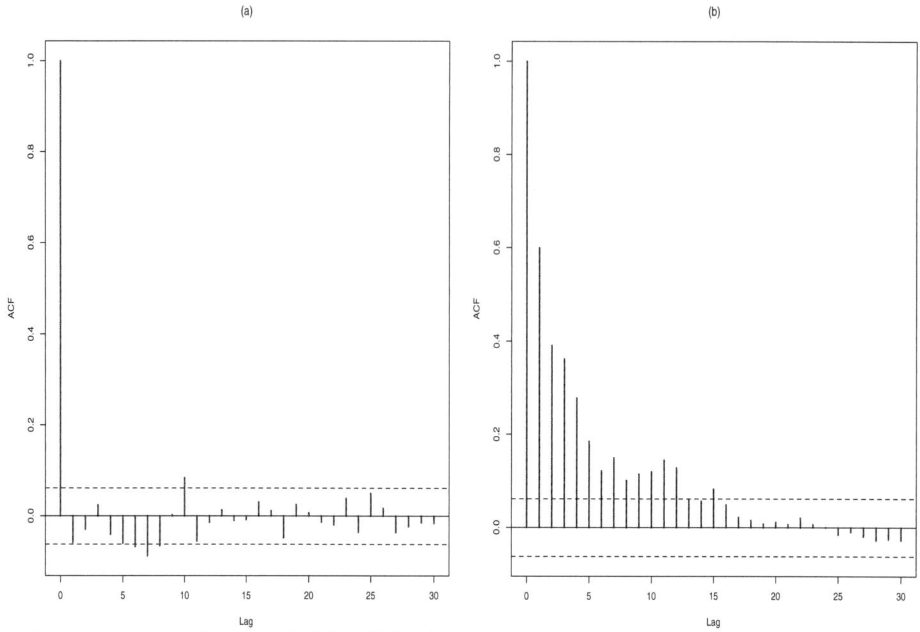

Figure 6.27 Simulated GARCH(1,1) process: 1000 observations with $\alpha_0 = 0.1$, $\alpha_1 = 0.7$, and $\beta_1 = 0.2$. (a) ACF of the series and (b) ACF of the squared series.

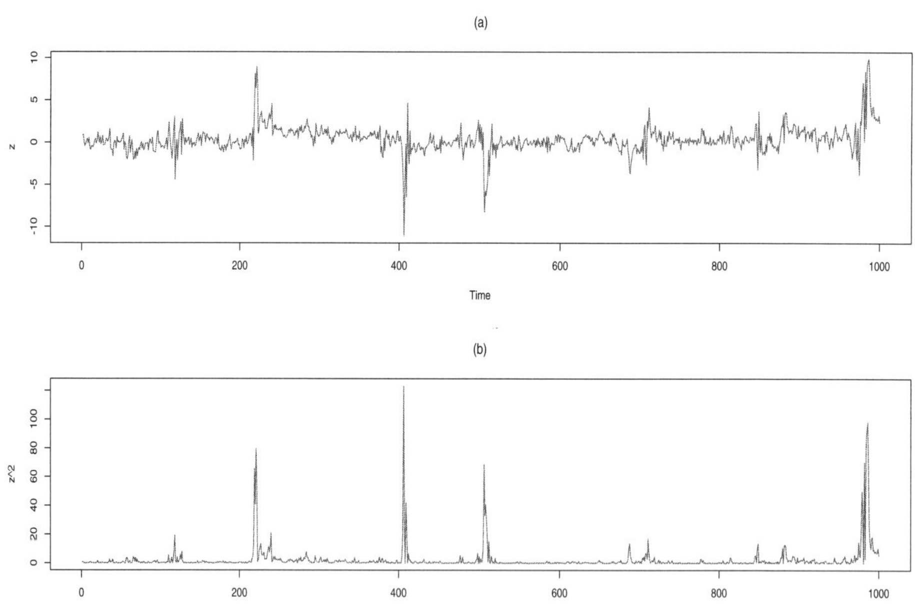

Figure 6.28 Simulated ARFIMA$(0, d, 0)$-GARCH$(1, 1)$ process: 1000 observations with $d = 0.4$, $\alpha_0 = 0.1$, $\alpha_1 = 0.7$, and $\beta_1 = 0.2$. (a) Series and (b) squared series.

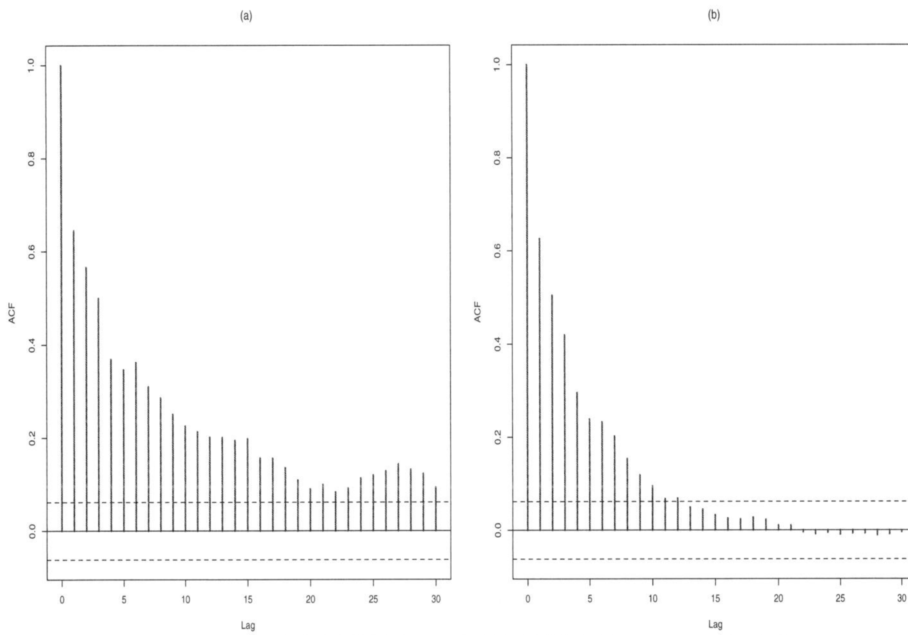

Figure 6.29 Simulated ARFIMA$(0, d, 0)$-GARCH$(1, 1)$ process: 1000 observations with $d = 0.4$, $\alpha_0 = 0.1$, $\alpha_1 = 0.7$, and $\beta_1 = 0.2$. (a) ACF of the series and (b) ACF of the squared series.

6.14 THRESHOLD AUTOREGRESSIVE MODELS

Nonlinearity is an extended phenomenon in different fields. In the previous sections we have examined heteroskedastic processes as tools for modeling financial time series. Nevertheless, there is a pleyade of other nonlinear models. One important example is the so-called *threshold time series models*. In this context, the process is assumed to have different regimes which are determined by, for instance, a function of the level of lagged values of the process.

In this section we briefly review two simple examples of threshold autoregressive process (TAR) and self-exciting threshold autoregressive process (SETAR).

A simple version of a threshold autoregressive process TAR(p) can be written as

$$y_t = \mu^{J_t} + \sum_{i=1}^{p} \phi_i^{J_t} y_{t-i} + \varphi^{J_t} \varepsilon_t,$$

where $J_t \in \{1, \ldots, J\}$ is a regime switching mechanism indicator and ε_t is an i.i.d. sequence with zero-mean and variance σ^2.

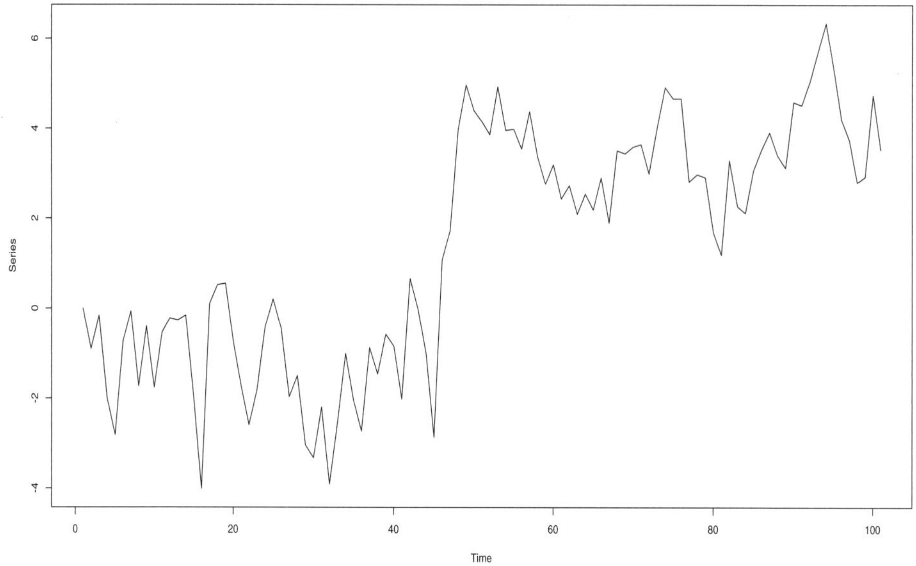

Figure 6.30 SETAR simulated time series.

A particular case of the above model is the self-exciting threshold autoregressive model (SETAR). Consider $p = 1$, $J = 2$ and this model can be written in terms of a thresholding variable z_t as follows

$$y_t = \begin{cases} \mu_1 + \phi_1 \, y_{t-1} + \varepsilon_t & \text{if} \quad z_{t-d} \leq r \\ \mu_2 + \phi_2 \, y_{t-1} + \varphi \, \varepsilon_t & \text{if} \quad z_{t-d} > r. \end{cases}$$

where r is a *threshold* parameter and d is a *delay* parameter. The variable z_t can be defined in terms of values of the process as, for example, $z_t = \beta_0 \, y_t + \beta_1 \, y_{t-1}$.

As an illustration of the features of a SETAR process, Figure 6.30 depicts the trajectory of a SETAR model with $\mu_1 = -1$, $\phi_1 = 0.2$, $\mu_2 = 1$, $\phi_2 = 0.7$, $\varphi = 1$ and $\text{Var}(\varepsilon_t) = 1$. Notice that the first half of the trajectory seems to be in one regime with mean -1 and the second half of the series seems to be in second regime which has mean 1. The sample ACF and sample PACF of this series are shown in Figure 6.31.

The R package tsDyn allows for the estimation of TAR and SETAR models. The output in this case is

```
> st=setar(Yt, m=1)
> summary(st)

Non linear autoregressive model
```

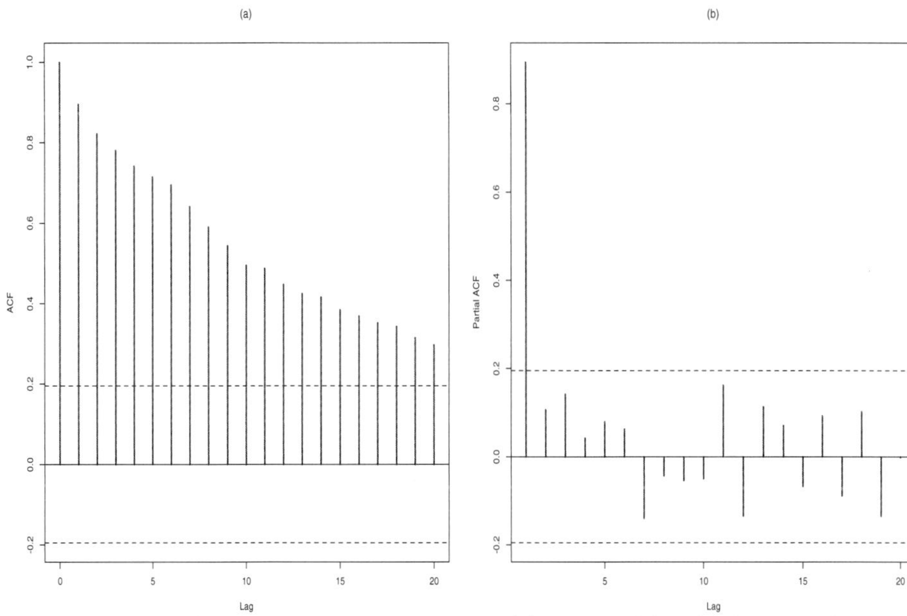

Figure 6.31 SETAR simulated time series. (b) Sample ACF, (b) Sample PACF.

```
SETAR model ( 2 regimes)

Coefficients:

Low regime:

   const.L     phiL.1
-1.0779184  0.2183248

High regime:

  const.H    phiH.1
1.1472667 0.6479871

Threshold:

-Variable: Z(t) = + (1) X(t)
-Value: 0.9398

Proportion of points in low regime: 71.72%       High regime: 28.28%

Residuals:

      Min          1Q     Median          3Q         Max
```

```
-3.654675 -0.685460  0.019659  0.744333  2.938611

Fit:

residuals variance = 1.099,  AIC = 19, MAPE = 531.6%

Coefficient(s):

          Estimate  Std. Error  t value  Pr(>|t|)
const.L  -1.07792     0.21597  -4.9910 2.678e-06 ***
phiL.1    0.21832     0.12389   1.7623   0.08121 .
const.H   1.14727     0.49396   2.3226   0.02231 *
phiH.1    0.64799     0.14221   4.5566 1.530e-05 ***
---

Signif. codes:  0 *** 0.001 ** 0.01 * 0.05 . 0.1   1

Threshold

Variable: Z(t) = + (1) X(t)

Value: 0.9398
```

Notice that the estimates produced by this package are close to the true parameter values.

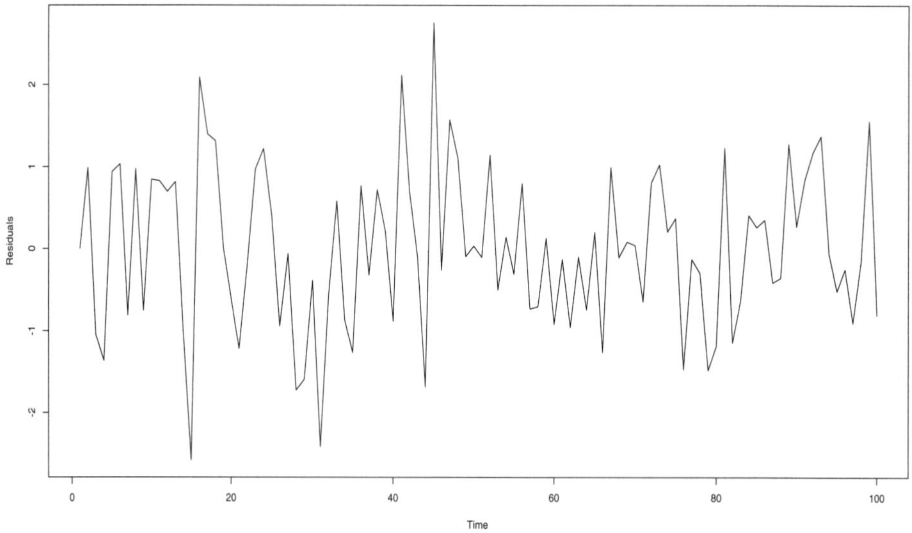

Figure 6.32 SETAR fitted model residuals.

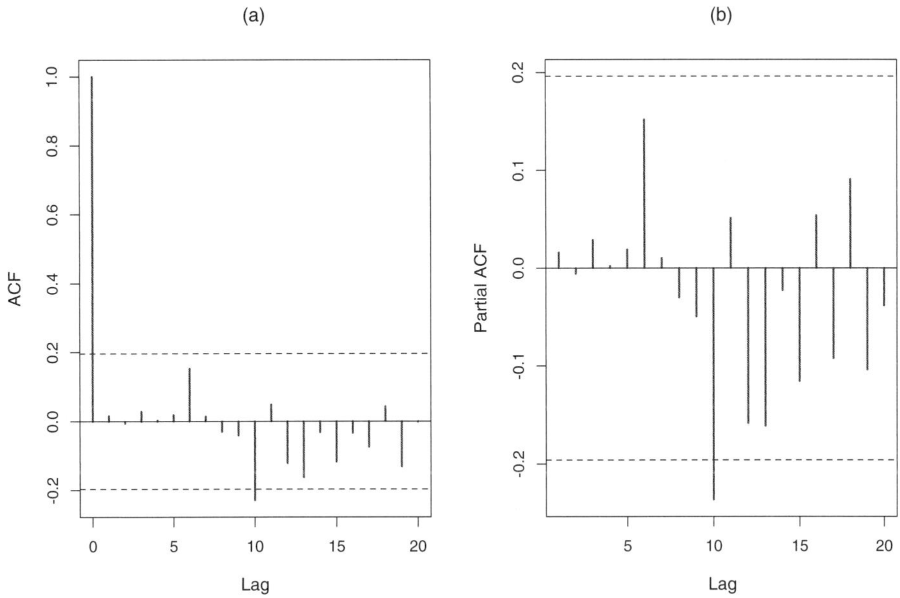

Figure 6.33 SETAR Residuals. (a) Sample ACF, (b) Sample PACF.

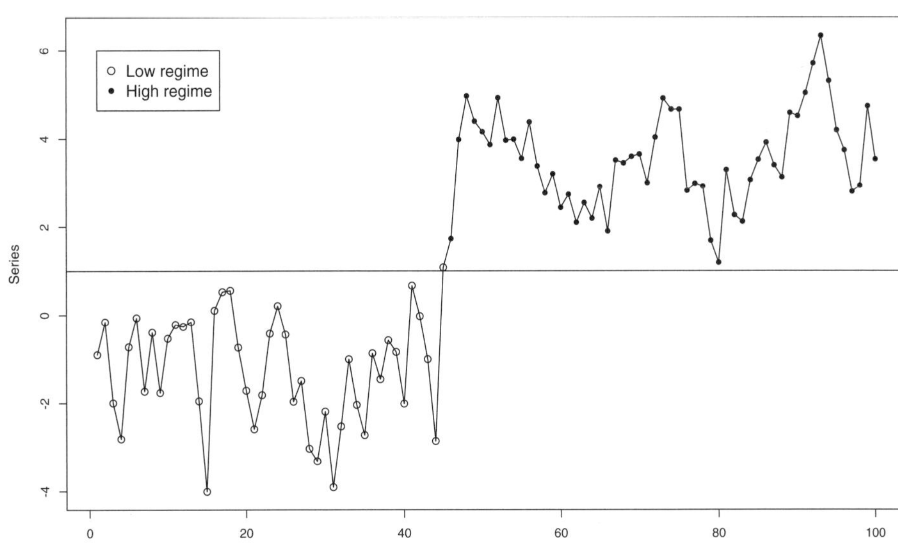

Figure 6.34 SETAR Fitting. Threshold and regimen classification.

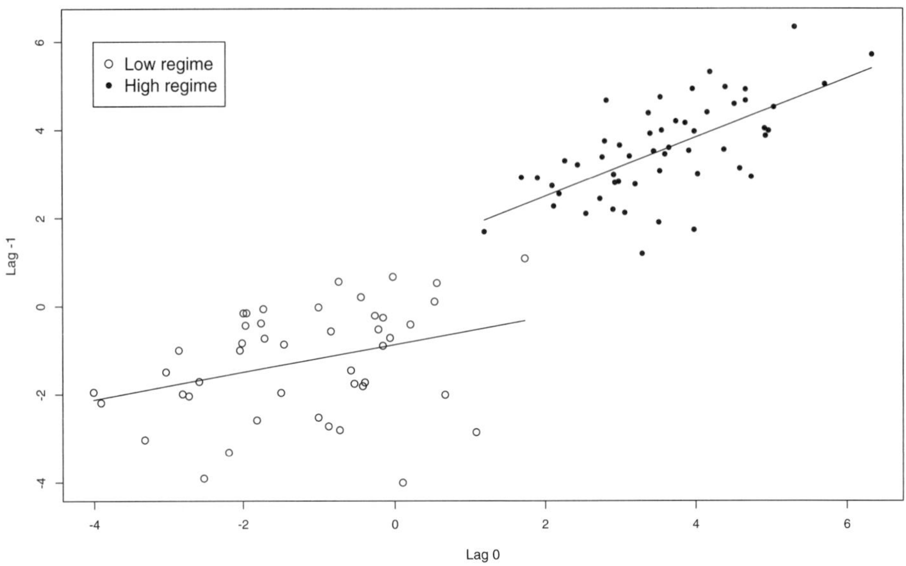

Figure 6.35 SETAR Fitting. Scatterplot of y_t and y_{t-1} along with empirical slope estimates for the two regimes.

6.15 BIBLIOGRAPHIC NOTES

Engle (1982) proposed the ARCH models to account for the stylized facts exhibited by many economic and financial time series. Based on this seminal work, a plethora of related models have been introduced. Among these methodologies we find the GARCH models [see, for example, Bollerslev (1986) and Taylor (1986)], the EGARCH models [see, for instance, Nelson (1991)], the *stochastic volatility processes* (SV) [see, for example, Harvey, Ruiz, and Shephard (1994)], the FIGARCH and FIEGARCH models [see, for instance, Baillie, Bollerslev, and Mikkelsen (1996) and Bollerslev and Mikkelsen (1996)], and the *long-memory generalized autoregressive conditionally heteroskedastic* (LMGARCH) models [see, for example, Robinson (1991), Robinson and Henry (1999) and Henry (2001)].

Most econometric models dealing with long-memory and heteroskedastic behaviors are nonlinear in the sense that the noise sequence is not necessarily independent. In particular, in the context of ARFIMA-GARCH models, the returns have long-memory and the noise has a conditional heteroskedasticity structure. These processes have received considerable attention; see, for example, Ling and Li (1997) and references therein.

A related class of interesting models is the extension of the ARCH(p) processes to the ARCH(∞) models to encompass the longer dependence observed in many squared financial series. The ARCH(∞) class was first introduced by Robinson (1991).

On the other hand, extensions of the stochastic volatility processes to the long-memory case have produced the LMSV models; see Harvey, Ruiz, and Shephard (1994), Ghysels, Harvey, and Renault (1996), Breidt, Crato, and de Lima (1998), and Deo and Hurvich (2003). Other estimation procedures for LMSV using state space systems can be found in Chan and Petris (2000) and Section 11 of Chan (2002). Furthermore, exact likelihood-based Bayesian estimation of LMSV is discussed in Section 4 of Brockwell (2004).

Threshold autoregressive processes (TAR) as well as self-exciting threshold autoregressive (SETAR) models are reviewed, for example, in Tsay (1989) and Tong (1990, 2011).

Problems

6.1 Suppose that the series $y_1, \ldots, y_{150}$ corresponds to the ARCH(1) process:

$$\begin{aligned} y_t &= \varepsilon_t \sigma_t, \\ \sigma_t^2 &= \alpha_0 + \alpha_1\, y_{t-1}^2, \end{aligned}$$

where ε_t is white noise $(0, 1)$.

(a) Assume that the MLE of α_1 es 0.32. Is this fitted process stationary?

(b) Build a 95% confidence interval for α_1.

6.2 Consider the model $y_t = \varepsilon_t y_{t-1}$ where ϵ_t is white noise $(0, \sigma^2)$

(a) Show that the conditional variance of y_t given y_{t-1} is $\sigma^2 y_{t-1}^2$

(b) Verify that under the assumption of second order stationarity if $\sigma^2 \neq 1$ then the variance of y_t is zero or infinite.

(c) Write this model as a Volterra expansion. What assumption on y_0 seems reasonable?

6.3 Explain briefly the following concepts.

(a) Return of a financial instrument.

(b) Conditional heteroskedasticity.

(c) Non linear process.

(d) Best lineal predictor.

6.4 Consider the following stationary process

$$\begin{aligned} x_t &= \theta\, y_{t-1} + y_t, \\ y_t &= (\alpha + \beta\, y_{t-1}^2)^{1/2}\, z_t, \end{aligned}$$

with $|\theta| < 1$, $\alpha > 0$, $0 \leq \beta < 1$ and $\{z_t\}$ is an i.i.d. N(0, 1) sequence.

(a) Determine the autocovariance function of $\{y_t\}$.

(b) Determine the autocovariance function of $\{x_t\}$.

Hint: Recall that if X and Y are random variables, then $E[g(X)] = E\{E[g(X)\,|\,Y]\}$.

6.5 Consider the ARCH(2) process that satisfies the equation

$$\begin{aligned} y_t &= \varepsilon_t\,\sigma_t, \\ \sigma_t^2 &= \alpha_0 + \alpha_1\,y_{t-1}^2 + \alpha_2\,y_{t-2}^2, \end{aligned}$$

where $\{\varepsilon_t\}$ is i.i.d. $(0,1)$. Let $\sigma_t^2(h) \equiv E[\sigma_{t+h}^2 | y_t, y_{t-1}, \ldots]$ the h-step predictor of the future volatility for $h \geq 1$.

(a) Verify that

$$\sigma_t^2(2) = \alpha_0 + \alpha_1\,\sigma_{t+1}^2 + \alpha_2\,y_t^2.$$

(b) Show that

$$\sigma_t^2(3) = \alpha_0\,(1+\alpha_1) + (\alpha_1^2 + \alpha_2)\,\sigma_{t+1}^2 + \alpha_1\,\alpha_2\,y_t^2.$$

6.6 Consider two ARCH(1) processes $\{y_t\}$ and $\{x_t\}$, independent, defined by

$$\begin{aligned} y_t &= \varepsilon_t\sigma_t, \\ \sigma_t^2 &= \alpha_0 + \alpha_1 y_{t-1}^2, \\ x_t &= \eta_t\nu_t, \\ \nu_t^2 &= \beta_0 + \beta_1 x_{t-1}^2, \end{aligned}$$

where $\{\varepsilon_t\}$ and $\{\eta_t\}$ are i.i.d. $N(0,1)$ sequences.

Define the new process $z_t = y_t \cdot x_t$.

(a) Show that $E(z_t) = 0$.

(b) Show that $E(z_t^2 | y_{t-1}, y_{t-2}, \ldots) = \sigma_t^2 \nu_t^2$.

(c) Verify that $E(z_t^2 z_{t+k}^2) = \left(\mu_y^2 + \frac{2\alpha_1^k \sigma^4}{1-\alpha_1^2}\right)\left(\mu_x^2 + \frac{2\beta_1^k \nu^4}{1-\beta_1^2}\right)$, where

$$\mu_y = \frac{\alpha_0}{1-\alpha_1} \quad \text{y} \quad \mu_x = \frac{\beta_0}{1-\beta_1}$$

$$\sigma^4 = \frac{\alpha_0^2(1+\alpha_1)}{(1-\alpha_1)(1-3\alpha_1^3)} \quad \text{y} \quad \nu^4 = \frac{\beta_0^2(1+\beta_1)}{(1-\beta_1)(1-3\beta_1^3)}$$

(d) Calculate $\mathrm{Cov}(z_t^2, z_{t+k}^2)$.

6.7 Consider the GARCH(1,1) process:

$$\begin{aligned} y_t &= \varepsilon_t\sigma_t, \\ \sigma_t^2 &= \alpha_0 + \alpha_1 y_{t-1}^2 + \beta_1\sigma_{t-1}^2 \end{aligned}$$

where $\{\varepsilon_t\}$ is i.i.d. $(0,1)$ Let $\sigma_t^2(h) = E(\sigma_{t+h}^2 | y_t, y_{t-1}, \dots)$ be the h-step volatility forecast.

(a) Show that for $h \geq 2$

$$\sigma_t^2(h) = \alpha_0 + (\alpha_1 + \beta_1)\sigma_t^2(h-1).$$

(b) Verify that the limit of $\sigma_t^2(h)$ as h increases satisfies

$$\lim_{h\to\infty} \sigma_t^2(h) = \frac{\alpha_0}{1-(\alpha_1+\beta_1)}.$$

6.8 Consider the *integrated* GARCH model IGARCH(1,1) defined by

$$\begin{aligned} y_t &= \varepsilon_t \sigma_t, \\ \sigma_t^2 &= \alpha_0 + \beta_1 \sigma_{t-1}^2 + (1-\beta_1) y_{t-1}^2, \end{aligned}$$

where ε_t is white noise $(0,1)$ and $\beta_1 \in (0,1)$. Let $\eta_t = y_t^2 - \sigma_t^2$.

(a) Show that η_t is white noise.

(b) Verify that y_t^2 satisfies

$$y_t^2 - y_{t-1}^2 = \alpha_0 + (1 - \beta_1 B)\eta_t.$$

(c) Based on the above, What mode satisfies y_t^2?

(d) Show that the ℓ-step variance predictor $\ell > 1$ is given by

$$\sigma_n^2(\ell) = \sigma_n^2(1) + (\ell - 1)\alpha_0.$$

6.9 Consider the following *exponential* GARCH, EGARCH$(1,0)$ model

$$\begin{aligned} y_t &= \varepsilon_t \sigma_t, \\ (1-\alpha B)\ln(\sigma_t^2) &= (1-\alpha)\alpha_0 + g(\varepsilon_{t-1}), \\ g(\varepsilon_{t-1}) &= \theta\varepsilon_{t-1} - \gamma[|\varepsilon_{t-1}| - E(|\varepsilon_{t-1}|)], \end{aligned}$$

where ϵ_t is white noise $N(0,1)$.

(a) Show that

$$(1-\alpha B)\ln(\sigma_t^2) = \begin{cases} \alpha_* + (\theta+\gamma)\varepsilon_{t-1} & \text{si} \quad \varepsilon_{t-1} \geq 0 \\ \alpha_* + (\theta-\gamma)\varepsilon_{t-1} & \text{si} \quad \varepsilon_{t-1} < 0, \end{cases}$$

where $\alpha_* = (1-\alpha)\alpha_0 - (\sqrt{2/\pi})\gamma$.

(b) Verify that

$$\sigma_t^2 = \sigma_{t-1}^{2\alpha} \exp(\alpha_*) \begin{cases} \exp[(\theta+\gamma)\frac{y_{t-1}}{\sqrt{\sigma_{t-1}^2}}] & \text{si} \quad y_{t-1} \geq 0 \\ \exp[(\theta-\gamma)\frac{y_{t-1}}{\sqrt{\sigma_{t-1}^2}}] & \text{si} \quad y_{t-1} < 0. \end{cases}$$

(c) What advantages has this model over a standard GARCH process?

6.10 Consider the stochastic volatility process $\{r_t\}$ given by

$$\begin{aligned} r_t &= \varepsilon_t \sigma_t, \\ \sigma_t &= \sigma \exp(\nu_t/2), \end{aligned}$$

where $\{\varepsilon_t\}$ is an independent and identically distributed sequence with zero-mean and unit variance and $\{\nu_t\}$ is a regular linear process satisfying

$$\nu_t = \sum_{j=0}^{\infty} \psi_j \eta_{t-j},$$

with $\sum_{j=0}^{\infty} \psi_j^2 < \infty$ and $\{\eta_t\}$ an independent and identically distributed sequence with zero-mean and unit variance, independent of the sequence $\{\varepsilon_t\}$. Show that the process r_t is strictly stationary and ergodic.

6.11 Assume that $\pi(B) = (1-B)^d$. Show that $\pi(B)\alpha_0 = 0$, where α_0 is any real constant and $d > 0$.

6.12 Show that the FIGARCH process may be written as

$$\theta(B)\sigma_t^2 = \omega + [\theta(B) - \phi(B)(1-B)^d] y_t^2, \tag{6.27}$$

where $\omega = \theta(B)\alpha_0$. What conditions must satisfy the polynomial $\theta(B)$ in order to ensure that ω is a positive constant?

6.13 Let $\lambda(d) = (1-B)^{-d}$ for $|d| < \frac{1}{2}$.

(a) Verify that the ARFIMA$(0,d,0)$-GARCH model may be written as

$$\lambda(d)\varepsilon_t(d) = c,$$

where c is a constant with respect to d. Note that the data $\{y_t\}$ do not depend on d.

(b) Let $\psi(B) = \sum_{j=0}^{\infty} \psi_j(d) B^j = (1-B)^{-d} = \lambda(d)^{-1}$. Show that

$$\sum_{j=0}^{\infty} \left[\frac{\partial}{\partial d}\psi_j(d)\right] \varepsilon_{t-j}(d) + \sum_{j=0}^{\infty} \psi_j(d) \left[\frac{\partial}{\partial d}\varepsilon_{t-j}(d)\right] = 0.$$

(c) Show that

$$\begin{aligned} \frac{\partial}{\partial d}\varepsilon_t(d) &= -\lambda(d) \sum_{j=0}^{\infty} \left[\frac{\partial}{\partial d}\psi_j(d)\right] \varepsilon_{t-j}(d) \\ &= -\lambda(d) \left[\frac{\partial}{\partial d}\lambda(d)\right] \varepsilon_t(d) \\ &= -\left[\frac{\partial}{\partial d}\log\lambda(d)\right] \varepsilon_t(d). \end{aligned}$$

(d) Verify that

$$\frac{\partial}{\partial d}\varepsilon_t(d) = -\sum_{j=1}^{\infty} \frac{1}{j}\varepsilon_{t-j}(d),$$

and prove that this expansion is well-defined.

6.14 Assume that the sequence $\{\epsilon_t\}$ in (6.7) corresponds to independent and identically distributed uniform random variables $U(-\sqrt{3}, \sqrt{3})$.

(a) Verify that $\{\epsilon_t\}$ is a sequence of zero-mean and unit variance random variables.

(b) Show that for this specification of $\{\epsilon_t\}$, the top Lyapunov exponent of the model in Example 6.1 is given by

$$\gamma = E[\log(\alpha_1 \epsilon^2 + \beta_1)] = 2\left[\log\sqrt{3\alpha_1 + \beta_1} + \sqrt{\frac{\beta_1}{3\alpha_1}}\arctan\sqrt{\frac{3\alpha_1}{\beta_1}} - 1\right].$$

Hint: The following formula could be useful:

$$\int \log(x^2 + a^2)\,dx = x\log(x^2 + a^2) + 2a\arctan\frac{x}{a} - 2x.$$

6.15 Consider the ARFIMA(p, d, q)-GARCH$(1,1)$ process:

$$\begin{aligned} y_t &= \sum_{j=0}^{\infty} \psi_j \varepsilon_{t-j}, \\ \varepsilon_t &= \epsilon_t \sigma_t, \\ \sigma_t^2 &= \alpha_0 + \alpha\varepsilon_{t-1}^2 + \beta\sigma_{t-1}^2, \end{aligned}$$

where ϵ_t is a random variable with density

$$f(\epsilon) = \frac{2}{\pi(1+\epsilon^2)^2},$$

for $-\infty < \epsilon < \infty$.

(a) Verify that the random variable ϵ satisfies

$$E(\epsilon) = 0,$$
$$\mathrm{Var}(\epsilon) = 1.$$

(b) Show that the top Lyapunov exponent in this case is given by

$$\gamma = 2\left[\log(\sqrt{\alpha} + \sqrt{\beta}) - \frac{\sqrt{\alpha}}{\sqrt{\alpha} + \sqrt{\beta}}\right]$$

(c) Verify whether the Lyapunov exponent γ is negative for $\alpha > 0$, $\beta > 0$, and $\alpha + \beta < 1$.

Hint: The following formula could be useful:

$$\int_0^\infty \log(a^2 + b^2x^2)\frac{dx}{(1+x^2)^2} = \frac{\pi}{2}\left[\log(a+b) - \frac{b}{a+b}\right],$$

for $a, b > 0$; see Gradshteyn and Ryzhik (2000, p. 557).

6.16 Consider the ARFIMA-GARCH process defined in Problem 6.15 where ϵ_t is a random variable with density

$$f(\epsilon) = \frac{1}{\pi\sqrt{1-\epsilon^2}},$$

for $\epsilon \in (-1, 1)$.

(a) Verify that ϵ is a zero-mean and unit variance random variable.

(b) Prove that the the top Lyapunov exponent in this case is

$$\gamma = 2\log\frac{\sqrt{\beta} + \sqrt{\alpha+\beta}}{2}.$$

(c) Show that the Lyapunov exponent γ is negative for $\alpha > 0$, $\beta > 0$, and $\alpha + \beta < 1$.

Hint: The following integrals could be useful:

$$\begin{aligned}\int \frac{dx}{\sqrt{1-x^2}} &= \arcsin x, \\ \int \frac{x^2 dx}{\sqrt{1-x^2}} &= \frac{1}{2}(\arcsin x - x\sqrt{1-x^2}),\end{aligned}$$

and

$$\int_0^1 \log(1+ax^2)\frac{dx}{\sqrt{1-x^2}} = \pi \log\frac{1+\sqrt{1+a}}{2},$$

for $a \geq -1$; see Gradshteyn and Ryzhik (2000, p. 558).

6.17 A definition of the FIEGARCH model is

$$\log(\sigma_t^2) = \omega + \phi(B)^{-1}(1-B)^{-d}\psi(B)g(\epsilon_{t-1}), \tag{6.28}$$

where $\phi(B) = 1 + \phi_1 B + \cdots + \phi_p B^p$, $\psi(B) = 1 + \psi_1 B + \cdots + \psi_q B^q$, and

$$g(\epsilon_t) = \theta\epsilon_t + \gamma[|\epsilon_t| - E(|\epsilon_t|)].$$

Another definition of a FIEGARCH process is

$$\phi(B)(1-B)^d \log(\sigma_t^2) = a + \sum_{j=1}^{q}(b_j|\epsilon_{t-j}| + \gamma_j\epsilon_{t-j}). \tag{6.29}$$

(a) Show that $\phi(B)(1-B)^d\omega = 0$.

(b) Starting from definition (6.28), prove that

$$\phi(B)(1-B)^d \log(\sigma_t^2) = -\psi(B)\gamma\mu + \theta\psi(B)\epsilon_{t-1} + \gamma\psi(B)|\epsilon_{t-1}|,$$

where $\mu = E(|\epsilon_1|)$.

(c) Show that by taking $b_j = \theta\psi_j$, $\gamma_j = \gamma\psi_j$, and $a = -\mu\gamma\sum_{j=1}^{q}\psi_j$, we obtain definition (6.29). Observe, however, that in model (6.29) we could release the parameters b_j and γ_j from the restriction $b_j\gamma = \gamma_j\theta$.

(d) Verify that by taking $\theta_j = b_j$, $\lambda_j = \gamma_j$, and $\alpha = a$, definition (6.16) is obtained.

6.18 Consider the FIEGARCH model described by equation (6.16) and assume that $\alpha = -\sum_{j=1}^{q}\theta_j\, E(|\epsilon_1|)$.

(a) Verify that

$$E[\alpha + \theta(B)|\epsilon_{t-1}| + \lambda(B)\epsilon_{t-1}] = 0.$$

(b) Show that conditional variance σ_t^2 may be formally written as

$$\sigma_t^2 = \exp\left\{\phi(B)^{-1}(1-B)^{-d}[\theta(B)(|\epsilon_{t-1}| - E\,|\epsilon_{t-1}|) + \lambda(B)\epsilon_{t-1}]\right\}.$$

(c) Show that the FIEGARCH process y_t may be formally written as

$$y_t = \epsilon_t \exp\left\{\tfrac{1}{2}\phi(B)^{-1}(1-B)^{-d}[\theta(B)(|\epsilon_{t-1}| - E\,|\epsilon_{t-1}|) + \lambda(B)\epsilon_{t-1}]\right\}.$$

(d) Consider a FIEGARCH$(0,d,1)$ where $\phi(B) = 1$, $(1-B)^{-d} = \sum_{j=0}^{\infty}\psi_j B^j$, $\theta(B) = \theta$, and $\lambda(B) = \lambda$. Show that σ_t^2 may be formally expressed as

$$\sigma_t^2 = \prod_{j=0}^{\infty} \exp\left[\theta\psi_j(|\epsilon_{t-1}| - E\,|\epsilon_{t-1}|) + \lambda\psi_j\epsilon_{t-1}\right].$$

(e) Under what conditions is the above infinite product well-defined?

6.19 Consider the following tree ring data from a location at Malleco, Chile, for the period from 1242 A.D. to 1975 A.D. This series is displayed in Figure 6.36.

A researcher proposes that the ring width during year t, say y_t, depends on the past as follows:

$$y_t = \begin{cases} \nu + z_t, & t = 1242 \\ \nu + \phi\, y_{t-1} + z_t, & t = 1243, 1244, \ldots, 1975 \end{cases}$$

with $\{z_t\}$ independent random variables with Normal$(0,\sigma)$ distribution. ν, ϕ ($|\phi| < 1$) and σ are parameters.

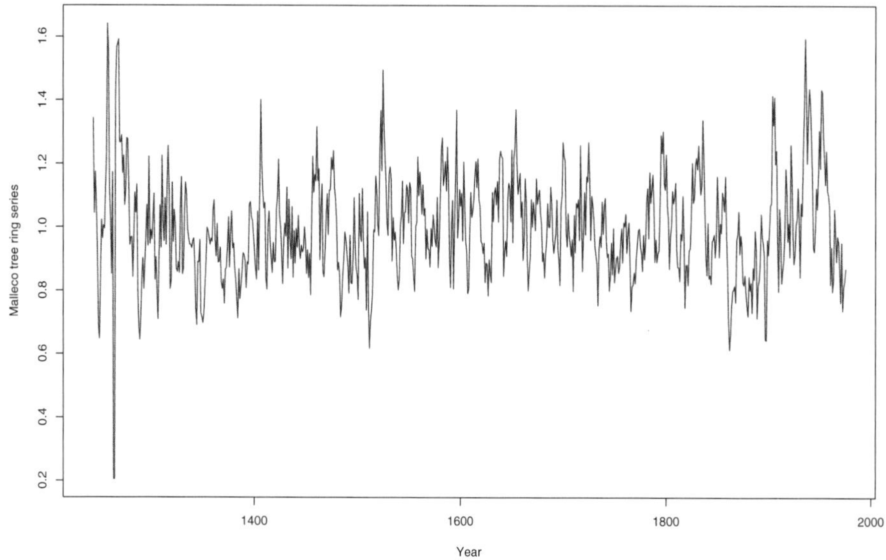

Figure 6.36 *Tree rings data at Malleco, Chile, for the period from 1242 A.D. to 1975 A.D.*

(a) Show that

$$y_{1242+k} = \frac{\nu\,(1-\phi^{k+1})}{1-\phi} + \sum_{j=0}^{k} \phi^j\, z_{1242+k-j}$$

(b) Show that

$$\mathrm{Cov}(y_t,\, y_{t+h}) = \sigma^2\, \phi^h \left(\frac{1-\phi^{2\,(t-1242+1)}}{1-\phi^2} \right)$$

for $t = 1242, \ldots, 1975$ and $h = 0, 1, 2, \ldots$.

6.20 Suppose that the process $\{x_t\}$ follows a TGARCH$(1,1)$ process defined as

$$\begin{aligned} x_t &= z_t \cdot \sigma_t \\ \sigma_t^2 &= \omega + (\alpha + \gamma \cdot \delta_{t-1})\, x_{t-1}^2 + \beta\, \sigma_{t-1}^2 \\ \delta_t &= \begin{cases} 1, & x_t < 0 \\ 0, & x_t \geq 0 \end{cases} \end{aligned}$$

with $\omega > 0$; $\alpha \geq 0$, $\gamma \geq 0$, $\beta \geq 0$, $0 \leq \alpha + \gamma + \beta < 1$ y $\{z_t\} \overset{\text{iid}}{\sim}$ Normal$(0,1)$. Find $E(x_t)$ and $\mathrm{Var}(x_t)$.

6.21 Consider the following GARCH(3, 2) process

$$x_t = z_t \cdot \sigma_t, \quad \{z_t\} \sim \text{WN}(0, 1)$$
$$\sigma_t^2 = \alpha_0 + \sum_{i=1}^{3} \alpha_i \, x_{t-i}^2 + \sum_{j=1}^{2} \beta_j \, \sigma_{t-j}^2,$$

with

$$\alpha_0 > 0, \quad \alpha_i \geq 0, \quad \beta_j \geq 0, \quad 0 \leq (\alpha_1 + \alpha_2 + \alpha_3 + \beta_1 + \beta_2) < 1.$$

Show that x_t^2 corresponds to an ARMA(3, 2) process and its noise is given by $\nu_t = \sigma_t^2(z_t^2 - 1)$.

6.22 Consider the $\{x_t\}$ process with mean μ given by:

$$x_t - \mu = \phi_1 \, (x_{t-1} - \mu) + \cdots + \phi_p \, (x_{t-p} - \mu) + z_t, \tag{6.30}$$

with $\{z_t\} \sim (0, \sigma^2)$. Show that (6.30) can be written as

$$\nabla \, x_t = \phi_0^* + \phi_1^* \, x_{t-1} + \phi_2^* \, \nabla \, x_{t-1} + \cdots + \phi_p^* \, \nabla \, x_{t-p+1} + z_t,$$

where $\phi_0^* = \mu \, (1 - \phi_1 - \cdots - \phi_p)$, $\phi_1^* = \sum_{i=1}^{p} \phi_i - 1$ y $\phi_j^* = -\sum_{i=j}^{p} \phi_i$ to $j = 2, \ldots, p$.

6.23 Let $\{x_1, \ldots, x_n\}$ observed values of a time series and $\hat{\rho}(h)$ sample autocorrelation function.

(a) If $x_t = a + b\,t$, where a y b are constants and $b \neq 0$, show that $h \geq 1$, $\hat{\rho}(h) \to 1$, when $n \to \infty$.

(b) If $x_t = c \, \cos(\omega \, t)$, where c y ω are constants ($c \neq 0$ y $\omega \in (-\pi, \pi]$), show that for any h, $\hat{\rho}(h) \to \cos(\omega \, h)$, when $n \to \infty$.

6.24 An ARMA model is fitted to a series of 500 observations collected sequentially in time. The results of the first 10 empirical autocorrelations is as follows:

k	1	2	3	4	5
$\hat{\rho}(k)$	−0.065	0.735	−0.061	0.386	−0.052
$\hat{\alpha}(k)$	−0.065	0.734	−0.002	−0.336	−0.005

k	6	7	8	9	10
$\hat{\rho}(k)$	0.238	−0.030	0.155	−0.033	0.071
$\hat{\alpha}(k)$	0.258	0.028	−0.162	−0.068	0.035

where the sample variance of the data is 2.708. Suppose that the residuals of model behave like a white noise of zero-mean and variance σ^2.

(a) Plot the ACF and PACF together with their empirical confidence bands. Based on these graphs, which model ARMA you find more suitable for this series?

(b) Find moment estimators of the parameters of the proposed model, including estimation of σ^2, and evaluate them according to the available information. If more than one possible value is available, choose the coefficient involving a causal and invertible model.

(c) Specify the asymptotic distribution of the autoregressive coefficients and/or moving-average model. Are they significantly different from zero at a 5% level?.

6.25 Consider a financial time series of 1500 observations. Two heteroskedastic models have been fitted to this series, a GARCH(1,1) along with an ARCH(2) process. The outputs from these fitted models are reported below. Which of the two models seems to better fit the series? Justify your answer.

Fitted Model 1

```
Title:
 GARCH Modeling

Call:
 garchFit(formula = ~garch(1, 1), data = y, trace = FALSE)

Mean and Variance Equation:
 data ~ garch(1, 1)
 [data = xx]

Conditional Distribution:
 norm

Coefficient(s):
        mu        omega       alpha1        beta1
3.2137e-05  9.1888e-07  1.8917e-01  7.1562e-01

Std. Errors:
 based on Hessian


Error Analysis:

        Estimate  Std. Error  t value Pr(>|t|)
mu      3.214e-05   6.677e-05    0.481     0.63
omega   9.189e-07   2.269e-07    4.050 5.11e-05 ***
alpha1  1.892e-01   3.047e-02    6.208 5.38e-10 ***
beta1   7.156e-01   4.429e-02   16.157  < 2e-16 ***
---
```

```
Signif. codes:  0 ***  0.001 **  0.01 *  0.05  .  0.1    1

Log Likelihood:
 6659.775    normalized:  4.43985


Standardised Residuals Tests:

                                Statistic p-Value
 Jarque-Bera Test   R    Chi^2  5.941889  0.0512548
 Shapiro-Wilk Test  R    W      0.9983002 0.1345606
 Ljung-Box Test     R    Q(10)  10.32037  0.4128496
 Ljung-Box Test     R    Q(15)  14.34846  0.4992822
 Ljung-Box Test     R    Q(20)  18.88156  0.5295365
 Ljung-Box Test     R^2  Q(10)  5.289371  0.8710286
 Ljung-Box Test     R^2  Q(15)  8.75091   0.8901772
 Ljung-Box Test     R^2  Q(20)  10.02904  0.9676424
 LM Arch Test       R    TR^2   7.27484   0.8389293

Information Criterion Statistics:
      AIC       BIC       SIC      HQIC
-8.874367 -8.860199 -8.874381 -8.869089
```

Fitted Model 2

```
Title:
 GARCH Modeling

Call:
 garchFit(formula = ~garch(2, 0), data = y, trace = FALSE)

Mean and Variance Equation:
 data ~ garch(2, 0)
 [data = xx]

Conditional Distribution:
 norm

Coefficient(s):
        mu        omega       alpha1       alpha2
3.5973e-05  5.4612e-06  2.5347e-01  1.6910e-01

Std. Errors:
 based on Hessian

Error Analysis:
```

```
        Estimate  Std. Error  t value Pr(>|t|)
mu     3.597e-05   6.977e-05    0.516    0.606
omega  5.461e-06   3.702e-07   14.752  < 2e-16 ***
alpha1 2.535e-01   4.267e-02    5.941 2.83e-09 ***
alpha2 1.691e-01   3.975e-02    4.254 2.10e-05 ***
---

Signif. codes:  0 *** 0.001 **  0.01 *  0.05 .   0.1   1

Log Likelihood:
 6631.342    normalized:  4.420894


Standardised Residuals Tests:

                                Statistic p-Value
 Jarque-Bera Test   R    Chi^2  32.41568  9.141626e-08
 Shapiro-Wilk Test  R    W      0.9954953 0.0001870913
 Ljung-Box Test     R    Q(10)  10.14591  0.427787
 Ljung-Box Test     R    Q(15)  14.62402  0.478823
 Ljung-Box Test     R    Q(20)  20.15955  0.4479896
 Ljung-Box Test     R^2  Q(10)  46.92196  9.756395e-07
 Ljung-Box Test     R^2  Q(15)  56.6825   9.289409e-07
 Ljung-Box Test     R^2  Q(20)  60.68203  5.582236e-06
 LM Arch Test       R    TR^2   56.86594  8.357072e-08

Information Criterion Statistics:
      AIC       BIC       SIC      HQIC
-8.836455 -8.822287 -8.836470 -8.831177
```

6.26 Consider the following stochastic volatility model

$$y_t = \varepsilon_t \sigma_t,$$
$$\sigma_t^2 = e^{v_t},$$

where $\{\varepsilon_t\}$ is a zero-mean and unit variance i.i.d. sequence and y$\{v_t\}$ corresponds to a Gaussian MA(1) process, that is,

$$v_t = \mu + \eta_t + \theta \eta_{t-1},$$

with $\eta_t \sim N(0, \sigma_\eta^2)$. Verify that:

(a) $E(y_t^2 | y_{t-1}, y_{t-2}, \ldots) = e^{v_t}$.

(b) $\log[E(y_t^2)] = \mu + \sigma_\eta^2(1 + \theta^2)/2$.

(c) $E\{\log[E(y_t^2 | y_{t-1}, y_{t-2}, \ldots]\} = \mu$.

6.27 Suppose that the sequence y_t satisfies a GARCH$(0, q)$ process defined as follows.

$$y_t = \varepsilon_t \sigma_t,$$
$$\sigma_t^2 = \alpha_0 + \beta_1 \sigma_{t-1}^2 + \cdots + \beta_q \sigma_{t-q}^2,$$

where $\{\varepsilon_t\}$ i.i.d. N(0,1). Show that:

(a) $E(\sigma_t^2 | y_{t-k}, y_{t-k-1}, \ldots) = \sigma_t^2$, for all $k \in \mathbb{Z}$.

(b) For $k \geq 1$.

$$E(y_{t+k}^2 | y_t, y_{t-1}, y_{t-2}, \ldots) = \sigma_{t+k}^2.$$

CHAPTER 7

PREDICTION

One of the fundamental aspects of the time series analysis is forecasting. Consequently, this chapter addresses the prediction of linear and nonlinear processes. Section 7.2 and Section 7.3 examine the formulation of one-step and multistep ahead predictors based on finite and infinite past. The innovations algorithm and approximate predictors are also described.

Forecasting future volatility is a crucial aspect in the context of heteroskedastic time series. Therefore, Section 7.4 discusses techniques for forecasting volatility for some of the models described in Chapter 6. Several illustrative applications are also discussed, including the prediction of ARMA, ARFIMA, GARCH and combinations of these processes. Building prediction bands are discussed in Section 7.5. Furthermore, these techniques are applied in Section 7.6 to the prediction of the S&P500 returns data introduced in Chapter 1.

Bibliographic notes are given in Section 7.7 and a list of problems is proposed at the end of this chapter. Additionally, some technical aspects such as vector spaces and the projection theorem are reviewed in Appendix A.

Time Series Analysis. First Edition. Wilfredo Palma.

7.1 OPTIMAL PREDICTION

Given a process $\{y_t : t \in \mathbb{Z}\}$, finding optimal predictors of y_{t+h} given the observed values $y_t, y_{t-1}, y_{t-2}, \ldots$ depends crucially on the definition of optimality. Suppose that $\widehat{y}_{t+h}$ denotes the h-step predictor of y_{t+h}. Typically, optimality here means that the variance of the prediction error $e_{t+h} = y_{t+h} - \widehat{y}_{t+h}$, $\text{Var}(e_{t+h})$, is minimal. If we are looking for *linear predictors*, that is,

$$\widehat{y}_{t+h} = \sum_{j=1}^{\infty} \alpha_j y_{t-j},$$

then the predictor that minimizes $\text{Var}(e_{t+h})$ is the conditional expectation

$$\widehat{y}_{t+h} = E(y_{t+h} | y_t, y_{t-1}, y_{t-2}, \ldots).$$

In what follows, $\mathcal{P}_t$ denotes the *infinite* past of the time series up to time t, $y_t, y_{t-1}, \ldots$. On the other hand, $\mathcal{F}_t$ denotes the *finite* past of the time series up to time t, $y_t, y_{t-1}, \ldots, y_1$. Thus, the previous predictor can be written as

$$\widehat{y}_{t+h} = E(y_{t+h} | \mathcal{P}_t).$$

7.2 ONE-STEP AHEAD PREDICTORS

Let $\{y_t\}$ be an invertible linear process with Wold representation

$$y_t = \sum_{j=0}^{\infty} \psi_j \varepsilon_{t-j}, \tag{7.1}$$

and AR(∞) expansion

$$y_t = \varepsilon_t + \sum_{j=1}^{\infty} \pi_j y_{t-j}, \tag{7.2}$$

where $\text{Var}(\varepsilon_t) = \sigma^2$. As described in the next subsections, one-step predictors of these processes are different depending whether we consider infinite or finite past.

7.2.1 Infinite Past

The best linear one-step predictor of y_{t+1} given its past $y_t, y_{t-1}, \ldots$ is given by

$$\widehat{y}_{t+1} = E[y_{t+1} | \mathcal{P}_t] = \sum_{j=1}^{\infty} \pi_j y_{t+1-j} = \sum_{j=1}^{\infty} \psi_j \varepsilon_{t+1-j},$$

with prediction error variance $E[y_{t+1} - \widehat{y}_{t+1}]^2 = \sigma^2$.

7.2.2 Finite Past

Notice that in practice we seldom have the full past, instead we only have a finite stretch of data $\{y_1, \ldots, y_t\}$, say. Under this circumstance, the best linear predictor of y_{t+1} based on its finite past is given by

$$\widetilde{y}_{t+1} = E[y_{t+1}|y_t, \ldots, y_1] = \phi_{t1}y_t + \cdots + \phi_{tt}y_1,$$

where $\phi_t = (\phi_{1t}, \ldots, \phi_{tt})'$ is the unique solution of the linear equation

$$\Gamma_t \phi_t = \gamma_t,$$

with $\Gamma_t = [\gamma(i-j)]_{i,j=1,\ldots,t}$ and $\gamma_t = [\gamma(1), \ldots, \gamma(t)]'$. The calculation of these coefficients can be carried out using the Durbin-Levinson algorithm described in Chapter 5. In particular, the prediction error variance of the one-step finite sample predictor $\widetilde{y}_{t+1}$,

$$\nu_t = E[y_{t+1} - \widetilde{y}_{t+1}]^2,$$

can be calculated by means of the recursive equations

$$\nu_t = \nu_{t-1}(1 - \phi_{tt}^2),$$

for $t \geq 1$, where ϕ_{tt} are called the *partial autocorrelation coefficients* and $\nu_0 = \gamma(0)$. Thus, ν_t may be written as

$$\nu_t = \gamma(0) \prod_{j=1}^{t} (1 - \phi_{jj}^2).$$

7.2.3 Innovations Algorithm

Another way to write the finite past predictors is based on expressing the current forecast in terms of the innovations of the previous predictors,

$$\widehat{y}_{t+1} = \theta_{t1}(y_t - \widehat{y}_t) + \theta_{t2}(y_{t-1} - \widehat{y}_{t-1}) + \cdots + \theta_{tt}(y_1 - \widehat{y}_1).$$

Note that the terms $\{y_t - \widehat{y}_t\}$ correspond to the innovations of the forecasting process, that is, $\text{Cov}(y_t - \widehat{y}_t, y_s - \widehat{y}_s) = 0$ for all t.

The coefficients $\theta_{i,j}$ can be calculated recursively by following the equations,

$$\begin{aligned}
\beta_1 &= \gamma(0) \\
\theta_{t,t-i} &= \frac{1}{\beta_i}\left[\gamma(t-1) + \sum_{j=1}^{i-1} \theta_{i,i-j}\theta_{t,t-j}\beta_j\right] \\
\beta_t &= \gamma(0) - \sum_{j=1}^{t-1} \theta_{t,t-j}^2 \beta_j.
\end{aligned}$$

EXAMPLE 7.1

Consider the MA(1) model

$$y_t = \varepsilon_t + \theta\varepsilon_{t-1},$$

where $\mathrm{Var}(\varepsilon_t) = \sigma^2$. In this case, by defining $\rho = \frac{\theta}{1+\theta^2}$ as the first-order autocorrelation of the process we can write

$$\begin{pmatrix} 1 & \rho & 0 & 0 & 0 & 0 & 0 \\ \rho & 1 & \rho & 0 & \cdots & 0 & 0 \\ 0 & \rho & 1 & \rho & 0 & \cdots & 0 \\ \vdots & & & & & & \vdots \\ 0 & 0 & \cdots & 0 & \rho & 1 & \rho \\ 0 & 0 & \cdots & 0 & 0 & \rho & 1 \end{pmatrix} \begin{pmatrix} \phi_{t1} \\ \phi_{t2} \\ \phi_{t3} \\ \vdots \\ \phi_{t,t-1} \\ \phi_{tt} \end{pmatrix} = \begin{pmatrix} \rho \\ 0 \\ 0 \\ \vdots \\ 0 \\ 0 \end{pmatrix}.$$

Solving this equation system successively yields

$$\phi_{11} = \rho,$$

$$\phi_{21} = \frac{\rho}{1-\rho^2}, \qquad \phi_{22} = \frac{\rho^2}{1-\rho^2},$$

$$\phi_{31} = \frac{\rho(1-\rho^2)}{1-2\rho^2}, \qquad \phi_{32} = -\frac{\rho^2}{1-2\rho^2}, \qquad \phi_{33} = \frac{\rho^3}{1-2\rho^2}.$$

On the other hand, a much simpler expression for the one-step predictor of this MA(1) process can be obtained by means of the Innovations Algorithm. In this case, an application of the method yields,

$$\begin{aligned} \widehat{y}_1 &= 0 \\ \widehat{y}_t &= \theta_{t1}(y_t - \widehat{y}_t), \end{aligned}$$

with $\theta_{t1} = \frac{\gamma(0)}{\beta_t}$ and $\theta_{ti} = 0$ for $i > 1$. Notice that

$$\theta_{t1} \to \theta,$$

as $t \to \infty$.

EXAMPLE 7.2

For a fractional noise FN(d), the partial autocorrelation function is given by

$$\phi_{tt} = \frac{d}{t-d}, \tag{7.3}$$

for $t \geq 1$. Hence, the prediction error variance is given by

$$\begin{aligned} \nu_t &= \gamma(0) \prod_{j=1}^{t} \left[1 - \left(\frac{d}{j-d} \right)^2 \right] \\ &= \gamma(0) \frac{\prod_{j=1}^{t} j \prod_{j=1}^{t} (j-2d)}{\left[\prod_{j=1}^{t} (j-d) \right]^2}. \end{aligned}$$

But, for any real α we have that

$$\prod_{j=1}^{t} (j-\alpha) = \frac{\Gamma(t+1-\alpha)}{\Gamma(1-\alpha)}.$$

Hence,

$$\nu_t = \gamma(0) \frac{\Gamma(t+1)\Gamma(t+1-2d)\Gamma(1-d)^2}{[\Gamma(t+1-d)]^2 \, \Gamma(1-2d)}.$$

Therefore, since from (2.25) the autocovariance function at zero lag is

$$\gamma(0) = \sigma^2 \frac{\Gamma(1-2d)}{[\Gamma(1-d)]^2},$$

we obtain the formula

$$\nu_t = \sigma^2 \frac{\Gamma(t+1)\Gamma(t+1-2d)}{[\Gamma(t+1-d)]^2},$$

for $t \geq 1$. Now, an application of expression (2.32) yields

$$\lim_{t \to \infty} \nu_t = \sigma^2.$$

Figure 7.1 displays the evolution of the partial autocorrelation coefficients ϕ_{tt} for $t = 1, \ldots, 20$ for a fractional noise FN(d) process with $d = 0.4$, an AR(1) model with $\phi_1 = 0.2$ and $\phi_2 = 0.4$, and a MA(1) process with $\theta_1 = 0.4$ and $\theta_2 = -0.2$.

Notice that for the fractional noise processes, $\phi_{tt} = \mathcal{O}(1/t)$ for large t. Despite the difficulty of finding explicit expressions for the partial autocorrelations for a general class of long-memory models, this rate of convergence to zero can be extended to any ARFIMA(p, d, q) process, as stated in the following result: Let $\{y_t\}$ be an ARFIMA(p, d, q) process with $0 < d < \frac{1}{2}$. Then, the partial autocorrelations ϕ_{tt} satisfy

$$|\phi_{tt}| \sim \frac{d}{t}, \qquad (7.4)$$

as $t \to \infty$.

Observe that the rate on t given by expression (7.4) does not depend on the value of the long-memory parameter d.

Figure 7.2 displays the evolution of the mean-squared prediction error ν_t for a fractional noise FN(d) with $d = 0.10$, $d = 0.30$, $d = 0.49$, $\sigma^2 = 1$, and $t = 1, \ldots, 40$.

As t increases, the effect of the remote past fades out and $\widetilde{y}_{t+1}$ becomes similar to $\widehat{y}_{t+1}$. In turn, the prediction error variance of the finite sample predictor ν_t becomes similar to the prediction error variance of the infinite past predictor, σ^2: If $\{y_t\}$ is a stationary process, then $\|\widehat{y}_{t+1} - \widetilde{y}_{t+1}\| \to 0$ and $\nu_t \to \sigma^2$ as $t \to \infty$.

Let $\delta_t = \|\widehat{y}_{t+1} - \widetilde{y}_{t+1}\|^2$ be the squared distance between the optimal predictor based on the full past and the best predictor based on the finite past. Then, we may write

$$\begin{aligned}
\|\widehat{y}_{t+1} - \widetilde{y}_{t+1}\|^2 &= \|(\widehat{y}_{t+1} - y_{t+1}) + (y_{t+1} - \widetilde{y}_{t+1})\|^2 \\
&= \|\widehat{y}_{t+1} - y_{t+1}\|^2 + \|y_{t+1} - \widetilde{y}_{t+1}\|^2 \\
&\quad -2\langle y_{t+1} - \widehat{y}_{t+1}, y_{t+1} - \widetilde{y}_{t+1}\rangle \\
&= \sigma^2 + \nu_t - 2\langle \varepsilon_{t+1}, y_{t+1} - \widetilde{y}_{t+1}\rangle.
\end{aligned}$$

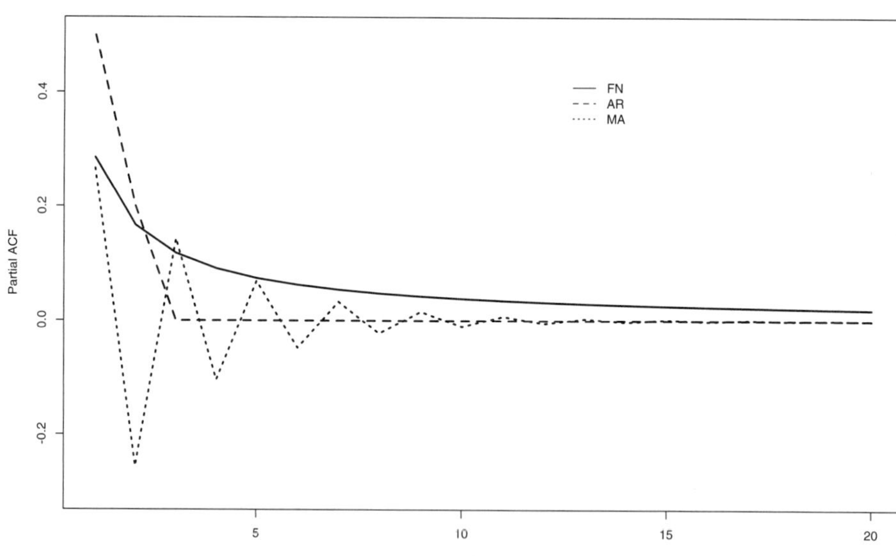

Figure 7.1 Evolution of the partial autocorrelation coefficients ϕ_{tt} for $t = 1, \ldots, 20$ for a fractional noise FN(d) process with $d = 0.4$ (heavy line), an AR(1) model with $\phi_1 = 0.2$ and $\phi_2 = 0.4$ (broken line), and a MA(1) process with $\theta_1 = 0.4$ and $\theta_2 = -0.2$ (dotted line).

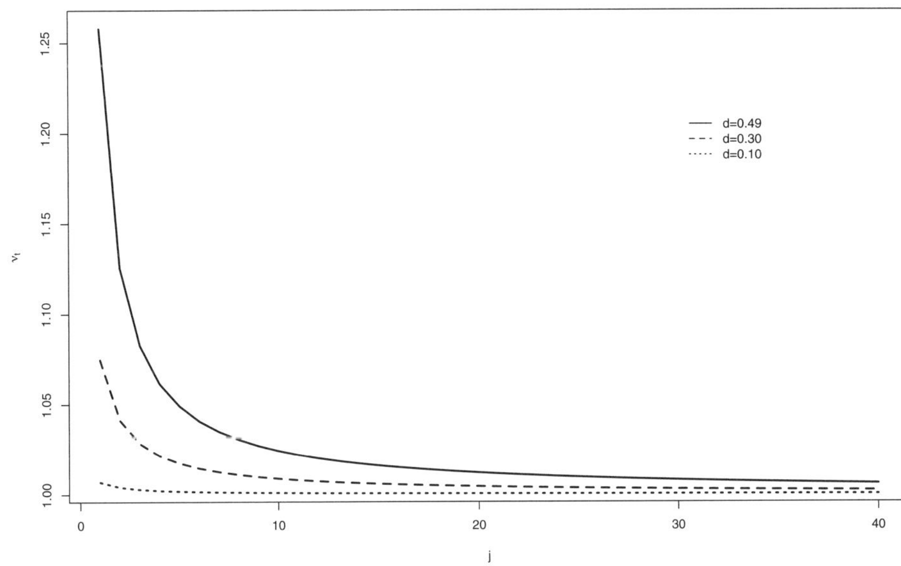

Figure 7.2 Evolution of the mean-squared prediction error ν_t for $t = 1, \ldots, 40$ for a fractional noise FN(d) process with $d = 0.10$, $d = 0.30$, $d = 0.49$, and $\sigma^2 = 1$.

Since $\langle \varepsilon_{t+1}, y_{t+1} \rangle = \sigma^2$ and $\langle \varepsilon_{t+1}, \widetilde{y}_{t+1} \rangle = 0$, we have

$$\delta_t = \nu_t - \sigma^2.$$

A precise rate at which ν_t converges to σ^2 as t increases is as follows: Let $\{y\}$ be an ARFIMA(p, d, q) process with unit variance noise and $0 < d < \frac{1}{2}$. Then,

$$\delta_t \sim \frac{d^2}{t}, \tag{7.5}$$

as $t \to \infty$.

7.2.4 An Approximate Predictor

Since the Durbin-Levinson algorithm for calculating the coefficients ϕ_{tj} is order $\mathcal{O}(n^2)$, for very large sample sizes it could be desirable a faster algorithm to obtain finite sample forecasts. One way to do this is approximating the regression coefficients ϕ_{tj} by π_j, based on the following fact: If $\{y_t\}$ is a stationary process, then $\phi_{tj} \to \pi_j$ as $t \to \infty$.

With this approximation, we introduce the finite sample predictor

$$\breve{y}_{t+1} = \sum_{j=1}^{t} \pi_j y_{t+1-j},$$

with prediction error variance

$$\mathrm{Var}(y_{t+1} - \check{y}_{t+1}) = \sigma^2 + r_t,$$

where $r_t = \mathrm{Var}(\sum_{j=t+1}^{\infty} \pi_j y_{t+1-j})$ or equivalently

$$r_t = \mathrm{Var}(\widehat{y}_{t+1} - \check{y}_{t+1}).$$

As expected, the prediction error variance of the approximate forecast is larger than the prediction error variance of $\widetilde{y}_{t+1}$. However, as t increases, these two predictors become similar: If $\{y_t\}$ is a stationary process with AR(∞) representation satisfying $\sum_{j=1}^{\infty} |\pi_j| < \infty$, then $r_t \to 0$ as $t \to \infty$.

For short-memory processes with autoregressive coefficients $\{\pi_j\}$ satisfying

$$|\pi_j| \sim c|\phi|^j,$$

for large j and positive constant c we have that

$$r_t \le c_1 |\phi|^{2t},$$

where $c_1 = \mathrm{Var}(y_0)[c|\phi|/(1-|\phi|)]^2$. Therefore, r_t converges to zero at an exponential rate.

However, for long-memory processes this rate is slower: If $\{y_t\}$ is a stationary and invertible process with AR(∞) and MA(∞) satisfying

$$\pi_j \sim \frac{j^{-d-1}}{\ell(j)\Gamma(-d)},$$
$$\psi_j \sim \frac{j^{d-1}\ell(j)}{\Gamma(d)},$$

as $j \to \infty$, with $0 < d < \frac{1}{2}$ and $\ell(\cdot)$ is a slowly varying function, then

$$r_t \sim \frac{d \tan(\pi d)}{\pi t}, \tag{7.6}$$

as $t \to \infty$.

Comparing expressions (7.5) and (7.6), we observe that for small values of d both terms δ_t and r_t behave similarly since

$$\frac{d \tan(\pi d)}{\pi t} \sim \frac{d^2}{t},$$

as $d \to 0$. On the contrary, when d approaches $\frac{1}{2}$, δ_t is bounded but r_t increases to infinity since $\tan(\pi/2) = \infty$.

7.3 MULTISTEP AHEAD PREDICTORS

7.3.1 Infinite Past

Let $\widehat{y}_t(h)$ be the best linear predictor of y_{t+h} based on the infinite past $\mathcal{F}_t$ for $h \geq 1$, which may be written as

$$\widehat{y}_t(h) = E[y_{t+h}|\mathcal{F}_t] = \sum_{j=0}^{\infty} \pi_j(h) y_{t-j} = \sum_{j=0}^{\infty} \psi_{j+h} \varepsilon_{t-j} = \sum_{j=h}^{\infty} \psi_j \varepsilon_{t+h-j}, \quad (7.7)$$

where the coefficients $\pi_j(h)$ for $j \geq 1$ are given by

$$\pi_j(h) = \sum_{i=0}^{h-1} \psi_i \pi_{j+h-i}.$$

The prediction error variance of $\widehat{y}_t(h)$, $\sigma^2(h) = E[y_{t+h} - \widehat{y}_t(h)]^2$, is

$$\sigma^2(h) = \sigma^2 \sum_{j=0}^{h-1} \psi_j^2. \quad (7.8)$$

7.3.2 Finite Past

The best linear predictor of y_{t+h} based on the finite past $\mathcal{P}_t$ is

$$\widetilde{y}_t(h) = \phi_{t1}(h) y_t + \cdots + \phi_{tt}(h) y_1,$$

where $\phi_t(h) = [\phi_{t1}(h), \ldots, \phi_{tt}(h)]'$ satisfies

$$\Gamma_t \phi_t(h) = \gamma_t(h),$$

with $\gamma_t(h) = [\gamma(h), \ldots, \gamma(t+h-1)]'$. Besides, the mean-squared prediction error is defined by

$$\sigma_t^2(h) = \|y_{t+h} - \widetilde{y}_t(h)\|^2.$$

Analogously to the one-step prediction case, we can use the approximate finite sample h-step ahead forecasts given by

$$\check{y}_t(h) = \sum_{j=0}^{t} \pi_j(h) y_{t-j},$$

with prediction error variance

$$\text{Var}[y_{t+h} - \check{y}_t(h)] = \sigma^2 + r_t(h),$$

where $r_t(h) = \mathrm{Var}[\sum_{j=t+1}^{\infty} \pi_j(h) y_{t-j}]$. For fixed h, this term behaves similarly to r_t, excepting a constant: Let $\{y_t\}$ be a stationary and invertible process with AR(∞) and MA(∞) satisfying

$$\begin{aligned} \pi_j &\sim \frac{j^{-d-1}}{\ell(j)\Gamma(-d)}, \\ \psi_j &\sim \frac{j^{d-1}\ell(j)}{\Gamma(d)}, \end{aligned}$$

as $j \to \infty$, with $0 < d < \frac{1}{2}$ and $\ell(\cdot)$ is a slowly varying function. If $\sum_{j=0}^{h-1} \psi_j \neq 0$, then

$$r_t(h) \sim \left(\sum_{j=0}^{h-1} \psi_j \right)^2 \frac{d \tan(\pi d)}{\pi t},$$

as $t \to \infty$.

7.4 HETEROSKEDASTIC MODELS

Consider a general heteroskedastic process $\{y_t : t \in \mathbb{Z}\}$ specified by the equations

$$y_t = \sum_{j=0}^{\infty} \psi_j \varepsilon_{t-j}, \tag{7.9}$$

$$\varepsilon_t = \epsilon_t \sigma_t, \tag{7.10}$$

$$\sigma_t^2 = f(\epsilon_{t-1}, \epsilon_{t-2}, \ldots, \eta_t, \eta_{t-1}, \ldots), \tag{7.11}$$

where f is function $\{\epsilon_t\}$ and $\{\eta_t\}$, which are sequences of independent and identically distributed random variables with zero-mean and unit variance and $\{\epsilon_t\}$ is independent of $\{\eta_t\}$.

Observe that this specification includes the ARFIMA-GARCH, the ARCH-type and the LMSV processes, among others, as shown in the following examples.

EXAMPLE 7.3

The conditional variance of the ARFIMA-GARCH(1, 1) model may be written as in (6.9),

$$f(\epsilon_{t-1}, \epsilon_{t-2}, \ldots, \eta_t, \eta_{t-1}, \ldots) = \alpha_0 \left[1 + \sum_{k=0}^{\infty} \prod_{j=1}^{k+1} (\alpha_1 \epsilon_{t-j}^2 + \beta_1) \right],$$

and $\eta_t = 0$ for all t.

■ EXAMPLE 7.4

The ARCH-type process is also included in specification (7.9)–(7.11) with the Volterra expansion

$$f(\epsilon_{t-1}, \epsilon_{t-2}, \dots) = \alpha_0 \sum_{k=0}^{\infty} \sum_{j_1,\dots,j_k=1}^{\infty} \alpha_{j_1}\alpha_{j_2}\cdots\alpha_{j_k}\epsilon_{t-j_1}^2\epsilon_{t-j_2}^2\cdots\epsilon_{t-j_k}^2,$$

and $\eta_t = 0$ for all t, cf., expression (6.14).

■ EXAMPLE 7.5

For the LMSV model introduced in Section 6.9, the conditional variance is given by

$$f(\eta_t, \eta_{t-1}, \dots) = \sigma^2 \exp\{\phi(B)^{-1}(1-B)^{-d}\theta(B)\eta_t\}.$$

Notice that in this case, the conditional variance does not depend directly on the sequence $\{\epsilon_t\}$ but it depends on random perturbations $\{\eta_t\}$ such as those appearing in (6.21).

■ EXAMPLE 7.6

The conditional variance for the FIEGARCH model may be written as

$$\begin{aligned} &f(\epsilon_{t-1}, \epsilon_{t-2}, \dots) \\ &\qquad = \exp\left\{\phi(B)^{-1}(1-B)^{-d}[\theta(B)(|\epsilon_{t-1}| - E\,|\epsilon_{t-1}|) + \lambda(B)\epsilon_{t-1}]\right\}; \end{aligned}$$

see Problem 6.18.

The following fact is fundamental for the formulation of prediction of the volatility techniques for the heteroskedastic models described by (7.9)–(7.11). For all $t \in \mathbb{Z}$ we have

$$E[\varepsilon_{t+h}^2|\mathcal{P}_t] = \begin{cases} E[\sigma_{t+h}^2|\mathcal{P}_t] & h \geq 1, \\ \varepsilon_{t+h}^2 & h < 1. \end{cases} \tag{7.12}$$

This can be verified as follows. First, notice that for $h < 1$ the result is trivial. For $h \geq 1$ observe that by the definition of ε_{t+h} we may write $E[\varepsilon_{t+h}^2|\mathcal{P}_t] = E[\epsilon_{t+h}^2\sigma_{t+h}^2|\mathcal{P}_t]$. But, since $\sigma_{t+h}^2 = f(\epsilon_{t+h-1}, \epsilon_{t+h-2}, \dots, \eta_{t+h}, \eta_{t+h-1}, \dots)$, σ_{t+h}^2 and ϵ_{t+h} are independent. Furthermore, ϵ_{t+h} is independent of $y_t, y_{t-1}, \dots$ for any $h \geq 1$. Thus,

$$E[\varepsilon_{t+h}^2|\mathcal{P}_t] = E[\epsilon_{t+h}^2]\,E[\sigma_{t+h}^2|\mathcal{P}_t].$$

Finally, by noting that $E[\epsilon_{t+h}^2] = 1$, the result is obtained.

7.4.1 Prediction of Returns

In the context of the heteroskedastic model (7.9), the forecasts of the returns can be obtained similarly to the cases discussed in Section 6.1. On the other hand, estimation the prediction error conditional variances is addressed in the next subsection. Practical illustrations of these techniques are provided later in this chapter.

7.4.2 Prediction of Volatility

Observe that from (7.7) the multistep ahead prediction error for $h \geq 1$ is given by

$$e_t(h) = y_{t+h} - \widehat{y}_t(h) = \varepsilon_{t+h} + \psi_1 \varepsilon_{t+h-1} + \cdots + \psi_{h-1}\varepsilon_{t+1}.$$

Therefore, the mean-squared prediction error is

$$\begin{aligned} E[e_t^2(h)|\mathcal{P}_t] &= E[\varepsilon_{t+h}^2|\mathcal{P}_t] + \psi_1^2\, E[\varepsilon_{t+h-1}^2|\mathcal{P}_t] + \\ &\quad \cdots + \psi_{h-1}^2\, E[\varepsilon_{t+1}^2|\mathcal{P}_t]. \end{aligned}$$

Thus, we may write

$$\begin{aligned} E[e_t^2(h)|\mathcal{P}_t] &= E[\sigma_{t+h}^2|\mathcal{P}_t] + \psi_1^2\, E[\sigma_{t+h-1}^2|\mathcal{P}_t] + \\ &\quad \cdots + \psi_{h-1}^2\, E[\sigma_{t+1}^2|\mathcal{P}_t]. \end{aligned}$$

Let $\sigma_t^2(h) = E[\sigma_{t+h}^2|\mathcal{P}_t]$, then the h-step prediction error conditional variance is

$$E[e_t^2(h)|\mathcal{P}_t] = \sum_{j=0}^{h-1} \psi_j^2 \sigma_t^2(h-j).$$

The calculation of the conditional variances $\sigma_t^2(j)$ depends on the specification of the heteroskedastic model. Some specific examples are discussed next.

■ EXAMPLE 7.7

Consider an ARCH(1) process and assume that we know the returns $\{y_n, y_{n-1}, \cdots\cdots\}$ and we have the equations $\sigma_n^2 = \alpha_0 + \alpha_1\, y_{n-1}^2$ and $\sigma_{n+1}^2 = \alpha_0 + \alpha_1\, y_n^2$. In order to forecast the conditional variance σ_{n+2}^2 we can write:

$$\hat{\sigma}_{n+2}^2 = E[\sigma_{n+2}^2|\mathcal{P}_n].$$

Thus,

$$\hat{\sigma}_{n+2}^2 = E[\alpha_0 + \alpha_1\, y_{n+1}^2|\mathcal{P}_n] = \alpha_0 + \alpha_1\, E(y_{n+1}^2|\mathcal{P}_n)$$

Since $\{y_t^2\}$ corresponds to an AR(1) process, we can write

$$y_{n+1}^2 = \alpha_0 + \alpha_1\, y_n^2 + \nu_{n+1}.$$

Now, by taking expectation we have

$$\begin{aligned} E(y_{n+1}^2\,|\mathcal{F}_n) &= \alpha_0 + \alpha_1\, E(y_n^2\,|\mathcal{F}_n) + E(\nu_{n+1}\,|\mathcal{F}_n) \\ &= \alpha_0 + \alpha_1\, y_n^2. \end{aligned}$$

On the other hand,

$$\widehat{\sigma}_{n+2}^2 = \alpha_0 + \alpha_1\,[\alpha_0 + \alpha_1\, y_n^2] = \alpha_0 + \alpha_1\,\alpha_0 + \alpha_1^2\, y_n^2.$$

Now, for predicting σ_{n+3}^2 we can write,

$$\widehat{\sigma}_{n+3}^2 = E[\alpha_0 + \alpha_1\, y_{n+2}^2|\mathcal{F}_n] = \alpha_0 + \alpha_1\, E(y_{n+2}^2|\mathcal{F}_n)$$

But,

$$y_{n+2}^2 = \alpha_0 + \alpha_1\, y_{n+1}^2 + \nu_{n+2}$$

so that by taking expectation on both sides

$$E(y_{n+2}^2\,|\mathcal{F}_n) = \alpha_0 + \alpha_1\, E(y_{n+1}^2\,|\mathcal{F}_n) + E(\nu_{n+2}\,|\mathcal{F}_n)$$

and then

$$\begin{aligned} E(y_{n+2}^2\,|\mathcal{F}_n) &= \alpha_0 + \alpha_1(\alpha_0 + \alpha_1\, y_n^2) + 0 \\ &= \alpha_0 + \alpha_1\,\alpha_0 + \alpha_1^2\, y_n^2 \end{aligned}$$

Thus,

$$\begin{aligned} \hat{\sigma}_{n+3}^2 &= \alpha_0 + \alpha_1\,\alpha_0 + \alpha_1^2\,\alpha_0 + \alpha_1^3\, y_n^2 \\ &= \alpha_0\,(1 + \alpha_1 + \alpha_1^2) + \alpha_1^3\, y_n^2. \end{aligned}$$

More generally, the h-step predictor is given by

$$\widehat{\sigma}_{n+h}^2 = \alpha_0 \left[\sum_{i=0}^{h-1} \alpha_1^i\right] + \alpha_1^h\, y_n^2 \tag{7.13}$$

Notice that the behavior of the predictor (7.13) as the forecasting horizon increases $h \to \infty$ is as follows

$$\widehat{\sigma}_{n+h}^2 \to_p \alpha_0 \left[\sum_{i=0}^{\infty} \alpha_1^i\right] = \frac{\alpha_0}{1-\alpha_1} = \mathrm{Var}(y_t).$$

EXAMPLE 7.8

The previous results established for ARCH(1) processes can be readily extended to the ARCH(p) model. In this case we have

$$\begin{aligned}\hat{\sigma}_n^2(1) &= \alpha_0 + \alpha_1\, y_n^2 + \cdots + \alpha_p\, y_{n+1-p}^2\\ \hat{\sigma}_n^2(2) &= \alpha_0 + \alpha_1\, \hat{\sigma}_n^2(1) + \alpha_2\, y_n^2 \cdots + \alpha_p\, y_{n+2-p}^2,\end{aligned}$$

and more generally,

$$\widehat{\sigma}_n^2(\ell) = \alpha_0 + \sum_{i=1}^{p} \alpha_i\, \widehat{\sigma}_n^2(\ell - i) \tag{7.14}$$

where $\widehat{\sigma}_n^2(\ell - i) = y_{n+\ell-i}^2$ para $\ell - i \leq 0$.

EXAMPLE 7.9

For the ARFIMA-GARCH(1, 1) we have

$$\sigma_{t+h}^2 = \alpha_0 + \alpha_1 \varepsilon_{t+h-1}^2 + \beta_1 \sigma_{t+h-1}^2.$$

Therefore, an application of (7.12) yields $\sigma_t^2(1) = \sigma_{t+1}^2$ and for $h \geq 2$,

$$\sigma_t^2(h) = \alpha_0 + (\alpha_1 + \beta_1)\sigma_t^2(h-1).$$

Solving this recursive equation we find the following solution for $h \geq 1$:

$$\sigma_t^2(h) = \alpha_0 \frac{1 - (\alpha_1 + \beta_1)^{h-1}}{1 - (\alpha_1 + \beta_1)} + (\alpha_1 + \beta_1)^{h-1} \sigma_{t+1}^2.$$

Since $0 \leq \alpha_1 + \beta_1 < 1$, we have that

$$\lim_{h \to \infty} \sigma_t^2(h) = \frac{\alpha_0}{1 - (\alpha_1 + \beta_1)},$$

where the term on the left hand of this equation corresponds to the variance of $\{e_t\}$.

EXAMPLE 7.10

For the general ARFIMA-GARCH(r, s) it is not hard to check that for $h > \max\{r, s\}$ we have

$$\sigma_t^2(h) = \alpha_0 + \sum_{j=1}^{r} \alpha_j \sigma_t^2(h-j) + \sum_{j=1}^{s} \beta_j \sigma_t^2(h-j),$$

and since $0 < \sum_{j=1}^{r} \alpha_j + \sum_{j=1}^{s} \beta_j < 1$,

$$\lim_{h \to \infty} \sigma_t^2(h) = \frac{\alpha_0}{1 - (\sum_{j=1}^{r} \alpha_j + \sum_{j=1}^{s} \beta_j)} = \mathrm{Var}(\varepsilon_t).$$

■ **EXAMPLE 7.11**

For the ARCH(∞) model, we have

$$\sigma_{t+h}^2 = \alpha_0 + \sum_{j=1}^{\infty} \alpha_j y_{t+h-j}^2 = \alpha_0 + \sum_{j=1}^{h-1} \alpha_j y_{t+h-j}^2 + \sum_{j=h}^{\infty} \alpha_j y_{t+h-j}^2.$$

Thus,

$$E[\sigma_{t+h}^2|\mathcal{F}_t] = \alpha_0 + \sum_{j=1}^{h-1} \alpha_j \, E[y_{t+h-j}^2|\mathcal{F}_t] + \sum_{j=h}^{\infty} \alpha_j y_{t+h-j}^2,$$

and then,

$$\sigma_t^2(h) = \alpha_0 + \sum_{j=1}^{h-1} \alpha_j \sigma_t^2(h-j) + \sum_{j=h}^{\infty} \alpha_j y_{t+h-j}^2.$$

Now, if $0 < \sum_{j=1}^{\infty} \alpha_j < 1$, then

$$\lim_{h\to\infty} \sigma_t^2(h) = \frac{\alpha_0}{1 - \sum_{j=1}^{\infty} \alpha_j} = \text{Var}(y_t).$$

7.5 PREDICTION BANDS

Given the forecasts $\widehat{y}_{t+h}$ and the estimated prediction error variances $\widehat{\sigma}_{t+h}^2$, we can build the approximate prediction bands

$$\left[\, \widehat{y}_{t+h} - z_\alpha \widehat{\sigma}_{t+h}, \quad \widehat{y}_{t+h} + z_\alpha \widehat{\sigma}_{t+h} \,\right]$$

where z_α corresponds to the $100\,(1-\alpha)$ percentile. For example, under Gaussianity, for the 90% prediction bands we can use $z_\alpha = 1.65$ while for 95% prediction bands this value is $z_\alpha = 1.96$.

■ **EXAMPLE 7.12**

Figure 7.3 displays a simulated Gaussian ARFIMA$(0, d, 0)$ process with zero-mean and unit variance white noise, long-memory parameter $d = 0.40$ and sample size $n = 1000$. The last 100 observations (dotted line) will be predicted using the estimated model.

The MLE of d calculated from the first 900 observations is $\widehat{d} = 0.4096$ with standard deviation $\widehat{\sigma}_d = 0.0251$. Besides, the estimated standard deviation of the white noise is $\widehat{\sigma} = 0.9735$. Figure 7.4 shows the observations from $t = 800$ to $t = 1000$, along with $\widehat{y}_{900}(h)$ forecasts with $h = 1, 2, \ldots, 100$ and 95% prediction bands. These multistep ahead predictors are based on the fitted model and on observations

$t = 1, 2, \dots, 900$. From Figure 7.4, notice that most of the future observations fall inside the prediction bands.

Figure 7.5 displays the theoretical and the empirical evolution of the prediction error standard deviation, from $t = 901$ to $t = 1000$. The *theoretical* prediction error standard deviation is based on the multistep prediction error variance formula (7.8) which yields

$$\widehat{\sigma}(h) = \widehat{\sigma}\sqrt{\sum_{j=0}^{h-1} \psi_j^2(\widehat{d})}, \tag{7.15}$$

for $h = 1, 2, \dots, 100$. Equation (7.15) provides only an approximation of the theoretical prediction error standard deviation at step h since we are dealing with a finite past of 900 observations. On the other hand, the empirical prediction error standard deviations are based on the Kalman filter output from a truncated state space representation with $m = 50$. Notice from this graph that the sample prediction error standard deviations are very close to their theoretical counterparts.

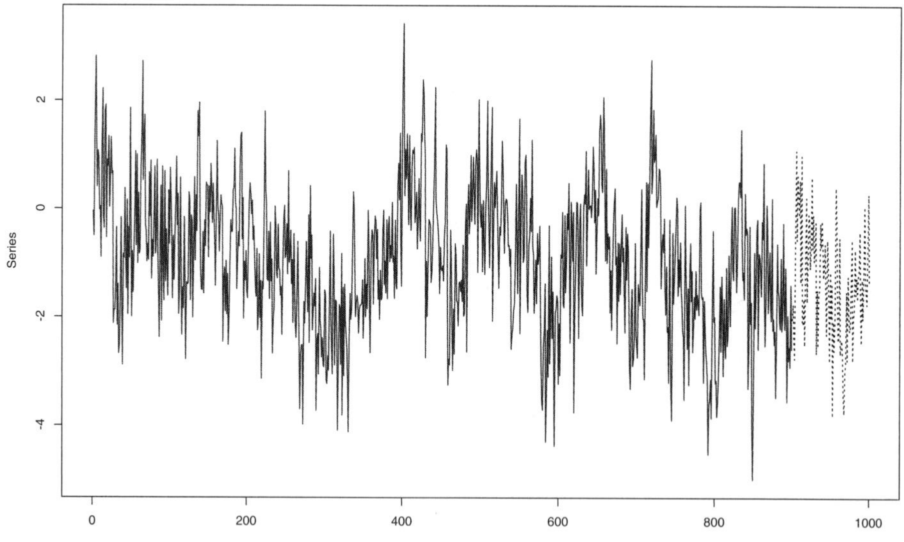

Figure 7.3 Simulated fractional noise process FN(d), with $d = 0.40$ and unit variance white noise. The dotted line indicates the last 100 observations that will be predicted using the estimated model.

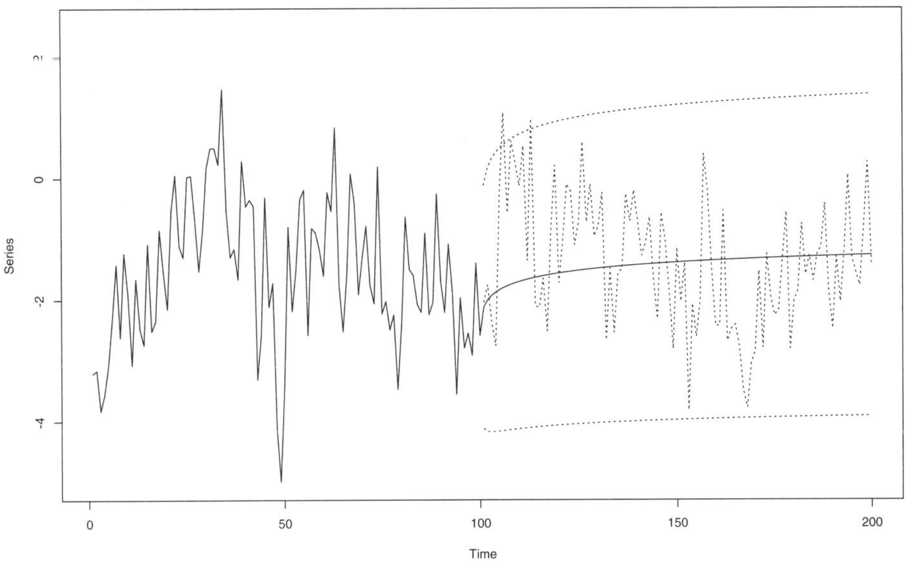

Figure 7.4 Simulated fractional noise process FN(d): Multistep forecasts of the last 100 values and 95% prediction bands.

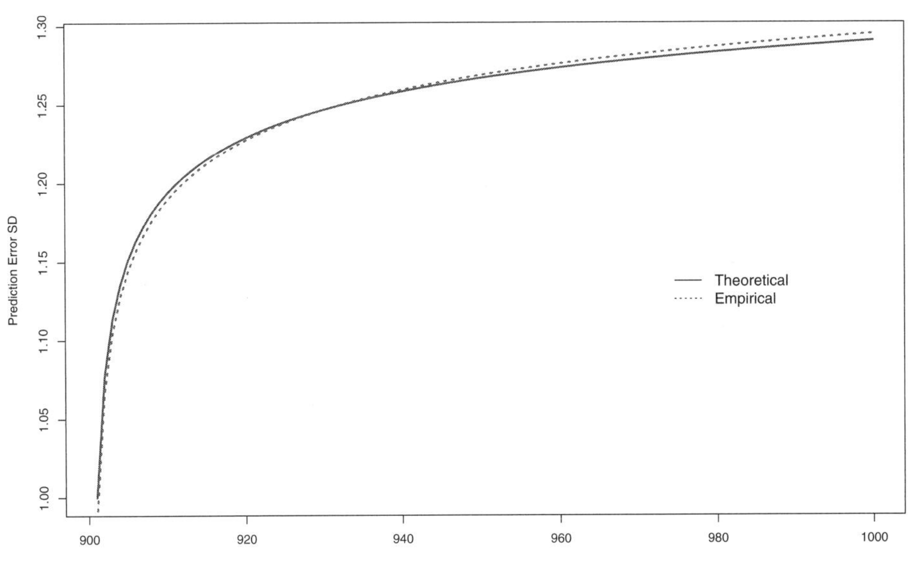

Figure 7.5 Simulated fractional noise process FN(d): Theoretical and empirical prediction error standard deviations of the last 100 observations.

EXAMPLE 7.13

We now illustrate the prediction of a short-memory process. Figure 7.6 shows a simulated trajectory of 1000 observations from an ARMA(1,1) process with parameters $\phi = 0.80$ and $\theta = -0.40$.

In order to illustrate the application of the forecasting methods discussed in this chapter, the series has been divided into two parts, the first 900 observations to fit the model and the remaining 100 observations for prediction.

The fitted model based on the 900 observations is reported below and the diagnostic plots are exhibited Figure 7.7, including the residuals, the ACF of residual and the Box-Ljung tests.

According to this set of graphs and tests, the fitting of this ARMA$(1,1)$ model seems adequate.

```
Coefficients:

         ar1      ma1  intercept
      0.8071  -0.3818     0.2215
s.e.  0.0329   0.0527     0.1085
```

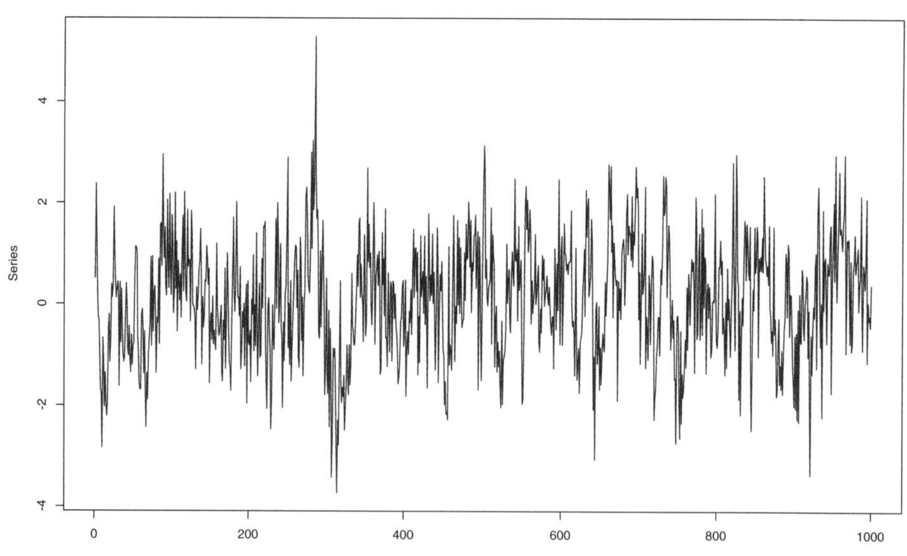

Figure 7.6 Example of prediction of an ARMA process. Simulated ARMA$(1,1)$ time series with $\phi = 0.8$ and $\theta = -0.4$.

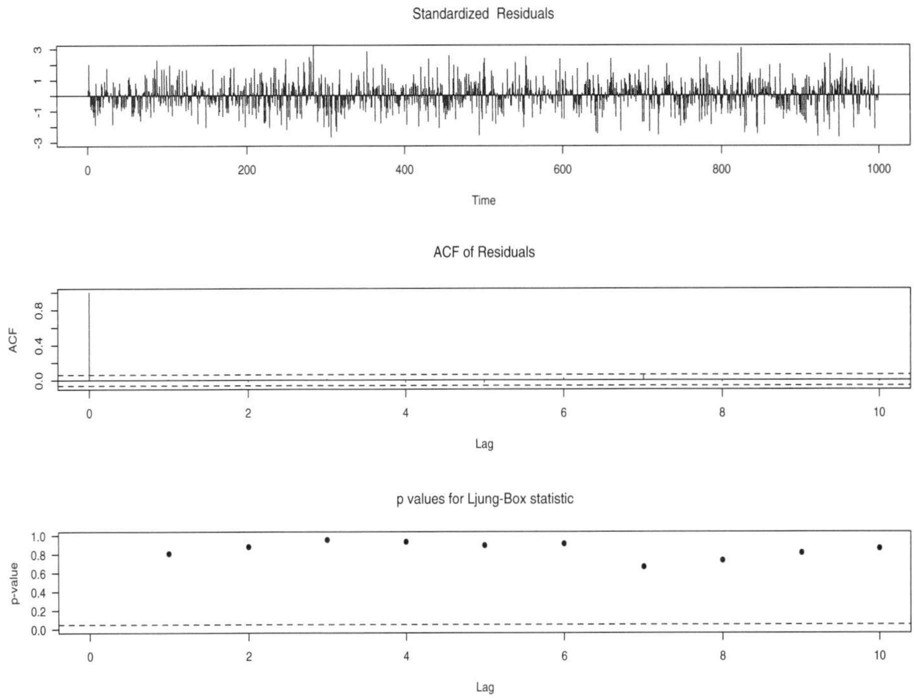

Figure 7.7 Example of prediction of an ARMA process. Fitted model diagnostics.

```
sigma^2 estimated as 1.04:  log likelihood = -1294.96,

aic = 2597.93
```

Figure 7.8 displays the series along the out-of-sample predictions and 95% prediction bands for observations from $t = 901$ to $t = 1000$. The predictions bands in this case are given by $\pm 2\sqrt{\nu_t}$.

Note that in this case the predictors converge to the mean of the process ($\mu = 0$) and that the prediction bands converge very fast to the corresponding to $\pm 2\sqrt{\mathrm{Var}(y_t)}$.

On the other hand, Figure 7.9 shows the last 100 values of the series along their corresponding forecasts and prediction bands. From this plot, observe that, as expected, most of the values lie inside of the prediction bands.

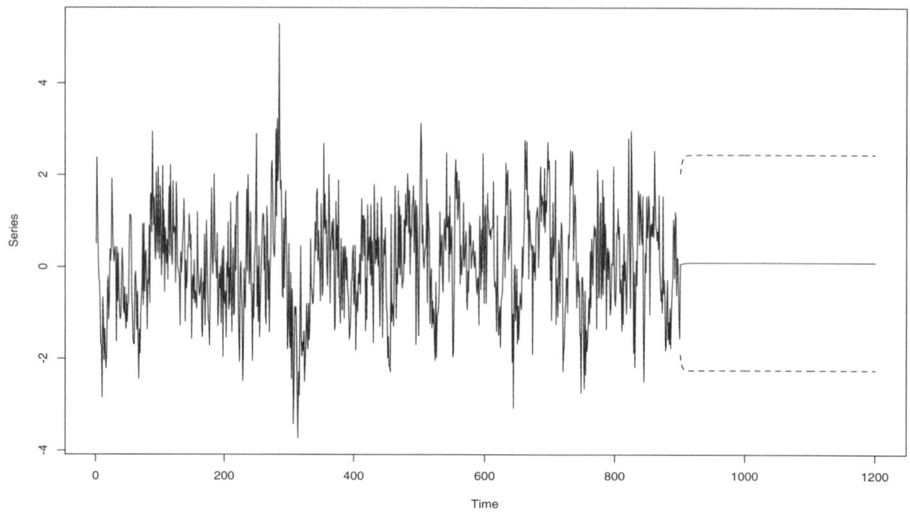

Figure 7.8 Example of prediction of an ARMA process. Time series, forecasts and predictions bands.

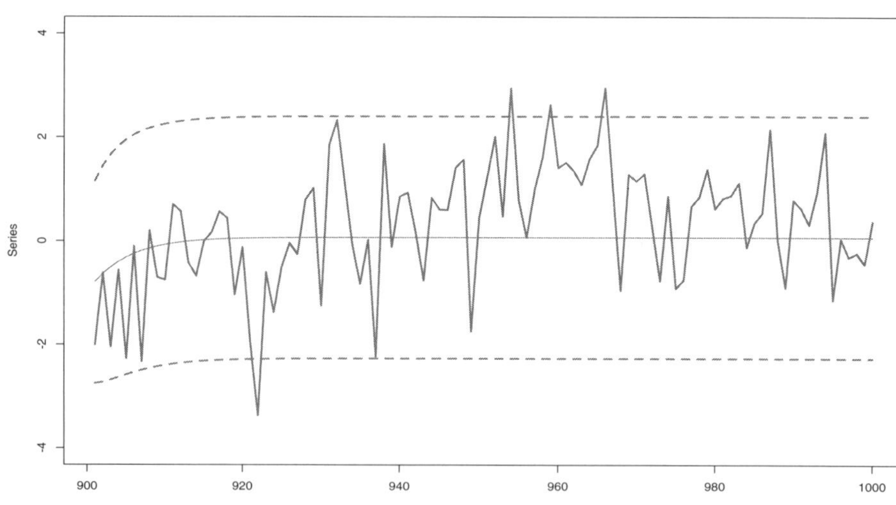

Figure 7.9 Example of prediction of an ARMA process. Last 100 observations, forecasts and predictions bands.

7.6 DATA APPLICATION

This section illustrates the application of prediction methods discussed in this chapter to a real-life time series. We consider the SP500 stock index introduced in Chapter 1. Recall that an AR(2)-GARCH(1, 1) model was fitted to these data in Section 6.11.

Figure 7.10 shows the SP500 returns while Figure 7.11 displays the volatility estimates for these data. Observe that the volatility estimates behave similarly to the variability of the returns. As expected, periods of high volatility produce large estimates while periods of low volatility generate small estimates. Figure 7.12 exhibits the corresponding estimates of the conditional standard deviations σ_t.

On the other hand, Figure 7.13 shows 100 out-of-sample forecasts of the SP500 returns along with 95% prediction bands.

In order to provide a comparison framework, we have also plotted 200 returns corresponding to the period $t = 15,905$ to $t = 16,104$. Notice that the predictors converge rapidly to the mean of the series, zero, and that the prediction bands increases as the forecasting horizon increases.

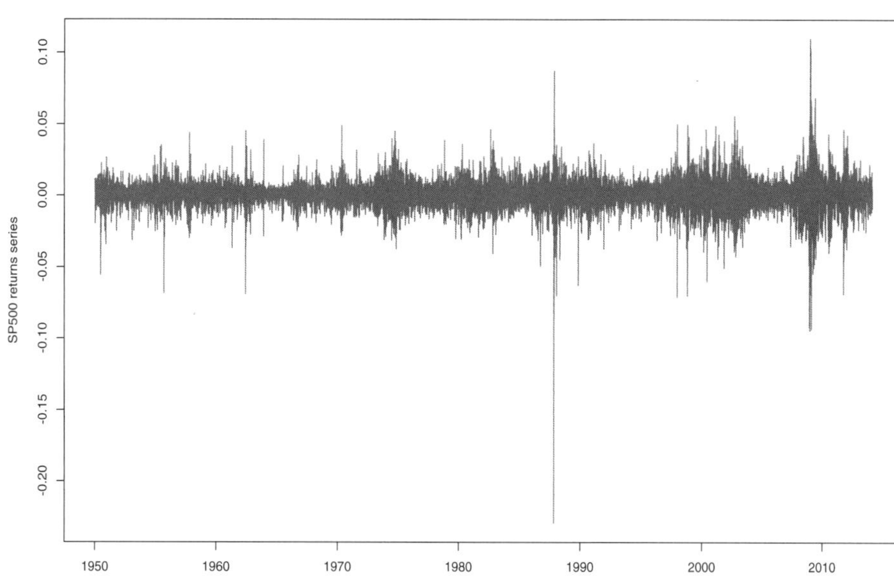

Figure 7.10 SP500 Returns data.

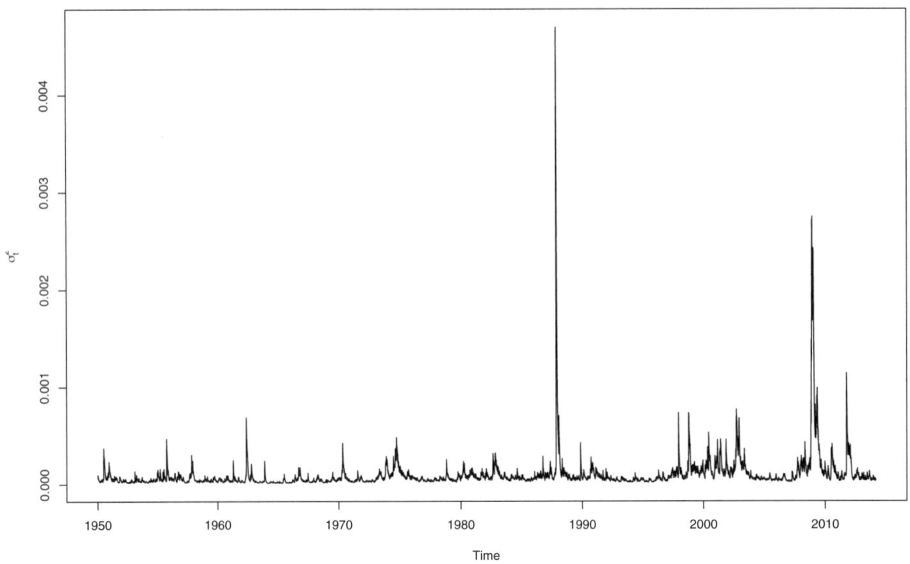

Figure 7.11 SP500 Returns data. Volatility estimates.

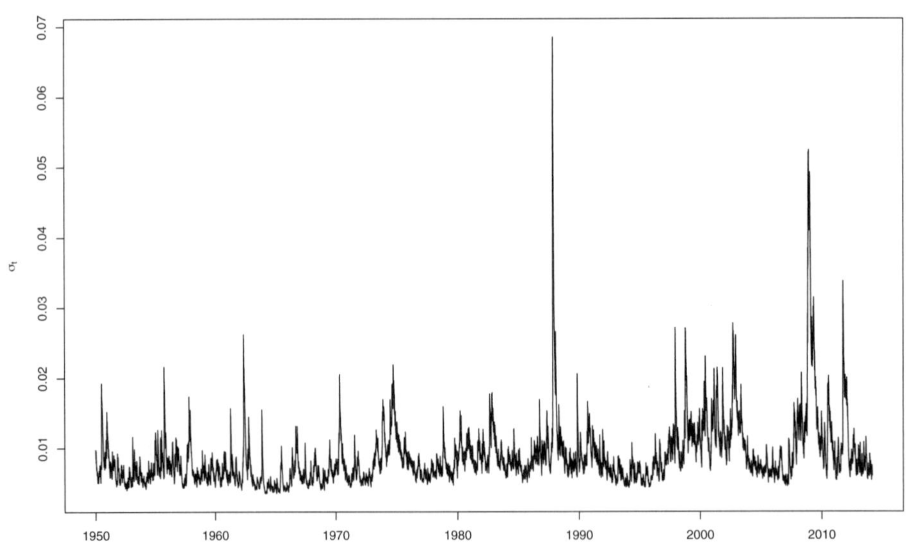

Figure 7.12 SP500 Returns data. Estimates of σ_t.

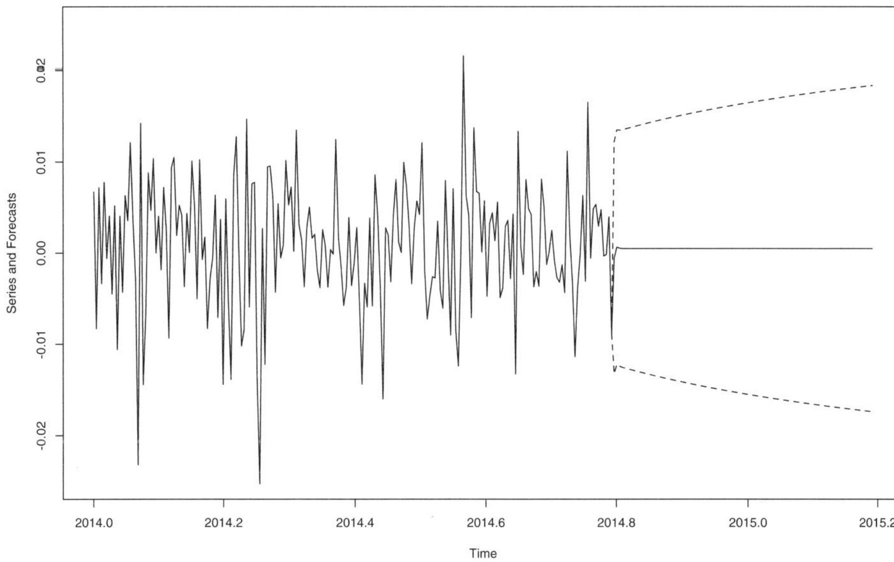

Figure 7.13 SP500 Returns data. Out-of-sample forecasts and prediction 95% bands.

7.7 BIBLIOGRAPHIC NOTES

The literature about prediction of stationary processes is extensive and spans several decades since the pioneering works on linear predictors by Kolmogorov, Wiener, and Wold, among others. The monographs by Rozanov (1967), Hannan (1970), and Pourahmadi (2001) offer excellent overviews of the theoretical problems involved in the prediction of linear processes. For a review of prediction methods in the context of long-memory processes; see, for example, Bhansali and Kokoszka (2003).

Optimal adaptive prediction with long-range-dependent models has been analyzed, for instance, by Ray (1993b), Tiao and Tsay (1994) and Basak, Chan, and Palma (2001).

Problems

7.1 Let $\{y_t : t \in \mathbb{Z}\}$ be an AR(2) process defined by

$$y_t = 0.2y_{t-1} + 0.6y_{t-2} + \varepsilon_t,$$

where $\{\varepsilon_t\}$ is a white noise sequence with zero-mean and variance $\sigma^2 = 2$.

(a) Find the best linear predictor of y_{n+2} based on the infinite past $y_n, y_{n-1}, \ldots$.

(b) Calculate the best linear predictor of y_{n+2} based on the finite past $y_n, y_{n-1}, \ldots, y_1$.

(c) Calculate the squared prediction error of two the previous forecast.

7.2 Show for the fractional noise process FN(d) we have

$$\phi_{tj} \approx \pi_j,$$

for all finite j as $t \to \infty$, where

$$\phi_{tj} = -\binom{n}{j} \frac{\Gamma(j-d)\Gamma(t-d-j+1)}{\Gamma(-d)\Gamma(t-d+1)},$$

and

$$\pi_j = \frac{\Gamma(j-d)}{\Gamma(j+1)\Gamma(-d)}.$$

7.3 Consider the following process $\{y_t\}$ with expected value μ for all time t, defined by the equation

$$y_t = \alpha + \phi\, y_{t-1} + z_t + (\theta + \eta)\, z_{t-1} + \theta\, \eta\, z_{t-2},$$

with $\alpha \in \mathbb{R}$, $|\phi| < 1$, $|\theta| < 1$, $|\eta| < 1$ y $\{z_t\} \sim \text{WN}(0, \sigma^2)$.

(a) Find an expression μ in terms of the coefficients of the process.

(b) Show this process can be written as

$$(1 - \phi\, B)\,[y_t - \mu] = \left[1 + (\theta + \eta)\, B + \theta\, \eta\, B^2\right] z_t$$

(c) Since $|\theta| < 1$ and $|\eta| < 1$, then the process $\{y_t - \mu\}$ is causal. Show that the coefficients ψ_k of its MA(∞) representation are

$$\psi_0 = 1, \quad \psi_1 = \phi + (\theta + \eta), \quad \psi_k = \phi^{k-2}\left[\phi^2 + (\theta + \eta)\, \phi + \theta\, \eta\right] \quad \text{for } k \geq 2.$$

(d) Given that $|\phi| < 1$, the the process $\{y_t - \mu\}$ is invertible. Show that the coefficients π_k of its AR(∞) representation are

$$\pi_0 = 1, \quad \pi_1 = +(\phi + \theta + \eta), \quad \pi_k = -(-\theta)^k + (\eta + \phi) \sum_{l=0}^{k-1} (-\theta)^l\, (-\eta)^{k-1-l} \quad \text{for } k \geq 2$$

(e) Find the autocovariance function of the process $\{y_t - \mu\}$.

(f) Determine the variance of prediction error h steps, $\sigma_t^2(h)$, under the assumption that all the past information is available up to time t.

7.4 If $y_t = z_t - \theta z_{t-1}$, where$|\theta| < 1$ and z_t is a white noise sequence with variance σ^2, verify that the best linear predictor of y_{t+1} based on y_j, $j \leq t$ is given by

$$\hat{y}_{t+1} = -\sum_{j=1}^{\infty} \theta^j\, y_{t+1-j},$$

and find its mean square error.

7.5 Consider the following formula for the coefficients of the best linear predictor from the Durbin-Levinson algorithm:

$$\phi_{tj} = \phi_{t-1,j} - \phi_{tt}\phi_{t-1,t-j},$$

for $j = 1, \ldots, n$. Let $\alpha_t = 1 - \sum_{j=1}^{t} \phi_{tj}$.

(a) Show that α_t satisfies the following recursive equation:

$$\alpha_t = (1 - \phi_{tt})\alpha_{t-1}.$$

(b) Verify that a solution to the above equation is

$$\alpha_t = \prod_{j=1}^{t}(1 - \phi_{jj}),$$

so that $\sum_{j=1}^{t} \phi_{tj} = 1 - \prod_{j=1}^{t}(1 - \phi_{jj})$.

7.6 Consider a stationary process $\{x_t\}$ given

$$x_t = z_t + \theta\, z_{t-s},$$

with $\{z_t\} \sim (0, \sigma^2)$ y $s \in \mathbb{N}$. Determine the coefficients $\theta_{n,j}$ of the one-step predictor of x_{n+1} given by

$$\hat{x}_{n+1} = \sum_{j=1}^{n} \theta_{n,j}(x_{n+1-j} - \hat{x}_{n+1-j}), \quad n = 1, 2, 3, \ldots$$

assuming that $\hat{X}_1 = 0$.

7.7 Let x_t be an ARMA(p, q) process. Show that σ^2 is the one-step prediction error variance and that $\gamma(0)$ is the infinite steps ahead prediction error variance. Furthermore, verify that $\gamma(0) \geq \sigma^2$.

7.8 A quarterly economic time series was modeled by $\nabla z_t = 0.5 + (1 - B + 0.5B^2)a_t$ with $\sigma_a^2 = 0.04$.

(a) Given $z_{48} = 130$, $a_{47} = -0.3$, $a_{48} = 0.2$, calculate and draw the predictions $\hat{Z}_{48}(l)$ for $l = 1, 2, \ldots, 12$

(b) Include 80% prediction bands in the graph.

7.9 Consider the MA(1) process where $\{\varepsilon_t\}$ is WN$(0, \sigma^2)$ y $|\theta| < 1$,

$$y_t = \varepsilon_t - \theta\, \varepsilon_{t-1}.$$

(a) Suppose that we want to find the best lineal predictor (BLP) of y_0 based on the infinite past $\{y_{-1}, y_{-2}, y_{-3}, \ldots\}$. Verify that in this case the BLP of y_0 is given by

$$\widehat{y}_0 = -\sum_{j=1}^{\infty} \theta^j y_{-j},$$

and that the prediction error variance is σ^2.

(b) Assume now that we only have the pst observations $\{y_{-1}, y_{-2}\}$ and we write the BLP of y_0 as

$$\widehat{y}_0 = \phi_1 y_{-1} + \phi_2 y_{-2}.$$

We know that in this case the vector $\phi = (\phi_1, \phi_2)'$ satisfies

$$\Gamma\,\phi = \gamma,$$

where Γ is the variance covariance matrix of $y = (y_{-1}, y_{-2})'$ y $\gamma = [\gamma(1), \gamma(2)]'$. Verify that

$$\phi = \frac{\rho}{1-\rho^2}(1,\ -\rho),$$

where $\rho = \frac{\gamma(1)}{\gamma(0)}$

7.10 Consider fractional noise a fractional noise process FN(d) where the coefficients of the best linear predictor are given by

$$\phi_{tj} = -\binom{n}{j}\frac{\Gamma(j-d)\Gamma(t-d-j+1)}{\Gamma(-d)\Gamma(t-d+1)},$$

and the coefficients of the infinite AR expansion are

$$\pi_j = \frac{\Gamma(j-d)}{\Gamma(j+1)\Gamma(-d)}.$$

Show that

$$\phi_{tj} \to \pi_j,$$

as $t \to \infty$.

7.11 Consider the AR(p) model

$$y_t + \phi_1 y_{t-1} - \cdots - \phi_p y_{t-p} = \varepsilon_t,$$

where $\{\varepsilon_t\}$ is a white noise sequence with variance σ^2. Verify that ν_t converges to σ^2 after p steps.

7.12 Find an expression for $\sigma_t^2(h)$, $h \geq 1$, for the ARFIMA-GARCH(r, s) model.

7.13 Show that for the ARFIMA-GARCH(1,1) model, the conditional variance at time $t+1$ may be written as

$$\sigma_{t+1}^2 = \frac{\alpha_0}{1-\beta-1} + \alpha_1 \sum_{j=0}^{\infty} \beta_1^j \varepsilon_{t-j}^2.$$

7.14 Verify the following result for the random variables x, y, z. If x is independent of y and z, then

$$E(xy|z) = E(x)\,E(y|z).$$

7.15 Show that for the fractional noise process, FN(d), we have

$$\delta_t \sim \frac{d^2}{t},$$

as $t \to \infty$.

7.16 Consider the h-step forecast for an ARCH(1) given by

$$\widehat{\sigma}^2_{n+h} = \alpha_0 \left[\sum_{i=0}^{h-1} \alpha_1^i \right] + \alpha_1^h\, y_n^2,$$

cf. (7.13). Prove that $\widehat{\sigma}^2_{n+h}$ increases with h.

7.17 Verify that the following expressions for the predictors of an ARCH(p) model hold:

$$\begin{aligned} \hat{\sigma}_n^2(1) &= \alpha_0 + \alpha_1\, y_n^2 + \cdots\cdots + \alpha_p\, y_{n+1-p}^2 \\ \hat{\sigma}_n^2(2) &= \alpha_0 + \alpha_1\, \widehat{\sigma}_n^2(1) + \alpha_2\, y_n^2 \cdots + \alpha_p\, y_{n+2-p}^2. \end{aligned}$$

Furthermore, show that for the general case we have,

$$\widehat{\sigma}_n^2(\ell) = \alpha_0 + \sum_{i=1}^{p} \alpha_i\, \widehat{\sigma}_n^2(\ell - i),$$

with $\widehat{\sigma}_n^2(\ell) = y_{n+\ell}^2$ for $\ell \leq 0$.

7.18 Consider the Hilbert space $\mathcal{L}_2$ and a subspace $\mathcal{M} \subseteq \mathcal{L}_2$. Show that the orthogonal projection of $y \in \mathcal{L}_2$ onto $\mathcal{M}$ is given by the conditional expectation

$$\widehat{y} = E(y|\mathcal{M}).$$

7.19 Consider the Hilbert space $\mathcal{H} = \overline{\text{sp}}\{e_t : t = 0, 1, 2, \dots\}$ where $\{e_t\}$ is an orthonormal basis, that is, $\langle e_t, e_s \rangle = 0$ for all $t \neq s$ and $\langle e_t, e_t \rangle = 1$ for all t.

(a) Let $x \in \mathcal{H}$, verify that this element may be written as

$$x = \sum_{t=0}^{\infty} \langle x, e_t \rangle e_t.$$

(b) Show that $\|x\|^2 = \sum_{t=0}^{\infty} \langle x, e_t \rangle^2$.

(c) Let $\mathcal{M} = \overline{\text{sp}}\{e_1, \ldots, e_N\}$ and let $\widehat{x}$ be the orthogonal projection of x on $\mathcal{M}$. Show that $\widehat{x} = \sum_{t=0}^{N} \langle x, e_t \rangle e_t$.

(d) Verify that $\|x - \widehat{x}\|^2 = \sum_{t=N+1}^{\infty} \langle x, e_t \rangle^2$.

7.20 Let $\{y_t : t \in \mathbb{N}\}$ be a sequence in a Hilbert space $\mathcal{H}$ such that $\sum_{t=1}^{\infty} \|y_t\| < \infty$. Show that $\sum_{t=1}^{\infty} y_t$ converges in $\mathcal{H}$.

7.21 Let $\mathcal{H}$ be a Hilbert space and suppose that $x, y \in \mathcal{H}$ are orthogonal vectors such that $\|x\| = \|y\| = 1$. Show that $\|\alpha x + (1-\alpha) y\| < 1$ for all $\alpha \in (0, 1)$. From this, what can you say about the set $\{y \in \mathcal{H} : \|y\| \leq 1\}$?

7.22 (Parallelogram law) Show that if $\mathcal{H}$ is an inner product space then

$$\|x + y\|^2 + \|x - y\|^2 = 2\|x\|^2 + 2\|y\|^2,$$

for all $x, y \in \mathcal{H}$.

7.23 Consider the following stochastic volatility process $\{r_t\}$ defined by

$$\begin{aligned} r_t &= \varepsilon_t \sigma_t, \\ \sigma_t &= \sigma \exp(\nu_t / 2), \end{aligned}$$

where $\{\varepsilon_t\}$ is an independent and identically distributed sequence with zero-mean and unit variance, and $\{\nu_t\}$ is a linear process:

$$\nu_t = \sum_{j=0}^{\infty} \psi_j \eta_{t-j},$$

with $\sum_{j=0}^{\infty} \psi_j^2 < \infty$ and $\{\eta_t\}$ is an independent and identically distributed sequence with zero-mean and unit variance, independent of the sequence $\{\varepsilon_t\}$.

(a) Show that the process r_t is stationary.

(b) Find the best linear predictor of y_{t+1} given the full past of the series.

CHAPTER 8

NONSTATIONARY PROCESSES

As discussed in Chapter 1, most real-life time series display a number of nonstationary features, including trends, seasonal behavior, explosive variances, trend breaks, among others. This chapter provides an overview of some of these problems and the time series analysis techniques developed to deal with them. In particular, we discuss the concepts of deterministic and stochastic trends as well as unit root procedures to test for a nonstationary explosive behavior. Autoregressive integrated moving-average processes are also briefly reviewed. These are very well-known models for dealing with integrated time series. On the other hand, techniques for modeling time-varying parameters are also discussed, focusing on the so-called locally stationary processes. In this case, the time series model is assumed to evolve very smoothly so that it can be locally approximated by stationary processes. This chapter also covers methodologies for handling abrupt structural changes as well as several examples and data applications.

Time Series Analysis. First Edition. Wilfredo Palma.

8.1 INTRODUCTION

On of the most common features in time series data is the presence of increasing or decreasing trends along with a number trends breaks. A basic question that arises is whether these trends are the result of a deterministic underlying pattern or it corresponds to the accumulation of random shocks over time. Of course, the observed time series may be the result of combinations of these two or other more complex data generation mechanisms. In what follows, we discuss two well known approaches to understand and model trends: deterministic and stochastic methodologies.

8.1.1 Deterministic Trends

Under the deterministic approach, the observed process is the result of a usually unknown underlying pattern $f(t)$ plus a noise ε_t,

$$y_t = f(t) + \varepsilon_t.$$

In order to estimate the trend, the function $f(t)$ can be written in terms of some parameter vector β. For instance, we may write

$$f(t) = \beta_0 + \beta_1 x_{t1} + \beta_2 x_{t2} + \cdots + \beta_p x_{tp},$$

where $x_{t1}, x_{t2}, \ldots, x_{tp}$ are deterministic covariates. In particular, by setting $x_{tj} = t^j$ we can generate a *polynomial trend* or *polynomial regression*. On the other hand, if $x_{tj} = \exp(ij)$ the resulting model corresponds to an *harmonic regression*.

Another way of specifying the function $f(t)$ is through local polynomials, so that the trend is flexible enough to capture short term movements in the data. In this nonparametric approach, the trend of process y_t is locally estimated as

$$\widehat{y}_t = \sum_{j=-m}^{m} w_j \, y_{t+j},$$

for $t = m+1, \ldots, N-m$. The optimal weights w_j are usually obtained by fitting cubic polynomials to the series y_t.

8.1.2 Stochastic Trends

If the underlying trend is assumed to be stochastic, the observed process y_t is usually understood as the result of a sequence of random shocks,

$$y_t = \varepsilon_1 + \varepsilon_2 + \cdots + \varepsilon_t,$$

where ε_t is a white noise or a sequence of independent random variables. Note that in this case, the differenced series satisfies

$$y_t - y_{t-1} = (1 - B) y_t = \Delta \varepsilon_t.$$

Thus, an stochastic trend process is more generally specified by

$$\Delta^d y_t = \varepsilon_t,$$

where d is a known differentiation order and ε_t is assumed to be a stationary process. Typically, a process satisfying this equation is referred to as *integrated process* of order d, I(d).

Assume that ε_t is a sequence of i.i.d. random variables with zero-mean and variance σ^2. The variance of an I(d) process is $\text{Var}(y_t) = t\,\sigma^2$. Thus, an integrated process possesses an explosive variability as t tends to infinity.

In contrast, under the same conditions, the variance of a process with deterministic trend is $\text{Var}(y_t) = \sigma$. Naturally, in this case the variability is not explosive.

In this sense, it is relevant to test whether the process possess a *unit root*, that is, it corresponds to an integrated process. This issue is discussed next.

8.2 UNIT ROOT TESTING

Consider the model

$$(1 - \phi B)y_t = \varepsilon_t,$$

where ϕ is an autoregressive parameter and ε_t is a zero-mean stationary sequence. As studied in previous chapters, if $|\phi| < 1$ then the process y_t can be expanded as

$$y_t = (1 - \phi B)^{-1}\varepsilon_t.$$

Thus, the process y_t is stationary. However, if $\phi = 1$ then y_t corresponds to a I(d) process with explosive variance. Therefore, the *unit root hypothesis* can be formally defined by $H_0 : \phi = 1$. A well known procedure for testing H_0 against the alternative hypothesis $H_1 : \phi < 1$ is the Dickey-Fuller statistic which is based on the least squares estimates

$$\begin{aligned}
\widehat{\phi} &= \frac{\sum_{t=1}^{n} y_t\, y_{t-1}}{\sum_{t=1}^{n} y_{t-1}^2} \\
\widehat{\sigma}^2 &= \frac{\sum_{t=1}^{n} (y_t - \widehat{\phi}\, y_{t-1})^2}{n-1}
\end{aligned}$$

Naturally, evidence in favor of the unit root hypothesis comes from an estimated value of ϕ close to one. Following this idea, the Dickey Fuller t-statistics is given by

$$\text{DF} = \frac{\widehat{\phi} - 1}{\sigma_{\widehat{\phi}}} = \frac{\sum_{t=1}^{n} \varepsilon_t\, y_{t-1}}{\widehat{\sigma}\sqrt{\sum_{t=1}^{n} y_{t-1}^2}}.$$

EXAMPLE 8.1

Consider the logarithm transformation of the SP500 series introduced in Chapter 1. An application of the Dickey-Fuller test to this series of log returns we get,

```
Augmented Dickey-Fuller Test

data:  z
Dickey-Fuller = -2.1531, Lag order = 25, p-value = 0.5135
alternative hypothesis: stationary.
```

Thus, according to this result, we cannot reject the unit root hypothesis for this series of log returns of the SP500 stock index at the 5% significance level.

EXAMPLE 8.2

When the Dickey-Fuller test is applied to the log returns of the IPSA stock index we obtain

```
Augmented Dickey-Fuller Test

data:  y
Dickey-Fuller = -1.6347, Lag order = 14, p-value = 0.733
alternative hypothesis: stationary,
```

so that based on this output, we cannot reject the unit root hypothesis for this series of log returns of the IPSA stock index at the 5% significance level.

8.3 ARIMA PROCESSES

An *autoregressive integrated moving-average* ARIMA(p, d, q) process y_t is defined by the equation

$$\phi(B)\Delta^d y_t = \theta(B)\varepsilon_t,$$

where $\phi(B)$ is an autoregressive polynomial of order p, $\theta(B)$ is an moving-average polynomial of order q, $\phi(B)$ and $\theta(B)$ have no common roots and ε_t is a white noise sequence. Note that the differenced process $\Delta^d y_t$ satisfies an ARMA(p, q) model.

Figure 8.1 to Figure 8.4 display 1000 simulated observations from ARIMA model

$$(1 - \phi B)(1 - B)^d y_t = \varepsilon_t - \theta\varepsilon_{t-1},$$

for different differentiation levels $d = 1$ and $d = 2$ as well as for distinct values of the autoregressive parameter ϕ. Note that these time series plots show apparent local trends. However, these are stochastic paths corresponding to

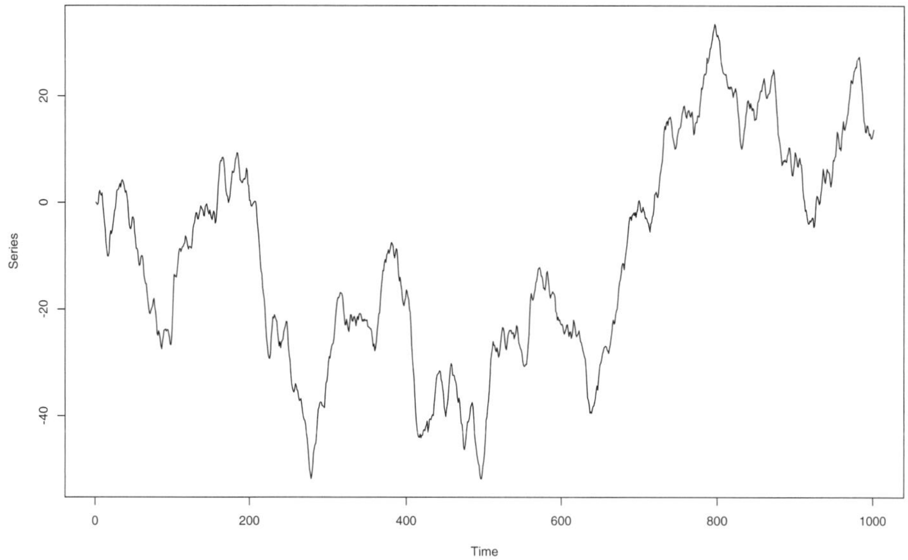

Figure 8.1 Simulated ARIMA(1, 1, 0) model with $\phi = 0.6$ and $\sigma^2 = 1$.

integrated processes. There is a clear difference in the paths of Figure 8.1 and Figure 8.2 due to the change of the parameter from positive to negative. Furthermore, these two cases are different from the series exhibited in Figure 8.3 and Figure 8.4. The second order integration represented in these plots reveals smooth but highly nonstationary paths.

After accounting for the differentiation level of the series, the estimation of a ARIMA model proceeds analogously to the estimation of an ARMA process. As an example, consider the ARIMA(1, 2, 1) exhibited in Figure 8.4. An application of the R function arima.mle produces the following parameter estimates

```
> fit

Call:
arima(x = y, order = c(1, 2, 1))

Coefficients:
         ar1     ma1
      0.2733  0.5750
s.e.  0.0409  0.0342

sigma^2 estimated as 0.9552: log likelihood = -1396.4, aic = 2798.8
```

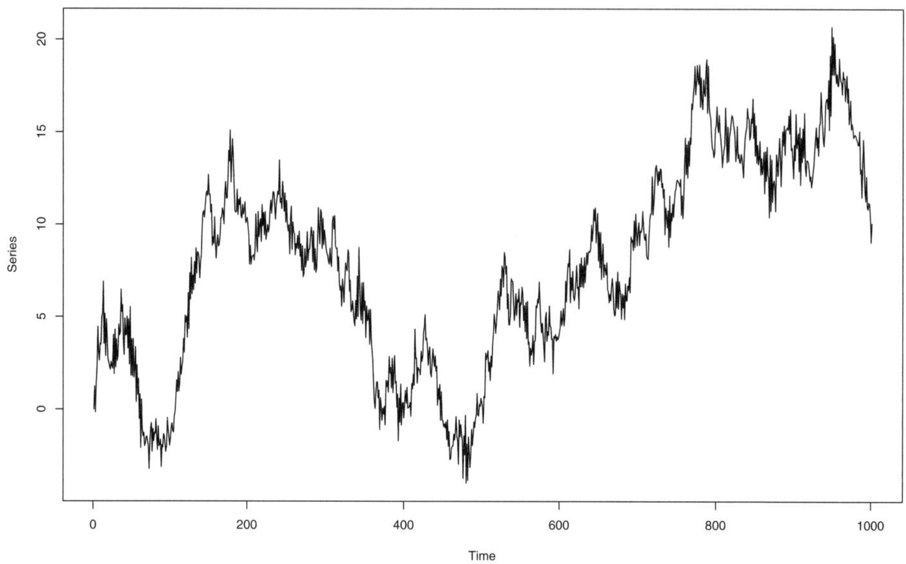

Figure 8.2 Simulated ARIMA$(1,1,0)$ model with $\phi = -0.6$ and $\sigma^2 = 1$.

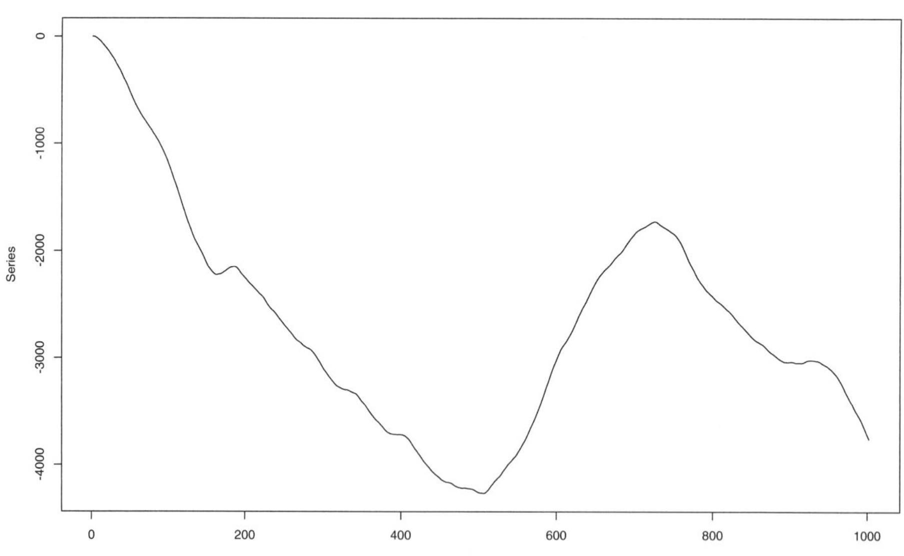

Figure 8.3 Simulated ARIMA$(1,2,0)$ model with $\phi = 0.2$ and $\sigma^2 = 1$.

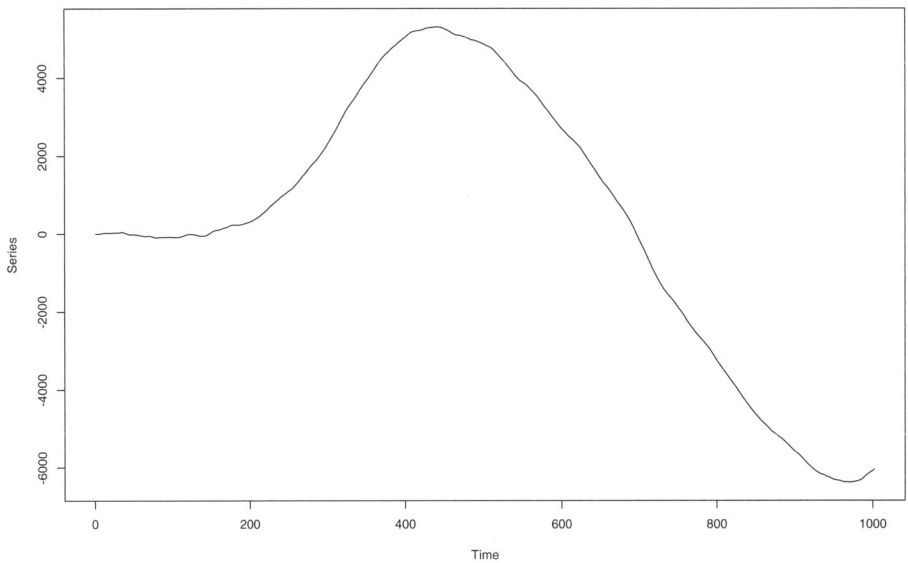

Figure 8.4 Simulated ARIMA(1, 2, 1) model with $\phi = 0.2$, $\theta = 0.6$ and $\sigma^2 = 1$.

8.4 LOCALLY STATIONARY PROCESSES

Another approach for modeling non stationary time series data is the class of locally stationary processes. Recall from Section 4.8 that a stationary process $\{y_t\}$ can be written in terms of a spectral representation as

$$y_t = \int_{-\pi}^{\pi} A(\lambda)\, e^{i\lambda t}\, dB(\lambda), \tag{8.1}$$

where $A(\lambda)$ is a transfer function and $B(\lambda)$ is an orthogonal increments process on $[-\pi, \pi]$ such that

$$\mathrm{Cov}[B(\lambda), B(\omega)] = \frac{\sigma^2}{2\pi}\delta(\lambda - \omega)d\lambda d\omega.$$

The representation (8.1) can be extended allowing the transfer function to evolve in time as follows,

$$y_{t,T} = \int_{-\pi}^{\pi} A^0_{t,T}(\lambda)\, e^{i\lambda t}\, dB(\lambda), \tag{8.2}$$

for $t = 1, \ldots, T$.

The transfer function $A^0_{t,T}(\lambda)$ of this class of nonstationary processes is assumed to change smoothly over time so that they can be locally approximated by stationary processes. Some examples are discussed below.

■ EXAMPLE 8.3

Consider the following time-varying version of the first-order moving-average process, denoted for simplicity as LSMA(1),

$$y_{t,T} = \sigma\left(\tfrac{t}{T}\right)\left[1+\theta\left(\tfrac{t}{T}\right)\varepsilon_{t-1}\right], \tag{8.3}$$

$t = 1, \ldots, T$, where $\{\varepsilon_t\}$ is a zero-mean and unit variance white noise sequence. The covariance structure of this model is,

$$\kappa_T(s,t) = \begin{cases} \sigma^2\left(\frac{t}{T}\right)\left[1+\theta^2\left(\frac{t}{T}\right)\right], & s=t, \\ \sigma\left(\frac{t}{T}\right)\sigma\left(\frac{t-1}{T}\right)\theta\left(\frac{t}{T}\right), & s=t-1, \\ \sigma\left(\frac{t}{T}\right)\sigma\left(\frac{t+1}{T}\right)\theta\left(\frac{t+1}{T}\right), & s=t+1, \\ 0 & \text{otherwise.} \end{cases}$$

In this case, the transfer function of the process is given by

$$A^0_{t,T}(\lambda) = A\left(\tfrac{t}{T},\lambda\right) = \sigma\left(\tfrac{t}{T}\right)\left[1+\theta\left(\tfrac{t}{T}\right)e^{i\lambda}\right]. \tag{8.4}$$

Furthermore, the time-varying spectral density is

$$f\left(\tfrac{t}{T},\lambda\right) = |A\left(\tfrac{t}{T},\lambda\right)|^2 = \sigma^2\left(\tfrac{t}{T}\right)\left[1+\theta^2\left(\tfrac{t}{T}\right)+2\theta\left(\tfrac{t}{T}\right)\cos\lambda\right]. \tag{8.5}$$

■ EXAMPLE 8.4

An extension of the previous model is the time-varying MA(∞) moving-average expansion

$$y_{t,T} = \sigma\left(\tfrac{t}{T}\right)\sum_{j=0}^{\infty}\psi_j\left(\tfrac{t}{T}\right)\varepsilon_{t-j}, \tag{8.6}$$

$t = 1, \ldots, T$, where $\{\varepsilon_t\}$ is a zero-mean and unit variance Gaussian white noise and $\{\psi_j(u)\}$ are coefficients satisfying

$$\psi_0(u) = 1, \qquad \sum_{j=0}^{\infty}\psi_j(u)^2 < \infty,$$

for all $u \in [0,1]$. This model will be denoted LSMA(∞) hereafter. The time-varying spectral density of (8.6) is

$$f_\theta(u,\lambda) = \sigma^2(u)|\sum_{j=0}^{\infty}\psi_j(u)e^{i\lambda j}|^2,$$

for $u \in [0,1]$ and $\lambda \in [-\pi,\pi]$. For simplicity, if $|\psi_j(u)| \leq K\exp(-aj)$ for $j \geq 1$ and $u \in [0,1]$ with K and a positive constants, model (8.6) will be called a *short-memory process*. On the other hand, if $|\psi_j(u)| \leq Kj^{d-1}$ for $u \in [0,1]$ and some $d \in (0,1/2)$, model (8.6) will be called a *long-memory process*. Another characterization is based on the spectral density. It is said that a LS process has *short memory* if its spectral density is bounded at $\lambda = 0$ for $u \in [0,1]$. On the other hand, the process has *long memory* if its spectral density is unbounded near the origin for $u \in [0,1]$.

■ EXAMPLE 8.5

Consider the LS autoregressive process LSAR(1) defined as

$$y_{t,T} = \phi(\tfrac{t}{T})y_{t-1,T} + \varepsilon_t, \tag{8.7}$$

for $T = 1,\ldots,T$. Suppose that $\phi(u) = \phi(0)$ for $u < 0$, and there exists a positive constant $K < 1$ such that $|\phi(u)| \leq K$ for $u < 1$. Thus, an expanded Wold expansion of this process is given by,

$$y_{t,T} = \sum_{j=0}^{\infty} \psi_j\,(t,T)\,\varepsilon_{t-j}, \tag{8.8}$$

where $\psi_0\,(t,T) = 1$ for all t,T, and for $j \geq 1$,

$$\psi_j\,(t,T) = \prod_{k=0}^{j-1} \phi(\tfrac{t-k}{T}). \tag{8.9}$$

From this, we conclude that the transfer function can be written as

$$A^0_{t,T}(\lambda) = 1 + \sum_{j=1}^{\infty}\prod_{k=0}^{j-1} \phi(\tfrac{t-k}{T})e^{i\lambda j}. \tag{8.10}$$

The spectral density of the limiting process is $f_\theta(u,\lambda) = \sigma(u)^2|1 - \phi(u)e^{i\lambda j}|^{-2}$. This process satisfies definition (8.2) and its spectral density is bounded at the origin for all u. Thus, this is a short-memory process.

■ EXAMPLE 8.6

Consider the LS autoregressive process LSAR(p)

$$y_{t,T} = \sum_{j=1}^{p} a_j(\tfrac{t}{T})y_{t-j,T} + \varepsilon_t,$$

for $T = 1,\ldots,T$. The spectral density of the limiting process is $f_\theta(u,\lambda) = \sigma(u)^2|1 - \sum_{j=1}^{p} a_j(u)e^{i\lambda j}|^{-2}$. This process satisfies definition (8.2). In this case, the spectral density is bounded at the origin under some regularity conditions on the roots of the polynomial $a(B) = 1 - \sum_{j=1}^{p} a_j B^j$. Thus, these LSAR($p$) processes have short memory.

■ EXAMPLE 8.7

Observe that a stationary fractional noise process (FN) with long-memory parameter d is given by

$$y_t = \sigma \sum_{j=0}^{\infty} \psi_j \varepsilon_{t-j}, \tag{8.11}$$

where $\psi_j = \frac{\Gamma(j+d)}{\Gamma(j+1)\Gamma(d)}$, where $\Gamma(\cdot)$ is the Gamma function. A nonstationary extension of this model is the LS fractional noise process (LSFN) with coefficients $\psi_j(u) = \frac{\Gamma[j+d(u)]}{\Gamma(j+1)\Gamma[d(u)]}$, where $d(\cdot)$ is a smoothly time-varying long-memory parameter. The covariances of a LSFN process are

$$\kappa_T(s,t) = \sigma\left(\tfrac{s}{T}\right)\sigma\left(\tfrac{t}{T}\right)\frac{\Gamma\left[1 - d\left(\frac{s}{T}\right) - d\left(\frac{t}{T}\right)\right]\Gamma\left[s - t + d\left(\frac{s}{T}\right)\right]}{\Gamma\left[1 - d\left(\frac{s}{T}\right)\right]\Gamma\left[d\left(\frac{s}{T}\right)\right]\Gamma\left[s - t + 1 - d\left(\frac{t}{T}\right)\right]},$$

for $s, t = 1, \ldots, T$, $s \geq t$. From this expression, and for large $s - t$ we have that

$$\kappa_T(s,t) \sim \sigma\left(\tfrac{s}{T}\right)\sigma\left(\tfrac{t}{T}\right)\frac{\Gamma\left[1 - d\left(\frac{s}{T}\right) - d\left(\frac{t}{T}\right)\right]}{\Gamma\left[1 - d\left(\frac{s}{T}\right)\right]\Gamma\left[d\left(\frac{s}{T}\right)\right]}(s-t)^{d\left(\frac{s}{T}\right) + d\left(\frac{t}{T}\right) - 1},$$

The spectral density of this process is given by

$$f_\theta(u,\lambda) = \frac{\sigma^2(u)}{2\pi}\left[2\sin\frac{\lambda}{2}\right]^{-2d_\theta(u)},$$

for $\lambda \in [-\pi, \pi]$. Thus, $f_\theta(u,\lambda) \sim \frac{\sigma^2(u)}{2\pi}|\lambda|^{-2d(u)}$, for $|\lambda| \to 0$. Consequently, $f_\theta(u,\lambda)$ has a pole at the origin and then this is a long-memory process.

■ EXAMPLE 8.8

Figure 8.5 exhibits a simulated locally stationary MA(1) model

$$y_t = \varepsilon_t + \theta_t\, \varepsilon_{t-1},$$

where ε_t is a Gaussian white noise sequence with zero-mean and unit variance, and the time-varying parameter θ_t evolves as

$$\theta_t = 0.9 - 1.8\,\frac{t}{n},$$

with $n = 400$. Observe that this parameter can be also written in terms of the rescaled time

$$\theta(u) = 0.9 - 1.8\,u,$$

for $u \in [0, 1]$.

The sample ACF of the generated process is provided in panel (a) of Figure 8.6. It is interesting to notice that this is just an heuristic exercise since this non stationary process does not have a well-defined ACF.

From this panel, we may consider that this time series is white noise. However, when we take a closer look at the sample ACF for the first half of the series, see Figure 8.6(b), it seems that this process behaves like a MA(1) model with positive first-order moving-average parameter θ.

A similar conclusion could be reached from panel (c) which shows the sample ACF of the second half of the observations. From this panel, we may think that the process is a MA(1) with negative parameter θ.

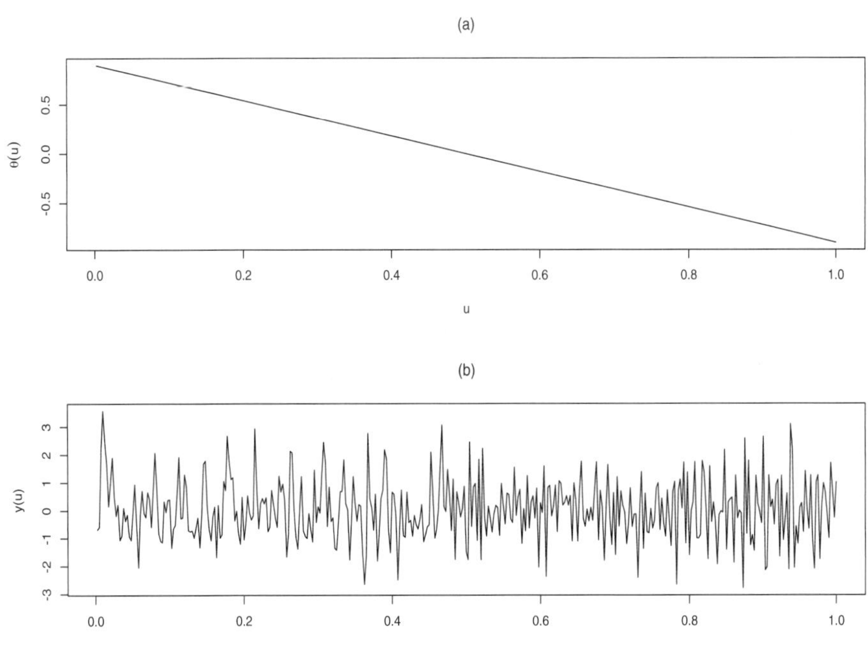

Figure 8.5 *Locally stationary MA(1) model with $\theta[u] = 0.9 - 1.8u$. (a) Evolution of $\theta(u)$, (b) Observed series y_t for $t = 1, \ldots, 400$.*

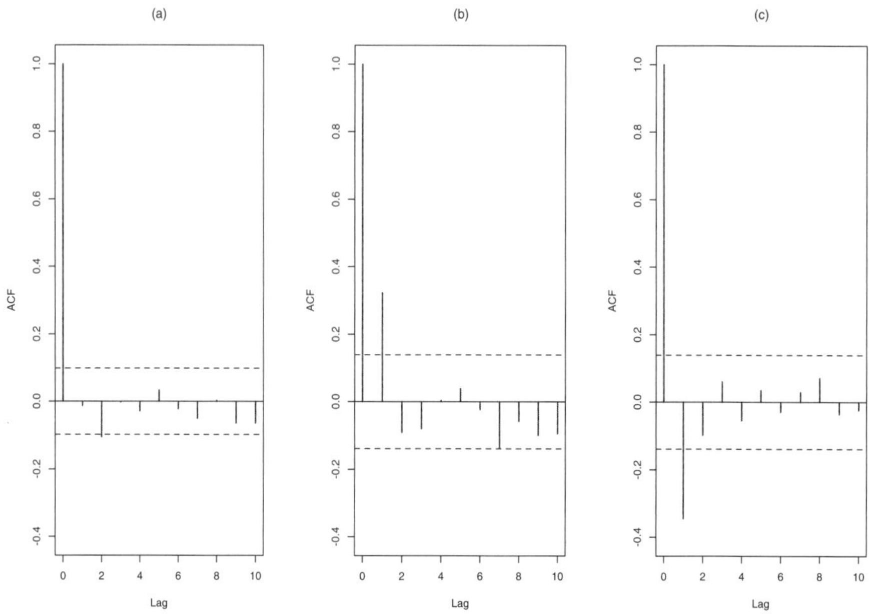

Figure 8.6 *Locally stationary MA(1) model with* $\theta[u] = 0.9 - 1.8u$. *(a) Sample ACF, (b) sample ACF of* y_t, *for* $t = 1, \ldots, 200$, *and (c) sample ACF of* y_t, *for* $t = 201, \ldots, 400$.

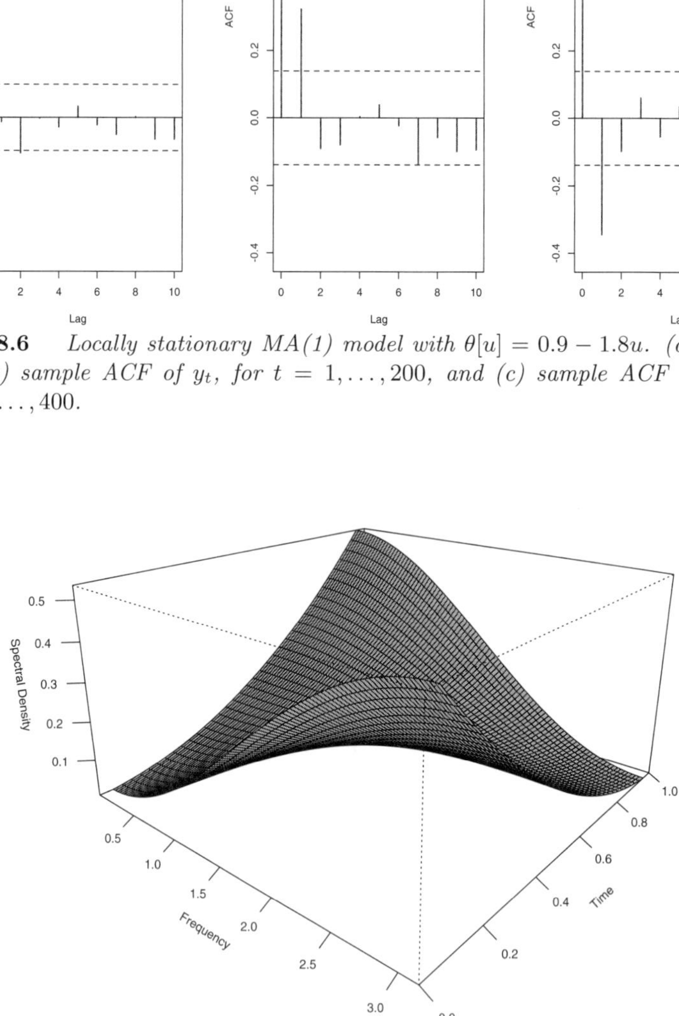

Figure 8.7 *Spectral density of the locally stationary MA(1) model with* $\theta(u) = 0.9 - 1.8\,u$.

■ **EXAMPLE 8.9**

To illustrate the use of locally stationary models in the context of strongly dependent time series, consider the following FN(d) process

$$y_t = \sigma_t (1 - B)^{d_t} \varepsilon_t,$$

where ε_t is a Gaussian white noise with zero-mean and unit variance. The evolution of the long-memory parameter d is in terms of the scaled time u given by

$$d(u) = 0.05 + 0.4\, u,$$

while the evolution of the standard deviation parameter σ is

$$\sigma(u) = 1 - 0.5\, u.$$

The evolutions of these two parameters are plotted in Figure 8.8 while the first 10 terms of the covariance matrix of this model is reported in Table 8.1. Notice that since this is not an stationary process, the covariance matrix is not longer Toeplitz. On the other hand, Figure 8.10 displays a simulated trajectory of this process with 1000 observations. From this figure, the variance of the series seems to decreases with time as expected from the specification of the model. Furthermore, Figure 8.11 exhibits the sample ACF for three blocks of observations of length 333 each. Note that the strength of the serial correlation seems to increase with time. This is due to the specification of the long-memory parameter d which increases from 0.05 to 0.45.

Table 8.1 Covariance Matrix of the LS FN model with $d(u) = 0.05 + 0.4\, u$ and $\sigma(u) = 1 - 0.5\, u$.

	1	2	3	4	5	6	7	8	9	10
1	0.92	0.12	0.10	0.09	0.09	0.09	0.09	0.09	0.10	0.10
2	0.12	0.84	0.16	0.12	0.11	0.11	0.11	0.11	0.12	0.12
3	0.10	0.16	0.77	0.19	0.15	0.13	0.13	0.14	0.14	0.15
4	0.09	0.12	0.19	0.71	0.22	0.17	0.16	0.16	0.17	0.18
5	0.09	0.11	0.15	0.22	0.66	0.25	0.21	0.20	0.20	0.22
6	0.09	0.11	0.13	0.17	0.25	0.63	0.29	0.25	0.25	0.27
7	0.09	0.11	0.13	0.16	0.21	0.29	0.61	0.34	0.32	0.33
8	0.09	0.11	0.14	0.16	0.20	0.25	0.34	0.62	0.41	0.42
9	0.10	0.12	0.14	0.17	0.20	0.25	0.32	0.41	0.68	0.57
10	0.10	0.12	0.15	0.18	0.22	0.27	0.33	0.42	0.57	0.91

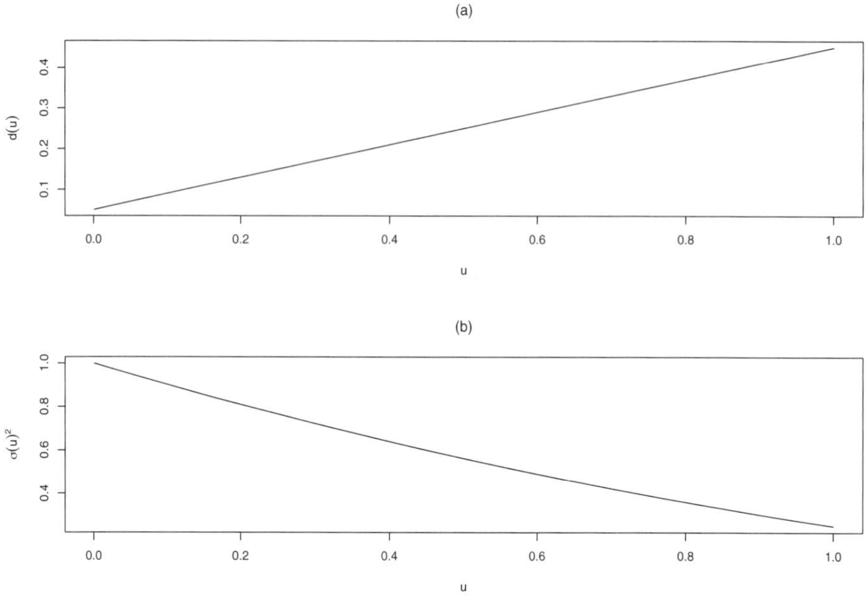

Figure 8.8 *Locally stationary FN(d) model specification with* $d(u) = 0.05 + 0.4\,u$ *and* $\sigma(u) = 1 - 0.5\,u$. *(a) Evolution of* $d(u)$. *(b) Evolution of the noise variance* $\sigma(u)^2$.

8.4.1 State-Space Representations

Given that the state space models described in Chapter 3 provide a very useful framework for the efficient calculation of estimates and forecasts, in this section we review the application of this representations to the case of locally stationary processes. Consider the following state space system,

$$\begin{aligned} x_{t+1,T} &= F_{t,T}x_{t,T} + V_{t,T}, \\ y_{t,T} &= G_{t,T}x_{t,T} + W_{t,T}, \end{aligned} \tag{8.12}$$

where $x_{t,T}$ is a state vector, $F_{t,T}$ is a state transition matrix, V_t is a state noise with variance $Q_{t,T}$, $y_{t,T}$ is the observation, $G_{t,T}$ is observation matrix and W_t is a observation noise with variance $R_{t,T}$.

The process (8.6) can be represented by the following infinite-dimensional state space system

$$\begin{aligned} x_{t+1,T} &= \begin{bmatrix} 0 \\ I_\infty \end{bmatrix} x_{t,T} + \begin{bmatrix} 1 & 0 & 0 & \cdots \end{bmatrix}' \varepsilon_{t+1}, \\ y_{t,T} &= \sigma(\tfrac{t}{T}) \left[1 \; \psi_1(\tfrac{t}{T}) \; \psi_2(\tfrac{t}{T}) \; \psi_3(\tfrac{t}{T}) \; \cdots \right] x_{t,T}, \end{aligned} \tag{8.13}$$

for $t = 1, \ldots, T$, $\mathrm{Var}(x_{t,T}) = I_\infty$, where $I_\infty = \mathrm{diag}\{1, 1, \ldots\}$, $R_{t,T} = 0$, $Q_{t,T} = (q_{ij})$ with $q_{ij} = 1$ if $i = j = 1$ and $q_{ij} = 0$ otherwise. In some

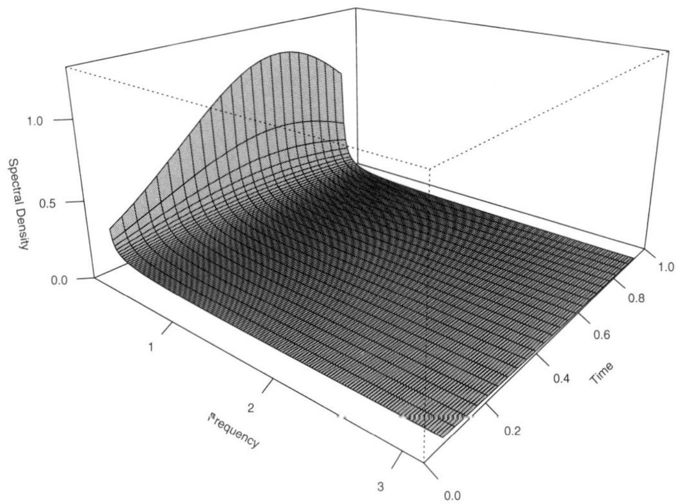

Figure 8.9 *Spectral Density of the locally stationary FN(d) model with* $d(u) = 0.05 + 0.4\,u$ *and* $\sigma(u) = 1 - 0.5\,u$.

cases, this state space representation may not be minimal. For instance, for LSAR(2) processes, the state space is 2-dimensional:

$$x_{t+1,T} = \begin{bmatrix} a_1(\frac{t}{T}) & a_2(\frac{t}{T}) \\ 0 & 1 \end{bmatrix} x_{t,T} + \varepsilon_{t+1}, \quad y_{t,T} = [1\ 0]\, x_{t,T}.$$

It is ussually more practical to approximate the model by,

$$y_{t,T} = \sigma\left(\tfrac{t}{T}\right) \sum_{j=0}^{m} \psi_j\left(\tfrac{t}{T}\right) \varepsilon_{t-j}, \tag{8.14}$$

for $t = 1, \dots, T$ and some positive integer m. A finite-dimensional state state system for (8.14) is given by

$$\begin{aligned} x_{t+1,T} &= \begin{bmatrix} 0 & 0 \\ I_m & 0 \end{bmatrix} x_{t,T} + \begin{bmatrix} 1 & 0 & \cdots & 0 \end{bmatrix}' \varepsilon_{t+1} \\ y_{t,T} &= \sigma(\tfrac{t}{T}) \left[1\ \psi_1(\tfrac{t}{T})\ \psi_2(\tfrac{t}{T})\ \psi_3(\tfrac{t}{T})\ \cdots \psi_m(\tfrac{t}{T})\right] x_{t,T}, \end{aligned} \tag{8.15}$$

for $t = 1, \dots, T$, where I_r denotes the $r \times r$ identity matrix hereafter. Let $r_m = \text{Var}[\sum_{j=m+1}^{\infty} \psi_j(u)\varepsilon_{t-j}]$ be the variance of the truncation error for approximating $\{y_{t,T}\}$ by the finite moving-average expansion (8.14). Then, the asymptotic magnitude of the truncation error when approximating (8.6) by

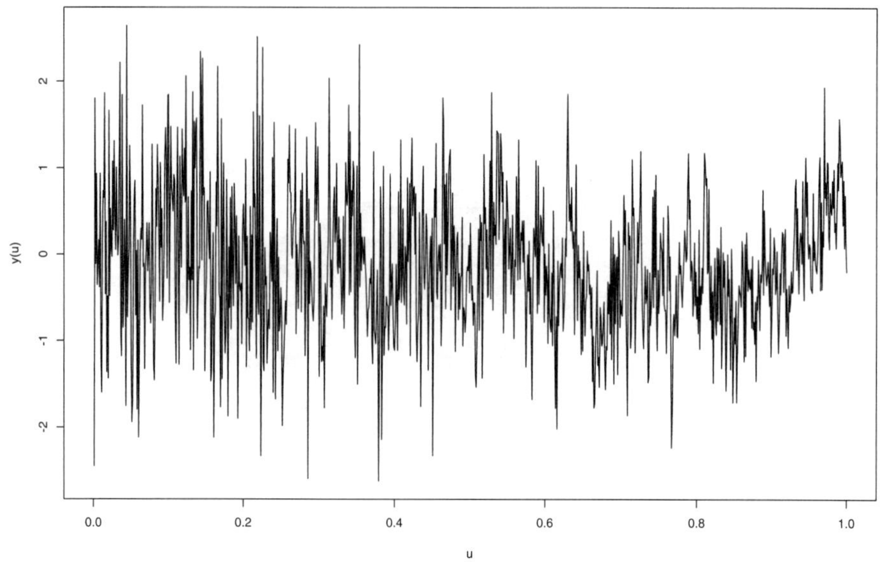

Figure 8.10 *Simulated locally stationary FN(d) model with $d(u) = 0.05 + 0.4\,u$ and $\sigma(u) = 1 - 0.5\,u$ and 1000 observations.*

(8.14) is, $r_m \sim \mathcal{O}(e^{-am})$ for a short-memory process and $r_m \sim \mathcal{O}(m^{2d-1})$ for a long-memory process, for large m, where $a > 0$ and $d = sup_u d(u) < 1/2$.

8.4.2 Whittle Estimation

Let $\theta \in \Theta$ be a parameter vector specifying model (8.2) where the parameter space Θ is a subset of a finite-dimensional Euclidean space. Given a sample $\{y_{1,T}, \ldots, y_{T,T}\}$ of the process (8.2) we can estimate θ by minimizing the Whittle log-likelihood function

$$\mathcal{L}_T(\theta) = \frac{1}{4\pi}\frac{1}{M}\int_{-\pi}^{\pi}\sum_{j=1}^{M}\left\{\log f_\theta(u_j,\lambda) + \frac{I_N(u_j,\lambda)}{f_\theta(u_j,\lambda)}\right\}d\lambda, \tag{8.16}$$

where $f_\theta(u,\lambda) = |A_\theta(u,\lambda)|^2$ is the time-varying spectral density of the limiting process specified by the parameter θ, $I_N(u,\lambda) = \frac{|D_N(u,\lambda)|^2}{2\pi H_{2,N}(0)}$ is a tapered periodogram with

$$D_N(u,\lambda) = \sum_{s=0}^{N-1} h\left(\frac{s}{N}\right) y_{[uT]-N/2+s+1,T}\; e^{-i\lambda s}, \quad H_{k,N} = \sum_{s=0}^{N-1} h\left(\frac{s}{N}\right)^k e^{-i\lambda s},$$

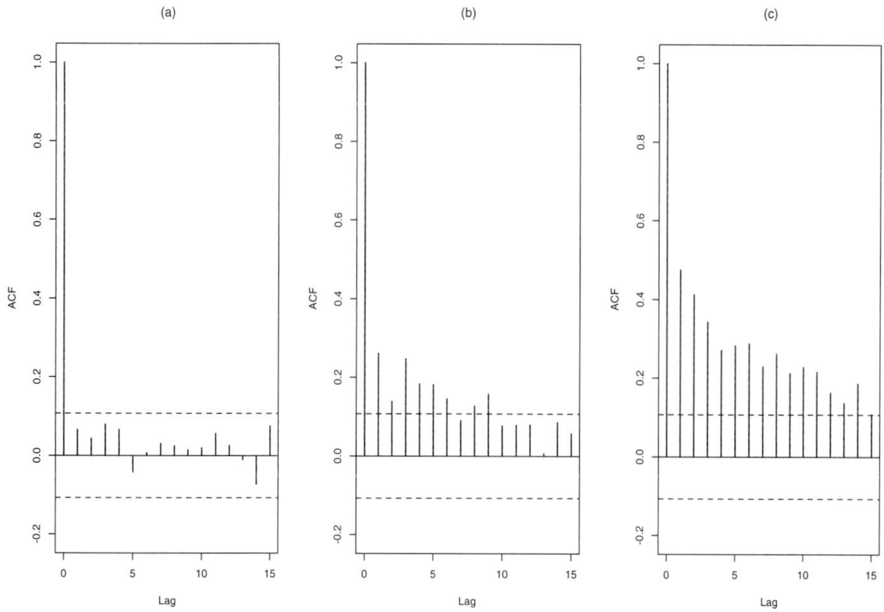

Figure 8.11 *Sample ACF for the simulated locally stationary FN(d) model with $d(u) = 0.05 + 0.4\,u$ and $\sigma(u) = 1 - 0.5\,u$ and 1000 observations. (a) sample ACF of y_t for $t = 1, \ldots, 333$, (b) sample ACF of y_t for $t = 334, \ldots, 666$ and (c) sample ACF of y_t for $t = 667, \ldots, 1000$.*

$T = S(M-1) + N$, $u_j = t_j/T$, $t_j = S(j-1) + N/2$, $j = 1, \ldots, M$ and $h(\cdot)$ is a data taper. Here, N is a block size, M denotes the number of blocks, and S is the shift between them.

The Whittle estimator of the parameter vector θ is given by

$$\widehat{\theta}_T = \arg\min \mathcal{L}_T(\theta), \tag{8.17}$$

where the minimization is over a parameter space Θ.

8.4.3 State Space Estimation

Consider the state space representation (8.13) of $y_{t,T}$. The Kalman filter equations can be used for estimating model parameters, state vectors, future observations and missing values. Let $\Delta_{t,T} = \mathrm{Var}(y_{t,T} - \widehat{y}_{t,T})$ be the prediction error variance and let $\Omega_{t,T} = \mathrm{Var}(x_{t,T} - \widehat{x}_{t,T}) = (\omega_{i,j}(t,T))$ be the state prediction error variance-covariance matrix. The Kalman recursive equations are as follows for the initial conditions $y_{0,T} = (0, 0, \ldots)$, $\widehat{x}_1 = E(x_1) =$

$(0, 0, \ldots)$ and $\Omega_{1,T} = E(x_1, x_1') = \{1, 1, \ldots\}$:

$$
\begin{aligned}
\Delta_{t,T} &= \sigma^2\left(\tfrac{t}{T}\right) \sum_{i,j=1}^{\infty} \psi_{i-1}\left(\tfrac{t}{T}\right) \omega_{i,j}(t, T)\, \psi_{j-1}\left(\tfrac{t}{T}\right), \\
\Theta_{t\,T}(i) &= \sigma\left(\tfrac{t}{T}\right) \sum_{j=1}^{\infty} \omega_{i-1,j}(t, T)\, \psi_{j-1}\left(\tfrac{t}{T}\right), \\
\omega_{t+1,T}(i,j) &= \omega_{t,T}(i+1, j+1) + q_{i,j} - \delta(t)\, \Theta_{t,T}(i)\, \Theta_{t,T}(j)/\Delta_{t,T}, \qquad (8.18) \\
\widehat{y}_{t,T} &= \sigma\left(\tfrac{t}{T}\right) \sum_{j=1}^{\infty} \psi_{j-1}\left(\tfrac{t}{T}\right) \widehat{x}_{t,T}(j), \\
\widehat{x}_{t+1,T}(i) &= \widehat{x}_{t,T}(i-1) + \Theta_{t,T}(i)(y_{t,T} - \widehat{Y}_{t,T})/\Delta_{t,T},
\end{aligned}
$$

where $\delta(t) = 1$ if observation $y_{t,T}$ is available and $\delta(t) = 0$ otherwise.

Let θ be the model parameter vector, then the log-likelihood function (up to a constant) can be obtained from (8.18),

$$
\mathcal{L}(\theta) = \sum_{t=1}^{T} \log \Delta_{t,T} + \sum_{t=1}^{T} \frac{\left(y_{t,T} - \widehat{y}_{t,T}\right)^2}{\Delta_{t,T}}.
$$

Hence, the exact MLE provided by the Kalman equations (8.18) is given by

$$
\widehat{\theta} = \arg\max_{\theta \in \Theta} \mathcal{L}(\theta),
$$

where Θ is a parameter space. Observe that the Kalman equations (8.18) can be applied directly to the general state space representation (8.12) or to the truncated representation (8.15), yielding in this case an approximate MLE.

■ EXAMPLE 8.10

Consider the following LSMA(2) process

$$
y_t = \varepsilon_t + \theta_1(t)\, \varepsilon_{t-1} + \theta_2(t)\, \varepsilon_{t-2},
$$

where the moving-average parameter evolve as

$$
\begin{aligned}
\theta_1(u) &= \alpha_0 + \alpha_1 u, \\
\theta_2(u) &= \beta_0 + \beta_1 u.
\end{aligned}
$$

Figure 8.12 displays a simulated trajectory of this process with 1000 observations and parameters $\alpha_0 = 0.2$, $\alpha_1 = 0.5$, $\beta_0 = 0.8$ and $\beta_1 = -0.6$. Additionally, the sample ACF of this process is exhibited in Figure 8.13. Given that this is not a stationary process, we consider heuristic estimates by blocks. Panel (a), first block of 333 observations, panel

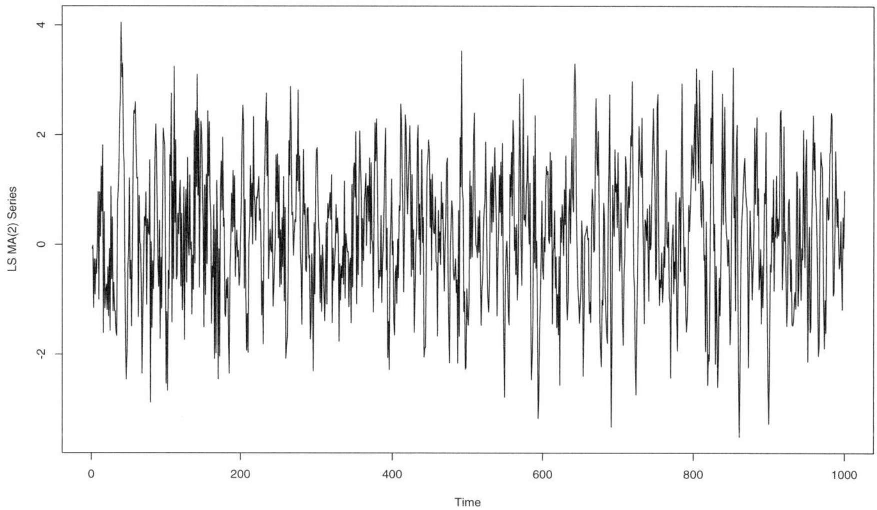

Figure 8.12 *Simulated LSMA(2) model with 1000 observations and parameters* $\alpha_0 = 0.2$, $\alpha_1 = 0.5$, $\beta_0 = 0.8$ *and* $\beta_1 = -0.6$.

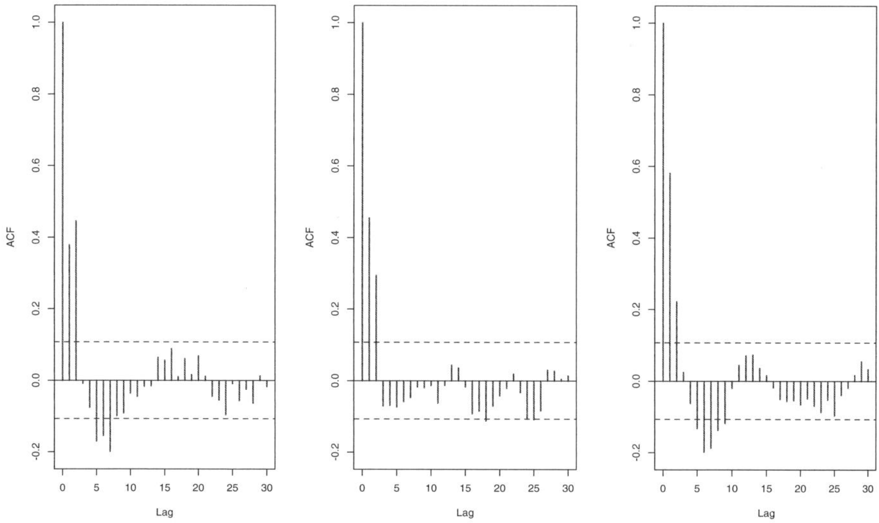

Figure 8.13 *Sample ACF of the simulated LSMA(2) model with 1000 observations and parameters* $\alpha_0 = 0.2$, $\alpha_1 = 0.5$, $\beta_0 = 0.8$ *and* $\beta_1 = -0.6$. *(a) first block of 333 observations, (b) second block of 333 observations, and (c) third block of 334 observations.*

Table 8.2 Whittle and Kalman Estimates for the LSMA(2) model parameters

Parameter	α_0	α_1	β_0	β_1
	0.2	0.5	0.8	-0.6
Whittle	0.2340	0.4209	0.7675	-0.6083
Kalfman	0.2352	0.4028	0.7739	-0.6097

(b), second block of 333 observations, and panel (c), third block of 334 observations. On the other hand, Table 8.2 reports the average parameter estimates for this model, based on 1000 repetitions and using both the Whittle and the Kalman methods.

Notice that both techniques produce estimates close to their theoretical counterparts.

■ EXAMPLE 8.11

Consider the following LSFN process

$$y_t = \sigma_t \,(1 - B)^{-d_t}\, \varepsilon_t,$$

where the parameters evolve as

$$\begin{aligned} \sigma(u) &= \alpha_0 + \alpha_1\, u, \\ d(u) &= \beta_0 + \beta_1\, u. \end{aligned}$$

Figure 8.14 displays a simulated trajectory of this LSFN process with 1000 observations a and time-varying parameters $\alpha_0 = 0.90$, $\alpha_1 = 0.10$, $\beta_0 = 0.15$ and $\beta_1 = 0.30$. Moreover, the sample ACF of this process is exhibited in Figure 8.15.

Analogously to the previous example, since this is a nonstationary process, we consider heuristic estimates by blocks. Panel (a), first block of 333 observations, panel (b), second block of 333 observations, and panel (c), third block of 334 observations.

On the other hand, Table 8.3 reports the average parameter estimates for this model, based on 1000 repetitions and using both the Whittle and the Kalman methods. From this table, we conclude that both approaches provide good parameter estimates in the LSFN example.

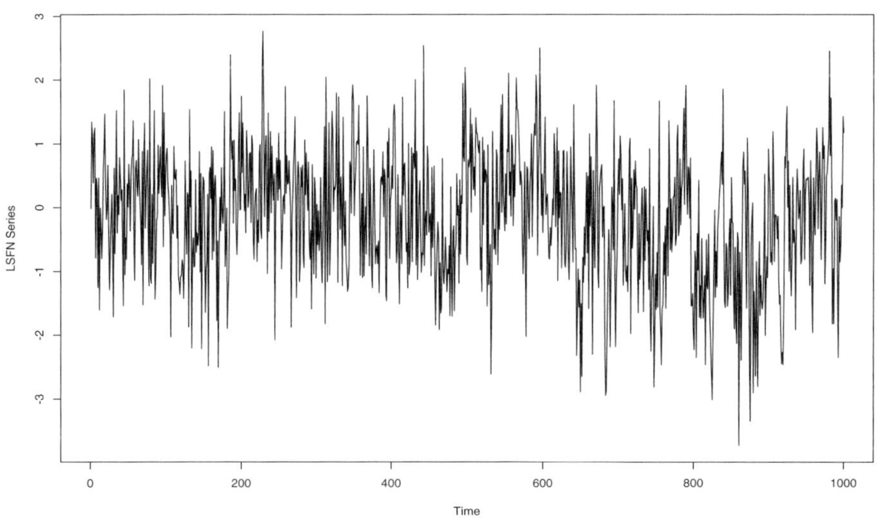

Figure 8.14 *Simulated LSFN model with 1000 observations and time-varying parameters* $\alpha_0 = 0.90$, $\alpha_1 = 0.10$, $\beta_0 = 0.15$ *and* $\beta_1 = 0.30$.

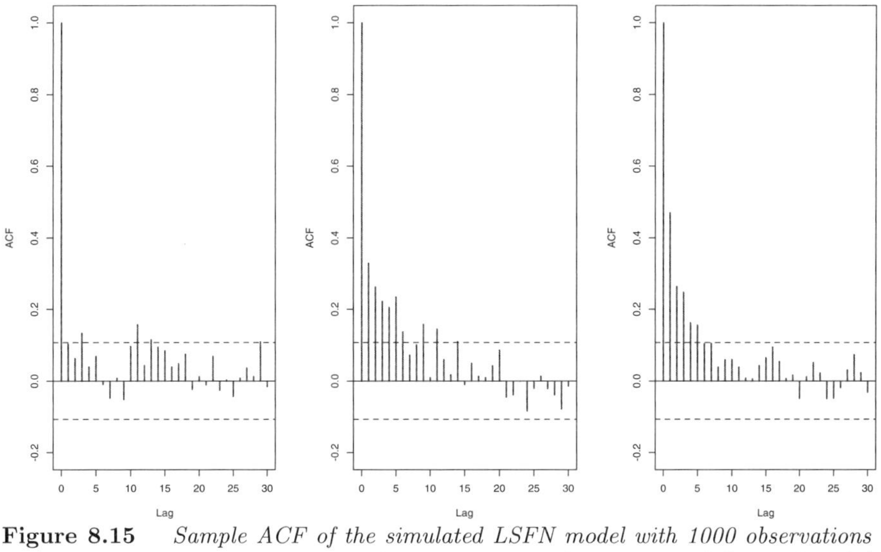

Figure 8.15 *Sample ACF of the simulated LSFN model with 1000 observations and time-varying parameters* $\alpha_0 = 0.90$, $\alpha_1 = 0.10$, $\beta_0 = 0.15$ *and* $\beta_1 = 0.30$. *(a) first block of 333 observations, (b) second block of 333 observations, and (c) third block of 334 observations.*

Table 8.3 Whittle and Kalman Estimates for the LSFN model parameters.

Parameter	α_0	α_1	β_0	β_1
	0.90	0.10	0.15	0.30
Whittle	0.9383	0.1053	0.1576	0.2491
Kalfman	0.9010	0.1826	0.1675	0.2431

8.4.4 Asymptotic Variance

Let θ_0 be the true value of the parameter θ, the Whittle estimator θ_n satisfies

$$\sqrt{n}(\theta_n - \theta_0) \to N\left[0, \Gamma(\theta_0)^{-1}\right],$$

as $n \to \infty$, where

$$\Gamma(\theta) = \frac{1}{4\pi}\int_0^1 \int_{-\pi}^{\pi} [\nabla \log f_\theta(u,\lambda)][\nabla \log f_\theta(u,\lambda)]' d\lambda du. \tag{8.19}$$

■ EXAMPLE 8.12

For an LSARFIMA$(1, d, 1)$ with polynomial evolution of its time-varying parameters $\phi(u)$, $\theta(u)$ and $d(u)$ we have the following formula for the asymptotic distribution of θ_T.

$$\Gamma = \begin{pmatrix} \Gamma_d & \Gamma_{d\,\phi} & \Gamma_{d\,\theta} & 0 \\ \Gamma_{\phi\,d} & \Gamma_\phi & \Gamma_{\phi\,\theta} & 0 \\ \Gamma_{\theta\,d} & \Gamma_{\theta\,\phi} & \Gamma_\theta & 0 \\ 0 & 0 & 0 & \Gamma_\sigma \end{pmatrix},$$

$$\Gamma_d = \frac{\pi^2}{6}\left[\frac{1}{i+j-1}\right]_{i,j=1,\ldots,P_d+1},$$

$$\Gamma_\phi = \int_0^1 \frac{1}{(1-[\phi(u)]^2)}\left[u^{i+j-2}\right]_{i,j=1,\ldots,P_\phi+1} du$$

$$\Gamma_\theta = \int_0^1 \frac{1}{(1-[\theta(u)]^2)}\left[u^{i+j-2}\right]_{i,j=1,\ldots P_\theta+1} du;$$

$$\Gamma_{d\,\phi} = -\int_0^1 \frac{\log[1+\phi(u)]}{\phi(u)}\left[u^{i+j-2}\right]_{i=1,\ldots,P_d+1;j=1,\ldots P_\phi+1} du$$

$$\Gamma_{d\,\theta} = \int_0^1 \frac{\log[1+\theta(u)]}{\theta(u)} \left[u^{i+j-2}\right]_{i=1,\ldots,P_d+1;j=1,\ldots P_\theta+1} du$$

$$\Gamma_{\phi\,\theta} = -\int_0^1 \frac{1}{1-\phi(u)\,\theta(u)} \left[u^{i+j-2}\right]_{i=1,\ldots,P_\phi+1;j=1,\ldots P_\theta+1} du$$

$$\Gamma_\sigma = 2\int_0^1 \frac{1}{[\sigma(u)]^2} \left[u^{i+j-2}\right]_{i,j=1,\ldots,P_\sigma+1} du.$$

EXAMPLE 8.13

Consider a LS-FN process where $d(u)$ and $\sigma(u)$ are given by

$$d(u) = \alpha_0 + \alpha_1 u + \cdots + \alpha_p u^p,$$
$$\sigma(u) = \beta_0 + \beta_1 u + \cdots + \beta_q u^q,$$

for $u \in [0,1]$. In this case the parameter vector is $(\alpha_0,\ldots,\alpha_p,\beta_0,\ldots,\beta_q)'$ and the matrix Γ given by (8.19) can be written as

$$\Gamma = \begin{pmatrix} \Gamma_\alpha & 0 \\ 0 & \Gamma_\beta \end{pmatrix},$$

where

$$\Gamma_\alpha = \left[\frac{\pi^2}{6\,(i+j+1)}\right]_{i,j=0,\ldots,p},$$

and

$$\Gamma_\beta = \left[\int_0^1 \frac{u^{i+j}\,du}{(\beta_0+\beta_1 u+\cdots+\beta_q u^q)^2}\right]_{i,j=0,\ldots,q}.$$

EXAMPLE 8.14

Consider a LS-FN process where $d(u)$ and $\sigma(u)$ are harmonic

$$d(u) = \alpha_0 + \alpha_1\cos(\lambda_1 u) + \cdots + \alpha_p\cos(\lambda_p u),$$
$$\sigma(u) = \beta_0 + \beta_1\cos(\omega_1 u) + \cdots + \beta_q\cos(\omega_q u),$$

for $u \in [0,1]$, where $\lambda_0 = 0$, $\lambda_i^2 \neq \lambda_j^2$ for $i,j = 0,\ldots,p$, $i \neq j$, $\omega_0 = 0$, and $\omega_i^2 \neq \omega_j^2$ for $i,j = 0,\ldots,q$, $i \neq j$. In this case $\theta = (\alpha_0,\ldots,\alpha_p,\beta_0,\ldots,\beta_q)'$ and Γ

$$\Gamma = \begin{pmatrix} \Gamma_\alpha & 0 \\ 0 & \Gamma_\beta \end{pmatrix},$$

where

$$\Gamma_\alpha = \frac{\pi^2}{12}\left[\frac{\sin(\lambda_i - \lambda_j)}{\lambda_i - \lambda_j} + \frac{\sin(\lambda_i + \lambda_j)}{\lambda_i + \lambda_j}\right]_{i,j=0,\dots,p},$$

$$\Gamma_\beta = \frac{\pi^2}{12}\left[\frac{\sin(\omega_i - \omega_j)}{\omega_i - \omega_j} + \frac{\sin(\omega_i + \omega_j)}{\omega_i + \omega_j}\right]_{i,j=0,\dots,q}.$$

EXAMPLE 8.15

Consider the LSARFIMA process defined by

$$\Phi(t/T, B)Y_{t,T} = \Theta(t/T, B)(1-B)^{-d(t/T)}\sigma(t/T)\varepsilon_t, \tag{8.20}$$

for $t = 1, \dots, T$, where for $u \in [0, 1]$,

$$\Phi(u, B) = 1 + \phi_1(u)B + \cdots + \phi_P(u)B^P$$

$$\Theta(u, B) = 1 + \theta_1(u)B + \cdots + \theta_Q(u)B^Q$$

Assume that $P = Q = 1$ in model (8.20) where $\sigma(u) = 1$ and $d(u)$, $\Phi(u, B)$, $\Theta(u, B)$ are specified by

$$\begin{aligned} d(u) &= \alpha_1 u, \\ \Phi(u, B) &= 1 + \phi(u)B, \qquad \phi(u) = \alpha_2 u, \\ \Theta(u, B) &= 1 + \theta(u)B, \qquad \theta(u) = \alpha_3 u, \end{aligned}$$

for $u \in [0, 1]$. In this case, $\theta = (\alpha_1, \alpha_2, \alpha_3)'$ and Γ from (8.19) can be written as

$$\Gamma = \begin{pmatrix} \gamma_{11} & \gamma_{12} & \gamma_{13} \\ \gamma_{21} & \gamma_{22} & \gamma_{23} \\ \gamma_{31} & \gamma_{32} & \gamma_{33} \end{pmatrix},$$

where

$$\begin{aligned} \gamma_{11} &= \frac{1}{2\alpha_1^3}\log\frac{1+\alpha_1}{1-\alpha_1} - \frac{1}{\alpha_1^2}, \quad |\alpha_1| < 1, \\ \gamma_{12} &= \frac{1}{(\alpha_1\alpha_2)^{3/2}}g(\alpha_1\alpha_2) - \frac{1}{\alpha_1\alpha_2}, \end{aligned}$$

with $g(x) = (\sqrt{x})$ for $x \in (0,1)$ and $g(x) = \arctan(\sqrt{-x})$ for $x \in (-1,0)$,

$$\gamma_{13} = \frac{1}{2\alpha_1}\left\{\left[\frac{1}{2} - \frac{1}{\alpha_1}\right] - \left[1 - \frac{1}{\alpha_1^2}\right]\log(1+\alpha_1)\right\},$$
$$\gamma_{22} = \frac{1}{2\alpha_2^3}\log\frac{1+\alpha_2}{1-\alpha_2} - \frac{1}{\alpha_2^2}, \quad |\alpha_2| < 1,$$
$$\gamma_{23} = \frac{1}{2\alpha_2}\left\{\left[1 - \frac{1}{\alpha_2^2}\right]\log(1+\alpha_2) - \left[\frac{1}{2} - \frac{1}{\alpha_2}\right]\right\},$$
$$\gamma_{33} = \frac{\pi^2}{18}.$$

8.4.5 Monte Carlo Experiments

In order to gain some insight into the finite sample performance of the Whittle estimator we report next a number of Monte Carlo experiments for the LSFN model

$$y_{t,T} = \sigma(t/T)\,(1-B)^{-d(t/T)}\,\varepsilon_t, \tag{8.21}$$

for $t = 1,\ldots,T$ with $d(u) = \alpha_0 + \alpha_1\,u$, $\sigma(u) = \beta_0 + \beta_1\,u$ and Gaussian white noise $\{\varepsilon_t\}$ with unit variance. Denote the parameter vector by $= (\alpha_0, \alpha_1)$ for $d(\cdot)$ and $= (\beta_0, \beta_1)$ for the noise scale $\sigma(\cdot)$.

The samples of the LSFN process are generated by means of the innovations algorithm. The Whittle estimates in these Monte Carlo simulations have been computed by using the cosine bell data taper

$$h(x) = \frac{1}{2}[1 - \cos(2\,\pi\,x)].$$

Table 8.4 reports the results from Monte Carlo simulations for several parameter values, based on 1000 replications. These tables show the average of the estimates as well as their theoretical and empirical standard deviations (SD) given by

$$\Gamma = \left[\frac{\pi^2}{6\,(i+j+1)}\right]_{i,j=0,1}, \qquad \Gamma = 2\left[\int_0^1 \frac{u^{i+j}\,du}{\sigma^2(u)}\right]_{i,j=0,1}. \tag{8.22}$$

Observe from this table that the estimated parameters are close to their true values. Besides, the empirical SD are close to their theoretical counterparts. These simulations suggest that the finite sample performance of the proposed estimators seem to be very good in terms of bias and standard deviations.

Table 8.4 Whittle maximum likelihood estimation for model (8.21): Sample size $T = 1024$, block size $N = 128$ and shift $S = 64$.

Parameters		Estimates		Theoretical SD		SD Estimates	
α_0	α_1	$\widehat{\alpha}_0$	$\widehat{\alpha}_1$	$\sigma(\widehat{\alpha}_0)$	$\sigma(\widehat{\alpha}_1)$	$\widehat{\sigma}(\widehat{\alpha}_0)$	$\widehat{\sigma}(\widehat{\alpha}_1)$
0.1000	0.2000	0.0896	0.2014	0.0490	0.0840	0.0534	0.1094
0.1500	0.2500	0.1261	0.2786	0.0490	0.0840	0.0515	0.1085
0.2000	0.2000	0.1853	0.2245	0.0490	0.0840	0.0743	0.1055
0.2000	0.2500	0.1877	0.2670	0.0490	0.0840	0.0573	0.0990
0.2500	0.2000	0.2627	0.2042	0.0490	0.0840	0.0772	0.1017
0.1000	0.2000	0.0755	0.2161	0.0490	0.0840	0.0533	0.1220
0.1500	0.2500	0.1428	0.2650	0.0490	0.0840	0.0689	0.0959
0.2000	0.2000	0.1812	0.2095	0.0490	0.0840	0.0476	0.1156
0.2000	0.2500	0.2030	0.2678	0.0490	0.0840	0.0475	0.0723
0.2500	0.2000	0.2546	0.1920	0.0490	0.0840	0.0510	0.1158
β_0	β_1	$\widehat{\beta}_0$	$\widehat{\beta}_1$	$\sigma(\widehat{\beta}_0)$	$\sigma(\widehat{\beta}_1)$	$\widehat{\sigma}(\widehat{\beta}_0)$	$\widehat{\sigma}(\widehat{\beta}_1)$
0.5000	0.5000	0.4862	0.5142	0.0270	0.0560	0.0201	0.0734
0.5000	0.5000	0.5018	0.5286	0.0270	0.0560	0.0353	0.0564
0.5000	0.5000	0.4701	0.5030	0.0270	0.0560	0.0255	0.0605
0.5000	0.5000	0.4879	0.5339	0.0270	0.0560	0.0135	0.0677
0.5000	0.5000	0.5020	0.4965	0.0270	0.0560	0.0295	0.0739
1.0000	-0.5000	0.9781	-0.4879	0.0380	0.0560	0.0255	0.0704
1.0000	-0.5000	0.9969	-0.5100	0.0380	0.0560	0.0406	0.0736
1.0000	-0.5000	1.0191	-0.5093	0.0380	0.0560	0.0438	0.0668
1.0000	-0.5000	1.0178	-0.4753	0.0380	0.0560	0.0376	0.0635
1.0000	-0.5000	1.0327	-0.4857	0.0380	0.0560	0.0365	0.0504

8.4.6 Data Application

Tree rings count is a usual procedure in studies of forest mass to determine growth and yield of both natural forests and forest plantations. These time series are also useful in paleoclimatology, as discussed in Chapter 1. Forest analysis can be implemented in species growing in temperate regions, where it is easy to identify the ring growth. In tropical climates, where there is little differentiation among seasons, growth rates are constant, making it difficult

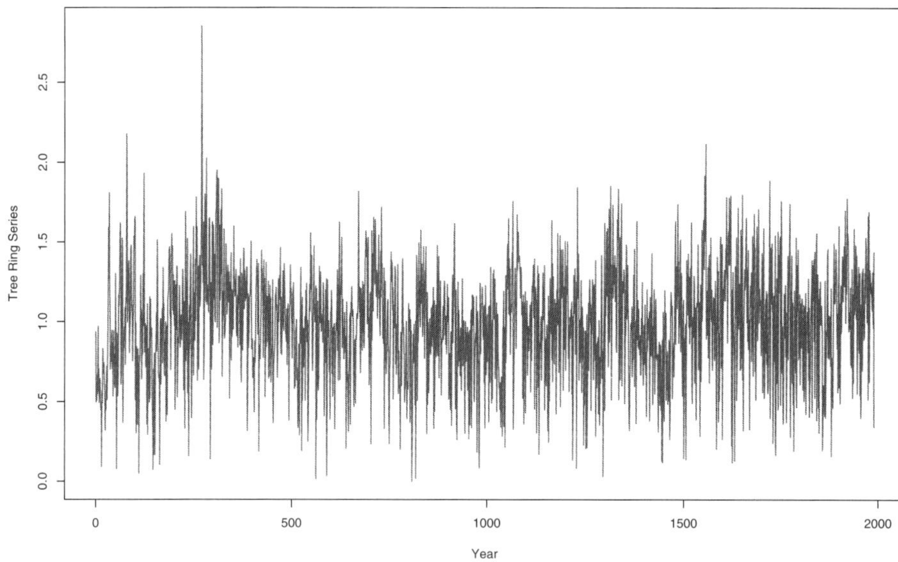

Figure 8.16 *Mammoth Creek Tree Ring Data.*

to clearly differentiate spring and winter wood. Consequently, this data set can be used as climate proxies and to indicate the chances of temperature and precipitation conditions in paleoclimatology.

Figure 8.16 displays annual tree-ring width of the Pinus Longaeva, measured at Mammoth Creek, Utah, from 0 AD to 1989 AD, cf. Chapter 1 and Appendix C.

Figure 8.17(a) shows the sample ACF of $x_{t,T}$, and the corresponding variances of the sample mean, that is *varplots*, are shown in panel (b). The dashed line corresponds to its expected behavior for a short-memory case with blocks of k observations, whereas the continuous line represents the expected behavior for a long-memory case. From both panels, this series seems to exhibit long-range dependence.

Moreover, a closer look at the sample ACF of the data reveals that the degree of persistence seems to vary over time. Indeed, Figure 8.18 shows the sample autocorrelation of three segments of the sample: observations 1 to 500, observations 750 to 1250 and observations 1490 to 1990. This figure provides information for arguing possible changes in the degree of dependence. This represents a clear evidence of a nonstationary process. Therefore, it seems that the data has a time-varying long-memory structure. Additionally, Figure 8.19 and Figure 8.20 depict two views of the time-varying periodogram of the data.

In order to handle these features, a locally stationary ARFIMA process is suggested. Figure 8.21 shows an heuristic estimator of the long-memory

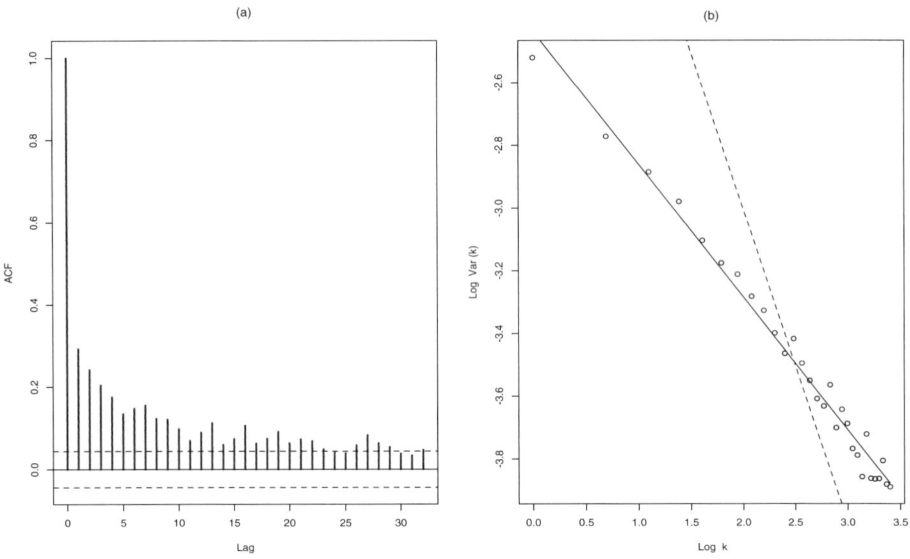

Figure 8.17 *Mammoth Creek Tree Ring Data. (a) Sample ACF, (b) Varplot.*

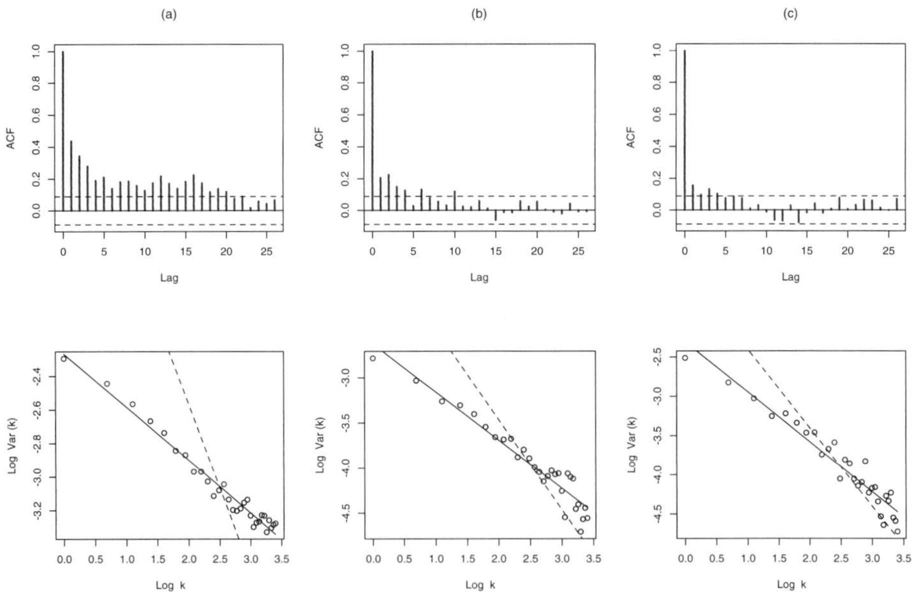

Figure 8.18 *Mammoth Creek Tree Ring Data. Sample ACF and Varplots: (a) Observations 1 to 500, (b) Observations 750 to 1250, (c) Observations 1490 to 1990.*

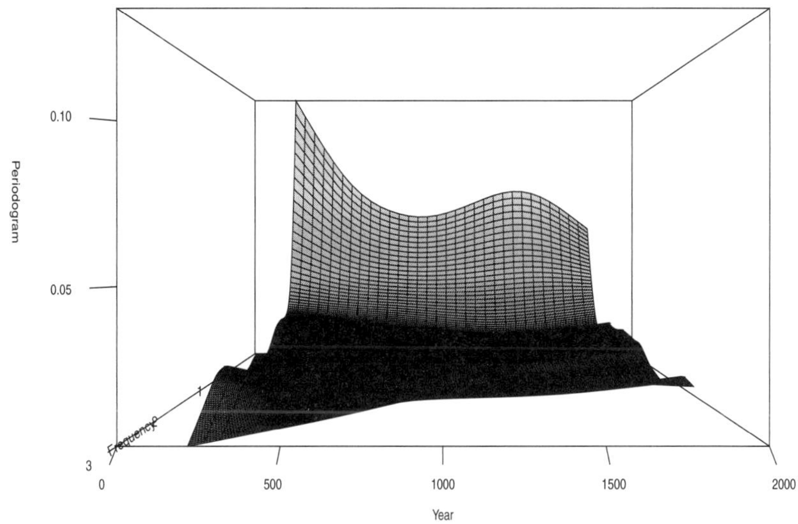

Figure 8.19 *Mammoth Creek tree ring data. Time-varying periodogram.*

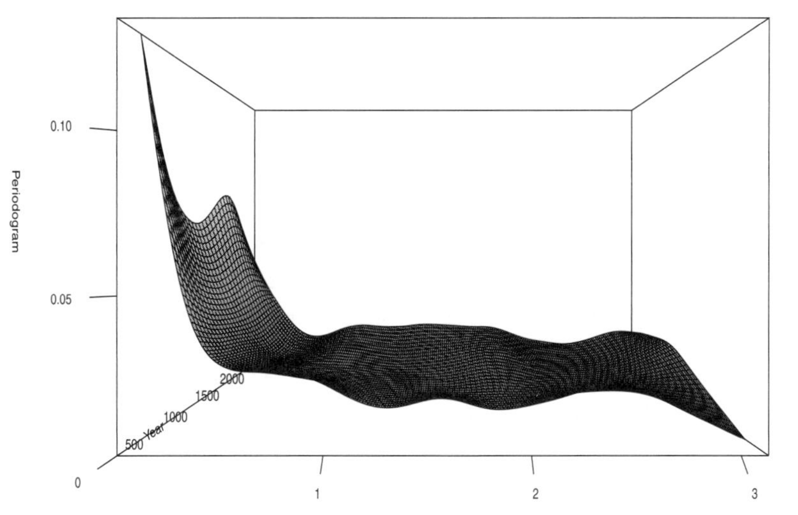

Figure 8.20 *Mammoth Creek tree ring data. Time-varying periodogram.*

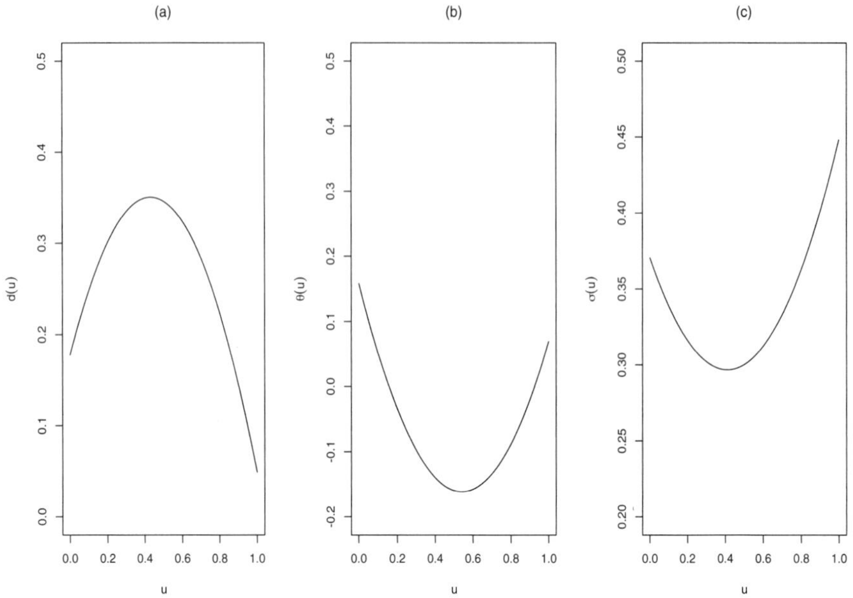

Figure 8.21 *Mammoth Creek tree ring data. Time-varying estimated parameters. (a) $d(u)$, (b) theta(u), (c) $\sigma(u)$.*

parameter and the variance of the noise scale along with stationary fractional noise and locally stationary fractional noise model estimates of these quantities. From this figure we suggest a linear and quadratic function for $d(u)$ and $\sigma(u)$ respectively, that is, a LSARFIMA$(1, d, 0)$ model with time-varying parameters given by,

$$d(u) = d_0 + d_1 u, \quad \theta(u) = \theta_0 + \theta_1 u, \quad \sigma(u) = \beta_0 + \beta_1 u + \beta_2 u^2. \tag{8.23}$$

Table 8.5 reports the parameter estimates using the Whittle method. The standard deviations and the t-tests have been obtained using (8.22) for $d(u)$, $\theta(u)$ and $\sigma(u)$, respectively. As we can observe in this table, the parameters (d_0, d_1) and $(\beta_0, \beta_1, \beta_2)$ are statistically significant at the 5% level.

The residuals of the model are plotted in Figure 8.22 along with the sample ACF, the partial ACF and the Ljung-Box tests. From these panels, it seems that there are no significant autocorrelations in the residuals. This conclusion is formally supported by the Ljung-Box tests. Consequently, the white noise hypothesis cannot be rejected at the 5% level.

As described in the previous data application, locally stationary processes are useful tools for modeling complex nonstationary time dependence structures.

Table 8.5 Tree Ring Data at Mammoth Creek, Utah. LSARFIMA$(1, d, 0)$ parameters estimated with the Whitle method.

Parameter	Estimate	SD	t-value
d_0	0.1769622	0.08445276	2.095399
d_1	0.8050242	0.39121972	2.057729
d_2	-0.9328672	0.37841028	-2.465227
θ_0	0.1593663	0.10967577	1.453067
θ_1	-1.1863239	0.50051007	-2.370230
θ_2	1.0963000	0.48266935	2.271327
β_0	0.3707303	0.01648537	22.488441
β_1	-0.3597924	0.07597368	-4.735751
β_2	0.4371512	0.07541373	5.796706

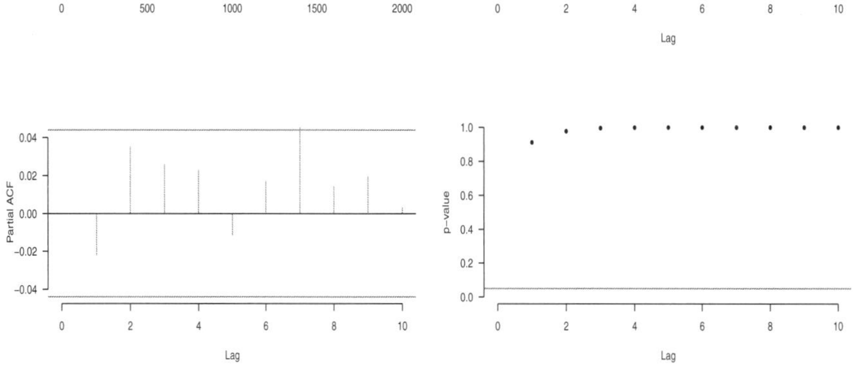

Figure 8.22 *Mammoth Creek tree ring data. Residual diagnostic of the model.*

8.5 STRUCTURAL BREAKS

As discussed throughout this book, real-life time series usually display structural changes. In the previous section, we discussed locally stationary processes where the model changes are gradual and continuous. On the other hand, in some ocassions the structural changes are rather abrupt. In these cases, there are other methodologies developed to deal with structural breaks. In what follows, we discuss a methodology for handling these nonstationarities by modeling the time series in terms of successive blocks of linear trends and seasonal models. Let y_t be a time series which can be described by

$$y_t = T_t + S_t + \varepsilon_t,$$

where T_t and S_t denote sequences of linear trends and seasonal components, and ε_t is an error sequence. More specifically, the linear trends are described by

$$T_t = \alpha_j + \beta_j\, t, \qquad \text{for} \qquad t \in (t_{j-1}, t_j]$$

where $t_0 = 0$ and $t_1, \ldots, t_m$ denote the times at which the trend breaks occur. On the other hand, the seasonal components are defined by

$$S_t = \begin{cases} \gamma_{ij} & \text{if} \quad t \,\text{is in season}\, i \,\text{and}\, t \in (s_{j-1}, s_j] \\ -\sum_{i=1}^{s-1} \gamma_{ij} & \text{if} \quad t \,\text{is in season}\, 0 \,\text{and}\, t \in (s_{j-1}, s_j] \end{cases}$$

where $s_0 = 0$ and $s_1, \ldots, s_p$ denote the times at which the seasonal breaks occur. The R package *breaks for additive seasonal and trend* bfast allows fot the estimation of the number of trend breaks m, the times $t_1, \ldots, t_m$, the number of seasonal breaks p, the times $s_1, \ldots, s_p$ along with the parameters α_j, β_j and γ_{ij}.

■ EXAMPLE 8.16

Consider the series of passenger enplanements introduced in Chapter 1. An application of the bfast methodology produces the following trend break decomposition shown in Figure 8.23.

```
  TREND BREAKPOINTS
 Confidence intervals for breakpoints
 of optimal 2-segment partition:

Call:
confint.breakpointsfull(object = bp.Vt, het.err = FALSE)

Breakpoints at observation number:
  2.5 % breakpoints 97.5 %
1    55          56     57
```

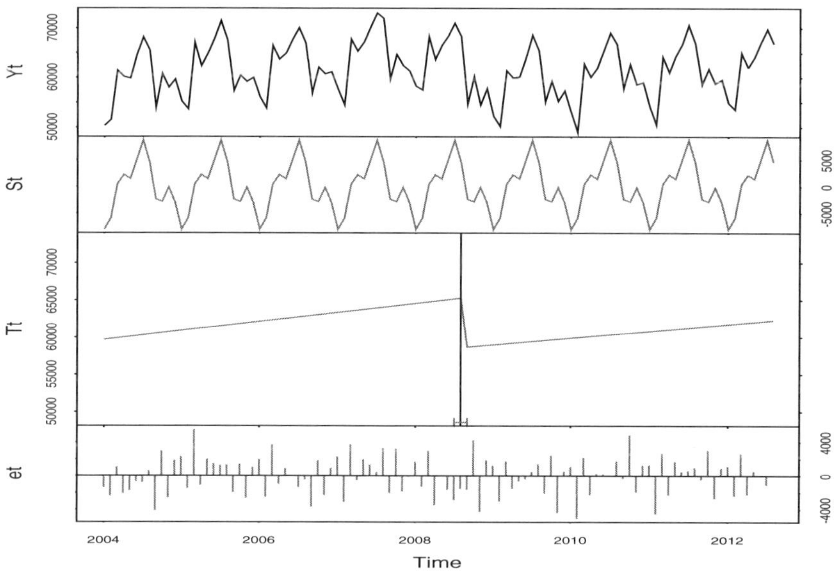

Figure 8.23 Breaks for Additive Seasonal and Trend Analysis of Passenger enplanements data.

```
Corresponding to breakdates:
  2.5 %   breakpoints 97.5 %
1 2008(7) 2008(8)     2008(9)

  SEASONAL BREAKPOINTS:  None
```

The analysis reported by this figure suggests that there was a trend change in the number of passenger enplanements around August 2008.

■ EXAMPLE 8.17

As another illustration of the bfast methodology consider the logarithm of the US employment in Arts, Entertainment and Recreation, for the period January 1990 to December 2012. In this case, the bfast method produces the following trend break decomposition.

```
  TREND BREAKPOINTS
 Confidence intervals for breakpoints
 of optimal 3-segment partition:

Breakpoints at observation number:
  2.5 % breakpoints 97.5 %
1   139          140    141
2   229          230    231
```

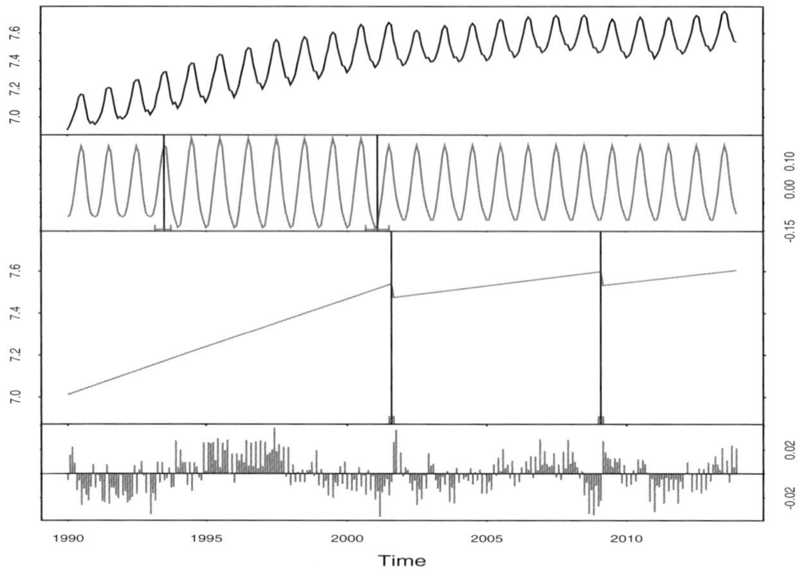

Figure 8.24 Breaks for Additive Seasonal and Trend Analysis of the logarithm of the US employment in Arts, Entertainment and Recreation, for the period January 1990 to December 2012.

```
Corresponding to breakdates:
  2.5 %   breakpoints 97.5 %
1 2001(7) 2001(8)     2001(9)
2 2009(1) 2009(2)     2009(3)

  SEASONAL BREAKPOINTS
 Confidence intervals for breakpoints
 of optimal 3-segment partition:

Breakpoints at observation number:
  2.5 % breakpoints 97.5 %
1    39            43     46
2   129           134    139

Corresponding to breakdates:
  2.5 %   breakpoints 97.5 %
1 1993(3) 1993(7)     1993(10)
2 2000(9) 2001(2)     2001(7)
```

According to the results, the analysis suggests structural changes in both, the linear trends and the seasonal components.

EXAMPLE 8.18

As an illustration of structural change in the serial dependence structure of the data consider the Nile river series described in Chapter 1. It has been noticed that the first 100 observations seems to have a different level of dependence than the rest of the series. In this example, we consider a fist block of 100 values, from 622 AD. to 721 AD., and a second block of 563 observations, from 722 AD. to 1284 AD. We do not consider the data after the year 1284 AD. in this study because in this period the series suffer a large number of data repetitions or missing data problems.

By means of the R package arfima, we compute the corresponding ARFIMA models to each block and obtain the following results.

```
> fit
Number of modes: 1

Call:
arfima(z = x.nile[1:100], order = c(0, 0, 0))

Coefficients for fits:
              Coef.1:    SE.1:
d.f           0.014179   0.0883828
```

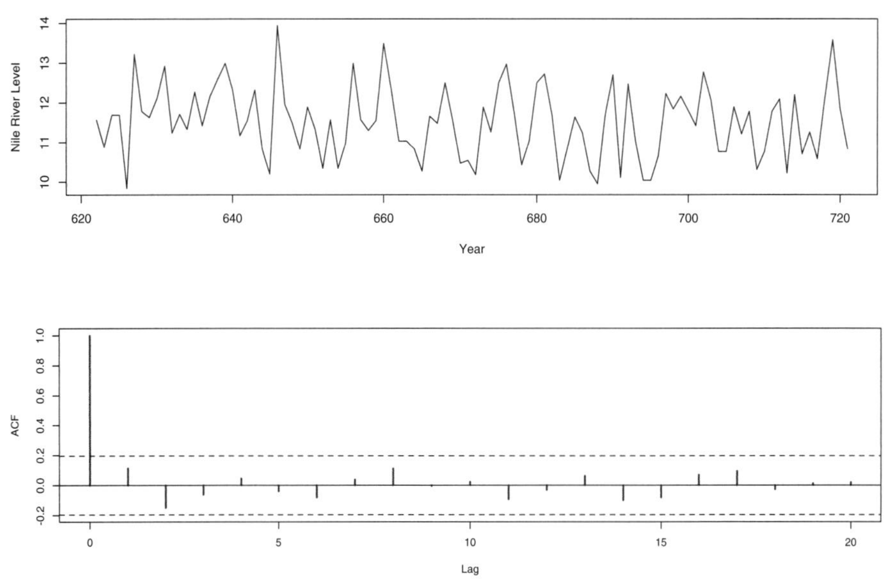

Figure 8.25 *Nile river levels from 622 A.D. to 721 A.D. and sample ACF.*

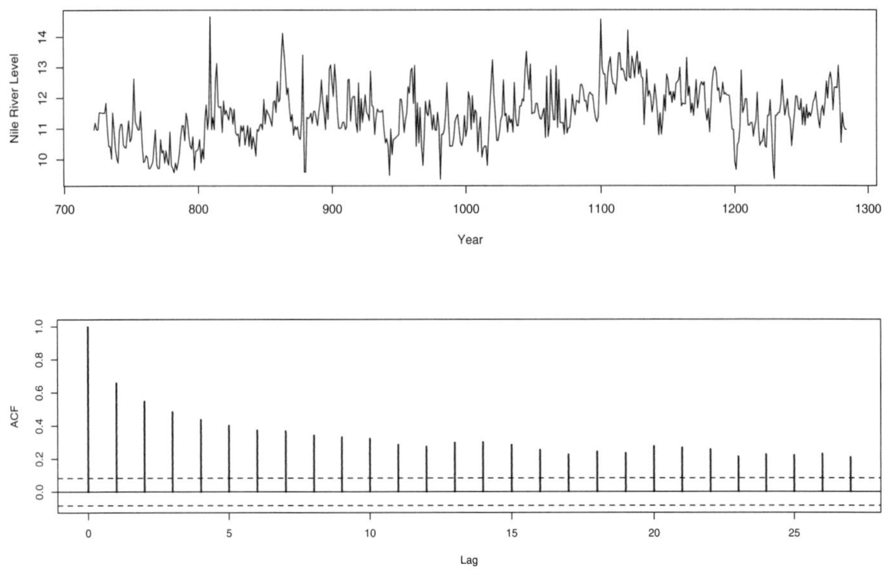

Figure 8.26 *Nile river levels from 722 A.D. to 1284 A.D. and sample ACF.*

```
Fitted mean    11.5133     0.0950155
logl           11.1362
sigma^2        0.808409
Starred fits are close to invertibility/stationarity boundaries


> fit
Number of modes: 1

Call:
arfima(z = x.nile[101:663], order = c(0, 0, 0))

Coefficients for fits:
             Coef.1:    SE.1:
d.f           0.44762    0.0302793
Fitted mean   11.4113    0.837629
logl          249.662
sigma^2       0.410754
Starred fits are close to invertibility/stationarity boundaries
```

As indicated by many studies, the first 100 observations displays almost null serial correlation which is reflected in the very low estimated parameter d. In fact, the following Ljung-Box white noise tests for lags

5, 10 and 20 indicate that this part of the Nile river series is compatible with a white noise model.

```
> Box.test(x.nile[1:100],lag=5,type="Ljung")

Box-Ljung test

data:  x.nile[1:100]
X-squared = 4.4495, df = 5, p-value = 0.4867

> Box.test(x.nile[1:100],lag=10,type="Ljung")

Box-Ljung test

data:  x.nile[1:100]
X-squared = 6.7533, df = 10, p-value = 0.7485

> Box.test(x.nile[1:100],lag=20,type="Ljung")

Box-Ljung test

data:  x.nile[1:100]
X-squared = 12.1105, df = 20, p-value = 0.9122
```

8.6 BIBLIOGRAPHIC NOTES

Techniques for estimating and forecasting ARIMA models are found in Box, Jenkins, and Reinsel (1994). Locally stationary processes have been playing an important role in time series analysis. They have provided a sound statistical methodology for modeling data exhibiting nonstationary features without resorting to data transformations, trend removals and other related techniques. The theory of LS processes is based on the principle that a nonstationary process can be locally approximated by a stationary one if the time variation of the model parameters is sufficiently smooth. The idea of developing techniques for handling directly nonstationary processes dates back to the sixties. For example, Priestley (1965), Priestley and Tong (1973), Tong (1973) and others developed the concept of evolutionary spectra. In the ninities, Dahlhaus (1996, 1997) provided a formal definition of a family of LS processes. There are several works on LS processes, including, Dahlhaus (2000), Jensen and Witcher (2000), Dahlhaus and Polonik (2006, 2009), Chandler and Polonik (2006), Palma and Olea (2010), Dette, Preuß, and Vetter (2011) and Palma, Olea, and Ferreira (2013), among others. Other classes of LS processes have been discussed for example by Wang, Cavanaugh, and Song (2001), Cavanaugh, Wang, and Davis (2003) and Last and Shumway (2008). The analysis of the asymptotic properties of the Whittle locally stationary estimates (8.17) is discussed, for example, in Dahlhaus (1997) and Palma and

Olea (2010). Furthermore, the R package LSTS allows for the estimation and prediction of LS models.

Problems

8.1 From the definition of the backshift operator B, find expressions for Bz_t and B^2z_t, if z_t is defined as:

(a) $z_t = \beta_0$,

(b) $z_t = \beta_0 + \beta_1 t$,

(c) $z_t = \beta_0 + \beta_1 x_t + \beta_2 t$,

(d) $z_t = \beta_1 x_t + \beta_2 y_t$, where β_0, β_1 and β_2 are constants, while x_t and y_t are time series.
Hint: The operator B is defined by the relationship $Bz_t = z_{t-1}$ for all t.

8.2 Let $\{x_t, t \in \mathbb{Z}\}$ a stationary stochastic process with autocorrelation function $\rho_x(k)$. Show that $\{y_t = (1 - B)x_t, t \in \mathbb{Z}\}$ is a stationary stochastic process and calculate $\rho_y(k)$ in terms of $\rho_x(k)$. If x_t is an ARMA(p,q) process, what can you say about the process ∇x_t?

8.3 From the definition of the operator ∇ ($\nabla = (1 - B)$), find expressions for ∇z_t and $\nabla^2 z_t$, if z_t is defined as in the previous exercise what is your guess about the general expression for $\nabla^d z_t$ if $d > 2$?

8.4 Consider the following processes z_t defined by

(i) $z_t = \beta_0 + \beta_1 \epsilon_t$,

(ii) $z_t = \beta_0 + \beta_1 t + \epsilon_t$,

(iii) $z_t = \beta_0^t e^{\epsilon_t}$, $\beta_0 > 0$

(iv) $z_t = \beta_0 + \epsilon_t + \beta_1 \epsilon_{t-1}$,

where β_0 and β_1 are constant and $\{\epsilon_t\}$ is a white noise process with zero-mean and variance σ_ϵ^2, define a new process y_t (as a function of z_t) that is stationary. Provide $E(y_t)$, $\mathrm{Var}(y_t)$ and $\mathrm{Cov}(y_t, y_{t+k})$ for $k = 1, 2, \ldots$.

8.5 Let $\{\varepsilon_t\}$ be a stationary process and $x_t = a + bt + \varepsilon_t$ with a and b constants.

(a) Show that ∇x_t is stationary.

(b) How would you obtain a stationary process stationary if the trend were quadratic?

8.6 Let $\{\epsilon_t\}$ be a sequence of independent random variables normally distributed, zero-mean and variance σ^2. Let a, b and c be constants Which of the following processes are stationary? For each statiuonary process calculate its expected value and its autocovariance function.

(a) $x_t = a + b\epsilon_t + c\epsilon_{t-1}$,

(b) $x_t = a + b\,\epsilon_0$,
(c) $x_t = \epsilon_1 \cos(ct) + \epsilon_2 \sin(ct)$,
(d) $x_t = \epsilon_0 \cos(ct)$,
(e) $x_t = \epsilon_t \cos(ct) + \epsilon_{t-1} \sin(ct)$,
(f) $x_t = \epsilon_t \epsilon_{t-1}$.

8.7 Let $x_t = a + b\,t$ for $t = 1, 2, \ldots$ with a and b constants. Show that the sample autocorrelations of this sequence $\hat{\rho}(k)$ satisfy $\hat{\rho}(k) \to 1$ as $n \to \infty$ for each fixed k.

8.8 Let S_t, $t = 0, 1, 2, \ldots$ a random walk with constant jump μ defined as $S_0 = 0$ and $S_t = \mu + S_{t-1} + x_t$, $t = 1, 2, \ldots$ where $x_1, x_2, \ldots$ are i.i.d. random variables with zero-mean and variance σ^2.

(a) Is the process $\{S_t\}$ stationary?
(b) Is the sequence $\{\nabla S_t\}$ stationary?

8.9 Consider a simple moving-average filter with weights $a_j = (2q+1)^{-1}$, $-q \le j \le q$.

(a) If $m_t = c_0 + c_1 t$, show that $\sum_{j=-q}^{j=q} a_j m_{t-j} = m_t$.
(b) If ε_t, $t = 0, \pm 1, \pm 2, \ldots$, are independent random variables with zero-mean and variance σ^2, show that the moving-average $A_t = \sum_{j=-q}^{j=q} a_j \varepsilon_{t-j}$ is *small* for large q in the sense that $E\,A_t = 0$ and $\mathrm{Var}(A_t) = \frac{\sigma^2}{2q+1}$.

8.10 Suppose that $m_t = c_0 + c_1 t + c_2 t^2$, $t = 0, \pm 1, \ldots$

(a) Show that

$$m_t = \sum_{i=-2}^{2} a_i m_{t+i} = \sum_{i=-3}^{3} b_i m_{t+i}, \; t = 0, \pm 1, \ldots$$

where $a_2 = a_{-2} = -\frac{3}{35}$, $a_1 = a_{-1} = \frac{12}{35}$, $a_0 = \frac{17}{35}$, and $b_3 = b_{-3} = -\frac{2}{21}$, $b_2 = b_{-2} = \frac{3}{21}$, $b_1 = b_{-1} = \frac{6}{21}$, $b_0 = \frac{7}{21}$.

(b) Suppose that $x_t = m_t + \varepsilon_t$ where $\{\varepsilon_t, t = 0, \pm 1, \ldots\}$ is a sequence of independent random variables normal, with zero-mean and variance σ^2. Consider $U_t = \sum_{i=-2}^{2} a_i x_{t+i}$ and $V_t = \sum_{i=-3}^{3} b_i x_{t+i}$.

(i) Calculate the mean and variance of U_t and V_t.

(ii) Find the correlation between U_t and U_{t+1} and between V_t and V_{t+1}.

(iii) Which of the two filtered series $\{U_t\}$ or $\{V_t\}$ would you expect to be have a smoother path?

8.11 Consider a zero-mean series y_t that satisfies an ARIMA(p, d, q) model. Please answer the following questions:

(a) Write the representation of y_t in terms of polynomials of the lag operator B.

(b) Consider the stationary part of y_t, is $x_t = (1-B)y_t$? What model is x_t?

(c) What are the required conditions for x_t to follow a causal and invertible ARMA(p,q). model?

(d) Derive the minimum means of predictors x_t using (i) squared error, the causal model representation, (ii) the invertible representation of the model and (iii) the difference equation representation model.

8.12 Let $\{x_t\}$ be the ARIMA(2,1,0) process

$$(1 - 0.8\,B + 0.25\,B^2)(1-B)x_t = z_t,$$

where $\{z_t\}$ is a WN$(0,1)$.

(a) Find the function $g(h) = P_n x_{n+h}$ for $h \geq 0$.

(b) Assuming that n is large, calculate $\sigma_n^2(h)$ for $h = 1, \ldots, 5$.

8.13 Show that the seasonal component S_t in the trend break model can be expressed as

$$S_t = \sum_{i=1}^{s-1} \gamma_{i,j}(d_{t,i} - d_{t,0}),$$

where the seasonal dummy variable $d_{t,i}$ satisfies $d_{t,i} = 1$ if t is in season i and 0 otherwise.

(a) Verify that if t is in season 0, then $d_{t,i} - d_{t,0} = 1$.

(b) Show that for all other seasons, $d_{t,i} - d_{t,0} = 1$ when t is in season $i \neq 0$.

8.14 Consider a LS-FN process where $d(u)$ and $\sigma(u)$ are specified by

$$\ell_1[\,d(u)\,] = \sum_{j=0}^{p} \alpha_j g_j(u), \qquad \ell_2[\,\sigma(u)\,] = \sum_{j=0}^{q} \beta_j h_j(u),$$

$(\alpha_0, \ldots, \alpha_p, \beta_0, \ldots, \beta_q)'$. Show that the matrix Γ in (8.19) is given by

$$\Gamma = \begin{pmatrix} \Gamma_\alpha & 0 \\ 0 & \Gamma_\beta \end{pmatrix},$$

$$\Gamma_\alpha = \frac{\pi^2}{6}\left[\int_0^1 \frac{g_i(u)\,g_j(u)}{[\ell_1'(d(u))]^2}\,du\right]_{i,j=0,\ldots,p},$$

$$\Gamma_\beta = 2\left[\int_0^1 \frac{h_i(u)\,h_j(u)}{[\sigma(u)\ell_2'(\sigma(u))]^2}\,du\right]_{i,j=0,\ldots,q}.$$

8.15 Consider the locally stationary ARFIMA$(0,d,1)$ model given by

$$y_{t,T} = \sigma\left(\tfrac{t}{T}\right)\left[1 - \theta\left(\tfrac{t}{T}\right)B\right](1-B)^{-d\left(\frac{t}{T}\right)}\varepsilon_t,$$

where $\theta(\cdot)$ is a smoothly varying moving-average coefficient satisfying $|\theta(u)| < 1$ for $u \in [0, 1]$. Verify that the covariances $\kappa_T(s,t)$ of this process are given by

$$\kappa_T(s,t) = \sigma\left(\tfrac{s}{T}\right)\sigma\left(\tfrac{t}{T}\right)\frac{\Gamma\left[1-d\left(\tfrac{s}{T}\right)-d\left(\tfrac{t}{T}\right)\right]\Gamma\left[s-t+d\left(\tfrac{s}{T}\right)\right]}{\Gamma\left[1-d\left(\tfrac{s}{T}\right)\right]\Gamma\left[d\left(\tfrac{s}{T}\right)\right]\Gamma\left[s-t+1-d\left(\tfrac{t}{T}\right)\right]}$$
$$\times\left[1+\theta\left(\tfrac{s}{T}\right)\theta\left(\tfrac{t}{T}\right)-\theta\left(\tfrac{s}{T}\right)\frac{s-t-d\left(\tfrac{t}{T}\right)}{s-t-1+d\left(\tfrac{s}{T}\right)}-\theta\left(\tfrac{t}{T}\right)\frac{s-t+d\left(\tfrac{s}{T}\right)}{s-t+1-d\left(\tfrac{t}{T}\right)}\right],$$

for $s,t = 1,\ldots,T$, $s \geq t$.

CHAPTER 9

SEASONALITY

Seasonal patterns arise in a great number of real-life time series data. For instance, this phenomenon occurs in revenue series, inflation rates, monetary aggregates, gross national product series, shipping data, and monthly flows of the Nile River; see Section 9.8 for specific references. Consequently, several statistical methodologies have been proposed to model this type of data including the Seasonal ARIMA (SARIMA) models, Gegenbauer autoregressive moving-average processes (GARMA), seasonal autoregressive fractionally integrated moving-average (SARFIMA) models, k-factor GARMA processes, and flexible seasonal fractionally integrated processes (flexible ARFISMA), among others.

In this chapter we review some of these statistical methodologies. A general long-memory seasonal process is described in Section 9.2. This section also discusses some large sample properties of the MLE and Whittle estimators such as consistency, central limit theorem, and efficiency. Calculation of the asymptotic variance of maximum-likelihood and quasi-maximum-likelihood parameter estimates is addressed in Section 9.4. The finite sample performance of these estimators is studied in Section 9.6 by means of Monte Carlo simulations while Section 9.7 is devoted to the analysis of a real-life data il-

Time Series Analysis. First Edition. Wilfredo Palma.

lustration of these estimation methodologies. Further reading on this topic are suggested in Section 9.8 and several problems are listed at the end of this chapter.

9.1 SARIMA MODELS

A simple seasonally integrated process with period s can be written as

$$(1 - B^s)y_t = \varepsilon_t,$$

where ε_t is a white noise sequence, or equivalently,

$$y_t = y_{t-s} + \varepsilon_t.$$

This basic model establishes that the observation at time t is the same as the observation at time $t - s$ except by an additive noise.

More generally, a SARIMA$(p, d, q) \times (P, D, Q)$ model with one seasonal component can be written as

$$\phi(B)\Phi(B^s)(1 - B)^d(1 - B^s)^D y_t = \theta(B)\Theta(B^s)\varepsilon_t,$$

where ε_t is a white noise sequence with zero-mean and variance σ^2 and the respective polynomials are given by

$$\begin{aligned}
\phi(B) &= 1 - \phi_1 B - \phi_2 B^2 - \cdots - \phi_p B^p, \\
\Phi(B^s) &= 1 - \Phi_1 B^s - \Phi_2 B^{2s} - \cdots - \Phi_P B^{Ps}, \\
\theta(B) &= 1 - \theta_1 B - \theta_2 B^2 - \cdots - \theta_q B^q, \\
\Theta(B^s) &= 1 - \Theta_1 B^s - \Theta_2 B^{2s} - \cdots - \Theta_Q B^{Qs}.
\end{aligned}$$

■ EXAMPLE 9.1

Figure 9.1 exhibits a simulated time series with 500 observations from a SARIMA$(1, 0, 1) \times (1, 1, 1)$ model with $\phi = -0.7$, $\Phi = 0.7$, $\theta = 0.4$, $\Theta = 0.2$ and $s = 12$. Note that in this case, apart from the seasonal behavior of the series, there is a random trend pattern generated by the seasonal differentiation.

On the other hand, Figure 9.2 shows a trajectory of a SARIMA$(1, 1, 1) \times (1, 1, 1)$ model with $\phi = -0.3$, $\Phi = -0.3$, $\theta = 0.4$, $\Theta = 0.2$ and $s = 12$.

Notice that the random trend is more extreme now as compared to the previous example. This could be expected from the fact that we have both standard differentiation $(1 - B)$ and seasonal differentiation $(1 - B^{12})$.

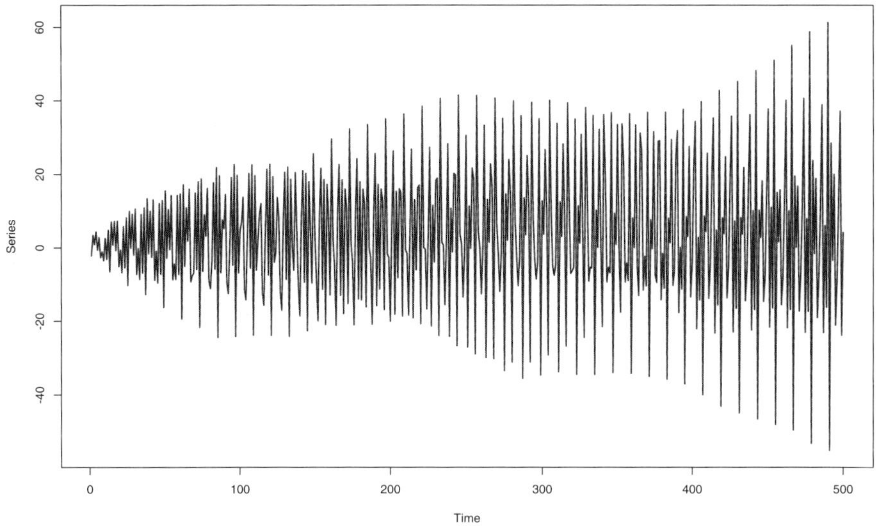

Figure 9.1 *Simulated 500 observations from a SARIMA model with* $\phi = -0.7$, $\Phi = 0.7$, $\theta = 0.4$, $\Theta = 0.2$, $s = 12$, $D = 1$ *and Gaussian white noise with unit variance.*

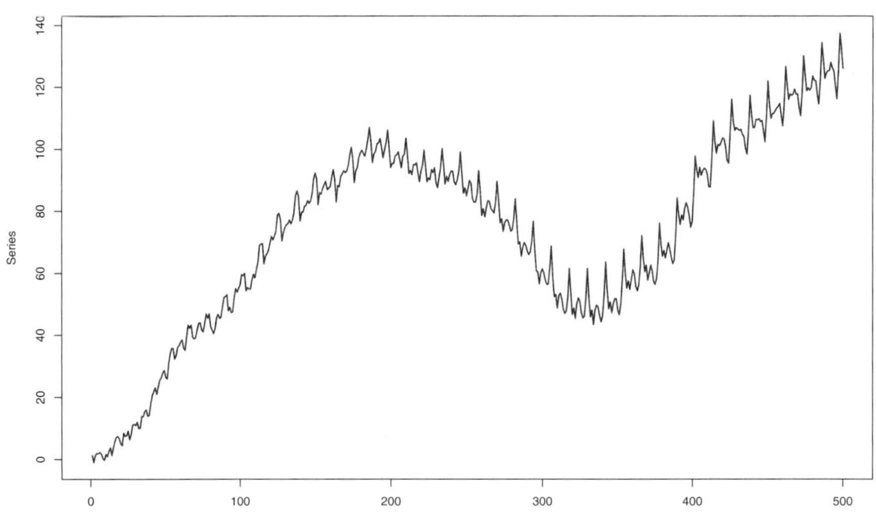

Figure 9.2 *Simulated 500 observations from a SARIMA model with* $\phi = -0.3$, $\Phi = -0.3$, $\theta = 0.4$, $\Theta = 0.2$, $s = 12$, $D = 1$, *and Gaussian white noise with unit variance.*

■ EXAMPLE 9.2

Figure 9.3 displays a sample of a SARIMA$(1, 0, 1) \times (1, 0, 1)$ model with $\phi = 0.5$, $\Phi = 0.6$, $\theta = 0.2$, $\Theta = 0.3$ and $s = 12$. Unlike the erratic pattern described by the two seasonal time series shown in the previous example, in this case the trajectory is not explosive. Additionally, the sample ACF of this series is plotted in Figure 9.4. Notice the cyclical pattern of this sample ACF.

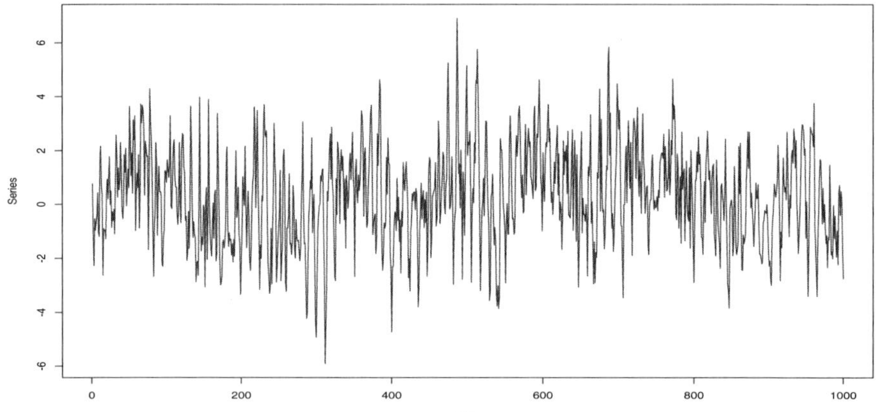

Figure 9.3 *Simulated 1000 observations from a SARIMA model with $\phi = 0.5$, $\Phi = 0.6$, $\theta = 0.2$, $\Theta = 0.3$, $s = 12$, and Gaussian white noise with unit variance.*

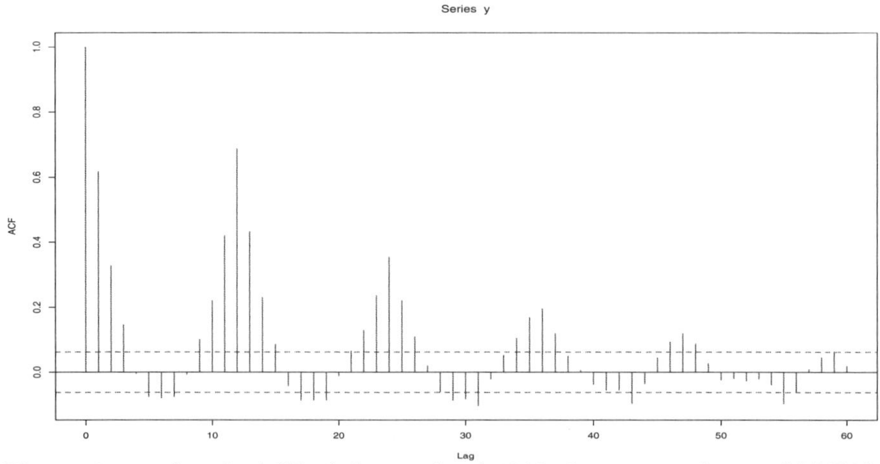

Figure 9.4 *Sample ACF of the simulated 1000 observations from a SARIMA model with $\phi = 0.5$, $\Phi = 0.6$, $\theta = 0.2$, $\Theta = 0.3$ and Gaussian white noise with unit variance.*

9.1.1 Spectral Density

The spectral density of the SARIMA$(p,0,q)\times(P,0,Q)$ process is given by

$$f(\lambda)=\frac{\sigma^2}{2\pi}\left|\frac{\theta(e^{i\,\lambda})\,\Theta(e^{i\,\lambda\,s})}{\phi(e^{i\,\lambda})\,\Phi(e^{i\,\lambda\,s})}\right|^2.$$

EXAMPLE 9.3

Figure 9.5 displays a sample of a SARIMA$(1,0,1)\times(1,0,1)$ model with $\phi=0.7$, $\Phi=0.6$, $\theta=0.4$, $\Theta=0.2$, and seasonal period $s=20$. The spectral density of this series is exhibited in Figure 9.6 while its periodogram is plotted in Figure 9.7.

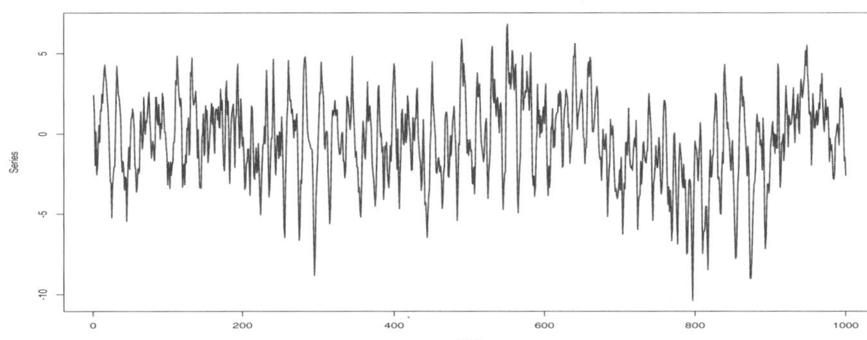

Figure 9.5 *SARIMA model with $\phi=0.7$, $\Phi=0.6$, $\theta=0.4$, $\Theta=0.2$, $s=20$, and Gaussian white noise with unit variance. Simulated series with 1000 observations.*

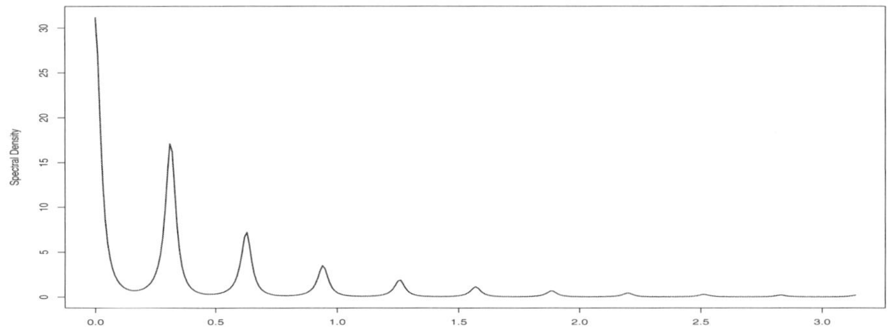

Figure 9.6 *SARIMA model with $\phi=0.7$, $\Phi=0.6$, $\theta=0.4$, $\Theta=0.2$, $s=20$, and Gaussian white noise with unit variance. Spectral Density.*

Furthermore, Figure 9.8 shows an smoothed version of the periodogram which uses a Daniell window.

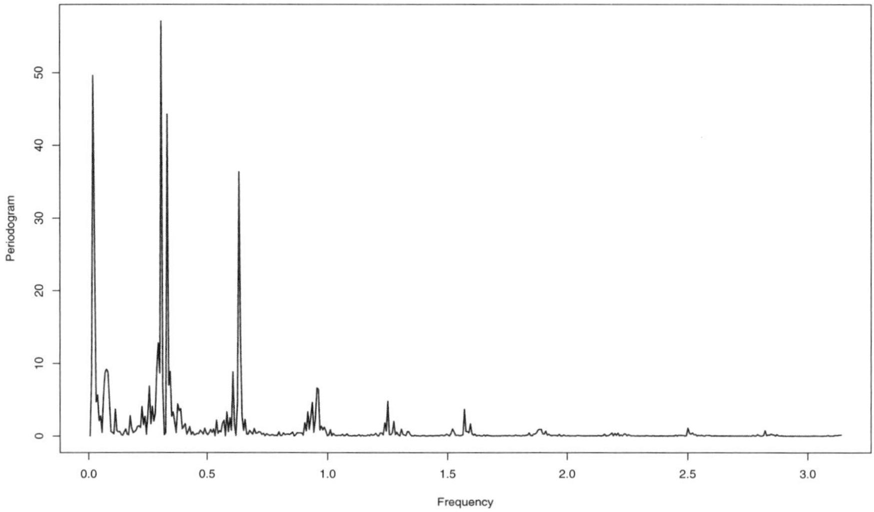

Figure 9.7 *SARIMA model with* $\phi = 0.7$, $\Phi = 0.6$, $\theta = 0.4$, $\Theta = 0.2$, $s = 20$, *and Gaussian white noise with unit variance. Periodogram.*

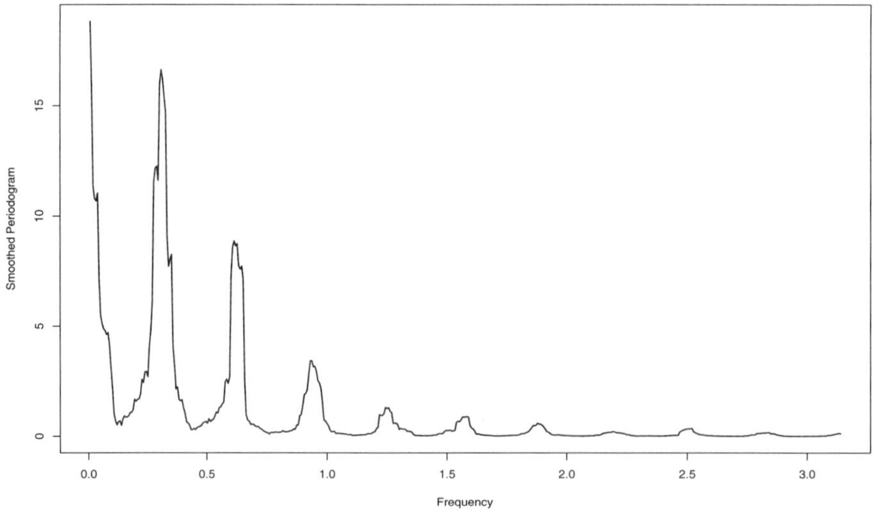

Figure 9.8 *SARIMA model with* $\phi = 0.7$, $\Phi = 0.6$, $\theta = 0.4$, $\Theta = 0.2$, $s = 20$, *and Gaussian white noise with unit variance. Smoothed periodogram using a Daniell kernel.*

9.1.2 Several Seasonal Components

The SARIMA models can also be extended to handle more than one seasonal period. For instance, we can write a SARIMA$(p,d,q) \times (P_1, D_1, Q_1)_{s_1} \cdots \times (P_m, D_m, Q_m)_{s_m}$ model with m seasonal component can be written as

$$\phi(B)\Phi_1(B^{s_1}) \cdots \Phi_m(B^{s_m}(1-B)^d(1-B^{s_1})^{D_1} \cdots (1-B^{s_m})^{D_m} \, y_t$$
$$= \theta(B)\Theta_1(B^{s_1}) \cdots \Theta_m(B^{s_m})\varepsilon_t,$$

where ε_t is a white noise sequence with zero-mean and variance σ^2 and

$$\begin{aligned} \phi(B) &= 1 - \phi_1 B - \phi_2 B^2 - \cdots - \phi_p B^p, \\ \Phi_i(B^{s_i}) &= 1 - \Phi_{i1} B^{s_i} - \Phi_{i2} B^{2s_i} - \cdots - \Phi_{iP} B^{Ps_i}, \\ \theta(B) &= 1 - \theta_1 B - \theta_2 B^2 - \cdots - \theta_q B^q, \\ \Theta_i(B^{s_i}) &= 1 - \Theta_{i1} B^{s_i} - \Theta_{i2} B^{2s_i} - \cdots - \Theta_{iQ} B^{Qs_i}. \end{aligned}$$

Moreover, the spectral density of this multiple seasonal components model is given by

$$f(\lambda) = \frac{\sigma^2}{2\pi} \left| \frac{\theta(e^{i\lambda})\,\Theta_1(e^{i\lambda s_1}) \cdots \Theta_m(e^{i\lambda s_m})}{\phi(e^{i\lambda})\,\Phi_1(e^{i\lambda s_1}) \cdots \Phi_m(e^{i\lambda s_m})} \right|^2.$$

9.1.3 Estimation

The estimation of SARIMA models are readily extended from the maximum likelihood techniques discussed in Chapter 5. In particular some of these methods are implemented in the statistical software R. Observe that the R package gsarima allows for the simulation and estimation of SARIMA models. In this case, the output is as follows,

```
ARIMA(1,0,1)(1,0,1)[12] with non-zero-mean

Coefficients:
         ar1     ma1    sar1    sma1  intercept
      0.5925  0.0515  0.5520  0.3466    -0.2413
s.e.  0.0589  0.0739  0.0493  0.0546     0.3138

sigma^2 estimated as 0.8816:  log likelihood=-683.25
AIC=1378.5   AICc=1378.67   BIC=1403.79
```

9.1.4 Estimator Performance

In this section we study the performance of maximum likelihood estimates of SARIMA models. As an illustration, consider the SARIMA$(1,0,1) \times (1,0,1)_s$ defined by

$$(1 - \phi B)(1 - \Phi B^s)y_t = (1 + \theta B)(1 + \Theta B^s)\varepsilon_t,$$

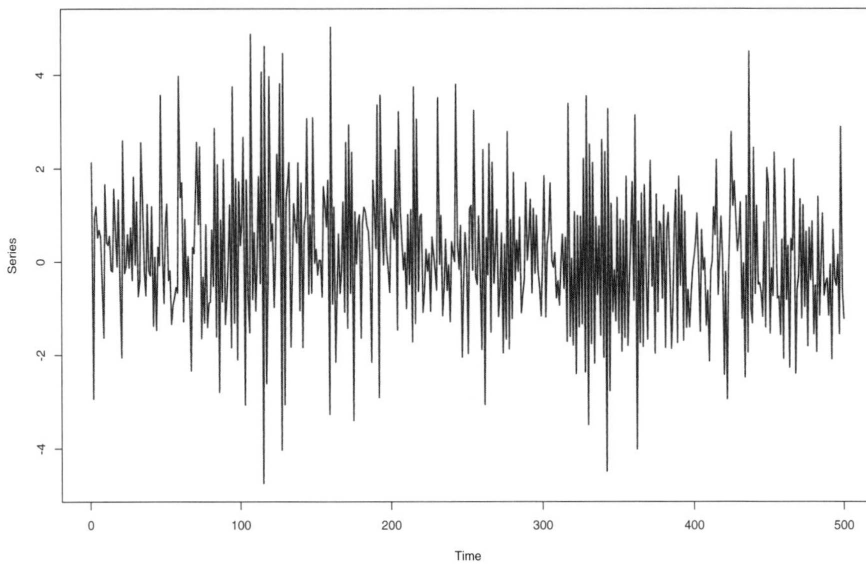

Figure 9.9 *Simulated 500 observations from a SARIMA model with* $\phi = -0.7$, $\Phi = 0.5$, $\theta = 0.2$, $\Theta = 0.3$, *and Gaussian white noise with unit variance.*

where ε_t is a Gaussian white noise sequence with zero-mean and unit variance. Figure 9.9 shows a trajectory of this seasonal model with with $\phi = -0.7$, $\Phi = 0.5$, $\theta = 0.2$ and $\Theta = 0.3$ while Figure 9.10 displays the sample ACF of this time series.

Table 9.1 reports the results from several simulations for different combinations of parameters ϕ, Φ, θ and Θ. The results are based on 1000 repetitions and time series with 500 observations. Notice that the average of the estimates are very close to their theoretical counterparts. Furthermore, Table 9.2 reports the estimates of the stander deviation of these maximum likelihood estimates. The first four columns of this table correspond to the average of estimates provided by the empirical Hessian matrix given by

$$H_n(\widehat{\theta}) = \nabla^2 \mathcal{L}_n(\widehat{\theta})$$

where $\mathcal{L}_n$ is the log-likelihood function, θ the vector of parameters of the seasonal model and $\widehat{\theta}$ the maximum likelihood estimator. Columns 5 to 8 of Table 9.2 report the empirical standard deviations of the ML estimators based on the 1000 Monte Carlo repetitions.

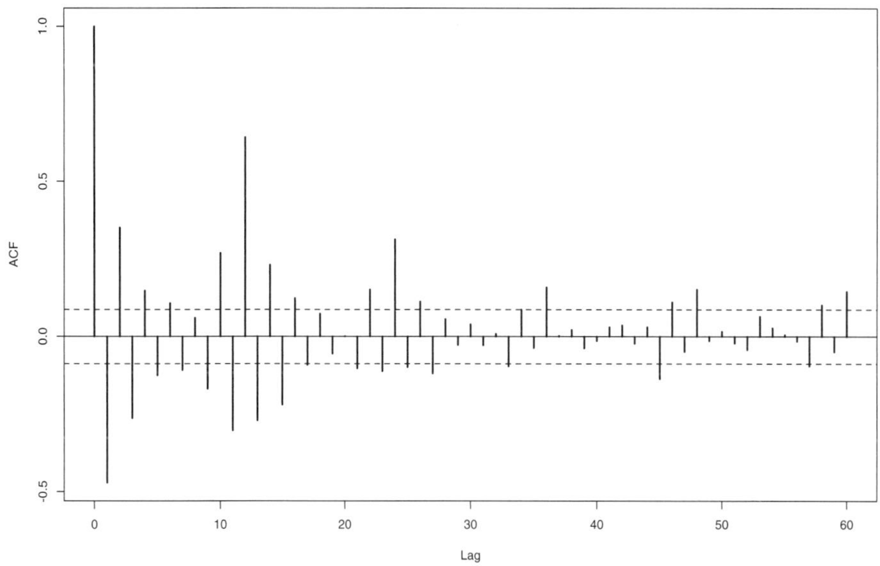

Figure 9.10 *Sample ACF of the simulated 500 observations from a SARIMA model with $\phi = -0.7$, $\Phi = 0.5$, $\theta = 0.2$, $\Theta = 0.3$, and Gaussian white noise with unit variance.*

Table 9.1 SARIMA Simulations with Sample Size $n = 500$ and Seasonal Period $s = 12$: Parameter Estimates.

ϕ	θ	Φ	Θ	$\widehat{\phi}$	$\widehat{\theta}$	$\widehat{\Phi}$	$\widehat{\Theta}$
0.5000	0.2000	0.6000	0.3000	0.4848	0.2021	0.5955	0.3008
0.3000	0.5000	0.3000	0.7000	0.2912	0.4797	0.3065	0.7010
0.3000	0.6000	-0.3000	0.5000	0.2914	0.6059	-0.2799	0.4833
0.7000	0.3000	0.5000	0.2000	0.6920	0.2981	0.4979	0.2024
-0.7000	0.2000	0.5000	0.3000	-0.6888	0.1939	0.4910	0.3059

9.1.5 Heating Degree Day Data Application

As an illustration of the versatility of SARIMA processes to model data exhibiting seasonal patterns as well as serial dependence we revisit the heating degree day data introduced in Chapter 1, see Figure 9.11. The periodogram of this time series is displayed in Figure 9.12. As expected from the nature of these data, the periodogram shows a peak at the frequency corresponding to a period $s = 12$. Consequently, the class of SARIMA processes is proposed for this time series. The selected model via AIC corresponds to a

Table 9.2 SARIMA Simulations with Sample Size $n = 500$ and Seasonal Period $s = 12$: Standard Deviation Estimates.

Hessian SD estimates				*Sample SD estimates*			
$\widehat{\sigma}(\widehat{\phi})$	$\widehat{\sigma}(\widehat{\theta})$	$\widehat{\sigma}(\widehat{\Phi})$	$\widehat{\sigma}(\widehat{\Theta})$	$\widehat{\sigma}(\widehat{\phi})$	$\widehat{\sigma}(\widehat{\theta})$	$\widehat{\sigma}(\widehat{\Phi})$	$\widehat{\sigma}(\widehat{\Theta})$
0.0628	0.0704	0.0477	0.0579	0.0627	0.0701	0.0492	0.0570
0.0640	0.0592	0.0531	0.0420	0.0657	0.0637	0.0552	0.0415
0.0563	0.0470	0.1806	0.1658	0.0576	0.0479	0.2027	0.1915
0.0393	0.0524	0.0624	0.0712	0.0385	0.0528	0.0598	0.0704
0.0568	0.0769	0.0571	0.0634	0.0606	0.0775	0.0563	0.0638

SARIMA$(1, 0, 0) \times (3, 0, 0)$ and the fitted model from the R function arima is as follows,

```
> fit

Call:

arima(x = y, order = c(1, 0, 0),

seasonal = list(order = c(3, 0, 0), period = 12))

Coefficients:

         ar1     sar1     sar2     sar3  intercept
      0.3623   0.3168   0.3245   0.3401   275.6695
s.e.  0.0521   0.0518   0.0499   0.0501    58.3368

sigma^2 estimated as 1408:

   log likelihood = -1859.91,  aic = 3731.83
```

Observe that all these coefficients are significant at the 5% level.

The residuals from this model are plotted in Figure 9.13. The sample ACF along with the Box-Ljung diagnostics test are reported in Figure 9.14. From these plots, it seems that the residuals do not have serial correlation. Figure Figure 9.15 correspond to a normal quantile-quantile plot. If the residual were normally distributed, then the dots should be close to the Gaussian quantile line (heavy line). Notice that their distribution seems to depart from normality in the tails.

Fitted values are plotted in Figure 9.16 along the observed values. These in-sample one-step predictions are very close to their true values. On the other hand, Figure 9.17 exhibits the out-of-sample forecasts up to 48 months ahead. Finally, 95% prediction bands for these forecasts are plotted in Figure 9.18.

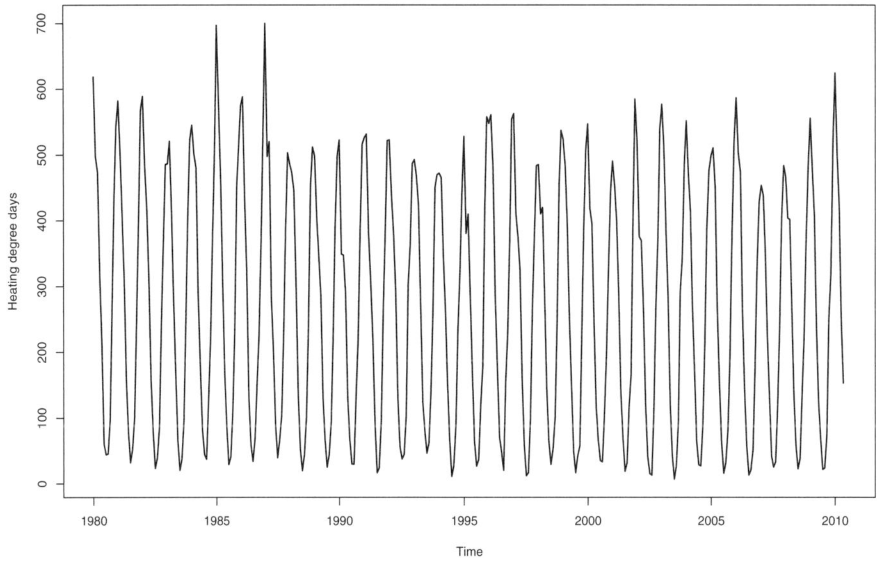

Figure 9.11 *SARIMA heating day degree application: Time series sata.*

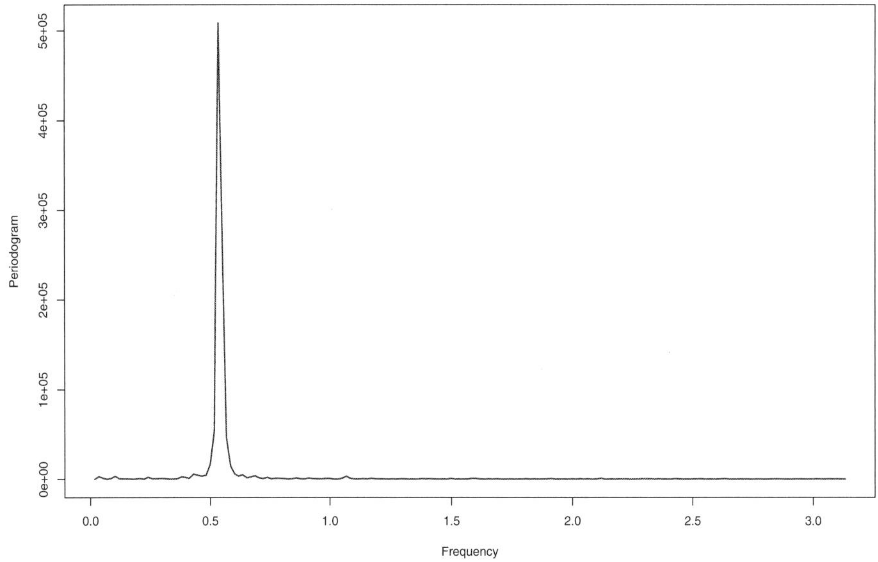

Figure 9.12 *SARIMA heating degree day data application: Periodogram.*

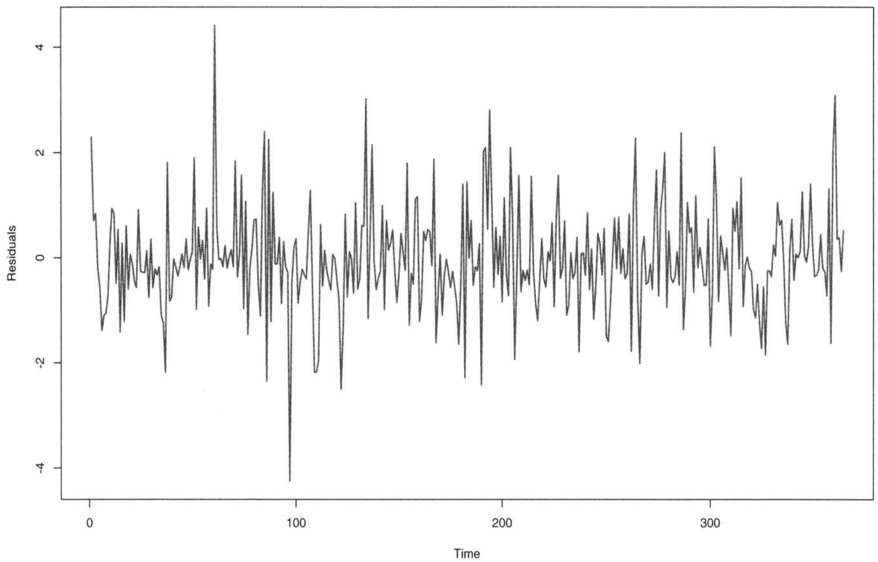

Figure 9.13 *SARIMA heating degree day data application: Residuals.*

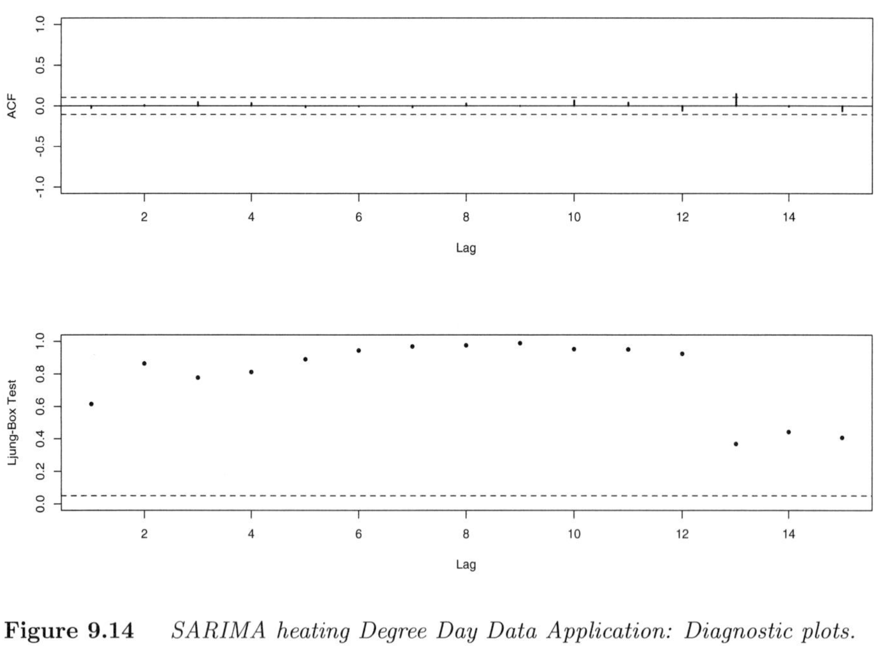

Figure 9.14 *SARIMA heating Degree Day Data Application: Diagnostic plots.*

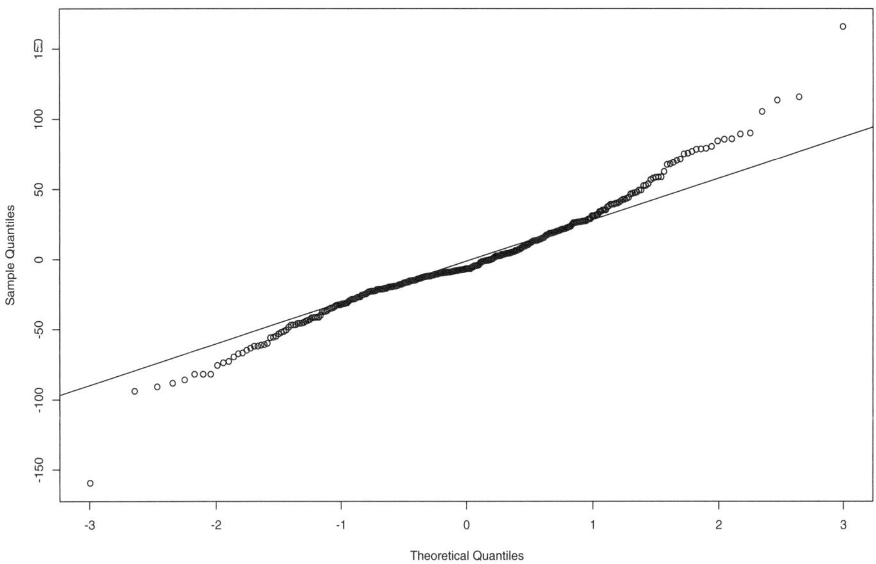

Figure 9.15 *SARIMA heating degree day data application: QQ plots.*

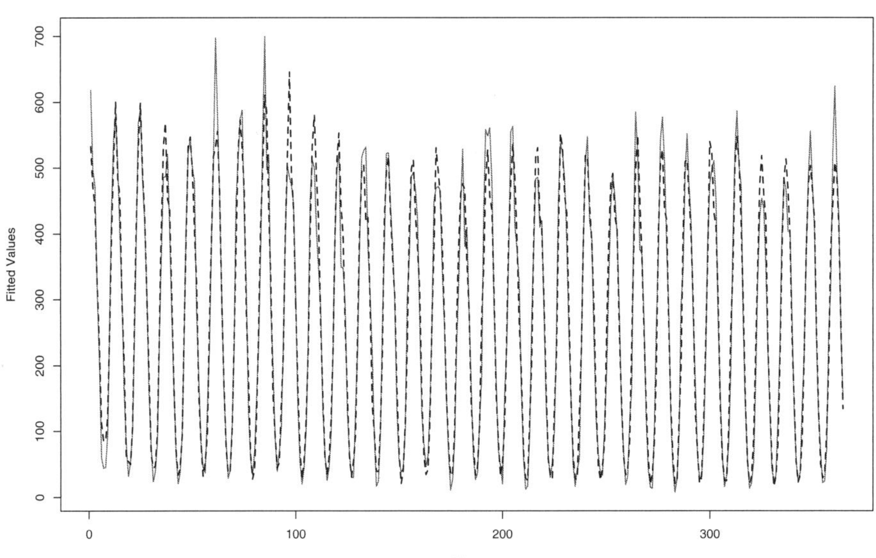

Figure 9.16 *SARIMA heating degree day data application: Fitted values.*

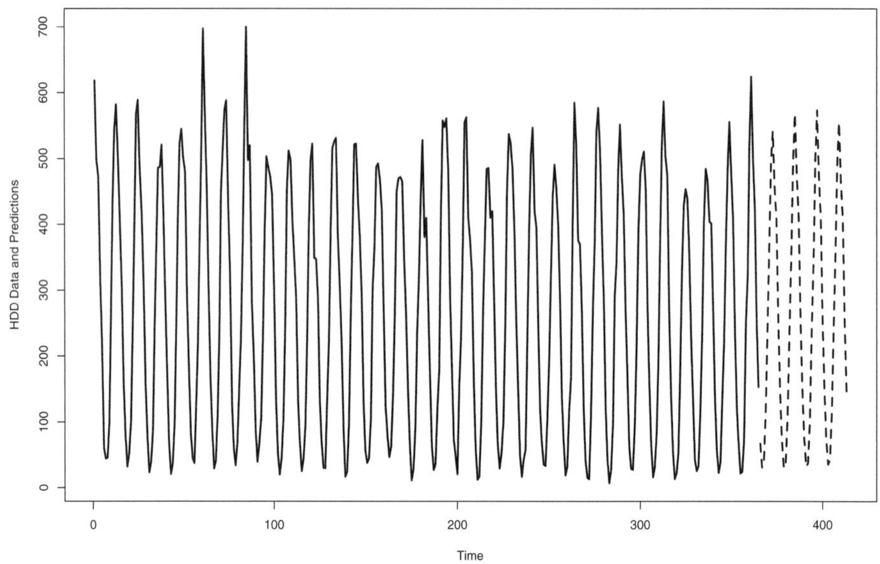

Figure 9.17 *SARIMA heating degree day data application: Out-of-sample predictions, 48 months ahead.*

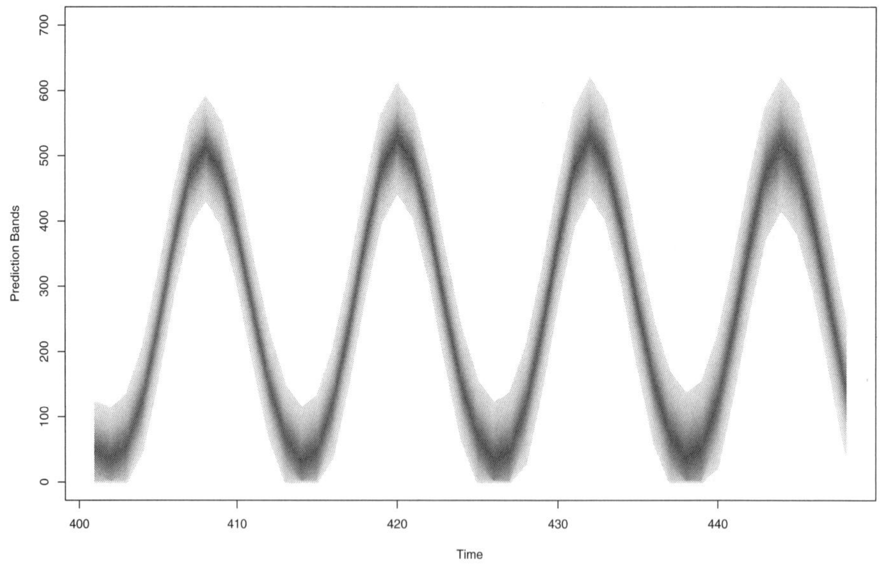

Figure 9.18 *SARIMA heating degree day data application: 95% Prediction bands for the out-of-sample forecasts, 48 months ahead.*

9.2 SARFIMA MODELS

In order to account for both seasonal and storming serial dependence, a motre general class ofseasonal long-memory processes may be specified by the spectral density

$$f(\lambda) = g(\lambda)|\lambda|^{-\alpha} \prod_{i=1}^{r} \prod_{j=1}^{m_i} |\lambda - \lambda_{ij}|^{-\alpha_i}, \tag{9.1}$$

where $\lambda \in (-\pi, \pi]$, $0 \leq \alpha, \alpha_i < 1$, $i = 1, ..., r$, $g(\lambda)$ is a symmetric, strictly positive, continuous, bounded function and $\lambda_{ij} \neq 0$ are poles for $j = 1, ..., m_i$, $i = 1, ..., r$. To ensure the symmetry of f, we assume that for any $i = 1, ..., r$, $j = 1, ..., m_i$, there is one and only one $1 \leq j' \leq m_i$ such that $\lambda_{ij} = -\lambda_{ij'}$.

As shown in the following examples, the spectral densities of many widely used models such as the seasonal ARFIMA process and the k-factor GARMA process satisfy (9.1).

Consider a seasonal ARFIMA model with multiple periods $s_1, \ldots, s_r$:

$$\phi(B) \prod_{i=1}^{r} \Phi_i(B^{s_i}) y_t = \theta(B) \prod_{i=1}^{r} \left[\Theta_i(B^{s_i})(1 - B^{s_i})^{-d_{s_i}}\right] (1 - B)^{-d} \varepsilon_t, \tag{9.2}$$

where $\phi(B)$, $\Phi_i(B^{s_i})$, $\theta(B)$, $\Theta_i(B^{s_i})$ are autoregressive and moving-average polynomials, for $i = 1, \ldots, r$.

The spectral density of the model described by (9.2) is given by

$$f_{s_1,\ldots,s_r}(\lambda) = \frac{\sigma^2}{2\pi} \frac{|\theta(e^{\imath\lambda})|^2}{|\phi(e^{\imath\lambda})|^2} |1 - e^{\imath\lambda}|^{-2d} \prod_{i=1}^{r} \frac{|\Theta_i(e^{\imath\lambda s_i})|^2 |1 - e^{\imath\lambda s_i}|^{-2d_{s_i}}}{|\Phi_i(e^{\imath\lambda s_i})|^2}.$$

Observe that this spectral density may be written as

$$f_{s_1,\ldots,s_r}(\lambda) = H(\lambda)|\lambda|^{-2d-2d_{s_1}-\cdots-2d_{s_r}} \prod_{i=1}^{r} \prod_{j=1}^{s_i} |\lambda - \lambda_{ij}|^{-2d_{s_i}},$$

which is a special case of (9.1) where

$$\begin{aligned} H(\lambda) &= \frac{\sigma^2}{2\pi} \frac{|\theta(e^{\imath\lambda})|^2}{|\phi(e^{\imath\lambda})|^2} \prod_{i=1}^{r} \frac{|\Theta_i(e^{\imath\lambda s_i})|^2}{|\Phi_i(e^{\imath\lambda s_i})|^2} \\ &\quad \times \frac{|\lambda|^{2d+2d_{s_1}+\cdots+2d_{s_r}} \prod_{i=1}^{r} \prod_{j=1}^{s_i} |\lambda - \lambda_{ij}|^{2d_{s_i}}}{|1 - e^{\imath\lambda}|^{2d} \prod_{i=1}^{r} |1 - e^{\imath\lambda s_i}|^{2d_{s_i}}}, \end{aligned}$$

and $\lambda_{ij} = 2\pi j/s_i$ for $i = 1, ..., r$, $j = 1, ..., [s_i/2]$, $\lambda_{ij} = 2\pi([s_i/2] - j)/s_i$ for $i = 1, ..., r$, $j = [s_i/2] + 1, \ldots, s_i$, $\alpha = 2d + 2d_{s_1} + \cdots + 2d_{s_r}$, and $\alpha_i = 2d_{s_i}$.

From Figure 9.19 and Figure 9.20 we may visualize the shape of the spectral density of a SARFIMA model for two sets of parameters. Figure 9.19 displays the spectral density of a SARFIMA$(0, d, 0) \times (0, d_s, 0)_s$ process with $d = 0.1$, $d_s = 0.3$, $s = 10$, and $\sigma^2 = 1$. On the other hand, Figure 9.20 shows the spectral density of a SARFIMA$(0, d, 0) \times (1, d_s, 1)_s$ process with $d = 0.1$, $d_s = 0.3$, $\Phi = -0.8$, $\Theta = 0.1$, $s = 10$, and $\sigma^2 = 1$. As expected, these plots have poles at the frequencies $\lambda = 2\pi j/s$, $j = 0, 1, \ldots, 5$.

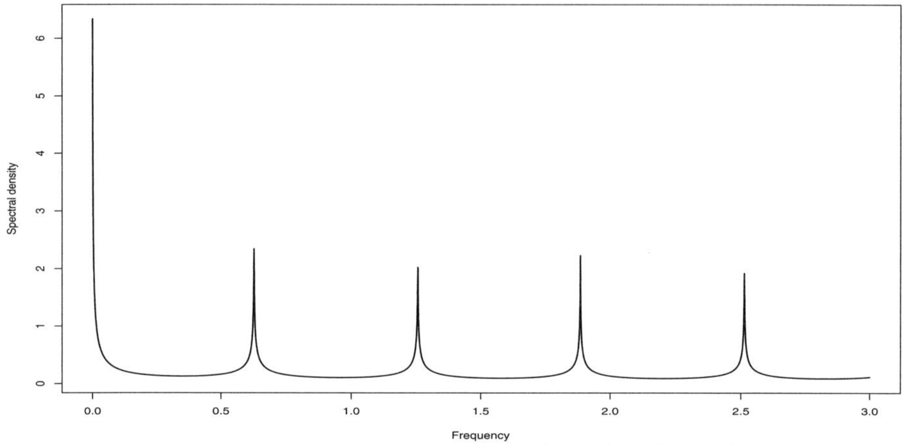

Figure 9.19 Spectral density of a SARFIMA$(0, d, 0) \times (0, d_s, 0)_s$ process with $d = 0.1$, $d_s = 0.3$, and $s = 10$.

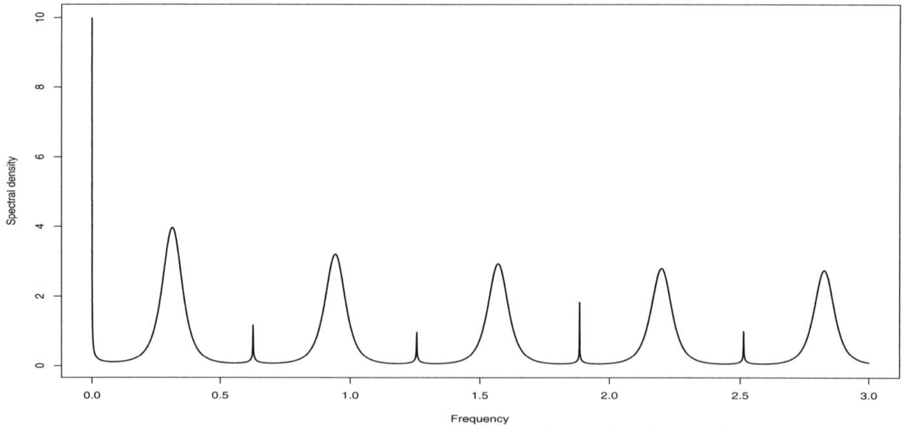

Figure 9.20 Spectral density of a SARFIMA$(0, d, 0) \times (1, d_s, 1)_s$ process with $d = 0.1$, $d_s = 0.3$, $\Phi = -0.8$, $\Theta = 0.1$ and $s = 10$.

9.3 GARMA MODELS

The spectral density of a k-factor GARMA process is given by

$$f(\lambda) = c|\theta(e^{i\lambda})|^2|\phi(e^{i\lambda})|^{-2}\prod_{j=1}^{k}|\cos\lambda - u_j|^{-d_j}, \tag{9.3}$$

where $c > 0$ is a constant, u_j are distinct values, $d_j \in (0, \frac{1}{2})$ when $|u_j| = 1$, and $d_j \in (0, 1)$ when $|u_j| \neq 1$.

For $|u_j| \leq 1$, we may write $u_j = \cos\lambda_j$ and this spectral density may be written in terms of (9.1) as follows:

$$f(\lambda) = H(\lambda)\prod_{j=1}^{k}|\lambda - \lambda_j|^{-d_j}|\lambda + \lambda_j|^{-d_j},$$

where

$$H(\lambda) = c|\theta(e^{i\lambda})|^2|\phi(e^{i\lambda})|^{-2}\prod_{j=1}^{k}\left|\frac{\cos\lambda - \cos\lambda_j}{\lambda^2 - \lambda_j^2}\right|^{-d_j}$$

is a strictly positive, symmetric, continuous function with

$$\lim_{\lambda\to\pm\lambda_\ell} H(\lambda) = c\frac{|\theta(e^{i\lambda_\ell})|^2}{|\phi(e^{i\lambda_\ell})|^2}\left|\frac{\sin\lambda_\ell}{2\lambda_\ell}\right|^{-d}\prod_{j\neq\ell}^{k}\left|\frac{\cos\lambda_\ell - \cos\lambda_j}{\lambda_\ell^2 - \lambda_j^2}\right|^{-d_j},$$

for $\lambda_\ell \neq 0$ and for $\lambda_\ell = 0$

$$\lim_{\lambda\to 0} H(\lambda) = 2^{d_\ell}c|\theta(1)|^2|\phi(1)|^{-2}\prod_{j\neq l}^{k}\left|\frac{1 - \cos\lambda_j}{\lambda_j^2}\right|^{-d_j}.$$

Observe that all these limits are finite and $H(\lambda)$ is a bounded function.

Figure 9.21 depicts the spectral density of a k-factor GARMA process with $k = 1$, $\lambda_1 = \pi/4$, and $d_1 = 0.1$. Notice that there is only one pole located at frequency $\pi/4$.

When the singularities λ_{ij} are known, the exact maximum-likelihood estimators of Gaussian time series models with spectral density (9.1) have the following large-sample properties. Let $\widehat{\theta}_n$ be the exact MLE and θ_0 the true parameter. Then, under some regularity conditions we have

(a) Consistency: $\widehat{\theta}_n \to \theta_0$ in probability as $n \to \infty$.

(b) Normality: $\sqrt{n}(\widehat{\theta}_n - \theta_0) \to N(0, \Gamma^{-1}(\theta_0))$, as $n \to \infty$, where $\Gamma(\theta) = (\Gamma_{ij}(\theta))$ with

$$\Gamma_{ij}(\theta) = \frac{1}{4\pi}\int_{-\pi}^{\pi}\left[\frac{\partial \log f_\theta(\lambda)}{\partial\theta_i}\right]\left[\frac{\partial \log f_\theta(\lambda)}{\partial\theta_j}\right]d\lambda, \tag{9.4}$$

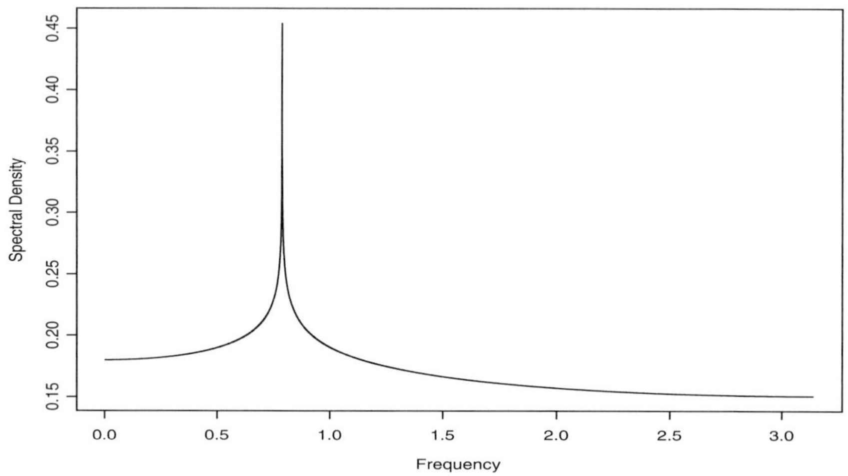

Figure 9.21 Spectral density of a k-factor GARMA process with $k = 1$, $\lambda_1 = \pi/4$, and $d_1 = 0.1$.

and $f_\theta(\lambda)$ is the spectral density (9.1).

(c) Efficiency: $\widehat{\theta}_n$ is an efficient estimator of θ_0.

When the location of the pole is unknown, we still can obtain similar large-sample properties for quasi-maximum-likelihood parameter estimates. Consider, for example, the class of long-memory seasonal models defined by the spectral density

$$f(\lambda) = \frac{\sigma^2}{2\pi} \left| 4 \sin \frac{\lambda + \omega}{2} \sin \frac{\lambda - \omega}{2} \right|^{-\alpha} |h(\lambda, \tau)|^2, \tag{9.5}$$

where $\theta = (\alpha, \tau)$ are the parameters related to the long-memory and the short-memory components, and ω denotes the unknown location of the pole.

Define the function

$$S(\theta, \omega) = \frac{1}{\widetilde{n}} \sum_{j=0}^{\widetilde{n}} \frac{I(\lambda_j)}{k(\lambda_j, \theta, \omega)}, \tag{9.6}$$

where $\widetilde{n} = [n/2]$, $I(\lambda_j)$ is the periodogram given by (4.6) evaluated at the Fourier frequency $\lambda_j = 2\pi j/n$ and

$$k(\lambda, , \theta, \omega) = \frac{2\pi}{\sigma^2} f(\lambda).$$

Consider the estimators $\widehat{\theta}_n$ and $\widehat{\omega}_n$ which minimize $S(\theta, \omega)$:

$$\begin{pmatrix} \widehat{\theta}_n \\ \widehat{\omega}_n \end{pmatrix} = \text{argmin}_{\Theta \times Q} S(\theta, \lambda_q),$$

where $Q = \{q : q = 0, 1, \ldots, \widetilde{n}\}$. Notice that ω belongs to a discrete set of frequencies $\lambda_0, \ldots, \lambda_{\widetilde{n}}$.

9.4 CALCULATION OF THE ASYMPTOTIC VARIANCE

Analytic expressions for the integral in (9.4) are difficult to obtain for an arbitrary period s. For a SARFIMA$(0, d, 0) \times (0, d_s, 0)_s$ model, the matrix $\Gamma(\theta)$ may be written as

$$\Gamma(\theta) = \begin{bmatrix} \dfrac{\pi^2}{6} & c(s) \\ & \\ c(s) & \dfrac{\pi^2}{6} \end{bmatrix}, \tag{9.7}$$

with $c(s) = (1/\pi) \int_{-\pi}^{\pi} \{\log|2\sin(\lambda/2)|\}\{\log|2\sin[s(\lambda/2)]|\} d\lambda$. An interesting feature of the asymptotic variance-covariance matrix of the parameter estimates (9.7) is that for a SARFIMA$(0, d, 0) \times (0, d_s, 0)_s$ process, the exact maximum-likelihood estimators $\widehat{d}$ and $\widehat{d}_s$ have the same variance.

An explicit expression for this integral can be given for $s = 2$. In this case,

$$\Gamma(\boldsymbol{\theta}) = \frac{\pi^2}{12} \begin{bmatrix} 2 & 1 \\ 1 & 2 \end{bmatrix}.$$

For other values of s, the integral may be evaluated numerically. For instance, Figure 9.22 shows the evolution of $\text{Var}(\widehat{d})$ as a function of the period s [see panel (a)] and the evolution of $\text{Cov}(\widehat{d}, \widehat{d}_s)$ as s increases [see panel (b)]. Both curves are based on the numerical evaluation of equation (9.7) and then inverting this matrix to obtain the asymptotic variance-covariance matrix of the parameters.

Observe that $\text{Var}(\widehat{d}_s)$, equivalently $\text{Var}(\widehat{d})$, starts at a value of $8/\pi^2$ and decreases to $6/\pi^2$ as $s \to \infty$. That is, for a very large period s, the asymptotic variance of $\widehat{d}_s$ is the same as the variance of $\widehat{d}$ from an ARFIMA$(0, d, 0)$ model.

9.5 AUTOCOVARIANCE FUNCTION

Finding explicit formulae for the ACF of a general seasonal model is rather difficult. However, we can obtain an asymptotic expression as the lag increases. Assume that $\alpha_1 > \alpha_2 \geq \cdots \geq \alpha_r$. Let $c_0, c_1, \ldots, c_{m_1}$ be constants. Then, for large lag h the autocovariance function $\gamma(h)$ satisfies

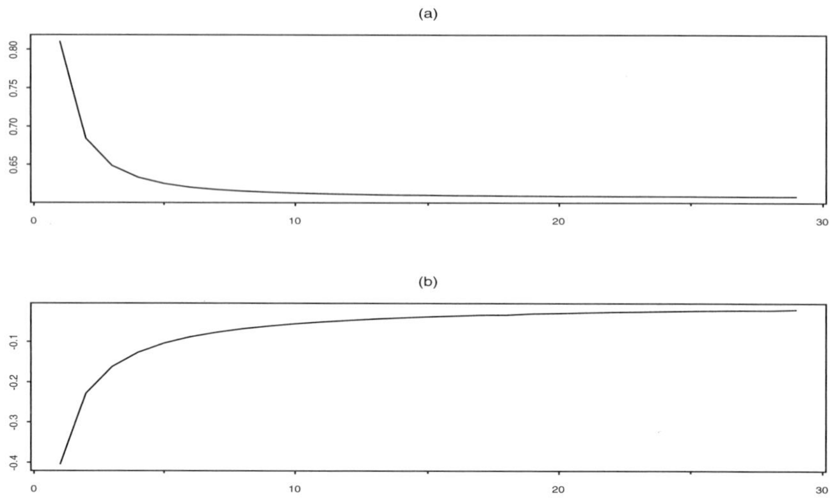

Figure 9.22 (a) Values of $\text{Var}(\widehat{d}_s)$ as a function of the period s and (b) values of $\text{Cov}(\widehat{d}, \widehat{d}_s)$ as a function of s.

(*a*) If $\alpha > \alpha_1$,

$$\gamma(h) \sim c_0 \, |h|^{\alpha - 1}.$$

(*b*) If $\alpha \leq \alpha_1$,

$$\gamma(h) \sim |h|^{\alpha_1 - 1} \left[c_0 \mathbf{1}_{\{\alpha = \alpha_1\}} + \sum_{j=1}^{m_1} c_j \cos(h\lambda_{1j}) \right].$$

Notice that the large lag behavior of the ACF depends on the maximum value of the exponents $\alpha, \alpha_1, \ldots, \alpha_r$.

For the SARFIMA process with $0 < d, d_{s_1}, \ldots, d_{s_r} < \frac{1}{2}$ and $d + d_{s_1} + \cdots + d_{s_r} < \frac{1}{2}$ the maximum exponent is always reached at zero frequency since $\alpha = d + d_{s_1} + \cdots + d_{s_r}$. Therefore in that case for large lag h the ACF behaves like

$$\gamma(h) \sim c_0 \, |h|^{2d + 2d_{s_1} + \cdots + 2d_{s_r} - 1}.$$

EXAMPLE 9.4

As an illustration of the shape of the autocovariance function of a seasonal long-memory process consider the SARFIMA$(0,d,0) \times (0,d_s,0)_s$ process described by the discrete-time equation

$$y_t = (1-B^s)^{-d_s}(1-B)^{-d}\varepsilon_t,$$

where $\{\varepsilon_t\}$ is a zero-mean and unit variance white noise.

In what follows, we plot the theoretical autocovariance function for three particular cases of this model from lag $h = 1$ to lag $h = 500$. The values of the ACF were calculated following the splitting method described in Subsection 5.10.11.

Figure 9.23 displays the theoretical ACF of a SARFIMA$(0,d,0) \times (0,d_s,0)_s$ process with parameters $d = 0.1$, $d_s = 0.3$, and $s = 12$.

Using the same method, Figure 9.24 shows the theoretical ACF of a SARFIMA $(0,d,0)\times(0,d_s,0)_s$ process with parameters $d = 0.3$, $d_s = 0.15$, and $s = 12$.

Finally, Figure 9.25 exhibits the theoretical ACF of a SARFIMA$(0,d,0)\times(0,d_s,0)_s$ process with parameters $d = 0.05$, $d_s = 0.44$, and $s = 24$.

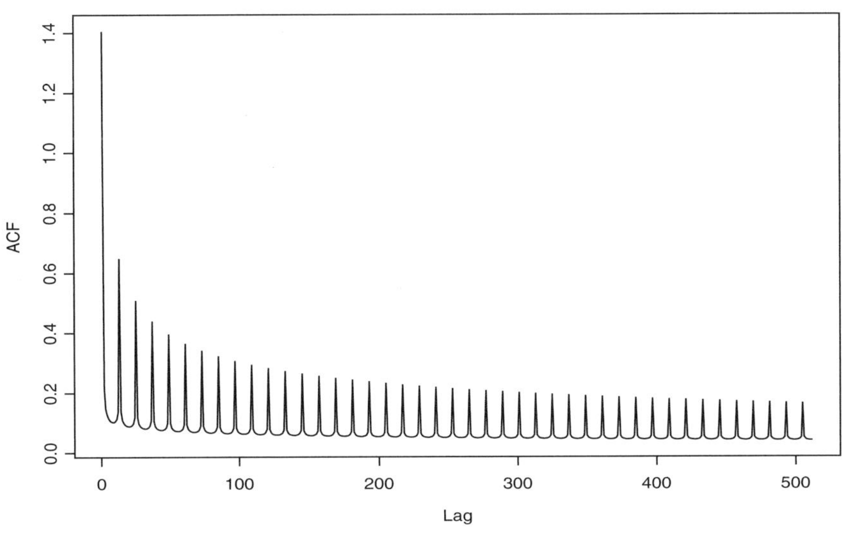

Figure 9.23 Autocovariance function of a SARFIMA$(0,d,0) \times (0,d_s,0)_s$ process with $d = 0.1$, $d_s = 0.3$, and $s = 12$.

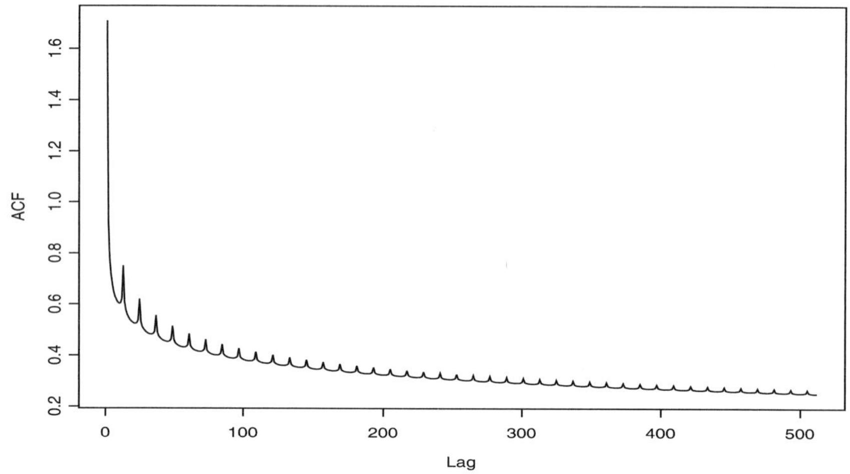

Figure 9.24 Autocovariance function of a SARFIMA$(0, d, 0) \times (0, d_s, 0)_s$ process with $d = 0.3$, $d_s = 0.15$, and $s = 12$.

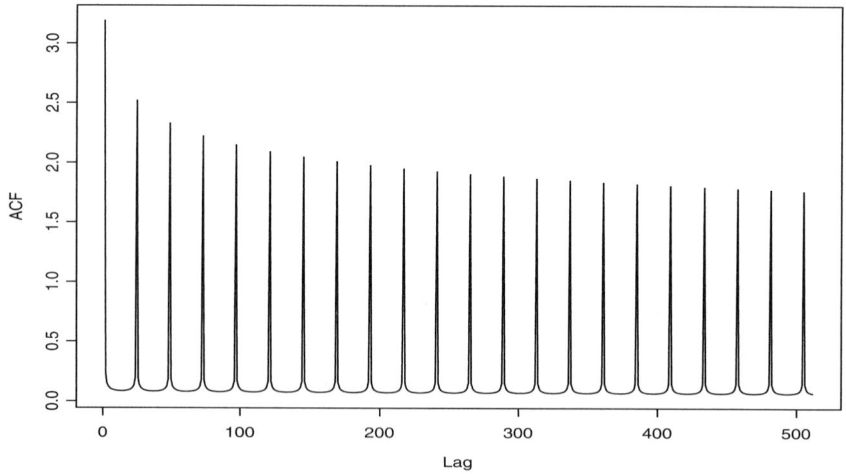

Figure 9.25 Autocovariance function of a SARFIMA$(0, d, 0) \times (0, d_s, 0)_s$ process with $d = 0.05$, $d_s = 0.44$, and $s = 24$.

9.6 MONTE CARLO STUDIES

In order to assess the finite sample performance of the ML estimates in the context of long-memory seasonal series, we show a number of Monte Carlo simulations for the class of SARFIMA models described by the difference equation

$$y_t - \mu = (1 - B^s)^{-d_s}(1 - B)^{-d}\varepsilon_t, \tag{9.8}$$

where μ is the mean of the series, and $\{\varepsilon_t\}$ are independent and identically distributed normal random variables with zero-mean and unit variance.

Table 9.3 to Table 9.6 report the results from the Monte Carlo simulations for the SARFIMA$(0, d, 0) \times (0, d_s, 0)_s$ process (9.8) with mean $\mu = 0$ assumed to be either *known* or *unknown* depending on the experiment, for different values of d, d_s, sample size n, seasonal period s. The white noise variance is $\sigma^2 = 1$ in all the simulations.

The finite sample performance of the MLE is compared to the Whittle estimate and the Kalman filter approach with truncation $m = 80$.

The results are based on 1000 repetitions, with seasonal series generated using the Durbin-Levinson algorithm with zero-mean and unit variance Gaussian noise.

Table 9.3 SARFIMA Simulations: Sample Size $n = 256$ and Seasonal Period $s = 6$

			Known Mean					
			Exact		Whittle		Kalman	
d	d_s		$\widehat{d}$	$\widehat{d}_s$	$\widehat{d}$	$\widehat{d}_s$	$\widehat{d}$	$\widehat{d}_s$
0.1	0.3	Mean	0.0945	0.2928	0.0590	0.2842	0.0974	0.3080
		S.D	0.0020	0.0018	0.0024	0.0039	0.0024	0.0026
0.2	0.2	Mean	0.1924	0.1901	0.1574	0.1602	0.2098	0.1909
		S.D	0.0023	0.0023	0.0034	0.0036	0.0022	0.0028
0.3	0.1	Mean	0.2924	0.0948	0.2591	0.0610	0.3046	0.1003
		S.D	0.0022	0.0022	0.0035	0.0024	0.0024	0.0023
			Unknown Mean					
			Exact		Whittle		Kalman	
			$\widehat{d}$	$\widehat{d}_s$	$\widehat{d}$	$\widehat{d}_s$	$\widehat{d}$	$\widehat{d}_s$
0.1	0.3	Mean	0.0806	0.2842	0.0590	0.2842	0.0749	0.2987
		S.D	0.0020	0.0020	0.0024	0.0039	0.0020	0.0030
0.2	0.2	Mean	0.1768	0.1812	0.1574	0.1601	0.1851	0.1799
		S.D	0.0027	0.0024	0.0034	0.0036	0.0018	0.0029
0.3	0.1	Mean	0.2755	0.0863	0.2591	0.0610	0.2800	0.0867
		S.D	0.0025	0.0022	0.0035	0.0024	0.0027	0.0022

Table 9.4 SARFIMA Simulations: Sample Size $n = 256$ and Seasonal Period $s = 10$

			Known Mean					
			Exact		Whittle		Kalman	
d	d_s		$\widehat{d}$	$\widehat{d}_s$	$\widehat{d}$	$\widehat{d}_s$	$\widehat{d}$	$\widehat{d}_s$
0.1	0.3	Mean	0.0955	0.2912	0.0599	0.2886	0.0991	0.3175
		S.D	0.0022	0.0016	0.0025	0.0032	0.0032	0.0022
0.2	0.2	Mean	0.1975	0.1916	0.1583	0.1640	0.2005	0.1963
		S.D	0.0022	0.0020	0.0034	0.0032	0.0027	0.0026
0.3	0.1	Mean	0.2947	0.0953	0.2601	0.0621	0.2978	0.1014
		S.D	0.0022	0.0022	0.0037	0.0023	0.0026	0.0023
			Unknown Mean					
			Exact		Whittle		Kalman	
			$\widehat{d}$	$\widehat{d}_s$	$\widehat{d}$	$\widehat{d}_s$	$\widehat{d}$	$\widehat{d}_s$
0.1	0.3	Mean	0.0806	0.2840	0.0599	0.2886	0.0725	0.3110
		S.D	0.0023	0.0017	0.0025	0.0032	0.0025	0.0025
0.2	0.2	Mean	0.1814	0.1837	0.1583	0.1640	0.1811	0.1894
		S.D	0.0025	0.0021	0.0034	0.0032	0.0030	0.0026
0.3	0.1	Mean	0.2781	0.0871	0.2601	0.0621	0.2698	0.0897
		S.D	0.0027	0.0022	0.0037	0.0023	0.0028	0.0022

Table 9.5 SARFIMA Simulations: Sample Size $n = 512$ and Seasonal Period $s = 6$

			Known Mean					
			Exact		Whittle		Kalman	
d	d_s		$\widehat{d}$	$\widehat{d}_s$	$\widehat{d}$	$\widehat{d}_s$	$\widehat{d}$	$\widehat{d}_s$
0.1	0.3	Mean	0.0995	0.2951	0.0803	0.2942	0.1057	0.3118
		S.D	0.0012	0.0010	0.0014	0.0016	0.0012	0.0013
0.2	0.2	Mean	0.1966	0.1977	0.1795	0.1839	0.1952	0.2021
		S.D	0.0013	0.0011	0.0017	0.0015	0.0014	0.0013
0.3	0.1	Mean	0.2962	0.0980	0.2811	0.0792	0.3060	0.0964
		S.D	0.0011	0.0011	0.0014	0.0013	0.0014	0.0012
			Unknown Mean					
			Exact		Whittle		Kalman	
			$\widehat{d}$	$\widehat{d}_s$	$\widehat{d}$	$\widehat{d}_s$	$\widehat{d}$	$\widehat{d}_s$
0.1	0.3	Mean	0.0919	0.2900	0.0803	0.2942	0.0870	0.3045
		S.D	0.0012	0.0010	0.0014	0.0016	0.0011	0.0013
0.2	0.2	Mean	0.1880	0.1923	0.1795	0.1839	0.1765	0.1943
		S.D	0.0014	0.0012	0.0017	0.0015	0.0011	0.0013
0.3	0.1	Mean	0.2878	0.0932	0.2811	0.0792	0.2849	0.0864
		S.D	0.0012	0.0012	0.0014	0.0013	0.0014	0.0012

Table 9.6 SARFIMA Simulations: Sample Size $n = 512$ and Seasonal Period $s = 10$

			Known Mean					
			Exact		Whittle		Kalman	
d	d_s		$\widehat{d}$	$\widehat{d_s}$	$\widehat{d}$	$\widehat{d_s}$	$\widehat{d}$	$\widehat{d_s}$
0.1	0.3	Mean	0.0979	0.2959	0.0768	0.3006	0.0995	0.3134
		S.D	0.0012	0.0009	0.0014	0.0016	0.0016	0.0014
0.2	0.2	Mean	0.1994	0.1957	0.1813	0.1832	0.2028	0.2007
		S.D	0.0012	0.0011	0.0016	0.0016	0.0014	0.0014
0.3	0.1	Mean	0.2963	0.0968	0.2801	0.0783	0.3070	0.0948
		S.D	0.0011	0.0012	0.0015	0.0014	0.0015	0.0013
			Unknown Mean					
			Exact		Whittle		Kalman	
			$\widehat{d}$	$\widehat{d_s}$	$\widehat{d}$	$\widehat{d_s}$	$\widehat{d}$	$\widehat{d_s}$
0.1	0.3	Mean	0.0896	0.2913	0.0768	0.3006	0.0799	0.3074
		S.D	0.0012	0.0009	0.0014	0.0016	0.0014	0.0014
0.2	0.2	Mean	0.1908	0.1908	0.1813	0.1832	0.1809	0.1931
		S.D	0.0013	0.0012	0.0016	0.0016	0.0015	0.0014
0.3	0.1	Mean	0.2876	0.0921	0.2801	0.0783	0.2890	0.0862
		S.D	0.0012	0.0012	0.0015	0.0014	0.0013	0.0012

The autocovariance function was computed by the convolution method of Subsection 5.10.11.

In order to explore the effect of the estimation of the mean we have considered two situations: *known mean* where the process is assumed to have zero-mean and *unknown mean* where the expected value of the process is estimated by the sample mean and then centered before the computations.

The exact MLE method has been implemented computationally by means of the Durbin-Levinson algorithm discussed in Chapter 5 with autocovariance calculated by the approach given in Subsection 5.10.11.

The Whittle method has been implemented by minimizing the following expression; see Section 5.8 and equation (9.6):

$$S(\theta) = \frac{2}{n} \sum_{k=1}^{[n/2]} \frac{I(\lambda_k)}{f_\theta(\lambda_k)},$$

where $\theta = (d, d_s)$ and $\lambda_k = 2\pi k/n$, with periodogram given by

$$I(\lambda_k) = \frac{1}{2\pi n} \left| \sum_{t=1}^{n} y_t e^{it\lambda_k} \right|^2,$$

Table 9.7 Asymptotic Standard Deviation of $\widehat{d}$ and $\widehat{d}_s$

n	$s = 6$	$s = 10$
256	0.0503	0.0503
512	0.0356	0.0356

see definition (4.6), and spectral density

$$f_\theta(\lambda) = \frac{1}{2\pi} \left|1 - e^{i\lambda}\right|^{-2d} \left|1 - e^{i\lambda s}\right|^{-2d_s}.$$

The approximate Kalman filter ML estimates are based on a finite state space representation of the truncated MA(∞) expansion described in Section 5.10.4.

From Table 9.3 to Table 9.6, it seems that for the *known mean* case the exact MLE and the Kalman methods display little bias for both sample sizes. On the other hand, the Whittle method presents a noticeable downward bias for both estimators $\widehat{d}$ and $\widehat{d}_s$.

The sample standard deviations of the estimates are close to their theoretical counterparts, reported in Table 9.7, for the three methods considered. However, the exact MLE seems to have slightly lower sample standard deviations than the other methods, for both long-memory parameters and both sample sizes. The theoretical values of the standard deviations of the estimated parameters given in Table 9.7 are based on formula (9.7).

In the *unknown mean* case, all the estimates seem to display a downward bias, which is stronger for the Whittle method. However, the bias displayed by this estimate is similar to the *known mean* case, since the Whittle algorithm is not affected by the estimation of the mean. Similarly to the previous case, the estimated standard deviations are comparable to the theoretical values, and the exact MLE displays slightly lower sample standard deviations than the other methods for most long-memory parameters and sample size combinations.

9.7 ILLUSTRATION

In this section we apply the maximum-likelihood estimation to the analysis of a time series consisting of *hyper text transfer protocol* (HTTP) requests to a World Wide Web server at the University of Saskatchewan. It has been reported that communication network traffic may exhibit long-memory behavior. The data analyzed here consist of the logarithm of the number of requests within one-hour periods.

The Internet traffic series is shown in Figure 9.26 while its sample autocorrelation function is displayed in Figure 9.27. Observe that the sample ACF

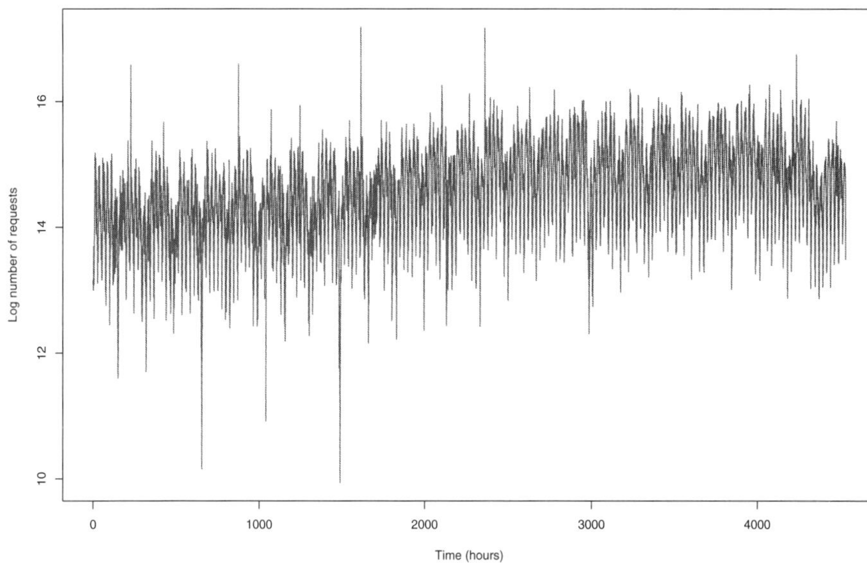

Figure 9.26 *Logarithm of HTTP requests time series data.*

decays slowly and exhibits a 24-hour periodicity. To account for these features we fit a SARFIMA model to this time series.

Table 9.8 reports the maximum-likelihood parameter estimates and the t-tests for the SARFIMA$(1,0,1) \times (0,d_s,0)_s$ with $s = 24$ process:

$$(1-\phi B)(y_t-\mu) = (1-\theta B)(1-B^s)^{-d_s}\varepsilon_t,$$

where ε_t is a white noise sequence with variance σ^2.

This model was selected by means of the Akaike's information criterion. From Table 9.8, notice that all the parameters included in the model are significant at the 5% level. The Student-t values reported on Table 9.8 are based on the numerical calculation of the inverse of the Hessian matrix, which approximates the asymptotic variance-variance matrix of the parameters.

Table 9.8 Log Internet Traffic Data: Maximum-Likelihood Estimation of the SARFIMA$(1,0,1) \times (0,d_s,0)_s$ Model

Parameter	d_s	ϕ	θ
Estimate	0.4456	0.8534	0.3246
Student-t	2.2558	7.5566	2.8623

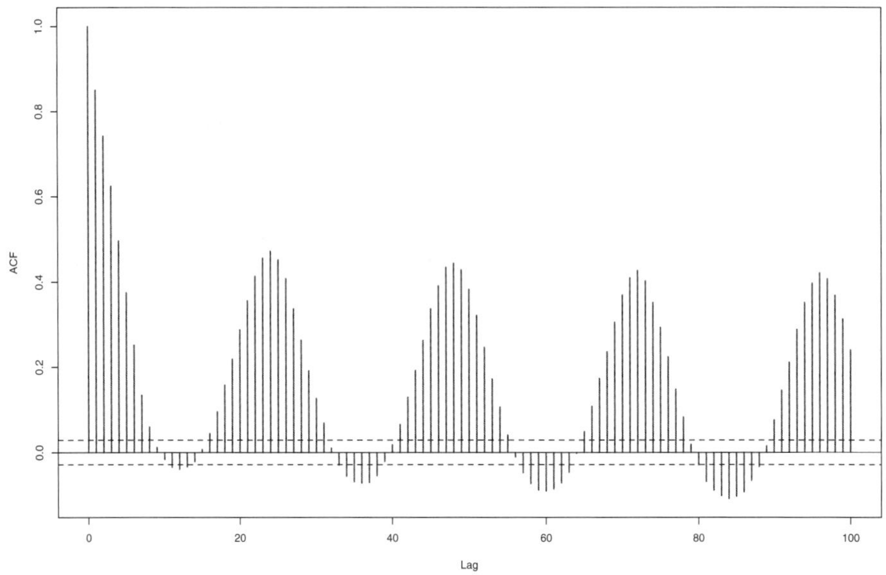

Figure 9.27 *Sample ACF of logarithm HTTP requests time series data.*

The standard deviation of the Internet traffic series is 0.6522 while the residual standard deviation is 0.2060. Thus, the fitted seasonal long-memory model explains roughly two thirds of the total standard deviation of the data.

9.8 BIBLIOGRAPHIC NOTES

Methods for estimating and forecasting SARIMA models are described in the book by Brockwell and Davis (1991) as well as in the monograph by Box, Jenkins, and Reinsel (1994). Long-range-dependent data with seasonal behavior have been reported fields as diverse as economics, physics, and hydrology. For example, inflation rates are studied by Hassler and Wolters (1995), revenue series are analyzed by Ray (1993a), monetary aggregates are considered by Porter-Hudak (1990), quarterly gross national product and shipping data are discussed by Ooms (1995), and monthly flows of the Nile River are studied by Montanari, Rosso, and Taqqu (2000).

Many statistical methodologies have been proposed to model this seasonal long-range-dependent data. For example, Abrahams and Dempster (1979) extend the fractional Gaussian noise process [see Mandelbrot and Van Ness (1968)] to include seasonal components.

On the other hand, Gray, Zhang, and Woodward (1989) propose the generalized fractional or Gegenbauer processes (GARMA), Porter-Hudak (1990) discusses (SARFIMA) models, Hassler (1994) introduces the flexible seasonal fractionally integrated processes (flexible ARFISMA), and Woodward, Cheng, and Gray (1998) introduce the k-factor GARMA processes.

Furthermore, the statistical properties of these models have been investigated by Giraitis and Leipus (1995), Chung (1996), Giraitis, Hidalgo, and Robinson (2001), and Palma and Chan (2005), among others. Finite sample performances of a number of estimation techniques for fractional seasonal models are studied in the papers by Reisen, Rodrigues, and Palma (2006a,b).

Problems

9.1 Let $y_t = \varepsilon_t + \varphi\, y_{t-s}$ be a SARIMA$(0,0,0) \times (1,0,0)_s$ model where the integer s corresponds to the seasonal period. Suppose that $|\varphi| < 1$ and that $\text{Var}(\varepsilon_t) = \sigma^2$.

(a) Show that the autocovariance function of this process is given by

$$\gamma(h) = \frac{\sigma^2}{1-\varphi^2}\varphi^{\left[\frac{h}{s}\right]},$$

where $[\cdot]$ denotes the integer function.

(b) Find the spectral density of this seasonal process.

9.2 Let $z_t = \sum_{j=1}^{n}(A_j \cos(\lambda_j t) + B_j \sin(\lambda_j t))$, where $t = 0, \pm 1, \ldots$ y $\lambda_1, \lambda_2, \ldots \lambda_n$ are positive constant and A_j, B_j are independent random variables, with zero-mean and variance $\sigma_j^2 = \text{Var}(A_j) = \text{Var}(B_j)$, $j = 1, \ldots, n$.

(a) Is this stationary process?

(b) Find the mean and autocovariance function of z_t

9.3 Consider the time series$\{y_t\}$:

$$\begin{aligned} y_t &= A \sin(\omega t) + x_t, \\ x_t &= a_0 + a_1\, t + a_2\, t^2 + \eta_t, \end{aligned}$$

where A is a random variable and $\{\eta_t\}$ is an integrated process of order 2, with $E(\eta_t) = 0$ for all t. That is, $(1 - B^2)\eta_t$ is a stationary process.

(a) Find the values of the frequency ω satisfying

$$(1 - B^2)y_t = (1 - B^2)x_t.$$

(b) Verify that for the values of ω found in part (a) we have that

$$z_t \equiv (1 - B^2)(1 - B)^2 y_t,$$

is a stationary process.

(c) Calculate the expected value $\mu = E(z_t)$.

(d) Is the process $\{z_t - \mu\}$ a white noise sequence?

9.4 Let $\{x_t\}$ be the time series defined by

$$x_t = A\cos(\pi t/3) + B\sin(\pi t/3) + y_t$$

where $y_t = z_t + 2.5z_{t-1}$, $\{z_t\} \sim \mathrm{WN}(0, \sigma^2)$, A and B are uncorrelated with zero-mean and variance ν^2, and z_t is uncorrelated with A and B for each t. Find the ACF of $\{x_t\}$.

9.5 Design a symmetric moving-average filter which removes seasonal components with period 3 and, simultaneously, it does not remove quadratic trends.

9.6 For which values of $\alpha, \alpha_1, \ldots, \alpha_r$ does the fractional seasonal model exhibits long memory?

9.7 Let s be a positive integer and $d \in (0, \frac{1}{2})$. Calculate the coefficients ψ_j in the expansion

$$(1 - B^s)^{-d} = \sum_{j=1}^{\infty} \psi_j B^j.$$

9.8 Let s be a positive integer and $d < 0$.

(a) Prove that the expansion

$$(1 - B^s)^d = \sum_{j=1}^{\infty} \pi_j B^j,$$

is absolutely convergent.

(b) Let z be a random variable such that $E(z^2) < \infty$. Show that

$$(1 - B^s)^d z = 0.$$

(c) Consider the equation

$$(1 - B^s)^d y_t = \varepsilon_t, \tag{9.9}$$

where $\{\varepsilon_t\}$ is a white noise sequence with finite variance. Prove that there is a stationary solution $\{y_t\}$ of equation (9.9).

(d) Show that the process $\{x_t\}$ defined as $x_t = y_t + z$ is also a stationary solution of equation (9.9).

9.9 Consider the SARFIMA$(0, 0, 0) \times (0, d_s, 0)_s$ process

$$y_t = (1 - B^s)^{-d_s} \varepsilon_t,$$

where ε_t is a white noise sequence with variance $\sigma = 1$. Let $\gamma_s(k)$ be the ACF of the process y_t and $\gamma(k)$ be the ACF of a fractional noise FN(d) process with unit noise variance. Verify that

$$\gamma_s(k) = \gamma(sk).$$

9.10 Consider the SARFIMA$(2, d, 2) \times (0, d_s, 0)_s$ process

$$(1 - \phi_1 B - \phi_2 B^2) y_t = (1 - \theta_1 B - \theta_2 B^2)(1 - B^s)^{-d_s} \varepsilon_t,$$

where ε_t is a white noise sequence with variance σ and

$$\boldsymbol{\theta} = (\phi_1, \phi_2, d, \theta_1, \theta_2, d_s, \sigma) = (0.2, 0.5, 0.3, -0.2, 0.5, 0.1, 1.3).$$

(a) Is this process stationary?
(b) Is this process invertible?

9.11 Calculate numerically the variance-covariance matrix $\Gamma(\theta)$ given in (9.7) for a SARFIMA$(0, d, 0) \times (0, d_s, 0)_s$ process with $\boldsymbol{\theta} = (d, d_s) = (0.1, 0.2)$ and $s = 12$.

9.12 Simulate a sample of 1000 observations from a Gaussian SARFIMA$(1, d, 1) \times (0, d_s, 0)_s$ process with parameters

$$\boldsymbol{\theta} = (\phi_1, \theta_1, d, d_s, \sigma) = (0.6, 0.3, 0.2, 0.2, 1).$$

9.13 Implement computationally the splitting method for calculating the theoretical ACF of a SARFIMA$(0, d, 0) \times (0, d_s, 0)_s$ process.

9.14 Write a state space system for the SARFIMA process described in the previous problem.

9.15 Calculate the MLE for the SARFIMA process in Problem 9.12 by means of the state space systems and the Whittle method.

9.16 Suppose that T is the variance-covariance matrix of a stationary Gaussian process $\{y_1, y_2, \ldots, y_n\}$ with fractional seasonal spectral density.

(a) Let $|T|$ be the Euclidean norm of T, that is,

$$|T| = [\mathrm{tr}(TT^*)]^{1/2}.$$

Show that

$$|T|^2 = \sum_{i=1}^{n} \sum_{j=1}^{n} \gamma^2(i - j) = n \sum_{j=0}^{n-1} \left(1 - \frac{j}{n}\right) \gamma^2(j).$$

(b) Verify that yields

$$\gamma(j) \leq K j^{\widetilde{\alpha}-1},$$

for $j \geq 1$, where $\widetilde{\alpha} = \max\{\alpha, \alpha_1, \ldots, \alpha_r\}$.

(c) Show that

$$|T|^2 \leq n \sum_{j=0}^{n-1} \gamma^2(j) \leq K n \sum_{j=1}^{n} j^{2\widetilde{\alpha}-2}.$$

(d) Verify that

$$\left|T\right|^2 \leq Kn^{2\widetilde{\alpha}},$$

and then

$$\left|T\right| \leq Kn.$$

9.17 Consider the GARMA process described by the spectral density

$$f(\lambda) = H(\lambda)|\lambda - \lambda_1|^{-d_1}|\lambda + \lambda_1|^{-d_1},$$

where $H(\lambda)$ is a $\mathcal{C}^\infty([-\pi, \pi])$ function. Verify that the ACF of this process satisfies the asymptotic expression

$$\gamma(h) \sim c_1 \cos(h\lambda_1)|h|^{2d_1-1},$$

as $|h| \to \infty$.

CHAPTER 10

TIME SERIES REGRESSION

In the previous chapters we have studied methods for dealing with serially dependent time series data but we have not discussed the problem of relating those time series to other covariates or trends. However, in many practical applications, the behavior of a time series may be related to the behavior of other data. A widely used approach to model these relationships is the linear regression analysis. Consequently, in this chapter we explore several aspects of the statistical analysis of linear regression models with serially dependent errors. A motivating example is discussed in Section 10.1. Some essential definitions about the model under study are given in Section 10.2. We then proceed to the analysis of some large sample properties of the least squares estimators (LSE) and the best linear unbiased estimators (BLUE). Aspects such as strong consistency, the asymptotic variance of the estimators, normality, and efficiency are discussed in Section 10.3 for the LSE and in Section 10.4 for the BLUE. Sections 10.5–10.7 present some important examples, including the estimation of the mean, the polynomial, and the harmonic regression. A real life data application to illustrate these regression techniques is presented in Section 10.8 while some references and further readings are discussed in the Section 10.9. This chapter concludes with a section of proposed problems.

Time Series Analysis. First Edition. Wilfredo Palma.

10.1 MOTIVATION

Consider the time series regression model

$$y_t = x_t \beta + \varepsilon_t,$$

where x_t is a regressor and ε_t is a stationary error sequence. A very simple example of this model is obtained under two essential assumptions: that the regression variable x_t is deterministic or *known* and that the error ε_t is a sequence of i.i.d. random variables with zero-mean and variance σ^2. Under this assumptions, the best estimate of the parameter β in the sense of unbiased and minimal variance corresponds to the least squares estimator (LSE) given by

$$\widehat{\beta} = \frac{\sum_{t=1}^{n} y_t x_t}{\sum_{t=1}^{n} x_t^2}.$$

Furthermore, the variance of this estimator is

$$\operatorname{Var}(\widehat{\beta}) = \frac{\sigma^2}{\sum_{t=1}^{n} x_t^2}.$$

On the other hand, if the error sequence is correlated, the variance of the LSE differs from the previous formula. For instance, assume that ε_t satisfies an AR(1) process with autoregressive parameter ϕ,

$$\varepsilon_t = \phi \varepsilon_{t-1} + e_t,$$

where the noise e_t is an i.i.d. sequence with zero-mean and variance $(1-\phi^2)\sigma^2$. Notice that given this parametrization of the variance of ε_t, we obtain that the variance of ε_t is σ^2, making both cases comparable. For $h = 1-n, \dots, n-1$ define

$$\alpha_h = \sum_{t=1}^{n-|h|} x_t x_{t+|h|},$$

so that we can write the variance of the LSE in this case as

$$\operatorname{Var}(\widehat{\beta}) = \frac{\sum_{h=1-n}^{n-1} \alpha_h \gamma(h)}{\sum_{t=1}^{n} x_t^2}.$$

Consequently, if r denotes the ratio between the two variances, we have that

$$r = \frac{1}{\sigma^2} \sum_{h=1-n}^{n-1} \alpha_h \gamma(h).$$

Consider the simple case where $x_t = 1$ for all $h = 1, \ldots, n$. Then, after some algebra, this quantity can be written as

$$r = \frac{1 + \phi - 2\phi^n}{(1 + \phi)(1 - \phi)^2}.$$

Since $|\phi| < 1$ we have that as the sample size increases to infinity

$$r = \frac{1}{(1 - \phi)^2}.$$

In turn, as ϕ approaches 1, r tends to infinity. Thus, as the level of serial dependence of the error sequence increases, the relative imprecision of the LSE increases as well. The lesson provided by this illustration is that serial dependence affects the quality of the estimates.

As an illustration, Figure 10.1 exhibits two examples of time series regressions for the linear trend with $\beta = 2$, $\sigma = 1$ with 200 observations. Figure 10.1(a) shows then case of independent Gaussian noise while Figure 10.1(b) displays the regression with AR(1) dependent Gaussian errors with autoregressive parameter $\phi = 0.95$. Notice that the values on panel (b) exhibit much more level of persistence, that is, the values tend to stay close to their respective neighbors.

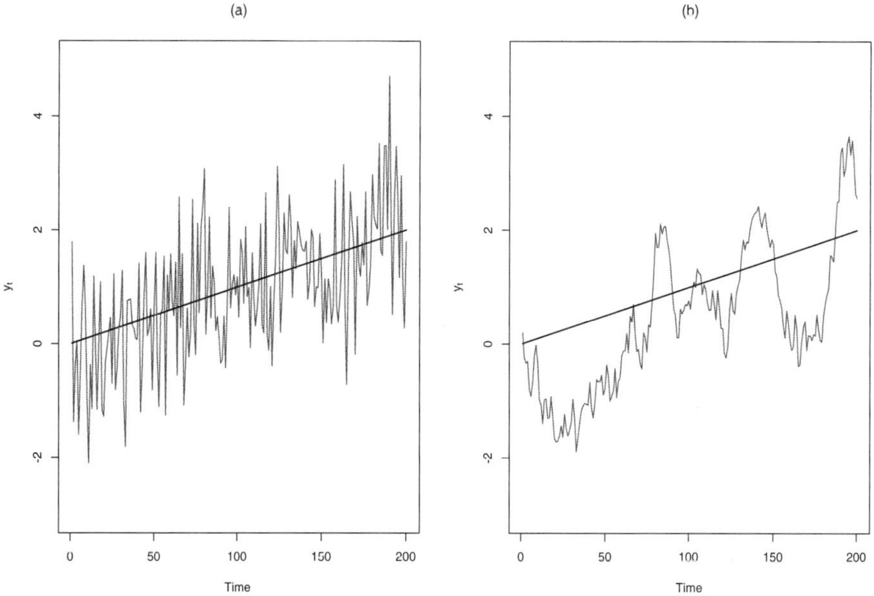

Figure 10.1 *Time series regressions. Linear trend with $\beta = 2$ and $\sigma = 1$. (a) Independent Gaussian noise. (b) AR(1) dependent Gaussian errors with autoregressive parameter $\phi = 0.95$.*

Table 10.1 Time series regression with constant and first-order autoregressive errors. Estimates of β for different values of ϕ.

ϕ	Independent	AR(1)	SD Independent	SD AR(1)	Ratio r
0.1000	2.0014	2.0014	0.0706	0.0782	1.2247
0.3000	2.0029	2.0040	0.0694	0.0939	1.8309
0.5000	1.9987	1.9989	0.0623	0.1069	2.9508
0.8000	2.0089	2.0274	0.0721	0.2154	8.9155

A similar exercise can be carried out considering the *best linear unbiased estimator* (BLUE) instead of the LSE.

To illustrate these issues, Table 10.1 and Table 10.2 report the results from the simulations of two regressions, a constant mean model

$$y_t = \beta + \varepsilon_t,$$

with $\beta = 2$, for $t = 1, \ldots, n$, and the linear trend model

$$y_t = \beta \frac{t}{n} + \varepsilon_t,$$

with $\beta = 2$, $n = 200$ and $t = 1, \ldots, n$. For the dependent error case, we consider the Gaussian AR(1) model with and $\sigma = 1$ and four different values for the autoregressive parameter ϕ, from a low level of dependence $\phi = 0.1$ to a high level of autocorrelation $\phi = 0.8$.

The results reported in both tables are based on 1000 repetitions. Notice that the least squares estimates of the regression parameter β seem to be unbiased in both tables and across the distinct values of ϕ. The effects of the error dependence can be observed in the levels of estimation error standard deviations. In particular, notice that in both Monte Carlo simulation sets the ratio r increases as the level of autocorrelation ϕ increases.

Table 10.2 Time series regression with linear trend and first-order autoregressive errors. Estimates of β for different values of ϕ.

ϕ	Independent	AR(1)	SD Independent	SD AR(1)	Ratio r
0.1000	2.0021	2.0024	0.1253	0.1384	1.2195
0.3000	1.9868	1.9823	0.1185	0.1611	1.8482
0.5000	1.9943	1.9904	0.1250	0.2153	2.9667
0.8000	1.9993	1.9996	0.1195	0.3527	8.7111

10.2 DEFINITIONS

Consider the linear regression model

$$y_t = x_t \boldsymbol{\beta} + \varepsilon_t, \tag{10.1}$$

for $t = 1, 2, \ldots$, where $x_t = (x_{t1}, \ldots, x_{tp})$ is a sequence of regressors, $\boldsymbol{\beta} \in \mathbb{R}^p$ is a vector of parameters, and $\{\varepsilon_t\}$ is a long-range-dependent stationary process with spectral density

$$f(\lambda) = |1 - e^{i\lambda}|^{-2d} f_0(\lambda), \tag{10.2}$$

where $f_0(\lambda)$ is a symmetric, positive, piecewise continuous function for $\lambda \in (-\pi, \pi]$ and $0 < d < \frac{1}{2}$.

The least squares estimator (LSE) of $\boldsymbol{\beta}$ is given by

$$\widehat{\boldsymbol{\beta}}_n = (X_n' X_n)^{-1} X_n' Y_n, \tag{10.3}$$

where X_n is the $n \times p$ matrix of regressors, $[X_n]_{ij} = x_{ij}$, $i = 1, \ldots, n$, $j = 1, \ldots, p$, and $Y_n = (y_1, \ldots, y_n)'$. The variance of the LSE is

$$\mathrm{Var}(\widehat{\boldsymbol{\beta}}_n) = (X_n' X_n)^{-1} X_n' \Gamma X_n (X_n' X_n)^{-1}, \tag{10.4}$$

where Γ is the variance-covariance matrix of $\{y_t\}$ with elements

$$\Gamma_{ij} = \int_{-\pi}^{\pi} f(\lambda) e^{i\lambda(i-j)} d\lambda. \tag{10.5}$$

On the other hand, the BLUE of $\boldsymbol{\beta}$ is

$$\widetilde{\boldsymbol{\beta}}_n = (X_n' \Gamma^{-1} X_n)^{-1} X_n' \Gamma^{-1} Y_n, \tag{10.6}$$

which has variance equal to

$$\mathrm{Var}(\widetilde{\boldsymbol{\beta}}_n) = (X_n' \Gamma^{-1} X_n)^{-1}. \tag{10.7}$$

10.2.1 Grenander Conditions

In order to analyze the large sample properties of the LSE and the BLUE, we introduce the following so-called Grenander conditions on the regressors. Let $X_n(j)$ be the jth column of the design matrix, X_n.

(1) $\|X_n(j)\| \to \infty$ as $n \to \infty$ for $j = 1, \ldots, p$.

(2) $\lim_{n\to\infty} \|X_{n+1}(j)\| / \|X_n(j)\| = 1$ for $j = 1, \ldots, p$.

(3) Let $X_{n,h}(j) = (x_{h+1,j}, x_{h+2,j}, \ldots, x_{n,j}, 0 \ldots, 0)'$ for $h \geq 0$ and $X_{n,h}(j) = (0, \ldots, 0, x_{1,j}, x_{2,j}, \ldots, x_{n+h,j})'$ for $h < 0$. Then, there exists a $p \times p$ finite matrix $R(h)$ such that

$$\frac{\langle X_{n,h}(i), X_n(j) \rangle}{\|X_{n,h}(i)\| \|X_n(j)\|} \to R_{ij}(h), \tag{10.8}$$

as $n \to \infty$ with $\langle x, y \rangle = x\bar{y}$ for complex numbers $x, y \in \mathbb{C}$ where $\bar{y}$ is the complex conjugate of y.

(4) The matrix $R(0)$ is nonsingular.

Notice first that under conditions (1)–(4), the matrix $R(h)$ may be written as

$$R(h) = \int_{-\pi}^{\pi} e^{ih\lambda}\, dM(\lambda), \tag{10.9}$$

where $M(\lambda)$ is a symmetric matrix function with positive semidefinite increments.

The asymptotic properties of the LSE estimates and the BLUE depend upon the behavior of $M_{jj}(\lambda)$ around frequency zero. Consequently, assume that for $j = 1, \ldots, s$, $M_{jj}(\lambda)$ suffers a jump at the origin, that is,

$$M_{jj}(0+) > M_{jj}(0), \qquad j = 1, \ldots, s,$$

and for $j = s+1, \ldots, p$, $M_{jj}(\lambda)$ is continuous at $\lambda = 0$, that is,

$$M_{jj}(0+) = M_{jj}(0), \qquad j = s+1, \ldots, p,$$

where $M_{jj}(0+) = \lim_{\lambda \to 0+} M_{jj}(\lambda)$.

Before stating the asymptotic behavior of these estimators, the following definitions and technical results are needed. Define the *characteristic function* of the design matrix by

$$M_{ij}^n(\lambda) = \int_{-\pi}^{\lambda} m_{ij}^n(\omega)\, d\omega,$$

where

$$m_{ij}^n(\lambda) = \frac{\langle \sum_{t=1}^n x_{ti} e^{-it\lambda}, \sum_{t=1}^n x_{tj} e^{-it\lambda} \rangle}{2\pi \|X_n(i)\| \|X_n(j)\|}.$$

Also, we define the function $\delta(\cdot)$ such that $\delta(x) = 1$ for $x = 0$ and $\delta(x) = 0$ for $x \neq 0$.

Notice that by Problem 10.14

$$\int_{-\pi}^{\pi} g(\lambda)\, dM^n(\lambda) \to \int_{-\pi}^{\pi} g(\lambda)\, dM(\lambda),$$

as $n \to \infty$ for any continuous function $g(\lambda)$ with $\lambda \in [-\pi, \pi]$.

10.3 PROPERTIES OF THE LSE

The large sample behavior of the LSE is studied in this section.

In what follows, the regressors are assumed to be nonstochastic. Let $\sigma_n = \lambda_{\min}(X_n' X_n)$ be the smallest eigenvalue of $X_n' X_n$ and let $\boldsymbol{\beta}$ be the true parameter. Under some conditions on σ_n, see Problem 10.17, the LSE is consistent. Finding the eigenvalues of the matrix $X_n' X_n$ may not be a simple task in some situations. Fortunately, there is a set of conditions that guarantees the strong consistency of the LSE which are simpler to verify, see Problem 10.18.

10.3.1 Asymptotic Variance

The asymptotic variance of the LSE is analyzed in this section. Consider now the following assumption. For any $\delta > 0$, there exists a positive constant c such that

$$\int_{-c}^{c} f(\lambda)\, dM_{jj}^{n}(\lambda) < \delta, \tag{10.10}$$

for every n and $j = s+1, \ldots, p$. With this additional condition the following result is obtained: If the Grenander conditions (1) to (4) are satisfied and that (10.10) holds. Define the $p \times p$ diagonal matrix

$$D_n = \operatorname{diag}(\|X_n(1)\| n^d, \ldots, \|X_n(s)\| n^d, \|X_n(s+1)\|, \ldots, \|X_n(p)\|).$$

Then,

(a) For $s = 0$,

$$D_n \operatorname{Var}(\widehat{\boldsymbol{\beta}}_n) D_n \to 2\pi R(0)^{-1} \int_{-\pi}^{\pi} f(\lambda)\, dM(\lambda) R(0)^{-1}, \tag{10.11}$$

as $n \to \infty$ if and only if condition (10.10) holds.

(b) For $s > 0$,

$$D_n^{-1}(X_n' X_n) \operatorname{Var}(\widehat{\boldsymbol{\beta}}_n)(X_n' X_n) D_n^{-1} \to 2\pi A, \tag{10.12}$$

as $n \to \infty$ where

$$A = \begin{pmatrix} B & 0 \\ 0 & C \end{pmatrix}, \tag{10.13}$$

and the elements of the $s \times s$ matrix B are given by

$$b_{ij} = f_0(0) \lim_{n \to \infty} n^{-2d} \int_{-\pi}^{\pi} |1 - e^{i\lambda}|^{-2d}\, dM_{ij}^{n}(\lambda),$$

for $i, j = 1, \ldots, s$ and the elements of the $(p-s) \times (p-s)$ matrix C are given by

$$c_{ij} = \int_{-\pi}^{\pi} f(\lambda)\, dM_{i+s, j+s}(\lambda).$$

10.3.2 Asymptotic Normality

Assume that the regression errors $\{\varepsilon_t\}$ corresponds to a strictly stationary process with spectral density satisfying (10.2) and Wold decomposition:

$$\varepsilon_t = \sum_{j=0}^{\infty} \psi_j \nu_{t-j}. \tag{10.14}$$

Thus, under some technical conditions, we have that

$$D_n^{-1}(X_n' X_n)(\widehat{\boldsymbol{\beta}}_n - \boldsymbol{\beta}) \to N_p(0, A),$$

as $n \to \infty$ where A is the variance-covariance matrix defined by (10.13).

10.4 PROPERTIES OF THE BLUE

Under some conditions, the BLUE satisfies

$$\widetilde{D}_n \operatorname{Var}(\widetilde{\boldsymbol{\beta}}_n)\widetilde{D}_n \to 2\pi \left[\int_{-\pi}^{\pi} f(\lambda)^{-1} dM(\lambda)\right]^{-1}, \tag{10.15}$$

as $n \to \infty$.

10.4.1 Efficiency of the LSE Relative to the BLUE

When $s > 0$, that is, $M(\lambda)$ displays a jump at the origin for some regressors, the LSE is not asymptotically efficient compared to the BLUE.

On the other hand, if $s = 0$ [that is, $M(\lambda)$ does not jump at zero frequency] then, under some conditions on the regressors, it can be established that the LSE is asymptotically efficient.

Assuming that $s = 0$ and conditions (1)–(5) and (10.10) hold, the LSE is efficient compared to the BLUE if and only if M increases at no more than p frequencies and the sum of the ranks of increases is p.

■ **EXAMPLE 10.1**

As an illustration of these theorems, consider the following trigonometric regression:

$$y_t = \boldsymbol{\beta} x_t + \varepsilon_t$$

where $x_t = \sum_{j=1}^{q} e^{i\lambda_j t}$, $\lambda_j \neq 0$ for $j = 1, \ldots, q$ are *known* frequencies and the error sequence has spectral density (10.2).

In this case,

$$\frac{\|X_n\|^2}{n} = \frac{1}{n}\sum_{t=1}^{n}\left(\sum_{j=1}^{q} e^{i\lambda_j t}\right)\left(\sum_{k=1}^{q} e^{-i\lambda_k t}\right) = \sum_{j=1}^{q}\sum_{k=1}^{q}\left(\frac{1}{n}\sum_{t=1}^{n} e^{i(\lambda_j - \lambda_k)t}\right).$$

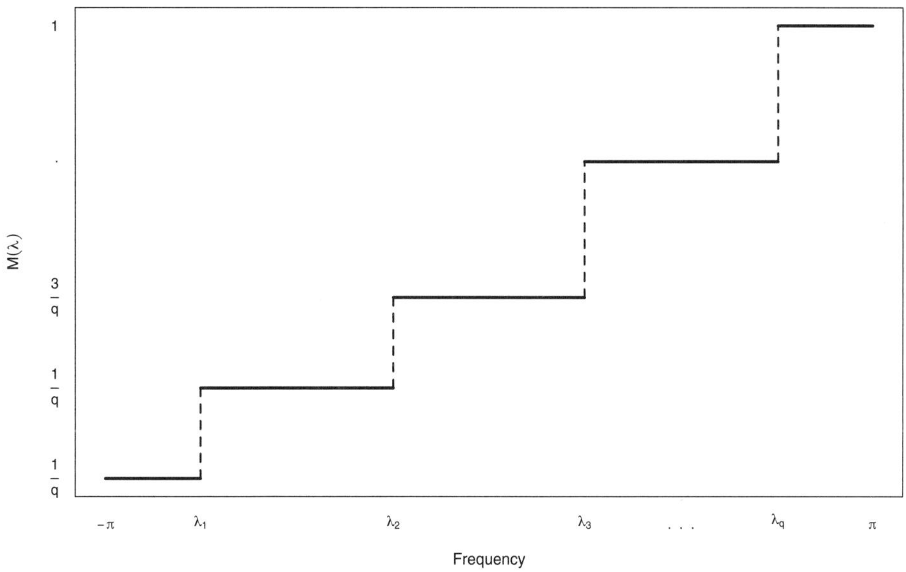

Figure 10.2 An example of trigonometric regression, function $M(\lambda)$.

Consequently, by Problem 10.16 we conclude that

$$\lim_{n\to\infty} \frac{\|X_n\|^2}{n} = \sum_{j=1}^{q}\sum_{k=1}^{q} \delta(\lambda_j - \lambda_k) = q.$$

Furthermore, $dM(\lambda) = (1/q)\sum_{j=1}^{q} \delta_D(\lambda - \lambda_j)\, d\lambda$, or equivalently,

$$M(\lambda) = \frac{1}{q}\int_{-\pi}^{\lambda} \sum_{j=1}^{q} \delta_D(\omega - \lambda_j)\, d\omega.$$

This function is displayed in Figure 10.2. Notice that $M(\lambda)$ exhibits q jumps located at frequencies $\lambda_1, \ldots, \lambda_q$. But, it does not jump at the origin. Therefore, $s = 0$ and $R(0) = 1$.

The asymptotic variance of the LSE satisfies

$$\begin{aligned} \lim_{n\to\infty} \|X_n\|^2 \operatorname{Var}(\widehat{\boldsymbol{\beta}}_n) &= \frac{2\pi}{q}\int_{-\pi}^{\pi} f(\lambda) \sum_{j=1}^{q} \delta_D(\lambda - \lambda_j)\, d\lambda \\ &= \frac{2\pi}{q}[f(\lambda_1) + \cdots + f(\lambda_q)]. \end{aligned}$$

On the other hand, the asymptotic variance of the BLUE is

$$\lim_{n\to\infty} \|X_n\|^2 \operatorname{Var}(\widetilde{\boldsymbol{\beta}}_n) = 2\pi q \left[\int_{-\pi}^{\pi} f(\lambda)^{-1} \sum_{j=1}^{q} \delta_D(\lambda - \lambda_j)\, d\lambda \right]^{-1}$$
$$= 2\pi q \left[f(\lambda_1)^{-1} + \cdots + f(\lambda_q)^{-1} \right]^{-1}.$$

The relative efficiency of the LSE compared to the BLUE is defined by

$$r(d) = \lim_{n\to\infty} \frac{\det \operatorname{Var}(\widetilde{\boldsymbol{\beta}}_n)}{\det \operatorname{Var}(\widehat{\boldsymbol{\beta}}_n)}. \qquad (10.16)$$

In this case we have

$$r(q) = \frac{q^2}{[f(\lambda_1)^{-1} + \cdots + f(\lambda_q)^{-1}][f(\lambda_1) + \cdots + f(\lambda_q)]}.$$

Thus, for $q = 1$, the LSE is asymptotically efficient and by Jensen's inequality, for $q \geq 2$ the LSE is not efficient.

Recalling that in this example $p = 1$, and since for $q = 1$, $M(\lambda)$ has only one jump at frequency λ_1 and consequently the LSE is efficient. On the other hand, for $q \geq 2$, $M(\lambda)$ has more than one jump and therefore the LSE is not efficient.

Consider, for instance, the frequencies $\lambda_j = \pi j/q$ for $j = 1, \ldots, q$. In this case we have

$$\lim_{q\to\infty} r(q) = \frac{\pi^2}{\int_0^{\pi} f(\lambda)\, d\lambda \int_0^{\pi} f(\lambda)^{-1}\, d\lambda}.$$

Figure 10.3 shows the behavior of $r(q)$ for a fractional noise model for different values of the long-memory parameter d and q.

For this model, we have that

$$\lim_{q\to\infty} r(q) = \frac{\Gamma(1-d)^2\Gamma(1+d)^2}{\Gamma(1-2d)\Gamma(1+2d)}.$$

As expected, for $d = 0$ the asymptotic relative efficiency is 1. On the other hand, for $d \to \frac{1}{2}$, the limiting relative efficiency is 0.

10.5 ESTIMATION OF THE MEAN

A simple but illustrative example of a linear regression model is

$$y_t = \mu + \varepsilon_t$$

where μ is the unknown mean of the process y_t. In this case, the LSE is

$$\widehat{\mu} = \frac{1}{n} \sum_{t=1}^{n} y_t = \frac{1}{n} \mathbf{1}' Y_n,$$

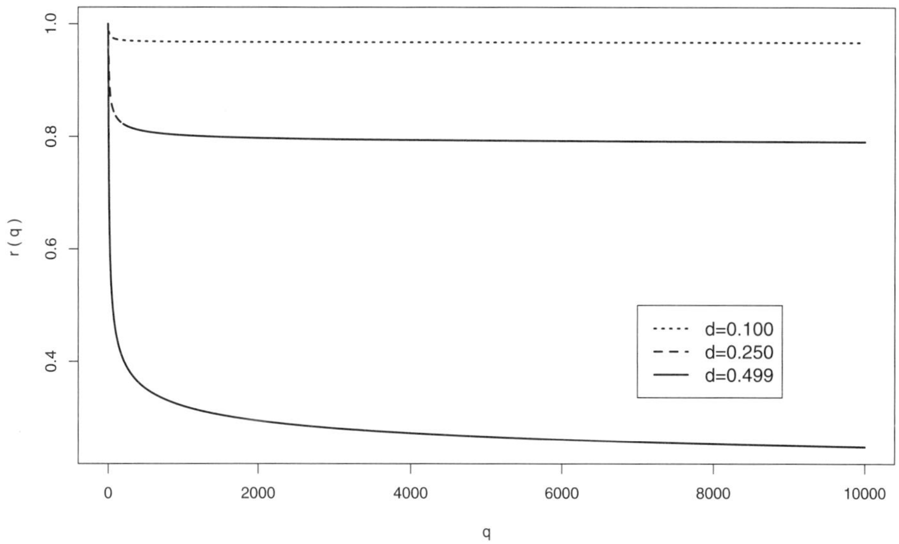

Figure 10.3 *An example of trigonometric regression: Relative efficiency.*

where $\mathbf{1} = (1, 1, \ldots, 1)'$ and the BLUE is

$$\widetilde{\mu} = (\mathbf{1}'\Gamma^{-1}\mathbf{1})^{-1}\mathbf{1}'\Gamma^{-1}Y_n.$$

In this case, $\sigma_n = n$ and then an application of Problem 10.18 establishes that the LSE is consistent.

10.5.1 Asymptotic Variance

On the other hand, the Grenander conditions are also satisfied: (1) $\|X_n\| = \sqrt{n} \to \infty$ as $n \to \infty$; (2) $\|X_{n+1}\|/\|X_n\| = 1 + 1/\sqrt{n} \to 1$ as $n \to \infty$; (3) $R_{11}(h) = \lim_{n\to\infty}[1 - |h|/n] = 1$ for all fixed h; and (4) $R(0) = 1$ is nonsingular. Furthermore, in this case

$$m^n(\lambda) = \frac{1}{2\pi n}\left|\sum_{t=1}^{n} e^{i\lambda t}\right|^2 = \frac{1}{2\pi n}\frac{|e^{in\lambda} - 1|^2}{|e^{i\lambda} - 1|^2} = \frac{1}{2\pi n}\frac{1 - \cos n\lambda}{1 - \cos\lambda},$$

which by Problem 10.15 tends to the Dirac functional operator $\delta_D(\lambda)$ as $n \to \infty$. Therefore,

$$M(\lambda) = \int_{-\pi}^{\lambda} \delta_D(\omega)\, d\omega = \mathbf{1}_{[0,\pi)}(\lambda).$$

Hence, $M(\lambda)$ has a jump at the origin as shown in Figure 10.4, and therefore $s = 1$. In this case $D_n = n^{d+1/2}$ and therefore

$$n^{1-2d}\,\mathrm{Var}(\widehat{\mu}) \to 2\pi b,$$

as $n \to \infty$ where

$$\begin{aligned} b &= \frac{1}{2\pi} f_0(0) \lim_{n\to\infty} n^{-1-2d} \int_{-\pi}^{\pi} |1 - e^{\imath\lambda}|^{-2-2d} |1 - e^{\imath n\lambda}|^2 d\lambda \\ &= \frac{\Gamma(1-2d)}{\Gamma(d)\Gamma(1-d)} f_0(0) \int_0^1 \int_0^1 |x-y|^{2d-1} dx\, dy \\ &= \frac{\Gamma(1-2d)}{d(1+2d)\Gamma(d)\Gamma(1-d)} f_0(0). \end{aligned}$$

For an ARFIMA(p, d, q) model $f_0(0) = \dfrac{\sigma^2}{2\pi} \dfrac{|\theta(1)|^2}{|\phi(1)|^2}$. Consequently,

$$\mathrm{Var}(\widehat{\mu}) \sim \frac{c_\gamma n^{2d-1}}{d(1+2d)} = \sigma^2 \frac{|\theta(1)|^2}{|\phi(1)|^2} \left[\frac{\Gamma(1-2d)}{d(1+2d)\Gamma(d)\Gamma(1-d)} \right] n^{2d-1}. \quad (10.17)$$

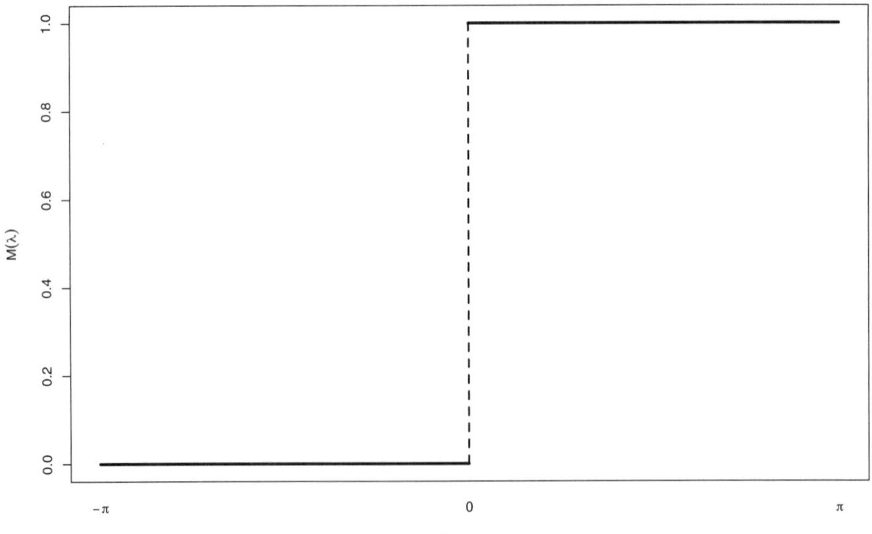

Figure 10.4 *Estimation of the mean: $M(\lambda)$.*

10.5.2 Relative Efficiency

Since in this case $M(\lambda)$ has a jump at zero frequency, the LSE (sample mean) is not asymptotically efficient as compared to the BLUE. However, we can analyze its relative efficiency with respect to the BLUE.

Following Adenstedt (1974), the BLUE of the mean of a fractional noise process with long-memory parameter d is given by

$$\widetilde{\mu} = \sum_{j=1}^{n} \alpha_j y_j,$$

where the coefficients α_j are

$$\alpha_j = \binom{n-1}{j-1} \frac{\Gamma(j-d)\Gamma(n+1-j-d)\Gamma(2-2d)}{\Gamma(n+1-2d)\Gamma(1-d)^2}.$$

The variance of this BLUE may be written as

$$\mathrm{Var}(\widetilde{\mu}) = \sigma^2 \frac{\Gamma(n)\Gamma(1-2d)\Gamma(2-2d)}{\Gamma(n+1-2d)\Gamma(1-d)^2} \sim \sigma^2 n^{2d-1} \frac{\Gamma(1-2d)\Gamma(2-2d)}{\Gamma(1-d)^2},$$

for large n. For an ARFIMA(p,d,q) process, the variance of the BLUE satisfies

$$\mathrm{Var}(\widetilde{\mu}) \sim \sigma^2 n^{2d-1} \frac{|\theta(1)|^2}{|\phi(1)|^2} \frac{\Gamma(1-2d)\Gamma(2-2d)}{\Gamma(1-d)^2}, \tag{10.18}$$

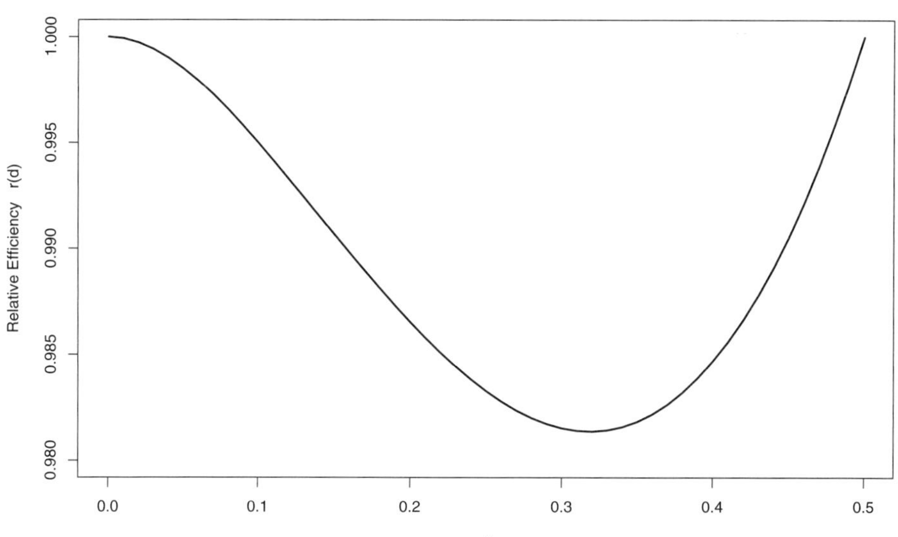

Figure 10.5 *Relative efficiency of the sample mean.*

for large n. Now, from (10.17) and (10.18) we conclude that

$$r(d) = \lim_{n\to\infty} \frac{\mathrm{Var}(\widetilde{\mu})}{\mathrm{Var}(\widehat{\mu})} = \frac{(1+2d)\Gamma(1+d)\Gamma(2-2d)}{\Gamma(1-d)}.$$

Figure 10.5 shows the relative efficiency $r(d)$ for $d \in (0, \frac{1}{2})$. The minimal relative efficiency is approximately 0.9813 reached at $d = 0.318$.

10.6 POLYNOMIAL TREND

Extending the previous example, consider the polynomial regression

$$y_t = \boldsymbol{\beta}_0 + \boldsymbol{\beta}_1 t + \cdots + \boldsymbol{\beta}_q t^q + \varepsilon_t,$$

where $\{\varepsilon_t\}$ satisfies (10.2). In this case, $p = q+1$, $X_n(i) = (1, 2^{i-1}, \ldots, n^{i-1})$ for $i = 1, \ldots, p$ and hence

$$X_n(i)' X_n(j) = \sum_{t=1}^{n} t^{i+j-2} = \frac{n^{i+j-1}}{i+j-1}[1+o(n)],$$

for $i, j = 1, \ldots, p$.

Thus, $D_{ii} = \dfrac{n^{2i-1+d}}{2i-1}[1+o(n)]$ and

$$m_{ij}^n(\lambda) = \frac{\sum_{t=1}^n \sum_{s=1}^n t^{i-1} s^{j-1} e^{\imath(s-t)\lambda}}{2\pi \|X_n(i)\| \|X_n(j)\|}.$$

This term converges to

$$m_{ij}(\lambda) = \frac{\sqrt{2i-1}\sqrt{2j-1}}{i+j-1}\delta_D(\lambda).$$

Therefore,

$$\begin{aligned} M_{ij}(\lambda) &= \int_{-\pi}^{\lambda} \frac{\sqrt{2i-1}\sqrt{2j-1}}{i+j-1}\delta_D(\lambda)\, d\lambda \\ &= \frac{\sqrt{2i-1}\sqrt{2j-1}}{i+j-1}\mathbf{1}_{(0,\pi)}(\lambda), \end{aligned}$$

and

$$R_i j(h) = \int_{-\pi}^{\pi} e^{\imath\lambda h}\, dM_{ij}(\lambda) = \frac{\sqrt{2i-1}\sqrt{2j-1}}{i+j-1},$$

which actually does not depend on the lag h. For instance, for $p=5$ we have that the matrix $R(h)$ is given by

$$R(h) = \begin{bmatrix} 1 & \frac{\sqrt{3}}{2} & \frac{\sqrt{5}}{3} & \frac{\sqrt{7}}{4} & \frac{\sqrt{9}}{5} \\ \frac{\sqrt{3}}{2} & 1 & \frac{\sqrt{15}}{4} & \frac{\sqrt{21}}{5} & \frac{\sqrt{27}}{6} \\ \frac{\sqrt{5}}{3} & \frac{\sqrt{15}}{4} & 1 & \frac{\sqrt{35}}{6} & \frac{\sqrt{45}}{7} \\ \frac{\sqrt{7}}{4} & \frac{\sqrt{21}}{5} & \frac{\sqrt{35}}{6} & 1 & \frac{\sqrt{63}}{8} \\ \frac{\sqrt{9}}{5} & \frac{\sqrt{27}}{6} & \frac{\sqrt{45}}{7} & \frac{\sqrt{63}}{8} & 1 \end{bmatrix}.$$

EXAMPLE 10.2

Consider an ARMA model

$$\phi(B)y_t = \theta(B)\varepsilon_t,$$

where ε_t is a white noise sequence with zero-mean and variance σ^2. The spectral density of this process is given by

$$f(\lambda) = \frac{\sigma^2}{2\pi}\left|\frac{\theta(e^{i\lambda})}{\phi(e^{i\lambda})}\right|^2.$$

In particular, at the origin we have

$$f(0) = \frac{\sigma^2}{2\pi}\left|\frac{\theta(1)}{\phi(1)}\right|^2.$$

Let C a square matrix given by

$$C = \int_{-\pi}^{\pi} f(\lambda) dM(\lambda),$$

so that

$$\begin{aligned} C_{ij} &= \int_{-\pi}^{\pi} f(\lambda) m_{ij}(\lambda)\, d\lambda \\ &= \frac{\sqrt{2i-1}\sqrt{2j-1}}{i+j-1}\int_{-\pi}^{\pi} f(\lambda)\delta_D(\lambda)\, d\lambda \\ &= f(0)\frac{\sqrt{2i-1}\sqrt{2j-1}}{i+j-1} \\ &= f(0)R_{ij}. \end{aligned}$$

Hence, the variance of the LSE satisfies

$$D_n \operatorname{Var}(\widehat{\boldsymbol{\beta}}_n) D_n \to 2\pi f(0)\, R(0)^{-1},$$

where D_n is a diagonal matrix with diagonal elements approximately $d_{ij} = \frac{n^{2i-1}}{2i-1}$. For simplicity, consider the case of a linear trend where the parameter to estimate is $(\boldsymbol{\beta}_0, \boldsymbol{\beta}_1)$. In this situation,

$$R(0) = \begin{bmatrix} 1 & \frac{\sqrt{3}}{2} \\ \frac{\sqrt{3}}{2} & 1 \end{bmatrix}.$$

and then

$$R(0)^{-1} = \begin{bmatrix} 4 & -2\sqrt{3} \\ -2\sqrt{3} & 4 \end{bmatrix}.$$

On the other hand,

$$D_n \sim \begin{bmatrix} n & 0 \\ 0 & \frac{n^3}{3} \end{bmatrix}.$$

Therefore, the large sample variance of the LSE satisfies

$$\operatorname{Var} \widehat{\boldsymbol{\beta}} \sim 2\pi\, f(0) \begin{bmatrix} \frac{4}{n^2} & -\frac{6\sqrt{3}}{n^4} \\ -\frac{6\sqrt{3}}{n^4} & \frac{36}{n^6} \end{bmatrix}.$$

or more precisely,

$$\operatorname{Var} \widehat{\boldsymbol{\beta}} \sim \sigma^2 \left| \frac{\theta(1)}{\phi(1)} \right|^2 \begin{bmatrix} \frac{4}{n^2} & -\frac{6\sqrt{3}}{n^4} \\ -\frac{6\sqrt{3}}{n^4} & \frac{36}{n^6} \end{bmatrix}.$$

10.6.1 Consistency

Notice that

$$\frac{\|x_n(j)\|^2}{n^\delta} = \frac{n^{2j-1-\delta}}{2j-1}[1 + o(n)].$$

Therefore, by taking $\delta = 1$ we have that

$$\liminf_{n\to\infty} \frac{\|x_n(j)\|^2}{n^\delta} > 0.$$

Furthermore, since $R(0)$ is nonsingular, we conclude by Problem 10.18 that the LSE is consistent.

10.6.2 Asymptotic Variance

Since $M(\lambda)$ has a jump at the origin, the asymptotic variance of $\widehat{\boldsymbol{\beta}}_n$ satisfies

$$D_n^{-1}(X_n' X_n) \operatorname{Var}(\widehat{\boldsymbol{\beta}}_n)(X_n' X_n) D_n^{-1} \to 2\pi B,$$

where

$$
\begin{aligned}
b_{ij} &= f_0(0) \lim_{n\to\infty} n^{-2d} \int_{-\pi}^{\pi} |1 - e^{i\lambda}|^{-2d} m_{ij}^n(\lambda)\, d\lambda \\
&= f_0(0) \lim_{n\to\infty} \frac{n^{-2d}}{\|X_n(i)\|\|X_n(j)\|} \sum_{t=1}^{n}\sum_{s=1}^{n} t^{i-1}s^{j-1} \int_{-\pi}^{\pi} \frac{|1-e^{i\lambda}|^{-2d}}{2\pi} e^{i(t-s)\lambda}\, d\lambda \\
&= f_0(0) \lim_{n\to\infty} \frac{n^{-2d}}{\|X_n(i)\|\|X_n(j)\|} \sum_{t=1}^{n}\sum_{s=1}^{n} t^{i-1}s^{j-1}\gamma(t-s),
\end{aligned}
$$

where $\gamma(\cdot)$ is the ACF of a fractional noise process FN(d) with $\sigma^2 = 1$. Therefore,

$$
\begin{aligned}
b_{ij} &= f_0(0)c_\gamma \lim_{n\to\infty} \frac{n^{-2d}}{\|X_n(i)\|\|X_n(j)\|} \sum_{t=1}^{n}\sum_{s=1}^{n} t^{i-1}s^{j-1}|t-s|^{2d-1} \\
&= f_0(0)c_\gamma \sqrt{(2i-1)(2j-1)} \\
&\times \lim_{n\to\infty} n^{-2d} \sum_{t=1}^{n}\sum_{s=1}^{n} \left(\frac{t}{n}\right)^{i-1} \left(\frac{s}{n}\right)^{j-1} \left|\frac{t}{n} - \frac{s}{n}\right|^{2d-1} \frac{1}{n^2} \\
&= f_0(0)c_\gamma \sqrt{(2i-1)(2j-1)} \int_0^1\int_0^1 x^{i-1}y^{j-1}|x-y|^{2d-1} dx\, dy \\
&= f_0(0)c_\gamma \sqrt{(2i-1)(2j-1)} \frac{B(i,2d)+B(j,2d)}{i+j+2d-1},
\end{aligned}
$$

see Problem 10.12 for finding the value of the double integral.

10.6.3 Relative Efficiency

Analogous to the estimation of the mean case, since $M(\lambda)$ has a jump at the origin, the LSE of the polynomial regression is not asymptotically efficient. The relative efficiency for $q = 1$ is given by

$$
r(d) = (9 - 4d^2) \left[\frac{(1+2d)\Gamma(1+d)\Gamma(3-2d)}{6\Gamma(2-d)}\right]^2.
$$

Figure 10.6 shows $r(d)$ for $d \in (0, \frac{1}{2})$. The minimal relative efficiency is $\frac{8}{9}$ reached at $d = 0.5$.

10.7 HARMONIC REGRESSION

Another important example is the harmonic regression

$$
y_t = \boldsymbol{\beta}_1 e^{i\lambda_1 t} + \boldsymbol{\beta}_2 e^{i\lambda_2 t} + \cdots + \boldsymbol{\beta}_q e^{i\lambda_q t} + \varepsilon_t,
$$

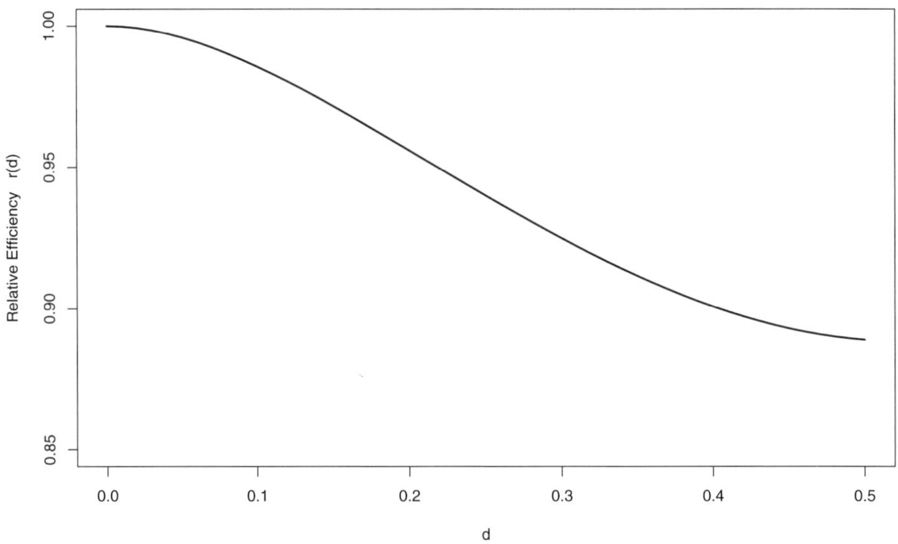

Figure 10.6 *Relative efficiency of the polynomial regression with $q = 1$.*

where $\{\varepsilon_t\}$ is a stationary process with spectral density (10.2) and $\lambda_j \neq 0$ for $j = 1, \ldots, q$ are known frequencies. In this case, $D_{ii} = \sqrt{n}$ for $i = 1, \ldots, q$, and by Problem10.15 we have that

$$\frac{\langle X_{n,h}(i), X_n(j)\rangle}{\|X_{n,h}(i)\|\|X_n(j)\|} = e^{\imath\lambda_j h}\frac{1}{n}\sum_{t=1}^{n} e^{\imath(\lambda_i-\lambda_j)t} \to e^{\imath\lambda_j h}\delta(\lambda_j - \lambda_i),$$

as $n \to \infty$. Hence, $R_{ij}(h) = e^{\imath\lambda_j h}\delta(\lambda_j - \lambda_i)$ and from equation (10.9) we conclude that $dM_{ij}(\lambda) = \delta(\lambda_j - \lambda_i)\delta_D(\lambda - \lambda_j)\, d\lambda$. Therefore, $M_{ij}(\lambda)$ does not have a jump at the origin for any $\lambda_j \neq 0$.

10.7.1 Consistency

Observe that $R(0) = I_n$ where I_n is the $n \times n$ identity matrix. Consequently, it satisfies the Grenander condition (4), that is, $R(0)$ is nonsingular. Besides,

$$\frac{\|X_n(j)\|^2}{n^\delta} = n^{1-\delta}.$$

Therefore, by taking $\delta = 1$ in Problem 10.18 we conclude that the LSE is consistent.

10.7.2 Asymptotic Variance

Since $R_{ij}(0) = \delta(i-j)$ and $D_{ii} = \sqrt{n}$ we have that

$$\int_{-\pi}^{\pi} f(\lambda)\, dM_{ij}(\lambda) = \int_{-\pi}^{\pi} f(\lambda)\delta_D(\lambda - \lambda_j)\, d\lambda = f(\lambda_j)\delta(i-j).$$

Consequently,

$$\lim_{n\to\infty} n\,\mathrm{Var}(\widehat{\boldsymbol{\beta}}_n) = 2\pi \begin{bmatrix} f(\lambda_1) & 0 & 0 & \cdots & 0 \\ 0 & f(\lambda_2) & 0 & \cdots & 0 \\ \vdots & & \ddots & & \vdots \\ 0 & \cdots & 0 & f(\lambda_{q-1}) & 0 \\ 0 & \cdots & 0 & 0 & f(\lambda_q) \end{bmatrix}.$$

10.7.3 Efficiency

As shown in Figure 10.7, $M_{jj}(\lambda)$ displays one jump of size one at frequency λ_j. Thus, $M(\lambda)$ increases at q frequencies each of rank one; so that the LSE is asymptotically efficient.

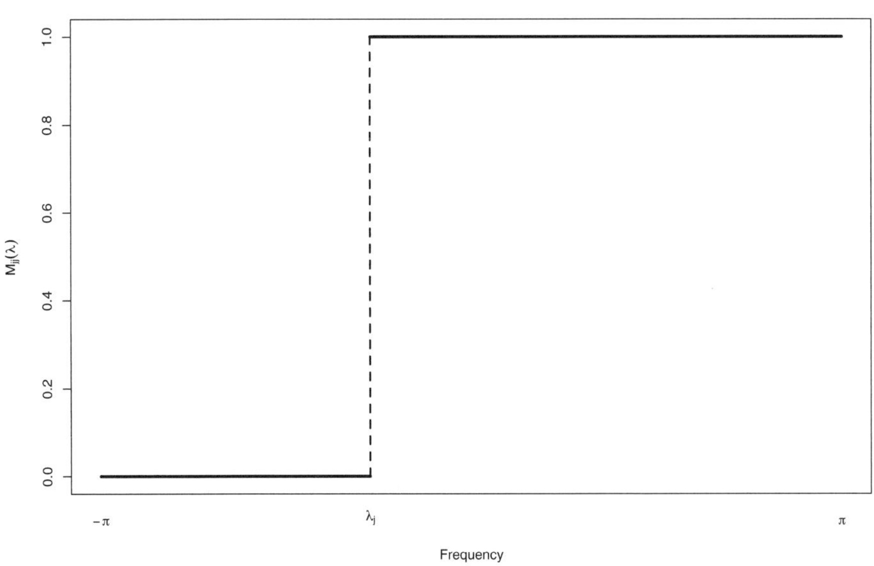

Figure 10.7 *Harmonic regression: $M_{jj}(\lambda)$.*

10.8 ILLUSTRATION: AIR POLLUTION DATA

As an illustration of the long-memory regression techniques discussed in this chapter consider the following air pollution data exhibited in Figure 10.8. This time series consists of 4014 daily observations of fine particulate matter with diameter less than 2.5 μm (PM2.5) measured in Santiago, Chile, during the period 1989–1999; see Section 10.9 for further details about these data.

In order to stabilize the variance of these data, a logarithmic transformation has been made. The resulting series is shown in Figure 10.9. This series displays a clear seasonal component and a possible downward linear trend. Consequently, the following model for the log-PM25 is proposed,

$$y_t = \beta_0 + \beta_1 t + \sum_{j=1}^{k} [a_j \sin(\omega_j t) + c_j \cos(\omega_j t)] + \varepsilon_t, \tag{10.19}$$

where $\varepsilon_t \sim (0, \sigma_\varepsilon^2)$.

An analysis of the periodogram of the detrended data and the ACF reveals the presence of three plausible seasonal frequencies, $\omega_1 = 2\pi/7$, $\omega_2 = 2\pi/183$, and $\omega_3 = 2\pi/365$.

The least squares fitting assuming uncorrelated errors is shown in Table 10.3. Observe that according to this table all the regression coefficients

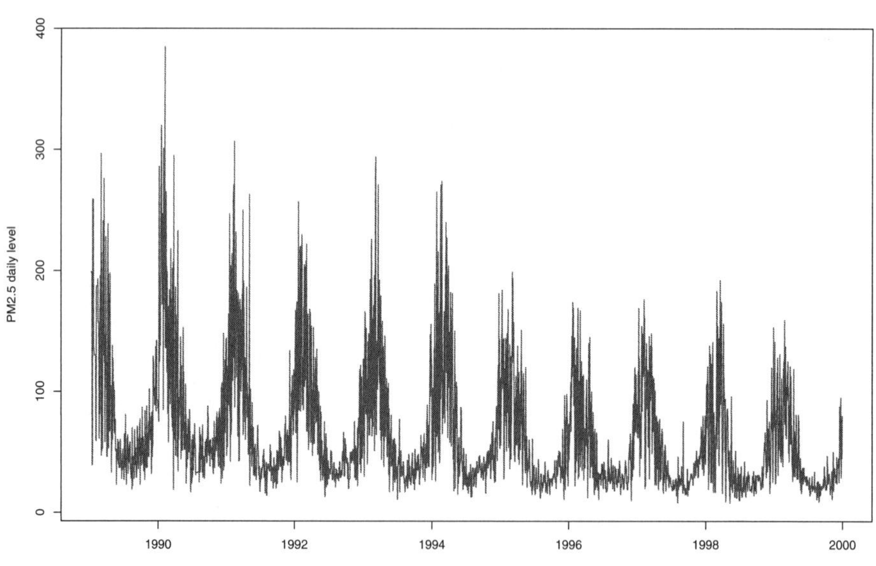

Figure 10.8 *Air pollution data: Daily PM2.5 measurements at Santiago, Chile, 1989 - 1999.*

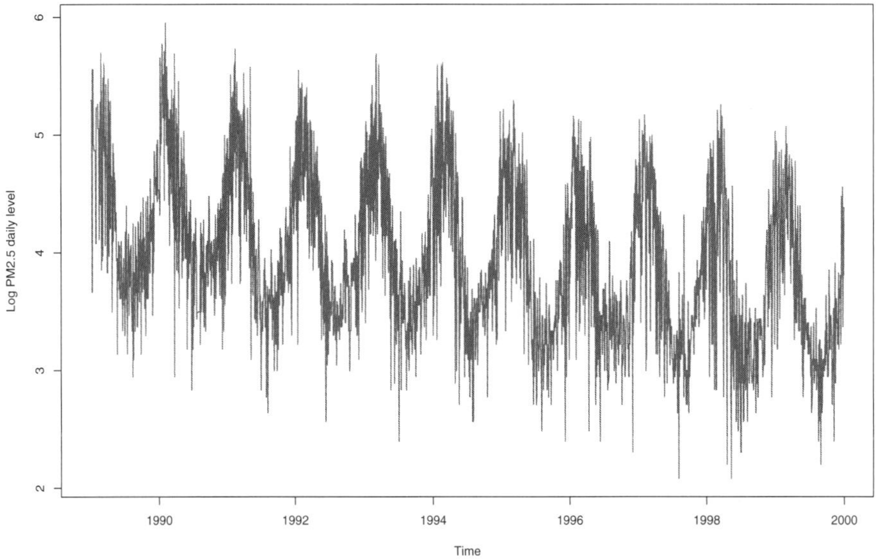

Figure 10.9 *Air pollution data: Log daily PM2.5 measurements at Santiago, Chile, 1989 - 1999.*

in model (10.19) are significant at the 5% level, excepting c_1. In particular, even though the LS estimate

of the linear trend coefficient β_1 is very small, $\widehat{\beta}_1 = -0.0002$, it is significant at the 5% level. The sample autocorrelation function of the residuals from the LSE fit is shown in Figure 10.10. As observed in this plot, the components of the autocorrelation function are significant even after 30 lags. Additionally, the variance plot (see Subsection 4.5.3) displayed in Figure 10.11 indicates the possible presence of long-range dependence in the data. As a result from these two plots, it seems that the disturbances ε_t in the linear regression model (10.19) may have long-memory correlation structure and the LSE fitting may not be adequate.

To account for the possible long-memory behavior of the errors, the following ARFIMA(p, d, q) model is proposed for the regression disturbances $\{\varepsilon_t\}$:

$$\phi(B)\varepsilon_t = \theta(B)(1 - B)^{-d}\eta_t,$$

where $\{\eta_t\}$ is a white noise sequence with variance σ^2. The model selected according to the Akaike's information criterion (AIC) is the ARFIMA$(0, d, 0)$, with $\widehat{d} = 0.4252$, $t_d = 34.37$, and $\widehat{\sigma}_\eta = 0.3557$. Table 10.4 shows the results from the least squares fit with ARFIMA errors. From this table, observe that the linear trend coefficient $\boldsymbol{\beta}_1$ is no longer significant at the 5% level. Similarly, the coefficients c_1 and c_2 are not significant at that level.

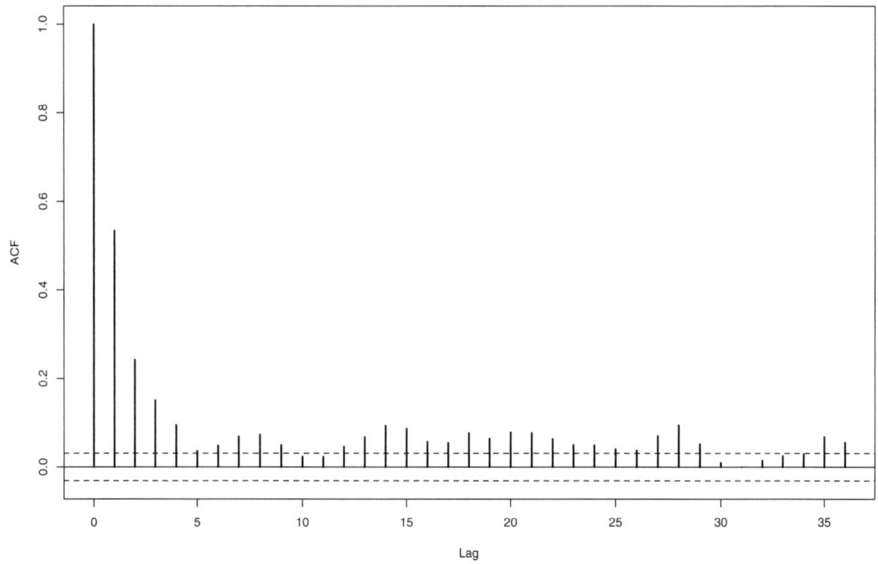

Figure 10.10 *Air pollution data: Sample autocorrelation function of the residuals from the least squares fit.*

Table 10.3 Air Pollution Data: Least Squares Fit

	Coefficient	Standard Deviation	t-stat	P-value
β_0	4.3148	0.0124	347.6882	0.0000
β_1	-0.0002	0.0000	-35.5443	0.0000
a_1	0.0775	0.0088	8.8556	0.0000
c_1	-0.0083	0.0088	-0.9446	0.3449
a_2	0.1007	0.0087	11.5086	0.0000
c_2	-0.0338	0.0088	-3.8590	0.0001
a_3	0.4974	0.0088	56.6914	0.0000
c_3	0.4479	0.0088	51.1620	0.0000

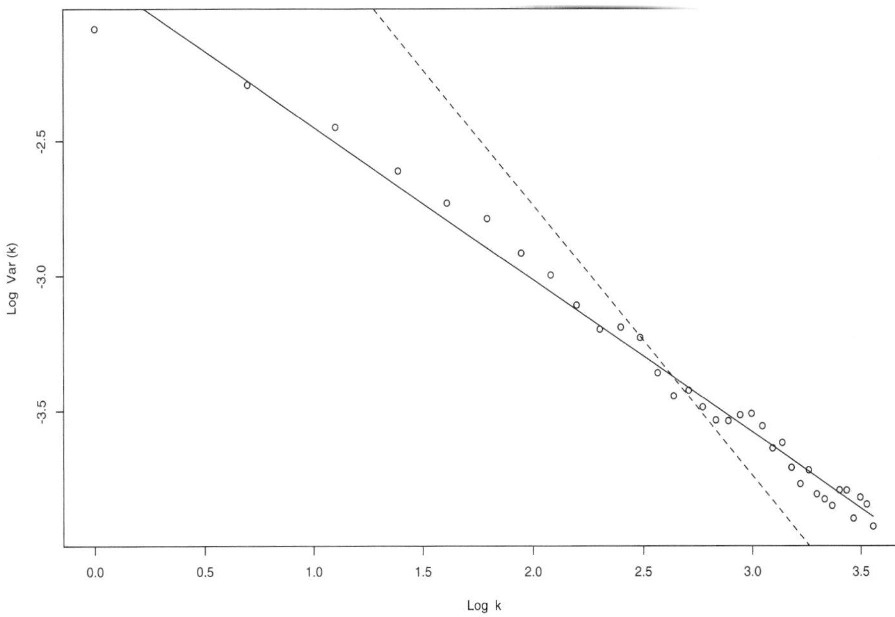

Figure 10.11 *Air pollution data: Variance plot of the residuals from the LSE fit.*

Table 10.4 Air Pollution Data: Least Squares Fit with Long-Memory Errors

	Coefficient	Standard Deviation	t-stat	P-value
β_0	4.3148	0.5969	7.2282	0.0000
β_1	-0.0002	0.0003	-0.7405	0.4590
a_1	0.0775	0.0084	9.1837	0.0000
c_1	-0.0083	0.0084	-0.9803	0.3270
a_2	0.1007	0.0363	2.7713	0.0056
c_2	-0.0338	0.0335	-1.0094	0.3129
a_3	0.4974	0.0539	9.2203	0.0000
c_3	0.4479	0.0447	10.0267	0.0000

10.9 BIBLIOGRAPHIC NOTES

Chapter 7 of Grenander and Rosenblatt (1957), Section VII.4 of Ibragimov and Rozanov (1978), and Chapter VIII of Hannan (1970) are excellent references for the analysis of regression with correlated errors. However, most of their results do not apply directly to strongly dependent processes. On the other hand, there is an extensive literature about regression data with long-memory disturbances; see, for example, Künsch (1986), Yajima (1988, 1991), Dahlhaus (1995), Sibbertsen (2001), and Choy and Taniguchi (2001), among others. Yajima (1991) established many asymptotic results for least squares error estimates and best linear unbiased estimates. As described in this chapter and according to Yajima, the convergence rates for the variance of these estimates as the sample size increases, depend on the structure of the characteristic function of the design matrix. The Grenander conditions were introduced by Grenander (1954). Many of the LSE and BLUE results discussed in this chapter concerning regression with long-memory errors are due to Yajima (1988, 1991). The air pollution data discussed in Section 10.8 are provided by the Ambient Air Quality Monitoring Network (MACAM in Spanish, www.sesma.cl) in Santiago, Chile. Further details about the fine particulate matter data and other ambient variables can be found in Iglesias, Jorquera, and Palma (2006).

Problems

10.1 Consider the first-order autoregressive model: $y_{t+1} = \phi y_t + \epsilon_{t+1}$ where ϵ_t is white noise with zero-mean and variance σ^2.

(a) Verify that $y_{t+k} = \phi^k y_t + \sum_{j=0}^{k} \phi^j \epsilon_{t+k-j}$

(b) Based on part a) verify that the best linear predictor of y_{t+k} given $\{y_t, y_{t-1}...\}$ is $\hat{y}_{t+k} = \phi^k y_t$

(c) Show that the prediction error $e_{t+k} = y_{t+k} - \hat{y}_{t+k}$ can be written as $e_{t+k} = \sum_{j=0}^{k} \phi^j \epsilon_{t+k-j}$

(d) Calculate $\text{Var}(e_{t+k})$.

10.2 Consider the following trend break time series model defined by

$$
\begin{aligned}
E(y_t) &= \begin{cases} \beta_0 & t = 1, ..., k, \\ \beta_0 + \beta_1 x_t & t = k+1, ..., n. \end{cases} \\
\text{Var}(y_t) &= \begin{cases} \sigma^2 & t = 1, ..., k, \\ \nu\sigma^2 & t = k+1, ..., n, \end{cases}
\end{aligned}
$$

where ν is a positive constant.

(a) Write this model as $y = x\,\beta + \varepsilon$, specifying the matrix x and $\text{Var}(\varepsilon) = V$.

(b) Verify that the estimator $\tilde{\beta} = (x'x)^{-1}x'y$ has greater variance than the estimator $\widehat{\beta} = (x'V^{-1}x)^{-1}x'V^{-1}y$.

10.3 Suppose that you have a simple regression model of the form $y_t = \beta t + \gamma x_t + \varepsilon_t$, where ε_t is a Gaussian white noise, t indicates time and x_t is a stationary time series. Show that the least squares estimator of γ for this model is exactly the same least squares estimator γ for model $y_t^* = \gamma x_t^* + \nu_t$, where y_t^* and x_t^* are the regression residuals of regression of y_t and x_t at time t.

10.4 Consider the process defined by the equation

$$y_t = \epsilon_t + \theta_1 \epsilon_{t-1} + \theta_2 \epsilon_{t-2},$$

where ε_t is a white noise sequence with zero-mean and variance σ^2.

(a) Calculate the ACF of y_t.

(b) Obtain the best linear predictors $\mathrm{P}_{[\{y_1\}]}y_2$, $\mathrm{P}_{[\{y_1,y_2\}]}y_3$

(c) Calculate the best linear predictors $\mathrm{P}_{[\{y_2\}]}y_3$ y $\mathrm{P}_{[\{y_2\}]}y_1$ and then obtain $\mathrm{corr}(e_3, e_1)$ where $e_3 = y_3 - \mathrm{P}_{[\{y_2\}]}y_3$ y $e_1 = y_1 - \mathrm{P}_{[\{y_2\}]}y_1$.

10.5 Consider the process $z_t = \beta + \epsilon_t$, where ϵ_t is and i.i.d. sequence with zero-mean and variance σ^2.

a) Find the least squares estimator (LSE) of β, $\hat{\beta}$.

b) Show that the k-step predictor, $\tilde{z}_{n+k} = E(z_{n+k}|z_1, ..., z : n)$, is $\tilde{z}_{n+h} = \beta$.

c) Given that β is an unknown parameter, you dicide to used the LSE to predict the value z_{n+k} by means of the formula $\hat{z}_{n+k} = \hat{\beta}$. Show that the mean squared error of this predictor is $E(\hat{z}_{n+k} - z_{n+k})^2 = \sigma^2(1 + \frac{1}{n})$.

10.6 Consider the time series y_t defined as

$$y_t = a + b\,t + w_t,$$

with $\{w_t\} \sim (0,\, \sigma^2)$. Show that applying a moving-average filter, the variance of the resulting process z_t defined as

$$z_t = \frac{1}{2\,k+1} \sum_{j=-k}^{k} y_{t-j}$$

is $1/(2\,k+1)$ parts of the variance of the series y_t.

10.7 Assume that x_j denotes the jth column of the $n \times p$ design matrix X_n, that is, $x_{tj} = t^{j-1}$ and let $D_n = \mathrm{diag}(\|x_1\|_n, \|x_2\|_n, \ldots, \|x_p\|_n)$ where $\|x_j\|_n = [\sum_{t=1}^{n} x_{tj}^2]^{1/2}$.

(a) Show that

$$\lim_{n\to\infty} D_n^{-1} X_n' X_n D_n^{-1} = M,$$

where $M = (m_{ij})_{i,j=1,\ldots,p}$ with $m_{ij} = \dfrac{\sqrt{2i-1}\sqrt{2j-1}}{i+j-1}$.

(b) Verify that

$$\lim_{n\to\infty} n^{-2d} D_n^{-1} X_n' \Gamma X_n D_n^{-1} = 2\pi f_0(0) H,$$

where

$$h_{ij} = \frac{\sqrt{2i-1}\sqrt{2j-1}\,\Gamma(1-2d)}{\Gamma(d)\Gamma(1-d)} \int_{-1}^{1}\int_{-1}^{1} x^{i-1} y^{j-1} |x-y|^{2d-1}\, dx\, dy.$$

10.8 Consider the harmonic regression

$$y_t = \alpha_1 \sin(\lambda_0 t) + \alpha_2 \cos(\lambda_0 t) + \varepsilon_t,$$

for $t = 1, \ldots, n$ where ε_t is a stationary long-memory process with spectral density (10.2).

(a) Show that y_t may be written as

$$y_t = \boldsymbol{\beta}_1 e^{i\lambda_0 t} + \boldsymbol{\beta}_2 e^{-i\lambda_0 t} + \varepsilon_t,$$

and find expressions for α_1 and α_2 in terms of $\boldsymbol{\beta}_1$ and $\boldsymbol{\beta}_2$.

(b) Verify that the LSE of α_1 and α_2 are

$$\widehat{\alpha}_1 = \frac{1}{2i}(i\widehat{\boldsymbol{\beta}}_2 + \widehat{\boldsymbol{\beta}}_1),$$

and

$$\widehat{\alpha}_2 = \frac{1}{2i}(i\widehat{\boldsymbol{\beta}}_2 - \widehat{\boldsymbol{\beta}}_1),$$

where $\widehat{\boldsymbol{\beta}}_1$ and $\widehat{\boldsymbol{\beta}}_2$ are the LSE of $\boldsymbol{\beta}_1$ and $\boldsymbol{\beta}_2$, respectively.

(c) Show that

$$\lim_{n\to\infty} n \operatorname{Var}(\widehat{\alpha}_1) = \lim_{n\to\infty} n \operatorname{Var}(\widehat{\alpha}_2) = \pi f(\lambda_0).$$

10.9 Consider the linear regression model

$$y_t = \boldsymbol{\beta} t^p e^{i\lambda_0 t} + \varepsilon_t,$$

where $p \in \{0, 1, 2, \ldots\}$ and ε_t is a stationary long-memory process with spectral density (10.2).

(a) Is the LSE of $\boldsymbol{\beta}$ consistent?

(b) Are the Grenander conditions satisfied in this case?

(c) Is the LSE asymptotically normal?

(d) Is the LSE asymptotically efficient in this case?

10.10 Let $h(\lambda) \geq 0$ and define the $n \times n$ matrix Γ with elements

$$\gamma_{ij} = \int_{-\pi}^{\pi} h(\lambda) e^{\imath(i-j)\lambda} d\lambda.$$

(a) Verify that Γ is symmetric, that is, $\gamma_{ij} = \overline{\gamma_{ji}}$.

(b) Show that Γ is positive semidefinite, that is, for any $x \in \mathbb{C}^n$

$$x^* \Gamma x \geq 0.$$

(c) Observe that since Γ is symmetric, it may be written as

$$\Gamma = UDU^*,$$

where U is a nonsingular $n \times n$ matrix and $D = \text{diag}(d_i)$ with $d_i \geq 0$ for $i = 1, \ldots, n$. Using this fact and the Cauchy-Schwartz inequality

$$|x^* y| \leq \|x\| \|y\|,$$

show that for all $u, v \in \mathbb{C}^n$

$$2|u^* \Gamma v| \leq u^* \Gamma u + v^* \Gamma v. \tag{10.20}$$

(d) Verify that

$$\int_{-\pi}^{\pi} 2\pi h(\lambda) m_{ij}^n(\lambda)\, d\lambda = \sum_{t=1}^{n} \sum_{s=1}^{n} \frac{\langle x_{ti}, x_{sj} \rangle}{\|x_n(i)\| \|x_n(j)\|} \gamma_{ij} = u^* \Gamma v$$

where $u = x_n(i)/\|x_n(i)\|$ and $v = x_n(j)/\|x_n(j)\|$.

(e) Using (10.20) show that

$$\left| \int_{-\pi}^{\pi} h(\lambda) m_{ij}^n(\lambda)\, d\lambda \right| \leq \int_{-\pi}^{\pi} h(\lambda) m_{ii}^n(\lambda)\, d\lambda + \int_{-\pi}^{\pi} h(\lambda) m_{jj}^n(\lambda)\, d\lambda.$$

(f) Deduce that for any $f(\lambda) \geq 0$,

$$\left| \int_A f(\lambda)\, dM_{ij}(\lambda) \right| \leq \int_A f(\lambda)\, dM_{ii}(\lambda) + \int_A f(\lambda)\, dM_{jj}(\lambda),$$

cf., Yajima (1991, p. 162).

10.11 Consider the following *trend break* regression

$$y_t = \boldsymbol{\beta}_0 + \boldsymbol{\beta}_1 x_t + \varepsilon_t,$$

for $t = 1, \ldots, n$, where the covariate x_t is defined by

$$x_t = \begin{cases} 0 & \text{if} \quad t \leq \dfrac{n}{2}, \\ 1 & \text{if} \quad \dfrac{n}{2} < t \leq n. \end{cases}$$

and for simplicity we assume that n is even. Let $X_n = [X_n(1), X_n(2)]$ the design matrix.

(a) Show that $\|X_n(1)\| = \sqrt{n}$ and $\|X_n(2)\| = \sqrt{n/2}$.

(b) Following expression (10.8) verify that

$$\frac{\langle X_{n,h}(1), X_n(2)\rangle}{\|X_{n,h}(1)\|\|X_n(2)\|} \to \frac{1}{\sqrt{2}},$$

as $n \to \infty$ and that

$$R(h) = \begin{bmatrix} 1 & \frac{1}{\sqrt{2}} \\ \frac{1}{\sqrt{2}} & 1 \end{bmatrix}.$$

(c) Are the Grenander conditions fulfilled in this case?

(d) Prove that the matrix $M(\lambda)$ may be written as

$$M(\lambda) = \delta_D(\lambda) \begin{bmatrix} 1 & \frac{1}{\sqrt{2}} \\ \frac{1}{\sqrt{2}} & 1 \end{bmatrix}.$$

(e) Let $\widehat{\boldsymbol{\beta}}_n$ be the LSE of $\boldsymbol{\beta} = (\boldsymbol{\beta}_1, \boldsymbol{\beta}_2)'$. Is this estimator consistent?

(f) Find an expression for the asymptotic variance of $\widehat{\boldsymbol{\beta}}_n$.

10.12 Show that

$$\int_0^1 \int_0^1 x^\alpha y^\beta |x-y|^\gamma \, dx \, dy = \frac{B(\alpha+1,\gamma+1) + B(\beta+1,\gamma+1)}{\alpha+\beta+\gamma+2},$$

for $\alpha > -1$, $\beta > -1$ and $\gamma > -1$, where $B(\cdot,\cdot)$ is the beta function. Hint: Try the change of variables $x = uv$ and $y = v$.

10.13 Consider the liner regression model

$$y_t = \boldsymbol{\beta} t^\alpha + \varepsilon_t,$$

for $t = 1, 2, \ldots, n$, where α is known and $\{\varepsilon_t\}$ is a long-memory stationary process with spectral density satisfying

$$f(\lambda) \sim c_f |\lambda|^{-2d},$$

as $|\lambda| \to 0$ with $0 < d < \frac{1}{2}$. Let $\widehat{\boldsymbol{\beta}}_n$ be the LSE of $\boldsymbol{\beta}$ and let $X_n = (1, 2^\alpha, \ldots, n^\alpha)'$.

(a) Verify that $\|X_n\|^2 \sim \dfrac{n^{2\alpha+1}}{2\alpha+1}$, for $\alpha > -\frac{1}{2}$.

(b) Show that if $\alpha > d - \frac{1}{2}$, then $\widehat{\boldsymbol{\beta}}_n$ is strongly consistent.

(c) Prove that

$$\frac{\langle X_{n,h}, X_n \rangle}{\|X_{n,h}\| \|X_n\|} \to R(h),$$

as $n \to \infty$, where $R(h) = 1$ for all $h \in \mathbb{Z}$.

(d) Are the Grenander conditions satisfied in this case?

(e) Show that the variance of $\widehat{\boldsymbol{\beta}}_n$ satisfies

$$\operatorname{Var}(\widehat{\boldsymbol{\beta}}_n) \sim \frac{4\pi c_f (2\alpha+1)^2}{\Gamma(d)\Gamma(1-d)} \frac{B(\alpha+1, 2d)}{2\alpha+2d+1} n^{2d-1-2\alpha},$$

as $n \to \infty$.

(f) Assume that the disturbances $\{\varepsilon_t\}$ are independent and identically distributed. Is the LSE $\widehat{\boldsymbol{\beta}}_n$ asymptotically normal?

10.14 Verify that under assumption (3), $M^n(\lambda)$ converges to $M(\lambda)$, that is,

$$\int_{-\pi}^{\pi} g(\lambda)\, dM^n(\lambda) \to \int_{-\pi}^{\pi} g(\lambda)\, dM(\lambda),$$

as $n \to \infty$ for any continuous function $g(\lambda)$ with $\lambda \in [-\pi, \pi]$.

10.15 Let the function $f_n(\cdot)$ be defined for $n \geq 1$ as

$$f_n(\lambda) = \frac{1}{2\pi n} \left| \sum_{t=1}^{n} e^{i\lambda t} \right|^2.$$

Show that the sequence of functions $\{f_n(\lambda)\}$ converges to the Dirac operator $\delta_D(\lambda)$ as $n \to \infty$. That is, for any continuous function $g(\lambda)$, $\lambda \in [-\pi, \pi]$ we have that

$$\int_{-\pi}^{\pi} f_n(\lambda) g(\lambda)\, d\lambda \to \int_{-\pi}^{\pi} g(\lambda)\delta_D(\lambda)\, d\lambda = g(0),$$

as $n \to \infty$.

10.16 Let the sequence of functions $\{f_n\}$ be defined by

$$f_n(\lambda) = \frac{1}{n} \sum_{t=1}^{n} e^{i\lambda t}.$$

Show that $f_n(\lambda) \to \delta(\lambda)$ as $n \to \infty$.

10.17 Consider the linear model (10.1) where the sequence of disturbances $\{\varepsilon_t\}$ is a stationary process with spectral density satisfying (10.2) and f_0 is a bounded function. If the following two conditions hold

(a) $n^{-2d}\sigma_n \to \infty$ as $n \to \infty$,

(b) $\sum_{n=p+2}^{\infty} \sigma_{n-1}^{-1} n^{2d-1} \log^2 n < \infty$,

then $\widehat{\boldsymbol{\beta}}_n \to \boldsymbol{\beta}$ almost surely as $n \to \infty$.

10.18 Consider the linear model (10.1) where the sequence of disturbances $\{\varepsilon_t\}$ is a stationary process with spectral density satisfying (10.2) and f_0 is a bounded function. Suppose that the Grenander conditions (3, with $h = 0$) and (4) hold. If

$$0 < \liminf_{n\to\infty} \frac{\|x_n(j)\|^2}{n^\delta},$$

for some $\delta > 2d$ and $j = 1, \ldots, p$, then $\widehat{\boldsymbol{\beta}}_n \to \boldsymbol{\beta}$ almost surely as $n \to \infty$.

CHAPTER 11

MISSING VALUES AND OUTLIERS

This chapter examines the effects of two relevant data problems of the statistical time series analysis. Many real-life time series data are incomplete or may exhibit observations with a magnitude that seems to be very different from the others. Both of these situations may produce important consequences in the statistical analysis, model fitting and prediction.

Regarding the first problem, notice that most of the methodologies studied so far assume complete data. However, in many practical situations only part of the data may be available. As a result, in these cases we are forced to fit models and make statistical inferences based on partially observed time series. In this chapter we explore the effects of data gaps on the analysis of long-memory processes and describe some methodologies to deal with this situation. As in many other areas of statistics, there are several ways to deal with incomplete time series data. A fairly usual approach, especially when there are only a small number of missing values, is to replace them by zero or the sample mean of the series. Another approach is to use some imputation technique such as cubic splines and other interpolation methods, including, for example, the repetition of previous values. In this chapter we discuss some of these techniques. Special attention is focused on the method of integrating out

Time Series Analysis. First Edition. Wilfredo Palma.

the missing values from the likelihood function. As discussed in Section 11.1, this approach does not introduce artificial correlation in the data and avoids some dramatic effects on parameter estimates produced by some imputation techniques. The definition of an appropriate likelihood function to account for unobserved values is studied in Section 11.2. In particular, it is shown that if the distribution of the location of the missing data is independent of the distribution of the time series, then the statistical inferences can be carried out by ignoring the missing observations. This section also discusses the computation of the likelihood function for a general class of time series models and modifications to the Kalman filter equations to account for missing values. Sections 11.3 and 11.4 are dedicated to investigate the effects of data gaps on estimates and predictions, respectively, establishing some general theoretical results. A number of illustrations of these results are discussed in Section 11.2. Estimation of missing values via interpolation is discussed in Section 11.5.

The second problem can be treated by means of specific techniques. For instance, outliers and intervention analysis are also discussed in this chapter. Section 11.7 describes some of the methodologies for modeling outliers. Additionally, it discusses the problem of interventions in the time series in the context of both known and unknown location. Suggestions for further reading on this topic are given in Section 11.8. This chapter concludes with a list of proposed problems.

11.1 INTRODUCTION

Missing data is an important problem for the analysis of time series. In particular, the parameter estimates may suffer serious distortions if the missing values are not appropriately treated. This point is illustrated by Table 11.1 which displays the maximum-likelihood estimates of the parameter ϕ of an AR(1) process, along with their standard deviations. The results are based on 1,000 simulations with $\phi = 0.8$ with sample size of 1,000 observations, for different percentages of missing values. Three maximum-likelihood estimates are considered, one based on the full sample (*full*), one based on the series with imputed missing values using a spline interpolation method (*imputed*), and one based only on the available data (*NAs*). Observe from Table 11.1 that the imputation method changes dramatically the estimated parameter ϕ. Besides, the standard deviations are greatly increased when compared to the MLE based on the full sample. On the other hand, when the missing values are adequately accounted for (MLE with NAs), the estimates are quite close to the true value of the first-order autoregressive parameter and the standard deviations are only marginally increased when compared to the full sample MLE. A similar situation is observed regarding the estimation of σ^2 which is reported in Table 11.2.

As discussed at the beginning of this chapter, there are many ways to deal with missing data in the context of time series. For example, the missing

Table 11.1 Maximum-likelihood estimates of ϕ for simulated AR(1) processes, based on 500 observations.[a]

%		$\widehat{\phi}$			SD	
Missing	Full	Imputed	NAs	Full	Imputed	NAs
10	0.7934	0.7119	0.7933	0.0275	0.0326	0.0280
20	0.7943	0.6309	0.7956	0.0266	0.0364	0.0312
30	0.7930	0.5519	0.8552	0.0298	0.0385	0.0310

[a] The three estimates correspond to the full, imputed, and incomplete data, respectively.

Table 11.2 Maximum-likelihood estimates of σ^2 for simulated AR(1) processes, based on 500 observations.[a]

%		$\widehat{\sigma^2}$			SD	
Missing	Full	Imputed	NAs	Full	Imputed	NAs
10	0.9966	1.2013	0.9970	0.0628	0.1100	0.0659
20	0.9929	1.3065	0.9943	0.0609	0.1365	0.0714
30	0.9967	1.3305	1.0136	0.0686	0.1831	0.0868

[a] The three estimates correspond to the full, imputed, and incomplete data, respectively.

values can be *imputed*, the data can be *repeated* as in the Nile River series, *replaced* by zeroes or other values, *ignored* or *integrated out*, among many other techniques. In the next section, we study some techniques for obtaining a likelihood function based only on the available data.

11.2 LIKELIHOOD FUNCTION WITH MISSING VALUES

An important issue when defining the likelihood function with missing data is the distribution of the *location* of the missing values. For example, if there is *censoring* in the collection of the data and any value above or below a given threshold is not reported, then the location of those missing observations will depend on the distribution of the time series. In other situations, the location may not depend on the values of the stochastic process but it may depend on time. For example, daily economical data may miss Saturday and Sunday observations, producing a systematic pattern of missing data. If the distribution of the location of missing data does not depend on the distribution of the time series—*observed* and *unobserved*—values, then maximum-likelihood estimations and statistical inferences can be carried out by ignoring the exact location of the missing values as described below.

11.2.1 Integration

Let the *full data* be the triple $(y_{\mathrm{obs}}, y_{\mathrm{mis}}, \eta)$, where y_{obs} denotes the observed values, y_{mis} represents the unobserved values, and η is the location of the missing values. In this analysis, η may be regarded as a random variable. The observed data is then given by the pair (y_{obs}, η), that is, the available information consists of the observed values and the pattern of the missing observations. The distribution of the observed data may be obtained by integrating the missing data out of the joint distribution of $(y_{\mathrm{obs}}, y_{\mathrm{mis}}, \eta)$:

$$\begin{aligned} f_\theta(y_{\mathrm{obs}}, \eta) &= \int f_\theta(y_{\mathrm{obs}}, y_{\mathrm{mis}}, \eta)\, dy_{\mathrm{mis}} \\ &= \int f_\theta(\eta | y_{\mathrm{obs}}, y_{\mathrm{mis}}) f_\theta(y_{\mathrm{obs}}, y_{\mathrm{mis}})\, dy_{\mathrm{mis}}. \end{aligned}$$

If the distribution of η does not depend on $(y_{\mathrm{obs}}, y_{\mathrm{mis}})$, then

$$\begin{aligned} f_\theta(y_{\mathrm{obs}}, \eta) &= \int f_\theta(\eta) f_\theta(y_{\mathrm{obs}}, y_{\mathrm{mis}})\, dy_{\mathrm{mis}} \\ &= f_\theta(\eta) \int f_\theta(y_{\mathrm{obs}}, y_{\mathrm{mis}})\, dy_{\mathrm{mis}}. \end{aligned}$$

Furthermore, if the location η does not depend upon the parameter θ, then

$$f_\theta(y_{\mathrm{obs}}, \eta) = f(\eta) \int f_\theta(y_{\mathrm{obs}}, y_{\mathrm{mis}})\, dy_{\mathrm{mis}}.$$

In this case, the likelihood function for θ satisfies

$$\mathcal{L}(\theta | y_{\mathrm{obs}}, \eta) \quad \propto \quad \int f_\theta(y_{\mathrm{obs}}, y_{\mathrm{mis}})\, dy_{\mathrm{mis}} = f_\theta(y_{\mathrm{obs}}) = \mathcal{L}(\theta | y_{\mathrm{obs}}).$$

Thus, maximum-likelihood estimates do not depend on the location of the missing data. Observe that this result is still valid if η depends upon time.

11.2.2 Maximization

An alternative way of dealing with unobserved data is through the maximization of the likelihood function $\mathcal{L}(\theta | y_{\mathrm{obs}}, y_{\mathrm{mis}})$ with respect to y_{mis}. Under this approach, let the function $\widetilde{\mathcal{L}}$ be defined by

$$\widetilde{\mathcal{L}}(\theta | y_{\mathrm{obs}}) = \max_{y_{\mathrm{mis}}} \mathcal{L}(\theta | y_{\mathrm{obs}}, y_{\mathrm{mis}}).$$

Consider, for instance, the class of random variables $y = (y_1, \ldots, y_n)' \in \mathbb{R}^n$, with zero-mean and joint density given by

$$f_\theta(y) = |\Sigma|^{-1/2} h(y' \Sigma^{-1} y), \tag{11.1}$$

where h is a positive real function and Σ is a symmetric definite positive matrix. Several well-known distributions are described by (11.1), including the multivariate normal, mixture of normal distributions and multivariate t, among others.

For this family of distributions we find that

$$\begin{aligned}\mathcal{L}(\theta|y_{\text{obs}}, y_{\text{mis}}) &= |\Sigma|^{-1/2} h[(y_{\text{obs}}, y_{\text{mis}})'\Sigma^{-1}(y_{\text{obs}}, y_{\text{mis}})] \\ &= |A|^{1/2} h[(y_{\text{obs}}, y_{\text{mis}})' A (y_{\text{obs}}, y_{\text{mis}})],\end{aligned}$$

where

$$A = \begin{bmatrix} A_{11} & A_{12} \\ A_{21} & A_{22} \end{bmatrix} = \Sigma^{-1} = \begin{bmatrix} \Sigma_{11} & \Sigma_{12} \\ \Sigma_{21} & \Sigma_{22} \end{bmatrix}^{-1},$$

with $\Sigma_{11} = \text{Var}[y_{\text{obs}}]$, $\Sigma_{22} = \text{Var}[y_{\text{mis}}]$, $\Sigma_{12} = \text{Cov}[y_{\text{obs}}, y_{\text{mis}}] = \Sigma'_{21}$.

Thus,

$$\frac{\partial \mathcal{L}(\theta|y_{\text{obs}}, y_{\text{mis}})}{\partial y_{\text{mis}}} = 2|A|^{1/2} h'(s)[y'_{\text{obs}} A_{12} + y'_{\text{mis}} A_{22}] = 0,$$

where $s = (y_{\text{obs}}, y_{\text{mis}}) A (y_{\text{obs}}, y_{\text{mis}})' \geq 0$. Consequently,

$$\widehat{y}_{\text{mis}} = -A_{22}^{-1} A_{21} y_{\text{obs}} = \Sigma_{21}\Sigma_{11}^{-1} y_{\text{obs}},$$

is a critical point of $\mathcal{L}(\theta|y_{\text{obs}}, y_{\text{mis}})$ as a function of y_{mis}. Furthermore, the Hessian is

$$H(\widehat{y}_{\text{mis}}) = 2|A|^{1/2} h'(s) A_{22},$$

which is negative definite if $h'(s) < 0$. This occurs for several distributions. For example, in the Gaussian case, $h(s) = \exp(-s^2/2)$, and hence its derivative satisfies

$$h'(s) = -se^{-s^2/2} < 0,$$

for any $s > 0$. Therefore, in this situation $\widehat{y}_{\text{mis}}$ is indeed a maximum and

$$\begin{aligned}\widetilde{\mathcal{L}}(\theta|y_{\text{obs}}) &= |\Sigma|^{-1/2} h(y'_{\text{obs}}\Sigma_{11}^{-1} y_{\text{obs}}) \\ &= |\Sigma_{22} - \Sigma_{21}\Sigma_{11}^{-1}\Sigma_{12}|^{-1/2} |\Sigma|_{11}^{-1/2} h(y'_{\text{obs}}\Sigma_{11}^{-1} y_{\text{obs}}) \\ &= |\Sigma_{22} - \Sigma_{21}\Sigma_{11}^{-1}\Sigma_{12}|^{-1/2} \mathcal{L}(\theta|y_{\text{obs}}).\end{aligned}$$

Hence, by defining $g(\theta) = |\Sigma_{22} - \Sigma_{21}\Sigma_{11}^{-1}\Sigma_{12}|^{-1/2}$, we conclude that

$$\widetilde{\mathcal{L}}(\theta|y_{\text{obs}}) = g(\theta)\mathcal{L}(\theta|y_{\text{obs}}).$$

Thus, the maximization of $\widetilde{\mathcal{L}}(\theta|y_{\text{obs}})$ with respect to θ may give a different result than the maximization of $\mathcal{L}(\theta|y_{\text{obs}})$.

11.2.3 Calculation of the Likelihood Function

Naturally, the actual calculation of the likelihood function depends on the joint distribution of the process. For example, for the class of distributions described by (11.1), the log-likelihood function is given by

$$\mathcal{L}(\theta) = -\tfrac{1}{2}\log\det\Sigma + \log h(y'\Sigma^{-1}y). \tag{11.2}$$

The matrix Σ can be diagonalized as in Chapter 5. That is, since Σ is assumed to be definite positive and symmetric, there is a lower triangular matrix L with ones in the diagonal and a diagonal matrix $D = \text{diag}\{d_1, \ldots, d_n\}$ such that

$$\Sigma = L'DL.$$

Define $e = Ly$. Hence, $y'\Sigma^{-1}y = e'De$ and $\det\Sigma = \det D$. Consequently,

$$\mathcal{L}(\theta) = -\frac{1}{2}\sum_{t=1}^{n}\log d_t + \log h\left(\sum_{t=1}^{n}\frac{e_t^2}{d_t}\right).$$

As discussed in the next section, in the presence of data gaps the likelihood function of this class of distributions may be calculated from the output of the Kalman filter equations.

11.2.4 Kalman Filter with Missing Observations

Here we analyze with greater detail the state space approach for dealing with missing observations that was outlined in Subsection 3.4.4. Consider the state space system

$$\left\{\begin{aligned} x_{t+1} &= Fx_t + H\varepsilon_t, \\ y_t &= Gx_t + \varepsilon_t, \end{aligned}\right.$$

where x_t is the state, y_t is the observation, F and G are system matrices, and the observation noise variance is $\text{Var}(\varepsilon_t) = \sigma^2$.

The following theorem summarizes the modifications to the Kalman filter equations in order to account for the missing observations.

Let $\widehat{x}_t$ be the projection of the state x_t onto $\mathcal{P}_{t-1} = \{\text{observed } y_s : 1 \leq s \leq t-1\}$ and let $\Omega_t = E[(x_t - \widehat{x}_t)(x_t - \widehat{x}_t)']$ be the state error variance, with $\widehat{x}_1 = 0$ and $\Omega_1 = E[x_1 x_1']$. Then, the state predictor $\widehat{x}_t$ is given by the following recursive equations for $t \geq 1$:

$$\begin{aligned}
\Delta_t &= G\Omega_t G' + \sigma^2, \\
K_t &= (F\Omega_t G' + \sigma^2 H)\Delta_t^{-1}, \\
\Omega_{t+1} &= \begin{cases} F\Omega_t F' + \sigma^2 HH' - \Delta_t K_t K_t', & y_t \text{ observed}, \\ F\Omega_t F' + \sigma^2 HH', & y_t \text{ missing}, \end{cases} \\
\nu_t &= \begin{cases} y_t - G\widehat{x}_t, & y_t \text{ observed}, \\ 0, & y_t \text{ missing}, \end{cases} \\
\widehat{x}_{t+1} &= F\widehat{x}_t + K_t \nu_t, \\
\widehat{y}_{t+1} &= G\widehat{x}_{t+1}.
\end{aligned}$$

Remark 11.1. Notice that the equation linking the state and the observation estimation (B.51) is the same, with or without missing data.

Based on the modified Kalman equations, the log-likelihood function of a stochastic process with missing data and probability distribution (11.1) is given by

$$\mathcal{L}(\theta) = -\frac{1}{2}\sum \log \Delta_t + \log h\left(\sum \frac{\nu_t^2}{\Delta_t}\right),$$

where the sum is over all the observed values of y_t.

■ **EXAMPLE 11.1**

Let y_t be a stationary process with autocorrelations ρ_k and variance σ_y^2 and suppose that we have observed the values y_1, y_2, y_4, ..., but the value y_3 is missing. By applying the modified Kalman equations described above we conclude that $\Delta_1 = \sigma_y^2$, $\Delta_2 = \sigma_y^2(1-\rho_1^2)$, $\Delta_3 = \sigma_y^2(1-\rho_2^2)$, $\Delta_4 = \sigma_y^2(1-\rho_1^2-\rho_2^2-\rho_3^2-2\rho_1\rho_2\rho_3)/(1-\rho_2^2)$, ... Thus, the magnitude of the jump in the prediction error variance from Δ_2 to Δ_3 is $\rho_2^2 - \rho_1^2$.

11.3 EFFECTS OF MISSING VALUES ON ML ESTIMATES

In Section 11.1 we discussed some of the effects of imputation on ML estimates. To further illustrate the effects of repeating values to fill data gaps consider the following example involving the Nile River time series, which is exhibited in Figure 11.1.

Figure 11.1(a) displays the original data. From this plot, it seems that during several periods the data was repeated year after year in order to complete the series. Figure 11.1(b) shows the same series, but without those repeated values (filtered series).

Table 11.3 shows the fitted parameters of an ARFIMA$(0,d,0)$ using an AR(40) approximation along the modified Kalman filter equations of Subsection 11.2.4 for both the original and the filtered data.

Table 11.3 Nile River Data Estimates

Series	$\widehat{d}$	$t_{\widehat{d}}$	$\widehat{\sigma}$
Original data	0.499	25.752	0.651
Filtered data	0.434	19.415	0.718

Observe that for the original data, the estimate of the long-memory parameter is 0.5, indicating that the model has reached the nonstationary boundary. On the other hand, for the filtered data, the estimate of d is inside the stationary region.

Thus, in this particular case, the presence of data irregularities such as the replacement of missing data with repeated values induces nonstationarity. On the other hand, when the missing data is appropriately taken care of, the resulting model is stationary.

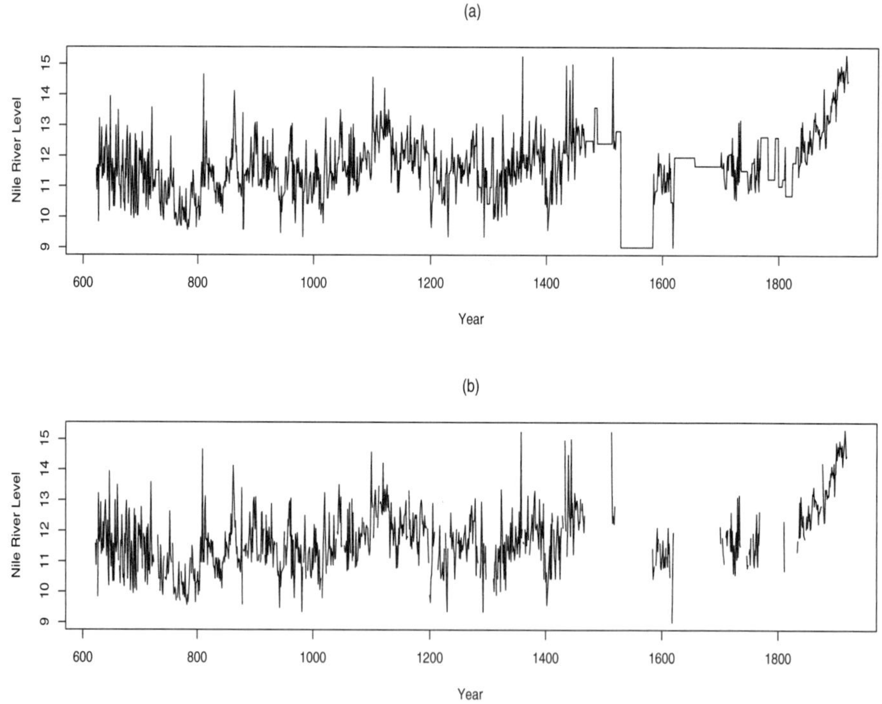

Figure 11.1 *Nile river data (A.D. 622 to A.D. 1921). (a) Original data and (b) filtered data.*

11.3.1 Monte Carlo Experiments

Table 11.4 displays the results from Monte Carlo simulations of Kalman maximum-likelihood estimates for a fractional noise ARFIMA$(0, d, 0)$ with missing values at random.

The sample size is $n = 400$ and the Kalman moving-average (MA) truncation uses $m = 40$. The long-memory parameters are $d = 0.1, 0.2, 0.3, 0.4$, and $\sigma = 1$. The number of missing values is 80 (20% of the sample) which have been selected randomly for each sample. The sample mean and standard deviations are based on 1000 repetitions.

From Table 11.4, notice that the bias is relatively low for all the values of the long-memory parameter. On the other hand, the sample standard deviation of the estimates seems to be close to the expected values. The expected standard deviation is $\sqrt{6/400\pi^2} = 0.0390$ for the full sample and $\sqrt{6/320\pi^2} = 0.04359$ for the incomplete data case.

Table 11.5 shows the results from Monte Carlo simulations of Kalman maximum-likelihood estimates for ARFIMA$(1, d, 1)$ processes with missing values at random. The sample size of the full data is $n = 400$ and the parameters are $d = 0.3$, $\phi = -0.5$, $\theta = 0.2$, and $\sigma^2 = 1$. The incomplete data have 80 missing observations.

From Table 11.5, the estimates of the long-memory parameter d seem to be slightly downward biased for both full and incomplete data.

On the other hand, the sample standard deviations are slightly higher than expected, for the estimates of d. The theoretical standard deviations reported in this table are based on formula (5.28), with $n = 400$ for the full data sample and $n = 320$ for the incomplete data case.

11.4 EFFECTS OF MISSING VALUES ON PREDICTION

In this section we turn our attention to the evolution of the one-step mean-squared prediction error, $E[y_t - \widehat{y}_t]^2$, for ARFIMA models, during and after a

Table 11.4 Maximum-Likelihood Estimates for Simulated FN(d) Processes Based on 1000 Observations for Different Values of the Long-Memory Parameter d

	Full Data		Incomplete Data	
d	Mean	SD	Mean	SD
0.1	0.0913	0.0387	0.0898	0.0446
0.2	0.1880	0.0396	0.1866	0.0456
0.3	0.2840	0.0389	0.2862	0.0484
0.4	0.3862	0.0388	0.3913	0.0508

Table 11.5 Finite Sample Behavior of Maximum-Likelihood Estimates with Missing Values.[a]

	Full Data			Incomplete Data		
	d	ϕ	θ	d	ϕ	θ
Mean	0.2559	-0.5102	0.1539	0.2561	-0.5086	0.1639
SD	0.0873	0.0729	0.1394	0.1070	0.0862	0.1795
Theoretical SD	0.0790	0.0721	0.1305	0.0883	0.0806	0.1459

[a] ARFIMA$(1, d, 1)$ model with sample size $n = 400$ and parameters $d = 0.3$, $\phi = -0.5$, and $\theta = 0.2$. The incomplete data have 80 missing observations selected randomly.

block of missing data. For simplicity, the analysis will be conducted by taking into account the full past instead of the finite past of the time series.

EXAMPLE 11.2

In what follows we illustrate the effects of missing values on the error variance of *in-sample* and *out-of-sample* predictors.

Figure 11.2 to Figure 11.4 show the evolution of the one-step prediction error variance for a fractional noise process with $d = 0.45$ and $\sigma = 1$, from $t = 1$ to $t = 200$.

In Figure 11.2, the heavy line represents the evolution for the full sample (no missing values) and the broken line indicates the evolution in the presence of a data gap from $t = 20$ to $t = 50$. It is clear that the prediction error variance increases from the beginning of the data gap up to $t = 50$ and then decays rapidly to 1.

The effect of missing values on out-of-sample forecasts can be analyzed with the help of Figure 11.3 and Figure 11.4. Unlike Figure 11.2 where the data from $t = 101$ to $t = 200$ were available, Figure 11.3 and Figure 11.4 depict the evolution of the prediction error variance in the case where the last observation is made a time $t = 100$ and no new data are available from $t = 101$ to $t = 200$. Naturally, this situation is similar to observing values from $t = 1$ to $t = 100$ and then predicting the future values from $t = 101$ to $t = 200$.

Observe in Figure 11.3 that the prediction error variance after $t = 100$ is not greatly affected by the presence of the data gap. On the other hand, a data gap of the same length but located closer to the end of the series may have a slightly greater effect on the variance of the forecasts, as suggested by Figure 11.4.

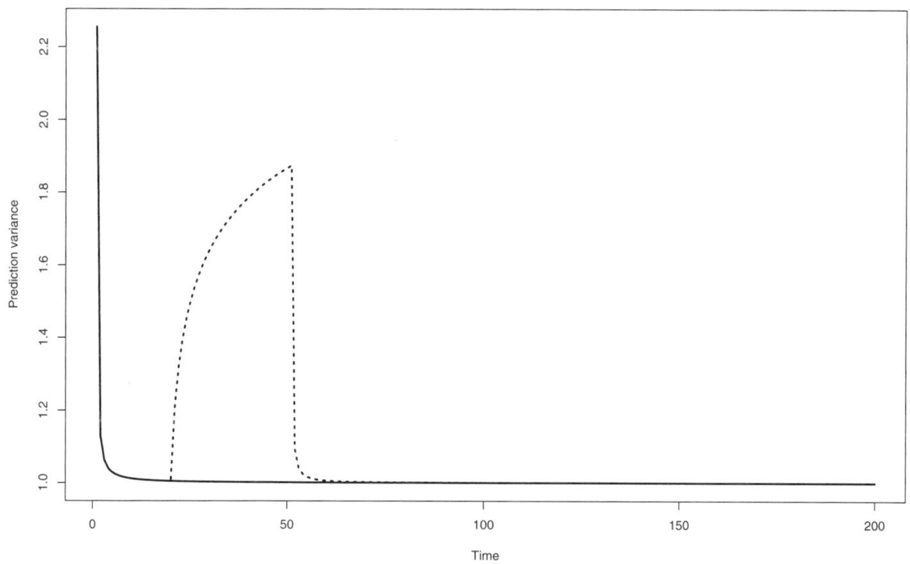

Figure 11.2 *Prediction error variance for a fractional noise process with* $d = 0.45$, $\sigma^2 = 1$, *and 200 observations. Heavy line: full sample. Broken line: Data gap from* $t = 20$ *to* $t = 50$.

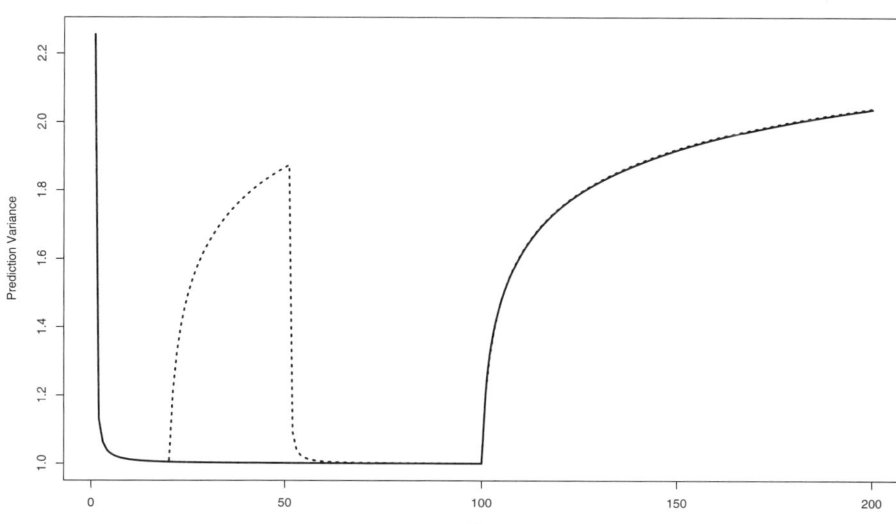

Figure 11.3 *Prediction error variance for a fractional noise process with* $d = 0.45$, $\sigma^2 = 1$, *and 200 observations. Heavy line: Data gap from* $t = 100$ *to* $t = 200$. *Broken line: Data gaps from* $t = 20$ *to* $t = 50$ *and from* $t = 100$ *to* $t = 200$.

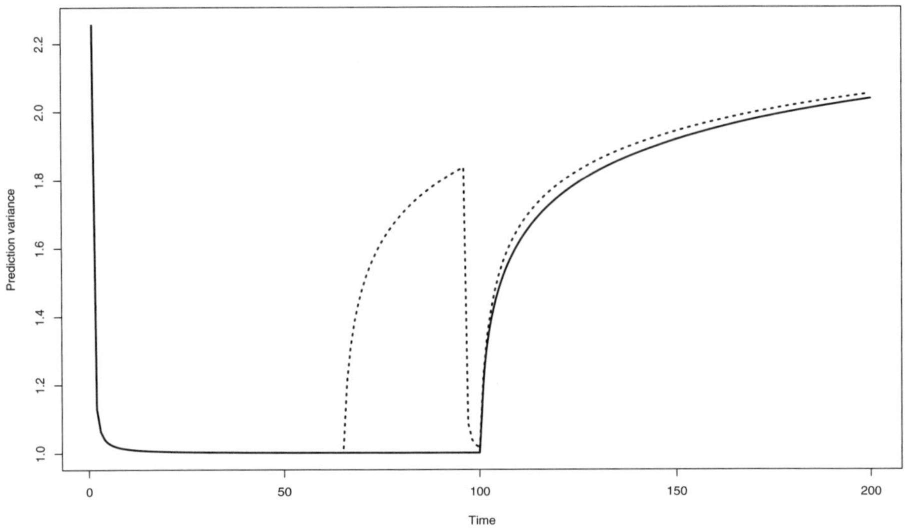

Figure 11.4 *Prediction error variance for a fractional noise process with $d = 0.45$, $\sigma^2 = 1$, and 200 observations. Heavy line: Data gap from $t = 100$ to $t = 200$. Broken line: Data gaps from $t = 65$ to $t = 95$ and from $t = 100$ to $t = 200$.*

11.5 INTERPOLATION OF MISSING DATA

We now focus our attention on the problem of finding estimates for the missing values of a long-memory linear process. To this end, let $\{y_t : t \in \mathbb{Z}\}$ be a stationary process with spectral density $f(\lambda)$ and autoregressive representation

$$\sum_{j=0}^{\infty} \pi_j y_{t-j} = \varepsilon_t, \tag{11.3}$$

where $\pi_0 = 1$ and $\{\varepsilon_t\}$ is a white noise sequence with variance σ^2.

Let $\mathcal{M} = \{y_{\text{obs}}\}$ be the vector space generated by the observed series and let y_{mis} be a missing value. According to the projection theorem of Chapter 1, the best linear interpolator of y_{mis} based on the observed data y_{obs} is given by

$$\widetilde{y}_{\text{mis}} = E(y_{\text{mis}}|\mathcal{M}).$$

In particular, if $y_{\text{mis}} = y_0$ and the full past and full future is available, that is, $y_{\text{obs}} = \{y_t, t \neq 0\}$, then the interpolator may be expressed as

$$\widetilde{y}_{\text{mis}} = -\sum_{j=1}^{\infty} \alpha_j (y_j + y_{-j}), \tag{11.4}$$

where the coefficients α_j are given by

$$\alpha_j = \frac{\sum_{i=0}^{\infty} \pi_i \pi_{i+j}}{\sum_{i=0}^{\infty} \pi_i^2},$$

for $j \geq 1$. In this case, the interpolation error variance is

$$\sigma_{\text{int}}^2 = 4\pi^2 \left[\int_{-\pi}^{\pi} \frac{d\lambda}{f(\lambda)} \right]^{-1}. \tag{11.5}$$

On the other hand, if $y_{\text{mis}} = y_0$ and the full past and part of the future is available, that is, $y_{\text{obs}} = \{y_t, t \neq 0, t \leq n\}$, then the best linear interpolator of y_0 may be written as

$$\widetilde{y}_{\text{mis}} = \widehat{y}_0 - \sum_{j=1}^{n} \beta_{j,n}(y_j - \widehat{y}_j),$$

where the coefficients $\beta_{j,n}$ are given by

$$\beta_{j,n} = \frac{\sum_{i=0}^{n-j} \pi_i \pi_{i+j}}{\sum_{i=0}^{n} \pi_i^2},$$

for $j = 1, 2, \ldots, n$ and $\widehat{y}_j$ is the best linear predictor of y_j based on $\{y_t, t < 0\}$. In this case, the interpolation error variance is

$$\sigma_{\text{int}}^2(n) = \frac{\sigma^2}{\sum_{i=0}^{n} \pi_i^2}. \tag{11.6}$$

Notice that

$$\sigma_{\text{int}}^2(n) \to \frac{\sigma^2}{\sum_{i=0}^{\infty} \pi_i^2},$$

as $n \to \infty$. But, recalling that the spectral density of process (11.3) is given by

$$f(\lambda) = \frac{\sigma^2}{2\pi} \left| \sum_{j=0}^{\infty} \pi_j e^{\imath\lambda j} \right|^{-2},$$

we conclude that

$$\begin{aligned} \int_{-\pi}^{\pi} \frac{d\lambda}{f(\lambda)} &= \frac{2\pi}{\sigma^2} \int_{-\pi}^{\pi} \left| \sum_{j=0}^{\infty} \pi_j e^{\imath\lambda j} \right|^2 d\lambda, \\ &= \frac{2\pi}{\sigma^2} \sum_{i=0}^{\infty} \sum_{j=0}^{\infty} \pi_i \pi_j \int_{-\pi}^{\pi} e^{\imath\lambda(i-j)} d\lambda \\ &= \frac{(2\pi)^2}{\sigma^2} \sum_{i=0}^{\infty} \pi_i^2. \end{aligned}$$

Therefore,

$$4\pi^2 \left[\int_{-\pi}^{\pi} \frac{d\lambda}{f(\lambda)} \right]^{-1} = \frac{\sigma^2}{\sum_{i=0}^{\infty} \pi_i^2}.$$

Hence, as expected,

$$\sigma_{\text{int}}^2(n) \to \sigma_{\text{int}}^2,$$

as $n \to \infty$.

Finally, if y_{obs} is a finite observed trajectory of a stationary time series, then we may use the projection theorem discussed in Chapter 1 directly to calculate the best linear interpolator of y_{mis} by means of the formula:

$$\widetilde{y}_{\text{mis}} = E(y_{\text{mis}} y_{\text{obs}}')[E(y_{\text{obs}} y_{\text{obs}}')]^{-1} y_{\text{obs}}. \tag{11.7}$$

Naturally, the calculation of this general expression implies obtaining the inverse of the variance-covariance matrix of the observed data and this procedure could be computationally demanding in the long-memory case.

■ EXAMPLE 11.3

Consider the fractional noise process FN(d). For this model we have that

$$\alpha_j = \frac{\Gamma(j-d)\Gamma(1+d)}{\Gamma(j+d+1)\Gamma(-d)},$$

for $j = 1, 2, \ldots$ Therefore, the best linear interpolator of y_0 based on the full past and full future is given by

$$\widetilde{y}_0 = \frac{d\Gamma(1+d)}{\Gamma(1-d)} \sum_{j=1}^{\infty} \frac{\Gamma(j-d)}{\Gamma(j+d+1)} (y_j + y_{-j}),$$

and its interpolation error variance equals

$$\sigma_{\text{int}}^2(d) = \sigma^2 \frac{\Gamma(1+d)^2}{\Gamma(1+2d)}. \tag{11.8}$$

Figure 11.5 displays the coefficients α_j for $j = 1, 2, \ldots, 40$ and $d = 0.4$. Observe that they approach zero rapidly. However, this decay rate may be much faster for short-memory processes. For instance, for an AR(1) model, $\alpha_j = 0$ for $j > 1$; see Problem 11.11.

The evolution of $\sigma_{\text{int}}(d)$ as d moves from 0 to 0.5 is depicted in Figure 11.6. In this case, $\sigma_{\text{int}}(0) = 1$ and $\sigma_{\text{int}}(0.5) = 0.8862$.

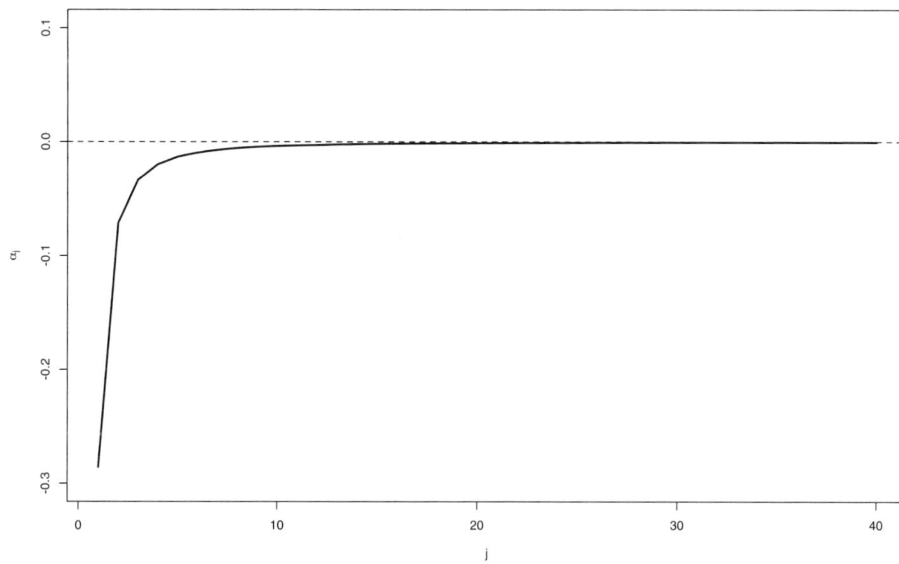

Figure 11.5 *Example 11.3: Behavior of the interpolation coefficients α_j for a fractional noise process with $d = 0.4$ and $j = 1, 2, \ldots, 40$.*

■ EXAMPLE 11.4

Extending the previous example, let $\{y_t : t \in \mathbb{Z}\}$ be an ARFIMA$(1, d, 0)$ process

$$(1 + \phi B)(1 - B)^d y_t = \varepsilon_t, \tag{11.9}$$

where $|\phi| < 1$, $d \in (-1, \frac{1}{2})$ and $\{\varepsilon_t\}$ is a white noise process with variance σ^2. Let η_j be the coefficients of the differencing operator $(1 - B)^d$,

$$(1 - B)^d = \sum_{j=0}^{\infty} \eta_j B^j.$$

Consequently, the coefficients of the autoregressive expansion (11.9)

$$(1 + \phi B)(1 - B)^d = \sum_{j=0}^{\infty} \pi_j B^j$$

are given by

$$\pi_j = \begin{cases} 1, & j = 0, \\ \eta_j + \phi \eta_{j-1}, & j \geq 1. \end{cases}$$

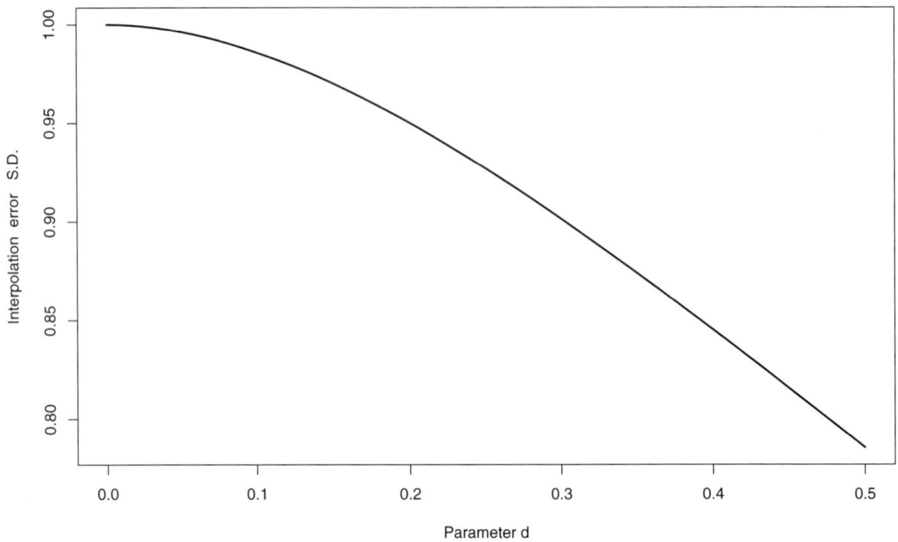

Figure 11.6 *Example 11.3: Interpolation error standard deviation of a fractional noise process based on the full past and full future, $\sigma_{\text{int}}(d)$ for different values of the long-memory parameter d.*

Hence,

$$\begin{aligned}\sum_{j=0}^{\infty}\pi_j^2 &= (1+\phi^2)\sum_{j=0}^{\infty}\eta_j^2 + 2\phi\sum_{j=0}^{\infty}\eta_j\eta_{j+1}\\ &= \left[1+\phi^2-\frac{2\phi d}{1+d}\right]\frac{\Gamma(1+2d)}{\Gamma(1+d)^2}.\end{aligned}$$

Finally, the interpolation error of y_0 based on $\{y_t, t \neq 0\}$ for this ARFIMA$(1,d,0)$ model is given by

$$\sigma_{\text{int}}^2 = \sigma^2\left[\frac{1+d}{1+\phi^2+d(1-\phi)^2}\right]\frac{\Gamma(1+d)^2}{\Gamma(1+2d)}.$$

■ EXAMPLE 11.5

Consider an ARFIMA(p,d,q) process with autocovariance function $\gamma_h(d,\phi,\theta)$ for lag h. By analyzing the interpolation error variances (11.5) and (11.6) we conclude that

$$\sigma_{\text{int}}^2(n) - \sigma_{\text{int}}^2 = \sigma^2\frac{\sum_{j=n+1}^{\infty}\pi_j^2}{\sum_{j=0}^{n}\pi_j^2\sum_{j=0}^{\infty}\pi_j^2}.$$

But, from (2.24) the AR(∞) coefficients of the ARFIMA process satisfy

$$\pi_j \sim \frac{\phi(1)}{\theta(1)\Gamma(-d)} j^{-d-1},$$

for large j. As a result,

$$\sigma^2_{\text{int}}(n) - \sigma^2_{\text{int}} \sim \left[\frac{\sigma^2 \phi(1)^2}{\theta(1)^2 \Gamma(-d)^2 \sum_{j=0}^{n} \pi_j^2 \sum_{j=0}^{\infty} \pi_j^2} \right] \sum_{j=n+1}^{\infty} j^{-2d-2}.$$

Now,

$$\sigma^2_{\text{int}}(n) - \sigma^2_{\text{int}} \sim \left[\frac{\sigma^2 \phi(1)^2}{(2d+1)\theta(1)^2 \Gamma(-d)^2 (\sum_{j=0}^{\infty} \pi_j^2)^2} \right] n^{-2d-1}.$$

Finally, by taking into account that

$$\sum_{j=0}^{\infty} \pi_j^2 = \frac{\gamma_0(-d, \theta, \phi)}{\sigma^2},$$

we conclude

$$\sigma^2_{\text{int}}(n) - \sigma^2_{\text{int}} \sim \left[\frac{\sigma^6 \phi(1)^2}{(2d+1)\theta(1)^2 \Gamma(-d)^2 \gamma_0(-d, \theta, \phi)^2} \right] n^{-2d-1}.$$

EXAMPLE 11.6

In order to illustrate the prediction and filtering procedures in the context of missing data we consider next the air pollution state space motivating example described in Chapter 3. Assume that there are two data gaps, one during the period $t = 351 - 400$ and another for the period $t = 701 - 800$. As in the case analyzed in Chapter 3, here we use the R package FKF to obtain predicted and filtered states.

Figure 11.7(a) depicts the predicted first state component while Figure 11.7(b) displays their estimated standard deviations. The corresponding plots of the second state component are shown in Figure 11.8. One the that hand, the filtered first state component along with their standard deviations are shown in Figure 11.9. Finally, Figure 11.10 exhibits the filtered second state component and the evolution of the standard deviations.

From these plots, notice that the standard deviations of predicted and filtered states increase rapidly as missing observations appear and then they decrease as the time series is actually observed.

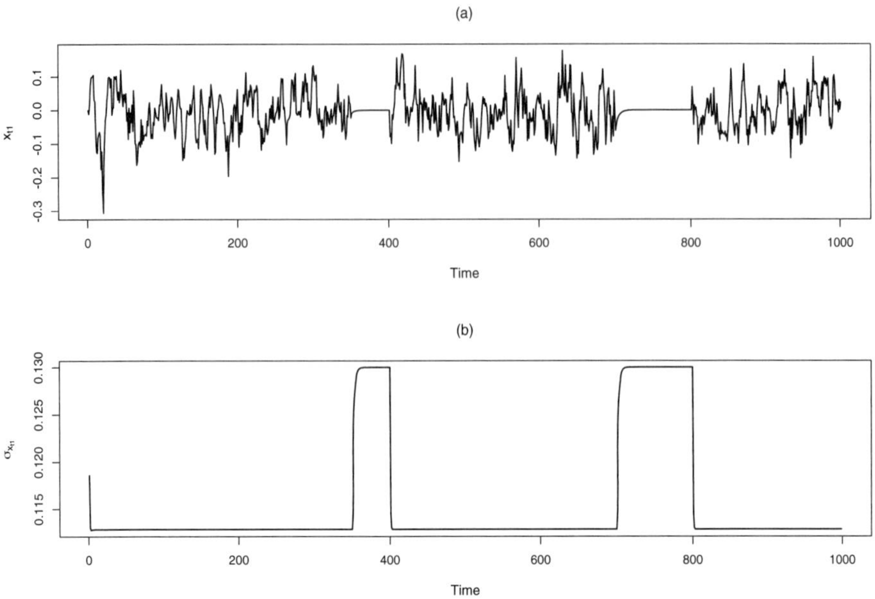

Figure 11.7 *Air pollution state space model example, predicted states, $t = 1000$. (a) predicted first state component. (b) prediction standard deviations.*

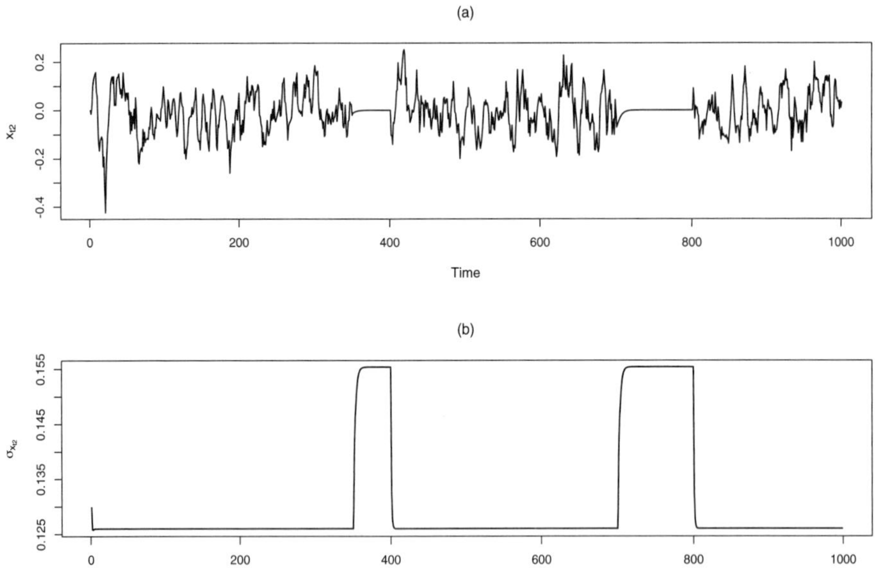

Figure 11.8 *Air pollution state space model example, predicted states, $t = 1000$. (a) predicted second state component. (b) prediction standard deviations.*

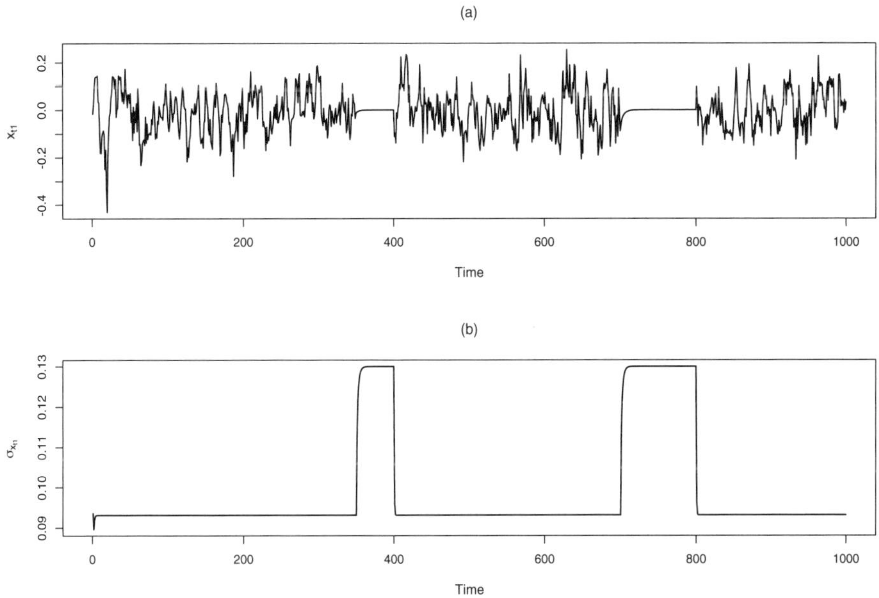

Figure 11.9 *Air pollution state space model example, filtered states, $t = 1000$. (a) filtered first state component. (b) filtering standard deviations.*

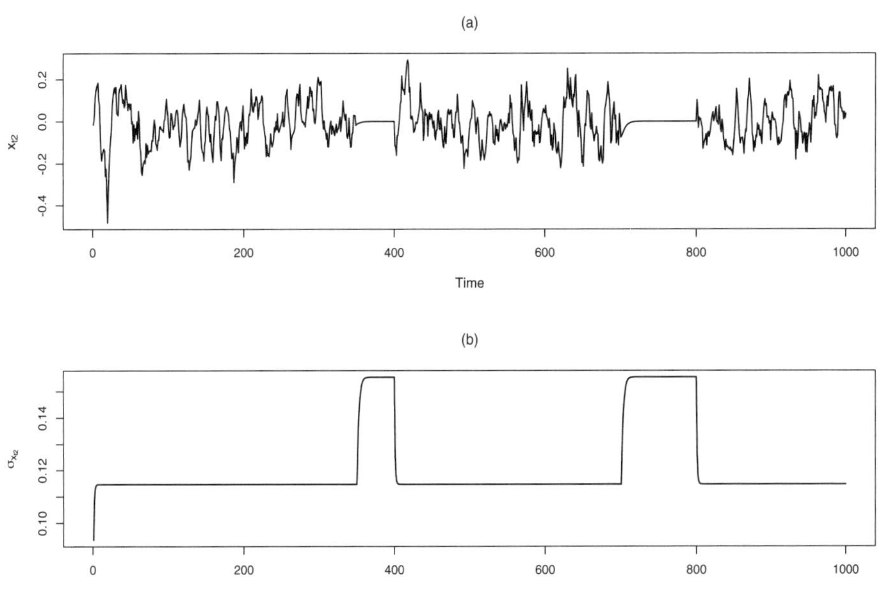

Figure 11.10 *Air pollution state space model example, filtered states, $t = 1000$. (a) filtered second state component. (b) filtering standard deviations.*

11.6 SPECTRAL ESTIMATION WITH MISSING VALUES

As discussed in previous chapters, spectral estimation is a fundamental tool for finding parameter estimates in time series. But, as expected missing values also affects the estimation of the spectral density. A simple technique for handling this problem in this context is assigning the mean value to each missing observation and correcting the number of observations to the effectively collected. Thus, the modified periodogram is given by

$$I(\lambda) = \frac{1}{2\pi n^*} \left| \sum_{j=1}^{n} y_j^* e^{\imath \lambda j} \right|^2 .$$

where

$$y_t^* = \begin{cases} y_t - \bar{y}, & \text{if } y_t \text{ is observed,} \\ 0 & \text{if } y_t \text{ is not observed,} \end{cases}$$

and n^* is the number of values actually observed. Another technique is based on the definition of the spectral density in terms of the autocovariances of the process,

$$f(\lambda) = \frac{1}{2\pi} \sum_{h=-\infty}^{\infty} \gamma(h) e^{\imath \lambda h}.$$

Thus, an estimator of the spectral density is

$$\widehat{f}(\lambda) = \frac{1}{2\pi} \sum_{h=1-n}^{n-1} \widehat{\gamma}(h) e^{\imath \lambda h}.$$

In order to estimate the autocovariance $\widehat{\gamma}(h)$ with incomplete data, the *Parzen* method suggests the following approach. Define the variable

$$a_t = \begin{cases} 1, & \text{if } y_t \text{ is observed,} \\ 0 & \text{if } y_t \text{ is not observed.} \end{cases}$$

Then, an estimate of the autocovariance is given by

$$\widehat{\gamma}(h) = \frac{\sum_{t=1}^{n-|h|} a_t \, (y_t - \bar{y}) a_{t+h} (y_{t+h} - \bar{y})}{\sum_{t=1}^{n-|h|} a_t \, a_{t+h}}$$

■ **EXAMPLE 11.7**

Consider a MA(1) model

$$y_t = \varepsilon_t + \theta \, \varepsilon_{t-1},$$

Table 11.6 Sample ACF of Full and Incomplete Data.

	$\gamma(0)$	$\gamma(1)$	$\gamma(0)$	$\gamma(1)$
	Full Sample		20% Missing	
Estimate	1.6331	0.7943	1.6318	0.7934
S.D.	0.0927	0.0698	0.0985	0.0822
	$\gamma(0)$	$\gamma(1)$	$\gamma(0)$	$\gamma(1)$
	Full Sample		40% Missing	
Estimate	1.6345	0.7934	1.6336	0.7912
S.D.	0.0914	0.0674	0.1102	0.1083

where ε_t is a normal white noise with zero-mean and unit variance. Table 11.6 exhibits two set of simulations of the estimation of $\gamma(0)$ and $\gamma(1)$, with a sample of 1000 observations, $\theta = 0.8$ and two scenarios, 20% and 40% of missing values. The estimates of the auto covariance function with missing observation are given by the Parzen method. Note that the ACF estimators seem to be unbiased, but the standard deviations increases as the number of missing observations increases.

Additionally, we can illustrate the estimation of the spectral density of this process which is given by

$$f(\lambda) = \frac{\sigma^2}{2\pi}(1 + \theta^2 + 2\,\theta\,\cos\lambda).$$

Figure 11.11 displays the theoretical spectral density of this MA(1) process (heavy line) along with the raw periodogram (gray line), the smoothed periodogram (broken line) and the Parzen based estimator (dotted line) in the case of complete data.

Notice that in this case the smoothed periodogram and the Parzen technique produce roughly similar results and follows the theoretical spectral density closely.

A similar plot is shown in Figure 11.12, where 40% of the data is missing. In this case, note that the raw periodogram (gray line) displays lower levels, as compared to the full sample situation. On the other hand, the smoothed periodogram seems to underestimate the true spectral density. However, the Parzen estimate follows closely their theoretical counterpart.

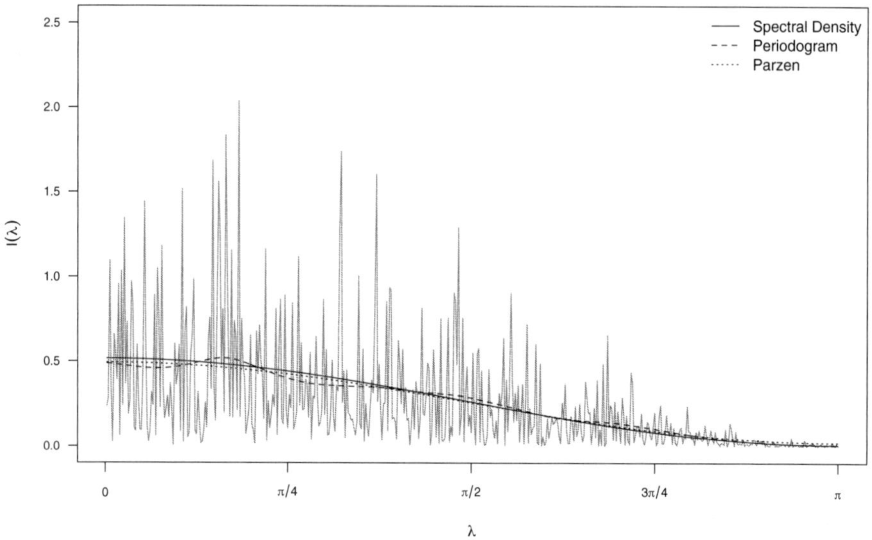

Figure 11.11 *Spectral estimation with complete data.*

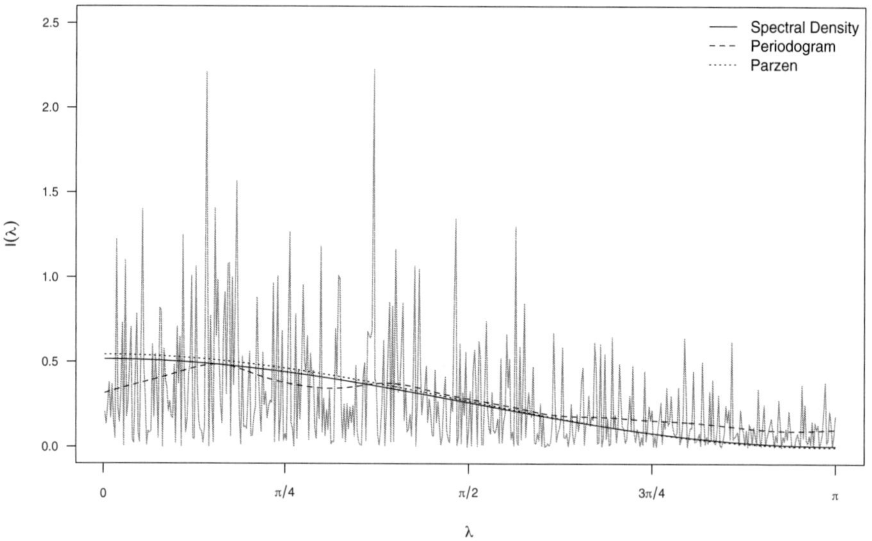

Figure 11.12 *Spectral estimation with incomplete data (40% of missing values).*

11.7 OUTLIERS AND INTERVENTION ANALYSIS

Generally speaking, an *outlier* in statistics is a value that is far beyond what is expected from its distribution. For example, a value equal to 20 in the case of a standard normal distribution $N(0,1)$ is far larger from the usual range of this random variable. In particular, the probability of observing a value of 20 or greater is practically zero (2.753624e-89). Outliers may cause serious problems in estimation, inference and forecasting of time series models. To illustrate this point, consider a very simple first-order autoregressive model,

$$y_t = \phi y_{t-1} + \varepsilon_t, \tag{11.10}$$

where $t = 1, \ldots, 200$ and ε_t is a Gaussian white noise sequence with zero-mean and unit variance. Figure 11.13 shows a realization of this process, assuming that the value of the input noise at time $T = 100$ is $\varepsilon_T = 20$.

The estimated AR(1) model with the original data is given by

```
Call:
arima(x = y1, order = c(1, 0, 0))

Coefficients:
         ar1  intercept
      0.6985    -0.1593
s.e.  0.0500     0.2491
```

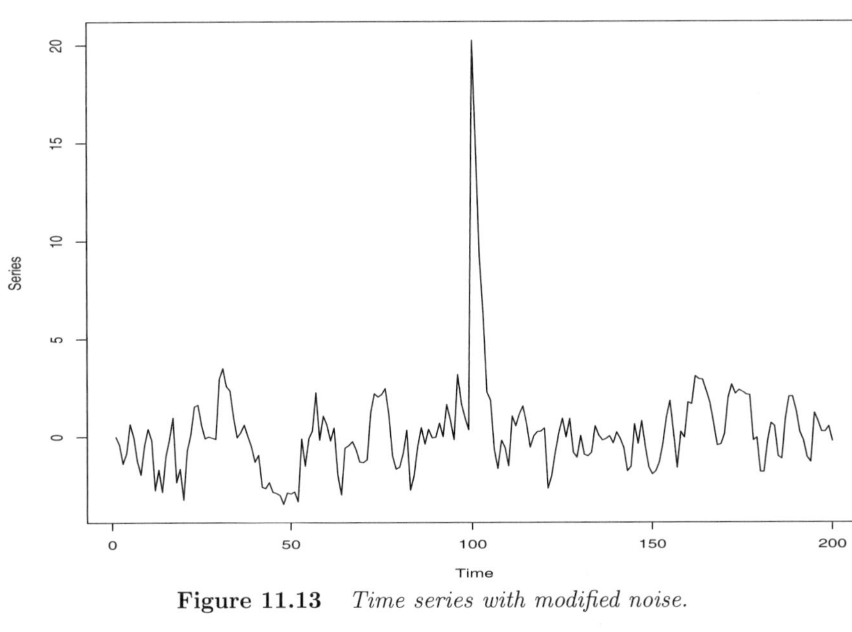

Figure 11.13 *Time series with modified noise.*

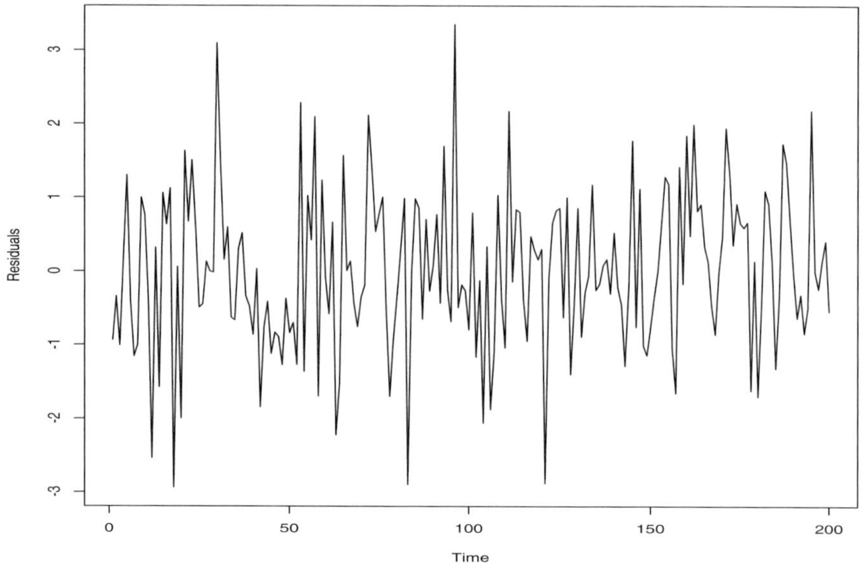

Figure 11.14 *Residuals of the original model.*

```
sigma^2 estimated as 1.154:  log likelihood = -298.46,  aic = 602.92
```

while the estimated model with the modified data is

```
fitCall:
arima(x = y2, order = c(1, 0, 0))

Coefficients:
         ar1  intercept
      0.6807     0.1710
s.e.  0.0512     0.3888

sigma^2 estimated as 3.148:  log likelihood = -398.79,  aic = 803.57
```

Note that even though the estimated parameter ϕ is similar in both cases, the estimated noise variances are quite different. This is not surprise since the variance of the original noise is 1.1615 while the modified noise is 3.1671. The impact on the residuals is quite dramatic if we compare Figure 11.14 and Figure 11.15. On the other hand, the effect of the prediction bands for these two models is quite important see Figure 11.16. Observe that the bands from the model fitted to the data with an outlier are much larger than the prediction bands from the original model.

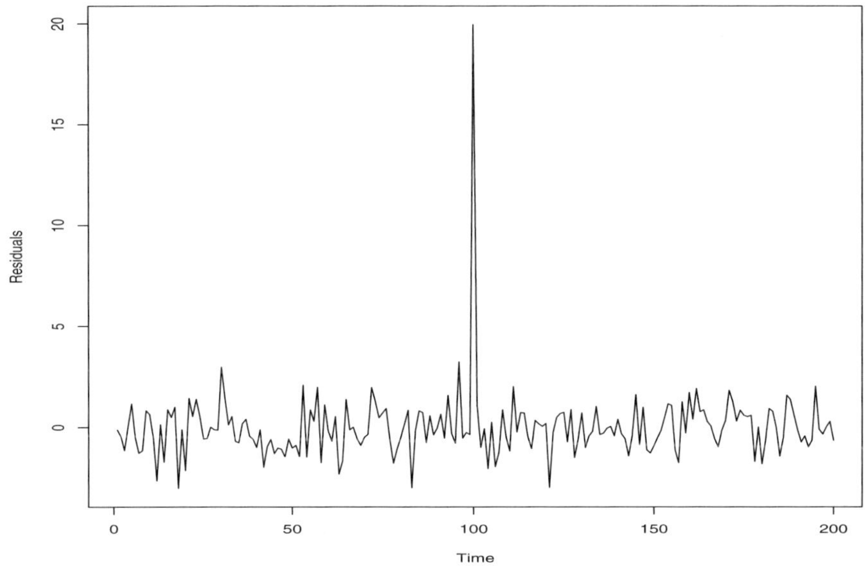

Figure 11.15 *Residuals from the modified noise model.*

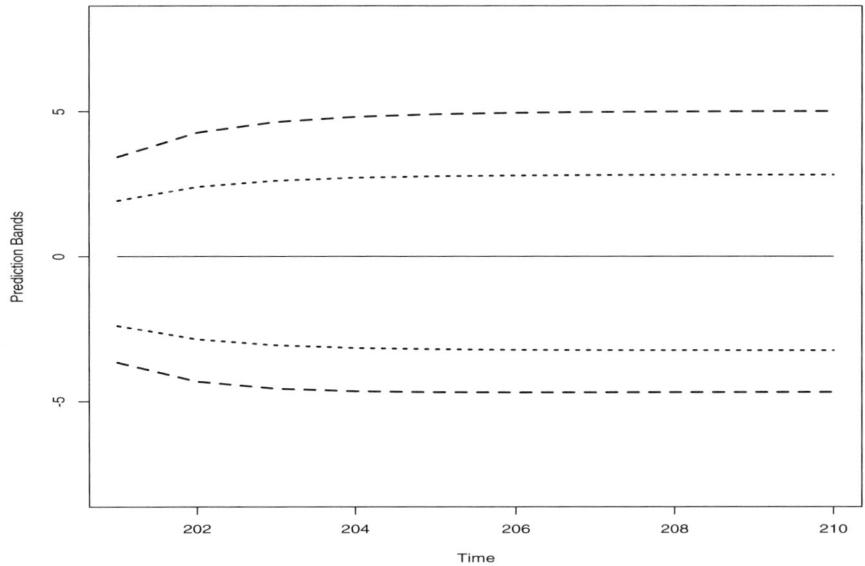

Figure 11.16 *95% prediction bands for 10 step ahead forecasts. Dotted line: Original model, Broken line: Model with one outlier.*

11.7.1 Methodology

When analyzing the different effects of one outlier on a time series we can classify some of them into four types: *Innovative Outlier* (IO), *Additive Outlier* (AO), *Temporary Change* (TC) and *Level Shift* (LS). These cases are briefly described next.

For simplicity, assume that the process under study corresponds to a seasonal ARMA (SARMA) model described by

$$y_t = \frac{\theta(B)}{\alpha(B)\phi(B)}\varepsilon_t,$$

where $\phi(B)$ and (B) are the autoregressive and moving-average polynomials, and $\alpha(B)$ is a seasonal polynomial. Suppose now that the observed process is a modified version of the original process y_t given by

$$x_t = y_t + \omega\,\psi(B)\,I_t(T),$$

where $I_t(T) = 1$ if $t = T$ and $I_t(T) = 1$, otherwise. In this context, T represents the location of the outlier, ω indicates the magnitude of the impact and the polynomial $\psi(B)$ corresponds to the structure of the perturbation. Among these events we can describe for instance an *innovational outlier* (IO) defined by

$$\psi(B) = \frac{\theta(B)}{\alpha(B)\phi(B)}.$$

Note that in this case the observed model is given by

$$x_t = \frac{\theta(B)}{\alpha(B)\phi(B)}\,[\varepsilon_t + \omega\,I_t(T)].$$

Thus, an innovational outlier model corresponds to a perturbation of the noise sequence at a specific time T. On the other hand, an *additive outlier* (AO) is defined by setting $\psi(B) = 1$, an *level shift* (LS) corresponds to

$$\psi(B) = \frac{1}{1-B},$$

while an *temporary change* model (TC) is obtained by setting

$$\psi(B) = \frac{1}{1-\delta\,B},$$

see Figure 11.17 for an illustration of these outlier structures.

11.7.2 Known time of the event

When the time of the event is known, the R package dynia allows for the estimation of the ARIMA model. In the case of model (11.10), the package output is as follows,

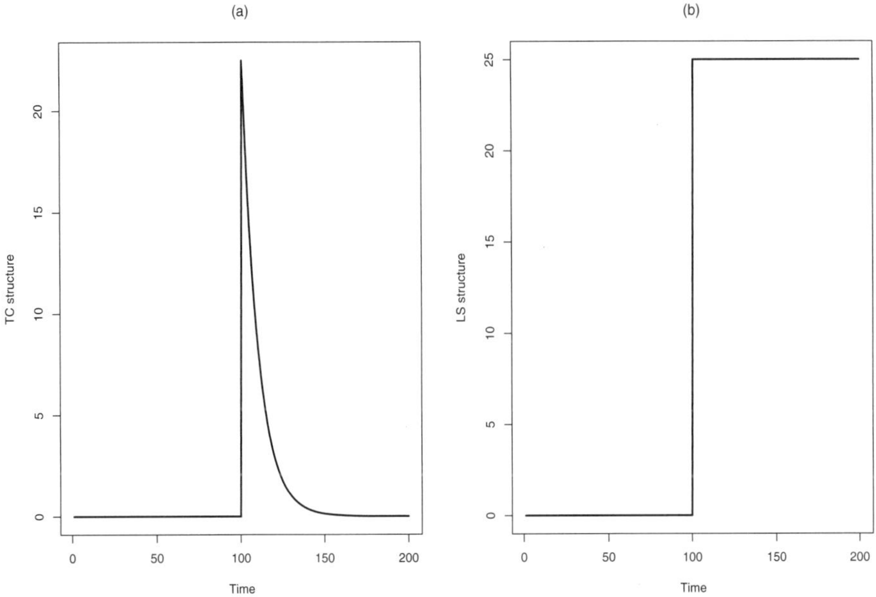

Figure 11.17 *TC and LS structures with $n = 200$, $T = 101$ and $\delta = 0.9$.*

```
$delta
[1] NA

$Int.Mod

Call:
arima(x = z, order = ..1, xreg = xreg2)

Coefficients:
         ar1  intercept    xreg2
      0.6882    -0.3406  20.1105
s.e.  0.0711     0.3630   1.1573

sigma^2 estimated as 1.324:  log likelihood = -156.25,  aic = 320.49

$`p-value`
NULL
```

That is, the impact is estimated as 20.1105 which is close to the true value of 20. At the same time, the estimated autoregressive parameter, $\widehat{\phi} = 0.6882$, is close to its true value $\phi = 0.7$.

11.7.3 Unknown time of the event

When the location of the outlier is unknown, the R package tsoutliers can estimate T, the parameters of the model as well as the magnitude of the jump.

In what follows, we illustrate the application of this methodology to an *Additive Outlier* (AO), *Temporary Change* (TC) and *Level Shift* (LS) and a combination of these cases. Before discussing the estimation of these different types of outliers, we show their effects on a simulated AR(1) series with $\phi = 0.8$ and $n = 200$ observations.

■ EXAMPLE 11.8

Consider the AR(1) process defined previously in (11.10) with $\phi = 0.8$, $\omega = 25$, $\delta = 0.9$ and $T = 101$. Figure 11.18 displays a simulated trajectory of this model while Figure 11.19 shows the effect of an innovation outlier at time $T = 100$ on this trajectory. On the other hand, the impact of an additive outlier is exhibited in Figure 11.20. The impact of a temporary change outlier is exhibited in Figure 11.21. Finally, the effect of a level shift outlier is displayed in Figure 11.22. Note in this last case, the effect of the event is change the mean of the series permanently.

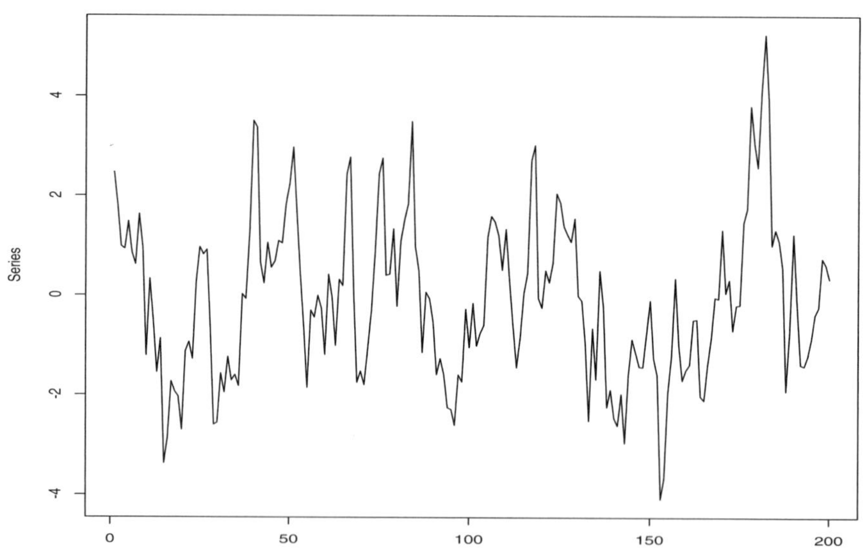

Figure 11.18 *Simulated AR(1) series with $\phi = 0.8$ and $n = 200$ observations.*

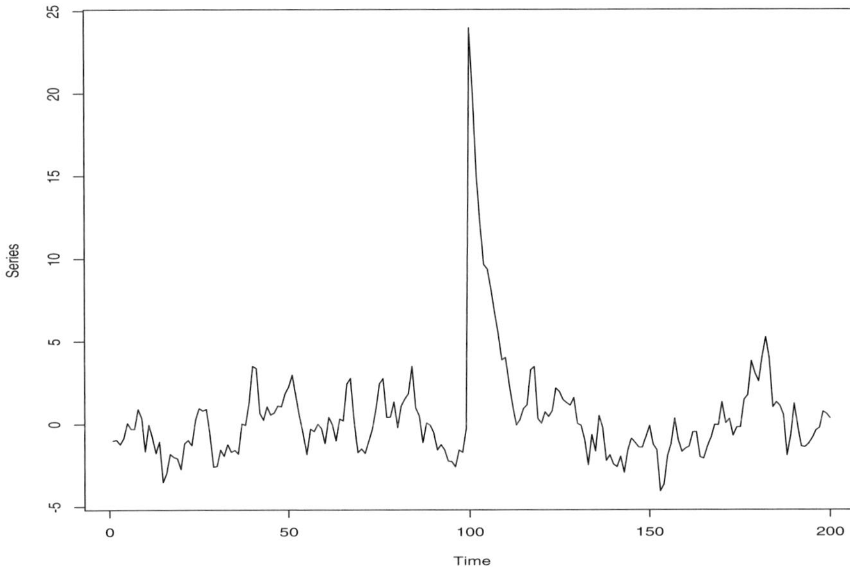

Figure 11.19 *Simulated AR(1) series with* $\phi = 0.8$, $n = 200$ *observations and one IO at time* $T = 100$.

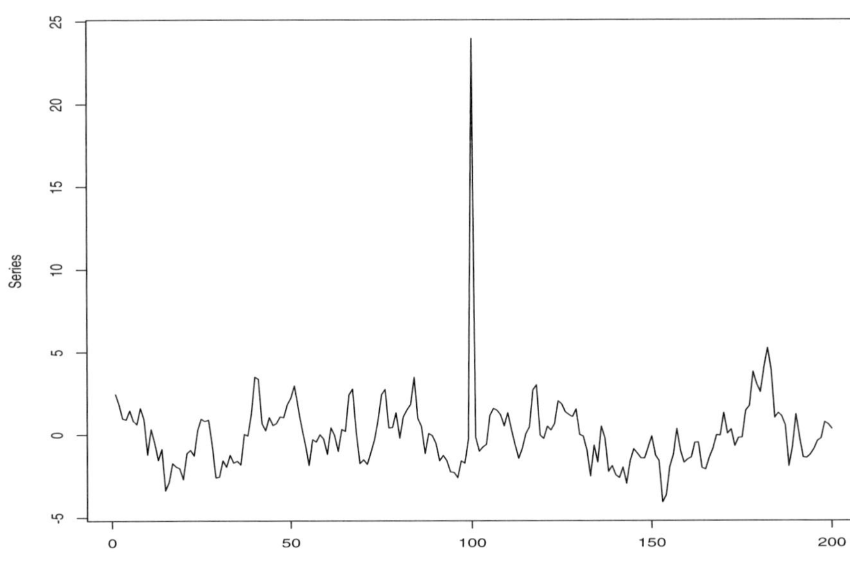

Figure 11.20 *Simulated AR(1) series with* $\phi = 0.8$, $n = 200$ *observations and one AO at time* $T = 100$.

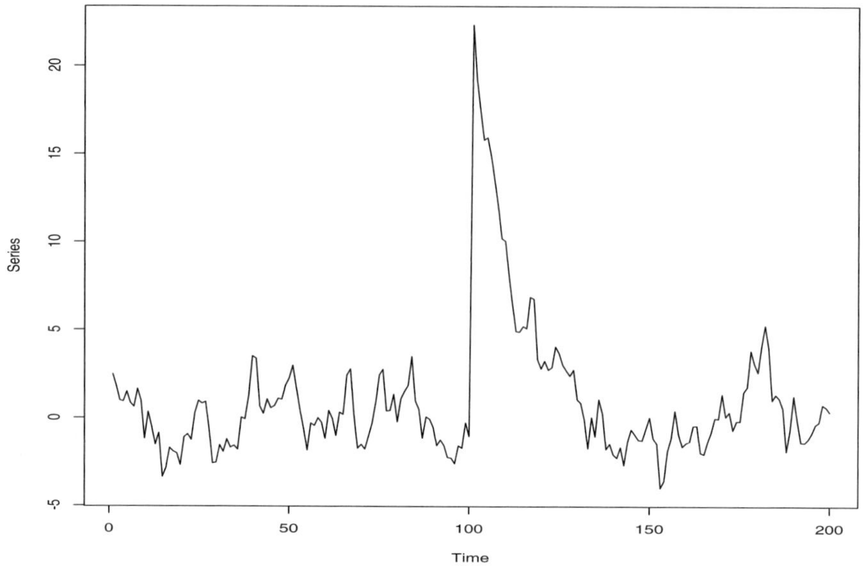

Figure 11.21 *Simulated AR(1) series with* $\phi = 0.8$, $n = 200$ *observations and one TC at time* $T = 101$.

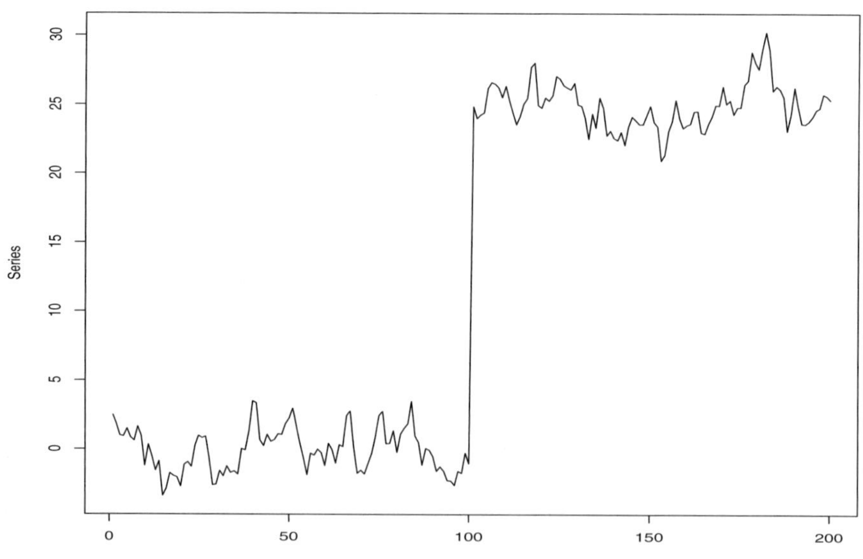

Figure 11.22 *Simulated AR(1) series with* $\phi = 0.8$, $n = 200$ *observations and one LS at time* $T = 101$.

EXAMPLE 11.9

Additive Outlier. Based on the illustrative model (11.10), we have generated an additive outlier of magnitude 20 at location $T = 100$. The series is displayed in Figure 11.23. Observe that the *tso* function of the *tsoutliers* package adequately estimates the time of the event as well as the parameter ϕ and the magnitude of the impact. A graphical summary of the fitting is provided by Figure 11.24.

```
> outlier.detection=tso(y,tsmethod="arima",
args.tsmethod=(list(order=c(1,0,0))),
types=c("AO", "NULL", "NULL"))

> outlier.detection

Series: y
ARIMA(1,0,0) with non-zero-mean

Coefficients:
         ar1  intercept    AO100
      0.6891     0.5061  19.2970
s.e.  0.0507     0.2296   0.8401
```

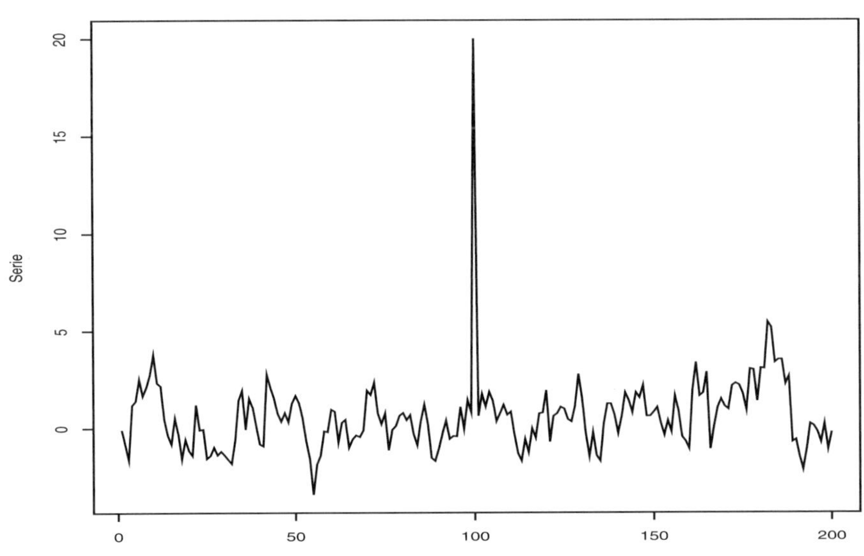

Figure 11.23 *Simulated AR(1) with $\phi = 0.7$, $n = 200$ and additive outlier AO at location $T = 100$ and magnitude 20.*

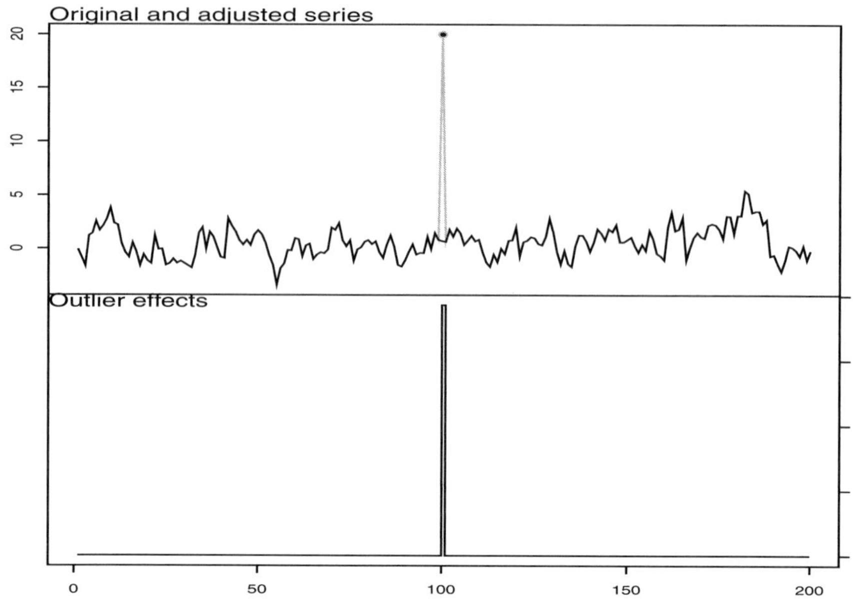

Figure 11.24 *Detection of AO structure with $n = 200$ and $T = 100$.*

```
sigma^2 estimated as 1.041:  log likelihood=-288.09

AIC=584.17   AICc=584.38   BIC=597.37

Outliers:
  type ind time coefhat tstat
1   AO 100  100    19.3 22.97
```

■ EXAMPLE 11.10

Temporary Change. Similar results are observed when the type of outlier is TC. The method estimate very well the location of the event, the autoregressive parameter ϕ and the magnitude of the impact. Figure 11.25 exhibits the fitting of the TC model as well as the effects of the outliers.

```
> outlier.detection=tso(y,tsmethod="arima",
args.tsmethod=(list(order=c(1,0,0))))
> outlier.detection
Call:
Series: y
ARIMA(1,0,0) with non-zero-mean

Coefficients:
```

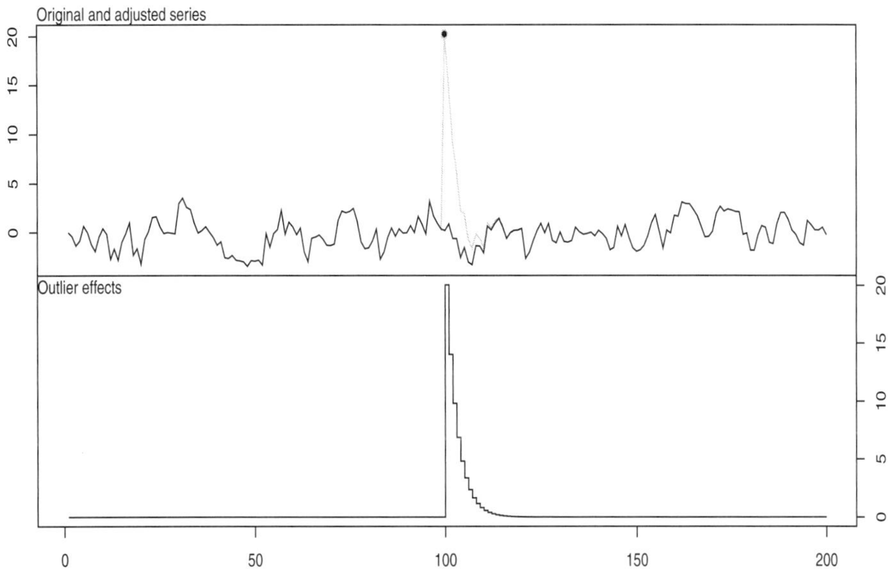

Figure 11.25 *Detection of TC structure with $n = 200$ and $T = 100$.*

```
          ar1  intercept    TC100
       0.6978    -0.1561  20.0454
s.e.   0.0500     0.2487   1.0756

sigma^2 estimated as 1.15:  log likelihood=-298.12
AIC=604.24   AICc=604.44   BIC=617.43

Outliers:
  type ind time coefhat tstat
1   TC 100  100   20.05 18.64
> plot(outlier.detection)
```

■ EXAMPLE 11.11

Level Shift. Consider the AR(1) time series with $\phi = 0.8$ and 200 observations shown in Figure 11.22. This process suffers a level change at time $T = 101$ of magnitude 25. Notice that the method detects adequately the location of the change and estimate very well the autoregressive parameter and the magnitude of the impact. The following is the output from the *two* function of the *tsoutliers* package. Moreover, Figure 11.26 exhibits a graphical analysis of the fitted level shift model.

```
> outlier.detection=tso(y,tsmethod="arima", args.tsmethod=
(list(order=c(1,0,0))),types=c("NULL", "LS", "NULL"))
```

```
> outlier.detection
Series: y

ARIMA(1,0,0) with non-zero-mean

Coefficients:
         ar1  intercept    LS101
      0.7723    -0.0918  25.1193
s.e.  0.0451     0.4190   0.5571

sigma^2 estimated as 1.057:  log likelihood=-289.81

AIC=587.62   AICc=587.83   BIC=600.82

Outliers:

  type ind time coefhat tstat
1   LS 101  101   25.12 45.09
```

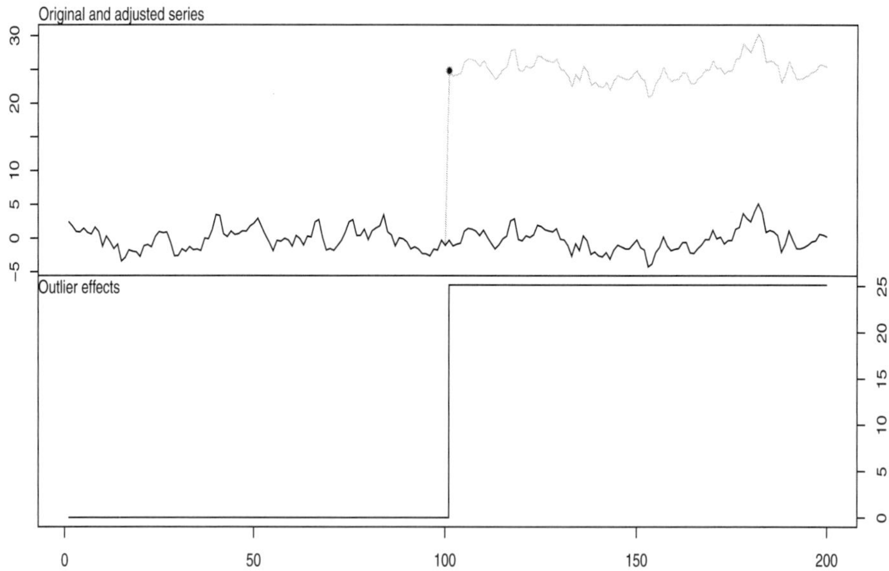

Figure 11.26 *Detection of LS structure with $n = 200$, $T = 101$ and $\delta = 0.9$.*

■ EXAMPLE 11.12

Combination of events. In this last illustration we consider the case where the process suffers two events, one temporary change at $T = 50$ and a level shift at $T = 150$. Figure 11.27 exhibits a simulated trajectory of this time series. The magnitudes of the first and second events are 20 and 30 respectively. Observe that the method appropriately estimate the location of these two events as well as their magnitudes. On the other hand, the estimate of the autoregressive structure is quite close to its theoretical counterpart. Figure 11.28 displays the fitting of the TC model as well as the effects of the outliers.

```
> outlier.detection=tso(y,tsmethod="arima",
args.tsmethod=(list(order=c(1,0,0))), types=c("LS", "TC", "NULL"))
> outlier.detection
Series: y
ARIMA(1,0,0) with non-zero-mean

Coefficients:
         ar1  intercept     TC50    LS151
      0.6796    -0.6013  20.1919  32.7470
s.e.  0.0528     0.2402   0.9621   0.4556

sigma^2 estimated as 0.918:  log likelihood=-275.55
```

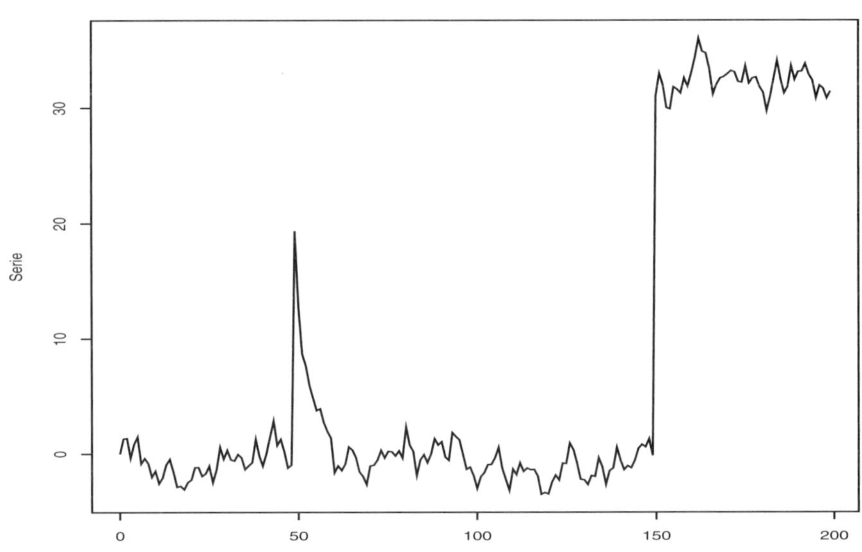

Figure 11.27 *Detection of LS structure with $n = 200$, $T = 101$ and $\delta = 0.9$.*

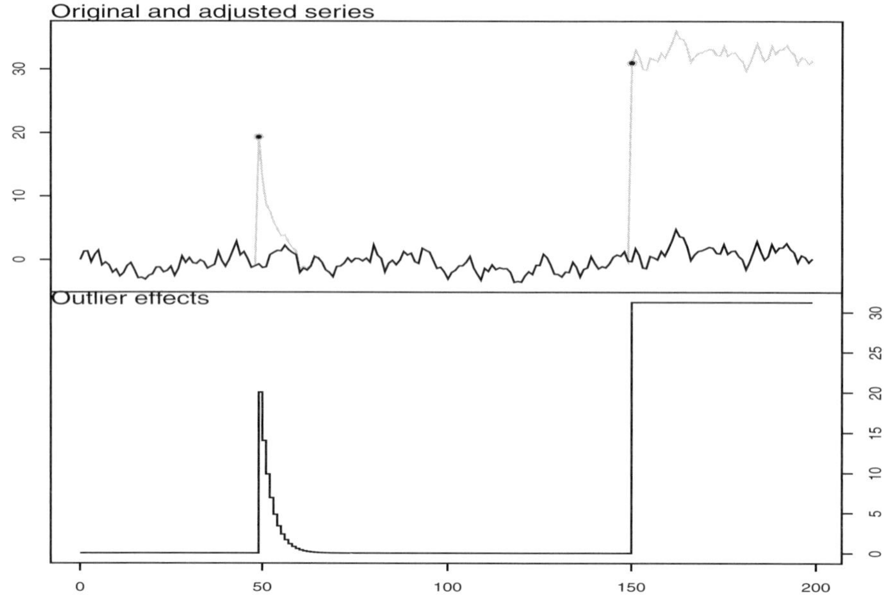

Figure 11.28 *Detection of a combination of two events, one TC outlier at $T = 50$ and LS at $T = 150$ with $n = 200$ and $\delta = 0.9$.*

```
AIC=561.09   AICc=561.4   BIC=577.58

Outliers:
  type ind time coefhat tstat
1   TC  50   49   20.19 20.99
2   LS 151  150   32.75 71.88
```

11.8 BIBLIOGRAPHIC NOTES

The time series literature on missing values is extensive. For instance, Jones (1980) develops a Kalman filter approach to deal with missing values in ARMA models. Ansley and Kohn (1983), Harvey and Pierse (1984), and Kohn and Ansley (1986) extend Jones' result to ARIMA processes. Further extensions of state space model techniques to ARFIMA models are considered by Palma and Chan (1997) and Palma (2000). Chapter 4 of Shumway and Stoffer (2011) discusses the Kalman filter modifications to account for missing data. An alternative approach for computing the maximum-likelihood estimates called the *expectation maximization (EM)* method to deal with data gaps is also discussed extensively in that book. Other methods for treating missing values are studied by Wilson, Tomsett, and Toumi (2003).

A good general reference to the problem of defining the likelihood function for incomplete data is the book by Little and Rubin (2002).

Estimation and interpolation of time series with missing values have been studied by many authors; see, for example, Cheng and Pourahmadi (1997), Bondon (2002), Damsleth (1980) and Pourahmadi (1989), among others. In addition, Yajima and Nishino (1999) address the problem of estimating the autocovariance function from an incomplete time series. A detailed study about interpolation of missing data appears in Chapter 8 of Pourahmadi (2001).

Finally, there is an extensive literature on detection and estimation of outliers, see for example Tsay, Peña, and Pankratz (2000), Chen and Liu (1993), Choy (2001) and Cai and Davies (2003), among others.

Problems

11.1 Let $\{y_t\}$ be a first-order autoregressive process such that

$$y_t = \phi y_{t-1} + \varepsilon_t,$$

where the white noise sequence $\{\varepsilon_t\}$ follows a standard normal distribution. Suppose that we have observed the values $\{y_1, y_2, y_3, y_5\}$ but y_4 is missing.

(a) Calculate the joint density of y_1, y_2, y_3, y_4, y_5, $f(y_1, y_2, y_3, y_4, y_5)$.

(b) Find z, the value that maximizes $f(y_1, y_2, y_3, y_4, y_5)$ with respect to y_4.

(c) Show that z corresponds to the *smoother* of y_4, that is,

$$z = E[y_4|y_1, y_2, y_3, y_5].$$

11.2 Suppose that $\{y_t : t \in \mathbb{Z}\}$ is a discrete-time stochastic process and we have observed the values $y_1, \dots, y_{m-1}, y_{m+1}, \dots, y_n$ but y_m has not been observed. Let f be the joint density of $y_1, \dots, y_n$ and let g be the unimodal conditional density of y_m given the observed values. Assume that the mode and the mean of g coincide. Let z be the value that maximizes f with respect to y_{mis}. Show that under these circumstances,

$$z = E[y_m|y_1, \dots, y_{m-1}, y_{m+1}, \dots, y_n].$$

11.3 Consider an invertible process with autoregressive decomposition, $y_t = \varepsilon_t - \sum_{j=1}^{\infty} \pi_j y_{t-j}$ and $\sigma = 1$. Let $e_t(1) = y_t - \widehat{y}_t$, where $\widehat{y}_t = P_{t-1} y_t$, P_{t-1} is the projection operator onto $\mathcal{H}_{t-1} = \{\text{observed } y_{t-1}, y_{t-2}, \dots\}$. Define the one-step prediction error by $\sigma_t^2(1) = \text{Var}[e_t(1)]$. Suppose that y_t is the first missing value in the time series. Since up to time t there are no missing observations, $\widehat{y}_t = \sum_{j=1}^{\infty} \pi_j y_{t-j}$, so that $y_t - \widehat{y}_t = \varepsilon_t$ and $\sigma_t^2(1) = \text{Var}[y_t - \widehat{y}_t] = \text{Var}[\varepsilon_t] = 1$.

(a) Show that since at time $t+1$, $\mathcal{H}_t = \mathcal{H}_{t-1}$ (no new information is available at time t) we have

$$\sigma_{t+1}^2(1) = \text{Var}[y_{t+1} - \widehat{y}_{t+1}] = \text{Var}[y_{t+1} - P_t y_{t+1}],$$

with

$$P_t y_{t+1} = -\pi_1 P_t y_t - \pi_2 y_{t-1} - \pi_3 y_{t-2} - \cdots$$

and

$$y_{t+1} - P_t y_{t+1} = \varepsilon_{t+1} - \pi_1(y_t - P_t y_t) = \varepsilon_{t+1} - \pi_1(y_t - P_{t-1} y_t).$$

(b) Verify that

$$\begin{aligned} \sigma_{t+1}^2(1) &= \pi_1^2 \operatorname{Var}[y_t - P_{t-1} y_t] + \operatorname{Var}[\varepsilon_{t+1}] \\ &= \pi_1^2 \sigma_t^2(1) + 1 = \pi_1^2 + 1 \geq \sigma_t^2(1). \end{aligned}$$

11.4 Implement computationally the modified Kalman filter equations of Subsection 11.2.4 for an fractional noise process.

11.5 Show that for $k \geq 1$,

$$\sigma_{t+k}^2(1) - 1 \geq \left\{ \frac{\pi_k}{\pi_{k+1}} \right\}^2 [\sigma_{t+k+1}^2(1) - 1].$$

11.6 Verify that for a fractional noise process, the one-step MSPE after a missing value is strictly decreasing. That is, for all $k \geq 1$,

$$\sigma_{t+k+1}^2(1) < \sigma_{t+k}^2(1).$$

11.7 Show that the one-step MSPE of an ARMA process with unit variance innovations converges to 1 at a exponential rate, that is,

$$|\sigma_{t+k}^2(1) - 1| \leq c a^{-k},$$

as $k \to \infty$ where $|a| < 1$.

11.8 Verify that the one-step MSPE of a fractional ARIMA process with unit variance innovations converges to 1 at a hyperbolic rate, that is,

$$|\sigma_{t+k}^2(1) - 1| \leq c k^{-\alpha},$$

as $k \to \infty$ with $c > 0$ and $\alpha = 2 - 2d > 0$.

11.9 Let $\widetilde{y}_t$ be the best linear interpolator of y_t based on $\{y_j, j \neq t\}$ and let $\widetilde{\sigma}^2$ be its error variance

$$\widetilde{\sigma}^2 = \operatorname{Var}(y_t - \widetilde{y}_t).$$

Let x_t be the standardized two-side innovation defined by

$$x_t = \frac{y_t - \widetilde{y}_t}{\widetilde{\sigma}^2}.$$

Let $\alpha(z) = 1 + \sum_{j=1}^{\infty} \alpha_j z^j + \sum_{j=1}^{\infty} \alpha_j z^{-j}$ where α_j are the coefficients appearing in the best linear interpolator (11.4) and $z \in \mathbb{C}$ with $|z| = 1$.

(a) Verify that x_t may be formally expressed as

$$x_t = \frac{1}{\widetilde{\sigma}^2} \alpha(B)\psi(B)\varepsilon_t,$$

where $\psi(B)$ is the Wold decomposition of the process (11.3).

(b) Prove that

$$\alpha(z) = c\pi(z)\pi(z^{-1}),$$

where

$$c = \frac{1}{\sum_{j=0}^{\infty} \pi_j^2},$$

π_j are the coefficients in the AR(∞) expansion (11.3) and $|z| = 1$.

(c) Based on the previous results, show that

$$x_t = \frac{c}{\widetilde{\sigma}^2} \pi(B^{-1})\varepsilon_t.$$

(d) Let f be the spectral density of y_t. Verify that the spectral density of the process x_t is given by

$$f_x(\lambda) = \frac{1}{(2\pi)^2 f(\lambda)}.$$

11.10 Let $\rho(k)$ be the autocorrelation of order k of a second-order stationary process. Verify that the coefficients α_j of the best linear interpolator of y_0 based on $\{y_t, t \neq 0\}$ satisfy the formula

$$\sum_{j=-\infty}^{\infty} \alpha_j \rho(k-j) = 0, \tag{11.11}$$

for any $k \in \mathbb{Z}$, $k \neq 0$.

11.11 Let $\{y_t : t \in \mathbb{Z}\}$ be the AR(1) process described by

$$y_t = \phi y_{t-1} + \varepsilon_t,$$

where $\{\varepsilon_t\}$ is a white noise sequence with variance σ^2.

(a) Show that for this AR(1) model, the best linear interpolator of y_0 with full past $\{y_t, t < 0\}$ and full future $\{y_t, t > 0\}$ is given by

$$\widetilde{y}_0 = \frac{\phi(y_1 + y_{-1})}{1 + \phi^2}.$$

(b) Prove that the interpolation error variance is

$$\sigma_{\text{int}}^2 = \frac{\sigma^2}{1+\phi^2}.$$

(c) Verify formula (11.11) in this case, with

$$\alpha_1 = \alpha_{-1} = -\frac{\phi}{1+\phi^2}.$$

11.12 Consider the MA(1) process $y_t = \varepsilon_t - \theta\varepsilon_{t-1}$ where $\{\varepsilon_t\}$ is a white noise process with variance σ^2 and $|\theta| < 1$.

(a) Show that the coefficients α_j of the best linear interpolator are given by

$$\alpha_j = \theta^j,$$

for $j \geq 0$ and then the best linear interpolator of y_0 based on $\{y_t, t \neq 0\}$ is

$$\widetilde{y}_0 = -\sum_{j=1}^{\infty} \theta^j (y_j + y_{-j}).$$

(b) Verify that in this case, formula (11.11) may be written as

$$\alpha_j - \left(\frac{1+\theta^2}{\theta}\right)\alpha_{j-1} + \alpha_{j-2} = 0,$$

for $k \in \mathbb{Z}$, $k \neq 0$.

(c) Show that interpolation error variance is given by $\sigma_{\text{int}}^2 = \sigma^2(1-\theta^2)$.

11.13 Prove that for a fractional noise process FN(d), the coefficients α_j of the best linear interpolator satisfy

$$\alpha_j \sim \frac{\Gamma(1+d)}{\Gamma(-d)} j^{-1-2d},$$

as $j \to \infty$.

11.14 Calculate the coefficients α_j of the best linear interpolator for the ARFIMA$(1, d, 0)$ discussed in Example 11.4. Recall that

$$\pi_j = \begin{cases} 1, & j = 0, \\ \eta_j + \phi\eta_{j-1}, & j \geq 1, \end{cases}$$

and that

$$\sum_{i=0}^{\infty} \eta_i \eta_{i+j} = \frac{\gamma_0(j)}{\sigma^2},$$

where $\gamma_0(j)$ is the autocovariance function of a fractional noise.

11.15 Let y_t be a stationary invertible process with AR(∞) representation $y_t = \varepsilon_t - \sum_{j=1}^{\infty} \pi_j y_{t-j}$ and MA(∞) decomposition $y_t = \sum_{j=1}^{\infty} \psi_j \varepsilon_{t-j}$, where the white noise sequence $\{\varepsilon_t\}$ has variance σ^2. Assume that the observations $y_0, ..., y_{m-1}$ are missing and define

$$\mathcal{H}_{k,m} = \{y_s, s < k, s \neq 0, 1, \dots, m-1\}.$$

Let $e(t, m, k)$ be the error of the best linear predictor of y_t given $\mathcal{H}_{k,m}$, that is, given all the available information before time k, and let $\sigma^2(t, m, k)$ be its variance. The following theorems characterize the behavior of $\sigma^2(t, m, k)$ during and after the data gap and specify the convergence rates.

Show that the variance $\sigma^2(k, m, k)$ satisfies

(a) For $k = 0, \dots, m$, $\sigma^2(k, m, k) = \sigma^2 \sum_{j=0}^{k} \psi_j^2$,

(b) For $k > m$, $\sigma^2(k, m, k) - \sigma^2 \leq \sigma_y^2 m^2 \max_{\{j \geq k-m\}} \pi_j^2$.

(c) $\lim_{k \to \infty} \sigma^2(k, m, k) = \sigma^2$.

11.16 Verify that for an ARFIMA model and fixed data gap length m we have

(a) $\sigma_y^2 - \sigma^2(k, m, k) \sim ck^{2d-1}$ for some constant $c > 0$ and large k, $k \leq m$,

(b) $\sigma^2(k, m, k) - \sigma^2 \sim ck^{-2d-2}$ for some constant $c > 0$ and large k, $k \gg m$.

11.17 Show that for an ARFIMA(p, d, q) process we have

$$\sigma^2(k, \infty, k) - \sigma^2 \sim \frac{d^2}{k},$$

as $k \to \infty$.

11.18 Show that for an ARFIMA(p, d, q) process the mean-squared prediction error of an isolated missing observation behaves as follows:

(a) for $k > 0$, $\sigma^2(k, 1, k) - \sigma^2 = \pi_k^2 \sigma^2(0, 1, k)$,

(b) for $k > 0$, if π_k is a monotonically decreasing sequence, then $\sigma^2(k, 1, k) - \sigma^2$ is a monotonically decreasing sequence converging to zero.

11.19 Show the missing values modifications of the state space Kalman recursive filter equations.

CHAPTER 12

NON-GAUSSIAN TIME SERIES

As discussed in the previous chapters, most real-life time series are not necessarily Gaussian. Moreover, in many situations it is not reasonable to approximate them by a real-valued continuos random variable. Counting data and positive observations are two examples where it is sometimes preferable to apply a specific non-Gaussian modeling approach. Counting data are usually the result of collecting the number of events occurred at a given time such as, for instance, the number of daily patients attending an hospital, the number of passenger enplanements, the number of animals of a certain species living in a specific geographic area, among others time series. On the other hand, positive data is often related to, for example, time between two events, useful life of a certain machine, height or weight of a person, among other physical measurements.

Generally speaking, there are two basic approaches to handle non-Gaussian data exhibiting serial dependence. These are the so-called *observation driven* models and the *parameter driven* processes. This chapter reviews these methodologies as well as techniques for handling zero-inflated observations.

Time Series Analysis. First Edition. Wilfredo Palma.

12.1 DATA DRIVEN MODELS

In this class of processes, the serial dependence observed in the data is usually defined in terms of its past values. Two important types of models belonging to this class are the so-called *integer-valued autoregressive* (INAR) processes and the *conditional distribution models*. In what follows we review some of these approaches.

12.1.1 INAR Processes

The basic idea of the INAR model is to take advantage of previously well established models in order to account for specific conditions of the data. For example, consider the case of a counting time series with a first-order autoregressive dependence. In this case, all the observations must be an integer number, so that the usual AR(1) model,

$$y_t = \alpha\, y_{t-1} + \varepsilon_t,$$

model is no longer adequate since it may produce non-integer values of y_t. Consequently, its definition can be modified via the *thinning* operation $\alpha \circ y_{t-1}$ to obtain an integer-valued process.

Let $\alpha \in [0, 1]$ and let z be a non-negative integer-valued random variable and let y_t be a sequence of non-negative integer-valued i.i.d. random variables with mean α and variance σ^2, independent of z. The thinning operation $\alpha \circ z$ is defined as

$$\alpha \circ z = \sum_{j=1}^{z} y_j.$$

Note that this expression generates a non-negative integer valued random variable. Based on the previous definition, we can write an INAR(1) process as follows,

$$y_t = \alpha \circ y_{t-1} + \varepsilon_t.$$

The ACF of y_t is $\rho(h) = \alpha^{|h|}$. Figure 12.1 displays a simulated trajectory of 200 observations from the INAR(1) model with $\alpha = 0.6$ and Poisson input noise. Moreover, Figure 12.2 shows the sample ACF and sample PACF of this process. It is interesting to notice that the first-order autoregressive dependence structure is fully reflected in these integer-valued observations.

On the other hand, Figure 12.3 exhibits a simulated INAR (1) process with $\alpha = 0.9$, Poisson input noise and 400 observations. As expected from the high level of positive dependence, note that the values of the process tend to stay at the same level of previous observations. This fact is reflected in Figure 12.4 which shows the sample ACF and PACF. The first-order autoregressive dependence is clearly observable in these two plots.

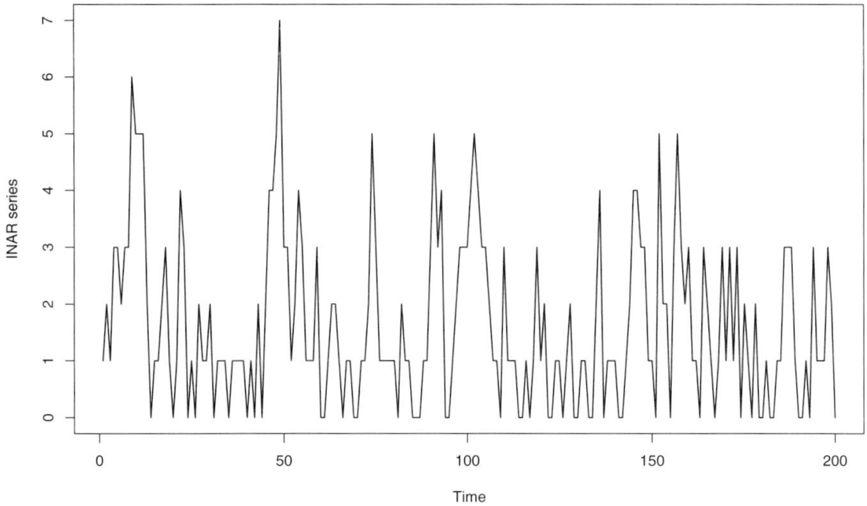

Figure 12.1 *Simulated INAR(1) process with $\alpha = 0.6$, $n = 200$, and Poisson input noise.*

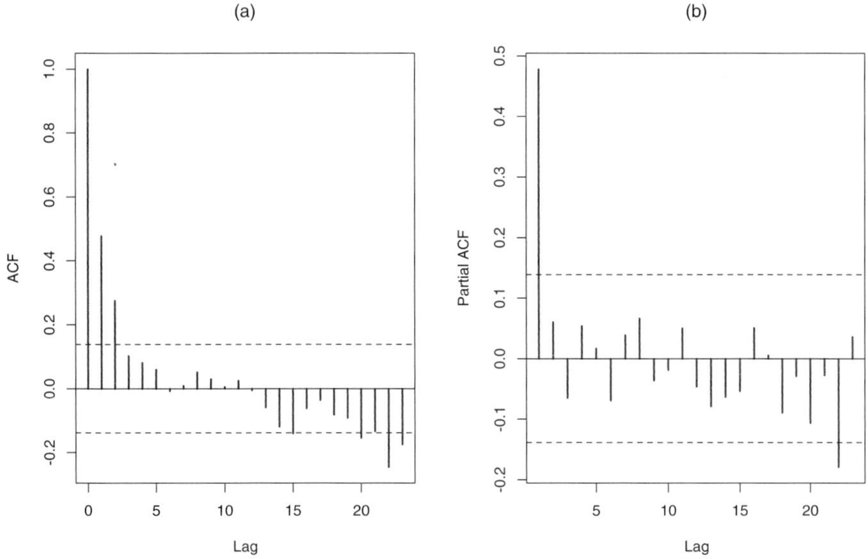

Figure 12.2 *(a) Sample ACF, and (b) sample PACF of the simulated INAR(1) process with $\alpha = 0.6$, $n = 200$ and Poisson input noise.*

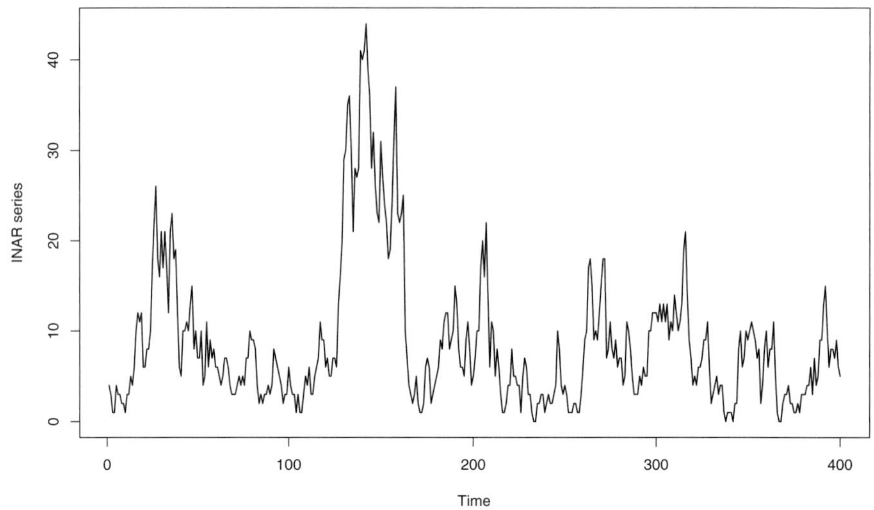

Figure 12.3 *Simulated INAR(1) process with $\alpha = 0.9$, $n = 400$ and Poisson input noise.*

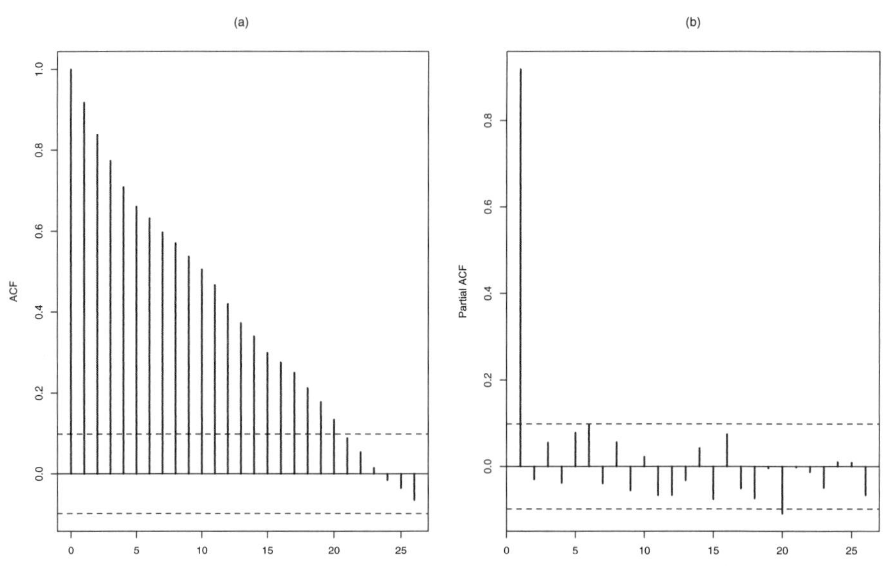

Figure 12.4 *(a) Sample ACF, and (b) sample PACF of the simulated INAR(1) process with $\alpha = 0.9$, $n = 400$ and Poisson input noise.*

The class of INAR models is versatile enough to handle additional exogenous variables such as trends or cycles. Consider for example the INAR(1) model with a time regressor described by

$$\begin{aligned} y_t &= \mu_t + z_t, \\ \mu_t &= \beta_0 + \beta_1 t, \\ z_t &= \alpha \circ z_{t-1} + \varepsilon_t. \end{aligned}$$

A particular application of the R package inarmix allows for the simulation and estimation of this model. Figure 12.5 shows a realization of INAR(1) process with $\alpha = 0.8$, $n = 200$ and linear trend $\mu_t = t - 0.1$. The output from *inarmix* is given below. Note that the estimated values are very close to their theoretical counterparts.

```
> oneclassfit <- inarmix(y~time,nclasses=1,id=subject,data=y)
> summary(oneclassfit)

 Call:
 inarmix(formula = y ~ time, nclasses = 1, id = subject, data = y)

                    Estimate        Std.Err       Z value      Pr(>z)
(Intercept) -5.975488e-01  5.472015e-05 -10920.08760 < 2.22e-16 ***
```

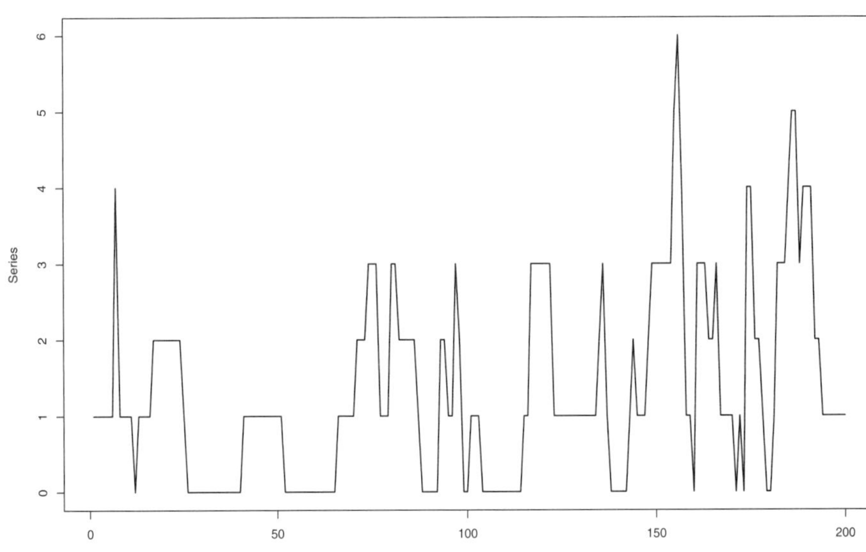

Figure 12.5 *Simulated INAR(1) process with $\alpha = 0.8$, $n = 200$, and linear trend $\mu_t = t - 0.1$.*

```
time          1.814660e+00 2.115357e-04  8578.50244 < 2.22e-16 ***
autocorr.     7.867933e-01 1.308610e-03   601.24361 < 2.22e-16 ***
scale         1.243723e+00 8.862073e-03    27.50184 < 2.22e-16 ***

 ------------------------------
log-likelihood: -211.5405
           BIC:  438.9760
           AIC:  429.0811
```

12.1.2 Conditional Distribution Models

The following class of models allows for the handling a large family of time series data distributions, including as a particular case, the normal distribution. We begin the discussion with a very simple example. Consider the Gaussian AR(1) model $y_t = \phi\, y_{t-1} + \varepsilon_t$, where ε_t is a sequence of independent normally distributed random variable with zero-mean and variance σ^2. Thus, we can write the model in terms of the conditional distribution specification,

$$\begin{aligned} y_t | y_{t-1} &\sim \mathrm{N}(\mu_t, \sigma^2), \\ \mu_t &= \phi\, y_{t-1}. \end{aligned}$$

Note that in this formulation of the AR(1) model, $E(y_t|y_{t-1}) = \mu_t$ and $\mathrm{Var}(y_t|y_{t-1}) = \sigma^2$. This simple model can be readily extended to an AR(p) process,

$$\begin{aligned} y_t | \mathcal{F}_{t-1} &\sim \mathrm{N}(\mu_t, \sigma^2), \\ \mu_t &= \phi_1\, y_{t-1} + \phi_2\, y_{t-2} + \cdots + \phi_p\, y_{t-p}. \end{aligned}$$

On the other hand, we can also allow for a different conditional distribution as well as a general *link function* $\ell(\cdot)$,

$$\begin{aligned} y_t | \mathcal{F}_{t-1} &\sim \mathrm{D}(\mu_t), \\ \ell(\mu_t) &= \mu_0 + \phi_1\, y_{t-1} + \phi_2\, y_{t-2} + \cdots + \phi_p\, y_{t-p}. \end{aligned}$$

Important examples of these more general processes are the Poisson distribution and the Negative Binomial distribution in the case of integer-valued time series and the Gamma and the log-Normal distributions in the case of positive data.

■ **EXAMPLE 12.1**

As a first illustration, Figure 12.6 displays a realization of a conditional Poisson AR(1) process with $\phi = 0.7$, $\mu_0 = 5.0$ and $n = 200$. Additionally, Figure 12.7 exhibits its sample ACF and sample PACF. As in the examples of INAR(1) models, in this case the first-order autoregressive dependence of the conditional model is also reflected in the observed values of the process.

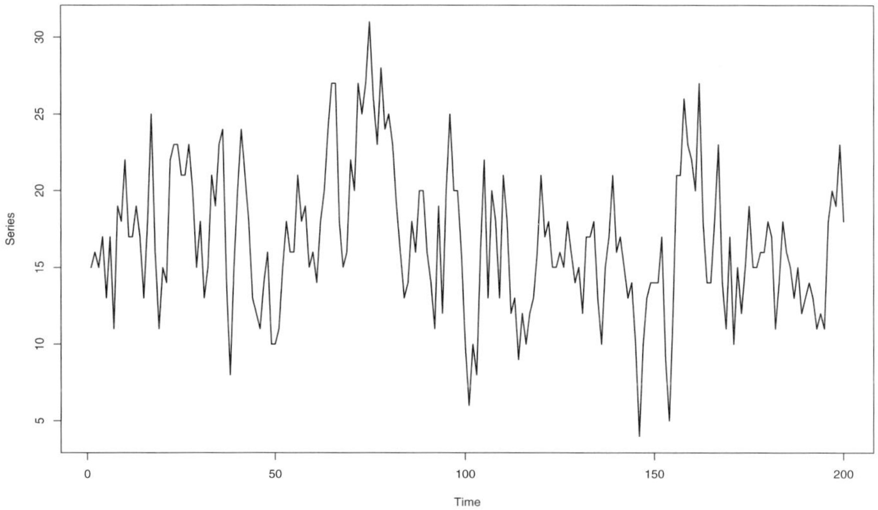

Figure 12.6 *Simulated conditional Poisson process with $\phi = 0.7$, $\mu_0 = 5$, and $n = 200$.*

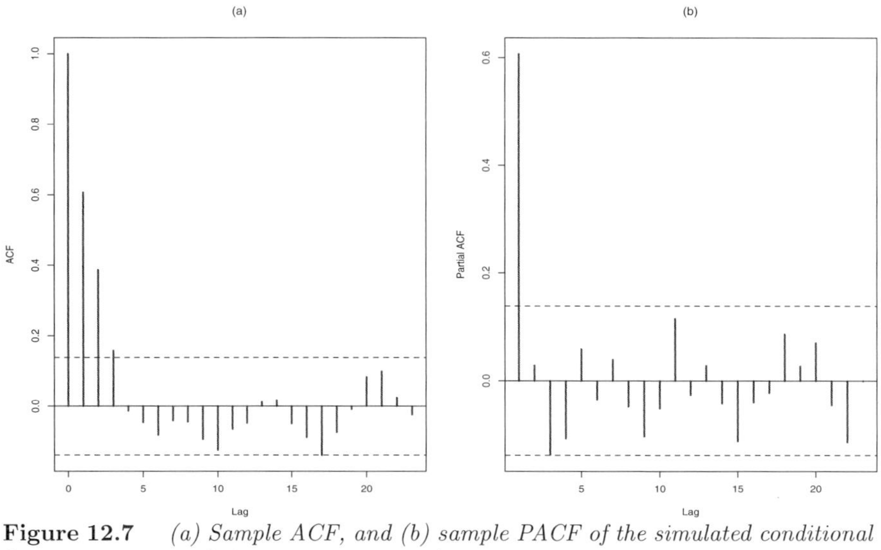

Figure 12.7 *(a) Sample ACF, and (b) sample PACF of the simulated conditional Poisson process with $\phi = 0.7$, $\mu_0 = 5$, and $n = 200$.*

EXAMPLE 12.2

The second case considers a Poisson conditional distribution but with negative first-order autocorrelation. Figure 12.8 displays a realization of a conditional Poisson AR(1) process with $\phi = -0.7$ and $n = 300$ observations. Additionally, Figure 12.9 exhibits its sample ACF and sample PACF. Notice that the negative first-order autoregressive dependence of the conditional model is reflected in the observed values of the process and on their sample ACF and PACF.

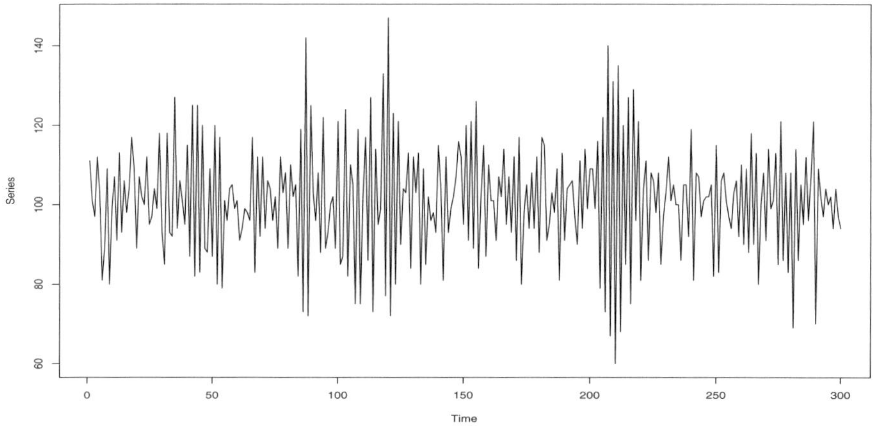

Figure 12.8 *Simulated conditional Poisson AR(1) process with $\phi = -0.7$ and $n = 300$.*

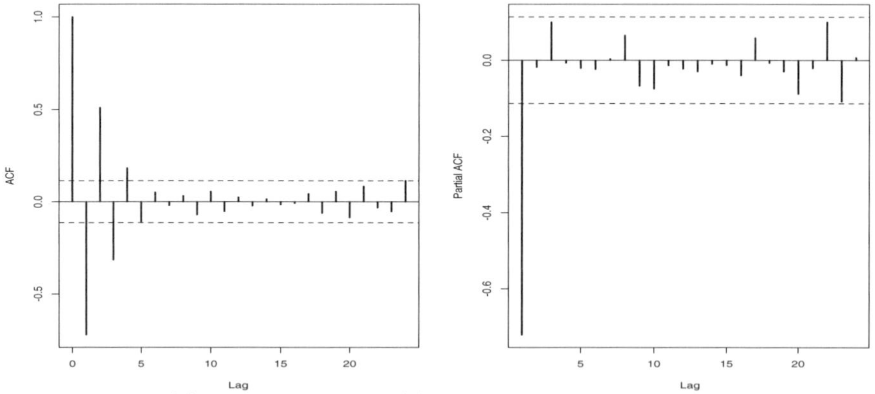

Figure 12.9 *(a) Sample ACF, and (b) sample PACF of the simulated Conditional Poisson AR(1) process with $\phi = -0.7$, and $n = 300$.*

EXAMPLE 12.3

This illustration considers a more complex serial dependence structure. Figure 12.10 displays a realization of a conditional Poisson ARMA$(1, 1)$ process with $\phi = 0.8$, $\theta = 0.6$ and $n = 300$ while Figure 12.11 exhibits its sample ACF and sample PACF. Observe that in this case, the ACF decays slowly to zero.

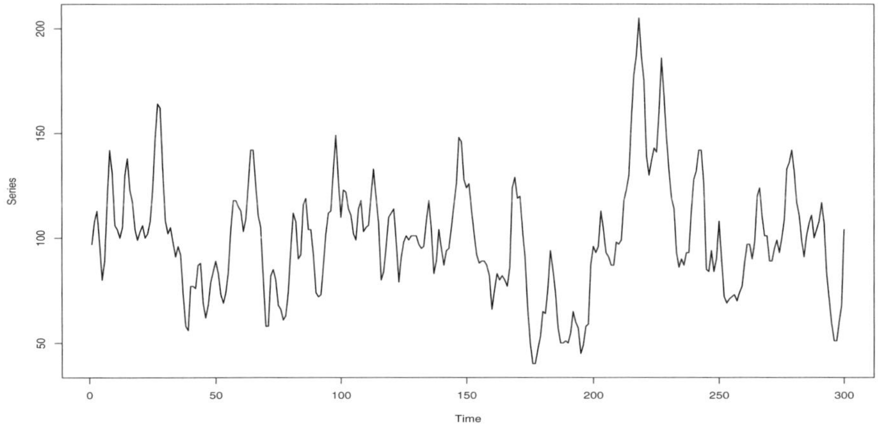

Figure 12.10 *Simulated conditional Poisson ARMA(1,1) process with* $\phi = 0.8$, $\theta = 0.6$ *and* $n = 300$.

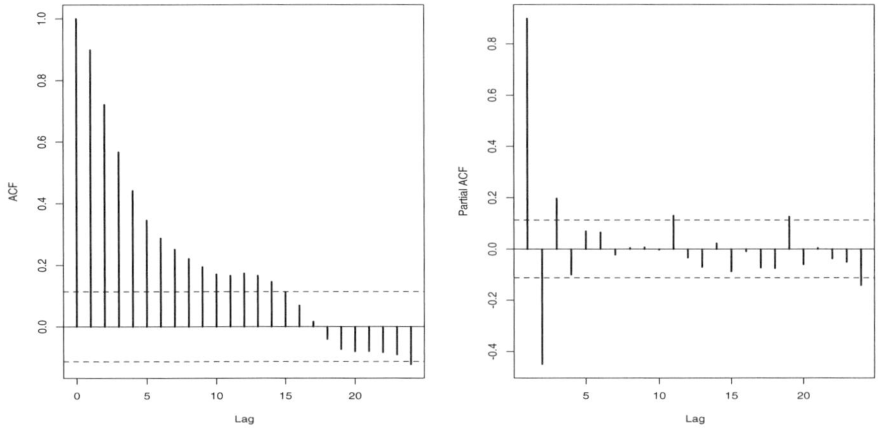

Figure 12.11 *(a) Sample ACF, and (b) sample PACF of the simulated conditional Poisson ARMA(1,1) process with* $\phi = 0.8$, $\theta = 0.6$, *and* $n = 300$.

■ EXAMPLE 12.4

The following example corresponds to an illustration with positive time series data. Figure 12.12 exhibits 200 simulated observations of a Gamma conditional distribution AR(1) model with autoregressive parameter $\phi = 0.7$, $\mu_0 = 3$ and *rate* 1. Notice that in this case the mean of the process is

$$E\, y_t = \frac{\mu_0}{1-\phi} = 10.$$

Thus, it is expected that this simulated series displays observations around that mean.

The sample ACF and sample PACF of this conditional Gamma process are shown in Figure 12.13. The AR(1) autocorrelation and partial autocorrelation structure of the model is clearly noticeable in these plots. In fact, as expected, the first component of the sample PACF has a value around 0.7.

Finally, Figure 12.14 shows the estimated density of the simulated series.

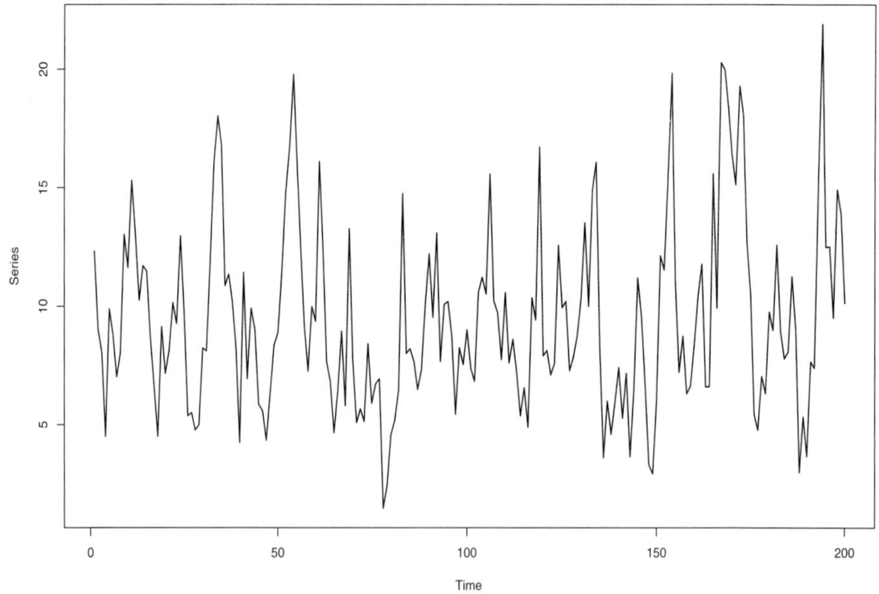

Figure 12.12 *Simulated conditional Gamma process with $\phi = 0.7$, $\mu_0 = 3$, rate = 1 and $n = 200$.*

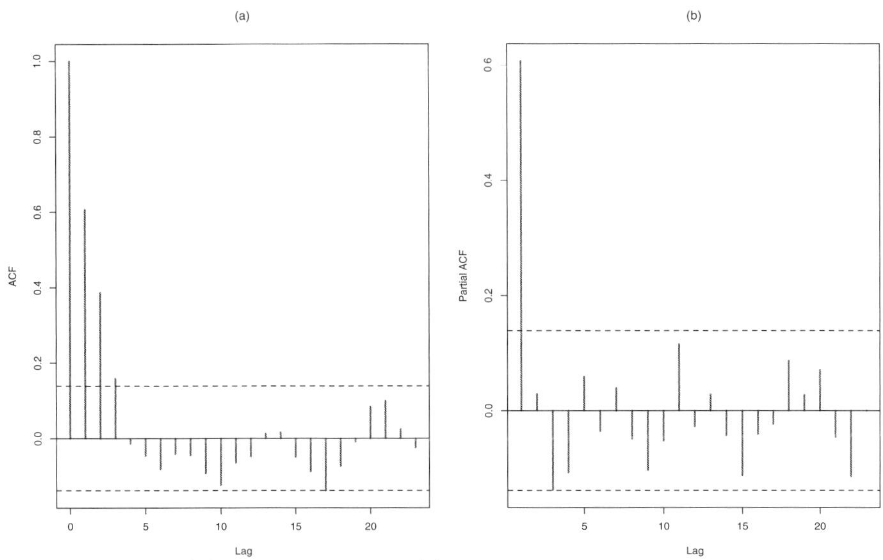

Figure 12.13 *(a) Sample ACF, and (b) sample PACF of the simulated conditional Gamma process with $\phi = 0.7$, rate $= 1$, and $n = 200$.*

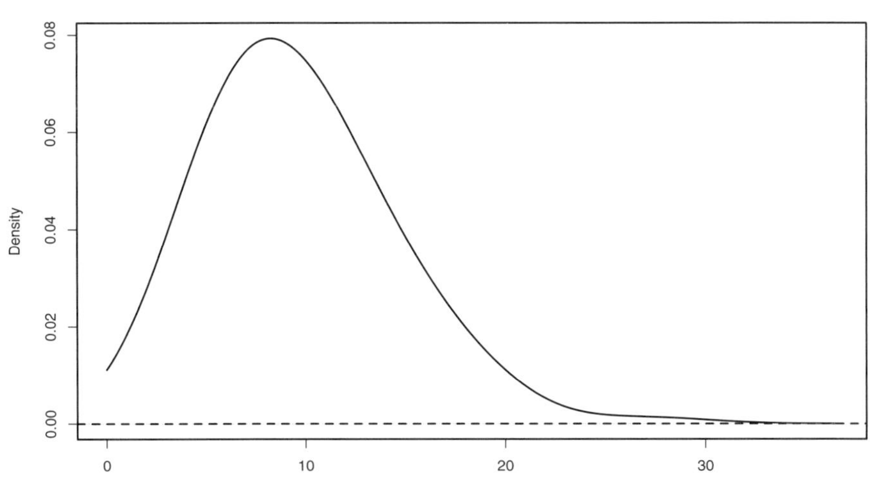

Figure 12.14 *Estimated density of the simulated Conditional Gamma process with $\phi = 0.7$, rate $= 1$, and $n = 200$.*

12.2 PARAMETER DRIVEN MODELS

The conditional distribution method can also be specified in terms of a latent process $\{\theta_t\}$, as for example,

$$\begin{aligned} y_t|\mathcal{F}_{t-1} &\sim \mathrm{D}(\theta_t), \\ \theta_t &= \theta_0 + \alpha\,\theta_{t-1} + \eta_t, \end{aligned}$$

where η_t is a white noise sequence. In this case, the *parameter* θ_t evolves according to an AR(1) process and its serial dependence is transferred to the *observed process* y_t.

As an illustration, consider the following ARMA(1, 1) parameter driven Poisson model defined by

$$\begin{aligned} y_t|\mathcal{F}_{t-1} &\sim \mathrm{Poisson}(\exp(\theta_t)), \\ \theta_t &= \phi\,\theta_{t-1} + \eta_t - \varphi\,\eta_{t-1}, \end{aligned}$$

with $\phi = 0.5$, $\varphi = 0.7$ and $\sigma^2 = 1$. Figure 12.15 exhibits a realization of this process with 500 observations. Figure 12.15(a) shows the series y_t while Figure 12.15(b) displays the underlying parameter process θ_t. Recall that in the context of a parameter driven modeling, the latent process θ_t corresponds to a *standard* ARMA model. The sample ACF and PACF of this process is exhibited in Figure 12.16.

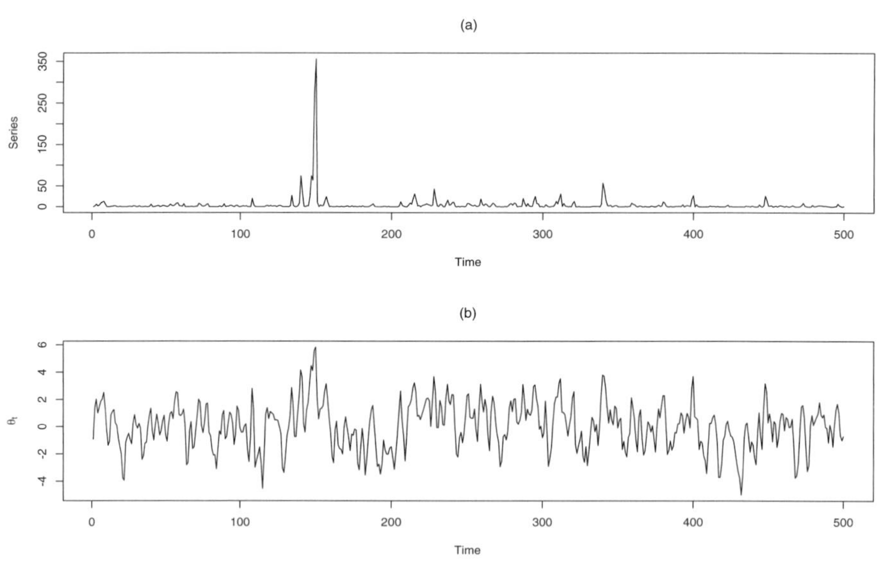

Figure 12.15 *Simulated parameter driven ARMA(1,1) Poisson process with $\phi = 0.5$, $\varphi = 0.7$, $\sigma^2 = 1$ and $n = 500$. (a) Time series, (b) evolution of θ_t.*

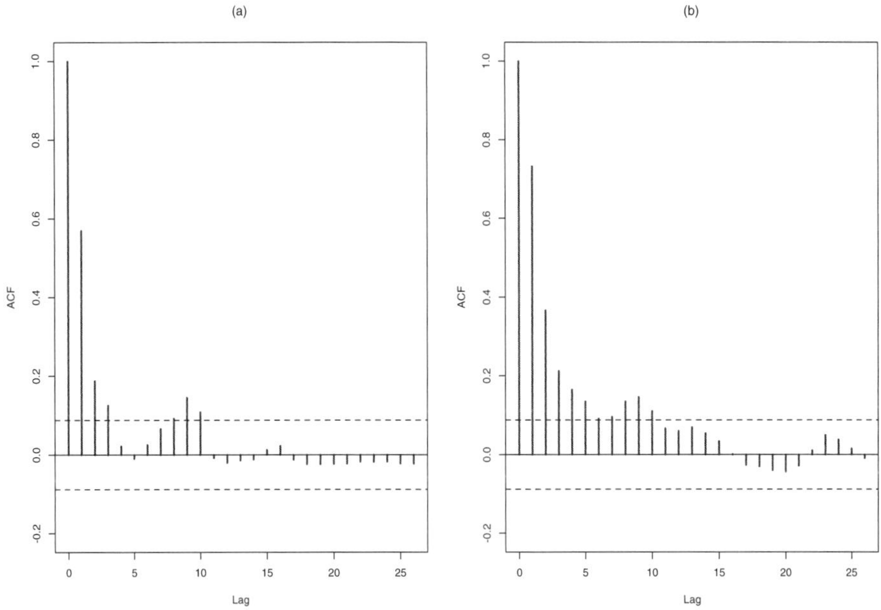

Figure 12.16 *(a) Sample ACF of the simulated parameter driven ARMA(1,1) Poisson process with $\phi = 0.5$, $\varphi = 0.7$, $\sigma^2 = 1$ and $n = 500$, (b) Sample ACF of θ_t.*

12.3 ESTIMATION

In order to provide a general framework for the estimation of different data contexts, let $G(\alpha, \beta)$ be a distribution corresponding to a discrete or continuous nonnegative random variable with mean α and variance β. Let g be a positive function; μ be a constant and $\{\pi_j\}_{j \geq 0}$ be a summable sequence of real numbers, that is $\sum_{j=0}^{\infty} |\pi_j| < \infty$. A conditional distribution process $\{y_t\}$ is defined as

$$y_t | \mathcal{F}_{t-1} \sim G(\lambda_t, g(\lambda_t)), \tag{12.1}$$

$$\lambda_t = \mu \sum_{j=0}^{\infty} \pi_j - \sum_{j=1}^{\infty} \pi_j y_{t-j}, \tag{12.2}$$

where $\mathcal{F}_t = \{y_t, y_{t-1}, ...\}$ represents the information up to time t. In addition, the conditional distribution function, G, may depend on other parameters besides π_j and μ. These parameters will be denoted by the vector η. Conditional on the information $\mathcal{F}_{t-1}$, y_t has distribution G with variance $\text{Var}[y_t|\mathcal{F}_{t-1}] = g(\lambda_t)$, which is a function of the conditional mean $\mathbf{E}[y_t|\mathcal{F}_{t-1}] = \lambda_t$. Model (12.1) - (12.2) can be written in different ways. For instance, if we define the sequence $\varepsilon_t = y_t - \mathbf{E}[y_t|\mathcal{F}_{t-1}] = y_t - \lambda_t$ then $\mathbf{E}(\varepsilon_t) = 0$ and, if $\mathbf{E}[g(\lambda_t)]$ is finite and constant, then $\{\varepsilon_t\}$ is an *innovation* process, that is, a zero-mean, uncorrelated sequence with finite constant variance.

Consider the model described by (12.1) - (12.2). Then, $\mathrm{Var}(\varepsilon_t) = \mathbf{E}[g(\lambda_t)]$, $\mathrm{Cov}(\varepsilon_t, \varepsilon_s) = 0$, for all $t \neq s$. If $\mathrm{Var}(\varepsilon_t)$ is a finite constant then $\{\varepsilon_t\}$ is an innovation process.

By replacing $\lambda_t = y_t - \varepsilon_t$ in (12.2) and since $\pi_0 = 1$, we can write the model as follows,

$$\Pi(B)(y_t - \mu) = \varepsilon_t,$$

where $\Pi(B) = \sum_{j=0}^{\infty} \pi_j B^j$, B is the backshift operator such that $By_t = y_{t-1}$. This representation is familiar in time series analysis with independent perturbations ε_t. For example, a conditional ARMA(p, q) can be written as

$$\Pi(B) = \Phi(B)\Theta(B)^{-1}$$

where $\Phi(B) = 1 - \phi_1 B - \cdots - \phi_p B^p$ and $\Theta(B) = 1 - \theta_1 B - \cdots - \theta_q B^q$.

Thus, model (12.1) - (12.2) with parametrization (12.3) satisfies,

$$\Phi(B)\,(y_t - \mu) = \Theta(B)\varepsilon_t, \tag{12.3}$$

$$\varepsilon_t | \mathcal{F}_{t-1} \sim (0, g(\lambda_t)). \tag{12.4}$$

To ensure the causality and invertibility of the filter (12.3), it is assumed that the polynomials $\Phi(B)$ and $\Theta(B)$ have no common roots, these are all located outside the unit circle. Note that in (12.3), even though the sequence $\{\varepsilon_t\}$ is not a strict white noise (independent), it is uncorrelated under some conditions. Therefore, the process $\{y_t\}$ and the ARMA model with independent input error sequence $\{\varepsilon_t\}$ have the same autocorrelation function

The setup provided by equations (12.3) - (12.4) is general enough to allow for modeling data with diverse distributions. For instance, the conditional distribution G, may belong to the exponential family with support in $\mathbb{R}_+$ such as Binomial, Gamma or Poisson distributions. In what follows, we discuss briefly these examples corresponding to continuous and discrete conditional distributions G, respectively.

(a) *Conditional Poisson:* Define the model as $y_t|\mathcal{F}_{t-1} \sim Poi(\lambda_t)$ where $\mathcal{Y} = \{0, 1, 2, \ldots\}$, η is null, and $g(\lambda_t) = \lambda_t$. In this case, $\mathbf{E}\{g(\lambda_t)\} = \mathbf{E}[\lambda_t] = \mu$.
(b) *Conditional Binomial:* Consider the model $y_t|\mathcal{F}_{t-1} \sim Bin(m, p_t)$ where n is fixed, $\mathcal{Y} = \{0, 1, 2, \ldots\}$, η is null, $\lambda_t = mp_t$ and $g(\lambda_t) = \lambda_t(m - \lambda_t)/m$. In this case, g is concave and bounded by $m/4$.
(c) *Conditional Gamma:* Let $y_t|\mathcal{F}_{t-1} \sim Gamma(\lambda_t/\beta, \beta)$, with $\eta = \beta > 0$, $\mathcal{Y} = (0, \infty)$, and $g(\lambda_t) = \beta\lambda_t$. For this distribution we have $\mathbf{E}\{g(\lambda_t)\} = \beta\mathbf{E}[\lambda_t]$.

Assume that the time series data $\{y_1, \ldots, y_n\}$ are generated by model (12.1)– (12.2) with parametrization (12.3). The vector of unknown parameters is denoted by θ. First, we estimate the mean of the processes. A simple estimator for the level of the process is the arithmetic mean, $\hat{\mu}_n = \frac{1}{n}\sum_{t=1}^{n} y_t$.

Once the mean μ is estimated by $\hat{\mu}_n$, the parameters δ and η may be estimated by using the maximum likelihood method. For computing the likelihood, we replace μ by $\hat{\mu}_n$. The *conditional* pseudo log-likelihood is given by

$$\mathcal{L}(\theta) = \sum_{t=2}^{n} \ell_t(\theta).$$

where $\ell_t(\theta) = \log f_\theta(y_t|\mathcal{F}_{t-1})$ and the contribution of the first observation, usually negligible for long time series, has been removed. Given that the conditional distribution G is a member of the exponential family we write:

$$f_\theta(y_t|\mathcal{F}_{t-1}) = a(\lambda_t, \eta)\psi(y_t) \exp\left\{\sum_{i=1}^{m} b_i(\lambda_t, \eta) R_i(y_t)\right\} \tag{12.5}$$

where the functions $a^*(\cdot)$ and $b_i(\cdot)$ depend on the information $\mathcal{F}_{t-1}$ only through λ_t and the functions ψ^* and R_i do not depend on δ and η. Then,

$$\ell_t(\theta) = C(\lambda_t) + \psi(y_t) + \sum_{i=1}^{m} b_i(\lambda_t) R_i(y_t), \tag{12.6}$$

In order to obtain the maximum likelihood estimator $(\hat{\delta}, \hat{\eta})$ and its precision, we need to calculate the *score* and the Hessian. The *score* is given by,

$$\frac{\partial \mathcal{L}(\theta)}{\partial \theta} = \left(\sum_{t=2}^{n} \frac{\partial \ell_t(\theta)}{\partial \theta}\right)$$

and the Hessian matrix is

$$\nabla^2 \mathcal{L}(\theta) = \left(\sum_{t=2}^{n} \frac{\partial^2 \ell_t(\theta)}{\partial^2 \theta_i \theta_j}\right)_{i,j=1,\ldots,p}.$$

EXAMPLE 12.5

Consider a conditional Poisson AR(1) model

$$y_t \sim \text{Poi}(\lambda_t)$$
$$\lambda_t = \mu_0 + \phi\, y_{t-1}$$

In this case the conditional distribution of y_t is

$$f(y_t|\lambda_t) = e^{-\lambda_t} \frac{\lambda_t^{y_t}}{y_t!}.$$

Consequently,

$$\ell_t(\theta) = -\lambda_t + y_t \log \lambda_t - \log y_t!,$$

where the parameter of interest is in this case $\theta = (\mu_0, \phi)$. The parameter space is given by $\Theta = (0, \infty) \times (-1, 1)$.

EXAMPLE 12.6

As an illustration of a time series model consisting of positive values, consider the conditional Gamma AR(1) model

$$y_t \sim \text{Gamma}(\lambda_t, \beta)$$
$$\lambda_t = \mu_0 + \phi\, \beta\, y_{t-1}$$

In this case the conditional distribution of y_t is

$$f(y_t|\lambda_t) = \frac{\beta^{\lambda_t}}{\Gamma(\lambda_t)} y_t^{\lambda_t - 1} e^{-\beta y_t}$$

Thus,

$$\ell_t(\theta) = \lambda_t \log(\beta) + \log \Gamma(\lambda_t) + (\lambda_t - 1)\log y_t - \beta y_t,$$

where $\theta = (\mu_0, \phi)$ is a two dimensional parameter vector.

EXAMPLE 12.7

An ACP(1,1) model is defined

$$\begin{aligned} y_t|\mathcal{F}_{t-1} &\sim \text{Poi}(\lambda_t) \\ \lambda_t &= \mu_0 + \alpha\, y_{t-1} + \beta\, \lambda_{t-1}. \end{aligned}$$

The mean of this process is given by

$$\mu = E\, y_t = \frac{\mu_0}{1 - \alpha - \beta},$$

while its variance is

$$\text{Var}\, y_t = \frac{\mu[1 - (\alpha + \beta)^2 + \alpha^2]}{1 - (\alpha + \beta)^2}.$$

Notice that even though the conditional distribution of y_t is equidispersed the distribution of the process is over dispersed since $\text{Var}\, y_t \geq E\, y_t$.

Figure 12.17 shows the evolution of the expected value of an ACP(1,1) process y_t as a function of $\alpha \in (0, 0.89)$ for $\beta = 0.1$ and $\mu_0 = 1$. On the other hand, the variance of this ACP(1,1) process is displayed in Figure 12.18. In addition, the ratio between the expected mean an variance of the ACP(1,1) process as a function of $\alpha \in (0, 0.89)$ is reported in Figure 12.19. Observe that in these three plots, the mean, variance and ratio increase substantially as the first-order autoregressive parameter α increases.

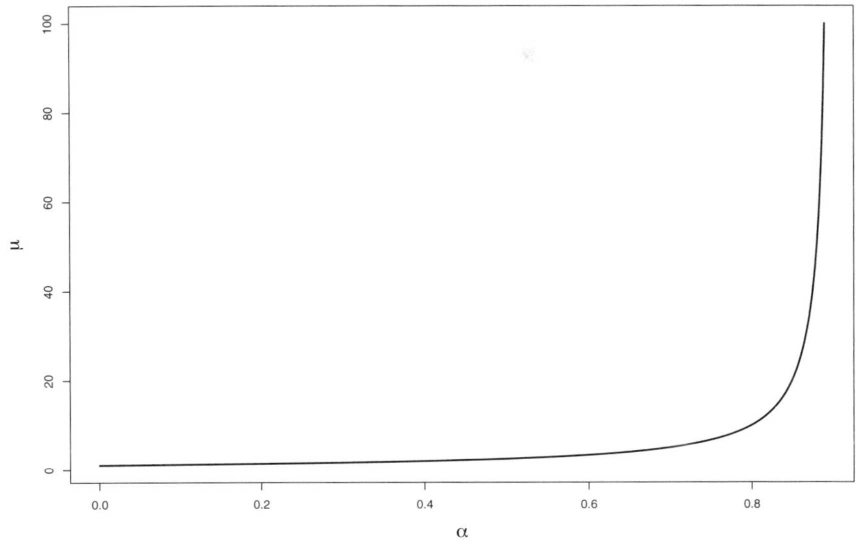

Figure 12.17 *Expected value of the ACP(1,1) process* y_t *as a function of* $\alpha \in (0, 0.89)$ *for* $\beta = 0.1$ *and* $\mu_0 = 1$.

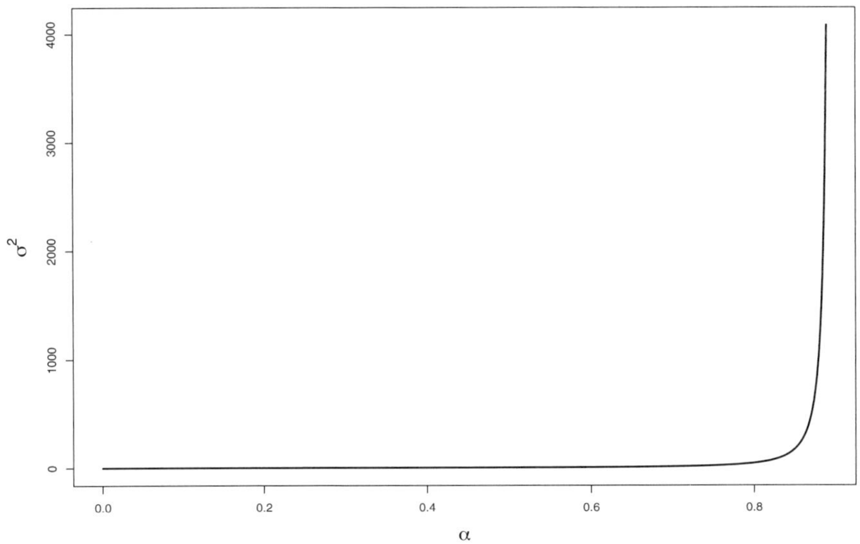

Figure 12.18 *Variance of the ACP(1,1) process* y_t *as a function of* $\alpha \in (0, 0.89)$ *for* $\beta = 0.1$ *and* $\mu_0 = 1$.

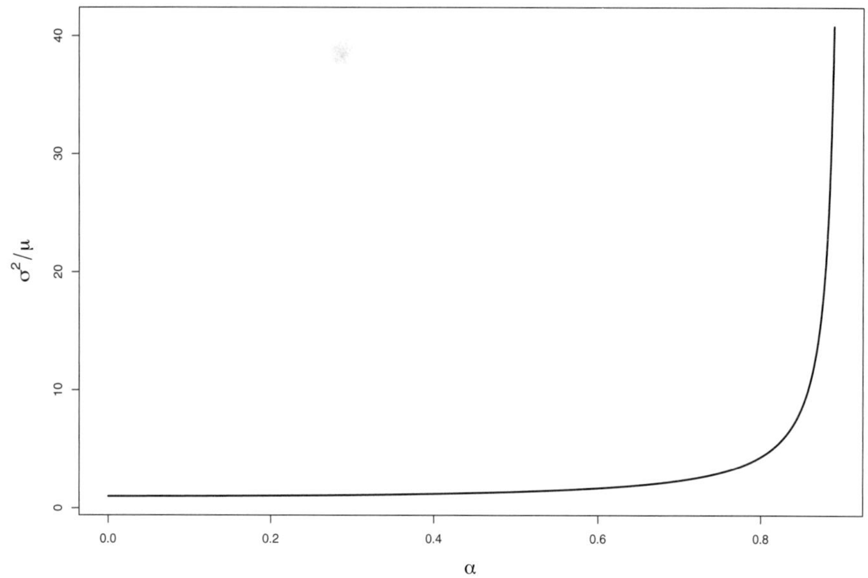

Figure 12.19 *Ratio between the expected mean an variance of the ACP(1,1) process y_t as a function of $\alpha \in (0, 0.89)$ for $\beta = 0.1$ and $\mu_0 = 1$.*

The autocorrelation structure of the ACP(1,1) model is given by

$$\rho(h) = (\alpha + \beta)^{|h|-1} \frac{\alpha\,[1 - \beta(\alpha + \beta)]}{1 - (\alpha + \beta)^2 + \alpha^2}$$

The R package ACP allows for the estimation and forecasting of conditional ARMA Poisson processes. Consider for example the ACP(1,1) model with the following parametrization,

$$\begin{aligned} y_t | \mathcal{F}_{t-1} &\sim \text{Poi}(\lambda_t) \\ \lambda_t &= a + b\, y_{t-1} + c\, \lambda_{t-1}. \end{aligned}$$

Figure 12.20 exhibits a simulated ACP(1,1) process with $a = 5$, $b = 0.6$, $c = 0.3$ and $n = 200$. Furthermore, Figure 12.21 shows the sample ACF and the sample PACF.

The estimated parameters from the ACP package are reported in the following output. Notice that even though in this example the program overestimate the intercept a, the estimators of b and c are close to their theoretical counterparts.

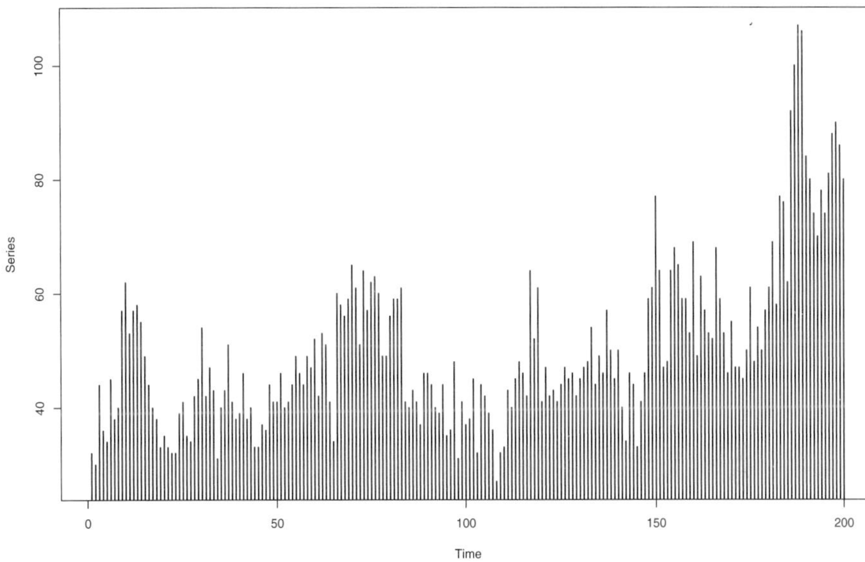

Figure 12.20 *Simulated ACP(1,1) process with* $a = 5$, $b = 0.6$, $c = 0.3$ *and* $n = 200$.

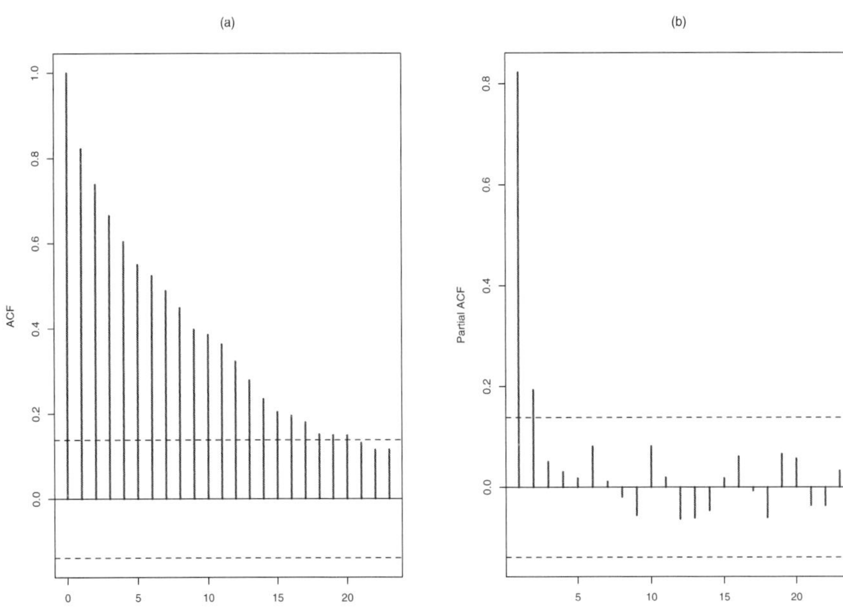

Figure 12.21 *(a) Sample ACF, and (b) PACF of the simulated ACP(1,1) process with* $a = 5$, $b = 0.6$, $c = 0.3$ *and* $n = 200$.

```
> summary(mod1)

Call:

acp.formula(formula = z ~ -1, data = zdata, p = 1, q = 1)

      Estimate    StdErr t.value    p.value
a    3.400014 1.599880  2.1252    0.03482 *
b 1 0.568413 0.065342  8.6990 1.322e-15 ***
c 1 0.368228 0.078570  4.6866 5.174e-06 ***

---

Signif. codes:  0 '***' 0.001 '**' 0.01 '*' 0.05 '.' 0.1 ' ' 1
```

Figure 12.22 displays the one-step forecasts generated by the ACP package. The heavy line corresponds to the observations while the broken line corresponds to the one=step predictors generated by the ACP(1, 1) model. Notice that the predicted values are close to the true values of the process. Observe also that one-step predictors have relatively less variation than the true observations. In particular, they usually do not reach the highest o lowest observed values.

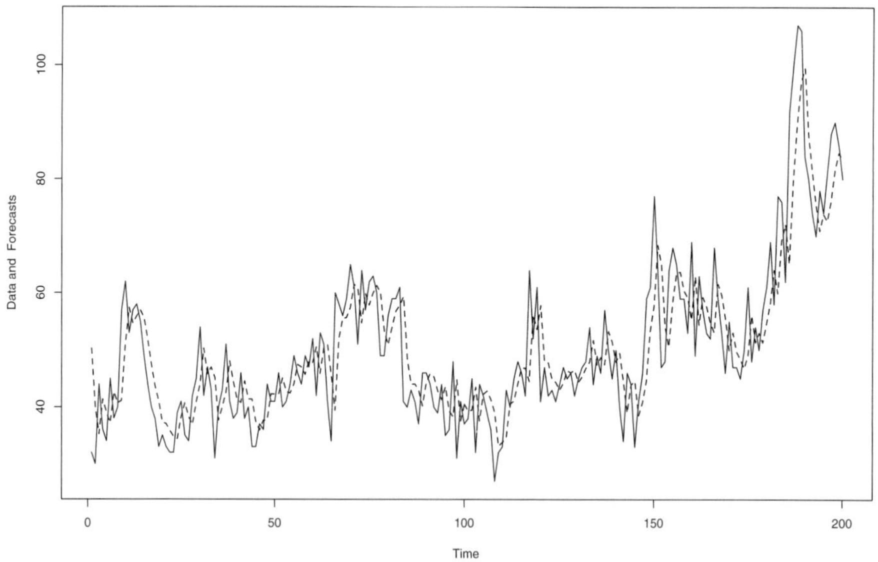

Figure 12.22 *Forecasts of the simulated ACP(1,1) process with $a = 5$, $b = 0.6$, $c = 0.3$ and $n = 200$. Heavy line: Observations, broken line: One=step predictors.*

12.3.1 Monte Carlo Experiments

In order to gain some insight on the finite sample performance of the quasi maximum likelihood estimator described above, the results from several Monte Carlo experiments are presented next.

In the following examples, we present first a simulated trajectory of the corresponding models and then we report the results for the parameter estimates.

EXAMPLE 12.8

Figure 12.23 displays a simulated trajectory of 500 observations of a conditional AR(1) Poisson model with $\phi = 0.7$ and $\mu_0 = 5$ while Figure 12.24 shows its sample ACF and sample PACF.

Table 12.1 shows the simulation results from a Poisson conditional distribution AR(1) model for different values of the parameters ϕ and μ_0, and two sample sizes $n = 500$ and $n = 1000$. The simulations results are based on 1000 repetitions. Observe that the estimated parameters are close to their true values.

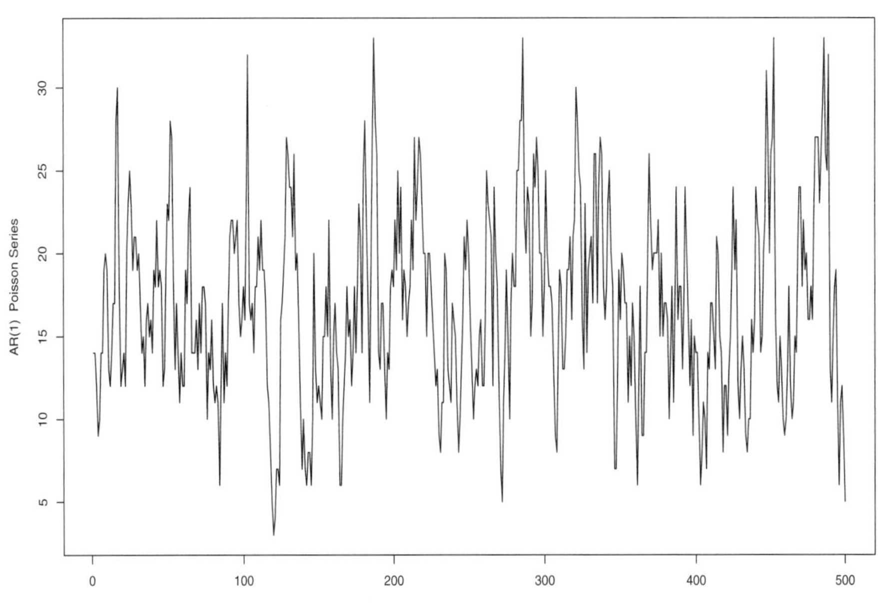

Figure 12.23 *Simulated conditional AR(1) Poisson model with* $\phi = 0.7$, $\mu_0 = 5$ *and* $n = 500$.

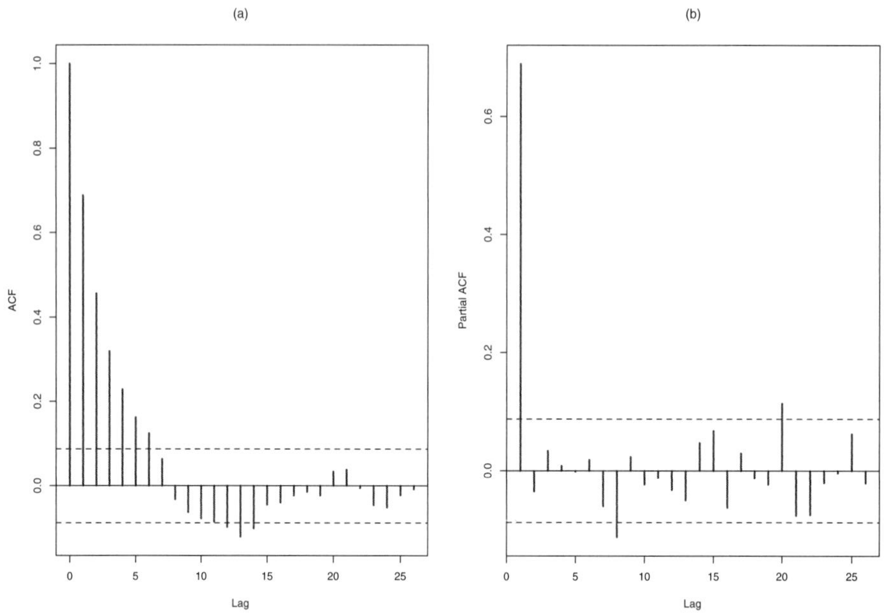

Figure 12.24 *(a) Sample ACF, and (b) PACF of the simulated conditional AR(1) Poisson model with* $\phi = 0.7$, $\mu_0 = 5$ *and* $n = 500$.

Table 12.1 Monte Carlo Experiments for Conditional AR(1) Poisson Models.

		$n = 500$			
ϕ	μ_0	$\hat{\phi}$	$\hat{\mu}_0$	$\widehat{\sigma}(\hat{\phi})$	$\widehat{\sigma}(\hat{\mu}_0)$
0.7	2	0.6961	2.0191	0.0371	0.2282
0.4	2	0.3922	2.0263	0.0485	0.1831
0.7	15	0.6926	15.3586	0.0351	1.7483
0.4	15	0.3915	15.2095	0.0485	1.2391
		$n = 1000$			
ϕ	μ_0	$\hat{\phi}$	$\hat{\mu}_0$	$\widehat{\sigma}(\hat{\phi})$	$\widehat{\sigma}(\hat{\mu}_0)$
0.7	2	0.6971	2.0172	0.0239	0.1550
0.4	2	0.3994	1.9987	0.0298	0.1082
0.7	15	0.6961	15.2008	0.0231	1.1204
0.4	15	0.3951	15.1268	0.0303	0.7357

EXAMPLE 12.9

As a another illustration of the estimation methodology of a non-Gaussian time series model, Figure 12.25 displays a simulated trajectory of 500 observations of a conditional ARMA(1,1) Poisson model with $\phi = 0.2$, $\theta = 0.6$ and $\mu_0 = 2$. Its sample ACF and sample PACF are shown in Figure 12.26. The mean of this process is given by

$$\mu = \frac{\mu_0}{1 - \phi - \theta},$$

which yields in this case $\mu = 10$. Observe that the values of this time series fluctuate around this mean.

Table 12.2 shows the simulation results from a Poisson conditional distribution ARMA$(1,1)$ model for different values of the parameters ϕ, θ, μ_0, and two sample sizes $n = 500$ and $n = 1000$. The simulations results are based on 1000 repetitions. Observe that the estimated parameters are close to their true values.

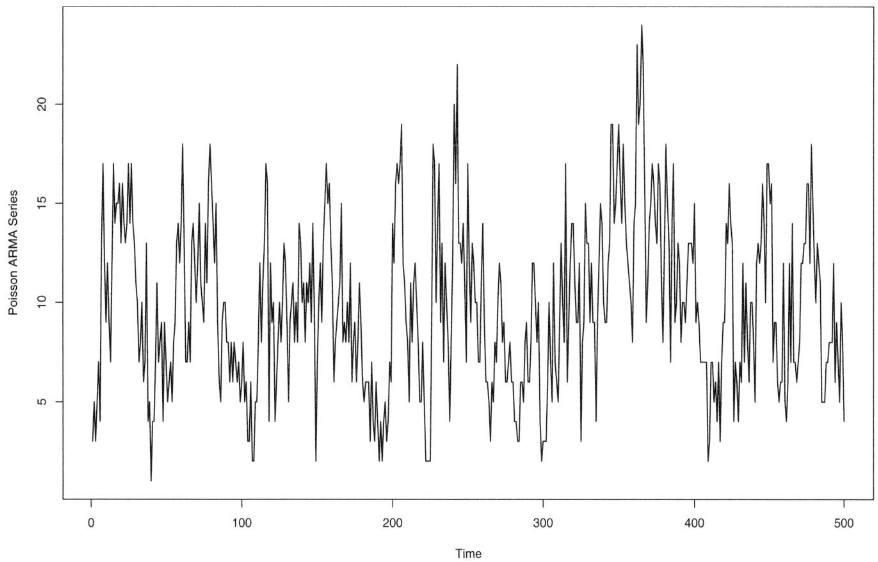

Figure 12.25 *Simulated conditional ARMA(1,1) Poisson model with* $\phi = 0.2$, $\theta = 0.6$, $\mu_0 = 2$ *and* $n = 500$.

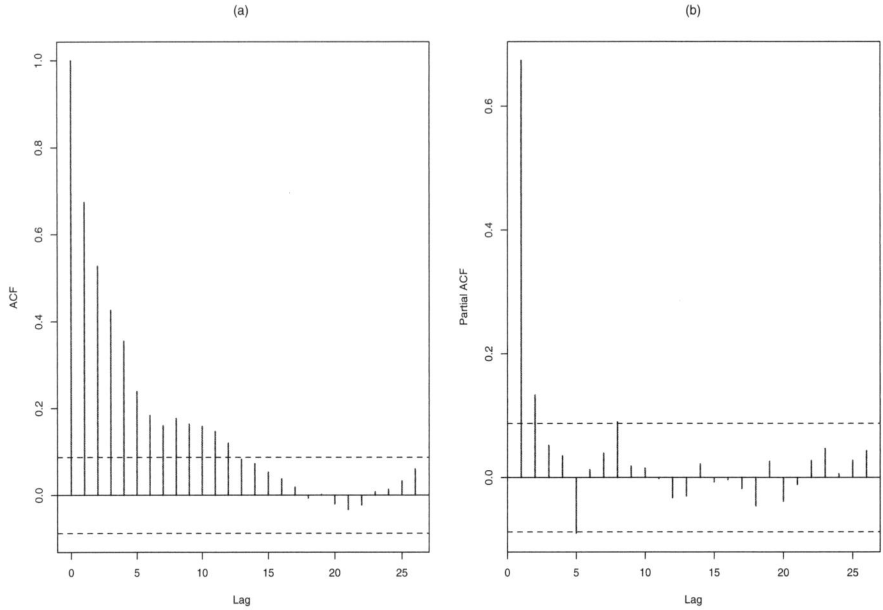

Figure 12.26 *Sample ACF and PACF of the simulated conditional ARMA(1) Poisson model with* $\phi = 0.2$, $\theta = 0.6$, $\mu_0 = 2$ *and* $n = 500$.

Table 12.2 Estimation of Conditional Poisson ARMA(1,1) models.

$n = 500$								
ϕ	θ	μ_0	$\hat{\phi}$	$\hat{\theta}$	$\hat{\mu}_0$	$\widehat{\sigma}(\hat{\phi})$	$\widehat{\sigma}(\hat{\theta})$	$\widehat{\sigma}(\hat{\mu}_0)$
0.7	0.1	2	0.6974	0.0924	2.0921	0.0513	0.0660	0.4079
0.1	0.7	2	0.1021	0.5870	3.1073	0.0425	0.2830	1.7308
0.6	0.2	2	0.5918	0.1983	2.0984	0.0502	0.0695	0.4553
0.2	0.6	2	0.1988	0.5660	2.3504	0.0452	0.1331	1.1086

$n = 1000$								
ϕ	θ	μ_0	$\hat{\phi}$	$\hat{\theta}$	$\hat{\mu}_0$	$\widehat{\sigma}(\hat{\phi})$	$\widehat{\sigma}(\hat{\theta})$	$\widehat{\sigma}(\hat{\mu}_0)$
0.7	0.1	2	0.6987	0.0974	2.0393	0.0334	0.0436	0.2709
0.1	0.7	2	0.1016	0.6569	2.4127	0.0274	0.1439	1.3249
0.6	0.2	2	0.5970	0.1974	2.0530	0.0336	0.0461	0.2832
0.2	0.6	2	0.2000	0.5862	2.1378	0.0306	0.0726	0.5684

■ **EXAMPLE 12.10**

Finally, Table 12.3 reports the estimation results from a set of Monte Carlo simulations of conditional Gamma ARMA(1,1) models with 500 observations, for different values of the parameters ϕ, θ and μ_0. Similarly to the previous illustrations, in this case the estimated parameters are close to their theoretical counterparts, even though the method seems to overestimate the parameter μ_0.

Table 12.3 Simulations of Conditional Gamma ARMA(1,1) models with 500 observations.

ϕ	θ	μ_0	$\hat{\phi}$	$\hat{\theta}$	$\hat{\mu}_0$	$\widehat{\sigma}(\hat{\phi})$	$\widehat{\sigma}(\hat{\theta})$	$\widehat{\sigma}(\hat{\mu}_0)$
0.5	0.3	2	0.4934	0.2936	2.1099	0.0612	0.4242	0.0439
0.3	0.6	2	0.2955	0.5982	2.1319	0.0547	0.6661	0.0381
0.4	0.4	5	0.3930	0.3966	5.2732	0.0687	1.1231	0.0425
0.6	0.3	3	0.5959	0.2927	3.3513	0.0609	0.8592	0.0512
0.2	0.5	2	0.1951	0.4822	2.1475	0.1265	0.7460	0.0388

12.3.2 Diagnostics

Non-Gaussian time series modeling involves both the experience of the analyst and the type of the problem under study. In practice, the nature of the variables defines the sample space. For example, counting processes lead naturally to discrete positive data. Besides, the distribution of data can be specified by means of tools such as histograms and $q-q$ plots. In some cases, the conditional distribution defines the form of $g(\lambda_t)$, for example, for a Poisson distribution, $g(\lambda_t) = \lambda_t$. But, in other situations we have some flexibility when defining $g(\lambda_t)$. The task of determining the conditional variance can be helped by observing the patterns of data and correlation structure from simulated time series. The sample autocorrelation function of both the observations and their squares, may give some clues about the underlying dependence structure of the data. Furthermore, the residuals $e_t = y_t - \hat{\lambda}_t$ can be used for assessing the goodness of fit, by checking the absence of correlation on the residuals.

12.3.3 Prediction

The distribution of the process $\{y_t\}$ conditional on the past information, $\mathcal{F}_{t-1}$, has mean λ_t. Therefore a natural one-step predictor of λ_t is $\hat{\lambda}_t$ which is based on (12.2),

$$\hat{\lambda}_t = \hat{\mu} \sum_{j=0}^{t-1} \hat{\pi}_j - \sum_{j=1}^{t-1} \hat{\pi}_j y_{t-j},$$

where each $\hat{\pi}_j$ depends on the parameter estimates $\hat{\delta}, \hat{\eta}$. Hence, the estimate conditional distribution is $y_t|\mathcal{F}_{t-1} \sim G(\hat{\lambda}_t, g(\hat{\lambda}_t))$ and construction of conditional prediction bands for one-step forecasts is based on the quantiles of this distribution.

12.4 DATA ILLUSTRATIONS

In what follows we illustrate the application of the techniques discussed in this chapter to the analysis of two real-life time series. The first case corresponds to the study IBM trading volume and the second application concerns the time series of glacial varves measurements discussed in Chapter 1.

12.4.1 IBM Trading Volume

Figure 12.27 exhibits the daily IBM trading volume in millions of transactions, from January 2, 2011 to December 31, 2014. The sample ACF and sample PACF of this time series are displayed in Figure 12.28.

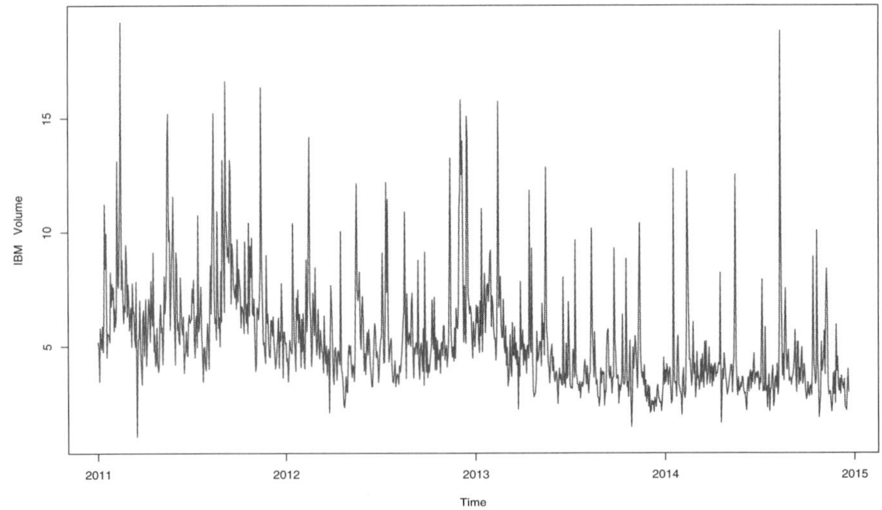

Figure 12.27 *Daily trading IBM volume in millions of transactions, from January 2, 2011 to December 31, 2014.*

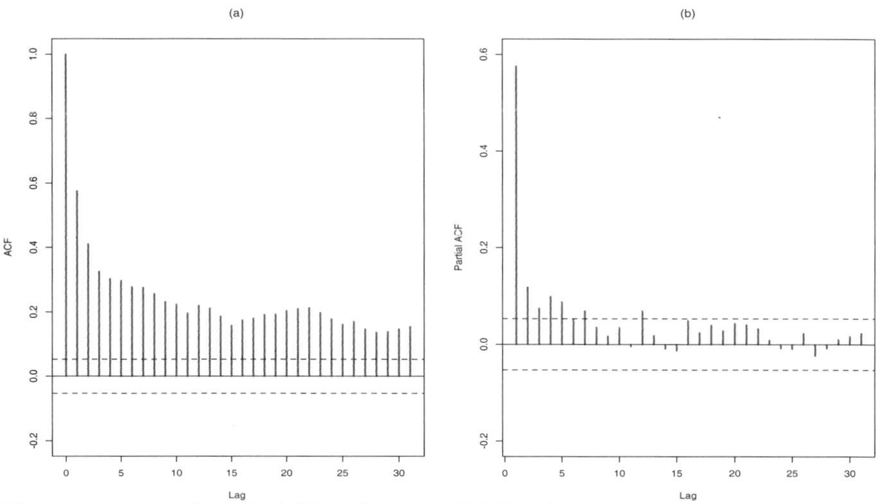

Figure 12.28 *Sample ACF and sample PACF of the daily trading IBM volume time series.*

The fitted ACP model for this time series is

```
Call:
acp.formula(formula = z ~ -1, data = zdata, p = 1, q = 1)

     Estimate   StdErr t.value    p.value
a    1.425442 0.263163  5.4166 7.606e-08 ***
b 1 0.529531 0.048736 10.8653 < 2.2e-16 ***
c 1 0.163728 0.080931  2.0231   0.04333 *
---
Signif. codes:  0 '***' 0.001 '**' 0.01 '*' 0.05 '.' 0.1 ' ' 1
```

On the other hand, the standardized residuals from this model are reported in Figure 12.29 along with the sample ACF. Notice that these residuals seem to be white noise. This hypothesis is formally tested by means of the Box-Ljung statistic,

```
> Box.test(e,lag=10,type="Ljung")

Box-Ljung test

data:  e
X-squared = 11.2483, df = 10, p-value = 0.3385
```

Furthermore, one-step forecasts are displayed in Figure 12.30. These predictions (broken line) seem to follows very closely the observe values (gray line).

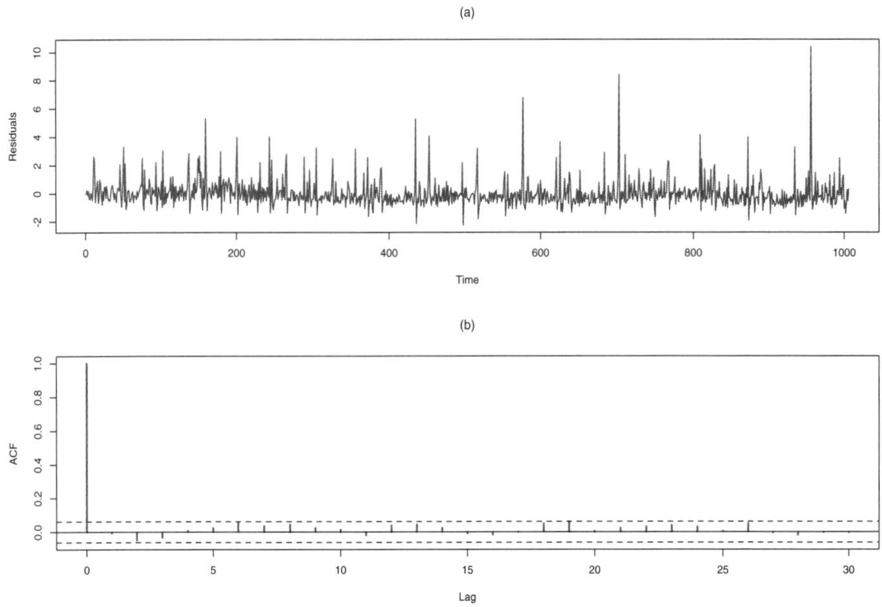

Figure 12.29 *(a) Residuals from the ACP(1,1) model of the IBM data. (b) Sample ACF of the residuals.*

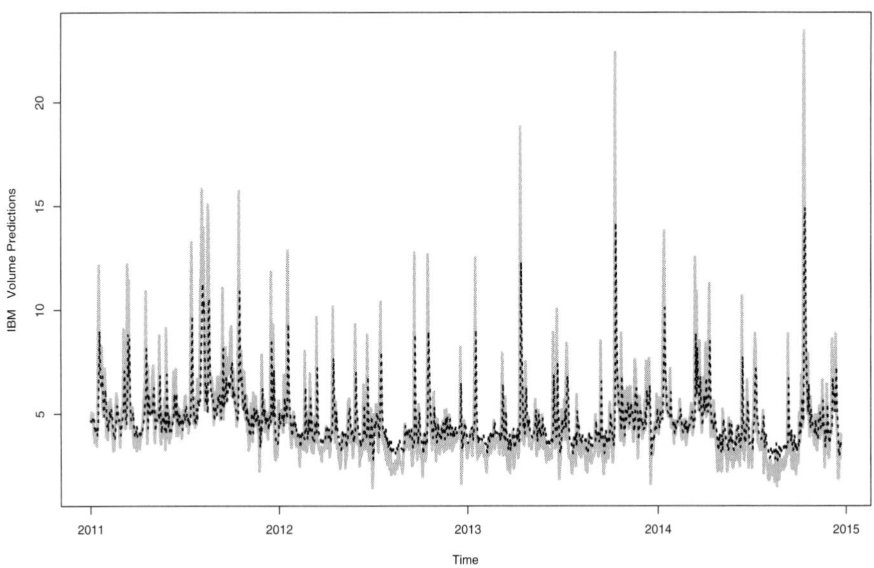

Figure 12.30 *One-step predictions from the ACP(1,1) model. IBM data (gray line) and One-step forecasts (broken line).*

12.4.2 Glacial Varves

As a second real-life data illustration of the techniques discussed in the previous sections for handling non-Gaussian time series, we consider a dataset containing measurements of the thicknesses of the yearly varves at one location in Massachusetts for the period 11,833–11,200 BC, see Figure 12.31.

As explained in Chapter 1, the analysis of this type of time series is highly relevant in the context of climatic and paleoclimatic studies.

The time series discussed here corresponds to a sequence of positive observations with an estimated density depicted in Figure 12.32. From this plot, it seems that the data could be fitted by a Gamma model. Compare, for example, with Figure 12.33 which shows the density of a sample of 1000 observations from a Gamma(λ, β) distribution with shape $\lambda = 1.88$ and rate $\beta = 0.068$.

According to this preliminary study, a conditional Gamma ARMA model is suggested for this time series. The selected model by the Akaike's criterion is an ARMA$(1, 1)$ process.

The sample autocorrelation structure of these data is exhibited in Figure 12.34.

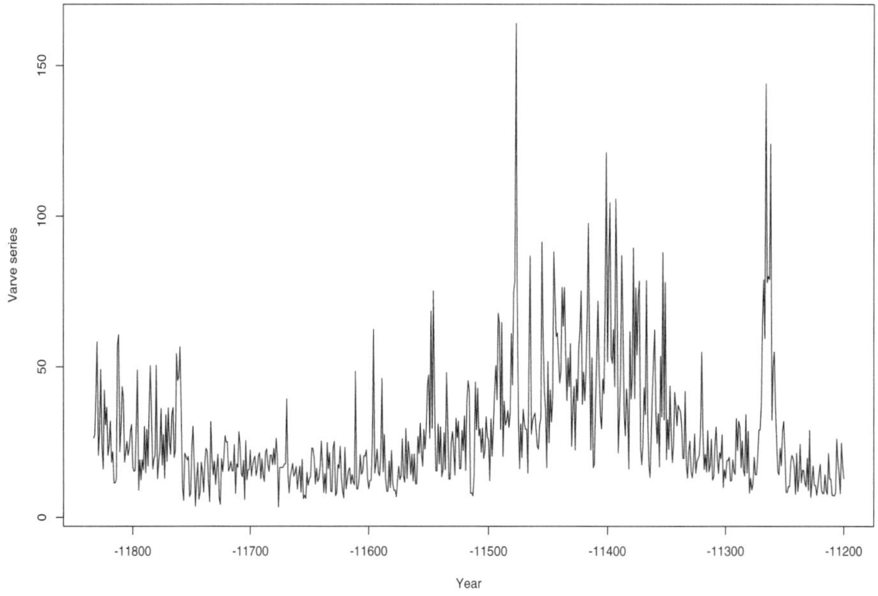

Figure 12.31 *Glacial varves time series data.*

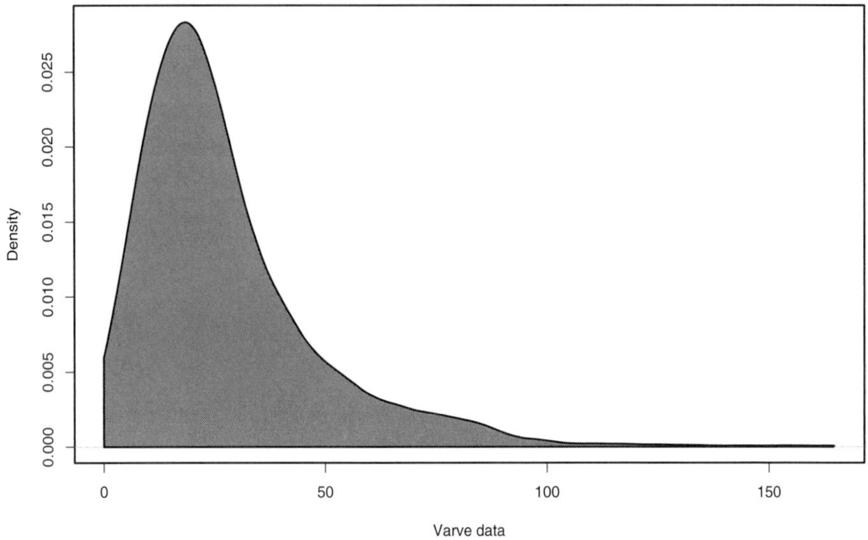

Figure 12.32 *Density of the glacial varves time series.*

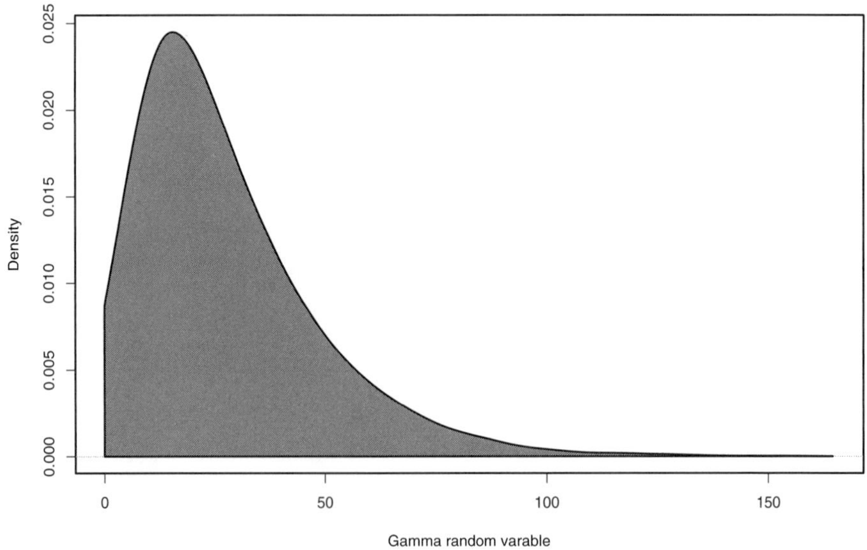

Figure 12.33 *Density of a Gamma*(λ, β) *distribution with shape* $\lambda = 1.88$ *and rate* $\beta = 0.068$.

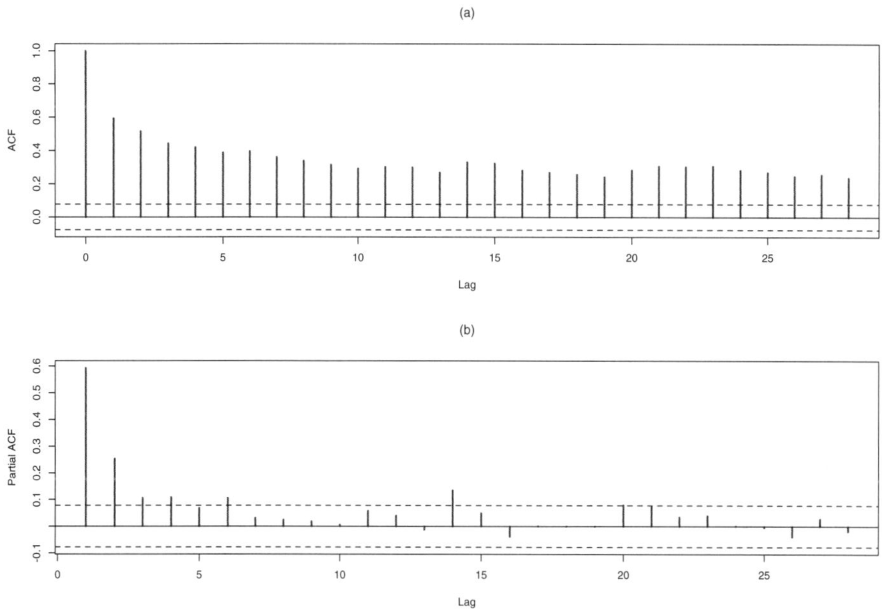

Figure 12.34 *(a) Sample ACF, and (b) sample PACF of the glacial varves.*

The fitted parameters are reported in Table 12.4 along with their standard errors and Student tests.

Figure 12.35 exhibits the sample ACF of the residuals and the Box-Ljung test suggests that they are compatible with the white noise hypothesis.

The one-step forecasts for this time series are displayed in Figure 12.36. Observe that the predictors are close to their observe values. Additionally, 95% prediction bands are provided in Figure 12.37. Notice that the predictions bands in this case are based on the 2.5% and 97.5% quantiles of the corresponding Gamma(λ_t, β) distribution.

```
> Box.test(e,lag=10,type="Ljung")
Box-Ljung test
X-squared = 11.2483, df = 10, p-value = 0.3385
```

Table 12.4 Gamma ARMA(1,1) estimates for the Glacial Varves Data.

	$\widehat{\mu}_0$	$\widehat{\phi}$	$\widehat{\theta}$
Parameter	0.2805	0.2762	0.6258
S.D.	0.0778	0.0298	0.0472
t-test	3.6054	9.2699	13.2588

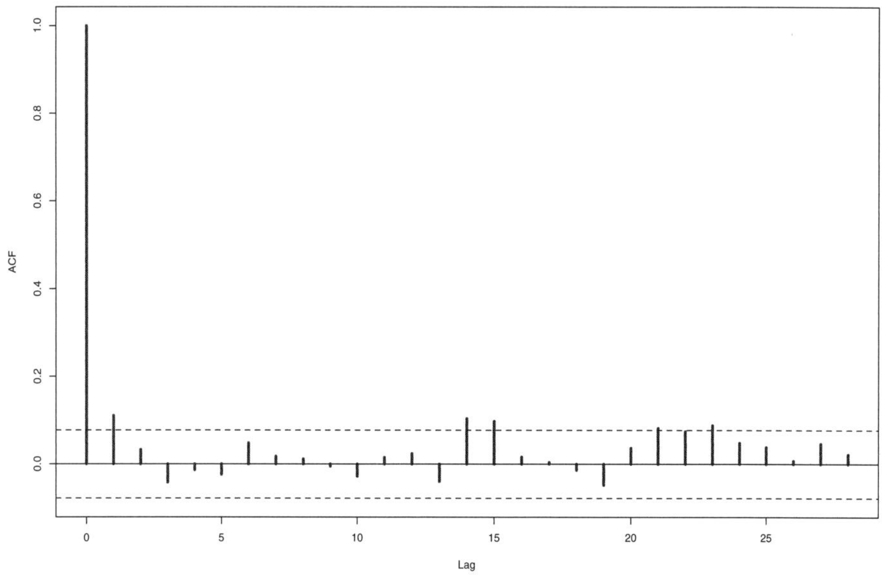

Figure 12.35 *Sample ACF of the residuals from the fitted glacial varves model.*

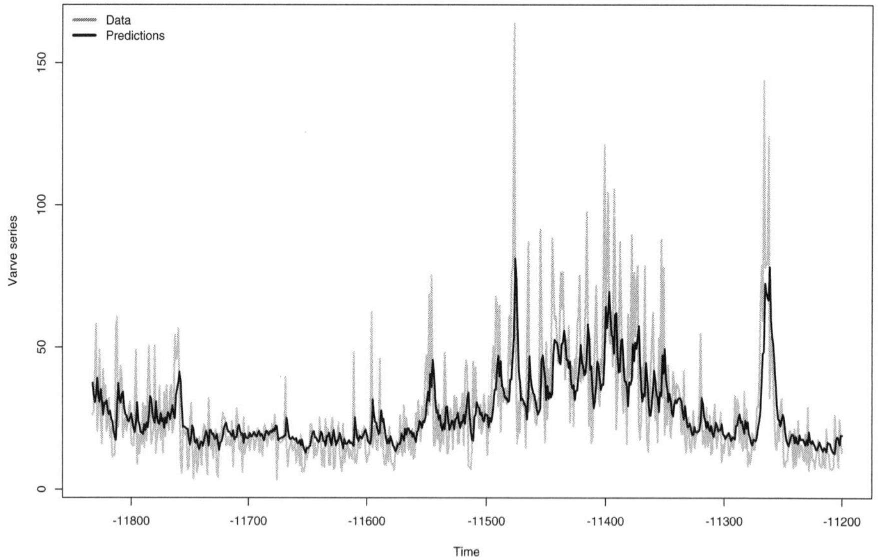

Figure 12.36 *One-step predictions of the glacial varves time series.*

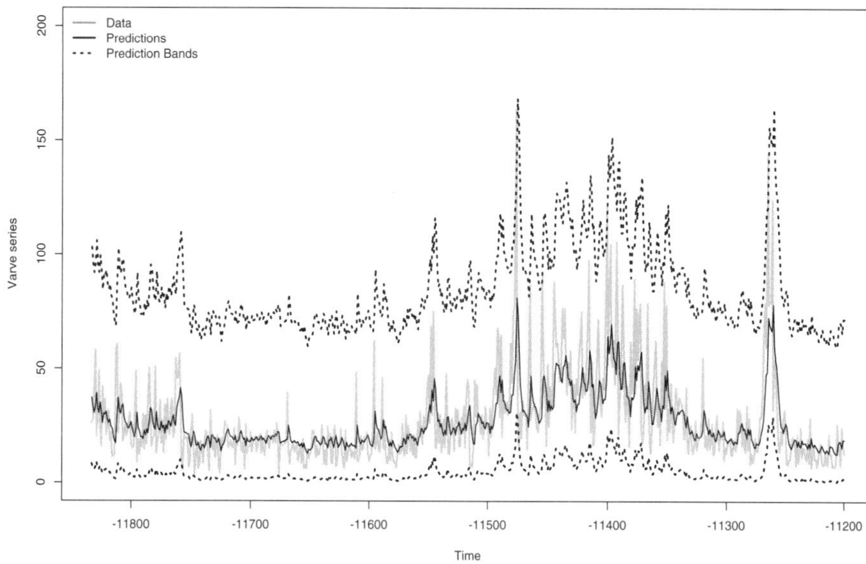

Figure 12.37 *Predictions bands of the glacial varves time series (95%).*

12.4.3 Voting Intentions

In this data illustration we consider the UK voting intention time series data introduced in Chapter 1. Figure 12.38 depicts the combined voting intentions of both the Conservative Party and the Labor Party. Note that the combined voting intention of these two political parties is expressed here as a proportion of the total.

On the other hand, Figure 12.39 displays the sample ACF and the PACF. Additionally, Figure 12.40 shows the estimated probability density of the data.

Based on the nature of the data as well as their dependence structure, the following conditional AR(p) Beta model is proposed

$$
\begin{aligned}
y_t|y_{t-1}, y_{t-2}, \dots &\sim \text{Beta}(a_t, b_t),\\
E(y_t|y_{t-1}, y_{t-2}, \dots) &= \mu_t,\\
\mu_t &= \mu_0 + \phi_1\, y_{t-1} + \phi_2\, y_{t-2} + \cdots + \phi_p\, y_{t-p},
\end{aligned}
$$

where the Beta distribution has probability density given by

$$f(y) = \frac{\Gamma(a_t + b_t)}{\Gamma(a_t)\Gamma(b_t)} y^{a_t - 1}(1 - y)^{b_t - 1}.$$

The mean of this Beta distribution is

$$\mu_t = \frac{a_t}{a_t + b_t}$$

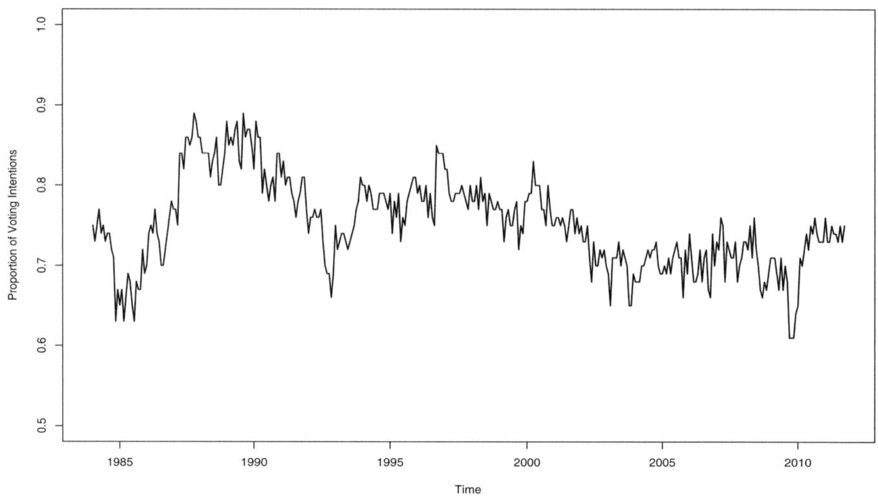

Figure 12.38 *Voting intention of the Conservative Party and the Labor Party time series.*

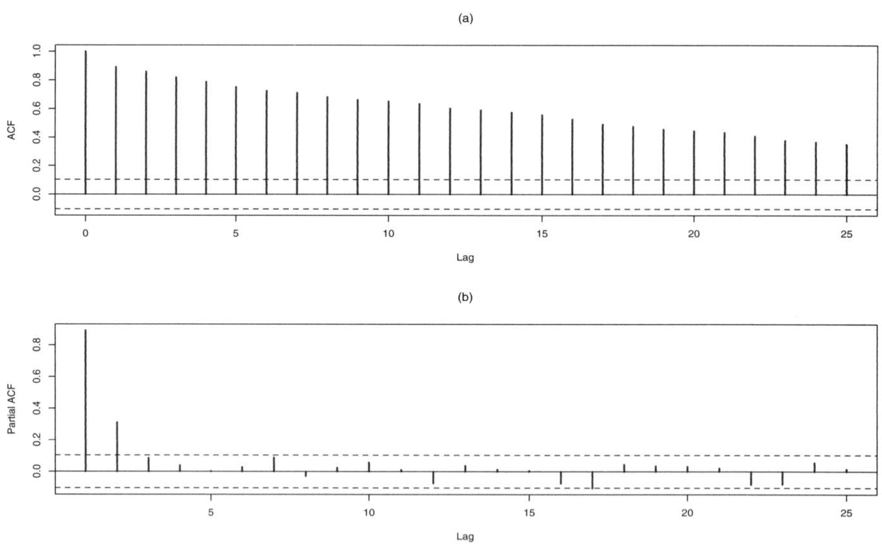

Figure 12.39 *(a) Sample ACF, (b) and PACF of the voting intention time series.*

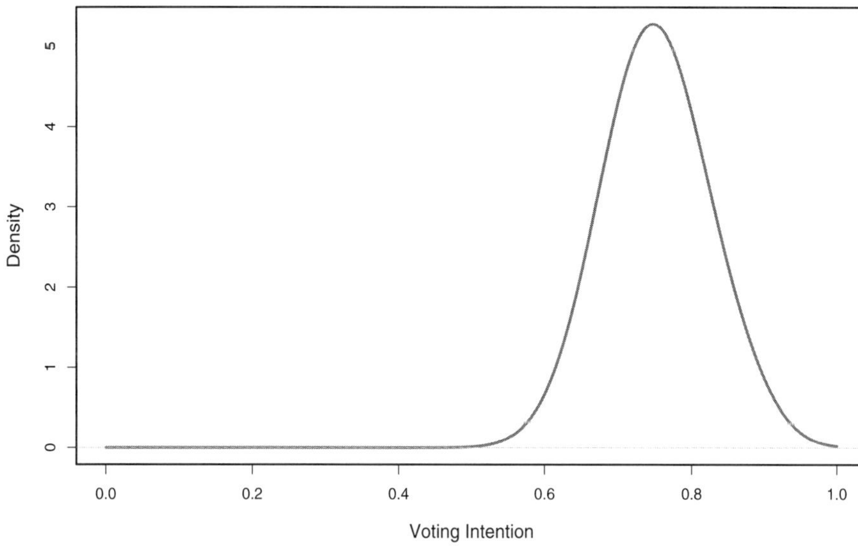

Figure 12.40 *Estimated probability density of the voting intention data.*

and the variance is

$$\sigma_t^2 = \frac{a_t\, b_t}{(a_t + b_t)^2(a_t + b_t + 1)}.$$

The mean of the series is 0.7520 and the standard deviation is 0.0563. The selected model according to the AIC is $p = 2$. The maximum likelihood parameter estimates are reported in Table 12.5, along with their estimated standard deviations and t-tests.

Figure 12.41 exhibits the one-step predictors while Figure 12.42 shows the residuals of the fitted Beta AR(2) model along with their sample ACF.

Table 12.5 Beta AR(2) estimates for the Voting Intention Data.

	$\widehat{\mu}_0$	$\widehat{\phi_1}$	$\widehat{\phi_2}$
Parameter	0.0664	0.6191	0.2950
S.D.	0.0181	0.0541	0.0541
t-test	3.6639	11.4436	5.4559

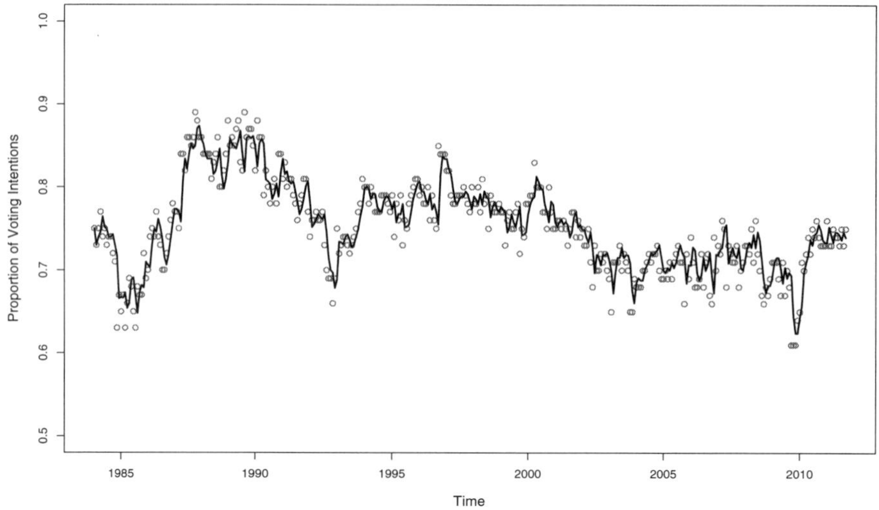

Figure 12.41 *Predicted voting intention of the Conservative Party and the Labor Party time series.*

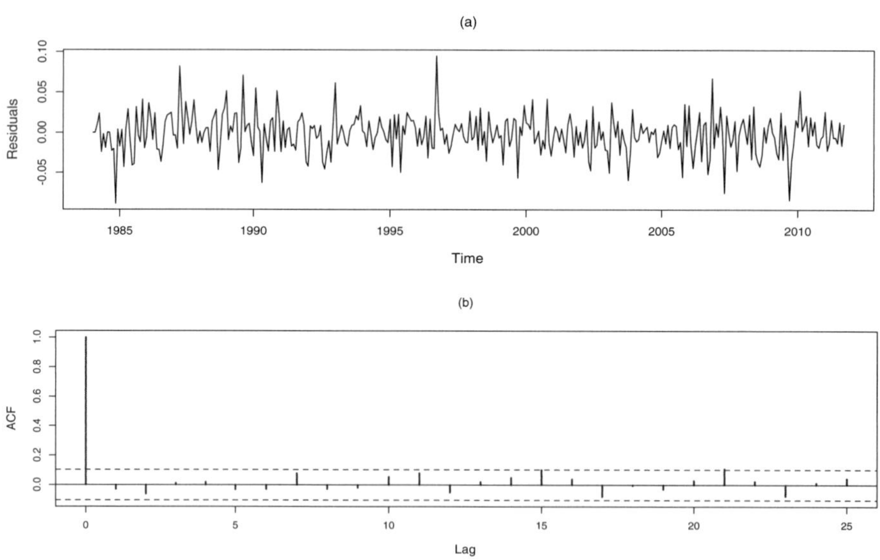

Figure 12.42 *Voting intention Beta model. (a) Residuals, and (b) sample ACF of the residuals.*

From these plots, the residuals seem to be a white noise sequence. The Box-Ljung test reported below indicates that this hypothesis is not rejected at the 5% significance level.

```
> Box.test(res,lag=10,type="Ljung")

Box-Ljung test

data:  res

X-squared = 6.3684, df = 10, p-value = 0.7834
```

12.5 ZERO-INFLATED MODELS

An important problem when modeling count time series data is that the presence of observations with zero counts may be greater than expected from, for instance, a Poisson model. Due to this situation, there are several so-called zero-inflated models (ZIM) that provide the flexibility of controlling the probability of zero counts. For example, a zero-inflated Poisson distribution is given by

$$\begin{aligned} f(y_t = 0|\lambda_t) &= \omega + (1-\omega)\, e^{-\lambda_t} \\ f(y_t = h|\lambda_t) &= (1-\omega)\, \frac{\lambda_t^{y_t}}{\Gamma(\lambda_t)}\, e^{-\lambda_t}, \qquad h \geq 1. \end{aligned}$$

In this model, the parameter ω controls the level of probability of zero counts in the time series. For $\omega = 0$, the model corresponds to the usual Poisson distribution. On the other hand, for $\omega = 1$ all the probability is given to the zero count. The mean of this zero-inflated Poisson distribution (ZIP) is

$$E(y_t) = (1-\omega)\, \lambda_t$$

and its variance is

$$\mathrm{Var}(y_t) = \lambda_t(1-\omega)\,(1+\lambda_t\, \omega).$$

Another model of interest in this context is the Negative Binomial. In this case, the probability is defined by

$$f(y_t|p_t) = \omega\, I_{\{y_t=0\}} + (1-\omega)\, \frac{\Gamma(k+y_t)}{\Gamma(k)\, y_t!}\, p_t^{y_t}\, (1-p_t)^k,$$

where the probability os success p_t is given by

$$p_t = \frac{k}{k+\lambda_t},$$

with k is an overdispersion coefficient and λ is an intensity parameter.

The conditional mean of this zero-inflated Negative Binomial (ZINB) distribution is

$$E(y_t|\lambda_t) = (1-\omega)\,\lambda_t$$

and its conditional variance is given by

$$\text{Var}(y_t|\lambda_t) = \lambda_t(1-\omega)\left(1+\lambda_t\,\omega+\frac{\lambda_t}{k}\right).$$

From this formula, we can see that the parameter $k>0$ allows for overdispersion of the distribution since as $k \to 0$, the variance increases. In fact, the variance to mean ratio is

$$\frac{\text{Var}(y_t|\lambda_t)}{E(y_t|\lambda_t)} = 1+\left(\omega+\frac{1}{k}\right)\lambda_t \geq 1,$$

since $\omega \geq 0$ and $k > 0$.

In order to account for serial correlation in the data, we can consider the latent process z_t that satisfies, for instance, an AR(p) model

$$z_t = \phi_1\, z_{t-1} + \phi_2\, z_{t-2} + \cdots + \phi_p\, z_{t-p} + \varepsilon_t,$$

where ε_t is a white noise sequence with zero-mean and variance σ^2. Additionally, let s_t be the state space vector that represents the process z_t. Regression variables can be included in the model via the link function

$$\log\lambda_t = \log w_t + x_t\,\beta + z_t$$

where $\log w_t$ is an offset variable and x_t is a set of regressors. Thus, we can write a Poisson-Gamma mixture model

$$\begin{aligned} s_t|s_{t-1} &\sim \text{N}(\Phi\, s_{t-1}, \Sigma), \\ u_t &\sim \text{Bernoulli}(\omega) \\ v_t &\sim \text{Gamma}(k, 1/k), \\ y_t|s_t, u_t, v_t &\sim \text{Poisson}((1-u_t)\, v_t\, \lambda_t). \end{aligned}$$

■ EXAMPLE 12.11

The R package ZIM provides a framework for simulating and estimating these zero-inflated time series models. As an illustration, consider the following model

$$z_t = \phi\, z_{t-1} + \varepsilon_t,$$

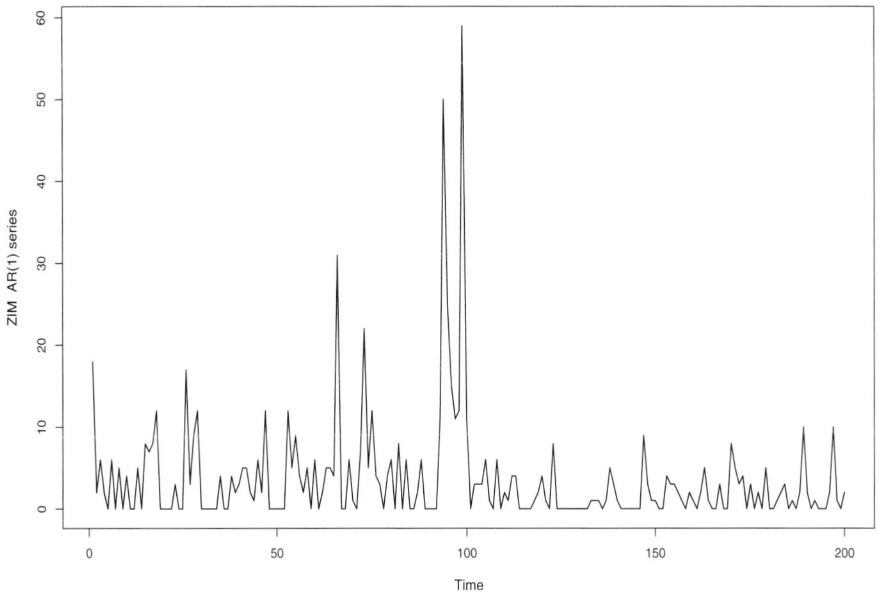

Figure 12.43 *Simulated Zero-inflated AR(1) model with $\phi = 0.4$ and 200 observations.*

where $\phi = 0.4$, $\sigma = 0.5$,

$$\log \lambda_t = \log w + x_t \beta + z_t$$

where the offset is constant $\log w = 2$, the regressor is

$$x_t = \begin{cases} 0 & \text{for} \quad t = 1, \ldots, 100, \\ 1 & \text{for} \quad t = 101, \ldots, 200. \end{cases}$$

In this case, the regression parameter $\beta = -1$ while the zero-inflation parameter ω is 0.3.

A simulated trajectory of 200 observations of this process is shown in Figure 12.43 while its sample ACF is displayed in Figure 12.44.

Figure 12.45 exhibits the evolution of the underlying state sequence s_t and its sample ACF is reported in Figure 12.46. Notice that in this simple case, the states s_t correspond to the z_t sequence.

On the other hand, the evolution of the Bernoulli sequence u_t in shown Figure 12.47 while the evolution of the Gamma sequence v_t is displayed in Figure 12.48.

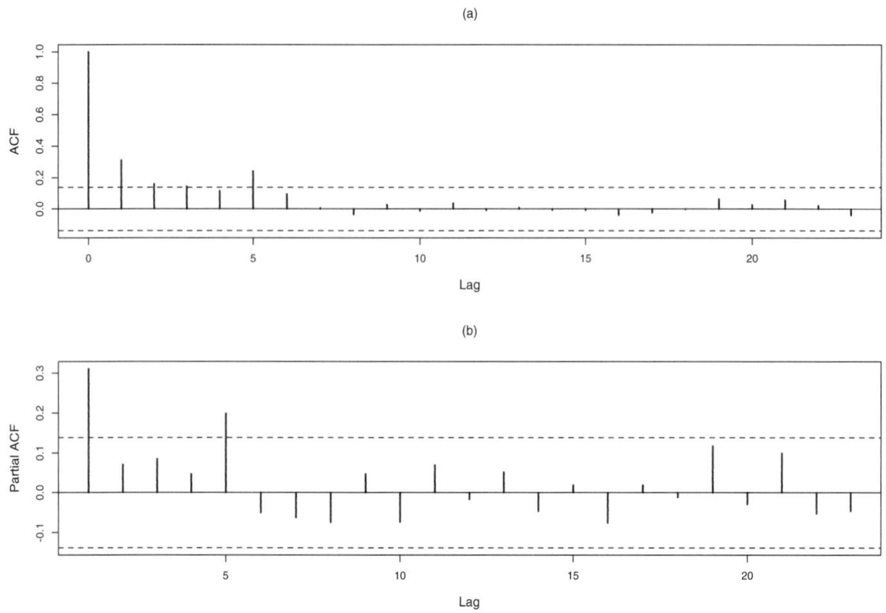

Figure 12.44 *Zero-inflated time series model. (a) Sample ACF of y_t, and (b) sample PACF of y_t.*

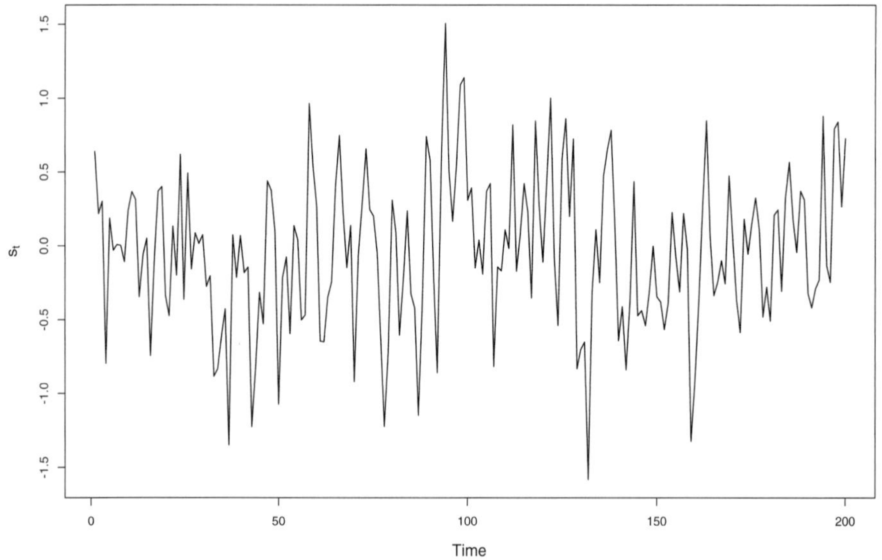

Figure 12.45 *Zero-inflated time series model: State sequence s_t.*

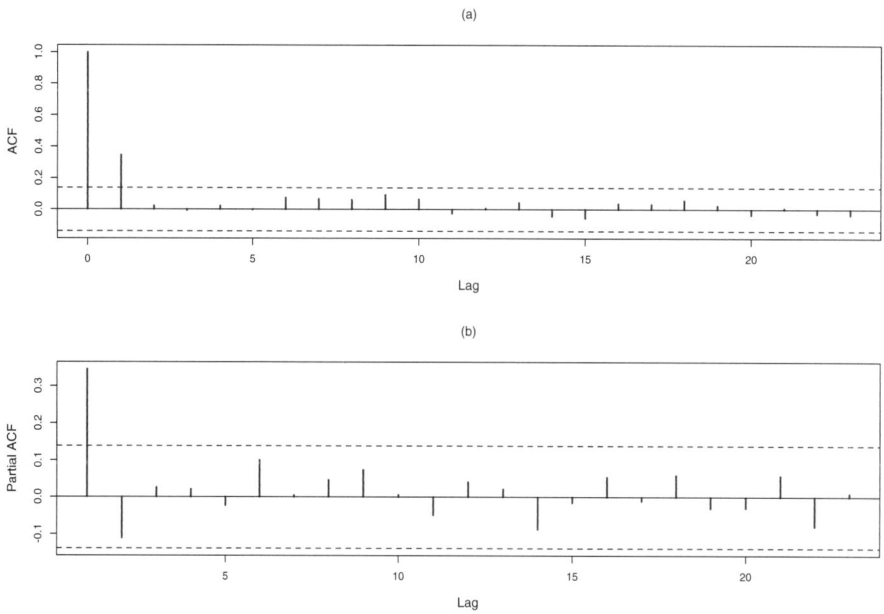

Figure 12.46 *Zero-inflated time series model. (a) Sample ACF of the state sequence s_t, (b) sample PACF of the state sequence s_t.*

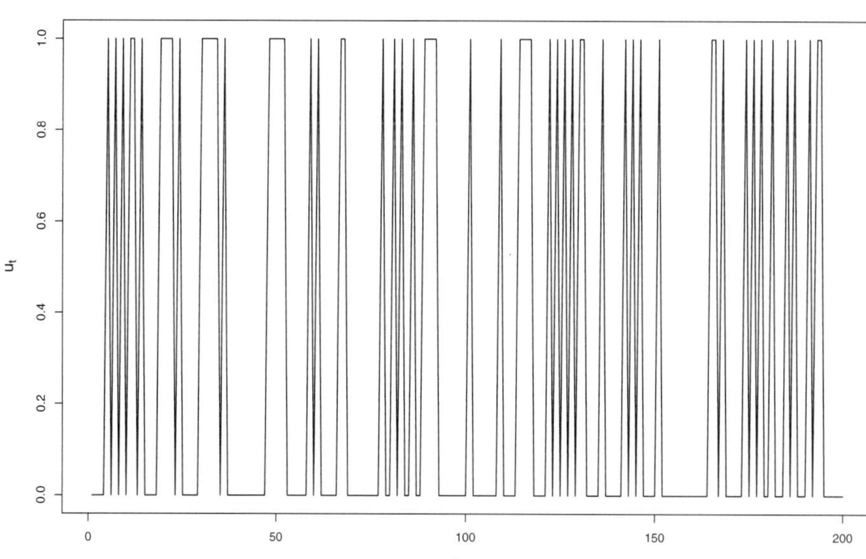

Figure 12.47 *Zero-Inflated Time Series Model: Bernoulli sequence u_t.*

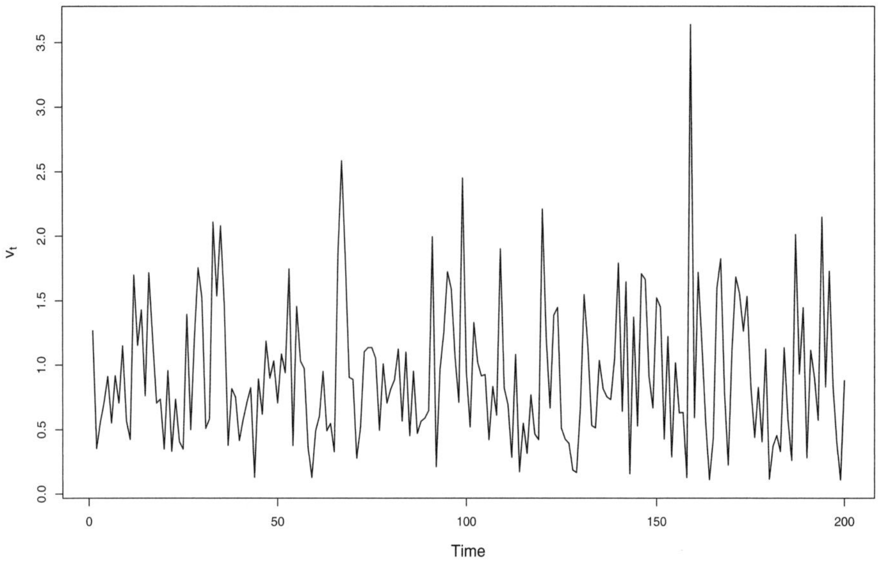

Figure 12.48 *Zero-Inflated Time Series Model: Gamma sequence v_t.*

The estimated parameters are reported in the following output:

```
> dzim(y~X,dist='zip')

Call:
dzim(formula = y ~ X,dist = "zip")
(Zero-inflation parameter taken to be 0.3016)

Coefficients (log-linear):
             Estimate Std. Error z value  Pr(>|z|)
(Intercept)  1.76327    0.21609  8.1600 3.349e-16 ***
X           -1.18216    0.26444 -4.4704 7.808e-06 ***

Coefficients (autoregressive):
    Estimate Std. Error z value Pr(>|z|)
ar1  0.17734    0.21326  0.8316   0.4057
---
Signif. codes: 0 '***' 0.001 '**' 0.01 '*' 0.05 '.' 0.1 ' ' 1

(Standard deviation parameter taken to be 0.8023)

Criteria for assessing goodness of fit
loglik: -441.858
aic: 893.716
bic: 910.2076
tic: 975.3338
```

Observe that the estimates are close to their theoretical counterparts. For example, the zero-inflation parameter ω is 0.3 and its estimate is 0.3016 while the regression parameter is -1 and its estimate is -1.18216. On the other hand, the estimate of the autoregressive parameter is lower than its true value.

12.6 BIBLIOGRAPHIC NOTES

The literature on non-Gaussian time series is vast. For instance, several models have been proposed to handle integer-valued time series or data with special features such as positivity. The idea of dividing these models into two groups: observation-driven and parameter-driven models was first suggested by Cox (1981). MacDonald and Zucchini (1997), Cameron and Trivedi (2013) and the review byMcKenzie (2003) provide an excellent overview of the literature in this area. In particular, observation-driven models for Poisson counts have been developed by Davis, Dunsmuir, and Streett (2003), see also Grunwald, Hyndman, Tedesco, and Tweedie (2000). Extensions to long-memory models have been studied by Palma and Zevallos (2011). The INAR models were introduced by McKenzie (1985) and Al-Osh and Alzaid (1987). Freeland and McCabe (2004) discuss forecasting of INAR processes. The R package inarmix is due to Henderson and Rathouz (2014). Kedem and Fokianos (2002) covers a number of regression time series methodologies, including models for count data. Methods for zero-inflated model are reviewed, for instance, by Yang, Cavanaugh, and Zamba (2014) and Yang, Zamba, and Cavanaugh (2013), among others.

Problems

12.1 Let z be a non-negative integer-valued random variable and consider the thinning operation $\circ$ defined in Section 12.1

(a) Show that

$$E(\alpha \circ z) = \alpha\, E(z).$$

(b) Verify that

$$E(\alpha \circ z)^2 = (\alpha^2 + \sigma^2)\, E(z).$$

(c) Show that

$$\text{Var}(\alpha \circ z)^2 = (\alpha^2 + \sigma^2)\, E(z) - \alpha^2\, (E\, z)^2.$$

12.2 Consider the following INAR(1) model.

$$y_t = \alpha \circ y_{t-1} + \varepsilon_t,$$

where $E\,y_t = \mu$ and $E\,\varepsilon_t = \mu_\varepsilon$. Assume that the process y_t is stationary.

(a) Show that

$$E\,y_t = \mu = \frac{\mu_\varepsilon}{1-\alpha}.$$

(b) Verify that

$$\operatorname{Var} y_t = (\alpha^2 + \sigma^2)\mu - \alpha^2\mu^2 + \sigma_\varepsilon^2.$$

12.3 Consider the conditional Gamma ARMA

$$y_t|\mathcal{F}_{t-1} \sim \text{Gamma}(\lambda_t, \beta),$$

where

$$\lambda_t = \mu_0 + \phi\,\beta\,y_{t-1} + \theta\,\lambda_{t-1},$$

with $\mu_0 > 1$, $0 \leq \phi, 1$, $0 \leq \theta, 1$ $\phi + \theta < 1$.

(a) Show that

$$E\,\lambda_t = \frac{\mu_0}{1-\phi-\theta}.$$

(b) Verify that

$$E\,y_t = \frac{\mu_0}{\beta(1-\phi-\theta)}.$$

(c) Show that λ_t can be written in terms of the infinite past of the observed process as follows

$$\lambda_t = \frac{\mu_0}{1-\theta} + \phi\beta \sum_{j=0}^{\infty} \theta^j\,y_{t-1-j}.$$

12.4 Consider the conditional parameter driven Poisson MA(1)

$$y_t|\lambda_t \sim \text{Poi}(\lambda_t),$$

with

$$\begin{aligned}\lambda_t &= \exp\theta_t \\ \theta_t &= \varepsilon_t + \varphi\,\varepsilon_{t-1},\end{aligned}$$

where ε_t is a Gaussian white noise sequence with zero-mean and variance σ^2.

(a) Verify that θ_t is a Gaussian sequence with zero-mean and variance

$$\operatorname{Var}\theta_t = (1+\varphi^2)\sigma^2.$$

(b) Show that λ_t is a Log-Normal sequence of random variables with

$$E\,\lambda_t = \exp[(1+\varphi^2)\sigma^2/2],$$

and

$$\mathrm{Var}\,\lambda_t = (\exp[(1+\varphi^2)\sigma^2] - 1)\,\exp[(1+\varphi^2)\sigma^2].$$

(c) Verify that the expected value of y_t is given by

$$E\,y_t = \exp[(1+\varphi^2)\sigma^2/2].$$

12.5 Suppose that $\sigma^2 \sim \mathrm{Gamma}(\alpha, \beta)$. An *overdispersed distribution* is obtained as

$$\mathcal{F} = \frac{\eta\sigma^2}{\chi^2_\eta},$$

where η is a parameter indicating the *degrees of freedom* of the χ^2 distribution. Verify that $\mathcal{F} \propto \mathrm{Fisher}(a, b)$ and find the parameters a and b.

APPENDIX A
COMPLEMENTS

As described in the preface, this text is intended for a general audience interested in time series analysis. Consequently, it focusses on methodologies and applications rather than on theoretical aspects of this discipline. Nevertheless, the following appendix attempts to provides some additional mathematical and statistical concepts for the analysis of stationary stochastic processes. Section A.1 begins with a description of the concepts of vector space, *norm* and then proceeds to the definition of *inner product* and *Hilbert spaces*. There is a strong motivation for introducing these spaces. Prediction, interpolation, and smoothing are key aspects of time series analysis, and they are usually obtained in terms of orthogonal projections onto some vector subspaces. A Hilbert space does not only guarantee the existence of the best linear predictor, interpolator, or smoother but also provides an explicit tool for obtaining them through the *projection theorem*. Several other fundamental concepts about linear processes such as *stationarity*, *singularity*, *regularity*, *causality*, *invertibility*, and *ergodicity* are discussed in Section A.2.

*

Time Series Analysis. First Edition. Wilfredo Palma.

A.1 PROJECTION THEOREM

Let $\mathcal{M}$ be a vector space. An *inner product* is a function $\langle \cdot, \cdot \rangle : \mathcal{M} \to \mathbb{R}$ such that for all $x, y, z \in \mathcal{M}$ and $\alpha, \beta \in \mathbb{R}$ it satisfies (i) $\langle \alpha x + \beta y, z \rangle = \alpha \langle x, z \rangle + \beta \langle y, z \rangle$, (ii) $\langle x, y \rangle = \langle y, x \rangle$, and (iii) $\langle x, x \rangle \geq 0$ and $\langle x, x \rangle = 0$ if and only if $x = 0$. The vector space $\mathcal{M}$ endowed with an inner product $\langle \cdot, \cdot \rangle$ is said to be an *inner product space.*

A *norm* on a vector space $\mathcal{M}$ is a function $\| \cdot \| : \mathcal{M} \to [0, \infty)$ satisfying for all $x, y \in \mathcal{M}$ and $\alpha \in \mathbb{R}$ (i) $\|x\| = 0$ if and only if $x = 0$, (ii) $\|\alpha x\| = |\alpha| \|x\|$, and (iii) $\|x + y\| \leq \|x\| + \|y\|$. The inner product $\langle \cdot, \cdot \rangle$ induces the norm $\|x\| = \sqrt{\langle x, x \rangle}$ in $\mathcal{M}$. The sequence $\{x_t\}$ in $\mathcal{M}$ is said to be a *Cauchy sequence* if and only if for all $\varepsilon > 0$ there is an integer n such that $\|x_t - x_s\| < \varepsilon$ for all $t, s > n$. The vector space $\mathcal{M}$ is *complete* if and only if every Cauchy sequence has a limit in $\mathcal{M}$. With these definitions we are ready to introduce the Hilbert spaces.

A *Hilbert* space is a complete inner product vector space. These spaces are particularly important for time series analysis mainly because of the following *projection theorem.*.

Let $\mathcal{M}$ be a closed subspace of the Hilbert space $\mathcal{H}$ and let $x \in \mathcal{H}$. Then, (a) there is a unique point $y \in \mathcal{M}$ such that $\|x - y\| = \inf_{z \in \mathcal{M}} \|x - z\|$ and (b) $y \in \mathcal{M}$ and $\|x - y\| = \inf_{z \in \mathcal{M}} \|x - z\|$ if and only if $y \in \mathcal{M}$ and $\langle x - y, z \rangle = 0$ for all $z \in \mathcal{M}$.

This results can be proved as follows. (a) By definition, there is a sequence $\{y_t\}$ in $\mathcal{M}$ such that $\|x - y_t\| \to \alpha$ where $\alpha = \inf_{z \in \mathcal{M}} \|x - z\|$. This is a Cauchy sequence since $(y_t + y_s)/2 \in \mathcal{M}$ and by the parallelogram law (see Problem 7.22) we have

$$\|(y_t - x) - (y_s - x)\|^2 + 4\|(y_t + y_s)/2 - x\|^2 = 2\|y_t - x\|^2 + 2\|y_s - x\|^2.$$

Thus, $\|y_t - y_s\|^2 = 2\|y_t - x\|^2 + 2\|y_s - x\|^2 - 4\alpha^2$, and then $\|y_t - y_s\| \to 0$ as $t, s \to \infty$. Since $\mathcal{H}$ is complete, there is $y \in \mathcal{H}$ such that $\|y - y_t\| \to 0$ as $t \to \infty$. Furthermore, since $\mathcal{M}$ is closed, we conclude that $y \in \mathcal{M}$. Now, by the continuity of the norm we have that $\|x - y\| = \lim_{t \to \infty} \|x - y_t\| = \alpha$. To show that this point is unique assume that there is another vector $z \in \mathcal{M}$ such that $\|x - z\| = \alpha$. By the parallelogram law we have

$$\|y - z\|^2 = 2\|y - x\|^2 + 2\|z - x\|^2 - 4\|(y + z)/2 - x\|^2 = 0,$$

so that $y = z$. (b) Suppose that $y \in \mathcal{M}$ and $\|x - y\| = \inf_{z \in \mathcal{M}} \|x - z\|$. If $z \in \mathcal{M}$, then $y + z \in \mathcal{M}$ and

$$\|x - y\|^2 \leq \|x - (y + z)\|^2 = \|(x - y) - z\|^2 = \|x - y\|^2 + \|z\|^2 - 2\langle x - y, z \rangle.$$

Therefore, $2\langle x - y, z \rangle \leq \|z\|^2$ for any $z \in \mathcal{M}$. For fixed z, $\lambda z \in \mathcal{M}$ for any $\lambda \geq 0$. Assuming that $\langle x - y, z \rangle \geq 0$ (otherwise take $-z$), we have

$0 \leq 2\lambda\langle x-y, z\rangle \leq \lambda^2\|z\|^2$, that is, $0 \leq 2\langle x-y, z\rangle \leq \lambda\|z\|^2$. Consequently, as $\lambda \to 0$ we have $\langle x-y, z\rangle = 0$. Conversely, consider a point $y \in \mathcal{M}$ satisfying $\langle x-y, z\rangle = 0$ for all $z \in \mathcal{M}$. Then, for any $z \in \mathcal{M}$ we have

$$\|x-z\|^2 = \|x-y\|^2 + \|y-z\|^2 \geq \|x-y\|^2.$$

Therefore, $\|x-y\| = \inf_{z\in\mathcal{M}} \|x-z\|$.

Given a subset M of a Hilbert space $\mathcal{H}$, the *span* of M, denoted by $\text{sp}M$, is the subspace generated by all finite linear combinations of elements of M and $\overline{\text{sp}}\, M$ denotes its closure in $\mathcal{H}$, that is, $\overline{\text{sp}}\, M$ contains all the limits of sequences in $\text{sp}M$. In what follows, we illustrate the concept of Hilbert space with several well-known examples.

■ EXAMPLE 1.1

Let n be a positive integer and consider the space $\mathbb{C}^n$ endowed with the Euclidean inner product

$$\langle x, y\rangle = x\overline{y} = \sum_{j=1}^{n} x_j \overline{y}_j,$$

where $x = (x_1, \ldots, x_n)'$, $y = (y_1, \ldots, y_n)'$, and $\overline{y}$ is the complex-conjugate of y. Then, $\mathbb{C}^n$ is a Hilbert space with norm $\|x\| = (\sum_{j=1}^{n} |x_j|^2)^{1/2}$.

■ EXAMPLE 1.2

Let $\{y_t : t \in \mathbb{Z}\}$ be a real, zero-mean stochastic process defined on a probability space $(\Omega, \mathcal{F}, \mathsf{P})$. Then, $\mathcal{L}_2(\Omega, \mathcal{F}, \mathsf{P})$ denotes the Hilbert space with inner product $\langle x, y\rangle = E(xy)$ and norm $\|y\| = \sqrt{E(y^2)}$, where $E(\cdot)$ stands for the expectation operator. This space will be simply called $\mathcal{L}_2$ throughout this book.

■ EXAMPLE 1.3

Let F be a distribution function and let $f(\lambda)$ and $g(\lambda)$ be two complex-valued functions with domain $[-\pi, \pi]$ such that

$$\int_{-\pi}^{\pi} |f(\lambda)|^2\, dF(\lambda) < \infty, \qquad \int_{-\pi}^{\pi} |g(\lambda)|^2\, dF(\lambda) < \infty.$$

Then,

$$\langle f, g\rangle_F = \int_{-\pi}^{\pi} f(\lambda)\, \overline{g(\lambda)}\, dF(\lambda)$$

is an inner product and the generated Hilbert space is denoted by $\mathcal{L}_2(F)$. Observe that by Hölder's inequality this inner product is well-defined since

$$|\langle f, g\rangle|^2 \leq \int_{-\pi}^{\pi} |f(\lambda)|^2 \, dF(\lambda) \int_{-\pi}^{\pi} |g(\lambda)|^2 \, dF(\lambda) < \infty.$$

EXAMPLE 1.4

As a particular case of Example 1.3, if $F(\lambda)$ corresponds to the Lebesgue measure over $[-\pi, \pi]$ given by $dF(\lambda) = d\lambda/2\pi$, then the resulting Hilbert space is denoted by $\mathcal{L}_2(d\lambda)$.

A.2 WOLD DECOMPOSITION

The autocovariance function of second-order stationary process, $\gamma(\cdot)$, can be written as

$$\gamma(h) = \int_{-\pi}^{\pi} e^{ih\lambda} \, dF(\lambda),$$

where the function F is right-continuous, non-decreasing, bounded over $[-\pi, \pi]$, and satisfies the condition $F(-\pi) = 0$. F is called the *spectral distribution* of $\gamma(\cdot)$. Furthermore, if

$$F(\lambda) = \int_{-\pi}^{\lambda} f(\omega) \, d\omega,$$

then $f(\cdot)$ is called the *spectral density* of $\gamma(\cdot)$.

A.2.1 Singularity and Regularity

Let $\mathcal{F}_t = \overline{\mathrm{sp}}\{y_s : s < t\}$ be the *past* of the process at time t and $\mathcal{F}_{-\infty} = \bigcap_{t=-\infty}^{\infty} \mathcal{F}_t$. The process is said to be *deterministic*, that is, *perfectly predictable*, if and only if $y_t \in \mathcal{F}_{-\infty}$ for all $t \in \mathbb{Z}$. In other words, a process is deterministic if and only if $\mathcal{F}_{-\infty} = \mathcal{F}_{\infty}$. These processes are also-called *singular*. On the other hand, a process is said to be *purely nondeterministic* or *regular* if and only if $\mathcal{F}_{-\infty} = \{0\}$.

A.2.2 Wold Decomposition

The following result, known as the *Wold decomposition theorem*, is a fundamental tool for analyzing stationary processes: Any stationary process is the sum of a regular process and a singular process; these two processes are orthogonal and the decomposition is unique.

According to the Wold representation theorem, a stationary purely nondeterministic process may be expressed as

$$y_t = \sum_{j=0}^{\infty} \psi_j \varepsilon_{t-j} = \psi(B)\varepsilon_t, \tag{A.1}$$

where $\psi_0 = 1$, $\sum_{j=0}^{\infty} \psi_j^2 < \infty$, $\{\varepsilon_t\}$ is a white noise sequence with variance σ^2. The *Wold expansion* (A.1) is unique and $\varepsilon_t \in \mathcal{F}_{t+1}$ for all $t \in \mathbb{Z}$.

Recall now that a complex-valued stochastic process $\{\varepsilon(\lambda) \colon \lambda \in [-\pi, \pi]\}$ is said to be a *right-continuous orthogonal-increment process* if it satisfies the following conditions: (i) $\|\varepsilon(\lambda)\| < \infty$, (ii) $\langle 1, \varepsilon(\lambda)\rangle = 0$, (iii) $\langle \varepsilon(\lambda_2) - \varepsilon(\lambda_1), \varepsilon(\omega_2) - \varepsilon(\omega_1)\rangle = 0$ for all disjoint intervals $(\lambda_1, \lambda_2]$ and $(\omega_1, \omega_2]$, and (iv) $\|\varepsilon(\lambda + \delta) - \varepsilon(\lambda)\| \to 0$, as $\delta \to 0_+$, where $\|\cdot\|$ is the $\mathcal{L}_2$ norm. The following result, usually called the *spectral representation theorem*, establishes that a stationary process can be written as an integral of a deterministic function with respect to a right-continuous orthogonal-increment sequence: Let $\{y_t\}$ be a zero-mean stationary process with spectral distribution F. Then, there exists a right-continuous orthogonal-increment process $\{\varepsilon(\lambda)\}$ such that $F(\lambda) = \|\varepsilon(\lambda) - \varepsilon(-\pi)\|^2$ for $\lambda \in [-\pi, \pi]$ and the process $\{y_t\}$ can be written as

$$y_t = \int_{-\pi}^{\pi} e^{\imath t\lambda}\, d\varepsilon(\lambda),$$

with probability one.

In time series analysis we often need to apply a *linear filter* $\{\psi_j : j \in \mathbb{Z}\}$ to a stationary process $\{x_t\}$, that is, $y_t = \sum_{j=-\infty}^{\infty} \psi_j x_{t-j} = \psi(B)x_t$. The following result provides general conditions on the linear filter $\{\psi_j \colon j \in \mathbb{Z}\}$ so that the filtered process $\{y_t\}$ is also stationary. Furthermore, it establishes conditions to ensure that this filtering procedure is reversible:

Let $\{x_t\}$ be a zero-mean stationary process with spectral representation

$$x_t = \int_{-\pi}^{\pi} e^{\imath\lambda t}\, d\varepsilon(\lambda)$$

and spectral distribution F. Let $\{\psi_j : j \in \mathbb{Z}\}$ be a sequence such that for $\lambda \in (-\pi, \pi]$

$$\sum_{j=-n}^{n} \psi_j e^{\imath\lambda j} \to \psi(e^{\imath\lambda}) = \sum_{j=-\infty}^{\infty} \psi_j e^{\imath\lambda j},$$

as $n \to \infty$ in the Hilbert space $\mathcal{L}(F)$ defined in Example 1.3. Then,

(a) The filtered process $\{y_t\}$ defined by

$$y_t = \sum_{j=-\infty}^{\infty} \psi_j x_{t-j}$$

is zero-mean stationary.

(b) The spectral distribution of $\{y_t\}$ is given by

$$F_y(\lambda) = \int_{-\pi}^{\lambda} |\psi(e^{-\imath\omega})|^2 \, dF(\omega).$$

(c) The process $\{y_t\}$ can be expressed as

$$y_t = \int_{-\pi}^{\pi} e^{\imath\lambda t} \psi(e^{-\imath\lambda}) \, d\varepsilon(\lambda).$$

(d) Assume that the sequence $\{\psi_j\}$ is such that $\psi(e^{-\imath\lambda}) \neq 0$ for $\lambda \in \Lambda$, where Λ^c has zero F-measure. Then, the process $\{x_t\}$ can be written as

$$x_t = \int_{-\pi}^{\pi} e^{\imath\lambda t} \pi(e^{-\imath\lambda}) \, d\varepsilon_y(\lambda),$$

where $\pi(e^{\imath\lambda}) = 1/\psi(e^{\imath\lambda})$ and $d\varepsilon_y(\lambda) = \psi(e^{-\imath\lambda}) \, d\varepsilon(\lambda)$.

The following is an another relevant result, especially in then study of the properties of long-memory processes for proving to prove stationarity, causality and invertibility.

Let $\{x_t\}$ be a zero-mean stationary process and let $\{\varphi_j : j \in \mathbb{Z}\}$ be an absolutely summable sequence. Then,

(a) The process $\{y_t\}$ defined by

$$y_t = \sum_{j=-\infty}^{\infty} \varphi_j x_{t-j} = \varphi(B) x_t$$

is zero-mean stationary.

(b) Suppose that the stationary process $\{x_t\}$ is given by

$$x_t = \sum_{j=-\infty}^{\infty} \eta_j \varepsilon_{t-j} = \eta(B) \varepsilon_t.$$

Then, we can write

$$y_t = \varphi(B)\eta(B)\varepsilon_t = \eta(B)\varphi(B)\varepsilon_t = \psi(B)\varepsilon_t,$$

where these equations are in the mean square sense and $\psi(B) = \eta(B)\varphi(B)$.

A.2.3 Causality

The process defined by (A.1) is said to be *causal* since the representation of y_t is based only on the present and past of the input noise, $\{\varepsilon_t, \varepsilon_{t-1}, \dots\}$. Causality is a key property for predicting future values of the process.

■ **EXAMPLE 1.5**

Stationarity and causality are two related concepts but not necessarily equivalent. To illustrate this point let $\{\varepsilon_t\}$ be a white noise sequence with $\text{Var}(\varepsilon_t) < \infty$ and consider, for instance, the process $y_t = \varepsilon_t + \varepsilon_{t+1}$. This process is stationary but not causal, since y_t depends on future values of the sequence $\{\varepsilon_t\}$. On the other hand, the process $y_t = \varepsilon_t + \varepsilon_{t-1}$ is causal and stationary.

A.2.4 Invertibility

A linear regular process (A.1) is said to be *invertible* if there exists a sequence of coefficients $\{\pi_j\}$ such that

$$\varepsilon_t = -\sum_{j=0}^{\infty} \pi_j y_{t-j}, \tag{A.2}$$

where this expansion converges in $\mathcal{L}_2$. From (A.2), and assuming $\pi_0 = -1$, the process $\{y_t\}$ may be expressed as

$$y_t = \varepsilon_t + \sum_{j=1}^{\infty} \pi_j y_{t-j}. \tag{A.3}$$

A.2.5 Best Linear Predictor

Let $\{y_t\}$ be an invertible process as in (A.3). The best linear prediction of the observation y_t based on its history $\overline{\text{sp}}\{y_t, y_{t-1}, y_{t-2}, \dots\}$ is given by

$$\widehat{y}_t = \sum_{j=1}^{\infty} \pi_j y_{t-j}. \tag{A.4}$$

Consequently, $\varepsilon_t = y_t - \widehat{y}_t$ is an innovation sequence, that is, an orthogonal process with zero-mean and constant variance representing the part of $\{y_t\}$ that cannot be linearly predicted from the past.

On the other hand, suppose that we only have a finite trajectory of the stationary process, $\{y_1, y_2, \dots, y_n\}$. The best linear predictor of y_{n+1} based on its finite past $\overline{\text{sp}}\{y_1, y_2, \dots, y_n\}$, $\widehat{y}_{n+1}$, satisfies the equations

$$\langle y_{n+1} - \widehat{y}_{n+1}, y_j \rangle = 0,$$

for $j = 1, \dots, n$. By writing the best linear predictor as

$$\widehat{y}_{n+1} = \phi_{n1} y_n + \phi_{n2} y_{n-1} + \cdots + \phi_{nn} y_1,$$

we conclude that the coefficients ϕ_{nj} satisfy the equations

$$\sum_{i=1}^{n} \phi_{ni} \langle y_{n+1-i}, y_j \rangle = \langle y_{n+1}, y_j \rangle,$$

for $j = 1, \ldots, n$. If $\langle y_i, y_j \rangle = \gamma(i-j)$, where $\gamma(\cdot)$ is an autocovariance function, then we have that

$$\sum_{i=1}^{n} \phi_{ni}\gamma(n+1-i-j) = \gamma(n+1-j).$$

A.2.6 Szegö-Kolmogorov Formula

The spectral measure of the purely nondeterministic process (A.1) is absolutely continuous with respect to the Lebesgue measure on $[-\pi, \pi]$ and has spectral density

$$f(\lambda) = \frac{\sigma^2}{2\pi} |\psi(e^{-i\lambda})|^2.$$

Hence, its autocovariance function $\gamma(\cdot)$ may be written as

$$\gamma(h) = \int_{-\pi}^{\pi} f(\lambda) e^{-i\lambda h}\, d\lambda. \tag{A.5}$$

An explicit relationship between the spectral density and the variance of the noise sequence $\{\varepsilon_t\}$ is given by the Szegö-Kolmogorov formula

$$\sigma^2 = \exp\left\{ \frac{1}{2\pi} \int_{-\pi}^{\pi} \log[2\pi f(\lambda)]\, d\lambda \right\}. \tag{A.6}$$

From this expression, observe that for a purely nondeterministic process, $\sigma^2 > 0$ or equivalently, $\int_{-\pi}^{\pi} \log f(\lambda)\, d\lambda > -\infty$.

A.2.7 Ergodicity

Let $\{y_t : t \in \mathbb{Z}\}$ be a strictly stationary process and consider the σ-algebra $\mathcal{F}$ generated by the sets $S = \{\omega\colon y_0(\omega) = [y_{t_1}(\omega), \ldots, y_{t_n}(\omega)] \in \mathcal{B}\}$ where $\mathcal{B}$ is a Borel set in $\mathbb{R}^n$. Let T be a *shift* operator $T\colon \Omega \to \Omega$ such that $TS = \{\omega\colon y_1(\omega) = [y_{t_1+1}(\omega), \ldots, y_{t_n+1}(\omega)] \in \mathcal{B}\}$. This operator is measure preserving since the process is strictly stationary. Let $T^{-1}S$ be the preimage of S, that is, $T^{-1}S = \{\omega\colon T(\omega) \in S\}$. A measurable set is said to be *invariant* if and only if $T^{-1}S = S$. Similarly, a random variable X is said to be *invariant* if and only if $X(\omega) = X(T^{-1}\omega)$ for all $\omega \in \Omega$.

Let $\mathcal{S}$ be the set containing all the *invariant* measurable events. The process $\{y_t : t \in \mathbb{Z}\}$ is said to be *ergodic* if and only if for all the events $S \in \mathcal{S}$, either $P(S) = 0$ or $P(S) = 1$. Loosely speaking, this means that the only invariant sets are the entire space or zero measure events.

The concept of *ergodic process* arises from the study of the long run behavior of the average of a stationary sequence of random variables. Consider, for example, the strictly stationary process $\{y_t\}$ and the average $\bar{y}_n = \sum_{t=1}^{n} y_t/n$.

Notice that the sequence $\{\bar{y}_n\}$ converges under some mild conditions. If $\{y_t\}$ is a strictly stationary process with $E(|y_t|) < \infty$, then there is an *invariant* random variable $\bar{y}$ such that $E(|\bar{y}|) < \infty$,

$$\lim_{n\to\infty} \bar{y}_n = \bar{y},$$

almost surely, and

$$E(\bar{y}) = E(y_t).$$

It is important to determine under what circumstances the limit $\bar{y}$ is a constant. For instance, as discussed in Example 1.6 below, if $\{y_t\}$ is a sequence of independent random variables with $E\,|y_t| < \infty$, then the law of the large numbers guarantees that the limit $\bar{y}$ is equal to $E(y_t)$. Nevertheless, if a process is such that all invariant random variables are constant with probability 1, by the above theorem we can conclude that the limit $\bar{y}$ is a constant with probability 1. Stationary sequences satisfying this requirement are called *ergodic processes*.

■ EXAMPLE 1.6

A simple example of a strictly stationary ergodic process is a sequence of independent and identically distributed random variables $\{\varepsilon_1, \varepsilon_2, \dots\}$ with $E(|\varepsilon_t|) < \infty$. In this case, if $\mu = E(\varepsilon_t)$ and

$$\bar{\varepsilon}_n = \frac{1}{n}\sum_{t=1}^{n} \varepsilon_t,$$

is the sample mean of size n, then by the strong law of large numbers we have that $\lim_{n\to\infty} \bar{\varepsilon}_n = \mu$.

The following result is important for the theoretical treatment of ergodic systems since it establishes that strict stationarity and ergodicity is preserved by measurable transformations. If $\{y_t : t \in \mathbb{Z}\}$ is a strictly stationary ergodic process and $\phi : \mathbb{R}^\infty \to \mathbb{R}$ is a measurable transformation such that $x_t = \phi(y_t, y_{t-1}, \dots)$, then $\{x_t\}$ is a strictly stationary ergodic process.

An important consequence is that a linear filter of an independent and identically distributed sequence is strictly stationary and ergodic as stated in the following result. If $\{\varepsilon_t\}$ be a sequence of independent and identically distributed random variables with zero-mean and finite variance. If $\sum_{j=0}^{\infty} \alpha_j^2 < \infty$, then

$$y_t = \sum_{j=0}^{\infty} \alpha_j \varepsilon_{t-j}, \tag{A.7}$$

is a strictly stationary ergodic process.

■ EXAMPLE 1.7

For a trivial example of nonergodic process consider $y_t = \varepsilon_t + \theta\varepsilon_{t-1} + \eta$ where $\{\varepsilon_t\}$ is a sequence of independent and identically distributed random variables with zero-mean and unit variance, and η is a random variable independent of $\{\varepsilon_t\}$ with $E(|\eta|) < \infty$. The process $\{y_t\}$ is strictly stationary but not ergodic since $\bar{y}_n \to \eta$ as $n \to \infty$ and η is an invariant random variable which is not necessarily a constant.

Ergodicity is an essential tool for carrying out statistical inferences about a stochastic process. Usually, only one trajectory of the process is available for analysis. Consequently, there is only one sample of the process at any given time t. Thanks to the ergodicity of the process we can make up for this lack of information by using the trajectory to estimate the moments of the distribution of a strictly stationary process $\{y_t : t \in \mathbb{Z}\}$. Specifically, assume that $E(|y_t^i y_s^j|) < \infty$ for positive integers i, j and define $\mu_{ij} = E(y_t^i y_s^j)$. Observe that this moment does not depend on t or s since the process is strictly stationary. Therefore, we may write $\mu_{ij} = E(y_0^i y_h^j)$, where $h = s - t$. The moment μ_{ij} can be estimated by the time-average

$$\widehat{\mu}_{ij}(n) = \frac{1}{n} \sum_{t=1}^{n-h} y_t^i y_{t+h}^j. \tag{A.8}$$

The following result, known as the *pointwise ergodic theorem for stationary sequences*, states that this estimate is strongly consistent for every fixed h If $\{y_t : t \in \mathbb{Z}\}$ is a strictly stationary and ergodic process such that $E(|y_t|^{i+j}) < \infty$, then $\widehat{\mu}_{ij}(n) \to \mu_{ij}$ almost surely as $n \to \infty$.

A.2.8 Fractional Brownian Motion

A Gaussian stochastic process $B(t)$ with continuous sample paths is called *Brownian motion* if it satisfies (1) $B(0) = 0$ almost surely, (2) $B(t)$ has independent increments, (3) $E[B(t)] = E[B(s)]$, and (4) $\mathrm{Var}[B(t) - B(s)] = \sigma^2|t - s|$.

On the other hand, a standard *fractional Brownian motion* $B_d(t)$ may be defined as

$$B_d(t) = c(d) \int_{-\infty}^{\infty} (t - s)_+^d - (-s)_+^d \, dB(s), \tag{A.9}$$

where

$$s_+ = \begin{cases} s & \text{if} \quad s \geq 0 \\ 0 & \text{if} \quad s < 0, \end{cases}$$

and

$$c(d) = \frac{\sqrt{\Gamma(2d + 2) \cos \pi d}}{\Gamma(d + 1)}.$$

For a precise definition of the stochastic integral in (A.9), see, for example, Taqqu (2003). The term *standard* in this definition corresponds to the fact that the process $B_d(t)$ has unitary variance at time $t = 1$, that is, $\mathrm{Var}[B_d(1)] = 1$. In addition, the standard fractional Brownian motion (A.9) satisfies for $d \in (-\frac{1}{2}, \frac{1}{2})$

$$E[B_d(t)] = 0, \tag{A.10}$$
$$\mathrm{Cov}[B_d(t), B_d(s)] = \tfrac{1}{2}\left(|t|^{2d+1} + |s|^{2d+1} - |t-s|^{2d+1}\right). \tag{A.11}$$

A real-valued stochastic process $\{y(t) : t \in \mathbb{R}\}$ is *self-similar* with index $H > 0$ if for any constant $c > 0$ and for any time $t \in \mathbb{R}$ it satisfies

$$y(ct) \sim c^H y(t),$$

where the symbol $\sim$ means that both terms have the same distribution. The self-similarity index H is called the *scaling exponent* of the process or the *Hurst exponent*, in honor of the British hydrologist H. E. Hurst who developed methods for modeling the long-range dependence of the Nile River flows.

Observe that from (A.10) and (A.11) we conclude that

$$B_d(t) \sim N(0, |t|^{2d+1}).$$

Hence, for any positive constant c we have

$$B_d(ct) \sim N(0, c^{2d+1}|t|^{2d+1}),$$

and then

$$B_d(ct) \sim c^{d+\frac{1}{2}} B_d(t).$$

Thus, the fractional Brownian motion $B_d(t)$ is a self-similar process with self-similarity index $H = d + \frac{1}{2}$.

A.3 BIBLIOGRAPHIC NOTES

An excellent introduction to Hilbert spaces can be found in chapters 1 and 2 of Conway (1990). Additionally, Chapter 9 of Pourahmadi (2001) discusses stationary stochastic processes in Hilbert spaces. A very readable proof of the Szegö-Kolmogorov formula is given in Hannan (1970). The concept of ergodicity is also described in Chapter IV of that book. A good overview of stochastic processes is given in the first chapter of Taniguchi and Kakizawa (2000). Carothers (2000) is a helpful reference on real analysis. The book by Stout (1974) is a very good source for convergence results, especially its Chapter 3 about limiting theorems for stationary random variables.

APPENDIX B

SOLUTIONS TO SELECTED PROBLEMS

This appendix provides solutions to some selected problems from different chapters throughout the book.

Problem 2.10. We can write

$$\gamma_x(k) = \begin{cases} 0, & |k| \neq n\,s \\ \frac{\phi^{|k|/s}\,\sigma^2}{1-\phi^2}, & |k| = n\,s \end{cases}$$

Consequently,

$$\rho_x(k) = \begin{cases} 0, & |k| \neq n\,s \\ \phi^{|k|/s}, & |k| = n\,s \end{cases}$$

*

Time Series Analysis. First Edition. Wilfredo Palma.

for $n \in \mathbb{N}_0$ and $s \in \mathbb{N}$. Finally, Analogously to the AR(1) process, we can write

$$\alpha(k) = \begin{cases} \phi, & k = s \\ 0, & k \neq s. \end{cases}$$

Problem 2.11. From these plots we can obtain some preliminary findings. Figure I seems to exhibit an exponential decay in both ACF graphs. It is a candidate for an ARMA process. Figure II corresponds to a AR(1) process with positive parameter. Figure III is an exponential decay in both ACF graphs, but the ACF decays faster. This series is a candidate for a MA(2) process. Figure IV is an exponential decay in both ACF graphs. It is a candidate for an ARMA process. Figure V seems to correspond to an AR(1) process with negative parameter. Figure VI shows no correlation structure, therefore it seems to indicate a white noise sequence. From this analysis we can conclude: (a) Figure I. (b) Figure II. (c) Figure III. (d) Figure IV.

Problem 2.70. We consider all possible cases: Case 1). $0 < t < s \Rightarrow |t| < |s|$

$$\begin{aligned}
\text{Cov}(B_0(s), B_0(t)) &= \frac{|t| + |s| - |s - t|}{2} \\
&= \frac{t + s - (s - t)}{2} \\
&= \frac{t + s - s + t}{2} \\
&= t \\
&= |t| \\
&= \min\{|t|, |s|\}
\end{aligned}$$

Case 2). $t < s < 0 \Rightarrow s - t < 0$ and $\Rightarrow |t| > |s|$

$$\begin{aligned}
\text{Cov}(B_0(s), B_0(t)) &= \frac{|t| + |s| - |s - t|}{2} \\
&= \frac{-t - s - (s - t)}{2} \\
&= \frac{-t - s - s + t}{2} \\
&= -s \\
&= |s| \\
&= \min\{|t|, |s|\}
\end{aligned}$$

Case 3). $t < 0 < s : |t| < |s| \Rightarrow 0 < |s| - |t| = s + t$

$$\begin{aligned}
\mathrm{Cov}(B_0(s), B_0(t)) &= \frac{|t| + |s| - |s - t|}{2} \\
&= \frac{-t + s - (s + t)}{2} \\
&= \frac{-t + s - s - t}{2} \\
&= -t \\
&= |t| \\
&= \min\{|t|, |s|\}
\end{aligned}$$

Case 4). $t < 0 < s : |t| > |s| \Rightarrow 0 < |t| - |s| = -t - s$

$$\begin{aligned}
\mathrm{Cov}(B_0(s), B_0(t)) &= \frac{|t| + |s| - |s - t|}{2} \\
&= \frac{-t + s - (-s - t)}{2} \\
&= \frac{-t + s + s + t}{2} \\
&= s \\
&= |s| \\
&= \min\{|t|, |s|\}
\end{aligned}$$

Then, we conclude that

$$\mathrm{Cov}(B_0(s), B_0(t)) = \frac{|t| + |s| - |s - t|}{2} = \min\{|t|, |s|\}$$

Problem 2.31. Part (a). The autocovariance function of the MA(∞) process

$$x_t = \sum_{j=0}^{\infty} \psi_j \varepsilon_{t-j}$$

is given by

$$\gamma_x(h) = \sum_{j=0}^{\infty} \psi_j \psi_{j+h}.$$

For the process y_t we have

$$\psi_0 = 1, \psi_j = \frac{1}{j}$$

Therefore,

$$\gamma_y(h) = \sum_{j=0}^{\infty} \psi_j \psi_{j+h} = 1 \cdot \frac{1}{h} + \sum_{j=1}^{\infty} \frac{1}{j} \frac{1}{j+h}.$$

The series can be written as a finite sum (see Hint), yielding

$$\gamma_y(h) = \frac{1}{h} + \frac{1}{h} \sum_{j=1}^{h} \frac{1}{j}. \tag{B.1}$$

In particular,

$$\gamma(0) = \sigma_\varepsilon^2 \left(1 + \sum_{j=1}^{\infty} \frac{1}{j^2} \right) = 1 + \frac{\pi^2}{6}$$

as $\sigma_\varepsilon^2 = 1$ by assumption.

Part (b). By the Hint it follows that

$$\frac{1}{h} \sum_{j=1}^{h} \frac{1}{j} = \frac{C}{h} + \frac{\log h}{h} + \frac{\mathcal{O}\left(\frac{1}{h}\right)}{h}$$

For $h \to \infty$ the first term goes to 0. The last term also as

$$\frac{\mathcal{O}\left(\frac{1}{h}\right)}{h} = \frac{\mathcal{O}\left(\frac{1}{h}\right)}{\frac{1}{h}} \frac{1}{h^2} \leq C \frac{1}{h^2} \to 0 \text{ for } h \to \infty.$$

Thus it remains $\frac{\log h}{h}$.

Part (c). For any $m > n > 0$ it holds

$$\sum_{h=m}^{n} \gamma(h) = \sum_{h=m}^{n} \frac{1}{h} \left(1 + \sum_{j=1}^{h} \frac{1}{j} \right) > \sum_{h=m}^{n} \frac{1}{h}.$$

Part (d). From the preceding inequality it follows that $\sum_{h=0}^{n} \gamma(h)$ is not a Cauchy sequence, thus it can not converge. Hence, by definition y_t is a long memory process.

Problem 2.24. For an ARFIMA$(1, d, 0)$ we have that $i = 0, j = 1, \psi(0) = 1, \xi_1 = (\phi_1(1 - \phi_1))^{-1}$ and $C = \frac{\gamma_0(h)}{\sigma^2} \left[\phi_1^2 \beta(h) + \beta(-h) - 1 \right]$. For an ARFIMA(0,d,1) we have $i = 1, j = 0, \psi(1) = \psi_1, \xi_0 = (1 - \psi_1)^{-1}$ and $C = \frac{\gamma_0(h)}{\sigma^2} \left[\beta'(h) + \beta'(-h) - 1 \right]$. Here $\beta(h) = F(d + h, 1, 1 - d + h, \phi_1)$, $\beta'(h) = F(d + h, 1, 1 - d + h, 1)$ and $F(.)$ denotes for the Gaussian hypergeometric function. Finally,

$$\gamma_0(h) = \sigma^2 \frac{\Gamma(1 - 2d)}{\Gamma(1 - d)\Gamma(d)} \frac{\Gamma(h + d)}{\Gamma(1 + h - d)},$$

where $\sigma^2$is the variance of the white noise process and $\Gamma(.)$ is the Gamma function.

Thus, the autocovariance function for an ARFIMA$(1, d, 0)$ is

$$\gamma(h) = \gamma_0(h) \frac{\phi_1^2 F(d+h, 1, 1-d+h, \phi_1) + F(d-h, 1, 1-d-h, \phi_1) - 1}{\phi_1(1-\phi_1)},$$

and for an ARFIMA(0,d,1) the ACF is given by

$$\gamma(h) = \gamma_0(h)[F(d+h, 1, 1-d+h, 1) + F(d-h, 1, 1-d-h, 1) - 1]/(1-\psi_1).$$

The results provided by Hosking(1981) are the autocovariance functions for $(1-B)^d y_t = a_t$ referred to a process $x_t = (1-\phi B)y_t$ in the case of and ARFIMA(1,d,0) and $(1-B)^d x_t = a_t$ are referred to a process $y_t = (1-\theta B)x_t$ in the case of ARFIMA(0,d,1), where B is the backward shift operator and $\{a_t\}$ is a white noise process.

Problem 2.20. Consider an stationary, causal and invertible ARFIMA(p, d, q) process

$$\phi(B)y_t = \theta(B)(1-B)^{-d}\varepsilon_t,$$

where $\varepsilon_t \sim \text{WN}(0, \sigma^2)$. Note that the model can be expressed as

$$y_t = (1-B)^{-d}\phi(B)^{-1}\theta(B)\varepsilon_t.$$

Then, the spectral density of the process may be written by

$$\begin{aligned} f(\lambda) &= \frac{\sigma^2}{2\pi}|1-e^{-i\lambda}|^{-2d}\frac{|\theta(e^{-i\lambda})|^2}{|\phi(e^{-i\lambda})|^2} \\ &= \frac{\sigma^2}{2\pi}\left[2\sin\frac{\lambda}{2}\right]^{-2d}\frac{|\theta(e^{-i\lambda})|^2}{|\phi(e^{-i\lambda})|^2}, \end{aligned}$$

and when $x \to 0$ then $\sin x \sim x$. Thus, the spectral density of the ARFIMA model is approximated to

$$f(\lambda) \sim \frac{\sigma^2}{2\pi}\frac{|\theta(1)|^2}{|\phi(1)|^2}|\lambda|^{-2d}, \qquad \text{when } |\lambda| \to 0.$$

Clearly, this spectral density may be also expressed as

$$f(\lambda) \sim |\lambda|^{-2d}\ell_2(1/|\lambda|), \qquad |\lambda| \to 0,$$

where

$$\ell_2(1/|\lambda|) = \frac{\sigma^2}{2\pi}\frac{|\theta(1)|^2}{|\phi(1)|^2}.$$

It is shown that the autocovariance function of the ARFIMA model is

$$\gamma(h) \sim c_\gamma |h|^{2d-1},$$

when $|h| \to \infty$ and where

$$c_\gamma = \frac{\sigma^2}{\pi} \frac{|\theta(1)|^2}{|\phi(1)|^2} \Gamma(1-2d)\sin(\pi d).$$

Note that c_γ is constant in h, i.e., may be specified as the slowly varying function $\ell_1(h) = c_\gamma$. On the other hand, the model presents the following asymptotic relationship

$$\psi_j \sim \frac{\theta(1)}{\phi(1)} \frac{j^{d-1}}{\Gamma(d)}, \qquad \text{when } j \to \infty,$$

then, we can expressed the asymptotic coefficient of the Wold expansion

$$\psi_j \sim j^{d-1}\ell_3(j), \qquad j \to \infty,$$

where

$$\ell_3(j) = \frac{\theta(1)}{\phi(1)} \frac{1}{\Gamma(d)},$$

is constant in j. Furthermore, the ARFIMA(p, d, q) model follows

$$\sum_{h=-\infty}^{\infty} |\gamma(h)| = \infty.$$

To verify this condition, we note that

$$\begin{aligned} \sum_{h=-\infty}^{\infty} |\gamma(h)| &\geq \sum_{h=m}^{\infty} |\gamma(h)| \\ &\geq \lim_{n\to\infty} \sum_{h=m}^{n} |\gamma(h)|, \end{aligned}$$

for $0 < m < n$, where m is a value big enough such that

$$\gamma(h) \sim c_\gamma |h|^{2d-1}, \qquad h \in [m, \infty).$$

Thus, for $d \in (0, \frac{1}{2})$

$$\lim_{n\to\infty} \sum_{h=m}^{n} |c_\gamma||h|^{2d-1} \sim \lim_{n\to\infty} \frac{|c_\gamma|}{2d} |n|^{2d} = \infty.$$

If $d \in (-1, 0)$, then

$$\sum_{h=-\infty}^{\infty} |\gamma(h)| = \gamma(0) \sum_{h=-\infty}^{\infty} |\rho(h)| = \gamma(0)\left(1 + 2\sum_{h=1}^{\infty} |\rho(h)|\right).$$

It follows that $|\rho(h)| < c|h|^{2d-1}$, with $c > 0$ constant. Then

$$\sum_{h=-\infty}^{\infty} |\gamma(h)| < \gamma(0)\left(1 + 2\sum_{h=1}^{\infty} c|h|^{2d-1}\right) = \gamma(0)\left(1 + 2c\sum_{h=1}^{\infty} \frac{1}{|h|^{1-2d}}\right) =$$

$$\gamma(0) + 2\gamma(0)c\sum_{h=1}^{\infty} \frac{1}{|h|^{1-2d}}.$$

If $1 - 2d \in (1, 3)$ the series is convergent, that is, $\sum_{h=1}^{\infty} \dfrac{1}{|h|^{1-2d}} < \infty$. Also we know that $\gamma(0) < \infty$, then we conclude

$$\sum_{h=-\infty}^{\infty} |\gamma(h)| < \infty.$$

Problem 2.23. The spectral density of the process $\phi(B)y_t = \theta(B)(1-B)^{-d}\varepsilon_t$ is given by $f(\lambda) = \frac{\sigma^2}{2\pi}\left[2\sin\frac{\lambda}{2}\right]^{-2d}\frac{|\theta(e^{-i\lambda})|^2}{|\phi(e^{-1\lambda})|^2}$, where $\phi(B)$ and $\theta(B)$ are respectively the autoregressive and moving-average polynomials of the backward shift operator B and $\{\varepsilon_t\}$ is a white noise process. As $|\lambda| \to 0$ those polynomials evaluated at the unit circle converge to $\phi(1)$ and $\theta(1)$ while $2\sin\frac{\lambda}{2} \to \lambda$. Thus, the limiting spectral density function is $f(\lambda) = \frac{\sigma^2}{2\pi}\frac{|\theta(1)|^2}{|\phi(1)|^2}|\lambda|^{-2d}$, which is the desired result.

Problem 2.25. For a stationary, causal and invertible FN(d) process we know that

$$\psi_k = \frac{\Gamma(k+d)}{\Gamma(k+1)\Gamma(d)}.$$

Using Stirling's approximation

$$\frac{\Gamma(n+\alpha)}{\Gamma(n+\beta)} \sim n^{\alpha-\beta}, \qquad n \to \infty,$$

it follows that

$$\psi_k \sim \frac{k^{d-1}}{\Gamma(d)}, \qquad k \to \infty.$$

Now, note that the coefficients correspond to

$$\pi_k = \frac{\Gamma(k-d)}{\Gamma(k+1)\Gamma(-d)},$$

Again, using Stirling's approximation

$$\pi_k \sim \frac{k^{-d-1}}{\Gamma(-d)}, \qquad k \to \infty.$$

Finally, the autocovariance function is defined by

$$\rho(k) = \frac{\Gamma(1-d)}{\Gamma(d)} \frac{\Gamma(k+d)}{\Gamma(1+k-d)},$$

and the limit of $\rho(k)$, when $k \to \infty$, is

$$\lim_{k\to\infty} \rho(k) = \frac{\Gamma(1-d)}{\Gamma(d)} \lim_{k\to\infty} \frac{\Gamma(k+d)}{\Gamma(1+k-d)} = \frac{\Gamma(1-d)}{\Gamma(d)} k^{2d-1}.$$

Problem 2.29. The autocovariance function of a fractional Gaussian noise is defined by

$$\gamma(h) = \frac{\sigma^2}{2} \left(|h+1|^{2d+1} - 2|h|^{2d+1} + |h-1|^{2d+1} \right),$$

where $\sigma^2 = \text{Var}(y_t)$. Note that $\gamma(k) = \gamma(-k)$, then we are just going to analyze the case when $k \geq 1$, because $k = 0$ is equivalent to the variance.

$$\begin{aligned}
\gamma(k) &= \frac{\sigma^2}{2} \left(|k+1|^{2d+1} - 2|k|^{2d+1} + |k-1|^{2d+1} \right) \\
&= \sigma^2 \left(\frac{|k+1|^{2H}}{2} - |k|^{2H} + \frac{|k-1|^{2H}}{2} \right).
\end{aligned}$$

where $H = d + \frac{1}{2}$. If $d \in (0, \frac{1}{2})$ implies that $H \in (\frac{1}{2}, 1)$, then the function $f(x) = x^{2H}$ is strictly convex so we can use

$$\begin{aligned}
\frac{(k+1)^{2H} + (k-1)^{2H}}{2} &= \frac{f(k+1) + f(k-1)}{2} \\
&\geq f\left(\frac{k+1+k-1}{2} \right) = f(k) = k^{2H}.
\end{aligned}$$

Thus, it follows that $\gamma(k) \geq 0$. If $d \in (-\frac{1}{2}, 0)$, then $H \in (0, \frac{1}{2})$, in this case the function $f(k) = k^{2H}$ is strictly concave. Then we have that

$$\begin{aligned}
\frac{(k+1)^{2H} + (k-1)^{2H}}{2} &= \frac{f(k+1) + f(k-1)}{2} \\
&\leq f\left(\frac{k+1+k-1}{2} \right) = f(k) = k^{2H},
\end{aligned}$$

In this case $\gamma(k) \leq 0$.

Problem 2.30. If η_j are the coefficients of an FN(d) process, then

$$\begin{aligned}
\varphi_j &= \eta_j - \eta_{j-1} \\
&= \frac{\Gamma(j+d)}{\Gamma(d)\Gamma(j+1)} - \frac{\Gamma(j-1+d)}{\Gamma(d)\Gamma(j)} \\
&= \frac{(j-1+d)\Gamma(j-1+d)}{\Gamma(d)j\Gamma(j)} - \frac{\Gamma(j-1+d)}{\Gamma(d)\Gamma(j)} \\
&= \frac{\Gamma(j-1+d)}{\Gamma(d)\Gamma(j)}\Big[-\frac{j-1+d}{j} - 1\Big] \\
&= \eta_j\Big[\frac{j-1+d}{j} - 1\Big] \\
&= \eta_j\Big[\frac{j-1+d-j}{j}\Big] \\
&= \eta_j\frac{(d-1)}{j}.
\end{aligned}$$

Therefore,

$$\begin{aligned}
\varphi_j &\sim \lim_{j\to\infty} \eta_j \frac{(d-1)}{j} \\
&= \lim_{j\to\infty} \frac{\Gamma(j-1+d)}{\Gamma(j+1)}\frac{(d-1)}{\Gamma(d)} \\
&= \lim_{j\to\infty} \frac{\Gamma(j-1+d)}{\Gamma(j+1)}\frac{1}{\Gamma(d-1)} \\
&= \frac{1}{\Gamma(d-1)} \lim_{j\to\infty} \frac{\Gamma(j-1+d)}{\Gamma(j+1)} \\
&= \frac{j^{d-2}}{\Gamma(d-1)},
\end{aligned}$$

as $j \to \infty$.

Problem 2.71 The fractional Brownian motion satisfies for $d \in (-1/2, 1/2)$

$$\begin{aligned}
E[B_d(t)] &= 0 \\
\text{Cov}[B_d(t), B_d(s)] &= \frac{1}{2}(\mid t\mid^{2d+1} + \mid s\mid^{2d+1} - \mid t-s\mid^{2d-1}).
\end{aligned}$$

Then it follows that

$$B_d(t) \sim N(0, \mid t\mid^{2d+1}).$$

So that,

$$E[B_d(t+h) - B_d(t)] = 0,$$

and

$$\begin{aligned}
& \operatorname{Var}[B_d(t+h) - B_d(t)] \\
= \ & \operatorname{Var}[B_d(t+h)] + \operatorname{Var}[B_d(t)] - 2\operatorname{Cov}[B_d(t+h), B_d(t)] \\
= \ & |\, t+h \,|^{2d+1} + |\, t \,|^{2d+1} - 2[1/2(|\, t+h \,|^{2d+1} + |\, t \,|^{2d+1} - |\, h \,|^{2d+1})] \\
= \ & |\, h \,|^{2d+1}
\end{aligned}$$

On the other hand,

$$E[B_d(h) - B_d(0)] = 0$$

and

$$\begin{aligned}
& \operatorname{Var}[B_d(h) - B_d(0)] \\
= \ & \operatorname{Var}[B_d(h)] + \operatorname{Var}[B_d(0)] - 2\operatorname{Cov}[B_d(h), B_d(0)] \\
= \ & |\, h \,|^{2d+1} + |\, 0 \,|^{2d+1} - 2[1/2(|\, h \,|^{2d+1} + |\, 0 \,|^{2d+1} - |\, h \,|^{2d+1})] \\
= \ & |\, h \,|^{2d+1} .
\end{aligned}$$

Therefore for (3) we have

$$\begin{aligned}
B_d(t+h) - B_d(t) &\sim N(0, |\, h \,|^{2d+1}) \\
B_d(h) - B_d(0) &\sim N(0, |\, h \,|^{2d+1})
\end{aligned}$$

Hence $B_d(t+h) - B_d(t) \sim B_d(h) - B_d(0)$, as requested.

Problem 2.72. Given that fractional Brownian motion has stationary increments, meaning that

$$\begin{aligned}
B_d(t+h) - B_d(t) &\overset{D}{\sim} B_d(h) - B_d(0) && \text{(B.2)} \\
&= B_d(h) && \text{(B.3)}
\end{aligned}$$

have the same distribution. In particular, the p-th moments are equal

$$E[B_d(t+h) - B_d(t)]^p = EB_d(h)^p.$$

Since a fractional Brownian motion is self-similar with $H = d + 1/2$,

$$B_d(h \cdot 1)^p = |h|^{p(d+1/2)} B_d(1)^p, \tag{B.4}$$

which proves the claim. In particular for $p = 2$,

$$E\left[B_d(t+h) - B_d(t)\right]^2 = |h|^{2d+1} \cdot E\, B_d(1)^2. \tag{B.5}$$

Note that $E\, B_d(1)^2$ equals 1, as

$$1 = \operatorname{Var}(B_d(1)) = E\left(B_d(1) - E\, B_d(1)\right)^2 = EB_d(1)^2. \tag{B.6}$$

Thus dividing by h^2 we get

$$E\left[\frac{B_d(t+h)-B_d(t)}{h}\right]^2 = |h|^{2d-1}.$$

Problem 2.73. If $\sum_{j=1}^{\infty}[j(j+n)]^{-\delta} = \mathcal{O}(n^{-\delta})$ then

$$\left|\frac{1}{n^{-\delta}}\sum_{j=1}^{\infty}[j(j+n)]^{-\delta}\right| \leq \mathcal{C}.$$

Note that

$$\begin{aligned}\frac{1}{n^{-\delta}}\sum_{j=1}^{\infty}[j(j+n)]^{-\delta} &= \frac{1}{n^{-\delta}}\sum_{j=1}^{\infty}\left(\frac{1}{j}-\frac{1}{j+n}\right)^{\delta}\frac{1}{n^{\delta}} \\ &= \sum_{j=1}^{\infty}\left(\frac{1}{j}-\frac{1}{j+n}\right)^{\delta} \\ &\leq \sum_{j=1}^{\infty}\left(\frac{1}{j}\right)^{\delta}.\end{aligned}$$

Since $\delta \in (1,2)$ we have that $\sum_{j=1}^{\infty}\left(\frac{1}{j}\right)^{\delta} < \mathcal{C}$ converge to a constant $\mathcal{C} > 0$. Thus,

$$\left|\frac{1}{n^{-\delta}}\sum_{j=1}^{\infty}[j(j+n)]^{-\delta}\right| \leq \mathcal{C}.$$

Finally, as this condition is satisfied, we conclude that

$$\sum_{j=1}^{\infty}[j(j+n)]^{-\delta} = \mathcal{O}(n^{-\delta}).$$

Problem 3.5. Part a). Let $x_t = y_{t-1}$, hence the $AR(1)$ process can be written as

$$\begin{aligned}x_{t+1} &= \phi x_t + \varepsilon_t, \\ y_t &= \phi x_t + \varepsilon_t.\end{aligned}$$

For this system,

$$F = \phi, \quad G = \phi, \quad H = 1. \tag{B.7}$$

Thus the observability and controllability matrices are given by

$$\mathcal{O} = (G', F'G', F'^2G', \ldots)' = \begin{pmatrix} \phi \\ \phi^2 \\ \vdots \end{pmatrix},$$

and

$$\mathcal{C} = (H, FH, F^2H, \ldots) = (1, \phi, \phi^2, \ldots)$$

The system is controllable for all ϕ, as $\mathcal{C}$ has full rank 1 for every ϕ, it is observable only for $\phi \neq 0$. Thus, it is a minimal state space representation if $\phi \neq 0$.

Part b). A system is stable if $F^n x$ converges to zero in the norm of $\mathcal{H}$ for all $x \in \mathcal{H}$ for $n \to \infty$. This system is stable since

$$||F^n|| = |\phi^n| \to 0 \text{ for } n \to \infty \text{ as } |\phi| < 1.$$

It is exponentially stable as for $c > 1$ and $\alpha = -\log|\phi|$ it holds

$$|\phi^n| \leq ce^{-\alpha n}$$

for all n. Since $|\phi| < 1$, it holds $\alpha > 0$.

Part c). An $AR(1)$ process y_t has the Wold representation ($|\phi| < 1$)

$$(1 - \phi B)y_t = \varepsilon_t \Leftrightarrow y_t = (1 - \phi B)^{-1}\varepsilon_t = \sum_{j=0}^{\infty} \phi^j \varepsilon_{t-j}.$$

Thus the $MA(\infty)$ coefficients ψ_j equal

$$\psi_j = \phi^j, \quad j = 0, 1, 2, \ldots,$$

with $\{\psi_j\} \in \ell_2$.

The Hankel matrix is defined as $\mathcal{H} : \ell_2 \to \ell_2$ with

$$\mathcal{H} = \begin{pmatrix} \psi_1 & \psi_2 & \psi_3 & \cdots \\ \psi_2 & \psi_3 & \psi_4 & \cdots \\ \psi_3 & \psi_4 & \psi_5 & \cdots \\ \vdots & \vdots & \vdots & \ddots \end{pmatrix} = \begin{pmatrix} \phi^1 & \phi^2 & \phi^3 & \cdots \\ \phi^2 & \phi^3 & \phi^4 & \cdots \\ \phi^3 & \phi^4 & \phi^5 & \cdots \\ \vdots & \vdots & \vdots & \ddots \end{pmatrix} = \phi \cdot \begin{pmatrix} 1 & \phi & \phi^2 & \cdots \\ \phi & \phi^2 & \phi^3 & \cdots \\ \phi^2 & \phi^3 & \phi^4 & \cdots \\ \vdots & \vdots & \vdots & \ddots \end{pmatrix}$$

Part d). For $\phi \neq 0$ the Hankel matrix has rank 1, as every column $\mathcal{H}_k$, $k > 1$ can be expressed as a factor multiplied by the first column, i.e.

$$\mathcal{H}_k = \phi^(k-1)\mathcal{H}_1, \quad \forall k \geq 1.$$

Part d). For $\phi \neq 0$ the Hankel matrix has rank 1, as one can write x_t as

$$x_t = \mathcal{H}_1(\varepsilon_{t-1}, \varepsilon_{t-2}, \ldots)'$$

where $\mathcal{H}_1$ is the first row of $\mathcal{H}$.

Problem 3.7. The variance of the truncation error is of order $\mathcal{O}(1/m)$ for the autoregressive representation truncated in m, and is of order $\mathcal{O}(m^{2d-1})$ for the MA(∞) process, also truncated in m. This is an advantage of the autoregressive process. However, approximation by moving-average have also some advantages: the algorithm implementation and the analysis of the estimates theoretical properties are more simple.

Problem 3.9 Part a). Let F be a state matrix such that

$$F = \begin{bmatrix} 1 & 0 & 0 \\ 1 & 0 & 0 \\ 0 & 1 & 0 \end{bmatrix}$$

Thus we can define the following pattern for F^n.

$$\begin{aligned} F &= \begin{bmatrix} 1 & 0 & 0 \\ 1 & 0 & 0 \\ 0 & 1 & 0 \end{bmatrix} \\ F^2 &= \begin{bmatrix} 1 & 0 & 0 \\ 1 & 0 & 0 \\ 1 & 0 & 0 \end{bmatrix} \\ F^3 &= \begin{bmatrix} 1 & 0 & 0 \\ 1 & 0 & 0 \\ 1 & 0 & 0 \end{bmatrix} \\ \vdots &= \quad \vdots \end{aligned}$$

Now we can show that if $x = \begin{bmatrix} x_1 & x_2 & x_3 \end{bmatrix}'$, then

$$F^n x = \begin{bmatrix} x_1 & x_1 & x_1 \end{bmatrix}' \quad \forall\, n$$

Note that $F^n x \not\to 0$ when $n \to \infty$ which indicates that the system is not stable.

Part b). Let G and F be the observation and transition matrixes respectively

$$\begin{aligned} G &= \begin{bmatrix} 1 & 0 & 0 \end{bmatrix}, \\ F &= \begin{bmatrix} 1 & 0 & 0 \\ 1 & 0 & 0 \\ 0 & 1 & 0 \end{bmatrix}. \end{aligned}$$

The system is observable if the matrix $\mathcal{O}$ is full rank, where $\mathcal{O}$ is defined as it follows:

$$\mathcal{O} = \left(G', F'G', F'^2G', \ldots\right)'.$$

Then,

$$\begin{aligned} G' &= \begin{bmatrix} 1 \\ 0 \\ 0 \end{bmatrix} \\ F'G' &= \begin{bmatrix} 1 \\ 0 \\ 0 \end{bmatrix} \\ F'^2G' &= \begin{bmatrix} 1 \\ 0 \\ 0 \end{bmatrix} \\ \vdots &= \vdots \end{aligned}$$

That is, the observability matrix $\mathcal{O}$ is of the form

$$\mathcal{O} = \begin{pmatrix} 1 & 1 & 1 & \dots \\ 0 & 0 & 0 & \dots \\ 0 & 0 & 0 & \dots \end{pmatrix}'.$$

Is clear that the rank of $\mathcal{O}$ is one and therefore this matrix isn't full rank, meaning that the system is not observable.

Part c). Let H and F be the lineal and transition matrices respectively

$$\begin{aligned} H &= \begin{bmatrix} 1 \\ 0 \\ 0 \end{bmatrix}, \\ F &= \begin{bmatrix} 1 & 0 & 0 \\ 1 & 0 & 0 \\ 0 & 1 & 0 \end{bmatrix}. \end{aligned}$$

The system is said to be controllable if the matrix $\mathcal{C} = \big(H, FH, F^2H, \dots\big)$ is full rank.

$$H = \begin{bmatrix} 1 \\ 0 \\ 0 \end{bmatrix}$$
$$FH = \begin{bmatrix} 1 \\ 1 \\ 0 \end{bmatrix}$$
$$F^2H = \begin{bmatrix} 1 \\ 1 \\ 1 \end{bmatrix}$$
$$F^3H = \begin{bmatrix} 1 \\ 1 \\ 1 \end{bmatrix}$$
$$\vdots = \vdots$$

That is, $\mathcal{C}$ is given by

$$\mathcal{C} = \begin{pmatrix} 1 & 1 & 1 & 1 & \dots \\ 0 & 1 & 1 & 1 & \dots \\ 0 & 0 & 1 & 1 & \dots \end{pmatrix}.$$

Note that $\mathcal{C}$ is full rank, then the system is controllable.

Part d). Given that tho state space system is not stable, there is no guaranty that the mean of the state vector does not increases through time. Moreover, we do not know if the initial value effect will vanish over time. As the system is not observable, there we cannot determine its initial state from the observed process.

Problem 3.12. For this system we have that

$$F = \phi, \quad G = \theta, \quad H = 1.$$

Part a). A system is strongly stable if $F^n x$ converges to zero for all $x \in \mathcal{H}$ for $n \to \infty$. In the univariate case this means that $\phi^n x$ must converge to zero in the norm of $\mathcal{H}$. Thus

$$||\phi^n x|| = \phi^n ||x||$$

must converge to 0. This holds for $|\phi| < 1$, as $||x|| < \infty$ for every $x \in \mathcal{H}$. For exponential stability there must exist $c > 0$ and $\alpha > 0$ such that

$$||F^n|| \leq ce^{-\alpha n}.$$

This means that

$$|\phi^n| \leq ce^{-\alpha n} \Leftrightarrow |\phi| \leq ce^{-\alpha}$$

Setting $\alpha = -\log|\phi|$ and $c > 1$ gives the inequality for all n. As $|\phi| < 1$, it holds $\alpha > 0$.

Part b). The observability matrix of this system is given by

$$\mathcal{O} = (G', F'G', F'^2G', \ldots)' = \begin{pmatrix} \theta \\ \theta\phi \\ \theta\phi^2 \\ \vdots \end{pmatrix} = \theta \begin{pmatrix} 1 \\ \phi \\ \phi^2 \\ \vdots \end{pmatrix}$$

The rank of the matrix $\mathcal{O}$ equals 0, if and only if $\theta = 0$. This value of θ reduces the system to the trivial case $y_t = \varepsilon_t$ and the structure of x_t gets lost – i.e. it is not observable. For $\theta \neq 0$, however, the system is observable.

For controllability we have

$$\mathcal{C} = (H, FH, F^2H, \ldots) = \left(1, \phi, \phi^2, \ldots\right),$$

which is of full rank (equal 1) for all $\phi \in R$.

A system is minimal if and only if it is controllable and observable. This can be garanteed for this system iff $\theta \neq 0$.

Part c). To ensure that the Kalman recursions are stable, $F - HG = \phi - \theta = \Phi$ must have eigenvalues with modulus smaller than 1, or equivalently the polynomial $1 - \Phi z$ must be invertible for $|z| \leq 1$. This is the case for $|\phi - \theta| < 1$.

Solution to Problem 3.13. Part (a). Let $x \in \mathfrak{H}$, and

$$F^n x = F^{n-1} F x = \begin{pmatrix} 0 & 0 & 0 & \ldots & 0 & 1 \\ 0 & 0 & 0 & \ldots & 0 & 0 \\ \vdots & \vdots & & & & \vdots \\ \vdots & \vdots & & & & \vdots \\ 0 & 0 & 0 & \ldots & 0 & 0 \\ 0 & 0 & 0 & \ldots & 0 & 0 \end{pmatrix} \cdot \begin{pmatrix} 0 & 1 & 0 & 0 & \ldots & 0 \\ 0 & 0 & 1 & 0 & \ldots & 0 \\ 0 & 0 & 0 & 1 & \ldots & 0 \\ \vdots & \vdots & & & \ddots & \\ 0 & 0 & 0 & \ldots & 0 & 1 \\ 0 & 0 & 0 & \ldots & 0 & 0 \end{pmatrix} x \to 0$$

Hence F is stable.

Part (b). Consider the matrix norm induced by the sup norm is

$$\| F \|_\infty = \max_{1 \leq i \leq n} \sum_{j=1}^{n} | a_{ij} | = \max_{1 \leq i \leq n} \left[a_{i1} + a_{i2} + \ldots + a_{in}\right] = 1$$

Part (d). We can write the observability matrix as,

$$\mathcal{O} = (G', F'G', F'^2G', ...)'$$

$$= \begin{pmatrix} \psi_n & 0 & 0 & \cdots & \cdots & \cdots & 0 \\ \psi_{n-1} & \psi_{n-1} & 0 & & \cdots & \cdots & 0 \\ \psi_{n-2} & \psi_{n-2} & \psi_{n-2} & 0 & 0 & \cdots & 0 \\ \psi_{n-3} & \psi_{n-3} & \psi_{n-3} & \psi_{n-3} & 0 & \cdots & 0 \\ \vdots & \vdots & \psi_{n-4} & \psi_{n-4} & \psi_{n-4} & 0 \;\cdots & 0 \\ \vdots & \vdots & \vdots & \vdots & \vdots & \ddots & 0 \\ \psi_1 & \psi_1 & \psi_1 & \cdots & & \cdots & \psi_1 \end{pmatrix}'.$$

Hence rank$(\mathcal{O}) = n$ if $\psi_n \neq 0$.

Problem 3.14. Part a). Solving the system directly we have: $y_t = y_t$; $y_{t+1-p} = \phi_p y_{t-p} + ... + \phi_1 y_1 + \varepsilon_{t+1-p}$ which renders the AR(p) process.

Part b). We can write

$$G' = \begin{pmatrix} 0 & 0 & 0 & ... & 1 \end{pmatrix}',$$

$$F'G' = \begin{pmatrix} \phi_p & \phi_{p-1} & \phi_{p-2} & ... & \phi_1 \end{pmatrix}',$$

$$F^{2'}G' = \begin{pmatrix} \phi_{p-1} & \phi_{p-2} & \phi_{p-3} & ... & \sum_j \phi_{p-j}\phi_{p-j-1} \end{pmatrix},$$

and so on, for $j = 1, ..., n$ and we note that the observability matrix

$$\mathcal{O} = (G', F'G', F^{2'}G', ...)$$

is full rank and the system is therefore observable.

Part c). We begin by solving

$$H = \begin{pmatrix} 0 & 0 & 0 & ... & 1 \end{pmatrix}',$$

$$FH = \begin{pmatrix} 0 & 0 & 0 & ... & \phi_1 \end{pmatrix}',$$

$$F^2H = \begin{pmatrix} \phi_{p-1} & \phi_{p-2} & \phi_{p-3} & ... & \sum_j \phi_{p-j}\phi_{p-j-1} \end{pmatrix}',$$

and we note that the controllability matrix $C = (H, FH, F^2H, ...)$ is full rank and the system is controllable.

Problem 3.16. We have that,

$$\begin{aligned}\sum_{j=0}^{\infty} \|F^j z^j\| &= \sum_{j=0}^{\infty} |z^j| \|F^j\| \\ &= \sum_{j=0}^{\infty} |z|^j \|F^j\| \\ &\leq \sum_{j=0}^{\infty} |z|^j c e^{-\alpha j} \\ &\leq c \sum_{j=0}^{\infty} \left(|z|\, e^{-\alpha}\right)^j \\ &\leq c(1 - |z| e^{-\alpha})^{-1}.\end{aligned}$$

Note that the series converge, as $|z| \leq 1$ and if the values of α and c are positive, i.e. always occurs that $|e^{-\alpha}| < 1$.

Part b). Using the following decomposition

$$(I - zF)\left(I + zF + z^2F^2 + \ldots + z^nF^n\right) = I - z^{n+1}F^{n+1},$$

the limit is equal to

$$\begin{aligned}\lim_{n\to\infty} (I - zF)\left(I + zF + z^2F^2 + \ldots + z^nF^n\right) &= \lim_{n\to\infty} \left(I - z^{n+1}F^{n+1}\right), \\ (I - zF)\sum_{j=0}^{\infty} F^j z^j &= \lim_{n\to\infty} \left(I - z^{n+1}F^{n+1}\right).\end{aligned}$$

Then $\sum_{j=0}^{\infty} F^j z^j$ is $(I - zF)^{-1}$ if and only if

$$\lim_{n\to\infty} \left(I - z^{n+1}F^{n+1}\right) = I.$$

Now, we just need to verify the convergence given by

$$\begin{aligned}\lim_{n\to\infty} \|I - z^{n+1}F^{n+1} - I\| &= \lim_{n\to\infty} \|z^{n+1}F^{n+1}\| \\ &= \lim_{n\to\infty} |z|^{n+1} \|F^{n+1}\| \\ &\leq \lim_{n\to\infty} |z|^{n+1} c e^{-\alpha(n+1)} \\ &\leq 0.\end{aligned}$$

Which means that $\lim_{n\to\infty}\left(I - z^{n+1}F^{n+1}\right) = I$, i.e.,

$$(I - zF)\sum_{j=0}^{\infty} F^j z^j = I,$$

and consequently,

$$\sum_{j=0}^{\infty} F^j z^j = (I - zF)^{-1}.$$

Problem 4.5. We can write

$$\begin{aligned}
\alpha_j &= \frac{1}{2\pi}\int_{-\pi}^{\pi} f(\lambda)e^{-i\lambda j}d\lambda \\
&= \frac{1}{2\pi}\int_{-\pi}^{\pi} \sum_{k=-m}^{m} \alpha_k e_k(\lambda)e^{-i\lambda j}d\lambda \\
&= \frac{1}{2\pi}\int_{-\pi}^{\pi} \sum_{k=-m}^{m} \alpha_k e^{i\lambda k}e^{-i\lambda j}d\lambda \\
&= \frac{1}{2\pi}\int_{-\pi}^{\pi} \sum_{k=-m}^{m} \alpha_k \langle e_k, e_j\rangle d\lambda.
\end{aligned}$$

But,

$$\langle e_k, e_j\rangle = \begin{cases} 1, & \text{si } k = j \\ 0, & \text{otherwise.} \end{cases}$$

That is, the sum is

$$\sum_{k=-m}^{m} \alpha_k \langle e_k, e_j\rangle = \alpha_j.$$

Hence,

$$\frac{1}{2\pi}\int_{-\pi}^{\pi} \sum_{k=-m}^{m} \alpha_k \langle e_k, e_j\rangle d\lambda = \frac{1}{2\pi}\int_{-\pi}^{\pi} \alpha_j d\lambda = \alpha_j.$$

Furthermore,

$$\begin{aligned}
\| f(\lambda) \|^2 &= \langle f(\lambda), f(\lambda)\rangle \\
&= \left\langle \sum_{j=-m}^{m} \alpha_j e_j(\lambda), \sum_{k=-m}^{m} \alpha_k e_k(\lambda) \right\rangle \\
&= \sum_{j=-m}^{m} \sum_{k=-m}^{m} \alpha_j \alpha_k \langle e_j(\lambda), e_k(\lambda)\rangle.
\end{aligned}$$

However,

$$\langle e_k, e_j\rangle = \begin{cases} 1, & \text{si } k = j \\ 0, & \text{otherwise.} \end{cases}$$

Thus,

$$\| f(\lambda) \|^2 = \sum_{j=-m}^{m} \alpha_j^2$$

and then

$$\| f(\lambda) \| = \sqrt{\sum_{j=-m}^{m} \alpha_j^2}.$$

Problem 4.6. Assume that $x'\Gamma x > 0$ for $x \in \mathbb{R}^n$, $x \neq 0$ then Γ is positive definite. Note that because $\frac{1}{n} \log \det(\Gamma) \to \frac{1}{2\pi} \int_{-\pi}^{\pi} \log[2\pi f_\theta(\lambda)] \partial\lambda$ we have that $f_\theta(\lambda) > c$ with $c > 0$ implies that $\det(\Gamma) > 0$ for every $\lambda \in (-\pi, \pi]$ and the result is proved.

Problem 4.7. Since the function $g(\lambda)$ is continuos on $[-\pi, \pi]$ for any $\varepsilon > 0$ there is a trigonometric polynomial $\phi_k(\lambda) = \sum_{l=-k}^{k} c_l e^{i\lambda l}$ such that $| g(\lambda) - \phi_k(\lambda) | < \varepsilon$ for all $\lambda \in [-\pi, \pi]$. For this polynomial we have

$$\begin{aligned}
\int_{-\pi}^{\pi} \phi_k(\lambda) I_n(\lambda) d\lambda &= \int_{-\pi}^{\pi} \sum_{l=-k}^{k} c_l e^{i\lambda l} I_n(\lambda) d\lambda \\
&= \sum_{l=-k}^{k} c_l \int_{-\pi}^{\pi} e^{i\lambda l} I_n(\lambda) d\lambda \\
&= \sum_{l=-k}^{k} c_l w(l, n)
\end{aligned}$$

By taking limits as $n \to \infty$

$$\begin{aligned}
\lim_{n\to\infty} \int_{-\pi}^{\pi} \phi_k(\lambda) I_n(\lambda) d\lambda &= \lim_{n\to\infty} \sum_{l=-k}^{k} c_l w(l, n) \\
&= \sum_{l=-k}^{k} c_l \gamma(l) = \sum_{l=-k}^{k} c_l \int_{-\pi}^{\pi} e^{i\lambda l} f(\lambda) d\lambda \\
&= \int_{-\pi}^{\pi} \sum_{l=-k}^{k} c_l e^{i\lambda l} f(\lambda) d\lambda = \int_{-\pi}^{\pi} \phi_k(\lambda) f(\lambda) d(\lambda).
\end{aligned}$$

This is true for an arbitrary ε, therefore the result is obtained.

Problem 4.10. Provided that $\{y_t : t \in \mathbb{Z}\}$ is stationary and $E\,|\gamma(h)| < \infty$ the Fourier transform of the series is given by

$$F(h) = \int_{-\pi}^{\pi} e^{i\lambda h} f(\lambda) d\lambda$$

which corresponds the autocovariance of the process, $\gamma(h)$. On the other hand, these autocovariances are the Fourier coefficients of the expansion of $f(\lambda)$ scaled by 2π, such that $f(\lambda) = \sum_{h=-\infty}^{\infty} C_h e^{-i\lambda h}$ with $C_h = \gamma(h)/2\pi$.

Problem 4.12 According to the Wold representation, a stationary process may be expressed as:

$$y_t = \sum_{j=0}^{\infty} \psi_j \varepsilon_{t-j}$$

where $\psi_0 = 1, \sum_{j=0}^{\infty} \psi_j^2 < \infty$, ε is a white noise sequence with variance σ^2. Let

$$\begin{aligned}
\gamma(h) = E(y_t y_{t+h}) &= E(\sum_{j=0}^{\infty} \psi_j \varepsilon_{t-j} \sum_{i=0}^{\infty} \psi_i \varepsilon_{t+h-i}) \\
&= E(\sum_{j=0}^{\infty} \sum_{i=0}^{\infty} \psi_j \psi_i \varepsilon_{t-j} \varepsilon_{t+h-i}) = \sum_{i,j=0}^{\infty} \psi_j \psi_i E(\varepsilon_{t-j} \varepsilon_{t+h-i}) \\
&= \sigma^2 \sum_{j=0}^{\infty} \psi_j \psi_{j+h}
\end{aligned}$$

Similarly for $\gamma(h) = E(y_t, y_{t-h})$. Hence

$$\gamma(h) = \sigma^2 \sum_{j=0}^{\infty} \psi_j \psi_{j+|h|}.$$

Problem 4.16.

a) The periodogram is defined as

$$I(\lambda) = \frac{1}{2\pi n} \Big| \sum_{j=1}^{n} y_j e^{i\lambda j} \Big|^2,$$

for $\lambda \in [-\pi, \pi]$ and $|z| = \sqrt{z\overline{z}}$, where $\overline{z}$ is the complex conjugate of z.

Thus,

$$\begin{aligned}
\frac{1}{n}\left|\sum_{j=1}^{n} y_j e^{i\lambda j}\right|^2 &= \frac{1}{n}\sum_{j=1}^{n} y_j e^{i\lambda j}\overline{\sum_{m=1}^{n} y_m e^{i\lambda m}} \\
&= \frac{1}{n}\sum_{j=1}^{n} y_j e^{i\lambda j}\sum_{m=1}^{n} y_m e^{-i\lambda m} \\
&= \frac{1}{n}\sum_{j=1}^{n}\sum_{m=1}^{n} y_j y_m e^{i\lambda(j-m)} \\
&= \frac{1}{n}\sum_{j=1}^{n} y_j^2 + \frac{1}{n}\sum_{j\neq m} y_j y_m e^{i\lambda(j-m)}
\end{aligned}$$

As the functions $e^{i\lambda s}$ are ortogonal, i.e.

$$\int_{-\pi}^{\pi} e^{i\lambda k} e^{-i\lambda s} d\lambda = \begin{cases} 0 & k \neq s \\ 2\pi & k = s \end{cases},$$

it follows that the integral is different to zero, only for $k = -(j - m)$.

$$\begin{aligned}
\int_{-\pi}^{\pi} e^{i\lambda k} I(\lambda) d\lambda &= \frac{1}{2\pi}\int_{-\pi}^{\pi} e^{i\lambda k}\left(\frac{1}{n}\sum_{j=1}^{n} y_j^2 + \frac{1}{n}\sum_{j\neq m} y_j y_m e^{i\lambda(j-m)}\right) d\lambda \\
&= \frac{1}{2\pi}\int_{-\pi}^{\pi} e^{i\lambda k}\left(\frac{1}{n}\sum_{j=1}^{n} y_j y_{j+k} e^{-i\lambda k}\right) d\lambda \\
&= \frac{1}{n}\sum_{j=1}^{n} y_j y_{j+k}
\end{aligned}$$

Note that this sum is only different from 0 for $|k| < n$, as otherwise no observations are available. Replace now y_j by $y_j - \overline{y}$. This gives the result.

b) Define the random variable

$$\delta_t := (y_t - \overline{y}_n)(y_{t+k} - \overline{y}_n).$$

It holds

$$\mathbb{E}\delta_t = \gamma(k),$$

and by the strong law of large numbers

$$\frac{1}{n}\sum_{t=1}^{n} \delta_t \to \mathbb{E}\delta_t,$$

for $N \to \infty$.

Additionally we can write

$$\frac{1}{n} = \frac{n-k}{n}\frac{1}{n-k},$$

and as for every fixed k the term $\frac{n-k}{n}$ converges to 1 for $n \to \infty$, we get that

$$\lim_{n\to\infty} w(k,n) = \gamma(k).$$

Problem 5.5. The log-likelihood function of a zero-mean stationary Gaussian process $\{y_t\}$ is given by (see Expression 4.1, page 66)

$$\mathcal{L}(\theta) = -\frac{1}{2}\log\det\Gamma_\theta + \frac{1}{2}\mathbf{y}'\Gamma_\theta^{-1}\mathbf{y},$$

where $\mathbf{y} = (y_1,\ldots,y_n)'$ and $\Gamma_\theta = E\,\mathbf{y}\mathbf{y}'$ and θ is the parameter vector. The maximum likelihood estimator is obtained by maximizing the log-likelihood with respect to θ.

By the Durbin Levinson algorithm, the log-likelihood function can be expressed as

$$\mathcal{L}(\theta) = -\frac{1}{2}\sum_{t=1}^{n}\log\nu_{t-1} - \frac{1}{2}\sum_{t=1}^{n}\frac{e_t^2}{\nu_{t-1}},$$

where $e_t = y_t - \widehat{y}_t$ and ν_t satisfies

$$\begin{aligned}
\phi_{tt} &= (\nu_{t-1})^2\left[\gamma(t) - \sum_{i=1}^{t-1}\phi_{t-1,i}\gamma(t-i)\right]\\
\phi_{tj} &= \phi_{t-1,j} - \phi_{tt}\phi_{t-1,t-j}, j = 1,\ldots,t-1,\\
\nu_0 &= \sigma_y^2,\\
\nu_t &= \nu_{t-1}(1-\phi_{tt}^2), j = 1,\ldots,t-1.
\end{aligned}$$

Another expression for $\mathcal{L}(\theta)$ can be obtained by the state space approach,

$$\mathcal{L}(\theta) = -\frac{1}{2}\left[n\log 2\pi + \sum_{t=1}^{n}\log\Delta t + n\log\sigma^2 + \frac{1}{\sigma^2}\sum_{t=1}^{n}\frac{(y_t-\overline{y})^2}{\Delta t}\right].$$

Differentiating $\mathcal{L}(\theta)$ with respect to σ^2 (omitting the factor $-\frac{1}{2}$) gives

$$\frac{d}{d\sigma^2}\mathcal{L}(\theta) = \frac{n}{\sigma^2} - \left(\frac{1}{\sigma^2}\right)^2\sum_{t=1}^{n}\frac{(y_t-\overline{y})^2}{\Delta t}$$

Setting equal to zero and multiplying by $\sigma^4 > 0$ gives

$$\begin{aligned}\frac{n}{\sigma^2} - \left(\frac{1}{\sigma^2}\right)^2 \sum_{t=1}^{n} \frac{(y_t - \overline{y})^2}{\Delta t} &= 0 \\ \sum_{t=1}^{n} \frac{(y_t - \overline{y})^2}{\Delta t} &= n\sigma^2\end{aligned}$$

Thus the ML estimator for σ^2 equals

$$\widehat{\sigma}^2_{ML} = \frac{1}{n} \sum_{t=1}^{n} \frac{(y_t - \overline{y})^2}{\Delta t}.$$

Problem 5.8. Let y_t be a FN(d) process with $d \in (-1, \frac{1}{2})$. Thus,

$$y_t = (1 - B)^{-d} \varepsilon_t$$

with $\varepsilon_t \sim \text{WN}(0, \sigma^2)$. The process is stationary, causal and invertible. Furthermore, we know that in a process $FN(d)$

$$\widehat{y}_{t+1} = \phi_{t1} y_n + \cdots + \phi_{tt} y_1,$$

where the coefficients of the prediction are given by

$$\phi_{tj} = -\binom{t}{j} \frac{\Gamma(j-d)\Gamma(t-d-j+1)}{\Gamma(-d)\Gamma(t-d+1)}.$$

On the other hand, we need to determine ν_t defined by

$$\nu_t = \gamma(0) \prod_{j=1}^{t} (1 - \phi_{jj}^2).$$

In the particular case of fractional noise, we have that

$$\nu_t = \sigma^2 \frac{\Gamma(t+1)\Gamma(t+1-2d)}{[\Gamma(t+1-d)]^2}.$$

With this results, the log-likelihood function for a sample $(y_1, \ldots, y_n)$ can be expressed as

$$\begin{aligned}\mathcal{L}(d, \sigma^2) &= -\frac{1}{2} \sum_{t=1}^{n} \log \nu_{t-1} - \frac{1}{2} \sum_{t=1}^{n} \frac{(y_t - \widehat{y}_t)^2}{\nu_{t-1}} \\ &= -\frac{1}{2} \sum_{t=1}^{n} \log(\sigma^2 f(t,d)) - \frac{1}{2} \sum_{t=1}^{n} \frac{(y_t - \widehat{y}_t)^2}{\sigma^2 f(t,d)}, \\ &= -\frac{n}{2} \log(\sigma^2) - \frac{1}{2} \sum_{t=1}^{n} \log(f(t,d)) - \frac{1}{2} \sum_{t=1}^{n} \frac{(y_t - \widehat{y}_t)^2}{\sigma^2 f(t,d)},\end{aligned}$$

where

$$f(t,d) = \frac{\Gamma(t)\Gamma(t-2d)}{[\Gamma(t-d)]^2}.$$

Thus, we can maximize $\mathcal{L}(\theta)$ for σ^2, where the estimate of σ^2 is given by

$$\widehat{\sigma}^2(\widehat{d}) = \frac{1}{n}\sum_{t=1}^{n} \frac{(y_t - \widehat{y}_t)^2}{f(t,\widehat{d})}$$

Thus, the estimated of d can be obtained conditioning the likelihood function on $\widehat{\sigma}^2(d)$

$$\mathcal{L}(d,\widehat{\sigma}^2(d)) = -\frac{n}{2}\log[\widehat{\sigma}^2(d)] - \frac{1}{2}\sum_{t=1}^{n}\log(f(t,d)) - \frac{1}{2}.$$

The term $f(t,d)$ is despicable in the estimation of d, because, given de Stirling's approximation

$$f(t,d) = \frac{\Gamma(t)\Gamma(t-2d)}{[\Gamma(t-d)]^2} \approx t^d t^{-d} \approx 1,$$

that is, $\sum_{t=1}^{n}\log(f(t,d)) \approx 0$, then, maximum likelihood estimation is equivalent to maximize

$$\mathcal{L}_1(d,\widehat{\sigma}^2(d)) = -\frac{n}{2}\log[\widehat{\sigma}^2(d)].$$

In the Haslett and Raftery method, if $M = n$ we must maximize

$$\mathcal{L}_2(d) = \text{constant} - \frac{1}{2}n\log[\widehat{\sigma}_2^2(d)].$$

Note that if $M = n$, the numeric algorithm is of order $\mathcal{O}(n^2)$ just like the maximum likelihood case and $(y_t - \widehat{y}_t)^2$ is also equivalent, so we truncate in $M = n$. Then we have

$$\widehat{\sigma}_2^2(d) = \frac{1}{n}\sum_{t=1}^{n}\frac{(y_t - \widehat{y}_t)^2}{\upsilon_t},$$

where $\upsilon_t = \kappa\,\nu_{t-1}$. In this way

$$\begin{aligned}\upsilon_t &= \kappa\,\sigma^2\frac{\Gamma(t)\Gamma(t-2d)}{[\Gamma(t-d)]^2}\\ &= \kappa\,\sigma^2 f(t,d).\end{aligned}$$

where $\kappa \propto \sigma^{-2}$. Finally $\widehat{\sigma}_2(d) \propto (1/n)\sum_{t=1}^{n}(y_t - \widehat{y}_t)^2/f(t,\widehat{d})$. Thus,

$$\mathcal{L}_2(d) \propto \text{constant} - \frac{1}{2}n\log[\widehat{\sigma}^2(d)].$$

We conclude that the two methods $\mathcal{L}_1(d)$ y $\mathcal{L}_2(d)$ are equivalent when we maximize with respect to d.

Problem 5.14. Consider the following partition $(y_{n+1}, y_n, \ldots, y_2, y_1) = (y_{n+1}, \boldsymbol{y}_n)$. Then, the likelihood function can be expressed as

$$f(y_{n+1}, \boldsymbol{y}_n|\boldsymbol{\theta}) = (2\pi)^{-(n+1)}|\Gamma(\boldsymbol{\theta})|^{-1/2} \exp\left\{-\frac{1}{2}(y_{n+1}\,\boldsymbol{y}_n)\,\Gamma(\boldsymbol{\theta})^{-1}\,(y_{n+1}\,\boldsymbol{y}_n)^t\right\},$$

where the matrix $\Gamma(\boldsymbol{\theta})$ can be partitioned of the following way

$$\Gamma(\boldsymbol{\theta}) = \begin{bmatrix} \Gamma_{1,1}(\boldsymbol{\theta}) & \Gamma_{1,n}(\boldsymbol{\theta}) \\ \Gamma_{n,1}(\boldsymbol{\theta}) & \Gamma_{n,n}(\boldsymbol{\theta}) \end{bmatrix}.$$

Then, by properties of the Normal distribution, the conditional density is given by

$$y_{n+1}|\boldsymbol{y}_n, \boldsymbol{\theta} \sim N\left(\Gamma_{1,n}(\boldsymbol{\theta})\Gamma_{n,n}(\boldsymbol{\theta})^{-1}\boldsymbol{y}_n,\, \Gamma_{1,1}(\boldsymbol{\theta}) - \Gamma_{1,n}(\boldsymbol{\theta})\Gamma_{n,n}(\boldsymbol{\theta})^{-1}\Gamma_{n,1}(\boldsymbol{\theta})\right)$$

where the sub-matrices are

$$\begin{aligned}
\Gamma_{1,1}(\boldsymbol{\theta}) &= \gamma(0) \\
\Gamma_{1,n}(\boldsymbol{\theta}) &= [\gamma(1)\ \gamma(2)\ \ldots\ \gamma(n)] \\
\Gamma_{n,1}(\boldsymbol{\theta}) &= [\gamma(1)\ \gamma(2)\ \ldots\ \gamma(n)]^t \\
\Gamma_{n,n}(\boldsymbol{\theta}) &= \begin{bmatrix} \gamma(0) & \gamma(1) & \ldots & \gamma(n-1) \\ \gamma(1) & \gamma(0) & \ldots & \gamma(n-2) \\ \vdots & \vdots & \ddots & \vdots \\ \gamma(n-1) & \gamma(n-2) & \ldots & \gamma(0) \end{bmatrix}.
\end{aligned}$$

Once these matrices are defined, we can calculate de mean and variance of de conditional distribution. Note that $\Gamma_{n,n}(\boldsymbol{\theta})$ can be decompose in

$$\Gamma_{n,n}(\boldsymbol{\theta}) = LDL',$$

where

$$\begin{aligned}
D &= \operatorname{diag}(\nu_0, \ldots, \nu_{n-1}) \\
L &= \begin{bmatrix} 1 & & & & & \\ -\phi_{11} & 1 & & & & \\ -\phi_{22} & -\phi_{21} & 1 & & & \\ -\phi_{33} & -\phi_{32} & -\phi_{31} & & & \\ \vdots & \vdots & \vdots & & 1 & \\ -\phi_{n-1,n-1} & -\phi_{n-1,n-2} & -\phi_{n-1,n-3} & \ldots & -\phi_{n-1,1} & 1 \end{bmatrix},
\end{aligned}$$

and the coefficients ϕ_{ij} and ν_t are given by the equations of the durbin-levinson algorithm in section 4.1.2, and $\Gamma(\boldsymbol{\theta})^{-1} = L'D^{-1}L$. Thus, the conditional means is

$$\Gamma_{1,n}(\boldsymbol{\theta})\Gamma_{n,n}(\boldsymbol{\theta})^{-1}\boldsymbol{y}_n = \Gamma_{1,n}(\boldsymbol{\theta})L'D^{-1}L\boldsymbol{y}_n = (L\,\Gamma_{n,1}(\boldsymbol{\theta}))'D^{-1}L\boldsymbol{y}_n$$

The result of $L\,\Gamma_{n,1}(\boldsymbol{\theta})$ is

$$L\,\Gamma_{n,1}(\boldsymbol{\theta}) = \begin{bmatrix} \gamma(1) \\ \gamma(2) - \phi_{11}\gamma(1) \\ \gamma(3) - \phi_{21}\gamma(2) - \phi_{22}\gamma(1) \\ \vdots \\ \gamma(n) - \sum_{t=1}^{n-1} \phi_{n-1,t}\gamma(n-t) \end{bmatrix} = \begin{bmatrix} \phi_{11}\,\nu_0 \\ \phi_{22}\,\nu_1 \\ \phi_{33}\,\nu_2 \\ \vdots \\ \phi_{nn}\,\nu_{n-1} \end{bmatrix}.$$

Then, we have that

$$(L\,\Gamma_{n,1}(\boldsymbol{\theta}))'D^{-1} = \begin{bmatrix} \phi_{11} & \phi_{22} & \phi_{33} & \cdots & \phi_{nn} \end{bmatrix},$$

Thus, the product $(L\,\Gamma_{n,1}(\boldsymbol{\theta}))'D^{-1}L$ is

$$(L\,\Gamma_{n,1}(\boldsymbol{\theta}))'D^{-1}L =$$

$$\begin{bmatrix} \phi_{n1} & \phi_{n2} & \phi_{n3} & \cdots & \phi_{n,n-1} & \phi_{n,n} \end{bmatrix}.$$

Finally,

$$\begin{aligned} E(y_{n+1}|\boldsymbol{y}_n, \boldsymbol{\theta}) &= (L\,\Gamma_{n,1}(\boldsymbol{\theta}))'D^{-1}L\boldsymbol{y}_n \\ &= \begin{bmatrix} \phi_{n1} & \phi_{n2} & \cdots & \phi_{n,n} \end{bmatrix} \begin{bmatrix} y_n \\ y_{n-1} \\ \vdots \\ y_1 \end{bmatrix}. \\ &= \sum_{j=1}^{n} \phi_{nj} y_{n+1-j}. \end{aligned}$$

Moreover, note that

$$\begin{aligned} Var(y_{n+1}|\boldsymbol{y}_n, \boldsymbol{\theta}) &= \Gamma_{1,1}(\boldsymbol{\theta}) - \Gamma_{1,n}(\boldsymbol{\theta})\Gamma_{n,n}(\boldsymbol{\theta})^{-1}\Gamma_{n,1}(\boldsymbol{\theta}) \\ &= \gamma(0) - \Gamma_{1,n}(\boldsymbol{\theta})L'D^{-1}L\Gamma_{n,1}(\boldsymbol{\theta}) \\ &= \gamma(0) - (L\Gamma_{n,1}(\boldsymbol{\theta}))'D^{-1}L\Gamma_{n,1}(\boldsymbol{\theta}) \\ &= \nu_0 - \sum_{t=1}^{n} \phi^2 \nu_{t-1} \\ &= \nu_0 - \sum_{t=1}^{n} (\nu_{t-1} - \nu_t) \\ &= \nu_n. \end{aligned}$$

According to Durbin-Levinson algorithm we have that $\nu_n = \nu_{n-1}[1-\phi_{tt}^2]$ with $\nu_0 = \gamma(0)$. thus, recursively

$$\nu_n = \gamma(0)\prod_{j=1}^{n}(1-\phi_{jj}^2)$$

With this we can conclude that

$$Var(y_{n+1}|\boldsymbol{y}_n, \boldsymbol{\theta}) = \gamma(0)\prod_{j=1}^{n}(1-\phi_{jj}^2)$$

Problem 5.19. For the ARFIMA$(1, d, 1)$ process

$$\Gamma(d,\phi,\theta) = \begin{pmatrix} \frac{\pi^2}{6} & -\frac{\log(1+\phi)}{\phi} & \frac{\log(1+\theta)}{\theta} \\ -\frac{\log(1+\phi)}{\phi} & \frac{1}{1-\phi^2} & \frac{1}{1-\phi\theta} \\ \frac{\log(1+\theta)}{\theta} & \frac{1}{1-\phi\theta} & \frac{1}{1-\theta^2} \end{pmatrix}.$$

Since an ARFIMA$(1, d, 0)$ model is nested in the ARFIMA$(1, d, 1)$ model with $\theta = 0$, we can use this result and simply omit the third column and row of $\Gamma(d, \phi, \theta)$ to get the 2×2 matrix

$$\Gamma(d,\phi) = \begin{pmatrix} \frac{\pi^2}{6} & -\frac{\log(1+\phi)}{\phi} \\ -\frac{\log(1+\phi)}{\phi} & \frac{1}{1-\phi^2} \end{pmatrix}.$$

The asymptotic covariance of $\widehat{d}$ and $\widehat{\phi}$ equals the off-diagonal element of

$$\Gamma(d,\phi)^{-1} = \frac{1}{\det\Gamma(d,\phi)}\begin{pmatrix} \frac{1}{1-\phi^2} & \frac{\log(1+\phi)}{\phi} \\ \frac{\log(1+\phi)}{\phi} & \frac{\pi^2}{6} \end{pmatrix}$$

The determinant of $\Gamma(d, \phi)$ equals

$$\det\Gamma(d,\phi) = \frac{\pi^2}{6}\frac{1}{1-\phi^2} - \left(-\frac{\log(1+\phi)}{\phi}\right)^2 \tag{B.8}$$

As the determinant is a continuous function (summation and multiplication), the limit for $\phi \to 0$ equals

$$\lim_{\phi\to 0}\det\Gamma(d,\phi) = \frac{\pi^2}{6} - \left(\frac{\log(1+\phi)}{\phi}\right)^2$$

The second term is of the form $\frac{0}{0}$, thus by l'Hospitals rule

$$\lim_{\phi\to 0}\frac{\log(1+\phi)}{\phi} = \lim_{\phi\to 0}\frac{\frac{1}{1+\phi}}{1} = 1$$

it holds that

$$\lim_{\phi\to 0} \frac{1}{\det \Gamma(d,\phi)} = \frac{1}{\frac{\pi^2}{6} - 1} \tag{B.9}$$

Thus the limit of the off-diagonal element of the inverse yields

$$\frac{1}{\frac{\pi^2}{6} - 1} \lim_{\phi\to 0} \frac{\log(1+\phi)}{\phi} = \frac{1}{\frac{\pi^2}{6} - 1}$$

Since

$$\mathrm{corr}(\widehat{d}, \widehat{\phi}) := \frac{cov(\widehat{d}, \widehat{\phi})}{\sqrt{var(\widehat{d})}\sqrt{var(\widehat{\phi})}},$$

the factor $\frac{1}{\det \Gamma(d,\phi)}$ cancels out for all d and ϕ, and it remains

$$\mathrm{corr}(\widehat{d}, \widehat{\phi}) = \frac{\frac{\log(1+\phi)}{\phi}}{\sqrt{\frac{\pi^2}{6}}\sqrt{\frac{1}{1-\phi^2}}}.$$

The limit of $\frac{\log(1+\phi)}{\phi}$ for $\phi \to 0$ is of the form $\frac{0}{0}$, thus by l'Hospital

$$\lim_{\phi\to 0} \frac{\log(1+\phi)}{\phi} = \lim_{\phi\to 0} \frac{\frac{1}{1+\phi}}{1} = 1.$$

Finally we get

$$\lim_{\phi\to 0} \mathrm{corr}(\widehat{d}, \widehat{\phi}) = \frac{\lim_{\phi\to 0} \frac{\log(1+\phi)}{\phi}}{\sqrt{\frac{\pi^2}{6}} \lim_{\phi\to 0} \sqrt{\frac{1}{1-\phi^2}}} = \frac{1}{\frac{\pi}{\sqrt{6}}} = \frac{\sqrt{6}}{\pi} \tag{B.10}$$

Problem 5.20. We know that for the ARFIMA$(1, d, 0)$ process it holds that

$$\Gamma(d,\theta) = \begin{pmatrix} \frac{\pi^2}{6} & \frac{log(1+\theta)}{\theta} \\ \frac{log(1+\theta)}{\theta} & \frac{1}{1-\theta^2} \end{pmatrix}$$

The asymptotic correlation of $\hat{d}$ and $\hat{\theta}$ is given by the off diagonal element of the inverse of $\Gamma(d, \theta)$. The inverse is given by

$$\Gamma^{-1}(d,\theta) = \frac{1}{\det \Gamma(d,\theta)} \begin{pmatrix} \frac{1}{1-\theta^2} & -\frac{log(1+\theta)}{\theta} \\ -\frac{log(1+\theta)}{\theta} & \frac{\pi^2}{6} \end{pmatrix}$$

The correlation is defined as

$$\begin{aligned} \mathrm{corr}(\hat{d}, \hat{\theta}) &= \frac{cov(\hat{d}, \hat{\theta})}{\sqrt{var(\hat{d})}\sqrt{var(\hat{\theta})}} \\ &= -\frac{\frac{\log(1+\theta)}{\theta}}{\sqrt{\frac{\pi^2}{6}}\sqrt{\frac{1}{1-\theta^2}}} \end{aligned}$$

Making $\theta \to 0$ hence

$$\operatorname{corr}(\hat{d}, \hat{\theta}) = -\frac{\sqrt{6}}{\pi}$$

because the rule L'Hôpital

$$\lim_{\theta \to 1} \frac{\log(1+\theta)}{\theta} = \lim_{\theta \to 1} \frac{\frac{1}{1+\theta}}{1} = 1$$

Problem 6.4. Part (a) We have that

$$\begin{aligned}
E(y_t) &= E\left[E\left(y_t \mid y_{t-1}\right)\right] \\
&= E\left\{E\left[(\alpha + \beta\, y_{t-1}^2)^{1/2}\, z_t \mid y_{t-1}\right]\right\} \\
&= E\left\{(\alpha + \beta\, y_{t-1}^2)^{1/2} \cdot E\left(z_t \mid y_{t-1}\right)\right\} \\
&= E\left[(\alpha + \beta\, y_{t-1}^2)^{1/2} \cdot E\left(z_t\right)\right],
\end{aligned}$$

Since z_t has zero-mean,

$$E(y_t) = E\left[(\alpha + \beta\, y_{t-1}^2)^{1/2} \cdot 0\right] = E(0) = 0,$$

and

$$\begin{aligned}
E(y_t^2) &= E\left[E\left(y_t^2 \mid y_{t-1}\right)\right] \\
&= E\left\{E\left[(\alpha + \beta\, y_{t-1}^2)\, z_t^2 \mid y_{t-1}\right]\right\} \\
&= E\left\{(\alpha + \beta\, y_{t-1}^2) \cdot E\left(z_t^2 \mid y_{t-1}\right)\right\} \\
&= E\left[(\alpha + \beta\, y_{t-1}^2) \cdot E\left(z_t^2\right)\right],
\end{aligned}$$

Since z_t has unit variance,

$$\begin{aligned}
E(y_t^2) &= E\left[(\alpha + \beta\, y_{t-1}^2) \cdot 1\right] \\
&= \alpha + \beta\; E(y_{t-1}^2) \\
&= \alpha + \beta\; E(y_t^2).
\end{aligned}$$

Now, by stationarity, $E(y_t^2) = \frac{\alpha}{1-\beta}$. Finally, for $k > 0$

$$\begin{aligned}
E(y_t \cdot y_{t+k}) &= E\left[E\left(y_t \cdot y_{t+k} \mid y_{t+k-1}\right)\right] \\
&= E\left\{E\left[\left(\alpha + \beta\, y_{t-1}^2\right)^{1/2} z_t \cdot \left(\alpha + \beta\, y_{t+k-1}^2\right)^{1/2} z_{t+k} \mid y_{t+k-1}\right]\right\} \\
&= E\left[\left(\alpha + \beta\, y_{t-1}^2\right)^{1/2} z_t \cdot \left(\alpha + \beta\, y_{t+k-1}^2\right)^{1/2} \cdot E\left(z_{t+k} \mid y_{t+k-1}\right)\right] \\
&= E\left[\left(\alpha + \beta\, y_{t-1}^2\right)^{1/2} z_t \cdot \left(\alpha + \beta\, y_{t+k-1}^2\right)^{1/2} \cdot E\left(z_{t+k}\right)\right],
\end{aligned}$$

Since z_t has zero-mean,

$$E(y_t \cdot y_{t+k}) = E\left[\left(\alpha + \beta\, y_{t-1}^2\right)^{1/2} z_t \cdot \left(\alpha + \beta\, y_{t+k-1}^2\right)^{1/2} \cdot 0\right] \;= E(0) = 0.$$

Now, for $k < 0$,

$$\begin{aligned} E(y_t \cdot y_{t+k}) &= E\left[E\left(y_t \cdot y_{t+k} \,|\, y_{t+k-1}\right)\right] \\ &= E\left\{E\left[\left(\alpha + \beta\, y_{t-1}^2\right)^{1/2} z_t \cdot \left(\alpha + \beta\, y_{t+k-1}^2\right)^{1/2} z_{t+k} \,|\, y_{t-1}\right]\right\} \\ &= E\left[\left(\alpha + \beta\, y_{t-1}^2\right)^{1/2} \cdot \left(\alpha + \beta\, y_{t+k-1}^2\right)^{1/2} z_{t+k-1} \cdot E\left(z_t \,|\, y_{t-1}\right)\right] \\ &= E\left[\left(\alpha + \beta\, y_{t-1}^2\right)^{1/2} \cdot \left(\alpha + \beta\, y_{t+k-1}^2\right)^{1/2} z_{t+k-1} \cdot E\left(z_t\right)\right], \end{aligned}$$

Since z_t has zero-mean,

$$E(y_t \cdot y_{t+k}) = E\left[\left(\alpha + \beta\, y_{t-1}^2\right)^{1/2} \cdot \left(\alpha + \beta\, y_{t+k-1}^2\right)^{1/2} z_{t+k} \cdot 0\right] \;= E(0) = 0.$$

Therefore,

$$\begin{aligned} \gamma_y(k) = \operatorname{Cov}(y_t,\, y_{t+k}) &= E(y_t\, y_{t+k}) - E(y_t)\; E(y_{t+k}) \\ &= \begin{cases} \dfrac{\alpha}{1-\beta}, & k = 0 \\ 0, & k \neq 0 \end{cases} \end{aligned}$$

Pert (b). From part (a) we have that $\{y_t\} \sim \mathrm{WN}\left(0,\, \frac{\alpha}{1-\beta}\right)$ and given that $\{x_t\}$ has a MA(1) structure, the function of autocovariance is given by

$$\gamma_x(k) = \begin{cases} \dfrac{\alpha}{1-\beta}\left(1 + \theta^2\right), & k = 0 \\ \dfrac{\alpha}{1-\beta}\,\theta, & |k| = 1 \\ 0, & |k| > 1. \end{cases}$$

Problem 6.11. Since backshift operator B is such that $BY_t = y_{t-1}$, if $Y \equiv \alpha_0$ with $\alpha_0 \in$ then the backshift operator becomes $B\alpha_0 = \alpha_0 \Rightarrow B \equiv 1$ so if $\pi(B) = (1-B)^d \Rightarrow \pi(B)\alpha_0 = \pi(1)\alpha_0 = (1-1)^d\alpha_0 = 0$ if $d > 0$.

Problem 6.12. An ARCH(∞) model is defined as

$$\begin{aligned} y_t &= \sigma_t \varepsilon_t, \\ \varepsilon_t &\overset{iid}{\sim} \mathrm{WN}(0, \sigma_\varepsilon^2), \\ \sigma_t^2 &= \alpha_0 + \sum_{j=1}^{\infty} \alpha_j y_{t-j}^2, \end{aligned}$$

where $\alpha_0 > 0$ and $\alpha_j \geq 0$ for $j \geq 1$. If the coefficients α_j are given by an ARFIMA(p,d,q) representation, then the ARCH(∞) model is called a FIGARCH(p,d,q) model.

Let $\pi(B) = \phi(B)(1-B)^d\theta(B)^{-1}$, then

$$\pi(B)y_t^2 = \alpha_0 + \nu_t$$

with $\nu_t = y_t^2 - \sigma_t^2$.

Multiplying both sides by $\theta(B)$ gives

$$\phi(B)(1-B)^d y_t^2 = \omega + \theta(B)\nu_t,$$

where $\omega = \theta(B)\alpha_0$. Since α_0 is a constant, $\theta(B)\alpha_0 = \theta(1)\cdot\alpha_0$. To ensure that ω is positive, the MA polynomial must be positive at $z=1$ (or $\lambda = 0$), that is, $\theta(1) > 0$.

Problem 6.13.

(a) An ARFIMA$(0,d,0)$-GARCH process is defined by

$$\varepsilon_t(d) = y_t(1-B)^d$$

for $\mid d \mid < 1/2$.Hence

$$\begin{aligned}
\varepsilon_t(d) &= y_t(1-B)^d \\
(1-B)^{-d}\varepsilon_t(d) &= y_t = c \\
\lambda(d)^{-1}\varepsilon_t(d) &= c
\end{aligned}$$

where c is a constant with respect to d

(b) Part of a derivative with respect to d is

$$\begin{aligned}
&\frac{\partial}{\partial d}\lambda^{-1}(d)\varepsilon_t(d) + \lambda^{-1}(d)\frac{\partial}{\partial d}\varepsilon_t(d) = 0 \\
=\ &\frac{\partial}{\partial d}(1-B)^{-d}\varepsilon_t(d) + (1-B)^{-d}\frac{\partial}{\partial d}\varepsilon_t(d) = 0 \\
=\ &\sum_{j=0}^{\infty}[\frac{\partial}{\partial d}\psi_j(d)]\varepsilon_{t-j}(d) + \sum_{j=0}^{\infty}\psi_j(d)[\frac{\partial}{\partial d}\varepsilon_{t-j}(d)] = 0
\end{aligned}$$

(c) In considering the first part of the equation (b)

$$\begin{aligned}
\lambda(d)^{-1}\frac{\partial}{\partial d}\varepsilon_t(d) &= -\frac{\partial}{\partial d}[\lambda^{-1}(d)\varepsilon_t(d)] \\
\frac{\partial}{\partial d}\varepsilon_t(d) &= -\lambda(d)\frac{\partial}{\partial d}[\lambda^{-1}(d)\varepsilon_t(d)] \\
\frac{\partial}{\partial d}\varepsilon_t(d) &= -\lambda(d)\sum_{j=0}^{\infty}\psi_j(d)[\frac{\partial}{\partial d}\varepsilon_{t-j}(d)] \\
\frac{\partial}{\partial d}\varepsilon_t(d) &= -\lambda(d)[\frac{\partial}{\partial d}\lambda^{-1}(d)]\varepsilon_t(d) \\
\frac{\partial}{\partial d}\varepsilon_t(d) &= -[\frac{\partial}{\partial d}\log\lambda^{-1}(d)]\varepsilon_t(d)
\end{aligned}$$

(d) Using the result

$$\log(1+x) = \sum_{j=1}^{\infty}\frac{(-1)^{j+1}}{j}x^j$$

and for the part (c). Hence

$$\begin{aligned}
\frac{\partial}{\partial d}\varepsilon_t(d) &= -[\frac{\partial}{\partial d}\log\lambda^{-1}(d)]\varepsilon_t(d) \\
&= -[\frac{\partial}{\partial d}\log(1-B)^{-d}]\varepsilon_t(d) \\
&- -[\frac{\partial}{\partial d} - d\log(1-B)]\varepsilon_t(d) \\
&= -[-\log(1-B)]\varepsilon_t(d) \\
&= -[-\sum_{j=1}^{\infty}\frac{(-1)^{j+1}(-B)^j}{j}]\varepsilon_t(d) \\
&= -[-\sum_{j=1}^{\infty}\frac{(-1)^{2j+1}B^j}{j}]\varepsilon_t(d) \\
&= -\sum_{j=1}^{\infty}\frac{1}{j}B^j\varepsilon_t(d) \\
&= -\sum_{j=1}^{\infty}\frac{1}{j}\varepsilon_{t-j}(d)
\end{aligned}$$

Problem 6.14. a) For a sequence $\{\epsilon_t\}$of independent and identically distributed uniform random variables $U(-\sqrt{3},\sqrt{3})$, the mean of the process is given by $E[\frac{1}{2}(-\sqrt{3}+\sqrt{3})] = 0$ and the variance is $E[\frac{1}{12}(\sqrt{3}+\sqrt{3})^2] = \frac{12}{12} = 1$.

b) Using the formula provided in Gradshteyn and Rizhik (2000, p.236), and denoting $x = 1$ and $a = \sqrt{\frac{\beta_1}{3\alpha_1}}$ the result for the Lyapunov exponent is direct.

Problem 7.2. Note that

$$\begin{aligned}
\phi_{tj} &= -\binom{t}{j}\frac{\Gamma(j-d)\Gamma(t-d-j+1)}{\Gamma(-d)\Gamma(t-d+1)} \\
&= -\frac{\Gamma(t+1)}{\Gamma(j+1)\Gamma(t-j+1)}\frac{\Gamma(j-d)\Gamma(t-d-j+1)}{\Gamma(-d)\Gamma(t-d+1)} \\
&= -\frac{\Gamma(j-d)}{\Gamma(j+1)\Gamma(-d)}\frac{\Gamma(t+1)\Gamma(t-d-j+1)}{\Gamma(t-j+1)\Gamma(t-d+1)},
\end{aligned}$$

and, when $t \to \infty$, we have that as $t \to \infty$,

$$\begin{aligned}
&-\frac{\Gamma(j-d)}{\Gamma(j+1)\Gamma(-d)}\frac{\Gamma(t+1)\Gamma(t-d-j+1)}{\Gamma(t-j+1)\Gamma(t-d+1)} \\
&\approx -\frac{\Gamma(j-d)}{\Gamma(j+1)\Gamma(-d)}t^{1+j-1}t^{-d-j+1+d-1} \\
&= -\frac{\Gamma(j-d)}{\Gamma(j+1)\Gamma(-d)}t^{j}t^{-j} \\
&= -\frac{\Gamma(j-d)}{\Gamma(j+1)\Gamma(-d)} \\
&= \pi_j
\end{aligned}$$

Problem 7.3. Part (a). Applying expectation to (6.30) we have that

$$\begin{aligned}
E(y_t) &= \alpha + \phi\, E(y_{t-1}) + E(z_t) + (\theta+\eta)\, E(z_{t-1}) + \theta\,\eta\, E(z_{t-2}) \\
\mu &= \alpha + \phi\,\mu + 0 + (\theta+\eta)\, 0 + \theta\,\eta\, 0 \\
\mu &= \alpha + \phi\,\mu.
\end{aligned}$$

Consequently, by stationarity, $\mu = \frac{\alpha}{1-\phi}$.

Part (b). From Part (a) we have that

$$\alpha = \mu - \phi\,\mu$$

replacing in (6.30) we obtain,

$$y_t = \mu - \phi\,\mu + \phi\, y_{t-1} + z_t + (\theta+\eta)\, z_{t-1} + \theta\,\eta\, z_{t-2}.$$

Hence,

$$\begin{aligned}
[y_t - \mu] &= \phi\,[y_{t-1} - \mu] + z_t + (\theta+\eta)\, z_{t-1} + \theta\,\eta\, z_{t-2} \\
&= \phi\,[B\, y_t - B\,\mu] + z_t + (\theta+\eta)\, z_{t-1} + \theta\,\eta\, z_{t-2} \\
&= \phi\, B\,[y_t - \mu] + z_t + (\theta+\eta)\, z_{t-1} + \theta\,\eta\, z_{t-2}
\end{aligned}$$

Thus,

$$[y_t - \mu] - \phi\, B\, [y_t - \mu] = z_t + (\theta + \eta)\, z_{t-1} + \theta\, \eta\, z_{t-2},$$

and,

$$\begin{aligned}(1 - \phi\, B)\, [y_t - \mu] &= z_t + (\theta + \eta)\, B\, z_t + \theta\, \eta\, B^2\, z_t \\ (1 - \phi\, B)\, [y_t - \mu] &= \left[1 + (\theta + \eta)\, B + \theta\, \eta\, B^2\right] z_t.\end{aligned}$$

Pert (c). Since $|\phi| < 1$, we have that

$$\begin{aligned}[y_t - \mu] &= (1 - \phi\, B)^{-1} \left[1 + (\theta + \eta)\, B + \theta\, \eta\, B^2\right] z_t \\ &= \left(\sum_{k=0}^{\infty} \phi^k\, B^k\right) \left[1 + (\theta + \eta)\, B + \theta\, \eta\, B^2\right] z_t \\ &= \left(\sum_{k=0}^{\infty} \phi^k\, B^k\right) + \left(\sum_{k=1}^{\infty} (\theta + \eta)\, \phi^{k-1}\, B^k\right) + \left(\sum_{k=2}^{\infty} \theta\, \eta\, \phi^{k-2}\, B^k\right) \\ &= B^0 + [\phi + (\theta + \eta)]\, B^1 + \left(\sum_{k=2}^{\infty} \left[\phi^k + \phi^{k-1}\, (\theta + \eta) + \theta\, \eta\, \phi^{k-2}\right] B^k\right) \\ &= B^0 + [\phi + (\theta + \eta)]\, B^1 + \left(\sum_{k=2}^{\infty} \phi^{k-2} \left[\phi^2 + \phi\, (\theta + \eta) + \theta\, \eta\right] B^k\right),\end{aligned}$$

that is,

$$\psi_0 = 1, \quad \psi_1 = \phi + (\theta + \eta), \quad \psi_k = \phi^{k-2} \left[\phi^2 + \phi\, (\theta + \eta) + \theta\, \eta\right] \quad k \geq 2,$$

as requested.

Part (e). The autocovariance function of $\{y_t - \mu\}$ is given by

$$\gamma(k) = \sigma^2 \sum_{j=0}^{\infty} \psi_j\, \psi_{j+|k|}$$

From this expression we can write

$$\begin{aligned}\gamma(0) &= \sigma^2 \left\{1 + [\phi + (\theta + \eta)]^2 + \frac{[\phi^2 + (\theta + \eta)\, \phi + \theta\, \eta]^2}{\phi^4} \cdot \left(\frac{\phi^4}{1 - \phi^2}\right)\right\} \\ &= \frac{\sigma^2}{1 - \phi^2} \left\{(1 - \phi^2) + (1 - \phi^2)\, [\phi + (\theta + \eta)]^2 + [\phi^2 + (\theta + \eta)\, \phi + \theta\, \eta]^2\right\} \\ &= \frac{\sigma^2}{1 - \phi^2} \left\{(1 - \phi^2)\, \{1 + [\phi + (\theta + \eta)]^2\} + [\phi^2 + (\theta + \eta)\, \phi + \theta\, \eta]^2\right\}.\end{aligned}$$

$$\begin{aligned}
\gamma(1) &= \sigma^2 \left\{ [\phi + (\theta + \eta)] + [\phi + (\theta + \eta)] \, [\phi^2 + (\theta + \eta)\,\phi + \theta\,\eta] \right. \\
&\quad \left. + \frac{[\phi^2 + (\theta + \eta)\,\phi + \theta\,\eta]^2}{\phi^3} \cdot \left(\frac{\phi^4}{1 - \phi^2} \right) \right\} \\
&= \frac{\sigma^2}{1 - \phi^2} \left\{ (1 - \phi^2) \, [\phi + (\theta + \eta)] \, \{1 + [\phi^2 + (\theta + \eta)\,\phi + \theta\,\eta]\} \right. \\
&\quad \left. + [\phi^2 + (\theta + \eta)\,\phi + \theta\,\eta]^2 \, \phi \right\}
\end{aligned}$$

$$\begin{aligned}
\gamma(k) &= \sigma^2 \left\{ \phi^{|k|-2} \, [\phi^2 + (\theta + \eta)\,\phi + \theta\,\eta] + [\phi + (\theta + \eta)] \, \phi^{|k|-1} \, [\phi^2 + (\theta + \eta)\,\phi + \theta\,\eta] + \right. \\
&\quad \left. \frac{[\phi^2 + (\theta + \eta)\,\phi + \theta\,\eta]^2}{\phi^4} \cdot \left(\frac{\phi^4}{1 - \phi^2} \right) \phi^{|k|} \right\} \\
&= \frac{\sigma^2 \, \phi^{|k|-2}}{1 - \phi^2} \left\{ (1 - \phi^2) \, [\phi^2 + (\theta + \eta)\,\phi + \theta\,\eta] \right. \\
&\quad + (1 - \phi^2) \, [\phi + (\theta + \eta)] \, \phi \, [\phi^2 + (\theta + \eta)\,\phi + \theta\,\eta] \\
&\quad \left. + [\phi^2 + (\theta + \eta)\,\phi + \theta\,\eta]^2 \, \phi^2 \right\} \\
&= \frac{\sigma^2 \, \phi^{|k|-2}}{1 - \phi^2} \left\{ (1 - \phi^2) \, [\phi^2 + (\theta + \eta)\,\phi + \theta\,\eta] \, \{1 + [\phi + (\theta + \eta)] \, \phi\} \right. \\
&\quad \left. + [\phi^2 + (\theta + \eta)\,\phi + \theta\,\eta]^2 \, \phi^2 \right\}
\end{aligned}$$

Part (f). Under the assumption of infinite information, the variance of the h-step prediction error is

$$\sigma_t^2(h) = \sigma^2 \sum_{j=0}^{h-1} \psi_j^2.$$

From this formula we can write,

$$\begin{aligned}
\sigma_t^2(1) &= 1 \\
\sigma_t^2(2) &= 1 + [\phi + (\theta + \eta)]^2 \\
\sigma_t^2(h) &= 1 + [\phi + (\theta + \eta)]^2 + \frac{\left[\phi^2 + (\theta + \eta)\,\phi + \theta\,\eta\right]^2}{\phi^4} \left(\frac{\phi^4 - \phi^{2\,h}}{1 - \phi^2} \right), \qquad h \geq 2.
\end{aligned}$$

Problem 7.5. If $\alpha_t = (1-\phi_{tt})\alpha_{t-1}$ and $\alpha_0 = 1$, then

$$\begin{aligned}\alpha_t &= (1-\phi_{tt})(1-\phi_{t-1,t-1})\alpha_{t-2}\\ &= (1-\phi_{tt})(1-\phi_{t-1,t-1})....(1-\phi_{t-(t-1),t-(t-1)})\alpha_{t-t}\\ &= (1-\phi_{tt})(1-\phi_{t-1,t-1})....(1-\phi_{11})\alpha_0\\ &= \prod_{j=1}^{t}(1-\phi_{jj}).\end{aligned}$$

Then, by using $\alpha_t = 1-\sum_{j=1}^{t}\phi_{tj} \Rightarrow \sum_{j=1}^{t}\phi_{tj} = 1-\alpha_t = 1-\prod_{j=1}^{t}(1-\phi_{jj})$.

Problem 7.6. We have that

$$\gamma_x(k) = \begin{cases} (1+\theta^2)\,\sigma^2, & k=0\\ \theta\,\sigma^2, & |k|=s\\ 0, & \text{otherwise.}\end{cases}$$

As an extension of the process MA(1) we may write,

$$\hat{X}_{n+1} = \begin{cases} 0, & n=0,\dots,s-1\\ \theta_{n,\,s}\,(x_{n+1-s}-\hat{X}_{n+1-s}), & n=s,s+1,s+2,\dots\end{cases}$$

where

$$\theta_{n,\,s} = \frac{\theta\,\sigma^2}{\upsilon_{n-s}}$$

and

$$\upsilon_n = \begin{cases} (1+\theta^2)\,\sigma^2, & n=0,1,\dots,s-1\\ (1+\theta^2)\,\sigma^2-\theta_{n,\,s}\,\upsilon_{n-s}, & n=s,s+1,s+2,\dots\end{cases}$$

Problem 7.12. A stochastic process y_t is ARFIMA-GARCH(r,s) if it is a solution to

$$\begin{aligned}\Phi(B)y_t &= \Theta(B)(1-B)^{-d}\varepsilon_t,\\ \varepsilon_t &= \eta_t\sigma_t,\\ \sigma_t^2 &= \alpha_0+\sum_{j=1}^{r}\alpha_j\varepsilon_{t-j}^2+\sum_{k=1}^{s}\beta_k\sigma_{t-k}^2,\end{aligned}$$

where $\mathcal{F}_{t-1}$ is the set generated by the past observations $y_{t-1}, y_{t-2},\dots$, $\sigma_t^2 = E\left[y_t^2|\mathcal{F}_{t-1}\right]$ is the conditional variance of the process y_t, the GARCH coefficients α_j, β_k are positive, and $\sum_{j=1}^{r}\alpha_j+\sum_{k=1}^{s}\beta_k < 1$, and η_t are i.i.d. zero-mean and unit variance random variables.

Since for all $t\in\mathbb{Z}$ we have

$$E[\varepsilon_{t+h}^2|\mathcal{F}_t] = \begin{cases} E[\sigma_{t+h}^2|\mathcal{F}_t] & h\geq 1,\\ \varepsilon_{t+h}^2 & h<1.\end{cases}$$

Furthermore it holds

$$E[e_t^2(h)|\mathcal{F}_t] = \sum_{j=0}^{h-1} \psi_j^2 \sigma_t^2(h-j), \tag{B.11}$$

where $e_t^2(h)$ is the multistep ahead prediction error

$$e_t(h) := y_t - \widehat{y}(h) = \varepsilon_{t+h} + \psi_1 \varepsilon_{t+h-1} + \ldots + \psi_{h-1} \varepsilon_{t+1}.$$

For the ARFIMA-GARCH(r, s) model we have

$$\sigma_{t+h}^2 = \alpha_0 + \sum_{j=1}^{r} \alpha_j \varepsilon_{t+h-j}^2 + \sum_{k=1}^{s} \beta_k \sigma_{t+h-k}^2.$$

Taking $E(\cdot|\mathcal{F}_t)$ on both sides,

$$\begin{aligned} \sigma_t^2(h) &= \alpha_0 + \sum_{j=1}^{r} \alpha_j \, E(\varepsilon_{t+h-j}^2|\mathcal{F}_t) + \sum_{k=1}^{s} \beta_k \, E(\sigma_{t+h-k}^2|\mathcal{F}_t) \\ &= \alpha_0 + \sum_{j=1}^{h-1} \alpha_j \sigma_t^2(h-j) + \sum_{j=h}^{r} \alpha_j \varepsilon_{t+h-j}^2 + \sum_{k=1}^{h-1} \beta_k \sigma_t^2(h-k) + \sum_{k=h}^{s} \beta_k \sigma_{t+h-k}^2. \end{aligned}$$

Problem 7.13. For the ARFIMA-GARCH$(1, 1)$ we have

$$\sigma_{t+h}^2 = \alpha_0 + \alpha_1 \varepsilon_{t+h-1}^2 + \beta_1 \sigma_{t+h-1}^2$$

Therefore for $h = 1$

$$\begin{aligned} \sigma_{t+1}^2 &= \alpha_0 + \alpha_1 \varepsilon_t^2 + \beta_1 \sigma_t^2 \\ &= \alpha_0 + \alpha_1 \varepsilon_t^2 + \beta_1 L(\sigma_{t+1}^2) \\ \sigma_{t+1}^2 (1 - L\beta_1) &= \alpha_0 + \alpha_1 \varepsilon_t^2 \\ \sigma_{t+1}^2 &= \frac{\alpha_0}{1 - L\beta_1} + \alpha_1 \sum_{j=0}^{\infty} \beta_1^j \varepsilon_{t-j}^2 \\ &= \frac{\alpha_0}{1 - \beta_1} + \alpha_1 \sum_{j=0}^{\infty} \beta_1^j \varepsilon_{t-j}^2 \end{aligned}$$

Finally we have

$$\sigma_{t+1}^2 = \frac{\alpha_0}{1 - \beta_1} + \alpha_1 \sum_{j=0}^{\infty} \beta_1^j \varepsilon_{t-j}^2.$$

Problem 7.22. Since $\mathcal{H}$ is an inner product space then we have: $\|x \pm y\|^2 = \langle x \pm y, x \pm y \rangle$ where $\langle \cdot, \cdot \rangle$ denotes an inner product function. Thus, $\|x + y\|^2 +$

$\|x-y\|^2 = \sum_i x_i^2 + 2\sum_i x_i y_i + \sum_i y_i^2 + \sum_i x_i^2 - 2\sum_i x_i y_i + \sum_i y_i^2 = 2\sum_i x_i^2 + 2\sum_i y_i^2 =$
$2\langle x,x\rangle + 2\langle y,y\rangle = 2\|x\|^2 + 2\|y\|^2$ for $i = 1, ..., n$.

Problem 9.7. Consider $B^s \equiv \widetilde{B}$

$$\begin{aligned}(1-B^s)^{-d} &= (1-\widetilde{B})^{-d} = \sum_{j=0}^{\infty} \psi_j \widetilde{B}^j \\ &= \sum_{j=0}^{\infty} \psi_j B^{sj} = \sum_{k=0}^{\infty} \psi_{k/s} B^k.\end{aligned}$$

But, by writing $k = sj$ and $j \in: j = \dfrac{k}{s}$ we have,

$$(1-B^s)^{-d} = \sum_{k=0}^{\infty} \psi_{k/s} B^k.$$

Problem 9.8.

a) The fractional filter is given by

$$(1-B)^d = \sum_{j=0}^{\infty} \eta_j B^j,$$

where

$$\eta_j = \frac{\Gamma(j+d)}{\Gamma(j+1)\Gamma(d)} = \binom{d}{j}.$$

This expansion holds for $d < 1/2, d \neq 0, -1, -2, \ldots$.

The function $\eta(z) = (1-z)^{-d}$ is analytic in the open unit disk $\{z : |z| < 1\}$ and analytic in the closed unit disk $\{z : |z| \leq 1\}$ for negative d. Thus the Taylor expansion is given by

$$\eta(z) = \sum_{j=0}^{\infty} \eta_j z^j. \tag{B.12}$$

In the seasonal case we have B^s instead of B for some positive $s \in \mathbb{N}$. Define $\tilde{B}_s = B^s$ and $\tilde{z}_s = z^s$. As all convergence conditions are based on the open and closed unit disk, and the function $f(z) = z^s$ preserves the property of being in this set or not, we can reduce the case of B^s to $\tilde{B}_s$.

Consequently we have

$$(1 - z^s)^d = (1 - \tilde{z}_s)^d = \sum_{j=0}^{\infty} \eta_j \tilde{z}_s^j = \sum_{j=0}^{\infty} \eta_j z^{s \cdot j}. \tag{B.13}$$

And this representation converges absolutely, for the same reasons as the function $(1 - z)^d$ converges to $\eta(z)$.

b) By assumption z is an element of $\mathbb{L}^2$, but constant over time. Hence,

$$Bz = z. \tag{B.14}$$

Thus applying any well-defined filter $a(B)$ to a random variable z, is equivalent to multiplying z by $a(1)$. If the filter has roots for $B = 1$, it holds

$$a(1)z = 0. \tag{B.15}$$

The filter $(1 - B^s)^d$ equals zero for $B = 1$ and $d > 0$.

c) Following the same arguments as in a) a stationary $\mathbb{L}^2$-convergent solution is given by

$$y_t = \sum_{j=0}^{\infty} \eta_j B^{s \cdot j} \varepsilon_t = \sum_{j=0}^{\infty} \eta_j \varepsilon_{t-s \cdot j}, \tag{B.16}$$

where

$$\eta_j = \frac{\Gamma(j+d)}{\Gamma(j+1)\Gamma(d)}, \qquad -1 < d < 0. \tag{B.17}$$

d) By assumption y_t is a solution to

$$(1 - B^s)^d y_t = \varepsilon_t. \tag{B.18}$$

In part b) we showed that for any random variable z it holds $(1-B^s)^d z = 0$. Using the linearity of the backshift operator we get

$$(1-B^s)^d x_t = (1-B^s)^d (y_t + z) = (1-B^s)^d y_t + (1-B^s)^d z = \varepsilon_t + 0 = \varepsilon_t. \tag{B.19}$$

Thus x_t is also a solution to the stochastic difference equation.

Problem 9.9. A SARFIMA process $\{y_t\}$ has spectral density function given by

$$f_s(\lambda) = \frac{1}{2\pi} \left|1 - e^{is\lambda}\right|^{-2d_s} = \frac{1}{2\pi} \left[2 \sin\left(\frac{s\lambda}{2}\right)\right]^{-2d_s}$$

Since $\sin\frac{s\lambda}{2} \to \frac{s\lambda}{2}$, when $\lambda \to 0^+$, one has

$$f_s(\lambda) \approx \frac{1}{2\pi}(s\lambda)^{-2d_s}$$

for $0 < d_s < \frac{1}{2}$ Therefore,

$$\begin{aligned}
\gamma_s(k) &= \int_{-\pi}^{\pi} e^{-i\lambda k} f_s(\lambda) \mathrm{d}\lambda \\
&= \int_{-\pi}^{\pi} e^{-i\lambda k} \frac{1}{2\pi} \left[2\sin\left(\frac{s\lambda}{2}\right)\right]^{-2d_s} \mathrm{d}\lambda \\
&\approx \int_{-\pi}^{\pi} e^{-i\lambda k} \frac{1}{2\pi} (s\lambda)^{-2d_s} \mathrm{d}\lambda \\
&\approx \int_{-\pi}^{\pi} e^{-i\lambda k} f(s\lambda) \mathrm{d}\lambda \\
&\approx \gamma(sk) \quad as \quad \lambda \to 0^+
\end{aligned}$$

where $f(\lambda)$ is the spectral density function the a $FN(d)$ process with unit noise variance. In general we have, for any $s \in \mathcal{N}$

$$\begin{aligned}
\gamma_s(k) &= \int_{-\pi}^{\pi} e^{-i\lambda k} \frac{1}{2\pi} \left[2\sin\left(\frac{s\lambda}{2}\right)\right]^{-2d_s} \mathrm{d}\lambda \\
&= \int_{-\pi}^{\pi} e^{-i\lambda sk} \frac{1}{2\pi} \left[2\sin\left(\frac{\lambda}{2}\right)\right]^{-2d_s} \mathrm{d}\lambda \\
&= \gamma(sk)
\end{aligned}$$

where $\lambda \in (-\pi, \pi]$ and $0 < d_s < \frac{1}{2}$.

Problem 9.14. Consider a SARFIMA$(0, d, 0) \times (0, d_s, 0)_s$ process. This process is defined by

$$y_t - \mu = (1 - B^s)^{-d_s}(1 - B)^{-d}\varepsilon_t, \tag{B.20}$$

where μ is the mean of the series, and $\{\varepsilon_t\}$ is a white noise sequence with finite variance.

The SARFIMA$(0, d, 0) \times (0, d_s, 0)_s$ process has a MA(∞) representation

$$y_t - \mu = \sum_{j=0}^{\infty} \psi_j \varepsilon_{t-j} \tag{B.21}$$

where the coefficients ψ_j are given by the inverse Fourier transform

$$\psi_j = \frac{1}{2\pi} \int_{-\pi}^{\pi} (1 - e^{i\lambda s})^{-d_s}(1 - e^{i\lambda})^{-d} d\lambda, \tag{B.22}$$

or equivalently by the convolution of the two filters

$$\psi_j = \sum_{k=0}^{j} a_k b_{k-j} \text{ where } a_k \leftrightarrow (1-B)^d, b_n \leftrightarrow (1-B^s)^{-d_s}. \tag{B.23}$$

Chapter 4 describes a method to convert a Wold representation to state space form. Following this method a State space form is given by

$$\begin{aligned}
x_t &= [y(t|t-1), y(t+1|t-1), y(t+2|t-1), \dots]', &\text{(B.24)}\\
F &= \begin{pmatrix} 0 & 1 & 0 & 0 & \dots \\ 0 & 0 & 1 & 0 & \dots \\ 0 & 0 & 0 & 1 & \dots \\ \vdots & \vdots & \vdots & \ddots & \end{pmatrix}, &\text{(B.25)}\\
H &= [\psi_1, \psi_2, \psi_2, \dots]' &\text{(B.26)}\\
G &= [1, 0, 0, \dots]' &\text{(B.27)}\\
y_t &= Gx_t + \varepsilon_t. &\text{(B.28)}
\end{aligned}$$

where

$$y(t+j|t-1) := E(y_{t+j}|y_{t-1}, y_{t-2}, \dots). \tag{B.29}$$

Problem 10.6. We have that

$$\begin{aligned}
\text{Var}(y_t) &= \text{Cov}(y_t,\, y_t) = \text{Cov}(a+bt+w_t,\, a+bt+w_t)\\
&= \text{Cov}(w_t,\, w_t) = \gamma_w(0)\\
&= \sigma^2.
\end{aligned}$$

Moreover,

$$\begin{aligned}
z_t &= \frac{1}{2k+1} \sum_{j=-k}^{k} y_{t-j}\\
&= \frac{1}{2k+1} \sum_{j=-k}^{k} [a + b(t-j) + w_{t-j}]\\
&= \frac{1}{2k+1} \sum_{j=-k}^{k} a + \frac{1}{2k+1} \sum_{j=-k}^{k} bt - \frac{1}{2k+1} \sum_{j=-k}^{k} j + \frac{1}{2k+1} \sum_{j=-k}^{k} w_{t-j}\\
&= a + bt + \frac{1}{2k+1} \sum_{j=-k}^{k} w_{t-j}
\end{aligned}$$

Then,

$$
\begin{aligned}
\operatorname{Var}(z_t) &= \operatorname{Cov}(z_t,\, z_t) \\
&= \operatorname{Cov}\left(a + bt + \frac{1}{2\,k+1}\sum_{j=-k}^{k} w_{t-j},\; a + bt + \frac{1}{2\,k+1}\sum_{i=-k}^{k} w_{t-i}\right) \\
&= \operatorname{Cov}\left(\frac{1}{2\,k+1}\sum_{j=-k}^{k} w_{t-j},\; \frac{1}{2\,k+1}\sum_{i=-k}^{k} w_{t-i}\right) \\
&= \left(\frac{1}{2\,k+1}\right)^2 \sum_{j=-k}^{k}\sum_{i=-k}^{k} \operatorname{Cov}\left(w_{t-j}, w_{t-i}\right)
\end{aligned}
$$

$$
\begin{aligned}
&= \left(\frac{1}{2\,k+1}\right)^2 \sum_{j=-k}^{k}\sum_{i=-k}^{k} \gamma_w(j-i) \\
&= \left(\frac{1}{2\,k+1}\right)^2 \left[(2\,k+1)\,\gamma_w(0) + \sum_{j=-k}^{k-1}\sum_{i=j+1}^{k} \gamma_w(i-j) + \sum_{i=-k}^{k-1}\sum_{j=i+1}^{k} \gamma_w(j-i)\right] \\
&= \left(\frac{1}{2\,k+1}\right)^2 \left[(2\,k+1)\,\sigma^2\right] \\
&= \frac{\sigma^2}{2\,k+1}.
\end{aligned}
$$

Therefore,

$$
\frac{\operatorname{Var}(z_t)}{\operatorname{Var}(y_t)} = \frac{1}{2\,k+1}.
$$

Problem 10.7. The design matrix has elements of the following form: $x_{tj} = t^{j-1}$. Thus, the product

$$
\boldsymbol{x}_n'\boldsymbol{x}_n = \begin{bmatrix}
\sum_{t=1}^{n} x_{t1}^2 & \sum_{t=1}^{n} x_{t1}x_{t2} & \cdots & \sum_{t=1}^{n} x_{t1}x_{tp} \\
\sum_{t=1}^{n} x_{t1}x_{t2} & \sum_{t=1}^{n} x_{t2}^2 & \cdots & \sum_{t=1}^{n} x_{t2}x_{tp} \\
\vdots & \vdots & \ddots & \vdots \\
\sum_{t=1}^{n} x_{t1}x_{tp} & \sum_{t=1}^{n} x_{t2}x_{t2} & \cdots & \sum_{t=1}^{n} x_{tp}^2
\end{bmatrix}.
$$

Multiplying on both sides by D_n^{-1} we get

$$
D_n^{-1}\boldsymbol{x}_n'\boldsymbol{x}_n D_n^{-1} = W,
$$

where the elements of the matrix W are given by

$$
w_{ij} = \frac{\sum_{t=1}^{n} x_{ti}x_{tj}}{\|x_i\|_n \|x_j\|_n}.
$$

By replacing $x_{tj} = t^{j-1}$, the elements of the matrix W are

$$w_{ij} = \frac{\sum_{t=1}^{n} t^{i-1} t^{j-1}}{\left[\sum_{t=1}^{n} t^{2i-2}\right]^{1/2} \left[\sum_{t=1}^{n} t^{2j-2}\right]^{1/2}} = \frac{\sum_{t=1}^{n} t^{i+j-2}}{\left[\sum_{t=1}^{n} t^{2i-2}\right]^{1/2} \left[\sum_{t=1}^{n} t^{2j-2}\right]^{1/2}}$$

For large values of n we have,

$$\sum_{t=1}^{n} t^{i+j-2} = \frac{n^{i+j-1}}{i+j-1}.$$

Now,

$$\lim_{n\to\infty} \frac{\sum_{t=1}^{n} t^{i+j-2}}{\left[\sum_{t=1}^{n} t^{2i-2}\right]^{1/2} \left[\sum_{t=1}^{n} t^{2j-2}\right]^{1/2}}$$

$$= \lim_{n\to\infty} \frac{\dfrac{n^{i+j-1}}{i+j-1}}{\left[\dfrac{n^{2i-1}}{2i-1}\right]^{1/2} \left[\dfrac{n^{2j-1}}{2j-1}\right]^{1/2}} = \frac{\sqrt{2i-1}\sqrt{2j-1}}{i+j-1} = m_{ij}.$$

Part b). We have that,

$$\lim_{n\to\infty} D_n^{-1} (\boldsymbol{x}_n' \boldsymbol{x}_n) \operatorname{Var}(\widehat{\boldsymbol{\beta}}_n) (\boldsymbol{x}_n' \boldsymbol{x}_n) D_n^{-1} \to 2\pi B$$

where

$$b_{ij} = f_0(0) \lim_{n\to\infty} \frac{n^{-2d}}{\|\boldsymbol{x}_i\|_n \|\boldsymbol{x}_j\|_n} \sum_{t=1}^{n} \sum_{s=1}^{n} t^{i-1} s^{j-1} \gamma(t-s).$$

Note that,

$$\begin{aligned} &D_n^{-1} (\boldsymbol{x}_n' \boldsymbol{x}_n) \operatorname{Var}(\widehat{\boldsymbol{\beta}}_n) (\boldsymbol{x}_n' \boldsymbol{x}_n) D_n^{-1} \\ &= D_n^{-1} (\boldsymbol{x}_n' \boldsymbol{x}_n) (\boldsymbol{x}_n' \boldsymbol{x}_n)^{-1} \boldsymbol{x}_n' \Gamma \boldsymbol{x}_n (\boldsymbol{x}_n' \boldsymbol{x}_n)^{-1} (\boldsymbol{x}_n' \boldsymbol{x}_n) D_n^{-1} \\ &= D_n^{-1} \boldsymbol{x}_n' \Gamma \boldsymbol{x}_n D_n^{-1}. \end{aligned}$$

Now,

$$\lim_{n\to\infty} n^{-2d} D_n^{-1} (\boldsymbol{x}_n' \boldsymbol{x}_n) \operatorname{Var}(\widehat{\boldsymbol{\beta}}_n) (\boldsymbol{x}_n' \boldsymbol{x}_n) D_n^{-1} = \lim_{n\to\infty} n^{-2d} D_n^{-1} \boldsymbol{x}_n' \Gamma \boldsymbol{x}_n D_n^{-1} \to 2\pi\delta,$$

where

$$\delta_{ij} = f_0(0) \lim_{n\to\infty} n^{-2d} \frac{n^{-2d}}{\|\boldsymbol{x}_i\|_n \|\boldsymbol{x}_j\|_n} \sum_{t=1}^{n} \sum_{s=1}^{n} t^{i-1} s^{j-1} \gamma(t-s).$$

For a $FN(d)$ with $\sigma^2 = 1$ we can see by Stirling's approximation,

$$\gamma(h) = \frac{\Gamma(1-2d)}{\Gamma(1-d)\Gamma(d)} \frac{\Gamma(h+d)}{\Gamma(1+h-d)} \overset{h\to\infty}{\to} \frac{\Gamma(1-2d)}{\Gamma(1-d)\Gamma(d)} h^{2d-1}.$$

Moreover, we can write

$$\begin{aligned}
\delta_{ij} &= f_0(0) \lim_{n\to\infty} n^{-2d} \frac{n^{-2d}}{\|\boldsymbol{x}_i\|_n \|\boldsymbol{x}_j\|_n} \sum_{t=1}^{n}\sum_{s=1}^{n} t^{i-1}s^{j-1}\gamma(t-s) \\
&= f_0(0) \lim_{n\to\infty} n^{-2d} \frac{n^{-2d}}{\|\boldsymbol{x}_i\|_n \|\boldsymbol{x}_j\|_n} \sum_{t=1}^{n}\sum_{s=1}^{n} t^{i-1}s^{j-1} \frac{\Gamma(1-2d)}{\Gamma(1-d)\Gamma(d)} |t-s|^{2d-1} \\
&= f_0(0) \frac{\Gamma(1-2d)}{\Gamma(1-d)\Gamma(d)} \lim_{n\to\infty} n^{-2d} \frac{n^{-2d}}{\|\boldsymbol{x}_i\|_n \|\boldsymbol{x}_j\|_n} \sum_{t=1}^{n}\sum_{s=1}^{n} t^{i-1}s^{j-1}|t-s|^{2d-1} \\
&= f_0(0) \frac{\Gamma(1-2d)}{\Gamma(1-d)\Gamma(d)} \times \\
&\quad \lim_{n\to\infty} n^{i+j-1} \frac{n^{-2d}}{\|\boldsymbol{x}_i\|_n \|\boldsymbol{x}_j\|_n} \sum_{t=1}^{n}\sum_{s=1}^{n} \left(\frac{t}{n}\right)^{i-1} \left(\frac{s}{n}\right)^{j-1} \left|\frac{t}{n}-\frac{s}{n}\right|^{2d-1} \frac{1}{n^2}.
\end{aligned}$$

Note that

$$\|\boldsymbol{x}_i\|_n \|\boldsymbol{x}_j\|_n = \left[\sum_{t=1}^{n} t^{2i-2}\right]^{1/2} \left[\sum_{t=1}^{n} t^{2j-2}\right]^{1/2} \simeq \frac{n^{i+j-1}}{\sqrt{2i-1}\sqrt{2j-1}}.$$

Thus,

$$\begin{aligned}
\delta_{ij} &= f_0(0) \frac{\Gamma(1-2d)}{\Gamma(1-d)\Gamma(d)} \sqrt{2i-1}\sqrt{2j-1} \\
&\quad \lim_{n\to\infty} n^{-2d} \sum_{t=1}^{n}\sum_{s=1}^{n} \left(\frac{t}{n}\right)^{i-1} \left(\frac{s}{n}\right)^{j-1} \left|\frac{t}{n}-\frac{s}{n}\right|^{2d-1} \frac{1}{n^2} \times \\
&= f_0(0) \frac{\Gamma(1-2d)}{\Gamma(1-d)\Gamma(d)} \sqrt{2i-1}\sqrt{2j-1} \int_{-1}^{1}\int_{-1}^{1} x^{i-1}y^{j-1}|x-y|^{2d-1}\,dx\,dy.
\end{aligned}$$

Note that

$$\delta_{ij} = f_0(0) h_{ij}$$

Consequently,

$$\lim_{n\to\infty} n^{-2d} D_n^{-1} \boldsymbol{x}_n' \Gamma \boldsymbol{x}_n D_n^{-1} = 2\pi f_0(0) H = 2\pi\delta.$$

Problem 10.9. This time series linear model has one regressor

$$\mathbf{x}_n = \begin{pmatrix} 1^p e^{i\lambda_0 1} \\ 2^p e^{i\lambda_0 2} \\ \vdots \\ n^p e^{i\lambda_0 n} \end{pmatrix}.$$

Consequently (for $h \geq 0$)

$$\mathbf{x}_{n,h} = \begin{pmatrix} h^p e^{i\lambda_0 h} \\ (h+1)^p e^{i\lambda_0 (h+1)} \\ \vdots \\ n^p e^{i\lambda_0 n} \\ 0 \\ \vdots \\ 0 \end{pmatrix},$$

with h zeros at the end. For $p = 0$ and/or $\lambda_0 = 0$ see the solutions in the text for the harmonic regression and polynomial trend, respectively. Thus, from now on, we assume that $p > 0$ and $\lambda_0 \neq 0$.

Part a). Since the Grenander conditions are satisfied (see Part b), to prove the consistency of $\widehat{\beta}_n$ we have to check if

$$\liminf_{n\to\infty} \frac{||\mathbf{x}_n||^2}{n^\delta} > 0,$$

for some $\delta > 2d$. Following b) we have

$$\begin{aligned} \frac{||\mathbf{x}_n||^2}{n^\delta} &= \frac{\sum_{t=1}^n t^{2p}}{n^\delta} \\ &\sim \frac{1}{2p+1}\frac{n^{2p+1}}{n^\delta} \\ &= \frac{1}{2p+1} n^{2p+1-\delta}. \end{aligned}$$

Choose $\delta = 1 > 2d$ for all $d \in (0, 1/2)$, then as $p \geq 0$ the last expression is always greater than zero for all n.

Part b). The norm of $\mathbf{x}_n$ equals (complex inner product)

$$\begin{aligned} ||\mathbf{x}_n||^2 &= \sum_{t=1}^n x_{n,t}\overline{x_{n,t}} \\ &= \sum_{t=1}^n \left(t^p e^{i\lambda_0 n}\right)\overline{\left(t^p e^{i\lambda_0 n}\right)} \\ &= \sum_{t=1}^n t^{2p} > n^{2p}, \end{aligned}$$

and clearly diverges for $n \to \infty$.

The ratio $\frac{||\mathbf{x}_{n+1}||^2}{||\mathbf{x}_n||^2}$ satisfies

$$\begin{aligned}\lim_{n\to\infty}\frac{\sum_{t=1}^{n+1}t^{2p}}{\sum_{t=1}^{n}t^{2p}} &= \lim_{n\to\infty}\frac{\sum_{t=1}^{n}t^{2p}+(n+1)^{2p}}{\sum_{t=1}^{n}t^{2p}}\\ &= 1+\lim_{n\to\infty}\frac{(n+1)^{2p}}{\sum_{t=1}^{n}t^{2p}}.\end{aligned}$$

Since $\sum_{t=1}^{n}t^{2p}\sim n^{2p+1}$, the second term converges to 0, hence the ratio to 1.

Given that there is only one regressor, the matrix $R(h)$ is in fact univariate, that is, a function of h.

$$\begin{aligned}R(h) &= \lim_{n\to\infty}\frac{<\mathbf{x}_{n,h},\mathbf{x}_n>}{||\mathbf{x}_{n,h}||||\mathbf{x}_n||}\\ &= \lim_{n\to\infty}\frac{\sum_{t=1}^{n}\left((t+h)^p e^{i\lambda_0(t+h)}\right)\overline{\left(t^p e^{i\lambda_0 t}\right)}}{\left(\sum_{t=1}^{n-h}(t+h)^p e^{i\lambda_0(t+h)}(t+h)^p e^{-i\lambda_0(t+h)}\right)^{1/2}\left(\sum_{t=1}^{n}t^{2p}\right)^{1/2}}\\ &= \lim_{n\to\infty}\frac{e^{i\lambda_0 h}\sum_{t=1}^{n-h}(t+h)^p t^p}{\left(\sum_{t=1}^{n-h}(t+h)^{2p}\right)^{1/2}\left(\sum_{t=1}^{n}t^{2p}\right)^{1/2}}\\ &= e^{i\lambda_0 h}\lim_{n\to\infty}\frac{\sum_{t=1}^{n-h}[(t+h)t]^p}{\left(\sum_{t=1}^{n-h}(t+h)^{2p}\right)^{1/2}\left(\sum_{t=1}^{n}t^{2p}\right)^{1/2}}.\end{aligned}$$

For every fixed $h>0$ the right term converges to 1, thus

$$R(h)=e^{i\lambda_0 h},$$

which is nonsingular at $h=0$, as $R(0)=1$, for all $\lambda_0\in R$. Thus, the Grenander conditions are satisfied.

Part c). Since we have a univariate case, the $p\times p$ matrix is a real number.

$$D_n=\left(\sum_{t=1}^{n}t^{2p}\right)^{1/2}n^d\sim n^{p+1/2+d} \tag{B.30}$$

Part d). We have to calculate $M(\lambda)$ and show that it does not have a jump at the origin. The spectral representation of $R(h)$ is given by

$$R(h)=\int_{-\pi}^{\pi}e^{i\lambda h}dM(\lambda). \tag{B.31}$$

As $R(h)=e^{i\lambda_0 h}$ it holds

$$e^{i\lambda_0 h}=\int_{-\pi}^{\pi}e^{i\lambda h}dM(\lambda)\Rightarrow M(\lambda)=\delta(\lambda-\lambda_0)d\lambda, \tag{B.32}$$

where $\delta(\cdot)$ is the Dirac delta function.

Thus $M(\lambda)$ exhibits a jump at λ_0. Thus it has a jump at the origin only for $\lambda_0 = 0$. By assumption we excluded this case from the analysis. Conditions (1)-(4) are satisfied, see b). Condition (5) requires that for some $\delta > 1 - 2d$

$$\max_{1 \le t \le n} \frac{\mathbf{x}_{n,t}}{||\mathbf{x}_n||^2} = o(n^{-\delta}). \tag{B.33}$$

The maximum of $\mathbf{x}_n$ is achieved for $t = n$, and $||\mathbf{x}_n||^2 \sim n^{2p+1}$, thus

$$\max_{1 \le t \le n} \frac{x_t}{||\mathbf{x}_n||^2} \sim \frac{n^p}{n^{2p+1}} = \frac{1}{n}. \tag{B.34}$$

Dividing by $n^{-\delta}$ gives

$$\frac{1}{n} n^{\delta} = n^{\delta - 1}. \tag{B.35}$$

This expression tends to 0 only for $\delta < 1$. Choose $\delta = 1 - d$. Thus, condition (5) ($\delta > 1 - 2d$) and $\delta < 1$ is satisfied.

Assumption (10.11) states that for any $\delta > 0$ there exists a constant c such that

$$\int_{-c}^{c} f(\lambda) dM^n(\lambda) < \delta. \tag{B.36}$$

for every n.

The spectral density equals

$$f(\lambda) = |1 - e^{i\lambda}|^{-2d} f_0(\lambda), \text{ with } 0 < d < 1/2,$$

and $f_0(\lambda)$ is a well-behaved function (see equation (10.2)).

Under the assumption $\lambda_0 \neq 0$, and given $M(\lambda) = \delta(\lambda - \lambda_0)d\lambda$ we can choose $c = \lambda_0/2$ and get

$$\int_{-\lambda_0/2}^{\lambda_0/2} f(\lambda)\delta(\lambda - \lambda_0) d^n\lambda = 0.$$

Hence, the LSE is asymptotically efficient.

Problem 10.10.

(a) By definition we have

$$\begin{aligned} \gamma_{ij}(i-j) &= \int_{-\pi}^{\pi} h(\lambda) e^{i(i-j)\lambda} d\lambda \\ &= \int_{-\pi}^{\pi} h(\lambda) e^{-i(j-i)\lambda} d\lambda \\ &= \overline{\gamma_{ji}(-(i-j))} \end{aligned}$$

Therefore $\gamma_{ij} = \overline{\gamma_{ji}}$

(b) Let x a vector complex, then

$$\begin{aligned} x^*\Gamma x &= \sum_{i,j=1}^{n} \gamma(i-j) x_i x_j' \\ &= \sum_{i,j=1}^{n} \int_{-\pi}^{\pi} h(\lambda) e^{i(i-j)\lambda} \mathrm{d}\lambda x_i x_j' \\ &= \int_{-\pi}^{\pi} \left| \sum_{j=1}^{n} x_j e^{i\lambda j} \right|^2 h(\lambda) \mathrm{d}\lambda \geq 0 \end{aligned}$$

were $h(\lambda) \geq 0$ for hypothesis

(d) For definition $m_{ij}^n = m_{ij}^n(\lambda) = \frac{<\sum_{t=1}^n x_{ti} e^{-it\lambda}, \sum_{t=1}^n x_{tj} e^{it\lambda}>}{2\pi \|x_n(i)\| \|x_n(j)\|}$ Hence

$$\begin{aligned} \int_{-\pi}^{\pi} 2\pi h(\lambda) m_{ij}^n(\lambda) \mathrm{d}\lambda &= \int_{-\pi}^{\pi} 2\pi h(\lambda) \frac{< \sum_{t=1}^n x_{ti} e^{-it\lambda}, \sum_{t=1}^n x_{tj} e^{it\lambda} >}{2\pi \parallel x_n(i) \parallel \parallel x_n(j) \parallel} \mathrm{d}\lambda \\ &= \int_{-\pi}^{\pi} \frac{\sum_{t=1}^n \sum_{s=1}^n < x_{ti}, x_{sj} >}{\|x_n(i)\| \|x_n(j)\|} h(\lambda) e^{i(i-j)\lambda} \mathrm{d}\lambda \\ &= \frac{\sum_{t=1}^n \sum_{s=1}^n < x_{ti}, x_{sj} >}{\|x_n(i)\| \|x_n(j)\|} \int_{-\pi}^{\pi} h(\lambda) e^{i(i-j)\lambda} \mathrm{d}\lambda \\ &= \sum_{t=1}^{n} \sum_{s=1}^{n} \frac{< x_{ti}, x_{sj} >}{\|x_n(i)\| \|x_n(j)\|} \gamma_{ij} \\ &= \sum_{t=1}^{n} \sum_{s=1}^{n} \frac{x_{ti}}{\|x_n(i)\|} \frac{x_{sj}}{\|x_n(j)\|} \gamma_{ij} \\ &= u^*\Gamma v \end{aligned}$$

where $u = \frac{x_n(i)}{\|x_n(i)\|}$ and $v = \frac{x_n(j)}{\|x_n(j)\|}$

(e) The part (d) we have

$$\int_{-\pi}^{\pi} 2\pi h(\lambda) m_{ij}^n(\lambda) \mathrm{d}\lambda = u^*\Gamma v$$

Then

$$\begin{aligned} \left| \int_{-\pi}^{\pi} 2\pi h(\lambda) m_{ij}^n(\lambda) \mathrm{d}\lambda \right| &= |u^*\Gamma v| \leq 2 |u^*\Gamma v| \\ &\leq u^*\Gamma u + v^*\Gamma v \\ &= 2\pi \left[\int_{-\pi}^{\pi} h(\lambda) m_{ii}^n(\lambda) \mathrm{d}\lambda + \int_{-\pi}^{\pi} h(\lambda) m_{jj}^n(\lambda) \mathrm{d}\lambda \right] \end{aligned}$$

The inequality is using the part (c). Therefore

$$\left|\int_{-\pi}^{\pi} h(\lambda)m_{ij}^n(\lambda)\right| \leq \int_{-\pi}^{\pi} h(\lambda)m_{ii}^n(\lambda)\mathrm{d}\lambda + \int_{-\pi}^{\pi} h(\lambda)m_{jj}^n(\lambda)\mathrm{d}\lambda$$

(f) We have that $M^n(\lambda)$ converge weakly to $M(\lambda)$, that is,

$$\int_{-\pi}^{\pi} g(\lambda)\mathrm{d}M^n(\lambda) \to \int_{-\pi}^{\pi} g(\lambda)\mathrm{d}M(\lambda)$$

Hence if defined $h(\lambda) = f(\lambda) \geq 0$ then by the part (e). We have

$$\left|\int_A f(\lambda)\mathrm{d}M_{ij}(\lambda)\right| \leq \int_A f(\lambda)\mathrm{d}M_{ii}(\lambda) + \int_A f(\lambda)\mathrm{d}M_{jj}(\lambda)$$

where $A \subset [-\pi, \pi]$

Problem 10.16. For $\lambda = 0$ we have that $f_n(0) = 1$ for all n. On the other hand, for $\lambda \neq 0$ we have

$$\left|\frac{1}{n}\sum_{t=1}^{n} e^{i\lambda t}\right| = \frac{1}{n}\left|\frac{e^{i\lambda n} - 1}{e^{i\lambda} - 1}\right| \leq \frac{2}{n}\left|\frac{1}{e^{i\lambda} - 1}\right| \to 0,$$

as $n \to \infty$.

Problem 11.1. Part a). The joint probability density of y_1, y_2, y_3, y_4, y_5 with $|\phi| < 1$ is given by

$$f(y_1, y_2, y_3, y_4, y_5) = f(y_5|y_4)f(y_4|y_3)f(y_3|y_2)f(y_2|y_1)f(y_1),$$

where

$$\begin{aligned} y_t|y_{t-1} &\sim N(\phi y_{t-1}, 1), \qquad t = 2, 3, 4, 5 \\ y_1 &\sim N(0, (1 - \phi^2)^{-1}). \end{aligned}$$

Thus

$$\begin{aligned} f(y_1, y_2, y_3, y_4, y_5) &= (2\pi)^{-5/2}(1 - \phi^2)^{1/2} \times \\ & \exp\left\{-\frac{1}{2}\sum_{t=2}^{5}(y_t - \phi y_{t-1})^2 - \frac{1}{2}y_1^2(1 - \phi^2)\right\}. \end{aligned}$$

Part b). Note that maximize $f(y_1, \ldots, y_5)$ respect to y_4 is equivalent to maximize

$$Q(y_1, y_2, y_3, y_4, y_5) = -\frac{1}{2}\sum_{t=2}^{5}(y_t - \phi y_{t-1})^2 - \frac{1}{2}y_1^2(1 - \phi^2).$$

Thus, the value of y_4 that maximize Q is given by the equation

$$\frac{\partial Q}{\partial y_4} = -y_4 + \phi y_3 + \phi y_5 - \phi^2 y_4 = 0.$$

Finally , the value that maximizes the density is

$$z = \tilde{y}_4 = \frac{\phi}{1+\phi^2}(y_3 + y_5).$$

Part c). We must calculate the conditional distribution of $y_4|y_1, y_2, y_3, y_5$,

$$f(y_4|y_1,y_2,y_3,y_5) = \frac{f(y_1,y_2,y_3,y_4,y_5)}{f(y_1,y_2,y_3,y_5)} = \frac{f(y_1,y_2,y_3,y_4,y_5)}{\int_{-\infty}^{\infty} f(y_1,y_2,y_3,y_4,y_5)dy_4}.$$

Thus, we need to calculate the density

$$f(y_1, y_2, y_3, y_5) =$$

$$\begin{aligned}
&= \int_{\infty}^{\infty} (2\pi)^{-5/2}(1-\phi^2)^{1/2} \exp\left\{-\frac{1}{2}\sum_{t=2}^{5}(y_t - \phi y_{t-1})^2 - \frac{1}{2}y_1^2(1-\phi^2)\right\} dy_4 \\
&= (2\pi)^{-5/2}(1-\phi^2)^{1/2} \exp\left\{-\frac{1}{2}\sum_{t=2}^{3}(y_t - \phi y_{t-1})^2 - \frac{1}{2}y_1^2(1-\phi^2)\right\} \\
&\times \int_{\infty}^{\infty} \exp\left\{-\frac{1}{2}(y_5 - \phi y_4)^2 - \frac{1}{2}(y_4 - \phi y_3)^2\right\} dy_4 \\
&= (2\pi)^{-5/2}(1-\phi^2)^{1/2} \exp\left\{-\frac{1}{2}\sum_{t=2}^{3}(y_t - \phi y_{t-1})^2 - \frac{1}{2}y_1^2(1-\phi^2)\right\} \\
&\times \int_{\infty}^{\infty} \exp\left\{-\frac{y_5^2}{2} + \frac{2\phi y_5 y_4}{2} - \frac{\phi^2 y_4^2}{2} - \frac{y_4^2}{2} + \frac{2\phi y_4 y_3}{2} - \frac{\phi^2 y_3^2}{2}\right\} dy_4
\end{aligned}$$

$$= (2\pi)^{-5/2}(1-\phi^2)^{1/2}$$

$$\exp\left\{-\frac{1}{2}\sum_{t=2}^{3}(y_t - \phi y_{t-1})^2 - \frac{1}{2}y_1^2(1-\phi^2) - \frac{y_5^2}{2} - \frac{\phi^2 y_3^2}{2}\right\}$$

$$\times \int_{\infty}^{\infty} \exp\left\{\frac{2\phi y_5 y_4}{2} - \frac{\phi^2 y_4^2}{2} - \frac{y_4^2}{2} + \frac{2\phi y_4 y_3}{2}\right\} dy_4$$

$$= (2\pi)^{-4/2}\frac{(1-\phi^2)^{1/2}}{(1+\phi^2)^{1/2}} \times$$

$$\exp\left\{-\frac{1}{2}\sum_{t=2}^{3}(y_t - \phi y_{t-1})^2 - \frac{1}{2}y_1^2(1-\phi^2) - \frac{y_5^2}{2} - \frac{\phi^2 y_3^2}{2} + \frac{\phi^2(y_3+y_5)^2}{2(1+\phi^2)^2}(1+\phi^2)\right\}.$$

With this result, we can calculate the conditional density

$$f(y_4|y_1, y_2, y_3, y_5) =$$
$$(2\pi)^{-1/2}(1+\phi^2)^{1/2} \exp\left\{\frac{2\phi y_4 y_5}{2} - \frac{\phi^2 y_4^2}{2} - \frac{y_4^2}{2} + \frac{2\phi y_4 y_3}{2} - \frac{\phi^2 (y_3 + y_5)^2}{2(1+\phi^2)^2}(1+\phi^2)\right\}.$$

Finally, we have that

$$f(y_4|y_1, y_2, y_3, y_5) = (2\pi)^{-1/2}(1+\phi^2)^{1/2} \exp\left\{-\frac{1+\phi^2}{2}\left(y_4 - \frac{\phi(y_3+y_5)}{(1+\phi^2)}\right)^2\right\}.$$

Thus

$$E(y_4|y_1, y_2, y_3, y_5) = \frac{\phi(y_3+y_5)}{(1+\phi^2)}.$$

which is the same result obtained in part b).

Problem 11.3. By assumption y_t is invertible and has an AR(∞) representation

$$\varepsilon_t = y_t + \sum_{j=1}^{\infty} \pi_j y_{t-j} \Leftrightarrow y_t = \varepsilon_t - \sum_{j=1}^{\infty} \pi_j y_{t-j}.$$

As the filter $\pi(L) = \sum_{j=0}^{\infty} \pi_j L^j$ with $\pi_0 = 1$ converges in $\mathbb{L}^2$, it holds $\sum_{j=0}^{\infty} |\pi_j|^2 < \infty$. The best linear predictor of y_t given the infinite past until $t-1$ equals the projection of y_t on H_{t-1}. For linear processes this is equivalent to the conditional expectation

$$\widehat{y}_t = E(y_t|H_{t-1}) = E(\varepsilon_t|H_{t-1}) - E(\sum_{j=1}^{\infty} \pi_j y_{t-j}|H_{t-1}) = -\sum_{j=1}^{\infty} \pi_j y_{t-j}.$$

a) By definition

$$\sigma_{t+1}^2(1) = [y_{t+1} - \widehat{y}_{t+1}] = [y_{t+1} - P_t y_{t+1}].$$

As the t-th observation is missing $H_{t-1} \equiv H_t$ and consequently the projection operators are equal $P_t = P_{t-1}$.

Hence,

$$
\begin{aligned}
P_t y_{t+1} = P_{t-1} y_{t+1} &= E(\varepsilon_{t+1}|H_{t-1}) - E(\sum_{j=1}^{\infty} \pi_j y_{t+1-j}|H_{t-1}) \\
&= 0 + E(\pi_1 y_t|H_{t-1}) - E(\sum_{j=2}^{\infty} \pi_j y_{t+1-j}|H_{t-1}) \\
&= \pi_1 P_{t-1} y_t - \sum_{j=2}^{\infty} \pi_j y_{t+1-j} \\
&= \pi_1 P_t y_t - \sum_{j=2}^{\infty} \pi_j y_{t+1-j}
\end{aligned}
$$

Therefore we can write the one step ahead prediction error at time $t+1$ as (using again $P_t = P_{t-1}$)

$$
\begin{aligned}
y_{t+1} - P_t y_{t+1} &= \varepsilon_{t+1} - \sum_{j=1}^{\infty} \pi_j y_{t+1-j} - \left(\pi_1 P_t y_t - \sum_{j=2}^{\infty} \pi_j y_{t+1-j} \right) \\
&= \varepsilon_{t+1} - \pi_1 y_t - \pi_1 P_t y_t \\
&= \varepsilon_{t+1} - \pi_1 (y_t - P_{t-1} y_t),
\end{aligned}
$$

which gives the equality

$$e_{t+1}(1) = \varepsilon_{t+1} - \pi_1 e_t(1).$$

Note that ε_{t+1} is uncorrelated with $e_t(1) \in H_t$.

b) As ε_{t+1} and $e_t(1)$ have expectation zero and are uncorrelated, we can compute the variance by

$$
\begin{aligned}
\sigma_{t+1}^2(1) &= \varepsilon_{t+1} + \pi_1^2 e_t(1) \\
&= \sigma_\varepsilon^2 + \pi_1^2 \sigma_t^2(1) \\
&= 1 + \pi_1^2.
\end{aligned}
$$

As $\pi_1^2 \geq 0$, the prediction error variance for $t+1$ is greater or equal to the prediction error variance for t, given a missing value at t, i.e.

$$\sigma_{t+1}^2(1) \geq \sigma_t^2(1).$$

Problem 11.7. Note that we have the subspace $\mathcal{H}$ generated by $\{y_s, s < k, s \neq t\}$. Thus,

$$
\begin{aligned}
y_{t+k} &= \varepsilon_{t+k} - \sum_{j=1}^{\infty} \pi_j y_{t+k-j} \\
\mathcal{P}_{\mathcal{H}} y_{t+k} &= -\pi_1 y_{t+k-1} - \ldots - \pi_{k-1} y_{t+1} - \pi_k \mathcal{P}_{\mathcal{H}} y_t - \pi_{k+1} y_{t-1} - \ldots.
\end{aligned}
$$

Hence,

$$\operatorname{Var}(y_{t+k} - \widehat{y}_{t+k}) = \operatorname{Var}(\varepsilon_{t+k} - \pi_k(y_t - \mathcal{P}_{\mathcal{H}} y_t)) = \operatorname{Var}(\varepsilon_{t+k}) + \pi_k^2 \operatorname{Var}(y_t - \mathcal{P}_{\mathcal{H}} y_t),$$

where $\operatorname{Var}(\varepsilon_{t+k}) = 1$ and $\operatorname{Var}(y_t - \mathcal{P}_{\mathcal{H}} y_t) < 1$. Therefore,

$$|\sigma_{t+k}^2(1) - 1| = |1 + \pi_k^2 \operatorname{Var}(y_t - \mathcal{P}_{\mathcal{H}} y_t) - 1| = |\pi_k^2 \operatorname{Var}(y_t - \mathcal{P}_{\mathcal{H}} y_t)| < |\pi_k^2|.$$

But we know that $|\pi_k| < M(1+\delta)^{-k}$ with $0 < \delta < \epsilon$ y $M > 0$, and once specified we have that

$$|\pi_k^2| = |\pi_k|^2 < M^2[(1+\delta)^2]^{-k} < ca^{-k},$$

where $|a| < 1$ y $c = M^2$. Thus, we conclude that

$$|\sigma_{t+k}^2(1) - 1| \leq ca^{-k}.$$

Problem 11.9.

a) The interpolator of y_t given $\{y_j | j \neq t\}$ is given by

$$\tilde{y}_t = -\sum_{j=1}^{\infty} \alpha_j (y_j + y_{-j}) = \left(\sum_{j=1}^{\infty} \alpha_j B^j + \sum_{j=1}^{\infty} \alpha_j B^{-j} \right) y_t,$$

where

$$\alpha_j = \frac{\sum_{i=0}^{\infty} \pi_i \pi_{i+j}}{\sum_{i=0}^{\infty} \pi_i^2}, \text{ for } j \geq 1.$$

Hence, x_t can be rewritten to

$$\begin{aligned} x_t &= \frac{y_t - \tilde{y}_t}{\tilde{\sigma}^2} \\ &= \frac{1}{\tilde{\sigma}^2} \left(y_t - \left(\sum_{j=1}^{\infty} \alpha_j B^j + \sum_{j=1}^{\infty} \alpha_j B^{-j} \right) y_t \right) \\ &= \frac{1}{\tilde{\sigma}^2} \alpha(B) y_t \\ &= \frac{1}{\tilde{\sigma}^2} \alpha(B) \psi(B) \varepsilon_t, \end{aligned}$$

where $\psi(B)\varepsilon_t$ is the Wold decomposition of y_t.

b) The product of two absolutely convergent series equals

$$\left(\sum_{j=0}^{\infty} a_j \right) \left(\sum_{k=0}^{\infty} b_k \right) = \sum_{n=0}^{\infty} c_n$$

where

$$c_n = \sum_{\ell=0}^{n} a_\ell b_{n-\ell}.$$

To obtain the result we proceed by matching terms.

When multiplying $\pi(z)$ by $\pi(z^{-1})$ the coefficients of z^j are given by all π_i and π_k such that

$$z^i z^{-k} = z^j \Rightarrow i - k = j \Leftrightarrow i = k + j.$$

As this equation holds for all $k = 0, 1, 2, \ldots$, we must sum over all k to get the coefficients α_j, i.e.

$$\alpha_j = c \cdot \sum_{k=0}^{\infty} \pi_k \pi_{k+j}.$$

As $\alpha_0 = 1$, the constant must equal

$$c = \frac{1}{\sum_{k=0}^{\infty} \pi_k^2}.$$

c) By definition the filter $\pi(B)$ is the inverse of $\psi(B)$ inverse, that is

$$\pi(B)\psi(B) = 1.$$

Using b) we can write

$$\alpha(z)\psi(z) = c\pi(z)\pi(z^{-1})\psi(z) = c\pi(z^{-1})\pi(z)\psi(z) = c\pi(z^{-1}). \quad \text{(B.37)}$$

Replacing z by B proves the expression.

d) In a) we proved that

$$x_t = \frac{1}{\tilde{\sigma}^2}\alpha(B)y_t, \quad \text{(B.38)}$$

hence the spectral density of x_t is given by

$$f_x(\lambda) = \frac{1}{\tilde{\sigma}^4}|\alpha(e^{i\lambda})|^2 f(\lambda) \quad \text{(B.39)}$$

Additionally we have that

$$\alpha(z) = \pi(z)\pi(z^{-1})c \Leftrightarrow \frac{\alpha(z)\pi(z)}{\pi(z^{-1})} = c \quad \text{(B.40)}$$

and since $\psi(z)\pi(z) = 1$ we have that

$$\alpha(z)\psi(z) = \pi(z)c. \quad \text{(B.41)}$$

Therefore,

$$\begin{aligned}
f_x(\lambda) &= \frac{c^2}{\tilde{\sigma}^4}|\pi(e^{i\lambda})|^2\frac{\sigma^2}{2\pi} && \text{(B.42)}\\
&= \frac{c^2}{\tilde{\sigma}^4}|\frac{1}{c}\frac{\alpha(e^{i\lambda})}{\pi(e^{-i\lambda})}|^2\frac{\sigma^2}{2\pi} && \text{(B.43)}\\
&= \frac{1}{\tilde{\sigma}^4}|\frac{\alpha(e^{i\lambda})}{\pi(e^{-i\lambda})}|^2\frac{\sigma^2}{2\pi} && \text{(B.44)}\\
&= \frac{1}{\tilde{\sigma}^4}|\frac{\alpha(e^{i\lambda})}{\pi(e^{-i\lambda})}\frac{\psi(e^{-i\lambda})}{\psi(e^{-i\lambda})}|^2\frac{\sigma^2}{2\pi} && \text{(B.45)}\\
&= \frac{1}{\tilde{\sigma}^4}|\frac{\alpha(e^{i\lambda})\psi(e^{-i\lambda})}{\pi(e^{-i\lambda})}\frac{1}{\psi(e^{-i\lambda})}|^2\frac{2\pi}{\sigma^2}\frac{\sigma^2}{2\pi}\frac{\sigma^2}{2\pi} && \text{(B.46)}\\
&= \frac{1}{\tilde{\sigma}^4}\frac{2\pi}{\sigma^2}\frac{\sigma^2}{2\pi}\frac{\sigma^2}{2\pi} && \text{(B.47)}\\
&= \frac{1}{\tilde{\sigma}^4}\frac{1}{f_y(\lambda)}\frac{\sigma^2}{2\pi}\frac{\sigma^2}{2\pi} && \text{(B.48)}\\
&= \frac{1}{f_y(\lambda)}\frac{\sigma^2}{(2\pi)^2}. && \text{(B.49)}
\end{aligned}$$

Problem 11.15. Part (a). This is a standard result for the k steps ahead error variance; see Subsection 7.3. Part (b). Let $a(k,m) = \sum_{j=k-m+1}^{k} \pi_j e(k-j,m,k)$ and take $t = k$ in the AR(∞) decomposition and subtract from both sides the best linear predictor. Then all terms vanish, except for y_k and those associated with the missing observations, which yields the identity

$$e(k,m,k) = \varepsilon_k - a(k,m). \qquad \text{(B.50)}$$

By the orthogonality of ε_k to all previous observations,

$$\sigma^2(k,m,k) = \sigma^2 + \mathrm{Var}[a(k,m)],$$

for $m \geq 1$. Now, by noting that

$$\begin{aligned}
\|a(k,m)\| &\leq \sum_{j=k-m+1}^{k} |\pi_j|\|e(k-j,m,k)\|\\
&\leq \max_{\{j\geq k-m\}} |\pi_j| \sum_{j=k-m+1}^{k} \|e(k-j,m,k)\|\\
&\leq \max_{\{j\geq k-m\}} |\pi_j| m\sigma_y,
\end{aligned}$$

part (b) is obtained. Part (c) is a direct consequence of part (b) since $\pi_j \to 0$ as $j \to \infty$ and $\sigma^2(k,m,k) \geq \sigma^2$ for all k.

Problem 11.16. Part (a). Notice that for $k \leq m$ we have that $\sigma^2(k,m,k) - \sigma_y^2 = \sigma^2 \sum_{j=k}^{\infty} \psi_j^2$. Since $\psi_j \sim c_0 j^{d-1}$ for large j, $\sigma^2 \sum_{j=k}^{\infty} \psi_j^2 \sim c k^{2d-1}$ for large k. Part (b). Observe that from equation (B.50) we have for $k \gg m$

$$
\begin{aligned}
\pi_{k-m+1}^{-2}\{\sigma^2(k,m,k) - \sigma^2\} &= \operatorname{Var}\left[\sum_{j=k-m+1}^{k} \frac{\pi_j}{\pi_{k-m+1}} e(k-j,m,k)\right], \\
&= \operatorname{Var}\left[\sum_{j=k-m+1}^{k} e(k-j,m,k)\right. \\
&\quad + \left.\sum_{j=k-m+1}^{k} b_{kj} e(k-j,m,k)\right],
\end{aligned}
$$

where $b_{kj} = \pi_j/\pi_{k-m+1} - 1$ for $j = k-m+1, \ldots, k$. For m fixed and large k this term is bounded by

$$
|b_{kj}| \leq \frac{c_1}{k},
$$

for $j = k-m+1, \ldots, k$ and $c_1 > 0$. Since $\|e(k-j,m,k)\| \leq \sigma_y$, we conclude that

$$
\pi_{k-m+1}^{-2}[\sigma^2(k,m,k) - \sigma^2] = \operatorname{Var}\left[\sum_{j=k-m+1}^{k} e(k-j,m,k)\right] + \mathcal{O}\left(\frac{1}{k}\right),
$$

for large k. Now, by defining the following random variable:

$$
z_k = \sum_{j=0}^{m-1} y_j - E\left[\sum_{j=0}^{m-1} y_j | y_k, \ldots, y_m, y_{-1}, y_{-2}, \ldots\right],
$$

we have

$$
\pi_{k-m+1}^{-2}\left\{\sigma^2(k,m,k) - \sigma^2\right\} = \operatorname{Var}[z_k] + \mathcal{O}\left(\frac{1}{k}\right).
$$

But, $0 < \operatorname{Var}[z_\infty] \leq \operatorname{Var}[z_k] \leq \operatorname{Var}[z_m] < \infty$. Therefore $\sigma^2(k,m,k) - \sigma^2 \sim c_2 \pi_{k-m+1}^2 \sim c k^{-2d-2}$ for $k \gg m$.

Problem 11.19. The modifications of the state covariance matrix equation and the state prediction are obtained by noting that if y_t is missing, then $\mathcal{P}_t = \mathcal{P}_{t-1}$ and $\widehat{x}_{t+1} = E[x_{t+1}|\mathcal{P}_t] = E[Fx_t + H\varepsilon_t|\mathcal{P}_t] = F\,E[x_t|\mathcal{P}_t] + E[H\varepsilon_t|\mathcal{P}_t] = F\,E[x_t|\mathcal{P}_{t-1}] = F\widehat{x}_t$. Thus, $x_{t+1} - \widehat{x}_{t+1} = F(x_t - \widehat{x}_t) + H\varepsilon_t$ and $\Omega_{t+1} = F\Omega_t F' + HH'\sigma^2$. On the other hand, since y_t is missing, the innovation is null, so that $\nu_t = 0$. Besides, let the observation y_t be missing. Hence, by definition

$$\widehat{y}_t = E[y_t|\mathcal{P}_{t-1}].$$

From the state space system equation we have

$$y_t = Gx_t + \varepsilon_t,$$

and therefore,

$$\widehat{y}_t = E[Gx_t + \varepsilon_t|\mathcal{P}_{t-1}] = G\widehat{x}_t + E[\varepsilon_t|\mathcal{P}_{t-1}] = G\widehat{x}_t. \tag{B.51}$$

APPENDIX C

DATA AND CODES

The financial time series data described in Chapter 1 are available at the Yahoo Finance website www.yahoo.finance.com, including also the IPSA stock index studied in Chapter 6. The Nile river time series is available at StatLib website www.statlib.cmu.edu. Heating degree days data, passenger enplanements, UK voting data, are available at the Datamarket website www.datamarket.com.

The source of the dengue data is DengueNet of the World Health Organization. They correspond to the annual number of combined dengue fever, dengue hemorrhagic fever and dengue shock syndrome combined. These data are available at www.who.int/denguenet.

The air pollution data are provided by the Ambient Air Quality Monitoring Network, www.sesma.cl in Santiago, Chile.

Tree ring data such as the Mammoth Creek series are available at the National Climatic Data Center, www.ncdc.noaa.gov. They were reported by Graybill (1990).

*

The mineral deposits series composed by stalagmite layer thickness observations taken at Shihua Cave, Beijing, China is reported by Tan et al. (2003).

Glacial varves data are available at www.stat.pitt.edu/∼stoffer/tsa.html. Daily, monthly and annual gold and copper prices data are available from at www.cochilco.cl/english. This site also reports prices of other metals. World metals inventories are also available at this site.

In this book we have made use of the following R libraries, bfast, forecast, nlts, gsarima, glarma, inarmix, wavelets, acp, ZIM, polynom, xtable, arfima, fArma, see R Core Team (2014) for furthers details.

REFERENCES

M. Abrahams and A. Dempster. (1979). *Research on Seasonal Analysis.* Progress Report on the ASA/Census Project on Seasonal Adjustment. Department of Statistics, Harvard University, Boston, MA.

P. Abry, P. Flandrin, M. S. Taqqu, and D. Veitch. (2003). Self-similarity and long-range dependence through the wavelet lens. In P. Doukhan, G. Oppenheim, and M. S. Taqqu, editors, *Theory and Applications of Long-Range Dependence.* Birkhäuser, Boston, MA, pp. 527–556.

R. K. Adenstedt. (1974). On large-sample estimation for the mean of a stationary random sequence. *Annals of Statistics* 2, 1095–1107.

M. A. Al-Osh and A. A. Alzaid. (1987). First-order integer-valued autoregressive (inar (1)) process. *Journal of Time Series Analysis* 8, 261–275.

G. S. Ammar. (1998). Classical foundations of algorithms for solving positive definite Toeplitz equations. *Calcolo. A Quarterly on Numerical Analysis and Theory of Computation* 33, 99–113.

B. D. O. Anderson and J. B. Moore. (1979). *Optimal Filtering.* Prentice-Hall, New York.

C. F. Ansley and R. Kohn. (1983). Exact likelihood of vector autoregressive-moving average process with missing or aggregated data. *Biometrika* 70, 275–278.

M. Aoki. (1990). *State Space Modeling of Time Series*. Springer, Berlin.

R. T. Baillie, T. Bollerslev, and H. O. Mikkelsen. (1996). Fractionally integrated generalized autoregressive conditional heteroskedasticity. *Journal of Econometrics* 74, 3–30.

G. K. Basak, N. H. Chan, and W. Palma. (2001). The approximation of long-memory processes by an ARMA model. *Journal of Forecasting* 20, 367–389.

W. Bell and S. Hillmer. (1991). Initializing the Kalman filter for nonstationary time series models. *Journal of Time Series Analysis* 12, 283–300.

A. F. Bennett. (1992). *Inverse Methods in Physical Oceanography*. Cambridge Monographs on Mechanics. Cambridge University Press, Cambridge.

J. Beran. (1994). *Statistics for Long-Memory Processes* Vol. 61, *Monographs on Statistics and Applied Probability*. Chapman and Hall, New York.

S. Bertelli and M. Caporin. (2002). A note on calculating autocovariances of long-memory processes. *Journal of Time Series Analysis* 23, 503–508.

R. J. Bhansali and P. S. Kokoszka. (2003). Prediction of long-memory time series. In P. Doukhan, G. Oppenheim, and M. S. Taqqu, editors, *Theory and Applications of Long-Range Dependence*. Birkhäuser, Boston, MA, pp. 355–367.

T. Bollerslev. (1986). Generalized autoregressive conditional heteroskedasticity. *Journal of Econometrics* 31, 307–327.

T. Bollerslev and H. O. Mikkelsen. (1996). Modeling and pricing long memory in stock market volatility. *Journal of Econometrics* 73, 151–184.

P. Bondon. (2002). Prediction with incomplete past of a stationary process. *Stochastic Processes and their Applications* 98, 67–76.

G. E. P. Box, G. M. Jenkins, and G. C. Reinsel. (1994). *Time Series Analysis*. Prentice Hall, Englewood Cliffs, NJ.

G. E. P. Box and G. C. Tiao. (1992). *Bayesian Inference in Statistical Analysis*. John Wiley & Sons, Inc., New York.

F. J. Breidt, N. Crato, and P. de Lima. (1998). The detection and estimation of long memory in stochastic volatility. *Journal of Econometrics* 83, 325–348.

D. R. Brillinger and P. R. Krishnaiah, editors. (1983). *Time series in the frequency domain* Vol. 3, *Handbook of Statistics.* North-Holland Publishing Co., Amsterdam.

A. E. Brockwell. (2004). A class of generalized long-memory time series models. Technical Report 813. Department of Statistics, Carnegie Mellon University, Pittsburgh.

P. J. Brockwell and R. A. Davis. (1991). *Time Series: Theory and Methods.* Springer, New York.

P. J. Brockwell and R. A. Davis. (2002). *Introduction to time series and forecasting.* Springer Texts in Statistics. Springer-Verlag, New York.

Y. Cai and N. Davies. (2003). A simple diagnostic method of outlier detection for stationary Gaussian time series. *Journal of Applied Statistics* 30, 205–223.

A. C. Cameron and P. K. Trivedi. (2013). *Regression analysis of count data* Vol. 53. Cambridge University Press, .

R. Carmona, W. L. Hwang, and B. Torresani. (1998). *Practical time-frequency analysis* Vol. 9, *Wavelet Analysis and its Applications.* Academic Press, Inc., San Diego, CA.

N. L. Carothers. (2000). *Real Analysis.* Cambridge University Press, Cambridge.

F. Castanié, editor. (2006). *Spectral analysis.* Digital Signal and Image Processing Series. ISTE, London.

F. Castanié, editor. (2011). *Digital spectral analysis.* Digital Signal and Image Processing Series. ISTE, London; John Wiley & Sons, Inc., Hoboken, NJ.

J. E. Cavanaugh, Y. Wang, and J. W. Davis. (2003). Locally self-similar processes and their wavelet analysis. In *Stochastic processes: modelling and simulation*, Vol. 21, *Handbook of Statist.* North-Holland, Amsterdam, pp. 93–135.

N. H. Chan. (2002). *Time Series. Applications to Finance.* Wiley Series in Probability and Statistics. John Wiley & Sons, Inc., New York.

N. H. Chan and W. Palma. (1998). State space modeling of long-memory processes. *Annals of Statistics* 26, 719–740.

N. H. Chan and G. Petris. (2000). Recent developments in heteroskedastic time series. In W. S. Chan, W. K. Li, and H. Tong, editors, *Statistics and Finance: An Interface.* Imperial College Press, London, pp. 169–184.

G. Chandler and W. Polonik. (2006). Discrimination of locally stationary time series based on the excess mass functional. *Journal of the American Statistical Association* 101, 240–253.

C. Chen and L. M. Liu. (1993). Joint estimation of model parameters and outlier effects in time series. *Journal of the American Statistical Association* 88, 284–297.

R. Cheng and M. Pourahmadi. (1997). Prediction with incomplete past and interpolation of missing values. *Statistics & Probability Letters* 33, 341–346.

K. Choy. (2001). Outlier detection for stationary time series. *Journal of Statistical Planning and Inference* 99, 111–127.

K. Choy and M. Taniguchi. (2001). Stochastic regression model with dependent disturbances. *Journal of Time Series Analysis* 22, 175–196.

C. F. Chung. (1996). A generalized fractionally integrated autoregressive moving-average process. *Journal of Time Series Analysis* 17, 111–140.

J. B. Conway. (1990). *A Course in Functional Analysis*, Vol. 96, *Graduate Texts in Mathematics.* Springer, New York.

D. R. Cox. (1981). Statistical analysis of time series: some recent developments. *Scandinavian Journal of Statistics* pp. 93–115.

D. R. Cox. (1984). Long-range dependence: A review. In H. A. David and H. T. David, editors, *Statistics: An Appraisal.* Iowa State University Press, Ames, IA, pp. 55–74.

R. F. Curtain and H. Zwart. (1995). *An Introduction to Infinite-Dimensional Linear Systems Theory*, Vol. 21, *Texts in Applied Mathematics.* Springer, New York.

R. Dahlhaus. (1989). Efficient parameter estimation for self-similar processes. *Annals of Statistics* 17, 1749–1766.

R. Dahlhaus. (1995). Efficient location and regression estimation for long range dependent regression models. *Annals of Statistics* 23, 1029–1047.

R. Dahlhaus. (1996). Asymptotic statistical inference for nonstationary processes with evolutionary spectra. In *Athens Conference on Applied Probability and Time Series Analysis, Vol. II (1995)*, Vol. 115, *Lecture Notes in Statist.* Springer, New York, pp. 145–159.

R. Dahlhaus. (1997). Fitting time series models to nonstationary processes. *The Annals of Statistics* 25, 1–37.

R. Dahlhaus. (2000). A likelihood approximation for locally stationary processes. *The Annals of Statistics* 28, 1762–1794.

R. Dahlhaus and W. Polonik. (2006). Nonparametric quasi-maximum likelihood estimation for Gaussian locally stationary processes. *The Annals of Statistics* 34, 2790–2824.

R. Dahlhaus and W. Polonik. (2009). Empirical spectral processes for locally stationary time series. *Bernoulli* 15, 1–39.

E. Damsleth. (1980). Interpolating missing values in a time series. *Scandinavian Journal of Statistics. Theory and Applications* 7, 33–39.

R. A. Davis, W. Dunsmuir, and S. B. Streett. (2003). Observation-driven models for poisson counts. *Biometrika* 90, 777–790.

R. S. Deo and C. M. Hurvich. (2003). Estimation of long memory in volatility. In P. Doukhan, G. Oppenheim, and M. S. Taqqu, editors, *Theory and Applications of Long-Range Dependence.* Birkhäuser, Boston, MA, pp. 313–324.

H. Dette, P. Preuß, and M. Vetter. (2011). A measure of stationarity in locally stationary processes with applications to testing. *Journal of the American Statistical Association* 106, 1113–1124.

P. J. Diggle. (1990). *Time series* Vol. 5, *Oxford Statistical Science Series.* The Clarendon Press, Oxford University Press, New York.

G. S. Dissanayake, M. S. Peiris, and T. Proietti. (2014). State space modeling of gegenbauer processes with long memory. *Computational Statistics & Data Analysis.*

J. A. Doornik and M. Ooms. (2003). Computational aspects of maximum likelihood estimation of autoregressive fractionally integrated moving average models. *Computational Statistics & Data Analysis* 42, 333–348.

P. Doukhan, G. Oppenheim, and M. S. Taqqu, editors. (2003). *Theory and Applications of Long-Range Dependence.* Birkhäuser, Boston, MA.

J. Durbin. (1960). The fitting of time series models. *International Statistical Review* 28, 233–244.

J. Durbin and S. J. Koopman. (2001). *Time Series Analysis by State Space Methods*, Vol. 24, *Oxford Statistical Science Series.* Oxford University Press, Oxford.

K. Dzhaparidze. (1986). *Parameter estimation and hypothesis testing in spectral analysis of stationary time series.* Springer Series in Statistics. Springer-Verlag, New York.

R. F. Engle. (1982). Autoregressive conditional heteroscedasticity with estimates of the variance of United Kingdom inflation. *Econometrica* 50, 987–1007.

P. Flandrin. (1999). *Time-Frequency/Time-Scale Analysis*, Vol. 10, *Wavelet Analysis and Its Applications.* Academic, San Diego, CA.

R. Fox and M. S. Taqqu. (1987). Central limit theorems for quadratic forms in random variables having long-range dependence. *Probability Theory and Related Fields* 74, 213–240.

R. K. Freeland and B. McCabe. (2004). Forecasting discrete valued low count time series. *International Journal of Forecasting* 20, 427–434.

W. A. Fuller. (1996). *Introduction to statistical time series.* Wiley Series in Probability and Statistics: Probability and Statistics. John Wiley & Sons, Inc., New York.

E. Ghysels, A. C. Harvey, and E. Renault. (1996). Stochastic volatility. In *Statistical Methods in Finance*, Vol. 14, *Handbook of Statistics.* North-Holland, Amsterdam, pp. 119–191.

L. Giraitis, J. Hidalgo, and P. M. Robinson. (2001). Gaussian estimation of parametric spectral density with unknown pole. *Annals of Statistics* 29, 987–1023.

L. Giraitis, P. Kokoszka, R. Leipus, and G. Teyssière. (2003). Rescaled variance and related tests for long memory in volatility and levels. *Journal of Econometrics* 112, 265–294.

L. Giraitis and R. Leipus. (1995). A generalized fractionally differencing approach in long-memory modeling. *Matematikos ir Informatikos Institutas* 35, 65–81.

I. S. Gradshteyn and I. M. Ryzhik. (2000). *Table of Integrals, Series, and Products.* Academic, San Diego, CA.

C. W. J. Granger. (1964). *Spectral analysis of economic time series.* In association with M. Hatanaka. Princeton Studies in Mathematical Economics, No. I. Princeton University Press, Princeton, N.J.

C. W. J. Granger and R. Joyeux. (1980). An introduction to long-memory time series models and fractional differencing. *Journal of Time Series Analysis* 1, 15–29.

S. Grassi and P. S. de Magistris. (2014). When long memory meets the kalman filter: A comparative study. *Computational Statistics & Data Analysis* 76, 301–319.

H. L. Gray, N. F. Zhang, and W. A. Woodward. (1989). On generalized fractional processes. *Journal of Time Series Analysis* 10, 233–257.

D. A. Graybill. (1990). Pinus longaeva tree ring data. *Mammoth Creek, Utah,* National Climatic Data Center.

U. Grenander. (1954). On the estimation of regression coefficients in the case of an autocorrelated disturbance. *Annals of Mathematical Statistics* 25, 252–272.

U. Grenander and M. Rosenblatt. (1957). *Statistical Analysis of Stationary Time Series.* John Wiley & Sons, Inc., New York.

G. K. Grunwald, R. J. Hyndman, L. Tedesco, and R. L. Tweedie. (2000). Non-gaussian conditional linear ar (1) models. *Australian & New Zealand Journal of Statistics* 42, 479–495.

P. Hall. (1997). Defining and measuring long-range dependence. In *Nonlinear Dynamics and Time Series (Montreal, PQ, 1995)*, Vol. 11, *Fields Inst. Commun.* Amer. Math. Soc., Providence, RI, pp. 153–160.

J. D. Hamilton. (1994). *Time Series Analysis.* Princeton University Press, Princeton, NJ.

E. J. Hannan. (1970). *Multiple Time Series.* John Wiley & Sons, Inc., New York.

E. J. Hannan and M. Deistler. (1988). *The Statistical Theory of Linear Systems.* Wiley, New York.

A. C. Harvey. (1989). *Forecasting Structural Time Series and the Kalman Filter.* Cambridge University Press, Cambridge.

A. C. Harvey and R. G. Pierse. (1984). Estimating missing observations in economic time series. *Journal of the American Statistical Association* 79, 125–131.

A. C. Harvey, E. Ruiz, and N. Shephard. (1994). Multivariate stochastic variance models. *Review of Economic Studies* 61, 247–265.

U. Hassler. (1994). (Mis)specification of long memory in seasonal time series. *Journal of Time Series Analysis* 15, 19–30.

U. Hassler and J. Wolters. (1995). Long memory in inflation rates: International evidence. *Journal of Business and Economic Statistics* 13, 37–45.

J. Hasslett and A. E. Raftery. (1989). Space-time modelling with long-memory dependence: Assessing Ireland's wind power resource. *Journal of Applied Statistics* 38, 1–50.

W. K. Hastings. (1970). Monte Carlo sampling methods using Markov chains and their applications. *Biometrika* 57, 97–109.

S. Haykin, editor. (1979). *Nonlinear methods of spectral analysis* Vol. 34, *Topics in Applied Physics.* Springer-Verlag, Berlin-New York.

M. Henry. (2001). Averaged periodogram spectral estimation with long-memory conditional heteroscedasticity. *Journal of Time Series Analysis* 22, 431–459.

J. R. M. Hosking. (1981). Fractional differencing. *Biometrika* 68, 165–176.

I. A. Ibragimov and Y. A. Rozanov. (1978). *Gaussian Random Processes*, Vol. 9, *Applications of Mathematics*. Springer, New York.

P. Iglesias, H. Jorquera, and W. Palma. (2006). Data analysis using regression models with missing observations and long-memory: an application study. *Computational Statistics & Data Analysis* 50, 2028–2043.

G. M. Jenkins and D. G. Watts. (1968). *Spectral analysis and its applications*. Holden-Day, San Francisco, Calif.-Cambridge-Amsterdam.

M. Jensen and B. Witcher. (2000). Time-varying long memory in volatility: detection and estimation with wavelets. Technical report. EURANDOM.

R. H. Jones. (1980). Maximum likelihood fitting of ARMA models to time series with missing observations. *Technometrics* 22, 389–395.

R. E. Kalman. (1961). A new approach to linear filtering and prediction problems. *Transactions of the American Society of Mechanical Engineers* 83D, 35–45.

R. E. Kalman and R. S. Bucy. (1961). New results in linear filtering and prediction theory. *Transactions of the American Society of Mechanical Engineers* 83, 95–108.

B. Kedem and K. Fokianos. (2002). *Regression models for time series analysis*. Wiley Series in Probability and Statistics. John Wiley & Sons, Inc., Hoboken, NJ.

T. Kobayashi and D. L. Simon. (2003). Application of a bank of Kalman filters for aircraft engine fault diagnostics. Technical Report E-14088. National Aeronautics and Space Administration, Washington, DC.

R. Kohn and C. F. Ansley. (1986). Estimation, prediction, and interpolation for ARIMA models with missing data. *Journal of the American Statistical Association* 81, 751–761.

L. H. Koopmans. (1995). *The spectral analysis of time series* Vol. 22, *Probability and Mathematical Statistics*. Academic Press, Inc., San Diego, CA.

T. W. Körner. (1989). *Fourier analysis*. Cambridge University Press, Cambridge.

T. W. Körner. (1993). *Exercises for Fourier analysis*. Cambridge University Press, Cambridge.

H. Künsch. (1986). Discrimination between monotonic trends and long-range dependence. *Journal of Applied Probability* 23, 1025–1030.

M. Last and R. H. Shumway. (2008). Detecting abrupt changes in a piecewise locally stationary time series. *Journal of Multivariate Analysis* 99, 191–214.

N. Levinson. (1947). The Wiener RMS (root mean square) error criterion in filter design and prediction. *Journal of Mathematical Physics* 25, 261–278.

W. K. Li and A. I. McLeod. (1986). Fractional time series modelling. *Biometrika* 73, 217–221.

S. Ling and W. K. Li. (1997). On fractionally integrated autoregressive moving-average time series models with conditional heteroscedasticity. *Journal of the American Statistical Association* 92, 1184–1194.

R. J. A. Little and D. B. Rubin. (2002). *Statistical Analysis with Missing Data.* Wiley Series in Probability and Statistics. Wiley, Hoboken, NJ.

I. L. MacDonald and W. Zucchini. (1997). *Hidden Markov and other models for discrete-valued time series* Vol. 110. CRC Press, .

B. B. Mandelbrot and J. W. Van Ness. (1968). Fractional Brownian motions, fractional noises and applications. *SIAM Review* 10, 422–437.

E. McKenzie. (1985). Some simple models for discrete variate time series. *Journal of the American Water Resources Association* 21, 645–650.

E. McKenzie. (2003). Discrete variate time series. *Handbook of statistics* 21, 573–606.

N. Metropolis, A. W. Rosenbluth, A. H. Teller, and E. Teller. (1953). Equations of state calculations by fast computing machines. *Journal of Chemical Physics* 21, 1087–1092.

A. Montanari, R. Rosso, and M. S. Taqqu. (2000). A seasonal fractional ARIMA model applied to Nile River monthly flows at Aswan. *Water Resources Research* 36, 1249–1259.

D. B. Nelson. (1991). Conditional heteroskedasticity in asset returns: a new approach. *Econometrica* 59, 347–370.

M. Ooms. (1995). Flexible seasonal long memory and economic time series. Technical Report EI-9515/A. Econometric Institute, Erasmus University, Rotterdam.

W. Palma. (2000). Missing values in ARFIMA models. In W. S. Chan, W. K. Li, and H. Tong, editors, *Statistics and Finance: An Interface.* Imperial College Press, London, pp. 141–152.

W. Palma. (2007). *Long-Memory Time Series: Theory and Methods.* Wiley Series in Probability and Statistics. John Wiley & Sons, Inc., Hoboken, NJ.

W. Palma and N. H. Chan. (1997). Estimation and forecasting of long-memory processes with missing values. *Journal of Forecasting* 16, 395–410.

W. Palma and N. H. Chan. (2005). Efficient estimation of seasonal long-range-dependent processes. *Journal of Time Series Analysis* 26, 863–892.

W. Palma and R. Olea. (2010). An efficient estimator for locally stationary Gaussian long-memory processes. *The Annals of Statistics* 38, 2958–2997.

W. Palma, R. Olea, and G. Ferreira. (2013). Estimation and forecasting of locally stationary processes. *Journal of Forecasting* 32, 86–96.

W. Palma and M. Zevallos. (2011). Fitting non-gaussian persistent data. *Applied Stochastic Models in Business and Industry* 27, 23–36.

D. B. Percival and A. T. Walden. (2006). *Wavelet Methods for Time Series Analysis*, Vol. 4, *Cambridge Series in Statistical and Probabilistic Mathematics.* Cambridge University Press, Cambridge.

S. Porter-Hudak. (1990). An application of the seasonal fractionally differenced model to the monetary aggregates. *Journal of the American Statistical Association, Applic. Case Studies* 85, 338–344.

M. Pourahmadi. (1989). Estimation and interpolation of missing values of a stationary time series. *Journal of Time Series Analysis* 10, 149–169.

M. Pourahmadi. (2001). *Foundations of Time Series Analysis and Prediction Theory.* John Wiley & Sons, Inc., New York.

S. J. Press. (2003). *Subjective and Objective Bayesian Statistics.* Wiley Series in Probability and Statistics. John Wiley & Sons, Inc., Hoboken, NJ.

W. H. Press, S. A. Teukolsky, W. T. Vetterling, and B. P. Flannery. (1992). *Numerical Recipes in FORTRAN.* Cambridge University Press, Cambridge.

W. H. Press, S. A. Teukolsky, W. T. Vetterling, and B. P. Flannery. (2007). *Numerical Recipes.* Cambridge University Press, Cambridge.

M. B. Priestley. (1965). Evolutionary spectra and non-stationary processes. *Journal of the Royal Statistical Society. Series B. Statistical Methodology* 27, 204–237.

M. B. Priestley. (1981a). *Spectral Analysis and Time Series. Vol. 1.* Academic, London.

M. B. Priestley. (1981b). *Spectral Analysis and Time Series. Vol. 2.* Academic, London.

M. B. Priestley and H. Tong. (1973). On the analysis of bivariate non-stationary processes. *Journal of the Royal Statistical Society. Series B. Methodological* 35, 153–166, 179–188.

R Core Team. R: A Language and Environment for Statistical Computing. R Foundation for Statistical Computing Vienna, Austria(2014). URL `http://www.R-project.org/`.

G. Rangarajan and M. Ding, editors. (2003). *Procesess with Long-Range Correlations.* Springer, Berlin.

B. K. Ray. (1993a). Long-range forecasting of IBM product revenues using a seasonal fractionally differenced ARMA model. *International Journal of Forecasting* 9, 255–269.

B. K. Ray. (1993b). Modeling long-memory processes for optimal long-range prediction. *Journal of Time Series Analysis* 14, 511–525.

B. K. Ray and R. S. Tsay. (2002). Bayesian methods for change-point detection in long-range dependent processes. *Journal of Time Series Analysis* 23, 687–705.

V. A. Reisen, A. L. Rodrigues, and W. Palma. (2006a). Estimating seasonal long-memory processes: a Monte Carlo study. *Journal of Statistical Computation and Simulation* 76, 305–316.

V. A. Reisen, A. L. Rodrigues, and W. Palma. (2006b). Estimation of seasonal fractionally integrated processes. *Computational Statistics & Data Analysis* 50, 568–582.

C. P. Robert. (2001). *The Bayesian Choice.* Springer Texts in Statistics. Springer, New York.

C. P. Robert and G. Casella. (2004). *Monte Carlo Statistical Methods.* Springer Texts in Statistics. Springer, New York.

P. M. Robinson. (1991). Testing for strong serial correlation and dynamic conditional heteroskedasticity in multiple regression. *Journal of Econometrics* 47, 67–84.

P. M. Robinson and M. Henry. (1999). Long and short memory conditional heteroskedasticity in estimating the memory parameter of levels. *Econometric Theory* 15, 299–336.

Y. A. Rozanov. (1967). *Stationary Random Processes.* Holden-Day, San Francisco.

R. H. Shumway and D. S. Stoffer. (2011). *Time series analysis and its applications.* Springer Texts in Statistics. Springer, New York.

P. Sibbertsen. (2001). S-estimation in the linear regression model with long-memory error terms under trend. *Journal of Time Series Analysis* 22, 353–363.

F. Sowell. (1992). Maximum likelihood estimation of stationary univariate fractionally integrated time series models. *Journal of Econometrics* 53, 165–188.

W. F. Stout. (1974). *Almost Sure Convergence.* Academic, New York–London.

T. Subba Rao and M. M. Gabr. (1984). *An Introduction to Bispectral Analysis and Bilinear Time Series Models* Vol. 24, *Lecture Notes in Statistics.* Springer, New York.

M. Taniguchi and Y. Kakizawa. (2000). *Asymptotic Theory of Statistical Inference for Time Series.* Springer Series in Statistics. Springer, New York.

M. S. Taqqu. (2003). Fractional Brownian motion and long-range dependence. In P. Doukhan, G. Oppenheim, and M. S. Taqqu, editors, *Theory and Applications of Long-Range Dependence.* Birkhäuser, Boston, MA, pp. 5–38.

M. S. Taqqu, V. Teverovsky, and W. Willinger. (1995). Estimators for long-range dependence: an empirical study. *Fractals* 3, 785–788.

S. J. Taylor. (1986). *Modelling Financial Time Series.* John Wiley & Sons, Inc., New York.

G. Teyssière and A. Kirman, editors. (2007). *Long Memory in Economics.* Springer, Berlin.

G. C. Tiao and R. S. Tsay. (1994). Some advances in non linear and adaptive modelling in time series. *Journal of Forecasting* 13, 109–131.

L. Tierney. (1994). Markov chains for exploring posterior distributions. *Annals of Statistics* 22, 1701–1762.

H. Tong. (1973). Some comments on spectral representations of non-stationary stochastic processes. *Journal of Applied Probability* 10, 881–885.

H. Tong. (1990). *Nonlinear time series* Vol. 6, *Oxford Statistical Science Series.* The Clarendon Press Oxford University Press, New York.

H. Tong. (2011). Threshold models in time series analysis–30 years on. *Statistics and its Interface* 4, 107–118.

R. S. Tsay. (1989). Testing and modeling threshold autoregressive processes. *Journal of the American Statistical Association* 84, 231–240.

R. S Tsay. (2005). *Analysis of Financial Time Series* Vol. 543. John Wiley & Sons, Inc., Hoboken, NJ.

R. S. Tsay. (2013). *An introduction to analysis of financial data with R.* Wiley Series in Probability and Statistics. John Wiley & Sons, Inc., Hoboken, NJ.

R. S. Tsay, D. Peña, and A. E. Pankratz. (2000). Outliers in multivariate time series. *Biometrika* 87, 789–804.

D. Veitch and P. Abry. (1999). A wavelet-based joint estimator of the parameters of long-range dependence. *Institute of Electrical and Electronics Engineers. Transactions on Information Theory* 45, 878–897.

Y. Wang, J. E. Cavanaugh, and C. Song. (2001). Self-similarity index estimation via wavelets for locally self-similar processes. *Journal of Statistical Planning and Inference* 99, 91–110.

P. Whittle. (1951). *Hypothesis Testing in Time Series Analysis.* Hafner, New York.

P. S. Wilson, A. C. Tomsett, and R. Toumi. (2003). Long-memory analysis of time series with missing values. *Physical Review E* 68, 017103 (1)–(4).

W. A. Woodward, Q. C. Cheng, and H. L. Gray. (1998). A k-factor GARMA long-memory model. *Journal of Time Series Analysis* 19, 485–504.

Y. Yajima. (1985). On estimation of long-memory time series models. *Australian Journal of Statistics* 27, 303–320.

Y. Yajima. (1988). On estimation of a regression model with long-memory stationary errors. *Annals of Statistics* 16, 791–807.

Y. Yajima. (1989). A central limit theorem of Fourier transforms of strongly dependent stationary processes. *Journal of Time Series Analysis* 10, 375–383.

Y. Yajima. (1991). Asymptotic properties of the LSE in a regression model with long-memory stationary errors. *Annals of Statistics* 19, 158–177.

Y. Yajima and H. Nishino. (1999). Estimation of the autocorrelation function of a stationary time series with missing observations. *Sankhyā. Indian Journal of Statistics. Series A* 61, 189–207.

M. Yang, J. E. Cavanaugh, and G. Zamba. (2014). State-space models for count time series with excess zeros. *Statistical Modelling.*

M. Yang, G. Zamba, and J. E. Cavanaugh. (2013). Markov regression models for count time series with excess zeros: A partial likelihood approach. *Statistical Methodology* 14, 26–38.

I. G Zurbenko. (1986). *The spectral analysis of time series.* North-Holland Publishing Co., Amsterdam.

TOPIC INDEX

ACP model, 456
ACP, 458
APARCH, 222
AR-ARCH, 242
ARCH, 213, 215, 244, 252
ARCH(∞), 221, 252–253, 281
ARFIMA, 64, 183, 203, 219–220, 224
ARFIMA-GARCH, 220, 245, 252, 256–258, 276, 280, 292
ARFIMA-GARCH, 218–220
ARMA, 93, 114, 157, 162, 177, 180, 187, 193, 213, 219, 434, 436
ARMA-APARCH, 231
ARMA-GARCH, 231
Akaike's information criterion, 153, 363, 389
definition, 153
Bayes estimator, 181
Bayes theorem, 180
Bayesian methods, 180–181, 194, 253
Beta distribution, 473
Beta model, 473
Box-Cox transformation, 41
Box-Ljung test, 34
Cauchy sequence, 488
Cauchy-Schwartz inequality, 395
Cholesky decomposition, 156
Cramer representation, 138
Daniell window, 129
Dickey-Fuller test, 298
Durbin-Levinson algorithm, 157, 269, 273, 291, 359, 361
EGARCH, 252
Euclidean inner product, 489
Euclidean norm, 367
Euler' constant, 80
FIEGARCH, 222, 252, 258–259, 277
FIGARCH, 45, 221–222, 252, 256
FKF, 100
Fisher information, 183
Fourier coefficients, 146
GARCH, 213, 216, 218–220, 244, 252
GARMA, 337, 351, 364–365
Gamma distribution, 446

Time Series Analysis. First Edition. Wilfredo Palma.

Gibbs sampler, 182
Grenander conditions, 373, 379, 396–397
Hölder's inequality, 489
Hankel Matrix, 93
Hankel matrix, 89, 93, 96, 115, 118
Haslett-Raftery estimate, 162–163, 179, 193, 202
Hessian matrix, 183
Hilbert space, 293–294, 487–489
Hurst exponent, 497
Hurst parameter, 175
INAR, 445
Joseph effect, 70
Kalman filter, 89, 99, 114–116, 156, 161, 165, 168, 282, 359, 362, 400, 404–405, 407, 434, 436
Kalman gain, 98
Kalman prediction equations, 99
Kalman recursions, 100
Kalman recursive equations, 98
LMGARCH, 252
Lebesgue measure, 490
Lyapunov exponent, 219, 257–258
MSPE, 436
Markov chain Monte Carlo (MCMC), 180–181, 183, 194
Markov chain, 181–182, 204–205
Markovian processes, 64
Metropolis-Hastings algorithm, 181–182
Negative Binomial distribution, 446
Nile river, 6
Noah effect, 70
Parzen method, 418
Poisson distribution, 446
Riemann zeta function, 174
SARFIMA, 337, 351–352, 355–357, 359, 363–364, 366–367
SARMA, 424
SETAR, 247–248
Stirling's approximation, 78
Student distribution, 184
Szegö-Kolmogorov formula, 494, 497
TAR process, 247–248
TAR, 247–248
Toeplitz structure, 157
Toeplitz, 307
Volterra expansion, 210, 221, 277
Whittle estimate, 158–160, 168, 179, 193–194, 337, 359, 361–362, 367
Wiener process, 84
Wishart distribution, 204
Wold expansion, 46, 48, 79, 88–89, 93, 96, 165, 168, 241–242, 268, 376, 437, 490–491
ZIM, 477–478
ZINB, 478
ZIP, 477
acceptance rate, 182
additive outlier, 424
approximate MLE, 220
asymmetric power autoregressive conditionally heteroskedastic, 222
best linear interpolator, 410–411, 436–438
best linear unbiased estimator (BLUE), 153, 369, 373–374, 376, 378–379, 381
bfast, 326
causality, 487, 493
  causal process, 492
characteristic function, 374, 392
colored noise, 127
complete space, 488
conjugate prior, 181
consistent, 375
controllability, 94, 114
controllable, 94
cumulant, 168
cyclical behavior, 337
data gaps, 399
data
  Internet traffic, 362–363
  Monthly US employment, 6
  Nile River, 337, 364, 401, 405, 497
  Particulate matter PM2.5, 7
  Passenger Enplanements, 9
  SP500 daily stock, 2
  air pollution, 388–389, 392, 557
  copper price, 212
deterministic, 490
detrended fluctuation analysis, 171
discrete wavelet transform, 174
dynia R package, 424
efficiency, 376
ergodicity, 494–495

ergodic process, 494
expectation maximization (EM), 434
exponential random variable, 45
exponentially stable, 92
fGarch, 225
fast Fourier transform, 158
flexible ARFISMA, 337
fractional Brownian motion, 71, 87, 497
fractional Gaussian noise, 71–72
fractional noise, 66–68, 163, 177, 179, 185, 187, 193, 202, 270–272, 290, 292–293, 366, 378, 381, 385, 407–408, 412, 436, 438–439
full data, 402
generalized gamma function, 204
gsarima, 343
harmonic regression, 385, 394
heteroskedastic, 212, 241, 276
hyper text transfer protocol (HTTP), 362
hyperbolic decay, 64
hypergeometric function, 67
improper prior, 180
imputation, 405
inarmix, 445
indicator function, 137
inner product, 488–489
innovation sequence, 493
innovational outlier, 424
integer part function, 45
intermediate memory, 220
interpolation error variance, 411–412
invariant, 494
invertible, 493
least squares estimator (LSE), 369, 373–379, 381, 384–387, 389, 394, 396
level shift, 424, 431, 433
leverage
 leverage effect, 222
linear filter, 491
linear regression model, 169, 171, 173, 369, 376, 378, 382, 385, 388–389, 392, 394–395
long-memory (definition), 64
long-memory stochastic volatility model (LMSV), 222, 224, 253, 276
loss function, 181
maximum-likelihood estimate (MLE), 166, 178, 183–185, 187, 193, 220, 281, 337, 353, 359, 361–362, 367, 400
missing values, xiv, 89, 99, 114, 161, 399–402, 404–408, 410, 434–436, 439, 554
nlts, 211
norm, 488
observability matrix, 94, 515
observability, 94
observable, 94
observation driven, 441
orthonormal basis, 148, 293
outlier, 421
parallelogram law, 294, 488
parameter driven Poisson model, 452
parameter driven, 441
partial autocorrelation coefficients, 269
particulate matter, 7
perfectly predictable, 490
periodogram, 125, 145, 148, 159, 361, 388
persistence, 371
polynomial regression, 382
posterior distribution, 181–182, 204
posterior mean, 181
prediction error variance, 268
prediction, 267
prior distribution, 180
projection theorem, 410, 412, 488
proposal distribution, 181–183
psi function, 174
purely nondeterministic, 490
quadratic loss, 181
quasi-maximum-likelihood estimate (QMLE), 162–163, 222
random walk process, 110
raw periodogram, 128
reachability, 114
regular process, 93, 490, 493
relative efficiency, 378, 381–382, 385
rescaled range statistic (R/S), 170
reversibility condition, 205
scaling exponent, 497
seasonal long-memory processes, 351
seasonality, 337
self-similar process, 497

semiparametric estimate, 168
singular process, 490
slowly varying function, 64, 220, 274, 276
spectral density, 140, 411, 490
spectral distribution, 490
spectral representation theorem, 491
spectral representation, 138
splitting method, 177
standardized two-side innovation, 436
state space model, 115–116
state space models, 113
state space system, 46, 48, 89–90, 92–98, 114–118, 160, 164, 166–167, 253, 282, 362, 367, 404, 434
  exponentially stable, 92
  extended, 113
  minimal, 95, 115
  minimality, 95
  observation equation, 48, 92
  observation matrix, 48, 92
  state predictor, 111–112
  state smoother error variance, 99
  state smoother, 99
  strongly stable, 92
  weakly stable, 92
state space systems, xiv, 93
stationary distribution, 182, 205
stationary increments, 87
stochastic volatility model (SV), 222
stochastic volatility, 252–253, 256
strict stationarity, 44, 495
strict white noise, 45
strictly stationary process, 45
strong law of large numbers, 495
structural model, 114
stylized facts, 212
subspace, 489
temporary change, 424
thinning operation, 442, 483
threshold autoregressive model, 247–248
time-varying spectral density, 140
transfer function, 138
transition matrix, 204
trend break regression, 395
tsDyn, 248
tsoutlier, 426, 429
uniform prior, 180
variance plot, 389
volatility, 213
wavelet, 136
  Daubechies wavelets, 137, 145
  Haar wavelet system, 137
  Littlewood-Paley decomposition, 148
  discrete wavelet transform (DWT), 136
  multiresolution analysis, 137
  scaling function, 137
wavelets package, 137
weak convergence, 397
weakly stationary process, 45, 493
white noise, 127
zero-inflated Negative Binomial, 478
zero-inflated Poisson distribution, 477
zero-inflated models, 477

AUTHOR INDEX

Abrahams M., 364
Abry P., 145, 174
Adenstedt R. K., 381
Ammar G. S., 193
Anderson B. D. O., 114
Ansley C. F., 114, 434
Aoki M., 114
Baillie R., 252
Basak G., 289
Bell W., 114
Bennett A. F., 114
Beran J., 72
Bertelli S., 193
Bhansali R. J., 289
Bollerslev T., 252
Bondon P., 435
Box G. E. P., 194, 331, 364
Breidt F. J., 253
Brockwell A. E., 253
Brockwell P. J., 39, 72, 114, 193, 364
Bucy R. S., 114
Caporin M., 193
Carothers N. L., 497
Casella G., 194
Chan N. H., 193, 253, 289, 365, 434
Cheng Q. C., 365
Cheng R., 435
Choy K., 392
Chung C. F., 365
Conway J. B., 497
Cox D. R., 72
Cox D.R., 483
Crato N., 253
Curtain R. F., 114
Dahlhaus R., 193, 392
Damsleth E., 435
Davis R. A., 39, 72, 114, 193, 364
Davis R.A., 483
Deistler M., 114
Dempster A., 364
Deo R., 253
Dette H., 332
Diggle P. J., 39
Dissanayake G. S., 193

Doornik J. A., 193
Dunsmuir W., 483
Durbin J., 114, 193
Engle R., 252–253
Ferreira G., 332
Flandrin P., 145
Flannery B. P., 145
Fokianos K., 39, 483
Fox R., 194
Fuller W. A., 39
Ghysels E., 253
Giraitis L., 193, 365
Gradshteyn I. S., 258
Grassi S., 193
Gray H. L., 365
Graybill D. A., 557
Grenander U., 392
Grunwald G. K., 483
Hall P., 72
Hamilton J. D., 39
Hannan E. J., 114, 289, 392, 497
Harvey A. C., 114, 252–253, 434
Haslett J., 163, 179, 193
Hassler U., 364–365
Henry M., 252
Hillmer S., 114
Hosking J. R. M., 72
Hurst H. E., 170
Hurvich C. M., 253
Hyndman R. J., 483
Ibragimov I. A., 392
Iglesias P., 392
Jenkins G. M., 331, 364
Jones R. H., 114, 434
Jorquera H., 392
Kakizawa Y., 497
Kalman R. E., 114
Kedem B., 39, 483
Kobayashi T., 114
Kohn R., 114, 434
Kokoszka P. S., 193
Kokoszka P., 289
Kolmogorov A. N., 289, 494, 497
Koopman S. J., 114
Leipus R., 193
Levinson N., 193
Li W. K., 252
Ling S., 252
Little R. J. A., 435
Mandelbrot B. B., 364
Mikkelsen H. O., 252
Montanari A., 364
Moore J. B., 114
Nelson D. B., 252
Nishino H., 435
Olea R., 332
Ooms M., 193, 364
Palma W., 114, 193, 289, 332, 365, 392, 483
Peiris M. S., 193
Percival D. B., 145
Petris G., 253
Pierse R. G., 434
Porter-Hudak S., 364–365
Pourahmadi M., 289, 435, 497
Press S. J., 194
Press W. H., 145
Preuß P., 332
Proietti T., 193
Raftery A., 163, 179, 193
Ray B. K., 114, 289, 364
Reinsel G. C., 331, 364
Reisen V., 365
Renault E., 253
Robert C. P., 194
Robinson P. M., 252–253
Rodrigues A., 365
Rosenblatt M., 392
Rosso R., 364
Rozanov Y. A., 289, 392
Rubin D. B., 435
Ruiz E., 253
Ryzhik I. M., 258
Shephard N., 253
Shumway R. H., 434
Shumway R. J., 39, 72
Simon D. L., 114
Sowell F., 72
Stoffer D. S., 434
Stoffer D., 39, 72
Stout W. F., 497
Streett S.B., 483
Surgailis D., 365
Szegö G., 494, 497
Taniguchi M., 392, 497

Taqqu M. S., 72, 145, 194, 364, 497
Tedesco L., 483
Teukolsky S. A., 145
Teverovsky V., 194
Teyssière G., 193
Tiao G. C., 194, 289
Tierney L., 194
Tomsett A. C., 434
Tong H., 253
Toumi R., 434
Tsay R. S., 39, 114, 253, 289, 435
Tweedie R. L., 483
Van Ness J. W., 364
Veitch D., 145, 174
Vetter M., 332
Vetterling W. T., 145
Walden A. T., 145
Wiener N., 289
Willinger W., 194
Wilson P. S., 434
Wold H., 289
Woodward W. A., 365
Yajima Y., 194, 392, 395, 435
Zevallos M., 483
Zhang N. F., 365
Zwart H., 114

WILEY SERIES IN PROBABILITY AND STATISTICS
ESTABLISHED BY WALTER A. SHEWHART AND SAMUEL S. WILKS

The *Wiley Series in Probability and Statistics* is well established and authoritative. It covers many topics of current research interest in both pure and applied statistics and probability theory. Written by leading statisticians and institutions, the titles span both state-of-the-art developments in the field and classical methods.

Reflecting the wide range of current research in statistics, the series encompasses applied, methodological and theoretical statistics, ranging from applications and new techniques made possible by advances in computerized practice to rigorous treatment of theoretical approaches.

This series provides essential and invaluable reading for all statisticians, whether in academia, industry, government, or research.

† ABRAHAM and LEDOLTER · Statistical Methods for Forecasting
AGRESTI · Analysis of Ordinal Categorical Data, *Second Edition*
AGRESTI · An Introduction to Categorical Data Analysis, *Second Edition*
AGRESTI · Categorical Data Analysis, *Third Edition*
AGRESTI · *Foundations of Linear and Generalized Linear Models*
ALSTON, MENGERSEN and PETTITT (editors) · Case Studies in Bayesian Statistical Modelling and Analysis
ALTMAN, GILL, and McDONALD · Numerical Issues in Statistical Computing for the Social Scientist
AMARATUNGA and CABRERA · Exploration and Analysis of DNA Microarray and Protein Array Data
AMARATUNGA, CABRERA, and SHKEDY · Exploration and Analysis of DNA Microarray and Other High-Dimensional Data, *Second Edition*
ANDĚL · Mathematics of Chance
ANDERSON · An Introduction to Multivariate Statistical Analysis, *Third Edition*
* ANDERSON · The Statistical Analysis of Time Series
ANDERSON, AUQUIER, HAUCK, OAKES, VANDAELE, and WEISBERG · Statistical Methods for Comparative Studies
ANDERSON and LOYNES · The Teaching of Practical Statistics
ARMITAGE and DAVID (editors) · Advances in Biometry
ARNOLD, BALAKRISHNAN, and NAGARAJA · Records
* ARTHANARI and DODGE · Mathematical Programming in Statistics
AUGUSTIN, COOLEN, DE COOMAN and TROFFAES (editors) · Introduction to Imprecise Probabilities
* BAILEY · The Elements of Stochastic Processes with Applications to the Natural Sciences
BAJORSKI · Statistics for Imaging, Optics, and Photonics
BALAKRISHNAN and KOUTRAS · Runs and Scans with Applications
BALAKRISHNAN and NG · Precedence-Type Tests and Applications
BARNETT · Comparative Statistical Inference, *Third Edition*
BARNETT · Environmental Statistics
BARNETT and LEWIS · Outliers in Statistical Data, *Third Edition*

*Now available in a lower priced paperback edition in the Wiley Classics Library.
†Now available in a lower priced paperback edition in the Wiley–Interscience Paperback Series.

BARTHOLOMEW, KNOTT, and MOUSTAKI · Latent Variable Models and Factor Analysis: A Unified Approach, *Third Edition*
BARTOSZYNSKI and NIEWIADOMSKA-BUGAJ · Probability and Statistical Inference, *Second Edition*
BASILEVSKY · Statistical Factor Analysis and Related Methods: Theory and Applications
BATES and WATTS · Nonlinear Regression Analysis and Its Applications
BECHHOFER, SANTNER, and GOLDSMAN · Design and Analysis of Experiments for Statistical Selection, Screening, and Multiple Comparisons
BEH and LOMBARDO · Correspondence Analysis: Theory, Practice and New Strategies
BEIRLANT, GOEGEBEUR, SEGERS, TEUGELS, and DE WAAL · Statistics of Extremes: Theory and Applications
BELSLEY · Conditioning Diagnostics: Collinearity and Weak Data in Regression
† BELSLEY, KUH, and WELSCH · Regression Diagnostics: Identifying Influential Data and Sources of Collinearity
BENDAT and PIERSOL · Random Data: Analysis and Measurement Procedures, *Fourth Edition*
BERNARDO and SMITH · Bayesian Theory
BHAT and MILLER · Elements of Applied Stochastic Processes, *Third Edition*
BHATTACHARYA and WAYMIRE · Stochastic Processes with Applications
BIEMER, GROVES, LYBERG, MATHIOWETZ, and SUDMAN · Measurement Errors in Surveys
BILLINGSLEY · Convergence of Probability Measures, *Second Edition*
BILLINGSLEY · Probability and Measure, *Anniversary Edition*
BIRKES and DODGE · Alternative Methods of Regression
BISGAARD and KULAHCI · Time Series Analysis and Forecasting by Example
BISWAS, DATTA, FINE, and SEGAL · Statistical Advances in the Biomedical Sciences: Clinical Trials, Epidemiology, Survival Analysis, and Bioinformatics
BLISCHKE and MURTHY (editors) · Case Studies in Reliability and Maintenance
BLISCHKE and MURTHY · Reliability: Modeling, Prediction, and Optimization
BLOOMFIELD · Fourier Analysis of Time Series: An Introduction, *Second Edition*
BOLLEN · Structural Equations with Latent Variables
BOLLEN and CURRAN · Latent Curve Models: A Structural Equation Perspective
BONNINI, CORAIN, MAROZZI and SALMASO · Nonparametric Hypothesis Testing: Rank and Permutation Methods with Applications in R
BOROVKOV · Ergodicity and Stability of Stochastic Processes
BOSQ and BLANKE · Inference and Prediction in Large Dimensions
BOULEAU · Numerical Methods for Stochastic Processes
* BOX and TIAO · Bayesian Inference in Statistical Analysis
BOX · Improving Almost Anything, *Revised Edition*
* BOX and DRAPER · Evolutionary Operation: A Statistical Method for Process Improvement
BOX and DRAPER · Response Surfaces, Mixtures, and Ridge Analyses, *Second Edition*
BOX, HUNTER, and HUNTER · Statistics for Experimenters: Design, Innovation, and Discovery, *Second Editon*
BOX, JENKINS, REINSEL, and LJUNG · Time Series Analysis: Forecasting and Control, *Fifth Edition*
BOX, LUCEÑO, and PANIAGUA-QUIÑONES · Statistical Control by Monitoring and Adjustment, *Second Edition*
* BROWN and HOLLANDER · Statistics: A Biomedical Introduction
CAIROLI and DALANG · Sequential Stochastic Optimization

*Now available in a lower priced paperback edition in the Wiley Classics Library.
†Now available in a lower priced paperback edition in the Wiley–Interscience Paperback Series.

CASTILLO, HADI, BALAKRISHNAN, and SARABIA · Extreme Value and Related Models with Applications in Engineering and Science
CHAN · Time Series: Applications to Finance with R and S-Plus®, *Second Edition*
CHARALAMBIDES · Combinatorial Methods in Discrete Distributions
CHATTERJEE and HADI · Regression Analysis by Example, *Fourth Edition*
CHATTERJEE and HADI · Sensitivity Analysis in Linear Regression
CHEN · The Fitness of Information: Quantitative Assessments of Critical Evidence
CHERNICK · Bootstrap Methods: A Guide for Practitioners and Researchers, *Second Edition*
CHERNICK and FRIIS · Introductory Biostatistics for the Health Sciences
CHILÈS and DELFINER · Geostatistics: Modeling Spatial Uncertainty, *Second Edition*
CHIU, STOYAN, KENDALL and MECKE · Stochastic Geometry and Its Applications, *Third Edition*
CHOW and LIU · Design and Analysis of Clinical Trials: Concepts and Methodologies, *Third Edition*
CLARKE · Linear Models: The Theory and Application of Analysis of Variance
CLARKE and DISNEY · Probability and Random Processes: A First Course with Applications, *Second Edition*
* COCHRAN and COX · Experimental Designs, *Second Edition*
COLLINS and LANZA · Latent Class and Latent Transition Analysis: With Applications in the Social, Behavioral, and Health Sciences
CONGDON · Applied Bayesian Modelling, *Second Edition*
CONGDON · Bayesian Models for Categorical Data
CONGDON · Bayesian Statistical Modelling, *Second Edition*
CONOVER · Practical Nonparametric Statistics, *Third Edition*
COOK · Regression Graphics
COOK and WEISBERG · An Introduction to Regression Graphics
COOK and WEISBERG · Applied Regression Including Computing and Graphics
CORNELL · A Primer on Experiments with Mixtures
CORNELL · Experiments with Mixtures, Designs, Models, and the Analysis of Mixture Data, *Third Edition*
COX · A Handbook of Introductory Statistical Methods
CRESSIE · Statistics for Spatial Data, *Revised Edition*
CRESSIE and WIKLE · Statistics for Spatio-Temporal Data
CSÖRGŐ and HORVÁTH · Limit Theorems in Change Point Analysis
DAGPUNAR · Simulation and Monte Carlo: With Applications in Finance and MCMC
DANIEL · Applications of Statistics to Industrial Experimentation
DANIEL · Biostatistics: A Foundation for Analysis in the Health Sciences, *Eighth Edition*
* DANIEL · Fitting Equations to Data: Computer Analysis of Multifactor Data, *Second Edition*
DASU and JOHNSON · Exploratory Data Mining and Data Cleaning
DAVID and NAGARAJA · Order Statistics, *Third Edition*
DAVINO, FURNO and VISTOCCO · Quantile Regression: Theory and Applications
* DEGROOT, FIENBERG, and KADANE · Statistics and the Law
DEL CASTILLO · Statistical Process Adjustment for Quality Control
DEMARIS · Regression with Social Data: Modeling Continuous and Limited Response Variables
DEMIDENKO · Mixed Models: Theory and Applications with R, *Second Edition*

*Now available in a lower priced paperback edition in the Wiley Classics Library.
†Now available in a lower priced paperback edition in the Wiley–Interscience Paperback Series.

DENISON, HOLMES, MALLICK, and SMITH · Bayesian Methods for Nonlinear Classification and Regression
DETTE and STUDDEN · The Theory of Canonical Moments with Applications in Statistics, Probability, and Analysis
DEY and MUKERJEE · Fractional Factorial Plans
DILLON and GOLDSTEIN · Multivariate Analysis: Methods and Applications
* DODGE and ROMIG · Sampling Inspection Tables, *Second Edition*
* DOOB · Stochastic Processes
DOWDY, WEARDEN, and CHILKO · Statistics for Research, *Third Edition*
DRAPER and SMITH · Applied Regression Analysis, *Third Edition*
DRYDEN and MARDIA · Statistical Shape Analysis
DUDEWICZ and MISHRA · Modern Mathematical Statistics
DUNN and CLARK · Basic Statistics: A Primer for the Biomedical Sciences, *Fourth Edition*
DUPUIS and ELLIS · A Weak Convergence Approach to the Theory of Large Deviations
EDLER and KITSOS · Recent Advances in Quantitative Methods in Cancer and Human Health Risk Assessment
* ELANDT-JOHNSON and JOHNSON · Survival Models and Data Analysis
ENDERS · Applied Econometric Time Series, *Third Edition*
† ETHIER and KURTZ · Markov Processes: Characterization and Convergence
EVANS, HASTINGS, and PEACOCK · Statistical Distributions, *Third Edition*
EVERITT, LANDAU, LEESE, and STAHL · Cluster Analysis, *Fifth Edition*
FEDERER and KING · Variations on Split Plot and Split Block Experiment Designs
FELLER · An Introduction to Probability Theory and Its Applications, Volume I, *Third Edition,* Revised; Volume II, *Second Edition*
FITZMAURICE, LAIRD, and WARE · Applied Longitudinal Analysis, *Second Edition*
* FLEISS · The Design and Analysis of Clinical Experiments
FLEISS · Statistical Methods for Rates and Proportions, *Third Edition*
† FLEMING and HARRINGTON · Counting Processes and Survival Analysis
FUJIKOSHI, ULYANOV, and SHIMIZU · Multivariate Statistics: High-Dimensional and Large-Sample Approximations
FULLER · Introduction to Statistical Time Series, *Second Edition*
† FULLER · Measurement Error Models
GALLANT · Nonlinear Statistical Models
GEISSER · Modes of Parametric Statistical Inference
GELMAN and MENG · Applied Bayesian Modeling and Causal Inference from ncomplete-Data Perspectives
GEWEKE · Contemporary Bayesian Econometrics and Statistics
GHOSH, MUKHOPADHYAY, and SEN · Sequential Estimation
GIESBRECHT and GUMPERTZ · Planning, Construction, and Statistical Analysis of Comparative Experiments
GIFI · Nonlinear Multivariate Analysis
GIVENS and HOETING · Computational Statistics
GLASSERMAN and YAO · Monotone Structure in Discrete-Event Systems
GNANADESIKAN · Methods for Statistical Data Analysis of Multivariate Observations, *Second Edition*
GOLDSTEIN · Multilevel Statistical Models, *Fourth Edition*
GOLDSTEIN and LEWIS · Assessment: Problems, Development, and Statistical Issues
GOLDSTEIN and WOOFF · Bayes Linear Statistics

*Now available in a lower priced paperback edition in the Wiley Classics Library.
†Now available in a lower priced paperback edition in the Wiley–Interscience Paperback Series.

GRAHAM · Markov Chains: Analytic and Monte Carlo Computations
GREENWOOD and NIKULIN · A Guide to Chi-Squared Testing
GROSS, SHORTLE, THOMPSON, and HARRIS · Fundamentals of Queueing Theory, *Fourth Edition*
GROSS, SHORTLE, THOMPSON, and HARRIS · Solutions Manual to Accompany Fundamentals of Queueing Theory, *Fourth Edition*
* HAHN and SHAPIRO · Statistical Models in Engineering
HAHN and MEEKER · Statistical Intervals: A Guide for Practitioners
HALD · A History of Probability and Statistics and their Applications Before 1750
† HAMPEL · Robust Statistics: The Approach Based on Influence Functions
HARTUNG, KNAPP, and SINHA · Statistical Meta-Analysis with Applications
HEIBERGER · Computation for the Analysis of Designed Experiments
HEDAYAT and SINHA · Design and Inference in Finite Population Sampling
HEDEKER and GIBBONS · Longitudinal Data Analysis
HELLER · MACSYMA for Statisticians
HERITIER, CANTONI, COPT, and VICTORIA-FESER · Robust Methods in Biostatistics
HINKELMANN and KEMPTHORNE · Design and Analysis of Experiments, Volume 1: Introduction to Experimental Design, *Second Edition*
HINKELMANN and KEMPTHORNE · Design and Analysis of Experiments, Volume 2: Advanced Experimental Design
HINKELMANN (editor) · Design and Analysis of Experiments, Volume 3: Special Designs and Applications
HOAGLIN, MOSTELLER, and TUKEY · Fundamentals of Exploratory Analysis of Variance
* HOAGLIN, MOSTELLER, and TUKEY · Exploring Data Tables, Trends and Shapes
* HOAGLIN, MOSTELLER, and TUKEY · Understanding Robust and Exploratory Data Analysis
HOCHBERG and TAMHANE · Multiple Comparison Procedures
HOCKING · Methods and Applications of Linear Models: Regression and the Analysis of Variance, *Third Edition*
HOEL · Introduction to Mathematical Statistics, *Fifth Edition*
HOGG and KLUGMAN · Loss Distributions
HOLLANDER, WOLFE, and CHICKEN · Nonparametric Statistical Methods, *Third Edition*
HOSMER and LEMESHOW · Applied Logistic Regression, *Second Edition*
HOSMER, LEMESHOW, and MAY · Applied Survival Analysis: Regression Modeling of Time-to-Event Data, *Second Edition*
HUBER · Data Analysis: What Can Be Learned From the Past 50 Years
HUBER · Robust Statistics
† HUBER and RONCHETTI · Robust Statistics, *Second Edition*
HUBERTY · Applied Discriminant Analysis, *Second Edition*
HUBERTY and OLEJNIK · Applied MANOVA and Discriminant Analysis, *Second Edition*
HUITEMA · The Analysis of Covariance and Alternatives: Statistical Methods for Experiments, Quasi-Experiments, and Single-Case Studies, *Second Edition*
HUNT and KENNEDY · Financial Derivatives in Theory and Practice, *Revised Edition*
HURD and MIAMEE · Periodically Correlated Random Sequences: Spectral Theory and Practice
HUSKOVA, BERAN, and DUPAC · Collected Works of Jaroslav Hajek— with Commentary
HUZURBAZAR · Flowgraph Models for Multistate Time-to-Event Data

*Now available in a lower priced paperback edition in the Wiley Classics Library.
†Now available in a lower priced paperback edition in the Wiley–Interscience Paperback Series.

JACKMAN · Bayesian Analysis for the Social Sciences
† JACKSON · A User's Guide to Principle Components
JOHN · Statistical Methods in Engineering and Quality Assurance
JOHNSON · Multivariate Statistical Simulation
JOHNSON and BALAKRISHNAN · Advances in the Theory and Practice of Statistics: A Volume in Honor of Samuel Kotz
JOHNSON, KEMP, and KOTZ · Univariate Discrete Distributions, *Third Edition*
JOHNSON and KOTZ (editors) · Leading Personalities in Statistical Sciences: From the Seventeenth Century to the Present
JOHNSON, KOTZ, and BALAKRISHNAN · Continuous Univariate Distributions, Volume 1, *Second Edition*
JOHNSON, KOTZ, and BALAKRISHNAN · Continuous Univariate Distributions, Volume 2, *Second Edition*
JOHNSON, KOTZ, and BALAKRISHNAN · Discrete Multivariate Distributions
JUDGE, GRIFFITHS, HILL, LÜTKEPOHL, and LEE · The Theory and Practice of Econometrics, *Second Edition*
JUREK and MASON · Operator-Limit Distributions in Probability Theory
KADANE · Bayesian Methods and Ethics in a Clinical Trial Design
KADANE AND SCHUM · A Probabilistic Analysis of the Sacco and Vanzetti Evidence
KALBFLEISCH and PRENTICE · The Statistical Analysis of Failure Time Data, *Second Edition*
KARIYA and KURATA · Generalized Least Squares
KASS and VOS · Geometrical Foundations of Asymptotic Inference
† KAUFMAN and ROUSSEEUW · Finding Groups in Data: An Introduction to Cluster Analysis
KEDEM and FOKIANOS · Regression Models for Time Series Analysis
KENDALL, BARDEN, CARNE, and LE · Shape and Shape Theory
KHURI · Advanced Calculus with Applications in Statistics, *Second Edition*
KHURI, MATHEW, and SINHA · Statistical Tests for Mixed Linear Models
* KISH · Statistical Design for Research
KLEIBER and KOTZ · Statistical Size Distributions in Economics and Actuarial Sciences
KLEMELÄ · Smoothing of Multivariate Data: Density Estimation and Visualization
KLUGMAN, PANJER, and WILLMOT · Loss Models: From Data to Decisions, *Third Edition*
KLUGMAN, PANJER, and WILLMOT · Loss Models: Further Topics
KLUGMAN, PANJER, and WILLMOT · Solutions Manual to Accompany Loss Models: From Data to Decisions, *Third Edition*
KOSKI and NOBLE · Bayesian Networks: An Introduction
KOTZ, BALAKRISHNAN, and JOHNSON · Continuous Multivariate Distributions, Volume 1, *Second Edition*
KOTZ and JOHNSON (editors) · Encyclopedia of Statistical Sciences: Volumes 1 to 9 with Index
KOTZ and JOHNSON (editors) · Encyclopedia of Statistical Sciences: Supplement Volume
KOTZ, READ, and BANKS (editors) · Encyclopedia of Statistical Sciences: Update Volume 1
KOTZ, READ, and BANKS (editors) · Encyclopedia of Statistical Sciences: Update Volume 2
KOWALSKI and TU · Modern Applied U-Statistics
KRISHNAMOORTHY and MATHEW · Statistical Tolerance Regions: Theory, Applications, and Computation

*Now available in a lower priced paperback edition in the Wiley Classics Library.
†Now available in a lower priced paperback edition in the Wiley–Interscience Paperback Series.

KROESE, TAIMRE, and BOTEV · Handbook of Monte Carlo Methods
KROONENBERG · Applied Multiway Data Analysis
KULINSKAYA, MORGENTHALER, and STAUDTE · Meta Analysis: A Guide to Calibrating and Combining Statistical Evidence
KULKARNI and HARMAN · An Elementary Introduction to Statistical Learning Theory
KUROWICKA and COOKE · Uncertainty Analysis with High Dimensional Dependence Modelling
KVAM and VIDAKOVIC · Nonparametric Statistics with Applications to Science and Engineering
LACHIN · Biostatistical Methods: The Assessment of Relative Risks, *Second Edition*
LAD · Operational Subjective Statistical Methods: A Mathematical, Philosophical, and Historical Introduction
LAMPERTI · Probability: A Survey of the Mathematical Theory, *Second Edition*
LAWLESS · Statistical Models and Methods for Lifetime Data, *Second Edition*
LAWSON · Statistical Methods in Spatial Epidemiology, *Second Edition*
LE · Applied Categorical Data Analysis, *Second Edition*
LE · Applied Survival Analysis
LEE · Structural Equation Modeling: A Bayesian Approach
LEE and WANG · Statistical Methods for Survival Data Analysis, *Fourth Edition*
LEPAGE and BILLARD · Exploring the Limits of Bootstrap
LESSLER and KALSBEEK · Nonsampling Errors in Surveys
LEYLAND and GOLDSTEIN (editors) · Multilevel Modelling of Health Statistics
LIAO · Statistical Group Comparison
LIN · Introductory Stochastic Analysis for Finance and Insurance
LINDLEY · Understanding Uncertainty, *Revised Edition*
LITTLE and RUBIN · Statistical Analysis with Missing Data, *Second Edition*
LLOYD · The Statistical Analysis of Categorical Data
LOWEN and TEICH · Fractal-Based Point Processes
MAGNUS and NEUDECKER · Matrix Differential Calculus with Applications in Statistics and Econometrics, *Revised Edition*
MALLER and ZHOU · Survival Analysis with Long Term Survivors
MARCHETTE · Random Graphs for Statistical Pattern Recognition
MARDIA and JUPP · Directional Statistics
MARKOVICH · Nonparametric Analysis of Univariate Heavy-Tailed Data: Research and Practice
MARONNA, MARTIN and YOHAI · Robust Statistics: Theory and Methods
MASON, GUNST, and HESS · Statistical Design and Analysis of Experiments with Applications to Engineering and Science, *Second Edition*
McCULLOCH, SEARLE, and NEUHAUS · Generalized, Linear, and Mixed Models, *Second Edition*
McFADDEN · Management of Data in Clinical Trials, *Second Edition*
* McLACHLAN · Discriminant Analysis and Statistical Pattern Recognition
McLACHLAN, DO, and AMBROISE · Analyzing Microarray Gene Expression Data
McLACHLAN and KRISHNAN · The EM Algorithm and Extensions, *Second Edition*
McLACHLAN and PEEL · Finite Mixture Models
McNEIL · Epidemiological Research Methods
MEEKER and ESCOBAR · Statistical Methods for Reliability Data
MEERSCHAERT and SCHEFFLER · Limit Distributions for Sums of Independent Random Vectors: Heavy Tails in Theory and Practice
MENGERSEN, ROBERT, and TITTERINGTON · Mixtures: Estimation and Applications

*Now available in a lower priced paperback edition in the Wiley Classics Library.
†Now available in a lower priced paperback edition in the Wiley–Interscience Paperback Series.

MICKEY, DUNN, and CLARK · Applied Statistics: Analysis of Variance and Regression, *Third Edition*
* MILLER · Survival Analysis, *Second Edition*
MONTGOMERY, JENNINGS, and KULAHCI · Introduction to Time Series Analysis and Forecasting, *Second Edition*
MONTGOMERY, PECK, and VINING · Introduction to Linear Regression Analysis, *Fifth Edition*
MORGENTHALER and TUKEY · Configural Polysampling: A Route to Practical Robustness
MUIRHEAD · Aspects of Multivariate Statistical Theory
MULLER and STOYAN · Comparison Methods for Stochastic Models and Risks
MURTHY, XIE, and JIANG · Weibull Models
MYERS, MONTGOMERY, and ANDERSON-COOK · Response Surface Methodology: Process and Product Optimization Using Designed Experiments, *Third Edition*
MYERS, MONTGOMERY, VINING, and ROBINSON · Generalized Linear Models. With Applications in Engineering and the Sciences, *Second Edition*
NATVIG · Multistate Systems Reliability Theory With Applications
† NELSON · Accelerated Testing, Statistical Models, Test Plans, and Data Analyses
† NELSON · Applied Life Data Analysis
NEWMAN · Biostatistical Methods in Epidemiology
NG, TAIN, and TANG · Dirichlet Theory: Theory, Methods and Applications
OKABE, BOOTS, SUGIHARA, and CHIU · Spatial Tesselations: Concepts and Applications of Voronoi Diagrams, *Second Edition*
OLIVER and SMITH · Influence Diagrams, Belief Nets and Decision Analysis
PALMA · Time Series Analysis
PALTA · Quantitative Methods in Population Health: Extensions of Ordinary Regressions
PANJER · Operational Risk: Modeling and Analytics
PANKRATZ · Forecasting with Dynamic Regression Models
PANKRATZ · Forecasting with Univariate Box-Jenkins Models: Concepts and Cases
PARDOUX · Markov Processes and Applications: Algorithms, Networks, Genome and Finance
PARMIGIANI and INOUE · Decision Theory: Principles and Approaches
* PARZEN · Modern Probability Theory and Its Applications
PEÑA, TIAO, and TSAY · A Course in Time Series Analysis
PESARIN and SALMASO · Permutation Tests for Complex Data: Applications and Software
PIANTADOSI · Clinical Trials: A Methodologic Perspective, *Second Edition*
POURAHMADI · Foundations of Time Series Analysis and Prediction Theory
POURAHMADI · High-Dimensional Covariance Estimation
POWELL · Approximate Dynamic Programming: Solving the Curses of Dimensionality, *Second Edition*
POWELL and RYZHOV · Optimal Learning
PRESS · Subjective and Objective Bayesian Statistics, *Second Edition*
PRESS and TANUR · The Subjectivity of Scientists and the Bayesian Approach
PURI, VILAPLANA, and WERTZ · New Perspectives in Theoretical and Applied Statistics
† PUTERMAN · Markov Decision Processes: Discrete Stochastic Dynamic Programming
QIU · Image Processing and Jump Regression Analysis
* RAO · Linear Statistical Inference and Its Applications, *Second Edition*

*Now available in a lower priced paperback edition in the Wiley Classics Library.
†Now available in a lower priced paperback edition in the Wiley–Interscience Paperback Series.

RAO · Statistical Inference for Fractional Diffusion Processes
RAUSAND and HØYLAND · System Reliability Theory: Models, Statistical Methods, and Applications, *Second Edition*
RAYNER, THAS, and BEST · Smooth Tests of Goodnes of Fit: Using R, *Second Edition*
RENCHER and SCHAALJE · Linear Models in Statistics, *Second Edition*
RENCHER and CHRISTENSEN · Methods of Multivariate Analysis, *Third Edition*
RENCHER · Multivariate Statistical Inference with Applications
RIGDON and BASU · Statistical Methods for the Reliability of Repairable Systems
* RIPLEY · Spatial Statistics
* RIPLEY · Stochastic Simulation
ROHATGI and SALEH · An Introduction to Probability and Statistics, *Third Edition*
ROLSKI, SCHMIDLI, SCHMIDT, and TEUGELS · Stochastic Processes for Insurance and Finance
ROSENBERGER and LACHIN · Randomization in Clinical Trials: Theory and Practice
ROSSI, ALLENBY, and McCULLOCH · Bayesian Statistics and Marketing
† ROUSSEEUW and LEROY · Robust Regression and Outlier Detection
ROYSTON and SAUERBREI · Multivariate Model Building: A Pragmatic Approach to Regression Analysis Based on Fractional Polynomials for Modeling Continuous Variables
* RUBIN · Multiple Imputation for Nonresponse in Surveys
RUBINSTEIN and KROESE · Simulation and the Monte Carlo Method, *Second Edition*
RUBINSTEIN and MELAMED · Modern Simulation and Modeling
RUBINSTEIN, RIDDER, and VAISMAN · Fast Sequential Monte Carlo Methods for Counting and Optimization
RYAN · Modern Engineering Statistics
RYAN · Modern Experimental Design
RYAN · Modern Regression Methods, *Second Edition*
RYAN · Sample Size Determination and Power
RYAN · Statistical Methods for Quality Improvement, *Third Edition*
SALEH · Theory of Preliminary Test and Stein-Type Estimation with Applications
SALTELLI, CHAN, and SCOTT (editors) · Sensitivity Analysis
SCHERER · Batch Effects and Noise in Microarray Experiments: Sources and Solutions
* SCHEFFE · The Analysis of Variance
SCHIMEK · Smoothing and Regression: Approaches, Computation, and Application
SCHOTT · Matrix Analysis for Statistics, *Second Edition*
SCHOUTENS · Levy Processes in Finance: Pricing Financial Derivatives
SCOTT · Multivariate Density Estimation
SCOTT · Multivariate Density Estimation: Theory, Practice, and Visualization
* SEARLE · Linear Models
† SEARLE · Linear Models for Unbalanced Data
† SEARLE · Matrix Algebra Useful for Statistics
† SEARLE, CASELLA, and McCULLOCH · Variance Components
SEARLE and WILLETT · Matrix Algebra for Applied Economics
SEBER · A Matrix Handbook For Statisticians
† SEBER · Multivariate Observations
SEBER and LEE · Linear Regression Analysis, *Second Edition*
† SEBER and WILD · Nonlinear Regression

*Now available in a lower priced paperback edition in the Wiley Classics Library.
†Now available in a lower priced paperback edition in the Wiley–Interscience Paperback Series.

SENNOTT · Stochastic Dynamic Programming and the Control of Queueing Systems
* SERFLING · Approximation Theorems of Mathematical Statistics
SHAFER and VOVK · Probability and Finance: It's Only a Game!
SHERMAN · Spatial Statistics and Spatio-Temporal Data: Covariance Functions and Directional Properties
SILVAPULLE and SEN · Constrained Statistical Inference: Inequality, Order, and Shape Restrictions
SINGPURWALLA · Reliability and Risk: A Bayesian Perspective
SMALL and McLEISH · Hilbert Space Methods in Probability and Statistical Inference
SRIVASTAVA · Methods of Multivariate Statistics
STAPLETON · Linear Statistical Models, *Second Edition*
STAPLETON · Models for Probability and Statistical Inference: Theory and Applications
STAUDTE and SHEATHER · Robust Estimation and Testing
STOYAN · Counterexamples in Probability, *Second Edition*
STOYAN and STOYAN · Fractals, Random Shapes and Point Fields: Methods of Geometrical Statistics
STREET and BURGESS · The Construction of Optimal Stated Choice Experiments: Theory and Methods
STYAN · The Collected Papers of T. W. Anderson: 1943–1985
SUTTON, ABRAMS, JONES, SHELDON, and SONG · Methods for Meta-Analysis in Medical Research
TAKEZAWA · Introduction to Nonparametric Regression
TAMHANE · Statistical Analysis of Designed Experiments: Theory and Applications
TANAKA · Time Series Analysis: Nonstationary and Noninvertible Distribution Theory
THOMPSON · Empirical Model Building: Data, Models, and Reality, *Second Edition*
THOMPSON · Sampling, *Third Edition*
THOMPSON · Simulation: A Modeler's Approach
THOMPSON and SEBER · Adaptive Sampling
THOMPSON, WILLIAMS, and FINDLAY · Models for Investors in Real World Markets
TIERNEY · LISP-STAT: An Object-Oriented Environment for Statistical Computing
and Dynamic Graphics
TROFFAES and DE COOMAN · Lower Previsions
TSAY · Analysis of Financial Time Series, *Third Edition*
TSAY · An Introduction to Analysis of Financial Data with R
TSAY · Multivariate Time Series Analysis: With R and Financial Applications
UPTON and FINGLETON · Spatial Data Analysis by Example, Volume II: Categorical and Directional Data
† VAN BELLE · Statistical Rules of Thumb, *Second Edition*
VAN BELLE, FISHER, HEAGERTY, and LUMLEY · Biostatistics: A Methodology for the Health Sciences, *Second Edition*
VESTRUP · The Theory of Measures and Integration
VIDAKOVIC · Statistical Modeling by Wavelets
VIERTL · Statistical Methods for Fuzzy Data
VINOD and REAGLE · Preparing for the Worst: Incorporating Downside Risk in Stock Market Investments
WALLER and GOTWAY · Applied Spatial Statistics for Public Health Data

*Now available in a lower priced paperback edition in the Wiley Classics Library.
†Now available in a lower priced paperback edition in the Wiley–Interscience Paperback Series.

WEISBERG · Applied Linear Regression, *Fourth Edition*
WEISBERG · Bias and Causation: Models and Judgment for Valid Comparisons
WELSH · Aspects of Statistical Inference
WESTFALL and YOUNG · Resampling-Based Multiple Testing: Examples and Methods for *p*-Value Adjustment
* WHITTAKER · Graphical Models in Applied Multivariate Statistics
WINKER · Optimization Heuristics in Economics: Applications of Threshold Accepting
WOODWORTH · Biostatistics: A Bayesian Introduction
WOOLSON and CLARKE · Statistical Methods for the Analysis of Biomedical Data, *Second Edition*
WU and HAMADA · Experiments: Planning, Analysis, and Parameter Design Optimization, *Second Edition*
WU and ZHANG · Nonparametric Regression Methods for Longitudinal Data Analysis
YAKIR · Extremes in Random Fields
YIN · Clinical Trial Design: Bayesian and Frequentist Adaptive Methods
YOUNG, VALERO-MORA, and FRIENDLY · Visual Statistics: Seeing Data with Dynamic Interactive Graphics
ZACKS · Examples and Problems in Mathematical Statistics
ZACKS · Stage-Wise Adaptive Designs
* ZELLNER · An Introduction to Bayesian Inference in Econometrics
ZELTERMAN · Discrete Distributions—Applications in the Health Sciences
ZHOU, OBUCHOWSKI, and McCLISH · Statistical Methods in Diagnostic Medicine, *Second Edition*

*Now available in a lower priced paperback edition in the Wiley Classics Library.
†Now available in a lower priced paperback edition in the Wiley–Interscience Paperback Series.